Topics in Geobiology

Volume 55

Series Editor

Kathleen A. Ritterbush, Geology and Geophysics, University of Utah, Salt Lake City, USA

The Topics in Geobiology series covers the broad discipline of geobiology that is devoted to documenting life history of the Earth. A critical theme inherent in addressing this issue and one that is at the heart of the series is the interplay between the history of life and the changing environment. The series aims for high quality, scholarly volumes of original research as well as broad reviews. Geobiology remains a vibrant as well as a rapidly advancing and dynamic field. Given this field's multi-discipline nature, it treats a broad spectrum of geologic, biologic, and geochemical themes all focused on documenting and understanding the fossil record and what it reveals about the evolutionary history of life. The Topics in Geobiology series was initiated to delve into how these numerous facets have influenced and controlled life on Earth. Recent volumes have showcased specific taxonomic groups, major themes in the discipline, as well as approaches to improving our understanding of how life has evolved. Taxonomic volumes focus on the biology and paleobiology of organisms–their ecology and mode of life–and, in addition, the fossil record–their phylogeny and evolutionary patterns–as well as their distribution in time and space. Theme-based volumes, such as predator-prey relationships, biomineralization, paleobiogeography, and approaches to high-resolution stratigraphy, cover specific topics and how important elements are manifested in a wide range of organisms and how those dynamics have changed through the evolutionary history of life.

Comments or suggestions for future volumes are welcomed.

Kathleen A. Ritterbush, Geology & Geophysics, The University of Utah, Salt Lake City, UT, USA Email: k.ritterbush@utah.edu

Alexander Lukeneder
Editor

Late Triassic Konservat-Lagerstätten within the Carnian Pluvial Episode

150 Years of Scientific Excavations in Austria

Springer

Editor
Alexander Lukeneder
Department of Geology and Paleontology
Natural History Museum Vienna
Vienna, Austria

ISSN 0275-0120
Topics in Geobiology
ISBN 978-3-032-19776-4 ISBN 978-3-032-19777-1 (eBook)
https://doi.org/10.1007/978-3-032-19777-1

This Springer imprint is published by the registered company Springer Nature Switzerland AG
The registered company address is: Gewerbestrasse 11, 6330 Cham, Switzerland

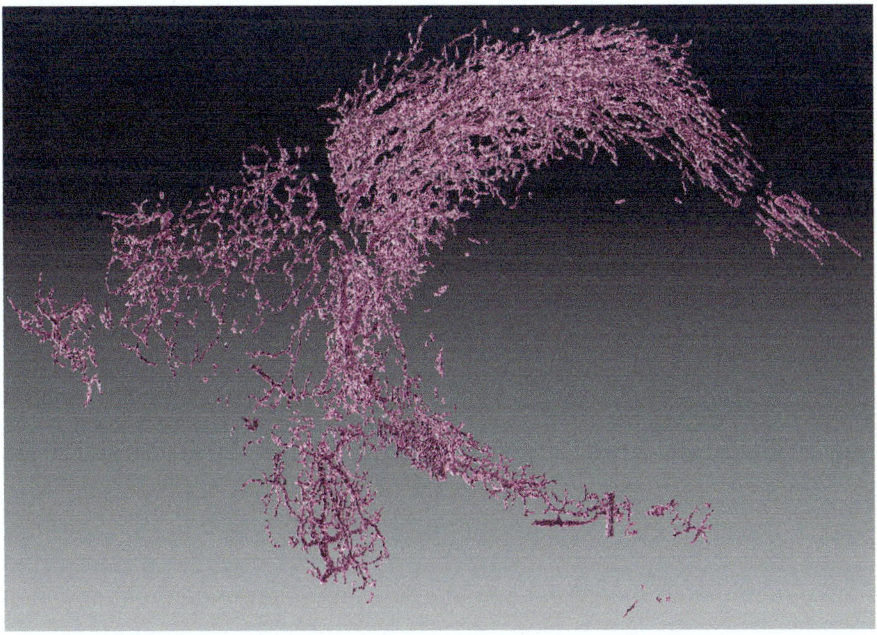

Micro-CT scan of a carbonized cephalic cartilage from the Late Triassic belemnoid *Phragmoteuthis polzbergensis*, NHMW 2012/0117/0028, from the Polzberg *Konservat-Lagerstätte* (Lower Austria). 3D-visualization of stained channel system by Petra Lukeneder, Valentin Blüml, Christina Kaurin, and Alexander Lukeneder (all Vienna). Specimen discovered by Alexander Lukeneder. Figure from Lukeneder et al. 2026, this volume, Chap. 8, Fig. 8.5a. Image by courtesy of Petra Lukeneder

We are just an advanced breed of monkeys
on a minor planet of a very average star.
But we can understand the Universe.
That makes us something very special.

—Stephen Hawking, British Astrophysicist
in 1988

Foreword

It is nearly 40 years since Alastair and Mike stumbled upon the realisation that the Carnian Stage of the Late Triassic had experienced a dramatic episode of climate change that had profoundly affected the Global biota (Simms and Ruffell, 1989). Initial evidence for the climate change was necessarily parochial and based on our own field observations—fluvial sandstones interrupting the arid red beds in southern England; a few caves penetrating outcrops of Carboniferous limestone in parts of England and Wales; and the obvious paleoenvironmental signal represented by the Raibl Beds of the Northern Calcareous Alps. Geochemical data in support of climate change was virtually non-existent and its ultimate cause was unclear, although we suspected increased CO_2 levels from volcanism (Simms and Ruffell, 1990).

Evidence for biotic change came initially from crinoids but a subsequent search through disparate papers revealed wider climatic effects and evidence for major biotic overturn in both the marine and terrestrial (Benton, 1986) realms. Paul Wignall was privy to some of the early discussions, since he shared an office with Alastair and Mike at Birmingham (Simms and Ruffell 2018), but he was not directly involved at that time. Focussing instead on his Ph.D. investigating the Upper Jurassic Kimmeridge Clay, he nonetheless recognised that there was a paleoclimatic component to that research. His paleoclimate credentials were strengthened hugely by subsequent work on Permo-Triassic boundary events and he became a valuable ally of the CPE, with Mike and Alastair happy to enlist him as the Third Man and a major contributor to our subsequent review of the climate evidence (Ruffell, Simms and Wignall, 2016)

Our role in identifying the climate signal is widely acknowledged but our contribution to recognising the biotic effects less so, despite the unambiguous title of our first paper: 'Synchroneity of climatic change and extinctions in the late Triassic' (Simms and Ruffell 1989). Others have since expanded considerably on this biotic aspect of the CPE (e.g., Dal Corso et al. 2020) but the main attributes were established by Simms et al. (1994). Although our work more than 30 years ago recognised the widespread biotic changes seen through the Carnian Stage, we were unaware of an additional fascinating element of this interval, that of the Early Carnian *Konservat-Lagerstätten* described in this volume. Previous publications relating to these *Lagerstätten* occur scattered through literature spanning more than a century, but by bringing together a diverse range of geoscientists to address multiple aspects of this particular example at a key time during Earth history, this volume can substantially increase awareness of it. Given the importance of the CPE in Earth history, the Austrian Triassic *Konservat-Lagerstätten* must now rank alongside others with a higher profile and long history of research: Ediacara and counterparts at other sites around the world; the Burgess Shale; and the Solenhofen Limestone.

The chapters in this new multi-author volume fall into three broad categories: documentary, covering the history of research here and an exploration of techniques; paleontological; and stratigraphic, encompassing paleontology, mineralogy and geochemistry.

In the first three chapters, Petra and Alexander Lukeneder provide an historical overview of the CPE. The 'signature' of the CPE in the Austrian succession was recognised long before the effect of climate change was even considered, and the importance of the Austrian Triassic in the global context of the CPE cannot be overestimated. They then explore the history of research into the Polzberg *Lagerstätte*, from its mining heyday around the turn of the nineteenth and twentieth centuries to a renaissance of interest in the past two decades that has led to publication of this volume, concluding in the third chapter with an overview of the almost 12,000 fossils from the *Polzberg Lagerstätte* held in Austrian museums and representing an invaluable data repository of the CPE.

In Chap. 4, Alexander Lukeneder demonstrates how the Polzberg deposits, once considered a local paleontological curiosity, actually represent part of a more extensive belt of *Konservat-Lagerstätten* in the Northern Calcareous Alps, and Chap. 19 provides an overview of the coeval Scheiblingbach *Konservat-Lagerstätt*e in the Northern Calcareous Alps of Styria. The study of both can provide new insights into Early Carnian marine ecosystems.

Chapters 5–11, 14 and 20 and 21, are paleontological descriptions of elements of the marine fauna. Some describe single species (Chavdar Rusev and Alexander Lukeneder on *Euestheria minuta*; Thomas Laville and Alexander Lukeneder on a new thylacocephalan; and Sebastian Stumpf and colleagues on a hybodontid shark) while others describe multiple specimens and species (Matúš Hyžný and colleagues on the decapod fauna; Alexander Lukeneder on bivalves; Petra Lukeneder and colleagues on belemnoids; Luke Alexander Parry and colleagues on polychaete worms; Luka Gale and Alexander Lukeneder on foraminifera; Andrea Tintori and colleagues on the fish fauna; and Alexander Lukeneder and colleagues on lungfish and coelacanth specimens). While the last of these addresses the role of new scanning technologies in the interpretation of old museum specimens, all feed into a broader narrative of raising awareness and increasing our understanding of marine ecosystems and taphonomic processes in the Polzberg region during the CPE.

The Polzberg *Lagerstätte* represents a largely marine environment, but three chapters explore aspects of the terrestrial flora. In Chap. 12, Evelyn Kustatscher and colleagues describe macrofloral remains preserved within the marine sediments at the Polzberg site, compare them with coeval floras from other locations and conclude that the apparent lack of hygrophytic taxa in the Polzberg flora reflects a taphonomic bias that favoured more robust elements. In Chap. 17, Guido Roghi and Alexander

Lukeneder demonstrate how floral changes, reflected in their palynological analyses, closely track the isotope curves and suggest that these curves are a reliable proxy for pulsed climate change. In Chap. 22, Evelyn Kustatscher and colleagues explore taphonomic effects on the preserved flora and some of the factors that underlie the Carnian 'amber spike'.

Four chapters are broadly stratigraphic in their approach, using a diverse range of analytical techniques to explore the record of the CPE in the Polzberg region. In Chap. 15, Alexander Lukeneder and colleagues describe the Großau drill core and how it provides stratigraphic correlation between the Polzberg and Scheiblingbach localities as well as supplying key core-logging information (magnetic susceptibility, gamma-ray, organic matter, etc.) that can be used as paleoenvironmental proxies and on which further studies might be based. The gamma ray results seem crude compared to geochemical changes described elsewhere in this volume, but comparable with the clay mineral stratigraphy from elsewhere. Could it be that, compared to elemental geochemistry, clay mineralogy and gamma-ray analyses are too insensitive to resolve paleoenvironmental change at the level of precision being sought?

In Chap. 13, Ibtissam Chraiki and colleagues use a multi-proxy mineralogical and geochemical approach to decipher paleoenvironmental changes through the CPE using bed-by-bed analysis redolent of how one strand of CPE research emerged (Roghi 2004) from the earlier global, broad-brush, approach. Their results show that clay mineralogy is insensitive to the change from marine to brackish conditions through the Reingraben Formation compared with the range of geochemical element ratios shown, and that the change from marine to brackish conditions, assumed to result from CPE runoff, occurs with no obvious change in outcrop lithology.

Jacopo Dal Corso and colleagues in Chap. 16 demonstrate how mercury concentration in sediments can be used as a proxy for volcanogenic input, with the timing of volcanic input in the Early Carnian Reifling Formation experiencing dilution (by way of C-isotopes) as terrigenous input occurred through the Middle Carnian Reingraben Formation. This confirms what would be expected as Wrangellia, and other, volcanic centres erupted in a megamonsoon, followed by the traditional CPE story of terrigenous input.

In Chap. 23, Johannes A. Weitz and co-workers focus on the organic components from the Lower Carnian succession Scheiblingbach ~30 km south-west of the Großau and Polzberg locations, comparing the organic-rich Göstling Formation with the low-organic content of the underlying carbonates of the Reifling Formation and overlying mudrocks of the Reingraben Formation. The time of Göstling Formation deposition appears be critical in the CPE transition, as waters became eutrophic during the change from marine-dominated to terrigenous inputs.

It is perhaps appropriate that the final chapter, by Mathias Harzhauser and colleagues, addresses possible astronomical constraints on climate and sediment deposition during the CPE. They combine two common approaches to CPE studies: (1) using lamina-by-lamina through bed-by-bed sediment cyclicity to elucidate (2) wider, possibly global, paleoclimate cyclicity through this interval. They demonstrate that short-period (e.g., as little as 0.2 ka) De Vries and Heinrich-Bond cycles dominate Carnian deposition through modulation of the Pangean Megamonsoon, the first time that sub-Milankovitch cycles have been detected in Mesozoic deposits. It suggests that future work may ultimately provide resolution of events in the CPE to the level of millenia.

The depth and diversity of the contributions in this volume add greatly to our knowledge of the CPE and how it developed. They provide a level of detail in many cases that was not available to us more than 30 years ago and encompass other aspects that were unexplored or yet to be discovered when the CPE was first proposed. Alexander Lukender and his colleagues should be commended for producing such a seminal volume and setting a benchmark for others working in this field. We feel honoured that an idea hatched almost 40 years ago, and considered crazy at the time by some, has led to such a substantial body of work as we see here.

Belfast, Northern Ireland — Alastair H. Ruffell
Tan-y-groes, United Kingdom — Michael J. Simms
Leeds, United Kingdom — Paul B. Wignall

References

Benton MJ (1986) More than one event in the Late Triassic extinction. Nature 321:857–859. https://doi.org/10.1038/321857a0

Dal Corso J, Bernardi M, Sun Y, Song H, Selfullah LJ, Preto N, Gianolla P, Ruffell A, Kustatcher E, Roghi G, Merico A, Hohn S, Schmidt AR, Marzolli A, Newton RJ, Wignall PB, Benton MJ (2020) Extinction and dawn of the modern world in the Carnian (Late Triassic). Sci Adv 6(38):eaba0099. https://doi.org/10.1126/sciadv.aba0099

Roghi G, (2004) Palynological investigations in the Carnian of the Cave del Predil area (Julian Alps, NE Italy. Rev Palbot Palyno 132:1–35

Ruffell A, Simms MJ, Wignall PB (2015) The Carnian Humid Episode of the late Triassic. Geol Mag 153:271–284. https://doi.org/10.1017/S0016756815000424

Simms MJ, Ruffell AH (1989) Synchroneity of extinction and climatic change in the late Triassic. Geology 17:265–268. https://doi.org/doi.org/10./1130/0091-7613(1989)017

Simms MJ, Johnson ALA, Ruffell A (1994) Biotic and climatic change in the Carnian (Triassic) of Europe and adjacent areas. In: Fraser NC, Dieter-Suess H (eds). In the Shadow of the Dinosaurs: Early Mesozoic Tetrapods. Cambridge University Press, 222–234. ISBN 0521452422, 448 p

Simms MJ, Ruffell A (2018) The Carnian Pluvial Episode: From discovery, through obscurity, to acceptance. Jl Geol Soc175:989–992. https://doi.org/10.1144/jgs2018-020

Preface

The idea to this special volume on the legacy of the Carnian Pluvial Episode (CPE) in Austria and the history of Late Triassic *Konservat-Lagerstätten* originated after years of intensive field work followed by publications on various taxonomic groups, on taphonomic processes resulting in, sometimes speculative, interpretations of paleoecological mechanisms, on trophic levels and food webs combined with climatological considerations. My intention was to span a bow from initial historical findings to the most recent results, from taxonomy to paleoecology, thereby also opening the field to the ancient climate. Knowing that I could not do that on my own, I attempted to assemble the most experienced and well-known experts from all over the world on the Carnian Pluvial Episode and Late Triassic inhabitants, marine and terrestrial, encompassing both paleozoology and paleobotany. This was a challenging but rewarding effort. Basically there were two motivations for this book. First, I wanted to expand, update, and, in essence, complete this book to be the reference and source for the Austrian Carnian Pluvial Episode and the Carnian *Konservat-Lagerstätten* formed during that period. Second, I wanted to describe the progress made in our knowledge concerning the Carnian Pluvial Episode and the resulting *Konservat-Lagerstätten* over the past 10 years.

Imagine you were one of the pioneers dealing with the formerly unknown Carnian Pluvial Episode—for example, Michael J. Simms, Alastair H. Ruffell and Paul B. Wignall in 1987—and no one believed in the existence of such a global humid episode. Today, this period has been widely noticed and broadly accepted by the scientific community. It is a great honour to be elected editor of this special volume on the Carnian Pluvial Episode of Austria and its linked *Konservat-Lagerstätten*. The book is meant to be the go-to source for the scientific community interested in the Carnian Pluvial Episode, Late Triassic climate changes and the incorporated treasures of *Konservat-Lagerstätten*. This severe global phase is marked by the aftermath of increased volcanic activity at the Wrangellia Large Igneous Province, enhanced humidification with periods of monsoon, the subsequent boost of terrigenous input into marginal seas and basins, as well as the co-occurring platform demise. Accordingly, the drastic environmental changes and adaptations of various marine and terrestrial organisms, along with their systematic and taxonomic classification, form the foundation for better understanding the trophic interactions and food chains within in Late Triassic paleobiota.

Although a remarkable number of scientists devoted their time and thoughts to the Carnian stage, the major knowledge and contributions on one of the most interesting periods of the Triassic, the Carnian Pluvial Episode, was generated only over the last 30 years, synonymic to the golden age of CPE studies. Many articles by colleagues and by the editor himself contributed essential knowledge and new data with conclusions and interpretations on the causes and aftermath of the Carnian Pluvial Episode. The incomplete list includes Michael J. Benton, Rainer Brandner, Jacopo Dal Corso, Piero Gianolla, Thomas Hornung, Leopold Krystyn, Heinz W. Kozur, Wolfram M. Kürschner, Evelyn Kustatscher, Jing Lu, Petra Lukeneder, Paolo Mietto, Charlotte S. Miller, Steven Mueller, James W. Ogg, Nereo Preto, Manuel Rigo, Guido Roghi, Alastair H. Ruffel, Wolfgang Schlager, Michael J. Simms, Yadong D. Sun, Yuki Tomimatsu, Paul B. Wignall, Peixin Zhang and He Zhao. Thus, the many authors from different institutions all over the world present a diversity of viewpoints, which may also include controversial hypotheses and theories.

This special volume on the Late Triassic Carnian Pluvial Episode would have been impossible without the enormous efforts and intense work on specific topics within and around this period and the associated *Konservat-Lagerstätten*. In addition, we greatly appreciate the support of numerous reviewers, who are named in the corresponding acknowledgements of the individual chapters.

In terms of content, I have structured the volume in main sections devoted to history, major collections, the Carnian Pluvial Episode, the Polzberg *Konservat-Lagerstätte*, the Großau drilling site, the Scheiblingbach locality, each with their distinct taxonomic groups of marine and terrestrial inhabitants. Each successive section is composed of specific chapters. The considerable number of chapters (24!), all authored by experts in the respective subjects, reflects the spectrum of Carnian Pluvial Episode workers currently active in academia. This ranges from junior scientists to senior scientists from many countries, with a high percentage of female authors and reviewers—the *who's who* within that scientific field. Participating scientists come from Austria, China, France, Germany, Ireland, Italy, Slovakia, Slovenia, the UK and the USA. A main aim was to be as *up to date* as possible. Accordingly, most of the data are presented for the first time, which is also reflected by the presentation of new fossil species. This collection of peer-reviewed papers presents reviews and compilations of the latest studies as well as fresh ideas and new data on various topics. Everyone, scientists and general public, who works on or is interested in climate changes and catastrophic episodes in Earth's history will find this collection an essential addition to their library. We included numerous chapters on the geochemistry and geophysics of different localities, a field that still offers vast opportunities for new research. Studying a wide field of taxonomic groups shows the possibility and implications of systematics on life histories, trophic interactions and paleocommunities. Furthermore, intense studies were conducted on biostratigraphy, biogeography, intraspecific variability, evolutionary trends, paleoecology, paleoclimatology, paleoceanography and many more. Finally, we added new information obtained from 3D-scanning and tomographic data obtained from micro computed tomography. Both methodologies provide the foundation for animation and help to resurrect ancient life for scientists as well as the general public.

The Foreword by *The Three Musketeers* of the Carnian Pluvial Episode—Michael J. Simms, Alastair H. Ruffel and Paul B. Wignall—is followed by the opening Preface along with the list of contributions and the names of the leading actors, the authors.

Chapter 1: The special volume on the Carnian Pluvial Episode of Austria is opened by Petra Lukeneder and Alexander Lukeneder review the history and concepts of the Carnian Pluvial Episode.

Chapter 2: In this chapter, Petra Lukeneder and Alexander Lukeneder with an overview of the discovery and legacy of the Polzberg *Konservat-Lagerstätte.*

Chapter 3: Petra Lukeneder and Alexander Lukeneder report on the fossil material of Late Triassic *Konservat-Lagerstätten* housed in the main Austrian institutions, namely the Natural History Museum Vienna, the University Vienna and the GeoSphere Austria and attributed collections.

Chapter 4: In the next section of the volume, the Austrian *Konservat-Lagerstätten* Belt is introduced by Alexander Lukeneder, referencing all known Carnian *Konservat-Lagerstätten* in Austria.

Chapter 5: Chavdar Rusev and Alexander Lukeneder describe the clam shrimp community of the Upper Triassic Polzberg *Konservat-Lagerstätte.*

Chapter 6: Matúš Hyžný and colleagues provide a synopsis of the decapod crustaceans from the Carnian deposits of Polzberg in Austria and Cave del Predil in Italy.

Chapter 7: Alexander Lukeneder highlights new data on bivalve mass occurrences deposited at numerous localities during the Carnian Pluvial Episode.

Chapter 8: Petra Lukeneder and colleagues present new data on the 3D reconstruction and microarchitecture of a reticulate belemnoid cephalic cartilage from the Late Triassic Polzberg and Cave del Predil *Konservat-Lagerstätten.*

Chapter 9: Luke Parry and colleagues provide new data on polychaete annelids from the Polzberg *Konservat-Lagerstätte.*

Chapter 10: The next section of the volume is devoted by Andrea Tintori and colleagues to the fish assemblage from the Early Carnian of Polzberg in Austria.

Chapter 11: Alexander Lukeneder and colleagues show first visualisations by 3D scanning and computer tomography of the Upper Triassic sarcopterygiids *Tellerodus* and *Coelacanthus* from the Polzberg *Konservat-Lagerstätte.*

Chapter 12: Evelyn Kustatscher and colleagues review the historical and more recently collected plant fossils of the Carnian Polzberg *Konservat-Lagerstätte* in Austria.

Chapter 13: Ibtissam Chraiki reviews the redox conditions and paleoenvironment of the Polzberg *Konservat-Lagerstätte* with the incorporated marine Polzberg paleobiota as evidence for the Carnian Pluvial Episode.

Chapter 14: Foraminifera from the Upper Triassic Carnian Pluvial Episode in Lower Austria are documented by Luka Gale and Alexander Lukeneder for the first time from the Carnian Pluvial Episode of Austria.

Chapter 15: Lukeneder and colleagues report the first data of a core drilling into the Carnian Pluvial Episode in Austria.

Chapter 16: Jacopo Dal Corso and colleagues present their work on carbon-isotope perturbation and mercury deposition in the Reifling Basin at 'Event 1' of the Carnian Pluvial Episode.

Chapter 17: Guido Roghi documents the palynology during the Austrian Carnian Pluvial Episode at numerous localities in Austria.

Chapter 18: Christian Pott and Alexander Lukeneder highlight a deltaic swamp forest as coastal wetland vegetation during the climax of the Carnian Pluvial Episode from Lunz am See in Austria.

Chapter 19: The marine Upper Triassic paleobiota from the Scheiblingbach *Konservat-Lagerstätte* in Austria are described for the first time by Alexander Lukeneder.

Chapter 20: Thomas Laville and Alexander Lukeneder note the first thylacocephalans from the Upper Triassic Scheiblingbach *Konservat-Lagerstätte* in Austria.

Chapter 21: Sebastian Stumpf and colleagues present the first hybodontiform record from the Upper Triassic Reifling Basin in Austria, which reveals the chondrichthyan role in Carnian food webs during the Carnian Pluvial Episode.

Chapter 22: Evelyn Kustatscher and colleagues present new data on the plant fossils of the Carnian of Scheiblingbach section in Austria.

Chapter 23: Johannes Weitz and colleagues provide a thorough description of the geochemistry and organic matter from the Scheiblingbach Carnian Pluvial Episode section.

Chapter 24: Mathias Harzhauser and colleagues propose a new potential modulation of a Late Triassic Pangean Megamonsoon by solar cycles tracing the Carnian Pluvial Episode of Austria.

This accumulation of highly diverse articles in the special volume on the Carnian Pluvial Episode in Austria commemorates the first successful historical excavations and initial point of research at the Polzberg *Konservat-Lagerstätte.* That early effort initiated the publication of a series of monographs on both invertebrate and vertebrate fossils. The excellent preservation of entire specimens, fragile structures and soft body parts within the entire paleocommunity at Großau, Lehen, Polzberg, Rehberg, Scheiblingbach, Steinbachgraben and other localities justifies the designation as classical *Konservat-Kagerstätten* within the Reifling Basin. All these sites are dominated by nektic ammonoids, belemnoids and actinopterygian fish inhabiting the water column, along with benthic bivalves, crustaceans and gastropods populating the sea floor.

This book enhances our understanding of the Carnian Pluvial Episode and serves as a reference for this important Late Triassic period. It presents a comprehensive review of the latest science on the geological and paleontological aspects of Earth during the Late Triassic epoch including the major Carnian Pluvial Episode. It reflects the latest knowledge on the Austrian Carnian Pluvial Episode and linked *Konservat-Lagerstätten* and contains new and unpublished taxa, data and results. Advances such as an explosion of new information about the Carnian Pluvial Episode, new technologies such as isotopic and mercury analysis, computed tomography combined with virtual paleontology, accompanied with the continuous discovery of new fossil taxa have given us the opportunity to present a comprehensive, *state-of-the-art* compilation on the Late Traissic Carnian Pluvial Episode and Carnian *Konservat-Lagerstätten* in Austria. As such it is intended for a variety of readers, certainly workers on the Carnian Pluvial Episode and *Konservat-Lagerstätten.*

Last but not least, I want to thank my children Lisa Marie and Felix Maximilian, who clearly were affected by the time-consuming and intense process during the creation and editorial work on this special volume, for their patience, inspiration and support.

Vienna, Austria
August 2026

Alexander Lukeneder

Acknowledgments

The projects on the Carnian Pluvial Episode and Carnian *Konservat-Lagerstätten* were financially supported by the Federal Government of Lower Austria (Department Science and Research; headquarters in St. Pölten, projects K3-F-964/001-2020 and K3-F-964/002-2023; Polzberg—*Konservat-Lagerstätte* von Weltruf and Drilling into Earth's History): the Austrian Academy of Sciences (headquarters in Vienna, project Polzberg Lukeneder 2021) represented by the National Committee for Geo/Hydro Sciences (Earth System Sciences Programme; the Society of Friends of the Natural History Museum Vienna (Freunde des Naturhistorischen Museums Wien; projects 'Fenster in die Karnische Krise', CPE Lukeneder 2023, Drilling into Earth's History); and the Natur- und UNESCO Geopark Steirische Eisenwurzen (Styrian Eisenwurzen Nature and UNESCO Geopark; headquarters in St Gallen, project Karnische Krise Lukeneder 2024). The projects were also supported by the communities of Lunz am See, Gaming, Göstling an der Ybbs (all Lower Austria) and Landl (Styria) as well as by the society 'Naturkundliche Gesellschaft Mostviertel' (Lower Austria) and the society 'Kultur- und Tourismusverein Gams bei Hieflau' (Styria).

Open access funding was financially supported by the Federal Government of Lower Austria (Department Science and Research), the Society of Friends of the Natural History Museum Vienna (Verein der Freunde des Naturhistorischen Museums Wien), the Natural History Museum Vienna, the Technical University of Leoben (Chair of Energy Geosciences), and the editor Alexander Lukeneder.

We are particularly grateful to Birgitt and Karl Aschauer (Waidhofen an der Ybbs), who made available abundant and exciting fossil material for scientific investigations.

The editor, Alexander Lukeneder, is personally and particularly grateful to the owners of the respective outcrop areas for sampling and digging permission. For the historical Polzberg site, thanks go to Franziska and Hermann Hofreiter as well as Eva and Karl Jagersberger (all Gaming), for the Großau drilling site to Felix Pöchhacker and family (Lunz am See), and for the Scheiblingbach locality to the Austrian Federal Forests Company (Österreichische Bundesforste; headquarters in Purkersdorf).

Contents

Contributors

Greta Alverà Department of Physics and Earth Science, University of Ferrara, Ferrara, Italy

Christian Aupissy UMR-CNRS IC2MP, Université de Poitiers, Poitiers, France

Valentin Blüml Department of Evolutionary Biology, University of Vienna, Vienna, Austria

Roberta Branz Museum of Nature South Tyrol, Bozen/Bolzano, Italy

Marcello Caggiati Department of Geosciences, University of Padova, Padova, Italy

Ibtissam Chraiki UMR-CNRS IC2MP, Université de Poitiers, Poitiers, France

Daoliang Chu State Key Laboratory of Geomicrobiology and Environmental Changes, China University of Geosciences, Wuhan, China

Davide Conedera Dipartimento di Geoscienze, Università di Padova, Padova, Italy

Jacopo Dal Corso State Key Laboratory of Geomicrobiology and Environmental Changes, China University of Geosciences, Wuhan, China

Davide Di Franco GeoSphere Austria, Vienna, Austria

Elea Dondoglio UMR-CNRS IC2MP, Université de Poitiers, Poitiers, France

Abderrazak El Albani UMR-CNRS IC2MP, Université de Poitiers, Poitiers, France

Anna El Koury UMR-CNRS IC2MP, Université de Poitiers, Poitiers, France

Claude Fontaine UMR-CNRS IC2MP, Université de Poitiers, Poitiers, France

Luka Gale Faculty of Natural Sciences and Engineering, Department of Geology, University of Ljubljana, Ljubljana, Slovenia;
Geological Survey of Slovenia, Ljubljana, Slovenia

Mohamed Ghnahalla Ecole Supérieure Polytechnique, Nouakchott, Mauritania

Piero Gianolla Department of Physics and Earth Science, University of Ferrara, Ferrara, Italy

Susanne Gier Department of Geology, University of Vienna, Vienna, Austria

Mathias Harzhauser Department of Geology and Paleontology, Natural History Museum Vienna, Vienna, Austria

Tastulek Haserbek State Key Laboratory of Geomicrobiology and Environmental Changes, China University of Geosciences, Wuhan, China

Matúš Hyžný Earth Science Institute, Slovak Academy of Sciences, Bratislava, Slovakia;
Department of Geology and Paleontology, Faculty of Natural Sciences, Comenius University, Bratislava, Slovakia

Christina Kaurin Department of Evolutionary Biology, University of Vienna, Vienna, Austria

Michael Klein 7Reasons Medien GmbH, Vienna, Austria

Evelyn Kustatscher Museum of Nature South Tyrol, Bozen/Bolzano, Italy;
Department of Natural History, Collection and Research Centre (SFZ), Tirolean State Museums, Hall in Tirol, Austria;
Department für Geo- und Umweltwissenschaften, Paläontologie und Geobiologie, Ludwig-Maximilians-Universität, München, Germany;
Bayerische Staatssammlung für Paläontologie und Geobiologie, München, Germany

Thomas Laville Faculty of Biology, Ludwig-Maximilians-Universität München, Planegg-Martinsried, Germany

Alexander Lukeneder Department of Geology and Paleontology, Natural History Museum Vienna, Vienna, Austria

Petra Lukeneder Department of Geology and Paleontology, Natural History Museum Vienna, Vienna, Austria;
Institute of Applied Geology, Department of Landscape, Water and Infrastructure, BOKU University, Vienna, Austria

Oleg Mandic Department of Geology and Paleontology, Natural History Museum Vienna, Vienna, Austria

David Misch Department Applied Geosciences and Geophysics, Technical University of Leoben, Leoben, Austria

Luke A. Parry Department of Earth Sciences, University of Oxford, Oxford, UK

Christian Pott LWL-Museum of Natural History, Westphalian State Museum with Planetarium, Münster, Germany

Guido Roghi Institute of Geosciences and Earth Resources—CNR, National Research Council (CNR), Padova, Italy

Chavdar Rusev iC Consulenten Zivitechniker GesmbH, Vienna, Austria

Reinhard F. Sachsenhofer Department Applied Geosciences and Geophysics, Technical University of Leoben, Leoben, Austria

Miriam Slodownik Botany Department, Trinity College Dublin, Dublin, Ireland

Haijun Song State Key Laboratory of Geomicrobiology and Environmental Changes, China University of Geosciences, Wuhan, China

Ksenija A. Stojanović Department of Applied Chemistry, University of Belgrade, Belgrade, Serbia

Sebastian Stumpf Department of Geology and Paleontology, Natural History Museum Vienna, Vienna, Austria;
Department of Paleontology, University of Vienna, Vienna, Austria

Li Tian State Key Laboratory of Geomicrobiology and Environmental Changes, China University of Geosciences, Wuhan, China

Andrea Tintori TRIASSICA Institute for Triassic Lagerstätten, Perledo, Italy

Johannes A. Weitz Department Applied Geosciences and Geophysics, Technical University of Leoben, Leoben, Austria

Viola Winkler Department of Central Research Laboratories, Natural History Museum Vienna, Vienna, Austria

Chapter 1
The Carnian Pluvial Episode – Historical Outline and Concepts

Petra Lukeneder and Alexander Lukeneder

Abstract The term "Carnian crisis" is known by different names, among them "Reingrabener Wende", "Carnian Pluvial Event" and "Carnian Pluvial Episode". They all refer to a brief period in the Late Triassic, characterized by an intense climate warming coupled with increased rainfall. This episode is still evident today in the fossil record marked by a terrestrial and marine faunal turnover. Multiple events in the early Carnian (Julian) to late Carnian (Tuvalian) are related to the coeval establishment of the Wrangellia Large Igneous Province, which released large quantities of volcanic gases into the atmosphere, whereby greenhouse gases (CO_2) triggered the global observed environmental changes and biological turnovers. Discrete short-term events resulted in long-term changes in terrestrial, freshwater and marine ecosystems. Research on this Triassic episode therefore also offers learning potential for recent climate crises. This article provides a short overview on the scientific history of the Carnian Pluvial Episode and discusses the past and recent interpretations of the Carnian Reingraben deposits in the Northern Calcareous Alps of Austria by previous authors.

Keywords Carnian crisis · Carnian Pluvial Episode · History of sciences · *Konservat-Lagerstätten* · Polzberg · Triassic

P. Lukeneder (✉) · A. Lukeneder
Department of Geology and Paleontology, Natural History Museum Vienna, Vienna, Austria
e-mail: petra.lukeneder@nhm.at

A. Lukeneder
e-mail: alexander.lukeneder@nhm.at

P. Lukeneder
Institute of Applied Geology, Department of Landscape, Water and Infrastructure, BOKU University, Vienna, Austria

A. Lukeneder (ed.), *Late Triassic Konservat-Lagerstätten within the Carnian Pluvial Episode*, Topics in Geobiology 55, https://doi.org/10.1007/978-3-032-19777-1_1

1.1 Early Mentions and Interpretations of the Carnian Pluvial Episode

Although the Triassic period is in general characterized by strong aridity (Miller et al. 2017), there was a brief humid episode marked by evidence for intense climate change with increased rainfall (Simms and Ruffell 1989; Kolar-Jurkovšek and Jurkovšek 2008; Mueller et al. 2016; Lukeneder and Lukeneder 2023; Lukeneder 2026, this volume), followed by a major marine and terrestrial faunal turnover (Preto et al 2019; Mazaheri-Johari et al. 2021). This humid climate pulse is known among others as the "Carnian Pluvial Event" (CPE) (Hornung et al. 2007a, b; Preto et al. 2008, 2019), which is directly linked to a floral and faunal turnover (Simms and Ruffel 1989) along the northern Tethys rim and is reported from many European localities and surrounding regions (Mueller et al. 2016). Heavy rainfall caused widespread and massive siliciclastic input into epicontinental basins and marginal Tethyan environments (Enos et al. 1998; Geyer 2002; Lehrmann et al. 2005; Hornung and Brandner 2005; Rigo et al. 2007; Hornung 2007b). This so-called conspicuous Carnian Pluvial Event/Episode (Simms and Ruffell 1989; Miller et al. 2017; Alverà et al. 2026; Lukeneder 2026, this volume) is known by several additional names: Middle Carnian Wet Intermezzo (MCWI; Kozur and Bachmann 2010; Ogg 2015), Carnian Humid Episode (Simms and Ruffell 1990; Ruffell and Simms 2016; Sun et al. 2016), Carnian Pluvial Phase (CPP; Mueller et al. 2016; Shi et al. 2017) and Carnian crisis (Hornung et al. 2007a). The here used term Carnian Pluvial Episode (Simms and Ruffell 2018) was actually coined and introduced to science long after its dicovery. High extinction rates had already been observed in Carnian sediments long before that term emerged. Schlager and Schöllnberger (1974) already pointed to a dark siliciclastic layer that abruptly interrupted a long phase of carbonate deposition in the deposits of the Northern Calcareous Alps (NCA). In the NCA this distinct climatic shift is known as "Reingrabener Wende" or Reingraben Turnover (Hornung and Brandner 2005), "Reingraben Event" (Schlager and Schöllnberger 1974; Hornung and Brandner 2005; Preto et al. 2019) and "Lunz Event" (Preto et al. 2019). Mueller et al. (2016) mention the term CPE as insufficient because this interval is not a distinct short-termed geological event but a longer phase of at least 1 Ma (Hornung et al. 2007b). Therefore, the term CPE shall be regarded as a junior synonym of the Reingraben Event (Hornung et al. 2007b; Mueller et al. 2016; The GGTS 2021). The well-known term "Carnian Pluvial Episode" was first mentioned and identified as a paleoclimatic event by Simms and Ruffel (1989). The enhanced humidification (documented by a change of the lithology; see also Lukeneder and Lukeneder 2023) combined with the deposition of volcanic material exhalated at the Wrangellia Large Igneous Province (LIP), triggered this global climate change (Simms and Ruffell 2018; Dal Corso et al. 2020; Hornung and Brandner 2005; Hornung et al. 2007a, b; Mueller et al. 2016; Lukeneder and Lukeneder 2021). Mueller et al. (2016) document a trend from dry climate during the Julian 1 to more humid and warmer conditions during the early Julian 2, resulting in the deposition of the Göstling Formation and the argillaceous Reingraben Formation (Lukeneder and Lukeneder 2021). Although the ultimate causes and complexity of the biotic crisis are still intensively studied, the drastic change to more humid climatic conditions was often given as the main reason (Hornung et al. 2007a).

Late Triassic localities comprising similar facies changes and preservational features are rare in the NCA of Austria. In addition to the localities already mentioned, the Reingraben Formation crops out in the Lunz region in Lower Austria (e.g., Rehberg, also known as Rechberg: Global Positioning System – GPS 47°51′37.60″N 15°3′59.30″E and Steinbachgraben near Göstling: GPS 47°48′21.91″N 14°56′59.13″E; Fig. 1.1) as well as in northern Styria (Scheiblingbachgraben near Großreifling: GPS 47°39′52.5″N 14°42′28.4″E). To the east, a particularly well-known occurrence is located in the region of Hinterbrühl (also: Brühl in Lower Austria) near Vienna.

Comparable paleontological Austrian sites (i.e. conservational deposits), although different in age, are known from Wiestal in Salzburg (Hornung et al. 2019) and Seefeld in Tyrol (Brandner and Poleschinski 1986). Both localities comprise deposits of the Norian Seefeld Formation (Lukeneder and Lukeneder 2023).

Fig. 1.1 The Lunz am See area is one of the hotspots of scientific research on the Carnian Pluvial Episode in the Northern Calcareous Alps of Austria. This region is well-known since the 19th century for its rich Triassic coal deposits and subsequent mining. Lake Lunzer See with the 1075 m high mount Maiszinken

1.2 Stratigraphic Position of the CPE

Precise biostratigraphic-dating for the CPE in the Northern Calcareous Alps (NCA) is mainly based on ammonoid zonation (Mojsisovics 1882; Krystyn 1978, 1991). For the CPE, a Julian age (Julian 1/IIc) was established by Hornung et al. (2007a). Lukeneder and Lukeneder (2023) give Julian 2 Ib, based on the *Austrotrachyceras austriacum* Zone (Fig. 1.2). Krystyn (1978) gives the stratigraphic scheme based upon conodontophorids (see also Mostler and Scheuring 1974). According to Kolar-Jurkovšek and Jurkovšek (2008), the stratigraphic development of the Northern Calcareous Alps (NCA) corresponds to Carnian "Raibl Beds" in the Northern Karavanke Mountains (NKM; see also Placer et al. 2002). Mueller et al. (2016) set the CPE onset at the Julian 1–2 boundary, lasting approx. until latest Julian or even early Tuvalian. Kohút et al. (2018) presented the maximum age of deposition of the Lunz Formation based on volcanic zircon dating. They dated the upper limit of the CPE and the Carnian stage in the Alpine-Carpathians realm, respectively, at 221.2 ± 1.6 Ma, which is much younger than the current age of the Carnian-Norian boundary at approximately 227 Ma. The CPE lasted for approximately 1.2–1.6 Ma (late Julian-early Tuvalian) (Miller et al. 2017; Mazaheri-Johari et al. 2021).

1.3 The Reingraben Formation and the Early Interpretations of the Polzberg Paleoenvironment

1.3.1 Starting Point at Polzberg

The *Konservat-Lagerstätte* Polzberg (Stur 1874; GPS: 47°53′5.90"N, 15° 4′27.70"E; ÖK 1:50,000, sheet 71 Ybbsitz, Ruttner and Schnabel 1988; sheet 72 Mariazell, Bauer and Schabel 1997; Austrian Map 2023; 712 m above sea level; Figs. 1.2 and 1.3) in Lower Austria is famous for its fossil-bearing sediments of the Carnian crisis (see also Lukeneder and Lukeneder 2021, 2022a, b, c; Lukeneder and Lukeneder 2023). The reconstruction of the Polzberg paleoenvironment as being dysoxic (Lukeneder and Lukeneder 2021) was by far not the first published interpretation. During the past 150 years, the Reingraben Formation has been interpreted in different ways. The type locality of the shaly Reingraben Formation is situated in the Raingraben Ravine (also: Reingraben Ravine, see also Tollmann 1976) east of Rohr am Gebirge (Lower Austria; GPS 47°53′22.6″N 15°44′52.8″E). It was discovered by Johann Nepomuk Czizek (1816–1895) and Dionys Stur (1827–1893) (Bittner 1882). Soon after its discovery (Stur 1874), the preserved Polzberg fauna attracted attention, although interpretations were rarely mentioned. Stur (1874) initially classified the Polzberg sediment as "Wengen shales" (=Wengerschiefer im Pölzberg), whereas Teller (1891) already assigned them to the Reingraben Formation (= Reingraben shales). Teller (1891) provided insights into the excavation process, but did not provide a clear interpretation of the environment. Abel (1906) was possibly the first

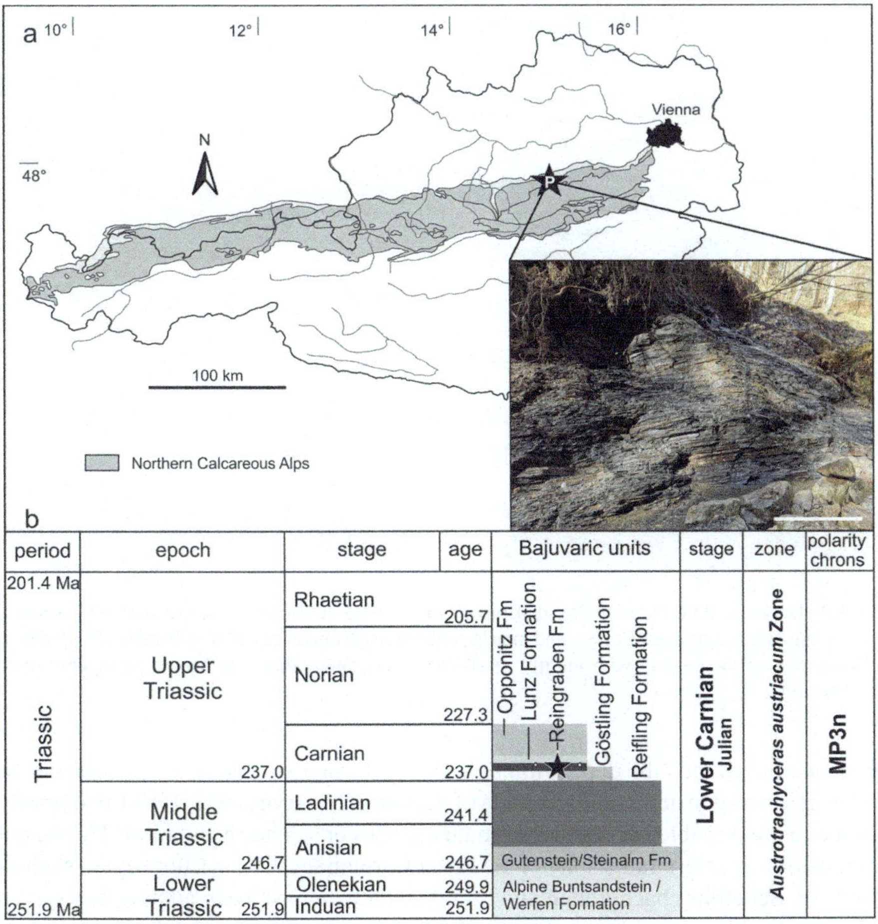

Fig. 1.2 **a** Austrian map with Northern Calcareous Alps marked. Black asterisk marks the position of the *Konservat-Lagerstätte* Polzberg (= Schindelberggraben ravine, Polzberg locality) near Lunz am See with indicated positions of new outcrops and historical adits. Enlargement shows main section from new excavation campaigns. Scale bar 1 m. **b** Stratigraphic chart with Bajuvaric units, formations and a lithological overview log of the Polzberg locality; calcareous deposits occur only in the older/lower part of the section. Adapted from Lukeneder and Lukeneder (2021, 2023)

to record the fauna in a more holistic manner and interpreted the Polzberg fossil site as a quiet spot of the ancient sea with shallow water in the Reifling Basin. Glaessner (1931), in turn, mentioned the Reingraben and Aonides shales of the Polzberg region as deposits of an inanimate, mud-filled bay with suboxic water. Glaessner (1931) concluded that benthonic life was not possible here and that pelagic animals died when entering this area. Based on the faunal elements, Glaessner (1931) interpreted the shaly deposits of the Reingraben Formation at Polzberg as autochthonous with rapid burial of the carcasses (Fig. 1.3). Krystyn (1991, 1998) took up the topic

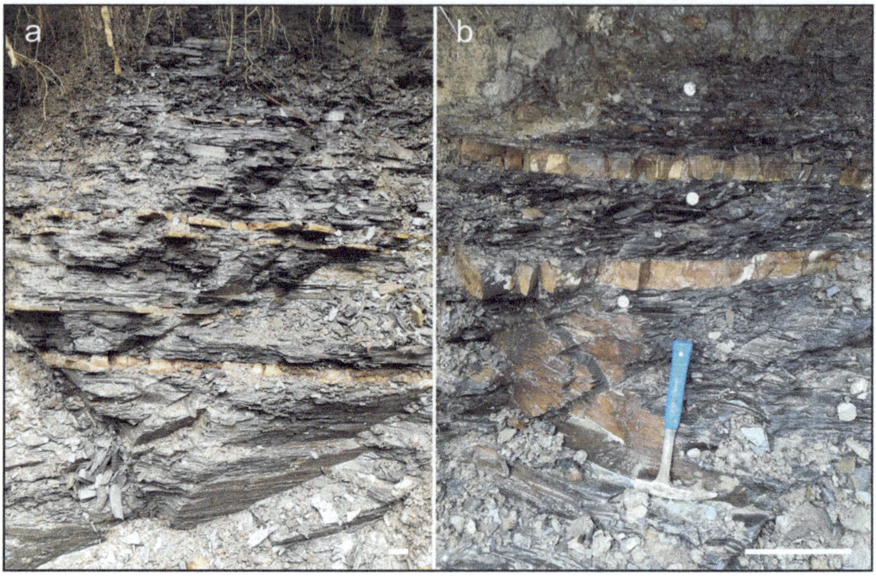

Fig. 1.3 Deposits of the Carnian Pluvial Episode in Polzberg. **a** Middle and upper part of the section comprising finely laminated clays and marls with mass occurrences of the bivalve *Peribositra*. **b** Upper part of the section with mainly fossil-free clays, intercalated by limestone layers. Scale bars 20 cm

again and interpreted the deposition as a result of water stratification and associated widespread oxygen deprivation in parts of the basin. Krystyn (1991, 1998) also briefly discussed the possible causes, e.g. climate changes or sea basin isolation. The faunal composition and the stratigraphy of the lower Carnian sections in Lunz region such as Polzberg, Scheiblingbachgraben and Steinbachgraben have been analyzed (Krystyn 1978, 1991). After a long break, Forchielli and Pervesler (2013) were the first to return to the Polzberg topic and focused on the arthropod group of the Thylacocephalia. They interpreted the lower part of the Reingraben Formation, with its predominantly nektonic fauna, as deposits of a deeper marine environment (Hornung and Brandner 2005; Hornung et al. 2007a; Forchielli and Pervesler 2013).

1.3.2 Hostile Environmental Conditions

The nutrient-rich claystones at Polzberg are characterized by a low taxonomic richness of distinct fossil groups, but by a high diversity when analyzing the entire assemblage (Fig. 1.4). Although only few groups such as the austrotrachyceratid ammonoids dominate, most of the taxonomic groups show relatively high numbers of individuals. The accompanying lack of bite marks in the coleoid proostraca supports the hypothesis for the absence of large predators in that near-bottom environment, which can be explained by the presence of a thicker and permanent low-oxygen layer. Low-oxygen conditions near the seafloor of the Late Triassic Reifling Basin are evident (see also Lukeneder and Lukeneder 2021). Benthic life is present only in the form of rare polychaete specimens, gastropods and bivalves, whereas mass occurrences of the most probably opportunistic bivalve *Peribositra* are limited to distinct layers in the middle part of the section. Notably, these bivalves were already embedded during juvenile stages and did not grow to adults. The mass occurrences of posidoniids could also point to occasional low-oxygen environments (McRoberts 2001; Hopkin and McRoberts 2005). The same interpretation was proposed for *Bositra buchi* (Oschmann 1994a, b; Röhl et al. 2001). Low-oxygen conditions are usually found associated with increased values of organic carbon. The amounts of TOC (Total Organic Carbon; weight %) vary between relatively low values of 0.3 and 1.4% (Lukeneder and Lukeneder 2021). Values of total sulphur (TS) content range between 1.8 and 0.3% (Lukeneder and Lukeneder 2021). Many of the argillaceous layers, especially in the younger part of the section, lack any fossil content. These macrofossil-free layers indicate inhospitable conditions at the relatively deep seafloor. In comparable *Konservat-Lagerstätten* the presence of juvenile bivalves and the contemporary absence of benthic fauna can be interpreted as inhospitable paleoenvironmental conditions near the seafloor (Hart et al. 2019). According to Forchielli and Pervesler (2013), who intensively studied the taphonomic pathways in *Konservat-Lagerstätten* such as in Polzberg, anoxia does not inhibit decay. Anoxia may, however, promote exceptional soft part preservation.

Hornung et al. (2007a) summarized the processes of platform demise, sea level fall and onset of warming in shallow-marine NW Tethyan environments. The humidification causes increasing freshwater influx, which elevates the nutrient content. This expands the oxygen-minimum zone (Hornung et al. 2007a). In particular, during the Julian 2 the expanded O_2-minimum zone filled almost the complete water column. Besides, only minor amounts of argillaceous material were deposited on the carbonate platforms, whereas the major depositional areas of the Reingraben Formation were located in basinal environments, where abundant argillaceous sediments were deposited (see Lukeneder 2026, Fig. 1.4).

1.4 Conclusions

The recent excavation campaigns at Polzberg (Lukeneder et al. 2024a, b) shed new light on this Late Triassic ecosystem, on its development and fossilization processes during the Carnian Pluvial Episode. The recent analyses of the Polzberg paleobiota embedded within dark laminated rocks of the Reingraben Formation allowed a precise reconstruction of this fossil ecosystem, influenced by climate change. Low resilience to these intense, rapid changes in environmental conditions during the CPE, ensured a rapid faunal turnover. Previous authors have pointed to the CPE as an analogue for the prediction of the effects of current global warming on marine and

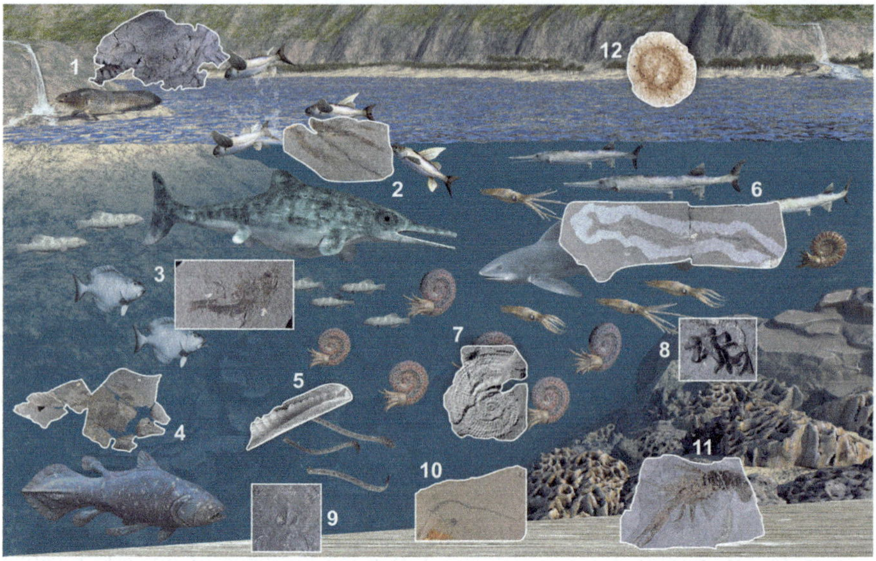

Fig. 1.4 Reconstruction of the Late Triassic paleoenvironment and vertebrate/invertebrate members of the early Carnian Polzberg paleobiota. Reconstruction is combined with fossil specimens in 1–10. **1** *Tellerodus sturii*, **2** *Thoracopterus telleri*, **3** *Nannolepis elegans*, **4** *"Coelacanthus" lunzensis*, **5** *Paragondolella indica*, **6** *Saurichthys calcaratus*, **7** *Austrotrachyceras minor*, **8** *Phragmoteuthis bisinuata* cartilage, **9** *Halobia rugosa*, **10** eunicid bristle worm, **11** *Platychela trauthi*, **12** *Patinasporites densus*. Figured members of paleobiota not to scale. Artwork by AL, final artwork by 7reasons; www.7reasons.net; adapted after Lukeneder and Lukeneder (2021)

terrestrial faunas. Simms and Ruffel (2018) understand the CPE as "*a critical stage in the development of significant elements in the modern biota and the corresponding ecosystems*". Studying the extinctions of the past can help us understand how modern ecosystems might respond to future disturbances. Predictions can then be made on the parts of a system particularly resilient to climate changes, based on estimated recovery rates. Simms and Ruffel (2018) themselves documented in detail the history of research from recognition to recent studies of the Carnian Pluvial Episode and its further perspectives.

Acknowledgments The authors are particularly grateful to the owners of the outcrop area, Franziska and Hermann Hofreiter, Eva and Karl Jagersberger (all Gaming), for sampling and digging permission. The work was done within the framework of and financially supported by projects of the Federal Government of Lower Austria (K3-F-964/001-2020 and K3-F-964/002-2023, Department Science and Research; headquarters in St. Pölten); the Austrian Academy of Sciences (headquarters in Vienna, project Polzberg Lukeneder 2021) represented by the National Committee for Geo/Hydro Sciences (Earth System Sciences Programme); and the Triassic project "Fenster in die Karnische Krise" of the Society of Friends of the Natural History Museum Vienna (Freunde des Naturhistorischen Museums Wien, CPE Lukeneder 2023). Wencke Wegner (SEM, microprobe) and Anton Englert (thin sections) (both Vienna) are acknowledged for technical support. The manuscript greatly benefited from valuable comments by the reviewers Lukas Plan (Vienna) and Oleg Mandic (Vienna), the production editor Ashok Arumairaj (Taramani), the in-house series editor Kathleen

A. Ritterbush (Utah), the publishing editor Simon Shah-Rohlfs and the production editor Antje Endemann (both Heidelberg).

Author Contribution AL initial idea for the present study. PL provided documentation. PL provided first draft of the paper. PL, AL contributed to writing the paper.

Funding Open access funding was enabled and the projects financially supported by the Federal Government of Lower Austria (K3-F-964/001–2020 and K3-F-964/002–2023, Department Science and Research; headquarters in St. Pölten); the Austrian Academy of Sciences (headquarters in Vienna, project Polzberg Lukeneder 2021) represented by the National Committee for Geo/Hydro Sciences (Earth System Sciences Programme); the Natural History Museum Vienna; and the Triassic project "Fenster in die Karnische Krise" of the Society of Friends of the Natural History Museum Vienna (Freunde des Naturhistorischen Museums Wien, CPE Lukeneder 2023) (to AL). The authors are responsible for the contents of this publication. The funders had no impact on conceptualization, design, data collection, analysis, decision to publish, or preparation of the manuscript.

Competing interests All authors stated that they have no competing interests.

Authors' contributions All authors stated that they have equally contributed to their chapters, read, edited and agreed to submit the final versions of their contributions.

Availability of data and materials All sources of information are provided in the text. Raw data related to the fossil material from Austria are available from the corresponding authors and Alexander Lukeneder upon request.

Additional information This article is part of the Special Volume: *Late Triassic Konservat-Lagerstätten within the Carnian Pluvial Episode. 150 Years of Scientific Excavations in Austria.*

Artificial Intelligence (AI) The chapter was produced without the use of AI.

Language The entire volume uses American English reviewed by a native scientist.

Received 2025.

Published online 2026.

References

Abel O (1906) Fossile Flugfische. Abh K K Geol Reichsanst 56:1–88

Alverà G, Dal Corso J, Chu D, Cruciani G, Roghi G, Caggiati M, Song H, Song H, Tian L, Du Y, Haserbek T, Gianolla P (2026) Episodic aridification at the onset of the Carnian Pluvial Episode in Western Tethys. Glob Planet Change 256:105192. https://doi.org/10.1016/j.gloplacha.2025.105192

Austrian Map (2023) Austrian Map Online. Bundesamt für Eich- und Vermessungswesen. https://www.bev.gv.at/

Bauer FK, Schnabel W (1997) Geologische Karte der Republik Österreich. Sheet Mariazell 72, 1:50.000. Geologische Bundesanstalt. https://doi.org/10.24341/tethys.37

Bittner A (1882) Die Geologischen Verhältnisse von Hernstein in Niederösterreich und der weiteren Umgebung. Wien. 1–309

Brandner R, Poleschinski W (1986) Stratigraphie und Tektonik am Kalkalpensüdrand zwischen Zirl und Seefeld in Tirol. Jber Mitt Oberrhein Geol Ver 68:67–92

Dal Corso J, Bernardi M, Sun Y, Song H, Seyfullah LJ, Preto N, Gianolla P, Ruffell A, Kustatscher E, Roghi G, Merico A, Hohn S, Schmidt AR, Marzoli A, Newton RJ, Wignall PB, Benton MJ (2020) Extinction and dawn of the modern world in the Carnian (Late Triassic). Sci Adv 6(18):eaba0099. https://doi.org/10.1126/sciadv.aba0099

Enos P, Jiayong W, Lehrmann DJ (1998) Death in Guizhou—Late Triassic drowning of the Yangtze carbonate platform: Sediment Geol 118:55–76

Forchielli A, Pervesler P (2013) Phosphatic cuticle in thylacocephalans: a taphonomic case study of (Arthropoda, Thylacocephala) from the Fossil-Lagerstätte Polzberg (Reingraben shales, Carnian, Upper Triassic, Lower Austria). Austr J Earth Sci 106(1):46–61

Geyer G (2002) Geologie von Unterfranken und angrenzenden Regionen: Klett-Perthes, 588 p

Glaessner MF (1931) Eine Crustaceenfauna aus den Lunzer Schichten Niederösterreichs. Jb KK Geol BA 81:467–486

Hart MB, Page KN, price GD, Smart CW (2019) Reconstructing the Christian Malford ecosystem in the Oxford Clay Formation (Callovian, Jurassic) of Wiltshire: exceptional preservation, taphonomy, burial and compaction. J Micropalaeontol 38:133–142. https://doi.org/10.5194/jm-38-133-2019

Hopkin EK, McRoberts CA (2005) A new Middle Triassic flat clam (Pterioida: Halobiidae) from the middle Anisian of North-Central Nevada, USA. J Paleontol 79(4):796–800. https://doi.org/10.1666/0022-3360(2005)079[0796:ANMTFC]2.0.CO;2

Hornung T, Brandner R (2005) Biochronostratigraphy of the Reingraben Turnover (Hallstatt Facies Belt): local black shale events controlled by regional tectonics, climatic change and plate tectonics. Facies 51:460–479

Hornung T, Brandner R, Krystin L, Joachimski MM, Keim L (2007a) Multistratigraphic constrains in the NW Tethyan "Carnian Crisis." Bull New Mexico Mus Nat Hist Sci 41:59–67

Hornung T, Krystin L, Brandner R (2007b) A Tethys-wide mid-Carnian (Upper Triassic) carbonate productivity crisis: Evidence for the Alpine Reingraben Event from Spiti (Indian Himalaya)? J Asian Earth Sci 30(2):285–302. https://doi.org/10.1016/j.jseaes.2006.10.001

Hornung T, Kogan I, Wolf G, Moosleitner G (2019) The Norian fish deposits of Wiestal ("Seefeld Member", Northern Calcareous Alps, Salzburg, Austria): Taxonomy and palaeoenvironmental implications. Austrian J Earth Sci 112:125–165. https://doi.org/10.17738/ajes.2019.000

Kohút M, Hofmann M, Havrila M, Linnemann U, Havrila J (2018) Tracking an upper limit of the "Carnian Crisis" and/or Carnian stage in the Western Carpathians (Slovakia). Int J Earth Sci 107:321–335. https://doi.org/10.1007/s00531-017-1491-8

Kolar-Jurkovšek T, Jurkovšek B (2008) New paleontological evidence of the Carnian strata in the Mežica area (Karavanke Mts, Slovenia): Conodont data for the Carnian Pluvial Event. Palaeogeogr Palaeoclimatol Palaeoecol 290:81–88. https://doi.org/10.1016/j.palaeo.2009.06.015

Kozur HW, Bachmann GH (2010) The Middle Carnian Wet Intermezzo of the Stuttgart Formation (Schilfsandstein). Germanic Basin. Palaeogeogr Palaeoclimatol Palaeoecol 290(1):107–119. https://doi.org/10.1016/j.palaeo.2009.11.004

Krystyn L (1978) Eine neue Zonengliederung im alpin-mediterranen Unterkarn. Schriftenr Erdwiss Komm Österr Akad Wiss 4:37–75

Krystyn L (1991) Die Fossillagerstatten der alpinen Trias. S. 23–78. In: Nagel D, Rabeder G (eds): Exkursionen im Jungpalaozoikum und Mesozoikum Österreichs. 1–119

Krystyn L (1998) Das Polzberger Fischlager. Österreichische Paläontologische Gesellschaft, 5. Jahrestagung (2.-4.Okt. 1998, Lunz am See), Field Trip Guide:12–14

Lehrmann DJ, Enos P, Payne JL, Montgomery P, Wei J, Yu Y, Xiao J, Orchard M (2005) Permian and Triassic depositional history of the Yangtze platform and Great Bank of Guizhou in the Nanpanjiang Basin of Guizhou and Guangxi, south China. Albertiana 33:149–168

Lukeneder A (ed) (2026) Late Triassic Konservat-Lagerstätten within the Carnian Pluvial Episode: 150 Years of Scientific Excavations in Austria. Top Geobiol 55:600 p. https://doi.org/10.1007/978-3-032-19777-1

Lukeneder A, Lukeneder P (2021) The Upper Triassic Polzberg palaeobiota from a marine *Konservat-Lagerstätte* deposited during the Carnian Pluvial Episode in Austria. Sci Rep 11:16644. https://doi.org/10.1038/s41598-021-96052-w

Lukeneder A, Lukeneder P (2022a) Taphonomic history and trophic interactions of an ammonoid fauna from the Upper Triassic Polzberg palaeobiota. Sci Rep 12:7455. https://doi.org/10.1038/s41598-022-11496-y

Lukeneder A, Lukeneder P (2023) New data on the marine Upper Triassic palaeobiota from the Polzberg *Konservat-Lagerstätte* in Austria. Swiss J Palaeontol 142:9. https://doi.org/10.1186/s13358-023-00269-3

Lukeneder A, Lukeneder P, Sachsenhofer RF, Roghi G, Rigo M (2024a) Multi-proxy record of the Austrian Upper Triassic Polzberg *Konservat-Lagerstätte* in light of the Carnian Pluvial Episode. Sci Rep 14:11194. https://doi.org/10.1038/s41598-024-60591-9

Lukeneder P, Fuchs D, Lukeneder A (2024b) Morphology, taxonomy and trophic interactions of rostrum-less coleoids from the Late Triassic Polzberg *Konservat-Lagerstätte* (Lower Austria). Swiss J Palaeontol 143:21. https://doi.org/10.1186/s13358-024-00319-4

Lukeneder P, Lukeneder A (2022b) Mineralized belemnoid cephalic cartilage from the late Triassic Polzberg *Konservat-Lagerstätte* (Austria). PLoS ONE 17(4):e0264595. https://doi.org/10.1371/journal.pone.0264595

Lukeneder P, Lukeneder A (2022c) Comment on "Triassic coleoid beaks and other structures from the Calcareous Alps revisited" by Doguzhaeva et al. (2022). Acta Palaeontol Pol 67(4):963–965.

Mazaheri-Johari M, Gianolla P, Mather TA, Frieling J, Chu D, Dal Corso J (2021) Mercury deposition in Western Tethys during the Carnian Pluvial Episode (Late Triassic). Sci Rep 11:17339. https://doi.org/10.1038/s41598-021-96890-8

McRoberts CA (2001) Paleobiology of the Triassic "Flat Clam" Halobia in oxygen-deficient marine facies. Geol Soc Amer 33(2):A-8

Miller CS, Peterse F, da Silva AC, Baranyi V, Reichart GJ, Kürschner WM (2017) Astronomical age constraints and extinction mechanisms of the Late Triassic Carnian crisis. Sci Rep:2557. https://doi.org/10.1038/s41598-017-02817-7

Mostler H, Scheuring BW (1974) Mikrofloren aus dem Langobard und Cordevol der Nördlichen Kalkalpen und das Problem des Beginns der Keupersedimentation im Germanischen Raum. Geol Paläont Mitt Innsbruck 4(4):1–35

Mojsisovics Ev (1882) *(1882):* Die Cephalopoden der mediterranen Trias. Abh Geol R-A 10:1–320

Mueller S, Krystyn L, Kürschner WM (2016) Climate variability during the Carnian Pluvial Phase—A quantitative palynological study of the Carnian sedimentary succession at Lunz am See, Northern Calcareous Alps, Austria. Palaeogeogr Palaeoclimatol Palaeoecol 441:198–211. https://doi.org/10.1016/j.palaeo.2015.06.008

Ogg JG (2015) The mysterious Mid-Carnian "Wet Intermezzo" global event. J Earth Sci 26(2):181–191. https://doi.org/10.1007/s12583-015-0527-x

Oschmann W (1994a) Adaptive pathways of benthic organisms in marine oxygen-controlled environments. N Jb Geol Paläontol Abh 191:393–444

Oschmann W (1994b) Der Kimmeridge Clay von Yorkshire als ein Beispiel eines fossilen Sauerstoff-kontrollierten Milieus. Beringeria 9:1–154

Placer L, Vrabec M, Trajanova M (2002) Kratek pregled geologije okolice Mežice. In: Horvat A, Košir A, Vreča P, Brenčič M (Eds), Vodnik po ekskurzijah, 1. Slovenski geološki kongres, Črna na Koroškem, 3–14

Preto N, Gianolla P, Roghi G (2008) The Carnian Pluvial Event: causes and effects. The Triassic climate, 44 p. Abstract book, Bolzano, June 3–7, 2008

Preto N, Bernadi M, Dal Corso J, Gianolla P, Kustatscher E, Roghi G, Rigo M (2019) The Carnian Pluvial Episode in Italy: History of the research and perspectives. BSPI 58(1):35–49

Rigo M, Preto N, Roghi G, Tateo F, Mietto P (2007) A rise in the carbonate Compensation Depth of western Tethys in the Carnian (Late Triassic): Deep-water evidence for the Carnian Pluvial Event: Palaeogeogr Palaeoclimatol Palaeo 246(2–4):31–182. https://doi.org/10.1016/j.palaeo.2006.09.013

Röhl HJ, Schmid-Röhl A, Oschmann W, Frimmel A, Schwark L (2001) The Posidonia Shale (Lower Toarcian) of SW-Germany: an oxygen-depleted ecosystem controlled by sea level and palaeoclimate. Palaeogeogr Palaeoclimatol Palaeoecol 165:27–52

Ruffell A, Simms MJ, Wignall PB (2016) The Carnian Humid Episode of the late Triassic: a review. Geol Mag 153(2):271–284

Ruttner A, Schnabel W (1988) Geologische Karte der Republik Österreich. Sheet Ybbsitz 71, 1:50.000. Geologische Bundesanstalt Wien

Schlager W, Schöllnberger W (1974) Das Prinzip stratigraphischer Wenden in der Schichtfolge der Nördlichen Kalkalpen. Austr J Earth Sci 66–67:165–193

Simms MJ, Ruffell AH (1989) Synchroneity of climatic change and extinctions in the Late Triassic. Geol 17(3):265–268. https://doi.org/10.1130/0091-7613(1989)017%3c0265:SOCCAE%3e2.3.CO;2

Simms MJ, Ruffell AH (1990) Climate and biotic change in the late Triassic. J Geol Soc 147(2):321–327. https://doi.org/10.1144/gsjgs.147.2.0321

Simms MJ, Ruffell AH (2018) The Carnian Pluvial Episode: From discovery, through obscurity, to acceptance. J Geol Soc 175(6):jgs2018–020. https://doi.org/10.1144/jgs2018-020

Shi Z, Preto N, Jiang H, Krystyn L, Zhang Y, Ogg JG, Jin X, Yuan J, Yang X, Du Y (2017) Demise of Late Triassic sponge mounds along the northwestern margin of the Yangtze Block, South China: Related to the Carnian Pluvial Phase? Palaeogeogr Palaeoclimatol Palaeoecol 474:247–263. https://doi.org/10.1016/j.palaeo.2016.10.031

Sun YD, Wignall PB, Joachimski MM, Bond DPG, Grasby SE, Lai XL, Wang LN, Zhang ZT, Sun S (2016) Climate warming, euxinia and carbon isotope perturbations during the Carnian (Triassic) Crisis in South China. Earth Planet Sci Lett 444:88–100. https://doi.org/10.1016/j.epsl.2016.03.037

Stur D (1874) Neue Aufschlüsse im Lunzer Sandsteine bei Lunz und ein neuer Fundort von Wengerschiefer im Pölzberg zwischen Lunzersee und Gaming. Verh K K Geol Reichsanst 1:271–273

Teller F (1891) Über den Schädel eines fossilen Dipnoers, Ceratodus Sturii nov. spec. aus den Schichten der oberen Trias der Nordalpen. Abh KK Geol Reichsanst 15:1–38

Tollmann A (1976) Analyse des Klassischen Nordalpinen Mesozoikums. Stratigraphie, Fauna und Fazies der Nördlichen Kalkalpen. Franz Deuticke Wien, 580 p

Chapter 2
Discovery and Legacy of the Polzberg *Konservat-Lagerstätte* (Carnian, Triassic)

Petra Lukeneder and Alexander Lukeneder

Abstract Since the 1870s the Polzberg *Konservat-Lagerstätte* (located between Gaming and Lunz am See, Lower Austria) has been known as a rich paleontological treasure trove. In 1874 the later director of the k. k. Geologische Reichsanstalt (today part of the GeoSphere Austria), Dionýs Štúr (1827–1893) himself noted this site as “very interesting”. Since then, the Late Triassic *Konservat-Lagerstätte* Polzberg has provided several thousand specimens of outstanding fossils, including different genera of fish, cephalopods (ammonites, coleoids), gastropods, but also several species of rarely physically preserved bristle worms. The first historical adit used exclusively for fossil mining at Polzberg (GPS: 47°53′23.31″ N 15°04′45.80″ E) was built and maintained by mine manager Joseph Haberfelner (1830–1913) for the k. k. Geologische Reichsanstalt (1885). Haberfelner also recovered the first *Belonorhynchus striolatus* Bronn at Polzberg, a fish that had previously been known only from the Southern Alps. Although the main goal of that adit was to collect fossil cephalopods, the important find of a fossil lungfish *Tellerodus sturii* (Teller, 1891) was made by Stur himself in 1885. 1909 marked the construction of another adit, also built by Haberfelner, for the Natural History Museum Vienna. Today, the museum has by far the most extensive collection of fossils from the Reingraben Formation of Polzberg. Based on historical maps, inventory books and diary entries from this period, the history of the Polzberg *Konservat-Lagerstätte* with its preserved flora and fauna, which is unique in Austria, is now being told.

Keywords *Konservat-Lagerstätte* · Polzberg · Discovery · Legacy · Triassic · Austria

P. Lukeneder (✉) · A. Lukeneder
Department of Geology and Paleontology, Natural History Museum Vienna, Vienna, Austria
e-mail: petra.lukeneder@nhm.at

A. Lukeneder
e-mail: alexander.lukeneder@nhm.at

P. Lukeneder
Institute of Applied Geology, Department of Landscape, Water and Infrastructure, BOKU University, Vienna, Austria

A. Lukeneder (ed.), *Late Triassic Konservat-Lagerstätten within the Carnian Pluvial Episode*, Topics in Geobiology 55, https://doi.org/10.1007/978-3-032-19777-1_2

2.1 Introduction

The first mention of the *Konservat-Lagerstätte* Polzberg (Fig. 2.1; GPS: 47°53′5.90″N, 15°4′27.70″E; Lower Austria. ÖK 1:50.000, sheet 71 Ybbsitz, Ruttner and Schnabel 1988; sheet 72 Mariazell, Bauer and Schnabel 1997; Austromap Online 2023; 712 m above sea level) can be found in Dionýs Štúrs' (also: Dionys Stur; Fig. 2.2a) report from 1874, in which new outcrops of the Lunz sandstones near Lunz am See and a new locality for "Wengen shales" at the "Pölzberg" north of Lake Lunzer See on the Seehof-Gaming line, in the so-called "Polzbergrotte", were described. This locality is known by several historical names, among them: Unter Polzberg (Stur 1874), Pölzberg (Stur 1874; 1886; 1888), Pölzlberggraben (notation on collection label of specimen at the Department of Palaeontology, University of Vienna, IPUW), Polzberg-Graben (Teller 1891; Krystyn 1991), Polzberggraben (Krystyn 1998), Schindelberggraben (Steindachner 1910, 1911; Glaessner 1931) (= Mount Schindlberg, 1066 m) and Polzberg (Krystyn 1978; Forchielli and Pervesler 2013) (see also Lukeneder and Lukeneder 2021).

This locality was originally discovered by mine manager Josef Haberfelner (also: Haberfellner) (Fig. 2.2b; 1830–1913). Over a period of 30 years Haberfelner managed the mines around Lunz am See which was an area of extensive coal mining. Additionally Haberfelner personally carried out intensive collecting activity. Over the years he collected fossils as well as minerals and entomological specimens such as beetles and butterflies. Haberfelner was a well-known personality from Lunz am See who was also involved in public education. An article in the Erlauftal-Bote from 12 January (1902) reports on a lecture on the geological history in Lunz, for which Haberfelner had provided a wealth of visual material (typical rocks and fossils) for demonstration purposes. Haberfelner also put together entire collections (rocks, fossils, beetles and butterflies) as needed, which could be purchased ("Transcript of price list of the collections of minerals, rocks, plant fossils and insects compiled by Josef Haberfelner, Marktgemeinde Lunz am See", see also Fig. 2.3). The Erlauftal-Bote (26 June 1904) reports on a meeting of the "Ybbstal teacher's club" at which participants were able to view the rich geological and mineralogical private collection of mining inspector Josef Haberfelner in Lunz am See. The article noted that the objects collected over 40 years were stacked up to the roof in the cellars. This underlines his tireless zeal for collecting and research.

On the occasion of Haberfelner's 80th birthday, paleontologist Friedrich Blaschke (1883–1911) published an article about Haberfelner (Blaschke 1910) and the artist Ernst Juch prepared a plaquette showing Haberfelner's portrait (Fig. 2.4a; see also figure in Illustriertes Wiener Extrablatt, 20 August 1910, p. 8). At his first sampling at Polzberg, Stur (1874) identified "*Ammonites Aon*" Münster (= *Trachyceras aon* Münster), *Estheria* sp., juvenile bivalves of halobiids (= posidoniid *Peribositra*) as fossils occurring within the so-called "Wengerschiefer" (old expression for Ladinian Wengen Shales). Traces of the coleoid cephalopod *Acanthotheutis bisinuata* Bronn (firstly described from Cave del Predil in northern Italy) were also present. Haberfelner, on the other hand, found an almost complete specimen of *Belonorhynchus*

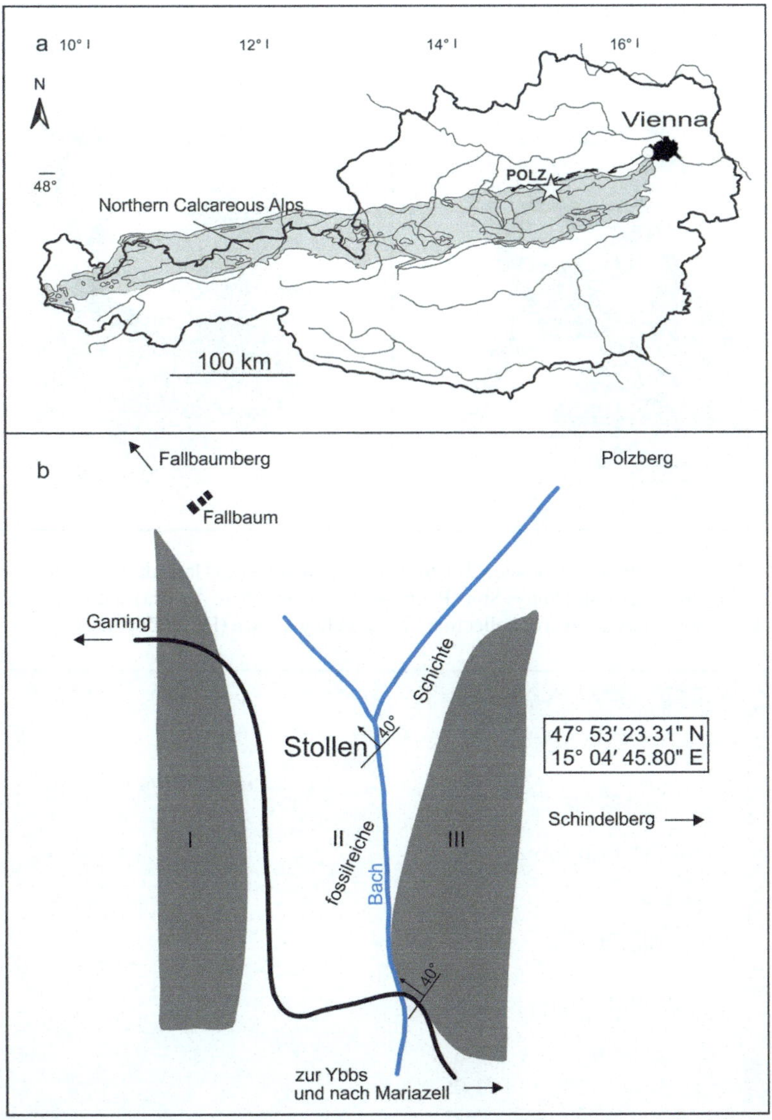

Fig. 2.1 **a** Austrian map with star POLZ marking the position of the Polzberg *Konservat-Lagerstätte* (= Schindelberggraben ravine, Polzberg locality). **b** Historical detail map (nominal scale approx. 1:12.500) of the adit of the GeoSphere Austria at Polzberg. The sketch is based on Friedrich Blaschkes notes as adapted and published in Glaessner (1931)

Fig. 2.2 Two collaborating scientists. **a** The Slovak paleontologist and later director of the former k. k. Geologische Reichsanstalt Dionys Stur (Photo Archive GeoSphere Austria) and **b** mine manager Josef Haberfelner whose extensive collections were widely known (Photo Zobodat)

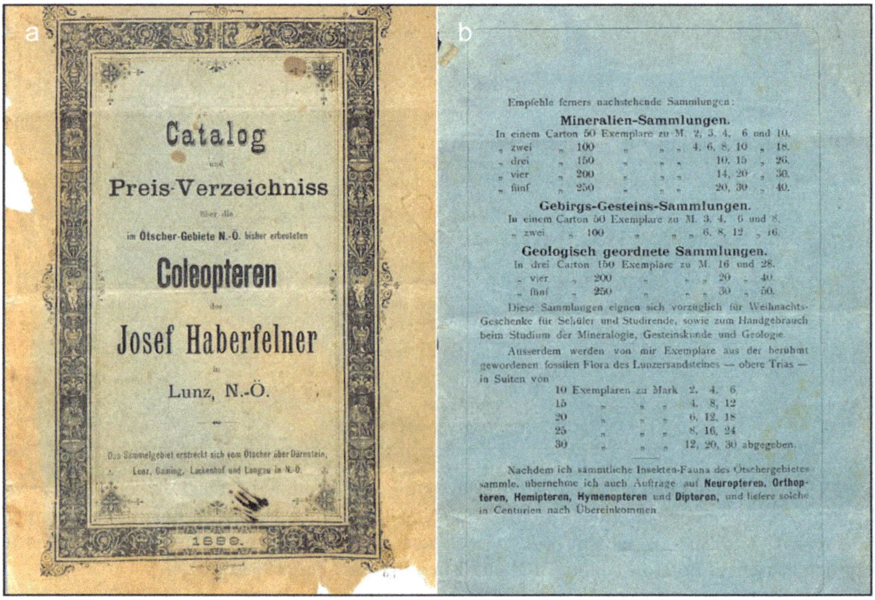

Fig. 2.3 **a** Haberfelner's catalogue of coelopterids from the Ötscher Mountain region in Lower Austria. **b** On the last page Haberfelner also offers collections of minerals and rocks for purchase

striolatus Bronn, previously known only from the Southern Alps. Stur furthermore identified remains of the plant *Voltzia Haueri* Stur. Dionys Stur (1874) already noted the presence of a collapsed tunnel at Polzberg, obviously dug by an "amateur in the hope to find coal". The heap of this tunnel was mentioned as being heavily covered with rocks of the "Wengen shales". The dilapidated tunnel was still visible at this point. In this tunnel, Haberfelner was able to find a fossil whose tail had a continuously radiating fin at the top and bottom, as is characteristic of the genus *Ceratodus* (Stur 1886). Teller (1891) later determined that this tail belonged to the coelacanthid *Coelacanthus lunzensis* Teller (see also Reis 1900). Haberfelner and Haberfelner (1902) again classified the fossil-bearing layers of the Polzberg *Konservat-Lagerstätte* as so-called "Wengen layers". They were described as slate clay, marl shale and thinly layered, mostly dark limestone, which is characterized by many and interesting "petrefacts" such as *Voltzia*, ammonites, fish, crabs and *Ceratodus* (today assigned to *Tellerodus*; see also Martin 1982). Haberfelner and Haberfelner (1902) mention a ringing sound for the black, siliceous limestone when struck.

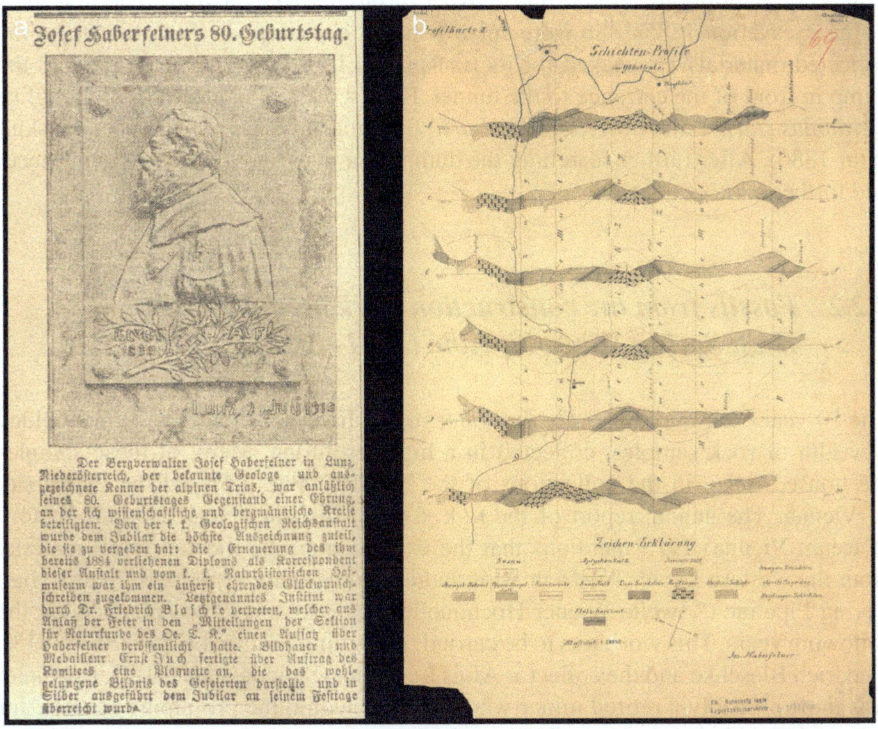

Fig. 2.4 **a** On the occasion of Haberfelner's 80th birthday the artist Ernst Juch prepared a plaquette showing a portrait of Haberfelner (Österreichische Nationalbibliothek, https://anno.onb.ac.at/). **b** Sections within Ybbstal region, drawn by Joseph Haberfelner. Wengen shales are indicated as the top layer of the Reifling limestone. Scale 1:25,530. *Credits* FA Rohstoffgeologie Lagerstättenarchivnummer L-1272/2 K

2.2 Historical Excavations at Polzberg

2.2.1 Excavation of the k. k. Geologische Reichsanstalt (1885)

During summer of 1885 Haberfelner was commissioned by the k. k. Geologische Reichsanstalt (today the Geological Survey is part of the GeoSphere Austria) to create a small tunnel for the occasional collection of fossils (mainly cephalopods) at the Polzberg Mountain. The exposed rocks were already classified as Reingraben shales (Stur 1886). Figure 2.4b shows a historical geological section through the Ybbsthal, drawn by Josef Haberfelner. Unfortunately, no maps or profile drawings from this Polzberg excavation have survived. Surprisingly, also no records are preserved in the field books of the involved geologists and paleontologists. Nevertheless, some information can be found in the historical literature.

Dionys Stur and Johann Kudernatsch mapped the region around Lunz am See (Fig. 2.5) including the area of the Polzberg *Konservat-Lagerstätte*. The first finds of the excavation 1885/1886 were small fish and cephalopods (Stur 1886). As the collected material was packed for its transport to Lunz am See, Stur searched the dump in front of the entrance to the tunnel. He discovered two teeth of the lungfish *Ceratodus sturii* Teller (= *Tellerodus sturii*) from the upper jaw and parts of the skull (Stur 1886). After further searching the dump, Stur also found the mandibular teeth and further parts of the skull.

2.2.2 Fossils from the construction work on the 2nd Vienna High Spring Water Pipeline (1901–1910)

The 10-year excavation work for the water supply line for the city of Vienna yielded a wealth of rock samples, collected in a high resolution. Many of these samples are housed today in the collections of the NHMW (Fig. 2.6) and the University of Vienna. The annual report of the k. k. Court Museum (today: Natural History Museum Vienna) 1913 mentions that the examination of collected fossil material from the western Lower Austrian Alps and along the route of the 2nd Vienna High Spring Pipeline (“Zweite Wiener Hochquellenwasserleitung”) was planned for the following years. This work was to be carried out by director Ernst Kittl (1854–1913), Friedrich Blaschke and the collector Alois Legthaler (Steindachner 1914). Legthaler was most probably a retired miner who collected within the pro-alpine Jurassic for Friedrich Trauth and the Natural History Museum Vienna before and during the First World War (see also Zapfe 1971; Trauth 1919). Legthaler was also mentioned as the “collector” from whom the NHMW bought 670 fossil specimens in 1900 (Harzhauser and Krenn 2023). Trauth (1948) summarizes in detail the geology and fossil content

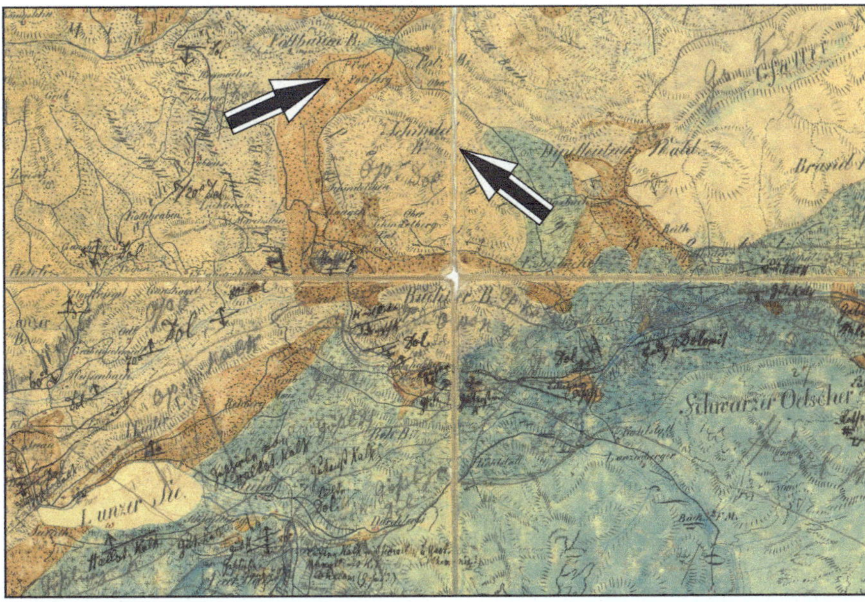

Fig. 2.5 Digitized original map of the Lunz region, by Dionys Stur and Johann Kudernatsch. ID M28150027, GEOKART – geologische Kartenwerke NÖ, GeoSphere Austria. Polzberg and Schindelberg indicated by white arrows

(e.g., Taf. XII, Fig. 2.4; *Trachyceras triadicum* = *Austrotrachyceras minor*; Krystyn 1991; Mojsisovics 1893a, 1893b) along the water pipeline.

2.2.3 Excavation by the Natural History Museum Vienna (1909)

During the excavation campaign by the Natural History Museum Vienna around 1909, Glaessner (1931) partially reproduced a written protocol prepared by Friedrich Blaschke. He stated that the new adit of the NHMW was built two meters next to the old adit of the k.k. Geologische Reichsanstalt (Glaessner 1931). Glaessner (1931) also included a detailed map made from a sketch by Blaschke (see also Fig. 2.1b). After eight meters of tunnel length, digging the tunnel was stopped and the fossil-rich layers were drilled. Josef Haberfelner was present during the entire construction work; Blaschke participated for two weeks and Friedrich Trauth (1883–1967) participated for one excavation week (Glaessner 1931). Blaschke was intensely involved in processing the Polzberg material and he meticulously documented the excavation work. However, his life and career ended very early. The newspaper "Der Schnee" reported on 1 April 1911 about an accident on the Bruderkogel in the Bösenstein group, where the club members Alfred Kernthaler, Friedrich Blaschke and Alexander

Fig. 2.6 The construction of Vienna's second spring water pipeline ("Hochquellenwasserleitung"; also: II. Kaiser-Franz-Josef-Hochquellenleitung) provided fossils from high-resoluted sampling. **a** Specimen of *Phragmoteuthis bisinuata* Bronn, collected by Josef Haberfelner. **b** Historical collection label from the NHMW referring to the construction of the second Vienna aqueduct, transcription and translation: *P. bisinuta* Bronn and tiny juvenile shells of *Halobia rugosa* Guembel. Age: Reingraben shales, Lower Carnian stage of the Upper Triassic. Locality: Lunz-Grubberg tunnel of the second Vienna aqueduct, approximately 540–560 m NNE of the tunnel entrance near Lunz, Lower Austria. Purchased from mine manager J. Haberfelner in Lunz. **c** Second, historical collection label (most probably older label), transcription and translation: The fossiliferous Reingraben shales (Coll. Haberfelner 1905 No. 57 with *Halobia rugosa* Gümbel, *Carnites floridus* Wulfen), approximately from the Sect. 540–560 m north from Lunz in the Lunzberg tunnel, at 540 fault, then Opponitz limestone, from 560 genuine Lunz layers; the remaining text is difficult to decipher. All scale bars 1 cm

Korani died in an avalanche (Unknown author 1911). Blaschke had been admitted as a volunteer at the Department of Geology and Paleontology since 1906. From 1910 on he held the status of assistant (Steindachner 1912; see also Zapfe 1971). Unfortunately the original documentation and sketches of the Polzberg excavation are not preserved. The historical, collapsed mines can still be guessed geomorphologically at GPS 47°53′23.31″ N, 15°4′45.80″ E (Lukeneder and Lukeneder 2021). The historical inventory books of the NHMW bear witness to the excavation and collection work (Fig. 2.7).

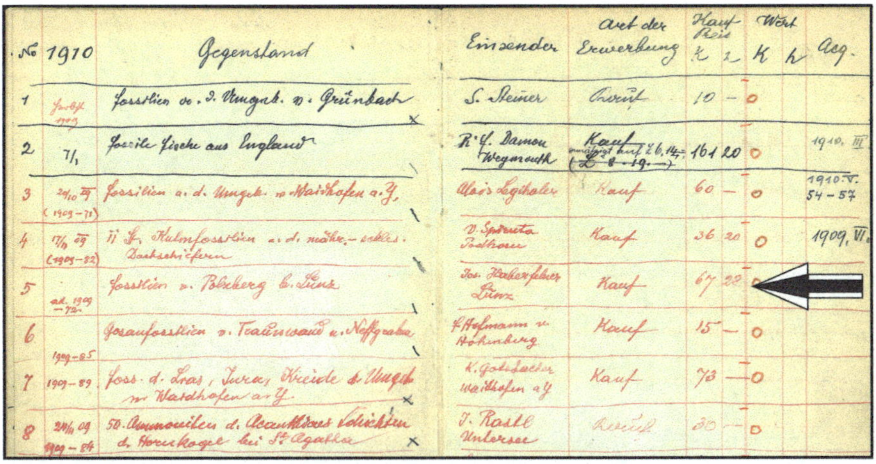

Fig. 2.7 Registration of fossils in an historical NHMW inventory book (1910) mentioning the acquisition of Polzberg fossils (indicated by white arrow). **a** Page that documents the purchase of Polzberg fossils from mine manager Josef Haberfelner. The price of the fossils was 67 Kronen and 28 Heller corresponding to a value of about 530 € (status: 29 May 2025)

2.3 Recent Paleontological Research at Polzberg

After a long period during which only individual, well-preserved Polzberg specimens were examined, the site and its fauna once again became the subject of increased investigations. Since 2005 the collectors Birgitt and Karl Aschauer have sampled the deposits in the area around the historical mine tunnels (Lukeneder and Lukeneder 2021). Examinations of the coleoids from Polzberg by Doguzhaeva et al. (2006; 2007ab; 2012) focused on selected well-preserved historical specimens, whereas Lukeneder et al. (2024c) undertook a holistic approach for describing the entire coleoid fauna. Forchielli and Pervesler (2013) were the first to re-examine the site with a focus on the arthropod fauna. The description of the new isopod species *Obtusotelson summesbergeri* Schädel et al. (2020) followed. More recently, the Polzberg fauna was extensively studied by Lukeneder and Lukeneder (2021; 2022a, b, c; Lukeneder et al. 2020; Lukeneder et al. 2023; Lukeneder et al. 2024a, b; Lukeneder et al. 2024c). As part of the polzberg project ("Polzberg – eine *Konservat-Lagerstätte* von Weltruf"), the new main section was recorded layer by layer and the fossil content was analyzed. During the field work (figs. 2.8 and 2.9) more than 5.300 fossils of marine taxa were collected and evaluated. These new data were enriched by examining the extensive historical material from austrian collections, including those of the Natural History Museum Vienna (NHMW) and the GeoSphere Austria (GBA). Table 2.1 collates the historical and recent excavation activities at Polzberg.

Fig. 2.8 Field work during the new excavation campaign at Polzberg (2021). **a** The authors Alexander Lukeneder and **b** Petra Lukeneder working at the new main section. The detailed record of the section and the bed-by-bed evaluation of the fossil content provided an accurate picture of the Late Triassic marine ecosystem

2.4 Conclusions

This article recalls the discovery and legacy of the exceptional Polzberg *Konservat-Lagerstätte* and the beginning of scientific excavations and fossil mining at Polzberg. The initial early scientific excavations were only possible due to the intensive and passionate work and support by the mine manager Josef Haberfelner from Lunz am See. Early explorers of the area, such as Dionys Stur and Friedrich Blaschke, recognized the potential of the deposits and their reports constitute the foundation for systematic excavations at Polzberg. This provided a solid foundation for the investigations of following scientists and for more recent research.

Fig. 2.9 Boxes with new material, excavated during the campaign 2021, ready for transport to the Natural History Museum Vienna. During the Polzberg project, about one ton of material was recovered bed-by-bed, examined and processed

Table 2.1 List of excavations and field work carried out at the Polzberg *Konservat-Lagerstätte*, starting with first sampling by Dionys Stur and Josef Haberfelner. Further excavations exclusively for fossil mining were conducted by the then k. k. Geologische Reichsanstalt and a few years later by the Natural History Museum Vienna. This was followed mainly by activities of fossil collectors and occasional sampling by the Department of Paleontology (University of Vienna). Since 2021 the sections have been systematically recorded and documented for the first time in the framework of a project of the NHMW

Year	Excavation
1874	Sampling by Dionys Stur and Josef Haberfelner
1885	k. k. Geologische Reichsanstalt (today: part of the GeoSphere Austria)
1901–1910	construction on the second Vienna high spring water pipeline
1909	Natural History Museum Vienna
around 2013	Collections of private fossil collectors; sampling of the Department of Paleontology (University of Vienna)
2021–2024	Polzberg project Lukeneder A and Lukeneder P (Natural History Museum Vienna)

Acknowledgments We are very grateful to the owner of the Polzberg section, Franziska and Hermann Hofreiter and Eva and Karl Jagersberger (all Gaming), for the digging permission during the whole project. We also want to thank Birgitt and Karl Aschauer (Waidhofen an der Ybbs) for providing many fossil specimens for our studies. Martin Maslo (Vienna) is thanked for access to the collections of the Department of Palaeontology (University of Vienna) and Davide Di Franco for access to the collections of the GeoSphere Austria (Vienna). Thomas Hofmann (Archive of the GeoSphere Austria) for support and permissions to use material of the GeoSphere. Bernhard Atzenhofer (Department of Mineral Resources of the GeoSphere Austria, Vienna) for providing profile sketches by Josef Haberfelner. Johann Sieberer (Lunz) provided archive material on Joseph Haberfelner. The work was done within the framework of and financially supported by projects of the Federal Government of Lower Austria (K3-F-964/001-2020 and K3-F-964/002-2023, Department Science and Research; headquarters in St. Pölten), the Austrian Academy of Sciences (headquarters in Vienna, project Polzberg Lukeneder 2021) represented by the National Committee for Geo/Hydro Sciences (Earth System Sciences Programme), and the Triassic project "Fenster in die Karnische Krise" of the Society of Friends of the Natural History Museum Vienna (Freunde des Naturhistorischen Museums Wien, CPE Lukeneder 2023). The manuscript greatly benefited from valuable comments by the reviewers Matija Križnar (Ljubljana) and Franz Ottner (Vienna), the production editor Ashok Arumairaj (Taramani), the in-house series editor Kathleen A. Ritterbush (Utah), the publishing editor Simon Shah-Rohlfs and the production editor Antje Endemann (both Heidelberg).

Author Contribution AL initial idea for the present study. PL provided documentation and first draft of the paper. PL, AL interpretation and writing the paper.

Funding Open access funding was enabled and the projects financially supported by the Federal Government of Lower Austria (K3-F-964/001-2020 and K3-F-964/002-2023, Department Science and Research; headquarters in St. Pölten); the Austrian Academy of Sciences (headquarters in Vienna, project Polzberg Lukeneder 2021) represented by the National Committee for Geo/Hydro Sciences (Earth System Sciences Programme); the Natural History Museum Vienna; and the Triassic project "Fenster in die Karnische Krise" of the Society of Friends of the Natural History Museum Vienna (Freunde des Naturhistorischen Museums Wien, CPE Lukeneder 2023) (to AL). The authors are responsible for the contents of this publication. The funders had no impact on concept.

Competing interests All authors stated that they have no competing interests.

Authors' contributions All authors stated that they have equally contributed to their chapters, read, edited and agreed to submit the final versions of their contributions.

Availability of data and materials All sources of information are provided in the text. Raw data related to the fossil material from Austria are available from the corresponding authors and Alexander Lukeneder upon request.

Additional information This article is part of the Special Volume: *Late Triassic Konservat-Lagerstätten within the Carnian Pluvial Episode. 150 Years of Scientific Excavations in Austria.*

Artificial Intelligence (AI) The chapter was produced without the use of AI.

Language The entire volume uses American English reviewed by a native scientist.

Received 2025.

Published online 2026.

References

Blaschke F (1910) Josef Haberfelner. Mitt Sek Naturk ÖTK 6:40–41

Bauer FK, Schnabel W (1997) Geologische Karte der Republik Österreich. Sheet Mariazell 72, 1:50.000. Geologische Bundesanstalt Wien. https://doi.org/10.24341/tethys.37

Bronn HG (1859) Nachtrag über die Trias-Fauna von Raibl. Neu Jb Mineral Geol Paläont 1859:39–45

Doguzhaeva LA, Mapes RH, Summesberger H, Mutvei H (2007a) The preservation of body tissues, shell, and mandibles in the ceratitid ammonoid *Austrotrachyceras* (Late Triassic), Austria. In Landman NH, Davis RA, Mapes RH (eds), Cephalopods present and past: new insights and fresh perspectives. Springer, pp 221–238. https://doi.org/10.1007/978-1-4020-6806-5_11

Doguzhaeva LA, Summesberger H (2012) Pro-ostraca of Triassic belemnoids (Cephalopoda) from Northern Calcareous Alps, with observations on their mode of preservation in an environment of northern Tethys which allowed for carbonization of non-biomineralized structures. N Jahrb Geol Paläontol Abh 266:31–38. https://doi.org/10.1127/0077-7749/2012/0266

Doguzhaeva LA, Summesberger H, Mutvei H (2006) An unique Upper Triassic coleoid from the Austrian Alps reveals pro-ostracum and mandibule ultrastructure. Acta Univ Carol Geol 49:69–82

Doguzhaeva LA, Summesberger H, Mutvei H, Brandstaetter F (2007b) The mantle, ink sac, ink, arm hooks and soft body debris associated with the shells in Late Triassic coleoid cephalopod *Phragmoteuthis* from the Austrian Alps. Palaeoworld 16:272–284. https://doi.org/10.1016/j.palwor.2007.08.002

Forchielli A, Pervesler P (2013) Phosphatic cuticle in thylacocephalans: A taphonomic case study of (Arthropoda, Thylacocephala) from the Fossil-Lagerstätte Polzberg (Reingraben shales, Carnian, Upper Triassic, Lower Austria). Austrian J Earth Sci 106:46–61

Glaessner MF (1931) Eine Crustaceenfauna aus den Lunzer Schichten Niederösterreichs. Jahrb K-K Geol BA Wien 81:467–486

Guembel CW Ritter von (1861) Geognostische Beschreibung des bayerischen Alpengebirges und seines Vorlandes. Justus Perthes, Gotha, 950 p

Haberfelner J, Haberfelner H (1902) Die Trias in den Alpen mit ihren kohlenführenden Lunzerschichten und deren bergmännische Bedeutung, pp 1–18

Harzhauser M, Krenn M (2023) The geological-paleontological collections of the Natural History Museum Vienna. An exhaustive evaluation of provenance context (1807 to 1918). Ann Naturhist Mus Wien Serie A 124:73–99

Krystyn L (1978) Eine neue Zonengliederung im alpin-mediterranen Unterkarn. Schrift Erdwiss Komm, Österr Akad Wiss 4:37–75

Krystyn L (1991) Die Fossillagerstätten der alpinen Trias. In Nagel D, Rabeder G (eds) Exkursionen im Jungpaläozoikum und Mesozoikum Österreichs. Österr Paläontol Ges Wien, pp 23–78

Krystyn L (1998) Das Polzberger Fischlager. Österreichische Paläontologische Gesellschaft, 5. Jahrestagung (2.-4.Okt. 1998, Lunz am See), Exkursionsführer 12–14

Lukeneder A (ed) (2026) Late Triassic Konservat-Lagerstätten within the Carnian Pluvial Episode: 150 years of scientific excavations in Austria. Top Geobiol 55:600 p. https://doi.org/10.1007/978-3-032-19777-1

Lukeneder A, Surmik D, Gorzelak P, Niedźwiedzki TB, Salamon MA (2020) Bromalites from the Upper Triassic Polzberg section (Austria); insights into trophic interactions and food chains of the Polzberg palaeobiota. Sc Rep 10:20545. https://doi.org/10.1038/s41598-020-77017-x

Lukeneder A, Lukeneder P (2021) The Upper Triassic Polzberg palaeobiota from a marine Konservat-Lagerstätte deposited during the Carnian Pluvial Episode in Austria. Sci Rep 11:16644. https://doi.org/10.1038/s41598-021-96052-w

Lukeneder A, Lukeneder P (2022a) Taphonomic history and trophic interactions of an ammonoid fauna from the Upper Triassic Polzberg palaeobiota. Sci Rep 12:7455. https://doi.org/10.1038/s41598-022-11496-y

Lukeneder A, Lukeneder P (2024) Ammonite taxonomy of the Carnian Polzberg *Konservat-Lagerstätte* in Austria. Jahrb Geol BA. (GeoSphere Austria) 162:27–55

Lukeneder A, Lukeneder P (2023) New data on the marine Upper Triassic palaeobiota from the Polzberg *Konservat-Lagerstätte* in Austria. Swiss J Palaeontol 142:9. https://doi.org/10.1186/s13358-023-00269-3

Lukeneder A, Lukeneder P, Sachsenhofer RF, Roghi G, Rigo M (2024a) Multi-proxy record of the Austrian Upper Triassic Polzberg *Konservat-Lagerstätte* in light of the Carnian Pluvial Episode. Sci Rep 14:11194. https://doi.org/10.1038/s41598-024-60591-9

Lukeneder P, Fuchs D, Lukeneder A (2024b) Morphology, taxonomy and trophic interactions of rostrum-less coleoids from the Late Triassic Polzberg *Konservat-Lagerstätte* (Lower Austria). Swiss J Palaeontol 143:21. https://doi.org/10.1186/s13358-024-00319-4

Lukeneder P, Lukeneder A (2022b) Mineralized belemnoid cephalic cartilage from the late Triassic Polzberg *Konservat-Lagerstätte* (Austria). PLoS ONE 17(4):e0264595. https://doi.org/10.1371/journal.pone.0264595

Lukeneder P, Lukeneder A (2022c) Comment on Triassic coleoid beaks and other structures from the Calcareous Alps revisited by Doguzhaeva et al. (2022). Acta Palaeontol Pol 67(4):963–965. https://doi.org/10.4202/app.01016.2022

Martin M (1982) Revision von *Tellerodus sturii* (TELLER) 1891. Verh Geol BA 2:21–31

Mojsisovics JAE (1893a) Die Cephalopoden der Hallstätter Kalke—Band I. Abh K-Kön Geol RA 6:835 p

Mojsisovics JAE (1893b) Die Cephalopoden der Hallstätter Kalke—Band II. Abh K-Kön Geol RA 6:524 p

Reis OM (1900) *Coelacanthus Lunzensis* Teller. Jahrb Geol BA 50:187–192. https://www.zobodat.at/pdf/JbGeolReichsanst_050_0187-0192.pdf

Ruttner A, Schnabel W (1988) Geologische Karte der Republik Österreich. Sheet Ybbsitz 71, 1:50.000. Geologische Bundesanstalt Wien.https://doi.org/10.24341/tethys.6

Schädel M, Eldijk T, Winkelhorst H, Reumer JWF, Haug J (2020) Triassic Isopoda: three new species from Central Europe shed light on the early diversity of the group. Bull Geosci 95:145–166

Steindachner F (1910) Notizen. Jahresbericht für 1909. Ann K-K Nat Hofmus 24:1–70. https://www.jstor.org/stable/41768087

Steindachner F (1911) Notizen. Jahresbericht für 1910. Ann K-K Nat Hofmus 25:1–54
Steindachner F (1912) Notizen. Jahresbericht für 1911. Ann K-K Nat Hofmus 26:1–55. https://www.jstor.org/stable/41768116
Steindachner F (1914) II. Musealarbeiten, Jahresbericht für 1913. Ann K-K Nat Hofmus 28:1–54
Stur D (1874) Neue Aufschlüsse im Lunzer Sandsteine bei Lunz und ein neuer Fundort von Wengerschiefer im Pölzberg zwischen Lunzersee und Gaming. Verh K-K Geol RA 1:271–273. https://www.zobodat.at/pdf/VerhGeolBundesanstalt_1874_0271-0273.pdf
Stur D (1886) Vorträge: Vorlage des ersten fossilen Schädels von *Ceratodus* aus den obertriadischen Reingrabner Schiefern von Pölzberg nördlich bei Lunz. Sitzung am 30. November, 381–383. https://www.zobodat.at/pdf/VerhGeolBundesanstalt_1886_0381-0383.pdf
Stur D (1888) Geologische Aufnahmen und Untersuchungen im Felde: Exkursion zum Pölzberg bei Lunz. Verh K-K Geol RA, Sitzung am 17. Jänner 1888
Teller F (1891) Über den Schädel eines fossilen Dipnoers, *Ceratodus Sturii* nov. spec. aus den Schichten der oberen Trias der Nordalpen. Abh K-K Geol RA 15:1–38
Trauth F (1919) Die Neuhauser Schichten, eine litorale Entwicklung des alpinen Bathonien. Verh K-K Geol BA 1919:333–339. https://www.zobodat.at/pdf/VerhGeolBundesanstalt_1919_0333-0339.pdf
Trauth F (1948) Geologie des Kalkalpenbereiches der Zweiten Wiener Hochquellenleitung. Abh Geol BA in Wien 26(1):1–99. https://www.zobodat.at/pdf/AbhGeolBA_26_0001-0099.pdf
Unknown author (1902) Lunz. Erlauftal-Bote, pp 1–12. Accessed 12 Jan 1902
Unknown author (1904) Lunz. Erlauftal-Bote, pp 1–10. Accessed 26 June 1904
Unknown author (1910) Josef Haberfelners 80. Geburtstag. Illustriertes Wiener Extrablatt, 1–16. Accessed 20 Aug 1910
Unknown author (1911) Der Schnee. Wochenschrift des Alpen-Skivereines 25:1–8
Wulfen FX (1793) Abhandlung vom Kärnthenschen Pfauenschweifigen Helmintholith oder dem Sogenannten Opalisirenden Muschelmarmor, pp 1–124
Zapfe H (1971) Catalogus Fossilium Austriae. Ein systematisches Verzeichnis aller auf österreichischem Gebiet festgestellten Fossilien. Springer, 140 p

Chapter 3
Fossil Material from the Carnian Pluvial Episode in Austrian Earth Science Collections

Petra Lukeneder and Alexander Lukeneder

Abstract Austria hosts several important collections (Natural History Museum Vienna, GeoSphere Austria, University Vienna) of fossils documenting the Carnian Pluvial Episode (CPE), with the Polzberg *Konservat-Lagerstätte* being the most prominent locality. As one of Austria´s paleontological highlights this site preserves abundant quantities of Triassic fossils, thousands of which are now held within the Austrian collections. Since its discovery around 1874, several unique finds from Polzberg have been made within the Carnian Reingraben Formation of this fossil site. These include the brackish lungfish *Ceratodus sturii* (Teller, 1891) and an almost complete specimen of the coleoid *Phragmoteuthis polzbergensis* (Lukeneder et al. 2024a; Lukeneder et al. 2024b). Polychaetes with soft tissue preservation represent another special feature of these deposits. Further localities in Austria where the Carnian Pluvial Episode is documented by collection material include the Großau section near Lunz am See (Lower Austria), the Steinbachgraben section near Göstling (Lower Austria), the Scheiblingbachgraben near Großreifling (Styria; all Austria). Cave del Predil in northern Italy (formerly known as "Raibl" as part of the Habsburg Empire) is particularly significant in relation to the Carnian Pluvial Episode. Many of the Polzberg taxa were first described from the almost contemporaneous Cave del Predil. Huge amounts of material from that location are available in several Austrian institutions. The value of these exclusive historical and recent findings also lies in their function as biodiversity repositories of historically important localities. They permit reconstructions of ancient trophic networks and help to document the biogeographical distribution of individual taxonomic groups.

P. Lukeneder (✉) · A. Lukeneder
Department of Geology and Paleontology, Natural History Museum Vienna, Vienna, Austria
e-mail: petra.lukeneder@nhm.at

A. Lukeneder
e-mail: alexander.lukeneder@nhm.at

P. Lukeneder
Institute of Applied Geology, Department of Landscape, Water and Infrastructure, BOKU University, Vienna, Austria

A. Lukeneder (ed.), *Late Triassic Konservat-Lagerstätten within the Carnian Pluvial Episode*, Topics in Geobiology 55, https://doi.org/10.1007/978-3-032-19777-1_3

Keywords Collections · History of sciences · Polzberg *Konservat-Lagerstätte* · Triassic

3.1 Introduction

In recent years, the Polzberg *Konservat-Lagerstätte* in Lower Austria, with its exceptionally preserved Late Triassic fossils, has been the subject of detailed studies. In addition to records and examinations of the main section (Lukeneder and Lukeneder 2023) by the use of modern methods (Lukeneder et al. 2020; Lukeneder and Lukeneder 2022a; Lukeneder and Lukeneder 2024; Lukeneder et al. 2024b; Lukeneder and Lukeneder 2026, this volume), the fossils from the extensive historical collections in Austrian institutions have also been brought into focus. Recent studies were therefore based on both newly excavated and historical material housed in the Austrian collections (Natural History Museum Vienna, GeoSphere Austria, University of Vienna, Lower Austrian State Collections) as well as material from private collections (Aschauer family, Waidhofen an der Ybbs, Lower Austria). The historical collections were enriched by newly excavated material (Lukeneder and Lukeneder 2021). The abundant fossils collected over many decades have enabled the reconstruction of ecological relationships and the assignment of previously enigmatic fossils to their taxonomic groups (Lukeneder and Lukeneder 2022b). As the recent project progressed, the importance of preserving and revising historical collections became increasingly clear. Specimens of previously undescribed polychaetes (Fig. 3.1a) were discovered within the paleobotanical collection and had therefore been misinterpreted as plant remains. The vertebrate members of the Polzberg palaeobiota include abundant fish such as the predatory *Saurichthys calcaratus* (Griffith 1977) (Fig. 3.1b). The Natural History Museum Vienna (NHMW) collection contains hundreds of coleoid remains, including specimens of the genus *Mojsisovicsteuthis* (Jeletzky 1965) (Fig. 3.1c) which have not yet been recognized as such (see also Lukeneder et al. 2024b). Specimens of the crustacean *Platychela trauthi* Glaessner 1931 (Fig. 3.2a; holotype stored at the NHM in Vienna; NHMW 1910/0015/0018; see also Lukeneder and Lukeneder 2023) are known from Polzberg, as are fish such as *Nannolepis elegans* (Griffith 1977) (Fig. 3.2b). To date, numerous invertebrate specimens (see for example Fig. 3.2c and d) remain undescribed.

3.2 Polzberg Fossils in Austrian Collections

The discovery and history of the Polzberg *Konservat-Lagerstätte* was documented in Lukeneder and Lukeneder (2026). Dionýs Štúr (also: Dionys Stur; 1827–1893) was the first to recognize the potential of this paleontological treasure (Stur 1874). During his work, he received active support from mine manager Josef Haberfelner (also: Haberfellner; 1830–1913) as well as from paleontologist Friedrich Blaschke

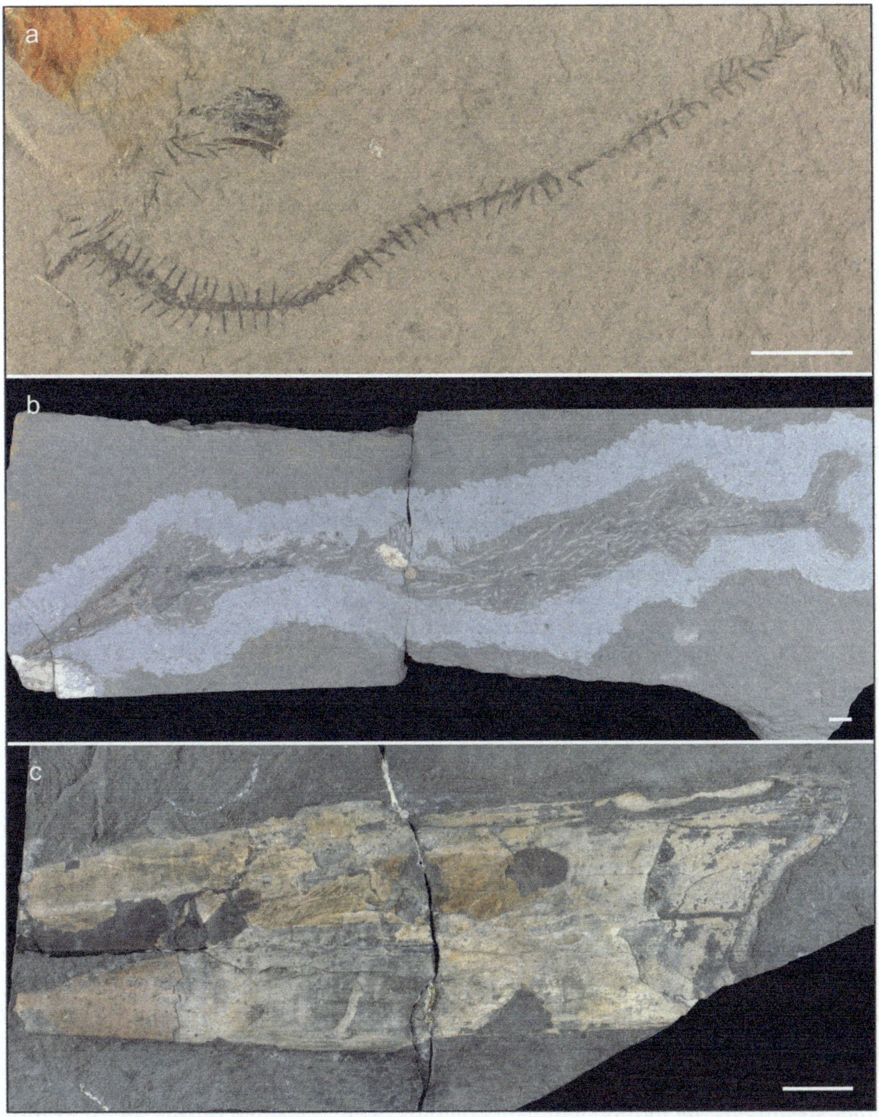

Fig. 3.1 **a** Eunicid polychaetes were discovered within the paleobotanical collection of the NHMW and in the collections of private collectors. **b** *Saurichthys calcaratus* (NHMW 2007z0170/0001). **c** Specimen of rostrum-less coleoid taxon *Mojsisovicsteuthis* from Polzberg (NHMW 2005z0005/0041), collected in 1908. All scale bars 1 cm

Fig. 3.2 Highlights from the Polzberg collection of the NHMW. **a** Holotype of *Platychela trauthi* (Glaessner 1931); NHMW 1910/0015/0018; lateral view, historical collection, exhibited in hall 8 NHM Vienna (Photo: Alice Schumacher, NHMW) **b** *Nannolepis elegans* (Griffith 1977) (NHMW 2007z0170/0148). **c** Indeterminate bivalve (NHMW 2012/0228/0000). **d** Large indeterminate gastropod. All scale bars 1 cm

(1883–1911), who also produced the only historical locality map of the *Konservat-Lagerstätte,* preserved in Glaessner (1931). The fossil fauna collected during early excavations at Polzberg, obtained from two mines which created exclusively for fossil sampling, is now housed at various Austrian institutions. The by far largest numbers of specimens, including holotypes and rare taxa, are housed at the Natural History Museum Vienna (NHMW), followed by the collection of GeoSphere Austria (formerly: Geological Survey of Austria, GBA; also known as k.k. Geologische Reichsanstalt from 1849 to 1918). The Polzberg collection at the NHMW (Fig. 3.3) consists of recently collected 5,290 specimens (as of 2021) and additional 6,397 specimens (see also Lukeneder and Lukeneder 2021, 2023) from historical collections. A larger number of specimens was bought from mine manager Haberfelner. Other specimens, including abundant coleoids, were collected during an excavation campaign in 1909 (Lukeneder and Lukeneder 2023). The most remarkable specimens are crustacean holotypes such as *Platychela trauthi* Glaessner 1931 (NHMW 1910/0015/0018), the isopod *Obtusotelson summesbergeri* (Schädel et al. 2020; NHMW 2020/0003/0001) and the arthropod group of the Thylacocephalia. Specimens of *Austriocaris carinata* Glaessner 1931 (including the holotype counterprint NHMW 1910/0015/0041) were already inventoried in 1910. Here as well, the new excavations and the revision of historical collection material has yielded new insights (Forchielli and Pervesler 2013; Laville et al. 2024).

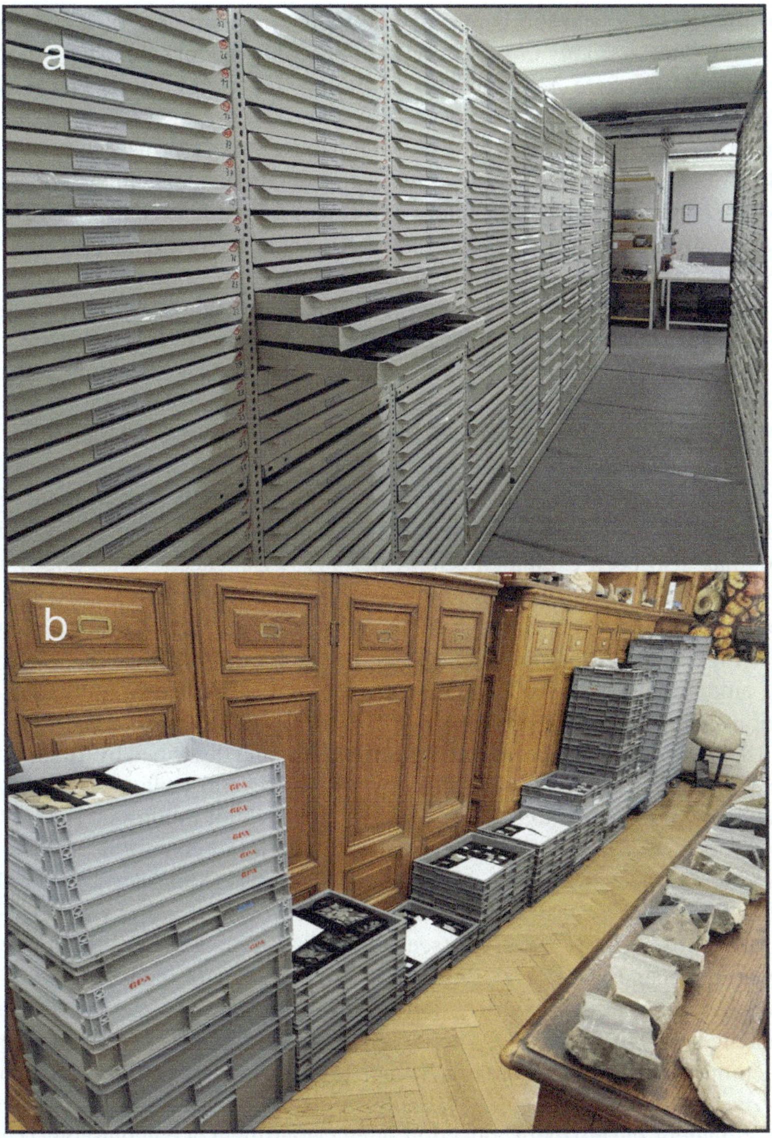

Fig. 3.3 **a** The Mesozoic collection of the Geological-Paleontological Department at the Natural History Museum Vienna includes several tens of drawers with Polzberg material. All material was reviewed and documented as part of the recent NHM Polzberg project. **b** The processing of the Polzberg fossils took several years. In particular, newly acquired fossils from the recent project were temporarily stored outside the collection for further processing

Both the NHMW and the University of Vienna (Department of Paleontology) own casts of the *Ceratodus sturii* holotype, collected by Dionys Stur (type specimen stored at the GeoSphere Austria). This specimen (Fig. 3.4a) can be viewed in the NHMW exhibition.

During the excavation work for the recent Polzberg project (project start in 2020), a remarkable coleoid fossil was discovered: an almost complete specimen of *Phragmoteuthis polzbergensis* Lukeneder, Fuchs and Lukeneder, 2024 (NHMW 2021/0123/0090; Fig. 3.4b and c). This has not only broadened our view on the coleoid family tree (first phragmoteuthid fossil with unambiguously ten arms), but also questioned the occurrence of *Phragmoteuthis bisinuata* (Bronn 1859) within the Polzberg subbasin (for details see also Lukeneder et al. 2024b). The holotype of this newly described species (NHMW 2006z0005/0021) is also stored at the NHMW collection. It includes a proostracum, differentiated arm hooks and a cephalic cartilage. In addition, the NHMW collection contains a wealth of figured originals such as *Gigantoperus telleri* (Abel 1906) (e.g., NHMW 2007z0170/0366) and *Thoracopterus niederisti* (Bronn 1858) (e.g., NHMW 2007z0170/0170, /0171, /0172, /0183; Fig. 3.4d), both figured in Griffith (1977). Bromalite material including fragments of ammonoids, fish and coleoids has also been examined (Glaessner 1931; Lukeneder et al. 2020) and is stored at the NHMW collection.

Overall, 282 Polzberg specimens are recorded in the collection of the GeoSphere Austria (Fig. 3.5a), including 46 specimens in the type collection. 184 specimens, mainly ammonoids (Fig. 3.5b), were donated by the Aschauer family. The type collection includes, among other taxa, all fragments of the *Ceratodus sturii* skull (GBA 1891/001/0002/01) discovered by Stur (1886) and described by Teller (1891). This extraordinary discovery was also reported in several contemporary newspapers (Unknown author 1886; Lobner 1927). During fossil mining in 1885/1886 at Polzberg, Josef Haberfelner found a specimen with a tail consisting of a continuously radiating fin, characteristic of the genus *Ceratodus* (Stur 1886). Teller (1891) attributed this tail to a coelacanthid specimen; *Coelacanthus lunzensis* (GBA 1891/001/0002/01 to 1891/001/0002/10) (see also Reis 1900). Another "*Coelacanthus*" *lunzensis* specimen (NHMW 2012/0228/3611), consisting of the basal part of a tail fragment, is stored in the historical collection of the NHMW (see also Lukeneder and Lukeneder 2023, Fig. 3.2f). The GeoSphere Austria also holds a specimen of the brachiopod *Cruratula* cf. *eudora* Bittner 1890 from the "Reifling Limestone of the Polzberggraben".

Unfortunately, the Polzberg collection at the Department of Paleontology (University of Vienna) consists of only a few inventoried specimens (for examples of specimen labels see Fig. 3.6); two of these (IPUW MFN 2318 = Inv. Nr. 1969 and IPUW MFN 18837 = Inv. Nr. 4938) are molds of *Ceratodus sturii* from the GBA collection. Emil Weinfurter (1904–1968) was a collector and self-taught paleontologist, who, in his will, bequeathed his extensive fossil collection to the Department of Palaeontology (University of Vienna) (Thenius 1968). Accordingly, today some Polzberg specimens are incorporated in the Emil Weinfurter collection (Fig. 3.6a), while others are distributed across the Triassic collections of the Department of Palaeontology (=Department of Palaeontology; IPUW) and the collection of the

Fig. 3.4 **a** Skull fragment of *Ceratodus sturii*, exhibited at the NHMW, cast of the original specimen from the GeoSphere Austria (Photo: Alice Schumacher, NHMW). **b** Ten-armed coleoid *Phragmoteuthis polzbergensis* (Lukeneder et al. 2024b) with preserved arm crown (NHMW 2021/0123/0090). **c** Enlargement of the hooked arm crown of *P. polzbergensis*, comprising nine preserved arms, indicating ten arms in the group of the Phragmoteuthida. **d** Specimen of *Thoracopterus polzbergensis* Bronn 1858 (NHMW 2007z0170/0172), figured original of Schultz (2013), Taf. 14, (Fig. 3.2). All scale bars 1 cm

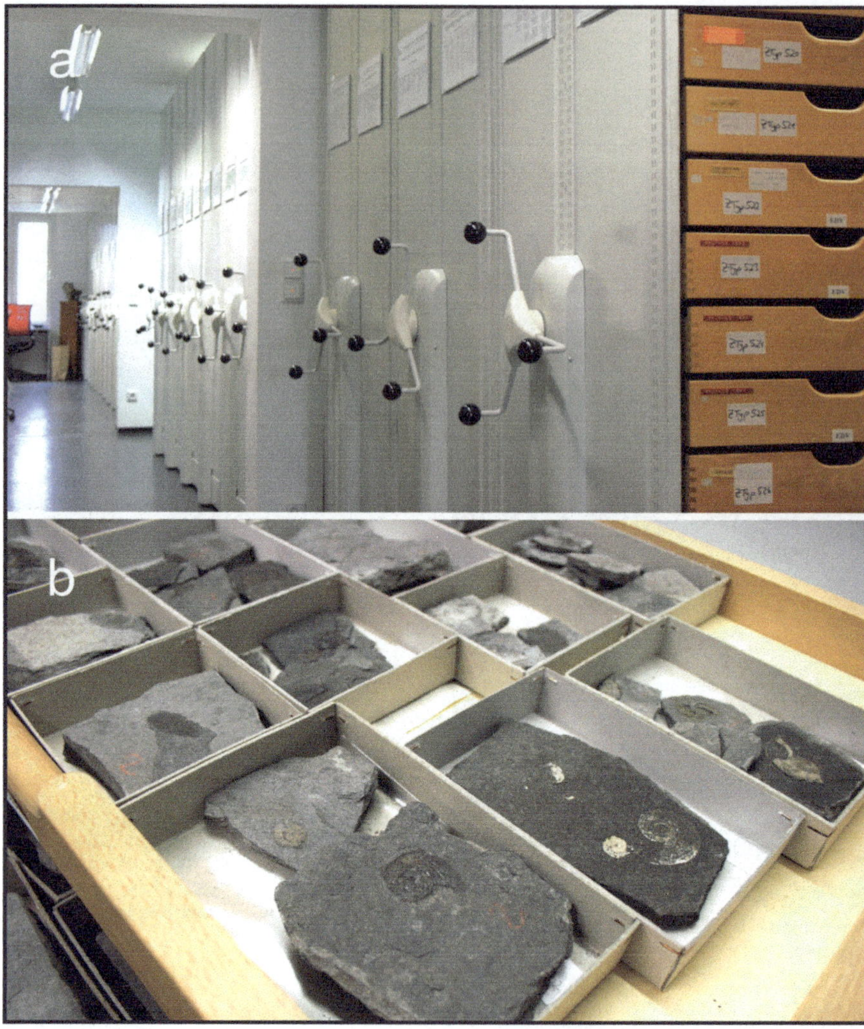

Fig. 3.5 **a** The Geosphere Austria's compaction system allows the storage of abundant fossils. The GBA Polzberg holotypes are kept separately from the remaining Polzberg fossils. **b** Ammonoids make up the vast majority of the Polzberg collection at the GBA. (Photos: Davide Di Franco, GeoSphere). Although the GeoSphere Austria collection contains by for fewer Polzberg fossils than the NHMW, it houses many holotype specimens

Department of Geology (IGUW). A personal visit to those collections yielded only 13 Polzberg specimens. As is the case with many other historical collections, the locality was often recorded simply as "Lunz" or "Lunz region," making later assignment to Polzberg difficult. Six inventoried specimens belong to the group Pisces, while the others are cephalopods (ammonoids) or rock samples. Several specimens in both the IPUW and the IGUW collections were donated by the paleontologist

Fig. 3.6 **a** The Weinfurter collection at the University of Vienna also includes Polzberg material. Individual Polzberg specimens are scattered across the various collection parts and are difficult to locate. Unfortunately in many cases only the locality "Lunz" is given. **b** A selection of Polzberg labels from the collection at the University of Vienna. Labels also document the involved persons such as Gustav Adolph von Arthaber. Note the different names "Polzberggraben", "Lunz" and "Pölzlberggraben" for one locality

Gustav Adolph von Arthaber (1864–1943) (see also notation on collection label in Fig. 3.6b).

There are certainly some smaller collections including Polzberg material that we could not list in detail here. The Lower Austrian State Collections are stored at Hart near St. Pölten. Four specimens (fish, ammonoids, rock sample) include the locality "Polzberg mine", although there are a few other specimens that may have originated from Polzberg but were only labeled "Lunz". Interestingly, some specimens are labeled with the designation "Collection Haberfelner at the Biological Station in Lunz am See" (Fig. 3.7a). Although the collection of the Montanuniversität in Leoben (Styria) contains an extensive collection from Lunz am See (including some Haberfelner specimens), no definitive Polzberg fossils could be identified.

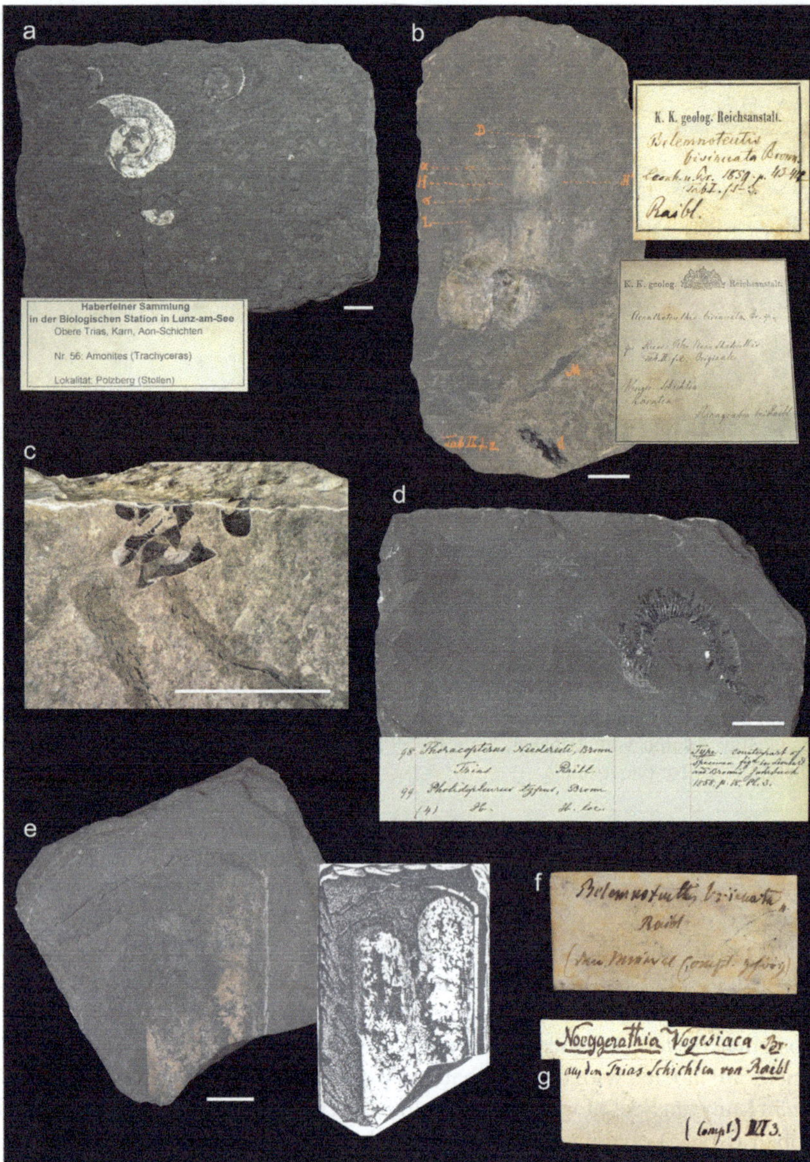
a
b
c
d
e
f
g
Haberfelner Sammlung
in der Biologischen Station in Lunz-am-See
Obere Trias, Karn, Aon-Schichten
Nr. 56: Amonites (Trachyceras)
Lokalität: Polzberg (Stollen)
K. K. geolog. Reichsanstalt.
Raibl.
Noeggerathia Vogesiaca
aus den Trias Schichten von Raibl

◂**Fig. 3.7** Fossils from Cave del Predil (northern Italy), formerly known as Raibl. **a** *Austrotrachyceras minor* from the Haberfelner collection of the Biological Station in Lunz am See (stored at the Lower Austrian State Collections at Hart near St. Pölten). **b** *Phragmoteuthis bisinuata* from Rinngraben ravine in Cave del Predil, collected by Dionys Stur. Red markings stem from measurements by Suess (1865). **c** Coleoid cephalic cartilage including hooks, assigned to the coleoid genus *Phragmoteuthis* (MFSN 53560) from the collection of the Museo Friulano di Storia Naturale in Udine (Italy). **d** Type of *Peltopleurus kneri* Woodward 1895 (NHMUK PV P 1099 a) from the collection of the Natural History Museum (NHMUK) in London (UK), including an enlarged entry from the historical inventory book, which indicates that a specimen of *Thoracopterus niederristi* from Raibl is also present in the collection (Photos: NHMUK). **e** Left: One type specimen of *Phragmoteuthis bisinuata* (Bronn 1859) from the collection of the University of Heidelberg (Germany) (GEOW-HD_OS-BRO-00002. Right: Copy of published drawing of the specimen by Heinrich Georg Bronn. **f** Historical collection label of *Phragmoteuthis bisinuata* specimen documenting the purchase of the specimens from the Mineralien Comptoir (backside of specimen GEOW-HD_OS-BRO-00003). **g** Historical collection label of the type specimen of *Noeggerathia vogesiaca* (GPIT-PV-75109) from the Triassic of Cave del Predil, stored at the Geological-Palaeontological Institute and Museum in Tübingen (Germany). The specimen was determined by "R. Florin" (most probably: Carl Rudolf Florin, 1894–1965) in 1926. All scale bars 1 cm

3.3 Fossils from the Reingraben Formation in the Lunz Region

Over time, deposits of the Reingraben Formation (=Reingraben Shales) have cropped out in many localities in the area around Lunz. The ammonoid *Carnites floridus* (Wulfen 1793) (GBA 2018/355/0001) is known from Sulzbachgraben (base of Pramelreith Mountain) near Lunz am See. Alexander Bittner (1850–1902) collected specimens of *Halobia rugosa* Gümbel, 1861 (GBA 2018/380/0003) near the chapel SW of the Labenbacher, near Wohlfahrtsschlag (Scheibbs region in Lower Austria). Dionys Stur also collected *Halobia haueri* Stur (=*Halobia rugosa* Gümbel, 1861) at the Klosterallee SW of Gutenstein (Lower Austria; GBA 2018/377/0001) and at the type locality of the Reingraben Formation, the Reingraben (also: Raingraben) near Rohr am Gebirge in Lower Austria (GBA 2018/378/0001). Reingraben shales with *Halobia haueri* are also known from the Klausgraben ravine near St. Anton an der Jeßnitz (GBA 2018/380/0002). *Orthoceras* sp. and *Halobia* sp. are reported from Kleinzell (GBA 2018/376/0017; Lower Austria). There are also specimens from the Saugraben and "Lake adit" ("See-Stollen") included in the Lower Austrian State Collections. Additional material was collected over the last years from the Großau section near Lunz am See, the Steinbachgraben section near Göstling (both Lower Austria), and the Scheiblingbachgraben near Großreifling (Styria).

3.4 Fossils from Cave del Predil (Northern Italy)

Cave del Predil, located in the Friuli-Julian Venetia region (northern Italy), was well known as "Raibl" during the Habsburg Empire and was an important center for zinc and lead mining. The so-called Raibl layers (consisting of Torer layers, dolomites, marls and calcareous layers, fossiliferous shales with fish and dark bituminous limestones, see also Konopasek, 2002) are almost contemporaneous to the Reingraben Shales in Austria. The Raibl fauna has been described numerous times and shows similarities to the Polzberg fauna. Overviews of the Raibl fauna and discussions on the paleoenvironment can be found in Bronn (1858), Kner (1865), Reuss (1858), Stur (1868), Wöhrmann (1893), Liebermann (1978) and Liebermann (1979). Fossils from the Carnian of Cave del Predil are now housed in many collections; among them, numerous specimens are stored at the Natural History Museum Vienna and GeoSphere Austria. Both institutions own material from the early examinations of the coleoid fauna by Eduard Suess (Fig. 3.7b). The Museo Friulano di Storia Naturale in Udine (Italy) owns a number of fossils from Cave del Predil as well, among them 11 coleoid specimens and coleoid cephalic cartilages (MFSN 53560; Fig. 3.7c). Thirty-four specimens from Cave del Predil can be accessed by an online search of the paleontological collection at the Natural History Museum (NHMUK) in London. They include the type of *Peltopleurus kneri* A. S. Woodward 1895 (Fig. 3.7d) as well as rock samples and minerals. The collection of the Institute of Applied Geology (IAG) at the BOKU University Vienna (formerly: University of Natural Resources and Life Sciences) holds a few additional specimens (not inventoried) from Cave del Predil, all of them part of the Crown Prince Rudolf Collection which dates back to the son of Empress Elisabeth (1837–1898), Rudolf (1858–1889), who was particularly interested in natural sciences. Crown Prince Rudolf was taught by geologist Christian Gottlob Ferdinand Ritter von Hochstetter (1829–1884), a participant of the Novara expedition (see also Harzhauser and Jovanovic-Kruspel 2023) and, from 1876, director of the Natural History Court Museum in Vienna. In July 1873, the Crown Prince visited the collections of the Carinthian State Museum with Ferdinand von Hochstetter; it is documented that he was particularly interested in the Raibl Triassic, among others. He was unable to visit Raibl himself because of the weather (Unknown author 1873). The type material of the coleoid *Phragmoteuthis bisinuata* (Bronn 1859) also comes from the Cave del Predil deposits and is located at the University of Heidelberg (Germany) (GEOW-HD_OS-BRO-00001; GEOW-HD_OS-BRO-00002; GEOW-HD_OS-BRO-00003; see also (Lukeneder et al. 2024b; Fig. 3.7e and f). Original material of Bronn (1858) can also be found at the Geological-Palaeontological Institute and Museum in Tübingen (Germany). The type specimen (GPIT-PV-75109) of *Noeggerathia vogesiaca* from Cave del Predil is stored there (Fig. 3.7g). At least a small collection of paleobotanical specimens is located at the Carinthian State Museum (Carinthia, Austria), while a considerable number of specimens (paleobotanical, fish, bivalves and cephalopods such as ammonoids and coleoids) from Cave del Predil are held within the collection of the Montanuniversität in Leoben (Styria).

3.5 Fossils from the Northalpine Raibl Beds

The Austrian collections contain fossils from numerous localities with deposits of the Carnian crisis. These sediments include the Northalpine Raibl Beds, which are exposed in the western regions of the Northern Calcareous Alps (Krainer et al. 2011). The dating of the Carnian (Julian-Tuvalian) Northalpine Raibl Beds ("Nordalpine Raibler Schichten"; see also Piller et al. 2004) is based on bivalves, ammonites and foraminifera (Tollmann 1976; Krainer et al. 2011). These beds in the region of the Karwendel and Mieming mountains comprise, among other fossils, fish teeth from Actinopterygii (*Saurichthys*, *Colobodus*? and indeterminate taxa) and lateral teeth of the small chondrichthyian *Palaeobates*. Additionally, reptile fragments ascribed to Eusauropterygia? and Prolacertiformes were reported (Krainer et al. 2011). A specimen of the coleoid *Aulacoceras inducens* Braun 1841 from the Raibl beds of the Schlernplateau, collected by Dionys Stur in 1868, is included in the collection of GeoSphere Austria.

3.6 Conclusions

Austria hosts numerous important collections of fossil material documenting the Carnian Pluvial Episode (CPE), with one of Austria´s paleontological highlights the most prominent Polzberg *Konservat-Lagerstätte*. The Polzberg *Konservat-Lagerstätte*, with its thousands of partly exceptionally preserved fossils, is unique in Austria. During the Carnian crisis, not only countless fish and cephalopods were embedded here, but in principle an entire ecosystem was preserved for 234 million years. An impressive total of 11,687 fossils (97.6%) of the Polzberg paleobiota are held within the Austrian collections of the Natural History Museum Vienna, 11,986 specimens in all Austrian collections (Table 3.1). This is largely thanks to the pioneering excavation activities of outstanding figures such as Dionys Stur (k. k. Geologische Reichsanstalt, today GeoSpere Austria) and mine manager Josef Haberfelner (Lunz am See), who recognized the importance of this locality and initiated the recovery of its remarkable biodiversity. The processing of all the historical material started soon after its discovery. Numerous articles published in recent years highlight the significance of Polzberg and its fossil fauna, and the use of modern methods yielded new perspectives on the material. The revision of the historical collection material has provided important insights into the Polzberg fauna. Especially regarding the coleoid *Mojsisovicsteuthis* specimens, it is noteworthy that almost exclusively historical material was revised (Lukeneder et al. 2024b). The almost contemporary deposits of the Carnian crisis are also found in Cave del Predil (Friuli Venezia Giulia, northern Italy). Due to the geographic and historical proximity to Austria, extensive material from this locality is also held within the Austrian collections, highlighting the broader biogeographical context of the Carnian Pluvial

Table 3.1 List of institutions with known Polzberg material. The number of specimens in the respective collections is given. Accordingly, the total number of Polzbergfossils in the Austrian collections is 11.986 specimens (as of 2025), excluding *Peribositra*

Institution/collection	Specimens n
Natural History Museum Vienna (NHMW)	11.687
GeoSphere Austria (GSA/GBA)	282
Institute of Paleontology, University of Vienna (IPUW)	13
Lower Austrian State Collections (LASC)	4
Total	11.986

Episode. The presence of abundant historical collection specimens from other localities with deposits of the Carnian crisis shows that these fossiliferous, fine-grained, almost black sediments have repeatedly attracted the attention of scientists and collectors. Despite decades of research, these specimens have often been studied only superficially. Lukeneder and Lukeneder (2021) demonstrated how a detailed analysis of historical collection specimens, in combination with material from recent excavations, enables the reconstruction of the Triassic ecosystem. The unique material from the historical adits is as significant as the samples collected bed-by-bed, further highlights the importance of historical collections and their preservation. This underscores the importance of historical collections, as well as their preservation and study. New methods and perspectives continue to advance research on historical deposits and collection material.

Acknowledgments The present work was done within the framework of and financially supported by projects of the Federal Government of Lower Austria (K3-F-964/001-2020 and K3-F-964/002-2023, Department Science and Research; headquarters in St. Pölten), the Austrian Academy of Sciences (headquarters in Vienna, project Polzberg Lukeneder 2021) represented by the National Committee for Geo/Hydro Sciences (Earth System Sciences Programme), and the Triassic project "Fenster in die Karnische Krise" of the society of friends of the Natural History Museum Vienna (Freunde des Naturhistorischen Museums Wien, CPE Lukeneder 2023). We are very grateful to Davide Di Franco and Martin Maslo (both Vienna) for access to the collections of the GBA (GeoSphere Austria) respectively the IPUW, and to Ronald Lintner and Fritz Egermann (both St. Pölten) for access to Lower Austrian State Collections. Martina Schmalholz and Manfred Vogt (both Heidelberg) are thanked for the access to Bronn's type material, Guiseppe Muscio (Udine) for sharing data on phragmoteuthids from the collection of the Museo Friulano di Storia Naturale, Udine (Italy), Dominik Kumertz (Leoben) for access to the extensive fossil collection of the Montanuniversität Leoben (Austria), and Alice Schumacher (Vienna) for specimen photography. The manuscript greatly benefited from valuable comments by the reviewers Davide Di Franco (Vienna) and Thomas Hofmann (Vienna), the production editor Ashok Arumairaj (Taramani), the in-house series editor Kathleen A. Ritterbush (Utah), the publishing editor Simon Shah-Rohlfs and the production editor Antje Endemann (both Heidelberg).

Author Contribution AL had the idea for the present study. PL, AL provided documentation. PL provided first draft of the paper. PL, AL contributed to writing the paper.

Funding Open access funding was enabled and the projects financially supported by the Federal Government of Lower Austria (K3-F-964/001-2020 and K3-F-964/002-2023, Department Science and Research; headquarters in St. Pölten); the Austrian Academy of Sciences (headquarters in

Vienna, project Polzberg Lukeneder 2021) represented by the National Committee for Geo/Hydro Sciences (Earth System Sciences Programme); the Natural History Museum Vienna; and the Triassic project "Fenster in die Karnische Krise" of the Society of Friends of the Natural History Museum Vienna (Freunde des Naturhistorischen Museums Wien, CPE Lukeneder 2023) (to AL). The authors are responsible for the contents of this publication. The funders had no impact on concept.

Competing interests All authors stated that they have no competing interests.

Authors' contributions All authors stated that they have equally contributed to their chapters, read, edited and agreed to submit the final versions of their contributions.

Availability of data and materials All sources of information are provided in the text. Raw data related to the fossil material from Austria are available from the corresponding authors and Alexander Lukeneder upon request.

Additional information This article is part of the Special Volume: *Late Triassic Konservat-Lagerstätten within the Carnian Pluvial Episode. 150 Years of Scientific Excavations in Austria.*

Artificial Intelligence (AI) The chapter was produced without the use of AI.

Language The entire volume uses American English reviewed by a native scientist.

Received 2025.

Published online 2026.

References

Abel O (1906) Fossile Flugfische. Jahrbuch K-K Geol RA 56, 1, Wien 1906. https://doi.org/10.5962/bhl.title.61007

Bittner A (1890) Brachiopoden der alpinen Trias. Abh K-K Geol RA 14:325 p

Braun CF (1841) Taxonomic names, in Beiträge zur Geognosie und Petrefacten-Kunde des südöstlichen Tirols vorzüglich der Schichten von St. Cassian. Beitr Petref-Kunde 4:1–152

Bronn HG (1858) Beiträge zur Triasischen Fauna und Flora der bituminösen Schiefer von Raibl, nebst Anhang über die Kurr'sche Sippe von Chiropteris aus dem Lettenkohle-Sandsteine. N Jahrb Min Geogn Geol Petrefactenk, 1–63

Bronn HG (1859) Nachtrag über die Trias-Fauna von Raibl. Neu Jb Mineral Geol Paläont 1859:39–45

Forchielli A, Pervesler P (2013) Phosphatic cuticle in thylacocephalans: A taphonomic case study of (Arthropoda, Thylacocephala) from the Fossil-Lagerstätte Polzberg (Reingraben shales, Carnian, Upper Triassic, Lower Austria). Austrian J Earth Sci 106:46–61

Glaessner MF (1931) Eine Crustaceenfauna aus den Lunzer Schichten Niederösterreichs. Jb Geol Bundesanst 81:467–486

Griffith J (1977) The upper Triassic fishes from Polzberg bei Lunz, Austria. Zool J Linn Soc 60(1):1–93. https://doi.org/10.1111/j.1096-3642.1977.tb00834.x

Guembel CW Ritter von (1861) Geognostische Beschreibung des bayerischen Alpengebirges und seines Vorlandes. Justus Perthes, Gotha, 950 p

Harzhauser M, Jovanovic-Kruspel S (2023) *Hochstetters* Auckland-Tagebuch. Erstes Neuseeland-Tagebuch des Geologen Ferdinand von Hochstetter: ein Aufschluss in Buchform. Naturhistorisches Museum Wien, Wien. ISBN 978–3–903096–48–6

Jeletzky JA (1965) Taxonomy and phylogeny of fossil Coleoidea (= Dibranchiata). Geol Surv Can Paper 65:76–78

Kner R (1865) Die Fische der bituminösen Schiefer von Raibl in Kärnthen. Sitzungsb d k. Akad d W Math Naturw Cl 1:152–197

Konopasek, R. (2002) Das Bergwerk in Raibl (Cave del Predil). Ein Rückblick auf die sechshundertjährige Geschichte des 1991 stillgelegten Blei-Zink-Bergbaues res montanarum 29/2002, 17–25

Krainer K, Lucas SC, Strasser M (2011) Vertebrate fossils from the Northalpine Raibl Beds, Western Northern Calcareous Alps, Tyrol (Austria). Austrian J Earth Sci 104(1):97–106

Laville T, Lukeneder P, Lukeneder A (2024) Revision of the thylacocephalan biota from the Upper Triassic Polzberg *Konservat-Lagerstätte*, Austria. Acta Palaeontol Pol 69:713–730. https://doi.org/10.4202/app.01180.2024

Liebermann HM (1978) Das Raibler Becken: eine paläogeographische Rekonstruktion aus dem südalpinen Karn von Raibl (Cave del Predil, Italien). Geol Paläont Mitt Innsbruck 7(4):7–20

Liebermann HM (1979) Die Bivalven- und Ostracodenfauna von Raibl und ihr stratigraphischer Wert. Verh Geol BA 2:85–131

Lobner R (1927) Unser Heimatboden im Wandel der Zeiten. St. Pöltner Zeitung 48:19

Lukeneder A (ed) (2026) Late Triassic Konservat-Lagerstätten within the Carnian Pluvial Episode: 150 years of scientific excavations in Austria. Top Geobiol 55:600 p. https://doi.org/10.1007/978-3-032-19777-1

Lukeneder A, Surmik D, Gorzelak P, Niedźwiedzki TB, Salamon MA (2020) Bromalites from the Upper Triassic Polzberg section (Austria); insights into trophic interactions and food chains of the Polzberg palaeobiota. Sci Rep 10:20545. https://doi.org/10.1038/s41598-020-77017-x

Lukeneder A, Lukeneder P (2021) The Upper Triassic Polzberg palaeobiota from a marine *Konservat-Lagerstätte* deposited during the Carnian Pluvial Episode in Austria. Sci Rep 11:16644. https://doi.org/10.1038/s41598-021-96052-w

Lukeneder A, Lukeneder P (2022a) Taphonomic history and trophic interactions of an ammonoid fauna from the Upper Triassic Polzberg palaeobiota. Sci Rep 12:7455. https://doi.org/10.1038/s41598-022-11496-y

Lukeneder A, Lukeneder P (2024) Ammonite taxonomy of the Carnian Polzberg *Konservat-Lagerstätte* in Austria. Jb Geol Bundesanst 162:27–55

Lukeneder A, Lukeneder P (2023) New data on the marine Upper Triassic palaeobiota from the Polzberg *Konservat-Lagerstätte* in Austria. Swiss J Palaeontol 142:9. https://doi.org/10.1186/s13358-023-00269-3

Lukeneder A, Lukeneder P, Sachsenhofer RF, Roghi G, Rigo M (2024a) Multi-proxy record of the Austrian Upper Triassic Polzberg Konservat-Lagerstätte in light of the Carnian Pluvial episode. Sci Rep 14:11194. https://doi.org/10.1038/s41598-024-60591-9

Lukeneder P, Fuchs D, Lukeneder A (2024b) Morphology, taxonomy and trophic interactions of rostrum-less coleoids from the Late Triassic Polzberg *Konservat-Lagerstätte* (Lower Austria). Swiss J Palaeontol 143:21. https://doi.org/10.1186/s13358-024-00319-4

Lukeneder P, Lukeneder A (2022c) Mineralized belemnoid cephalic cartilage from the late Triassic Polzberg *Konservat-Lagerstätte* (Austria). PLoS ONE, 17(4):e0264595. https://doi.org/10.1371/journal.pone.0264595

Lukeneder P, Lukeneder A (2026) Discovery and early scientific excavations at the Polzberg *Konservat-Lagerstätte.* In: Late Triassic *Konservat-Lagerstätten* within the Carnian Pluvial episode: 150 years of scientific excavations in Austria (ed Lukeneder A). Topics Geobiol 55:444. https://doi.org/10.1007/978-3-032-19777-1

Piller WE, Egger H, Erhart CW, Gross M, Harzhauser M, Hubmann B, van Husen D, Krenmayr H-G, Krystyn L, Lein R, Lukeneder A, Mandl GW, Rögl F, Roetzl R, Rupp C, Schnabel W, Schönlaub HP, Summesberger H, Wagreich M, Wessely G (2004) Die stratigraphische Tabelle von

Österreich 2004 (sedimentäre Schichtfolgen). Österreichische Akademie der Wissenschaften, Wien
Reis OM (1900) Coelacanthus lunzensis teller. Jb Geol Bundesanst 50:187–192
Reuss AE (1858) Ueber fossile Krebse aus den Raibler Schichten in Kaernthen. Beitr Palaeont Oesterr 1:1–6, Wien, Olmuetz
Schädel M, Eldijk T, Winkelhorst H, Reumer JWF, Haug J (2020) Triassic Isopoda: Three new species from Central Europe shed light on the early diversity of the group. Bull Geosci 95:145–166. https://doi.org/10.3140/bull.geosci.1773
Schultz O (2013) Catalogus fossilium austriae. Ein systematisches Verzeichnis aller auf österreichischem Gebiet festgestellten Fossilien. Band 3: Pisces. Verlag der Österreichischen Akademie der Wissenschaften, 57. ISBN 978–3–7001–7238–3
Stur D (1868) Beiträge zur Kenntniss der geologischen Verhältnisse der Umgegend von Raibl und Kaltwasser. Jahrb KK Geol RA 1:71–122
Stur D (1874) Neue Aufschlüsse im Lunzer Sandsteine bei Lunz und ein neuer Fundort von Wengerschiefer im Pölzberg zwischen Lunzersee und Gaming. Verh K-K Geol RA 1:271–273
Stur D (1886) Vorträge: vorlage des ersten fossilen Schädels von Ceratodus aus den obertriadischen Reingrabner Schiefern von Pölzberg nördlich bei Lunz. Sitzung am 30. 381–383
Suess E (1865) Über die Cephalopoden-Sippe Acanthoteuthis R. Wagu. Sitzungsber K Akad Wiss Math-Naturwiss Kl 51:225–244
Teller F (1891) Über den Schädel eines fossilen Dipnoers, Ceratodus Sturii nov. spec. aus den Schichten der oberen Trias der Nordalpen. Abh K-K Geol RA 15:1–38
Thenius E (1968) Emil Weinfurter. Austrian J Earth Sci 61:188–191
Tollmann A (1976) Analyse des klassischen nordalpinen Mesozoikums. Stratigraphie, Fauna und Fazies der Nördlichen Kalkalpen. Franz Deuticke Wien, 580 p
Unknown author (1886) Tagesbericht. Aus der geologischen reichsanstalt. Wiener Allgemeine Zeitung. Morgenblatt. Nr. 2430, 2
Unknown author (1873) Das Maria-Theresia Standbild in Klagenfurt und die Anwesenheit des Kronprinzen Rudolf in Kärnten vom 3. bis 19. Carinthia I 63:213–215
Wöhrmann S (1893) Die Raibler Schichten nebst kritischer Zusammenstellung ihrer Fauna. Jahrb KK Geol RA 4:617–768
Woodward AS (1895) Catalogue of the fossil fishes in the British museum (Natural History), Part III, 544
Wulfen FX (1793) Abhandlung vom Kärnthenschen Pfauenschweifigen Helmintholith oder dem Sogenannten Opalisirenden Muschelmarmor:124 p

Chapter 4
The Lower Carnian *Konservat-Lagerstätten* Belt in Austria

Alexander Lukeneder

Abstract A new geological term marked by the presence of frequent lower Carnian *Konservat-Lagerstätten* in the Northern Calcarous Alps of Austria—the Carnian *Konservat-Lagerstätten* Belt (CKLB)—is introduced. We examined several marine Upper Triassic localities and new fossil material obtained from bed-by-bed collections. All are of equal age and stem from the same characteristic fossiliferous interval at the basal well-laminated Reingraben Formation. Over 15,000 fossils in historical collections and new excavations of various marine taxa were studied and described in numerous papers. The fossiliferous layers from the Early Carnian (Julian 2, *Austrotrachyceras austriacum* Zone) feature excellent preservation of invertebrate, vertebrate and plant remains. Remarkable assemblages comprise ammonoids (*Austrotrachyceras, Paratrachyceras, Carnites*), coleoids (*Mojsisovicsteuthis, Phragmoteuthis*), bivalves (*Peribositra, Halobia*), gastropods (caenogastropods), arthropod thylacocephalans (*Atropicaris, Austriocaris, Paraostenia*), crustaceans with decapods (*Platychela, Antrimpos*), isopods (*Obtusotelson, Discosalaputium*) and branchiopods (*Euestheria*), polychaetes (*Palaeoaphrodite*, Eunicidae), coelacanth remains (*Coelacanthus*), frequent acytinopterygian fishes (*Peltopleurus*), a lungfish (*Tellerodus*), bromalites (coprolites, regurgitalites) along with frequent coniferopsid plant remains (*Voltzia,* indet plants) accumulated in layers. The lower Carnian paleobiotas were primary deposited in the normal marine Reifling Basin, characterized by a fluctuating basinal environment with intermittent influx of freshwater, accompanied by the enhanced presence of land plants. The most remarkable and unique feature of all these sections, deposited during the Carnian Pluvial Episode, is the developed formation of *Konservat-Lagerstätten.* These are charactzerized by the preservation of entire and fragile fossils, soft parts and thealmost complete paleobiota referring to the taphonomic history and the trophic food web of ancient paleocommunities. The prerequisites for high-quality preservation—calm and oxygen-depleted conditions—prevailed at the sea floor of the Reifling Basin. This triggered the deposition of the black, laminated, marly basal part of the Reingraben Formation, generating a zone termed the Carnian *Konservat-Lagerstätten* Belt.

A. Lukeneder (✉)
Department of Geology and Paleontology, Natural History Museum Vienna, Vienna, Austria
e-mail: alexander.lukeneder@nhm.at

A. Lukeneder (ed.), *Late Triassic Konservat-Lagerstätten within the Carnian Pluvial Episode*, Topics in Geobiology 55, https://doi.org/10.1007/978-3-032-19777-1_4

Keywords Carnian *Konservat-Lagerstätten* Belt · Paleobiota · Early Carnian · Late Triassic · Austria

4.1 Introduction

The paleobiotas of fossiliferous sites known as *Konservat-Lagerstätte* (Seilacher 1970; see also Krystyn 1991; Forchielli and Pervesler 2013) are characterized by a special mode of fossil preservation, making them precious sources of paleobiological information (Briggs 2014; marine conservation deposit in Klug et al. 2024a). Such conservation *Lagerstätten* provide unique insights into paleocommunities, food chains and dietary habits, as well as into trophic interactions between the inhabitants of marine ecosystems (Lukeneder et al. 2020, and references therein). New findings of additional fossil groups and the soft body preservation of numerous organisms shed light on such otherwise hidden Upper Triassic organisms. Middle to Upper Triassic occurrences exhibiting the preservation style of *Konservat-Lagerstätten* are rare (Bernardi et al. 2011; Furrer 2003, 2024; Pieroni and Furrer 2020; Klug et al. 2024a, b) and hence even more important for drawing conclusions about paleoenvironments along with their inhabitants and interactions. Middle Trissic Monte San Giorgio (S. Switzerland and N. Italy) and Upper Triassic Cave del Predil (formerly Raibl; N. Italy; Wöhrmann 1889, 1893; Lukeneder and Lukeneder 2022a, b) are in comparable numerous aspects (paleocommunity, paleoecology), though not the same exact age. The genesis of the Polzberg section is comparable, though not equal in age and taxa, to the Austrian Upper Triassic Seefeld Formation (Norian) fish accumulations from Seefeld (Tyrol; Brandner and Poleschinski 1986) and Wiestal (Salzburg; Hornung et al. 2019). Klug et al. (2024b) analyzed 45 Triassic conservation deposits and differentiated four main types of *Konservat-Lagerstätten*: Burgess type, Monte San Giorgio type, Holzmaden type and Solnhofen type. The main criteria were the facies and their fossil content of the distinct *Lagerstätte*. The type localities correspond to Paleozoic black shales, Triassic black shales, pyrite-rich black shales and platy limestones respectively. Though known for over 150 years, the Polzberg *Konservat-Lagerstätte* along with its formation processes and fossil assemblages are poorly understood and new taxa appear almost annually. The first upper Triassic locality termed as *Konservat-Lagerstätte* is located in Lower Austria (Fig. 4.1) and also known as Schindelberg or Pölzberg (Stur 1874, 1886) in historical collections, namely the Polzberg *Konservat-Lagerstätte*. It is associated with lower Carnian Reingraben Shales ("Reingrabner Schiefer", "*Trachyceras* Schiefer"). The unique feature of this *Konservat-Lagerstätte* is the basal few meters, comprising the fossiliferous Reingraben Formation with a uniquely preserved fauna (Trauth 1935; Krystyn 1991; Lukeneder and Lukeneder 2021). The paleontological sites in the region of Polzberg are known since the nineteenth century (Stur 1886; Teller 1891; Abel 1906). More recently, new paleontological data and faunal elements were published from the Polzberg *Lagerstätte* (Forchielli and Pervesler 2013; Griffith 1977; Doguzhaeva et al.

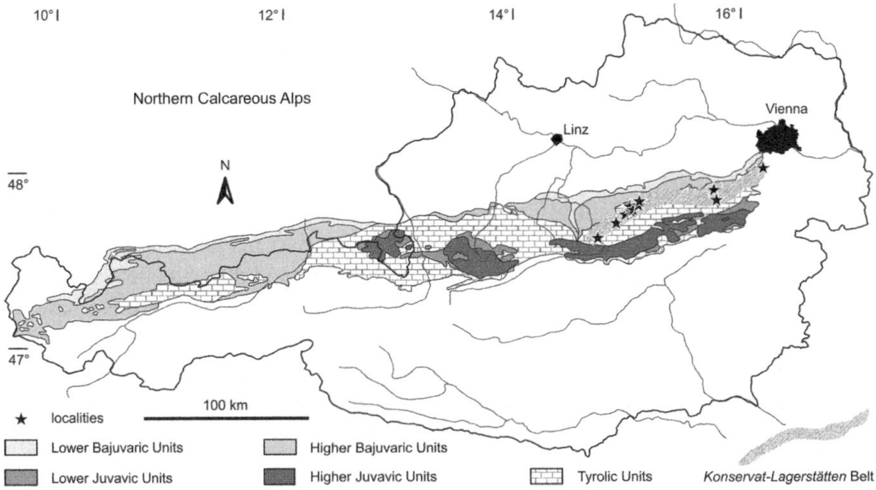

Fig. 4.1 Map of the marine Carnian *Konservat-Lagerstätten* Belt (CKLB; oblique shading) in the Northern Calcareous Alps of Styria and Lower Austria. Marked positions of the main *Konservat-Lagerstätten* localities (black asterisks) within the basal argillaceous Reingraben Formation

2007a, b; Doguzhaeva and Summesberger 2012; Schädel et al. 2020; Lukeneder et al. 2020, 2024b; Lukeneder and Lukeneder 2021, 2022a, b, 2024a, b; Lukeneder 2026).

Since the first papers (Stur 1874, 1886; Teller 1891) on fossils of the first-noted Triassic *Konservat-Lagerstätte*, namely the Polzberg *Konservat-Lagerstätte* in Lower Austria, numerous papers have been published on this specific Late Triassic fossil site (Seilacher 1970; Allison 1988; see also Krystyn 1991; Forchielli and Pervesler 2013). The pioneer period of sampling here started in the 1870s, resulting in a significant sampling period of over 150 years. Modern studies include paleoenvironmental analyses and investigations on the trophic food web (Lukeneder et al. 2020; Lukeneder and Lukeneder 2021), the taphonomy, trophic history and taxonomy of ammonoids (Lukeneder and Lukeneder 2022a, b, 2024a, b), and new findings of unique, mineralized belemnoid cartilage (Lukeneder and Lukeneder 2022b).

Fossil remnants from *Konservat-Lagerstätten* form valuable sources of paleobiological information. For the present contribution, excavations with bed-by-bed collecting were done during several field campaigns (60 days sampling) during four consectutive years between September 2020 and August 2025 (and still ongoing). The resulting details on the fauna and flora yielded new insights into the lower Carnian paleocommunities of the Northern Calcareous Alps in the north of Styria and Lower Austria (Lukeneder and Lukeneder 2021, 2023; Lukeneder et al. 2024a). This Special Volume (Lukeneder 2026) provides more details on distinct taxonomic groups (thylacocephalans, bristle worms, conodonts, gastropods, bivalves, fishes, phragmoteuthids, bromalites), the geochemistry, geophysics, lithological aspects and food chains and dietary habits, as well as trophic interactions between the inhabitants from the Upper Triassic marine ecosystems. These paleobiotas were deposited during the Carnian Julian 2 Ib (*Austrotrachyceras austriacum* Zone, *A. minor* biohorizon), within a period known as the worldwide Upper Triassic Carnian Pluvial Episode (CPE; Rigo et al. 2007; Lukeneder et al. 2012; Ruffell et al. 2015; Dal Corso et al. 2018; Simms and Ruffell 2018, 1990; Zhang et al. 2023, 2024, Lukeneder et al. 2024a; Alverà et al. 2026; Lukeneder 2026, this volume; Wet Intermezzo, Kozur and Bachmann 2010; Ogg 2015). The CPE features an enhanced humidification of the Carnian climate, resulting in the deposition of the fossiliferous Reingraben Formation at Scheiblingbach and Polzberg. As noted by Lukeneder and Lukeneder (2021) as well as Lukeneder et al. (2024b), such laminated deposits provide excellent conditions for the preservation of entire and soft body fossils, a prerequisite for *Konservat-Lagerstätten.*

The study presents insights into the biostratigraphic and geographic range of the unique lower Carnian *Konservat-Lagerstätten* Belt within the frame of an intraplatform basinal environment, the Reifling Basin.

4.2 Material and Methods

Historical excavation campaigns, conducted by Joseph Haberfelner, to obtain the Polzberg fossils were organized by the Geological Survey of Austria (GBA; since 2023 GeoSphere Austria) in 1885 and the Natural History Museum of Vienna (NHMW) in 1909 (Stur 1886; Teller 1891; Griffith 1977), resulting in 6397 fossil specimens. More recently, additional material (n 5290) was collected by the author (and team) and private collectors (Birgitt and Karl Aschauer; n 282 donated to GBA collections). The fossil fauna from early collecting at Polzberg is deposited in various Austrian institutions. The enormous amount of historical and recent material in collections from the comparable Scheiblingbachgraben (n 4602), Steinbachgraben (n 2650), Rehberg (n 437), Polzberg (11,687), Rad (n 124), and Hinterbrühl (n 47) localities encompasses approx 30,000 fossil remains. They are held in distinct collections of the NHMW, the GBA, and the Departments of Earth Sciences at the University of Vienna (DES; n 13), with smaller amounts in the Lower Austria State Collections (LASC; n 21). The by far highest numbers of specimens including holotypes and rare taxa are held at the NHMW, followed by the collection of the GBA.

4.3 Geologic Setting and Lithology

The Reifling Basin is an elongated, approximately 300–500 km long, NE to SW striking intraplatform basin. It is termed after Großreifling in northern Styria, in its narrow sense from its western border in St Gallen (Styria) to the eastern Rohr im Gebirge (Lower Austria). The Reifling Basin is framed to the north and south by Wetterstein Carbonate Platforms. It is characterized by the occurrence of specific lithological successions, here mainly the Lunzer Fazies (= Lunz Facies Zone LFZ) interdigitating with the Rohrer Fazies (= Rohr Facies Zone RFZ) to the east, and the Traunalpen Fazies (= Traunalpen Facies Zone TFZ) to the west (Spengler 1931; Thenius 1974; Tollmann 1976; Wessely 2006). Note that the LFZ reaches the westernmost areas of the Northern Calcareous AlpsNorthern Calcareous Alps the (NCA), descending below the Vienna Basin and finding its continuation in the Carpathian Mountains (Dobrá Voda Depression, pers. comm. Michael Moser). The Frankenfelser Fazies (Frankenfels Facies Zone) follows to the north and the Ötscher Fazies (Ötscher Facies Zone) to the south. Traunalpen, Frankenfels, Lunz and Rohr Facies Zones are subsumed into the main Hauptdolomit Fazies, whereas the southern Ötscher Fazies is intergrated to the main Dachstenkalk Fazies (Tollmann 1976).

Most important for the occurrence of the lower Carnian Reingraben Formation with the incorporated *Konservat-Lagerstätten* from Rad, Polzberg, Rehberg, Großau, Göstling an der Ybbs, and Scheilingbachgraben is the Lunz Facies Zone (Fig. 4.1). Bajuvaric nappes (Lunz Nappe and Sulzbach Nappe) and the northern Tyrolic Nappes (e.g., Reisalpen Nappe and Göller Nappe) are important in this respect. Within the LFZ the following lithological units occur and characterize the zone: Steinalm Formation (Anisian), Gutenstein Formation (Anisian), Reifling Formation (Anisian to Carnian), Raming Formation (Anisian to Carnian), Partnach Formation (Ladinian to Carnian), Göstling Formation (Carnian), Reingraben Formation (Carnian), Lunz Formation (Carnian), and Opponitz Formation (Carnian; Spengler 1959, 1963; Wagner 1970; Thenius 1974; Tollmann 1966, 1976; Oberhauser 1980; Faupl 2003; Piller et al. 2004; Wessely 2006).

After a strongly bioturbated top of the Göstling Formation, the overlying terrigeneous siliciclastic deposits of the Reingraben Formation (*A. austriacum* Zone, *A. minor* Subzone, Fig. 4.2) consist of a pronounced alternation of dark gray to black argillaceous marls ($CaCO_3$ 5–20%) intercalated by rare limestone/dolomite layers. No macro-bioturbation is visible. This lithological change marks the 'Reingrabener Wende' (Schlager and Schöllnberger 1974; = Reingraven Turnover, Mueller et al. 2016) within the humid phase of the Carnian Pluvial Episode (Ruffell et al. 2015). The bivalve *Peribositra* appears in all ontogenetic stages and, in layer-/rock-forming quantities, mostly in the articulated butterfly position lumachelles. The Late Triassic *Konservat-Lagerstätten* localities comprise invertebrates and vertebrates (fishes). Most of the fossils derive from the lowermost three calcareous meters of the Reingraben Formation, but macrofossil occur throughout the section. The paleobiota shows a nekton-dominated fauna with abundant fishes and cephalopods (Lukeneder and Lukeneder 2021, 2023; Lukeneder et al. 2020, 2024a).

The Upper Triassic outcrops at **Polzberg (Po)** are located on the western slope of Mount Schindelberg (1066 m), north of the Ois River, 4 km northeast of Lunz am See in Lower Austria (Stur 1874, 1886; Lukenerder and Lukeneder 2021; 1:50 000, geological map, sheet 71 Ybbsitz; Ruttner and Schnabel 1988 and sheet 72 Mariazell; Bauer and Schnabel 1997; Figs. 4.1 and 4.3). Within the Lunz Nappe, the Reifling Basin (Krystyn 1991; Krystyn and Lein 1996; Lukenerder and Lukeneder 2021)—an Upper Triassic intraplatform basin—is located between Polzberg and Großreifling. The exact position of the fossiliferous localities in the southern area of the Lunz Nappe within the lower part of the Reingraben Formation (Tollmann 1976; Piller

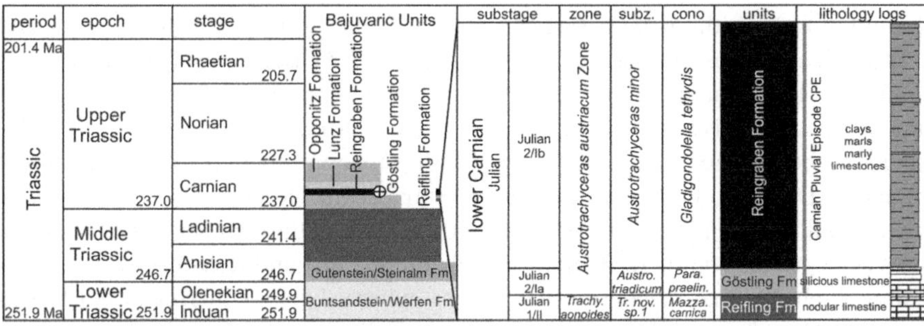

Fig. 4.2 Stratigraphic position of the Carnian *Konservat-Lagerstätten* Belt (CKLB) deposits during the lower Carnian. White circle: position of the CKLB in Austria. Compiled lithological units deposited during the Carnian Pluvial Episode (CPE). Austro *Autrotrachyceras*, Trachy *Trachyceras*, chrono chronostratigraphy, subz subzone, Fm Formation. Adapted after Lukeneder and Lukeneder (2021)

et al. 2004) was determined by GPS (global positioning system): N 47°53′4.98″, E 15°4′28.15″.

The **Rehberg section (Re)** is located near Lunz am See (N 47°51′37.60″, E 15°3′59.30; Lower Austria) on the northwestern slope of Mount Maißzinken (1057 m), situated 2.5 km south to the Polzberg site, 700 m west of the lake Lunzer See on the Sulzbach Nappe within the Lunz Nappe System (High-Bajuvaric System; Geo map 71 Ybbsitz; Ruttner und Schnabel 1988; Figs. 4.1 and 4.3). The section exhibits approx. 2.5 m of the Göstling Formation and, at the top, the argillaceous Reingraben Formation with characteristic entire fish and ammonite abundance.

The **Lehen section (Le)** is located near Lunz am See (N 47°50′1.26″, E 15°1′6.83″; Lower Austria) on the northern slope of Mount Großkopf (1147 m), situated 4.7 km southwest of the Rehberg site on the Sulzbach Nappe within the Lunz Nappe System (High-Bajuvaric System; Geo map 71 Ybbsitz; Ruttner und Schnabel 1988; Figs. 4.1 and 4.3). The section exhibits the topmost Reifling Formation, approx. 4 m of the Göstling Formation and, at the top, the argillaceous Reingraben Formation.

The **Großau section (Gro)** is located on the northeastern slope of Mount Großkopf (1147 m), 300 m southeast of the Ybbs River, in the Ypps valley connecting Göstling and Lunz am See (Figs. 4.1 and 4.3). The drilling site (N 47°49′24.40″, E 15°00′0.80″) is located 5 km southwest of the Lunz am See center and 5 km northeast of Göstling an der Ybbs in the southwest of Lower Austria (Trauth 1948; geological map, 71 Ybbsitz, Ruttner and Schnabel 1988). The southernmost tectonic element of the NCA in southeastern areas of Lower Austria around Grossau is the Lunz Nappe, occasionally interpreted as being a distinct nappe, namely the Sulzbach Nappe.

The **Steinbach section (St)** is located near Göstling an der Ybbs (N 47°48′22.20″, E 14°57′0.60″; Lower Austria), on the northern slope of Mount Großschöntaler Berg (881 m), situated 4 km southwest of the Großau drilling site on the Sulzbach Nappe within the Lunz Nappe System (High-Bajuvaric System; Geo map 71 Ybbsitz; Ruttner und Schnabel 1988; Figs. 4.1 and 4.3). The outcop is located within and under

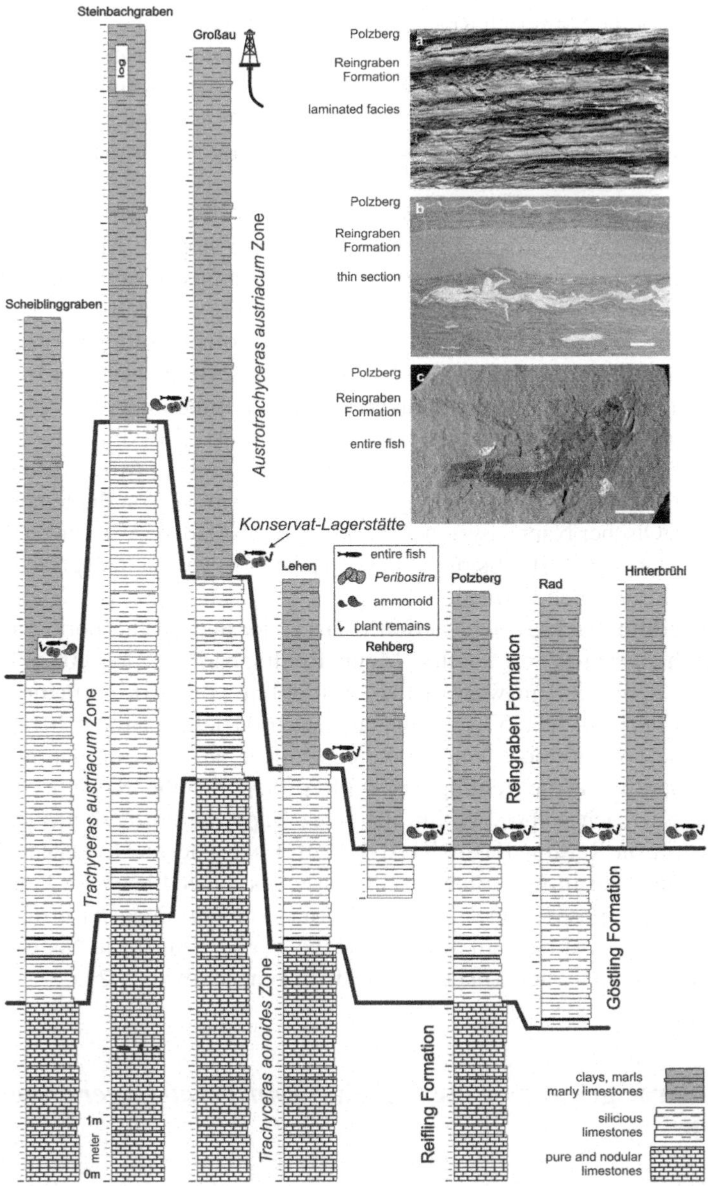

Fig. 4.3 Main localities of the marine Carnian *Konservat-Lagerstätten* Belt (CKLB), Northern Calcareous Alps of Styria and Lower Austria. Geographically arranged from west (Scheiblinggraben in Styria) to east (Hinterbrühl in Lower Austria). Position of the CKLB marked by fossil symbols at the basal argillaceous Reingraben Formation. **a** Laminated, argillaceous facies of the basal Reingraben Formation at the Polzberg locality (Lower Austria). **b** Thin section of the basal laminated argillaceous layers of the Reingraben Formation, note white remnants of crushed ammonoid shells (*Austrotrachyceras minor*). **c** Entire fish, *Nannolepis elegans* (NHMW 2007z0170/0148) from the lower Carnian Polzberg *Konservat-Lagerstätte*. Scale bars: **a** 50 mm, **b** 1 mm, **c** 10 mm

the lower end of the Steinbach River and at the transition into the bigger Ybbs River, and on both banks of the Ybbs River, accessible only at low water of the river. The section exhibits the Reifling Formation, approx. 10 m of the Göstling Formation and at the top the argillaceous Reingraben Formation with frequent intercalated calcareous layers. The calcareous layers are of allodapic character with transported fragments and fossils (e.g., foraminifera).

The **Rad section (Rad)** is located 2.5 km south of Kleinzell (N 47°57′7.90″, E 15°43′37.50″; Lower Austria) on the northwestern slope of Mount Hirschkogel (990 m), along a forest road on the tectonic Reisalpen Nappe of the Tyrolic System (Geofast map 74 Hohenberg; Moser and Pavlik 2013; Figs. 4.1 and 4.3). Single oolitic layers with bivalves, and foraminifera are present above the Göstling Formation. They were transported into the basin from the edge or slope of the basin, or from nearby platforms within the basal argillaceous Reingraben Formation.

The **Hinterbrühl locality (Hint)** is located 4 km west of Mödling (N 48°4′53.45″ and E 16°14′20.05″; Lower Austria) on the northern slope of Mount Schwarzkogel (406 m) along abandoned historical quarries. It is positioned on the tectonic Göller Nappe of the Ötscher Nappe System (Tyrolic System, Geo map 58 Baden, Schnabel 1997; Figs. 4.1 and 4.3). Historical fossil specimens mirror the presence of the argillaceous Reingraben Formation.

The **Scheiblingbachgraben-section (Sch)** is located on the northern slope of Mount Lärchkogel (1087 m), south of Mount Reiflingberg (712 m), 600 m southwest of the Enns River, 300 m west of the Großreifling center in northern Styria (N 47°39′52.50″, E 14°42′27.75″) (geological map, 100 Hieflau, Kreuss 2014; Figs. 4.1 and 4.3). The northernmost tectonic element of the NCA in northernmost Styria around Großreifling is the Reifling Unit (= Reilfinger Scholle, Krystyn et al. 2008), part of the Sulzbach Nappe. Within the Lunz Nappe in northernmost Styria and Lower Austria, the Reifling Basin (Krystyn 1991; Lukeneder and Lukeneder 2021)—an intraplatform basin during the Late Triassic—is located between Polzberg and Großreifling. It exposes Upper Triassic (Carnian, Julian) deposits of the Reifling Basin, which is part of the Lunz Nappe and the southern Sulzbach Nappe (Reifling Unit, NCA; Lukeneder and Lukeneder 2021; Lukeneder et al. 2024a).

4.4 Biostratigraphy of the Carnian *Konservat-Lagerstätten* Belt

The Göstling Formation is ascribed to the *A. triadicum* Subzone, whereas the Reingraben deposits belong to the higher *A. minor* Subzone (Julian 2, Fig. 4.2). The lowermost, fossiliferous part of the Reingraben shales is dated based on the occurrence of *Austrotrachyceras minor*, *Paratrachyceras haberfellneri* and the accompanying *Carnites floridus* as Julian 2 (Julian 2/Ib), similar to the biostratigraphic situation at Polzberg in Lower Austria (Lukeneder et al. 2024b). The presence of biostratigraphic index ammonoid members together with the mass occurrence of the benthic bivalve

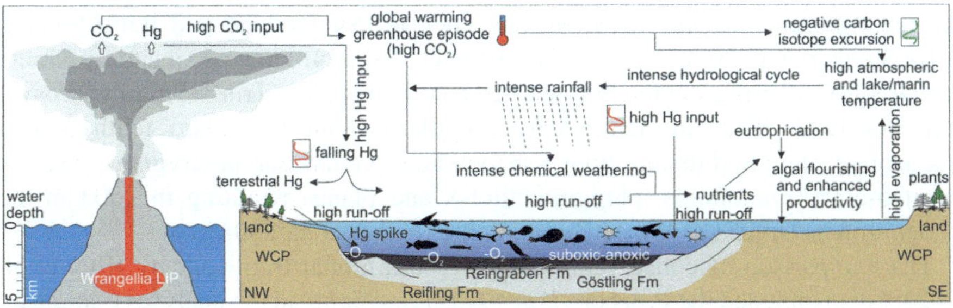

Fig. 4.4 Schematic environmental reconstruction and possible climate alteration in the lower Carnian Reifling Basin triggered by enhanced emission of CO_2 and Hg from the Wrangellia Large Igneous Province (LIP) during the Carnian Pluvial Episode in the lower Carnian (Julian). LIP Large Igneous Province, WCP Wetterstein Carbonate Platform, Fm Formation, Hg mercury, C carbon, O oxygen. Scheme adapted after Zhang et al. (2024)

Peribositra within the fossiliferous layers strengthen the biostratigraphical assignment as Julian age (Mueller et al. 2016; Lukeneder and Lukeneder 2021, 2022a, b; Lukeneder et al. 2024a). Concerning the ammonoid zonation and the specific duration of ammonoid zones in the late Carnian (Mueller et al. 2016; Rigo et al. 2018; Dal Corso et al. 2020), the analyzed sequence encompasses deposits of approximately 500 000 years (Fig. 4.4).

4.5 The Polzberg *Konservat-Lagerstätten* Type

Middle to Upper Triassic occurrences exhibiting the preservation style of *Konservat-Lagerstätten* (= conservation deposits) are rare (Bernardi et al 2011; Furrer 2003, 2024; Pieroni and Furrer 2020; Klug et al. 2024a, b). This makes them even more important for drawing conclusions about paleoenvironments along with their inhabitants and interactions. The genesis of the Polzberg section is comparable, though not contemporaneous in age and taxa, to the Austrian Upper Triassic Seefeld Formation (Norian) fish accumulations from Seefeld (Tyrol; adapt to Brandner and Poleschinski 1986) and Wiestal (Salzburg; Hornung et al. 2019). The most famous Triassic fossil *Konservat-Lagerstätte* scientifically excavated over a long history of over 100 years is the UNESCO World Heritage Site Monte San Giorgio at the border from S. Switzerland and N. Italy (Furrer 2003, 2024; Furrer and Vandelli 2014; Klug et al. 2024a, b). Triassic deposits from the Monte San Giorgio site span from the Olenekian Servino Formation, over the Anisian Bellano Formation, the San Salvadore Dolomite and the Besano Formation, over the Ladinian Meride Limestone (with interbedded subunits and black shale intervals) to the Carnian Pizzella Marls. Klug et al. (2024b) reviewed especially those units with black shales comprising the well-known conservation deposits such as in the fossiliferous Besano Formation (formerly Grenzbitumenzone), the younger fossil-bearing units Cava inferiore, Cava superiore, Cassina beds, and the Kalkschieferzone of Monte San Giorgio (Switzerland and Italy; Stockar 2010; Stockar et al. 2012). As noted by Klug et al. (2024b), the fossil sites at Monte San Giorgio still represent the best-studied record of Middle Triassic marine ecosystems worldwide. They are rich in marine reptiles (eosauropterygians, ichthyosaurs, placodonts), chondrichthyans, bony fish (actinopterygians, sarcopterygians), various invertebrates (ammonoids, phragmotethids), and plants, resulting in 2003 in the recognition as UNESCO World Heritage Site (Felber et al. 2004).

The comparison of conservation deposits in a global framwork by Klug et al. (2024b) also included the Polzberg *Konservat-Lagerstätte* in the analyses (variance–covariance principal component analysis) with respect to biostratigraphy, paleobiota, preservation mode, thickness, range and depositional characteristics. The authors concluded that the Polzberg (Reingraben Formation; not Lunz Fm as given in Klug et al. 2024b) conservation deposits plotted near real Black Shale *Lagerstätten* as the Holzmaden type (e.g., Oxford Clay, Posidonienschiefer), Monte San Giorgio type (Besano Formation, Anisisn, Middle Triassic), and the Paris Biota (Thaynes Formation, Olenekian, Early Triassic). Note, however, that the Polzberg (Reingraben Formation, Carnian, Late Triassic) and the Paris Biota (Thaynes Formation) are chracteristically more invertebrate dominated. From own investigations and years of excavation in the fossiliferous Reingraben Formation, the Polzberg *Konservat-Lagerstätte* does not represent a real Black Shale Lagerstätte of the Holzmaden type (Klug et al. 2024b) as noted for Posidonienschiefer or Oxford Clay. This is because Polzberg and contemporaneous *Lagerstätten* of the lower Carnian of Austria are not associated with shales, are mostly not black, and have less total and only a rare pyrite content. More likely the Polzberg *Konservat-Lagerstätten* type, and hence all sites within the Carnian *Konservat-Lagerstätten* Belt (CKLB), are an independant type of conservation deposits, the Polzberg type.

The Polzberg type of *Lagerstätte* appears in all contemporaneous Austrian localities within the lower Carnian *Konservat Lagerstätten* Belt (CKLB; Figs. 4.1 and 4.3) in a similar mode:

status	*Konservat-Lagerstätte* (conservation deposit)
epoch	Late Triassic
stage	Carnian
substage	lower Carnian
zonation	Julian 2/Ib
biostratigraphic range	basal *A. minor* Subzone, *Austrotrachyceras austriacum* Zone (Fig. 4.2)
age	aprox. 233–234 million years
duration approximately	200 000 to 500 000 years
Carnian Pluvial Episode	formed during humid cosmopolitan interval CPE
causes	strong volcanic activity, increased emission of CO_2 and Hg from the Wrangellia Large Igneous Province (LIP), followed by humidification with heavy rain and global warming with enhanced terrestrial input into marine basins (Fig. 4.4)

geological unit	Reingraben Formation, Northern Calcareous Alps
facies	laminated marls with allodapic limestone intercalations, partly lumachelles, increasing clay content
thickness of level	approx. 5 m
bioturbation	absent
paleoenvironment	marine intraplatform basin, Reilfing Basin
water body	normal marine
freshwater	with intermittant freshwater influx
sea floor	dysoxic to partly anoxic, partly stagnant conditions
sediment	laminated, no bioturbation, anoxic
paleobiota	dominated by invertebrates (Fig. 4.5)
invertebrates	dominated in the Julian 2/Ib by benthic bivalves, gastropods, arthropods, polychaetes, nektic ammonoids, coleoids (Fig. 4.6)
vertebrates	dominated by nektic carnivorous actinopterygian fishes, sarcopteryian fishes, rare chondrichthyans
trace fossils	frequent coprolites and regurgitalites containing ammonoids, coleoids and conodonts
plants	dominated by coniferopsids
palynology	rich terrestrial association, typical Carnian sporomorphs
taphonomy	exceptional preservation with entire and articlulated carcasses, in situ buccal masses, fine morphological details and soft parts, carbonized and phosphatized fossil remnants

Late Triassic *Konservat-Lagerstätten*

main features	Hinterbrühl	Rad	Polzberg	Rehberg	Lehen	Großau	Steinbach	Scheibling
entire fish	☆	☆	★	★	☆	★	★	★
entire coleoids	☆	☆	★	☆	☆	☆	☆	☆
entire crustacea	☆	☆	★	★	☆	★	★	★
ammo/soft parts	☆	☆	★	★	☆	★	★	★
ammo/buccal	☆	☆	★	★	☆	★	★	★
ceph/cartilage	☆	☆	★	★	☆	☆	★	★
polychaetes	☆	☆	★	☆	☆	☆	☆	☆
bivalve/butterfly	☆	☆	★	★	☆	★	★	★
bromalites	☆	☆	★	★	☆	★	★	★
plant remains	☆	☆	★	☆	☆	★	★	★
no bioturbation	☆	☆	★	★	☆	★	★	★
laminated facies	☆	☆	★	★	☆	★	★	★

Fig. 4.5 Characteristic invertebrate, vertebrate and plant members of the lower Carnian paleobiota of the Austrian Carnian *Konservat-Lagerstätten* Belt (CKLB), Northern Calcareous Alps. Main *Konservat-Lagerstätten* localities: black asterisks presence, gray asterisk supposed presence

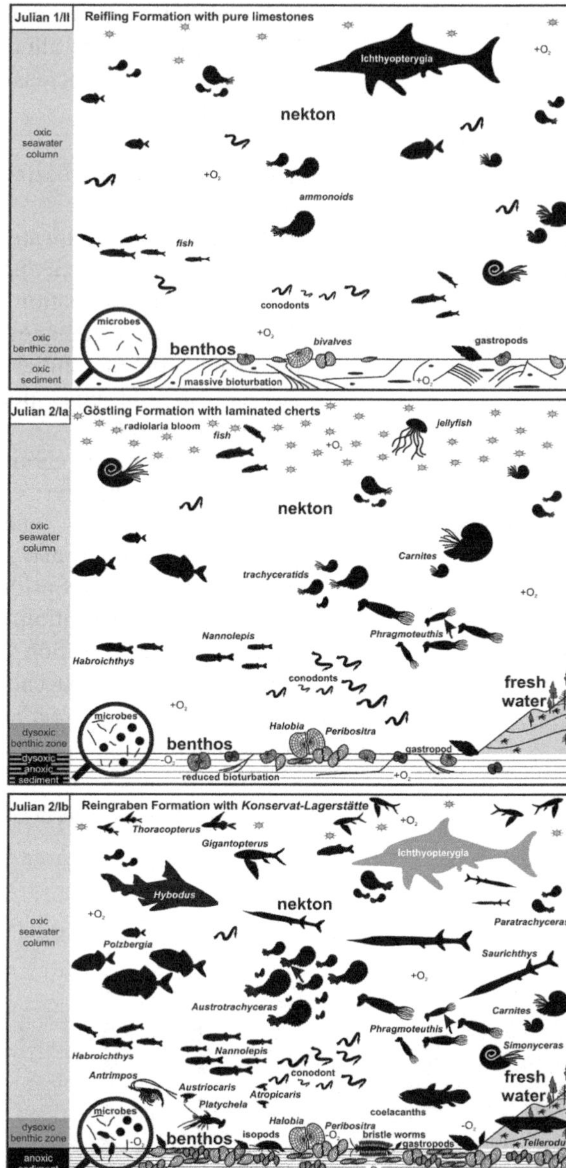

Fig. 4.6 Lower Carnian macrofaunal paleobiota (black) in the Reifling Basin based on direct evidence from the fossil record of all localities in the Carnian *Konservat-Lagerstätten* Belt. Direct evidence from fossils and bromalites. Ichthyopterygia not detected yet, most likely present. Figured members of paleobiota not to scale. Note the change in diversity and abundance during the lower Carnian and the adaptation of the ecosystem and paleocommunities during and after normal conditions **a** in the Julian 1/II (Reifling Formation), **b** the initial phase of the CPE in the Julian 2/Ia (Göstling Formation) to **c** the maximum in the Julian2/Ib (Reingraben Formation). Adapted after Lukeneder and Lukeneder (2021)

geochemistry	low total organic carbon (mean 0.92%) and sulfur values (mean 1.05%) carbonate content varies from 6 to 80%
pyrite	rare to absent
projects on the CPE	Drilling into the Earth's History (K3-F-964/002–2023, 2023–2026; Federal Government of Lower

	Austria, Freunde NHMW Trias); Polzberg—a *Konservat-Lagerstätte* of Weltruf (K3-F-964/001–2020; 2021–2022, Austrian Academy of Science, Federal Government of Lower Austria).
literature	Stur (1874, 1886), Teller (1891), Abel (1906), Glaessner (1931, 1965), Trauth (1935), Griffith (1977), Krystyn (1978, 1991), Doguzhaeva et al. (2007a, b), Doguzhaeva and Summesberger (2012), Hornung et al. (2007), Hornung (2008), Forchielli and Pervesler (2013), Schädel et al. (2020), Lukeneder and Lukeneder (2021, 2022a, b, 2023, 2024a, b), Lukeneder et al. (2020, 2024b), Laville et al. (2024), Lukeneder et al. (2023, 2024a, b), Lukeneder (2026 this volume).

4.6 Conclusions

A new term for the tectonic and geographic area of lower Carnian *Konservat-Lagerstätten* is introduced. The Carnian *Konservat-Lagerstätten* Belt (CKLB) is a unique feature within the Northern Calcareous Alps of Styria and Lower Austria, formed during the Carnian Pluvial Episode (CPE). Historical (1885 to 1909) and recent excavations (2015 to 2025) with bed-by-bed collecting from several lower Carnian localities detected the lower Carnian *Konservat-Lagerstätte* in all sections at the base of the lower Carnian (Julian 2/Ib) Reingraben Formation. Thousands of fossil remains from the Late Triassic (Early Carnian) provided new details and insights into a new paleocommunity characterizing the *Konservat-Lagerstätten.* The CKLB was deposited within the same marine intraplatform Reifling Basin. All sections and localities are equivalent in age and paleobiota based on primary investigation of the well-known Polzberg *Konservat-Lagerstätte.* The basal sediments of the Reingraben Formation were deposited at an epipelagic to upper mesopelagic position at the sea floor of the NE/SW elongated Reifling Basin. Thin-shelled posidoniid bivalves dominate in the lower Carnian deposits of the Reifling Basin, comprising the well-known fossiliferous Polzberg, the newly described Scheiblingbach, and the newly detected and described Steinbachgraben, Lehen, Rehberg, Rad, and Hinterbrühl *Konservat-Lagerstätten* localities, all located in the CKLB. Within lower Carnian deposits of the CKLB, ceratitid nektic/nektobenthic ammonoids and nektic actinopterygian fishes prevailed in the overlying oxygenated water column. The new fossil assemblages from the CKLB provide new insights into this lower Carnian marine ecosystem, deposited during the CPE. All *Konservat-Lagerstätten* within the CKLB localities feature an excellent preservation of entire carcasses and soft body as well as of fragile skeletons (nektic and benthic taxa). This points to unique paleoenvironmental conditions in the lower Carnian dysoxic bottom water at the marginal edges of the Reifling Basin. Future work will concentrate on detecting and analyzing more entire sections, confirming and strengthening the range assumed for the CKLB within the Lunz, Sulzbach, Göller, and Reisalpen nappes.

Acknowledgments The author is particularly grateful to the owners of the outcrop area, the Austrian Federal Forests Company (Österreichische Bundesforste), for sampling and digging permission. Especially Lukas Nöst (forest district Grossreifling) and Dominic Asztaller (forest dristrict Grossreifling) for ther advice and help during the field campaigns. I thank Petra Lukeneder (Gablitz), Veronika, N.I. Koukal (Vienna) and Justine and Umberto Uprimny (Pichlern) for their help with cleaning outcrops and sampling fossils. The work was done within the framework of and financially supported by projects of the Federal Government of Lower Austria (K3-F-964/001-2020 and K3-F-964/002-2023, Department Science and Research; headquarters in St. Pölten); the Austrian Academy of Sciences (headquarters in Vienna, project Polzberg Lukeneder 2021) represented by the National Committee for Geo/Hydro Sciences (Earth System Sciences Programme); the Triassic projects "Fenster in die Karnische Krise" and special budget for the realization of the special volume (Lukeneder CPE book 2025) of the Society of Friends of the Natural History Museum Vienna (Freunde des Naturhistorischen Museums Wien, CPE Lukeneder 2023); and the Natur- und UNESCO Geopark Steirische Eisenwurzen (St Gallen, project Karnische Krise Lukeneder 2024). The manuscript greatly benefited from valuable comments by Petra Lukeneder (Vienna) and an anonymous reviewer, the production editor Ashok Arumairaj (Taramani), the in-house series editor Kathleen A. Ritterbush (Utah), the publishing editor Simon Shah-Rohlfs and the production editor Antje Endemann (both Heidelberg).

Author Contribution AL designed research, provided documentation on fossil material and wrote the paper.

Funding Open access funding enabled the projects, which were additonally financially supported by the Federal Government of Lower Austria (K3-F-964/001-2020 and K3-F-964/002-2023, Department Science and Research; headquarters in St. Pölten); the Austrian Academy of Sciences (headquarters in Vienna, project Polzberg Lukeneder 2021) represented by the National Committee for Geo/Hydro Sciences (Earth System Sciences Programme); the Natural History Museum Vienna; and the Triassic project "Fenster in die Karnische Krise" of the Society of Friends of the Natural History Museum Vienna (Freunde des Naturhistorischen Museums Wien, CPE Lukeneder 2023); the Natural History Museum Vienna; and the Natur- und UNESCO Geopark Steirische Esienwurzen (St Gallen, project Karnische Krise Lukeneder 2024) (to AL). The authors are responsible for the contents of this publication. The funder had no impact on conceptualization, design, data collection, analysis, decision to publish, or preparation of the manuscript.

Competing interests All authors stated that they have no competing interests.

Authors' contributions The author states that he has read, edited and agreed to submit the final version of the contribution.

Availability of data and materials All sources of information are provided in the text. Raw data related to the fossil material from Austria are available from the corresponding authors and Alexander Lukeneder upon request.

Additional information This article is part of the Special Volume: *Late Triassic Konservat-Lagerstätten within the Carnian Pluvial Episode. 150 Years of Scientific Excavations in Austria.*

Artificial Intelligence (AI) The chapter was produced without the use of AI.

Language The entire volume uses American English reviewed by a native scientist.

Received 2025.

Published online 2026.

References

Abel O (1906) Fossile Flugfische. Abh K-K Geol RA 56:1–88.https://doi.org/10.5962/bhl.title.61007

Alverà G, Dal Corso J, Chu D, Cruciani G, Roghi G, Caggiati M, Song H, Song H, Tian L, Du Y, Haserbek T, Gianolla P (2026) Episodic aridification at the onset of the Carnian Pluvial Episode in Western Tethys. Glob Planet Change 256:105192. https://doi.org/10.1016/j.gloplacha.2025.105192

Allison PA (1988) Konservat-Lagerstätten: Cause and classification. Paleobiology 14:331–344. https://doi.org/10.1017/S0094837300012082

Bauer FK, Schnabel W (1997) Geologische Karte der Republik Österreich. Sheet Mariazell 72, 1:50.000. Geologische Bundesanstalt Wien

Bernardi M, Avanzini M, Bizzarini F (2011) Vertebrate fauna from the San Cassiano Formation (early Carnian) of the Dolomites region. Geo Alp 8:122–127

Brandner R, Poleschinski W (1986) Stratigraphie und Tektonik am Kalkalpensüdrand zwischen Zirl und Seefeld in Tirol. Jber Mitt Oberrhein Geol Ver 68:67–92. https://doi.org/10.1127/jmogv/68/1986/67

Briggs DEG (2014) Konservat-Lagerstätten 40 years on: the exceptional becomes *mainstream.* In: Laflamme M, Schiffbauer JD, Darroch SAF (eds) Reading and Writing of the Fossil Record: Preservational Pathways to Exceptional Fossilization. Paleontol Soc Papers 20:1–14. https://doi.org/10.1017/S1089332600002771

Dal Corso J, Bernardi M, Sun Y, Song H, Seyfullah LJ, Preto N, Gianolla P, Ruffell A, Kustatscher E, Roghi G, Merico A, Hohn S, Schmid, AR, Marzoli A, Newton RJ, Wignall PB, Benton MJ (2020) Extinction and dawn of the modern world in the Carnian (Late Triassic). Sci Adv 6(38):eaba0099. https://doi.org/10.1126/sciadv.aba0099

Dal Corso J, Ruffell A, Preto N (2018) The Carnian Pluvial Episode (Late Triassic): new insights into this important time of global environmental and biological change. J Geol Soc London 175:986–988. https://doi.org/10.1144/jgs2018-185

Doguzhaeva LA, Summesberger H (2012) Pro-ostraca of Triassic belemnoids (Cephalopoda) from Northern Calcareous Alps, with observations on their mode of preservation in an environment of northern Tethys which allowed for carbonization of non-biomineralized structures. Neues Jahrb. Geol Palaontol Abh 266:31–38. https://doi.org/10.1127/0077-7749/2012/0266

Doguzhaeva LA, Summesberger H, Mutvei H, Brandstaetter F (2007a) The mantle, ink sac, ink, arm hooks and soft body debris associated with the shells in Late Triassic coleoid cephalopod *Phragmoteuthis* from the Austrian Alps: Palaeoworld 16(4):272–284. https://doi.org/10.1016/j.palwor.2007.08.002

Doguzhaeva LA, Mapes RH, Summesberger H, Mutvei H (2007b) The preservation of body tissues, shell, and mandibles in the ceratitid ammonoid Austrotrachyceras (Late Triassic), Austria. In: Landman NH, Davis RA, Mapes RH (eds) Cephalopods present and past: new insights and fresh perspectives. Springer, Dordrecht, 221–238. https://doi.org/10.1007/978-1-4020-6806-5_11

Faupl P (2003) Historische Geologie. 2. Auflage, facultas, UTB Stuttgart, 271 p

Felber M, Furrer H, Tintori A (2004) The Triassic of Monte San Giorgio in the World Heritage List of UNESCO: an opportunity for science, the local people and tourism. Eclogae Geol Helv 97:1–2. https://doi.org/10.1007/s00015-004-1118-5

Forchielli A, Pervesler P (2013) Phosphatic cuticle in thylacocephalans: a taphonomic case study of (Arthropoda, Thylacocephala) from the *Fossil-Lagerstätte* Polzberg (Reingraben shales, Carnian, Upper Triassic, Lower Austria). AJES 106:46–61

Furrer HD (2003) Der Monte San Giorgio im Südtessin—vom Berg der Saurier zur Fossil-Lagerstätte internationaler Bedeutung. Neujahrsbl Natur Ges Zürich 206:1–64

Furrer H, Vandelli A (2014) Guide to the Museum of fossils from Monte San Giorgio Meride. Fondazione del Monte San Giorgio, 1–127

Furrer HD (2024) The history of palaeontological research and excavations at Monte San Giorgio. Swiss J Palaeontol 142:18. https://doi.org/10.1186/s13358-024-00314-9

Glaessner MF (1931) Eine Crustaceen fauna aus den Lunzer Schichten Niederösterreichs. Jahrb Geol BA 81:467–486

Glaessner MF (1965) Vorkommen fossiler Dekapoden (Crustacea) in Fisch-Schiefern. Senckenb Lethaea 46a:111–122

Griffith J (1977) The Upper Triassic fishes from Polzberg bei Lunz. Zool J Linn Soc 60:1–93

Hornung T (2008) The Carnian Crisis in the Tethyan Realm. 1–235, Dr. Müller, Saarbrücken

Hornung T, Krystyn L, Brandner RA (2007) A Tethys-wide mid-Carnian (Upper Triassic) carbonate productivity crisis: Evidence for the Alpine Reingraben Event from Spiti (Indian Himalaya)? J Asian Earth Sci 30(2):285–302. https://doi.org/10.1016/j.jseaes.2006.10.001

Hornung T, Kogan I, Wolf G, Moosleitner G (2019) The Norian fish deposits of Wiestal ("Seefeld Member", Northern Calcareous Alps, Salzburg, Austria)—taxonomy and palaeoenvironmental implications. Austrian J Earth Sci 112(2):125–165. https://doi.org/10.17738/ajes.2019.0008

Klug C, Scheyer TM, Klein N, Liu J, Stockar R eds (2024a) Special Issue: 100 years of scientific excavations at UNESCO World Heritage Site Monte San Giorgio and Global Research on Triassic Marine Lagerstätten. Swiss J Palaeontol 143 (Spec Issue):1–498, Springer, Basel. https://doi.org/10.1186/s13358-024-00328-3Merid

Klug C, Spiekman SNF, Bastiaans D, Scheffold B, Scheyser TM (2024b) The marine conservation deposits of Monte San Giorgio (Switzerland, Italy): the prototype of Triassic black shale Lagerstätten. Swiss J Palaeontol 143:11. https://doi.org/10.1186/s13358-024-00308-7

Kozur HW, Bachmann GH (2010) The middle Carnian Wet Intermezzo of the Stuttgart Formation (Schilfsandstein), Germanic Basin. Palaeogeogr Palaeoclimatol Palaeoecol 290:107–119. https://doi.org/10.1016/j.palaeo.2009.11.004

Kreuss O (2014) 100 Hieflau. Geologische Karte, 1: 50 000. Geofast, Geosphere Austria. https://doi.org/10.24341/tethys.101

Krystyn L (1978) Eine neue Zonengliederung des mediterranen Unterkarn. Schrift Erdwiss Komm ÖAW 4:37–75

Krystyn L (1991) Die Fossillagerstätten der alpinen Trias. In: Nagel D, Rabeder G (eds) Exkursionen im Jungpaläozoikum und Mesozoikum Österreichs. Österr Paläontol Ges Wien 1991:23–78

Krystyn L, Lein R (1996) Triassische Becken- und Plattformsedimente der östlichen Kalkalpen. Exkursionsführer SEDIMENT 96, 11. Sedimentologentreffen 1996:1–26

Krystyn L, Lein R, Richoz S (2008) Der Gamsstein: Werden und Vergehen einer Wettersteinkalk-Plattform. Exkursionsführer PANGEO 2008. J Alp Geol 49:157–172

Laville T., Lukeneder P, Lukeneder A (2024) Revision of the thylacocephalan biota from the Upper Triassic Polzberg Konservat-Lagerstätte, Austria. Acta Palaeontol Pol 69 (4): 713–730. https://doi.org/10.4202/app.1180.2024

Lukeneder A (ed) (2026) Late Triassic Konservat-Lagerstätten within the Carnian Pluvial Episode: 150 years of scientific excavations in Austria. Top Geobiol 55:600 p. https://doi.org/10.1007/978-3-032-19777-1

Lukeneder A, Lukeneder P (2021) The Upper Triassic Polzberg palaeobiota from a marine *Konservat-Lagerstätte* deposited during the Carnian Pluvial Episode in Austria. Sci Rep 11:16644. https://doi.org/10.1038/s41598-021-96052-w

Lukeneder P, Lukeneder A (2022) Comment on "Triassic coleoid beaks and other structures from the Calcareous Alps revisited" by Doguzhaeva et al. (2022). Acta Palaeontol Pol 67(4):963–965. https://doi.org/10.4202/app.01016.2022

Lukeneder A, Lukeneder P (2022a) Taphonomic history and trophic interactions of an ammonoid fauna from the Upper Triassic Polzberg palaeobiota. Sci Rep 12:7455. https://doi.org/10.1038/s41598-022-11496-y

Lukeneder P, Lukeneder A (2022b) Mineralized belemnoid cephalic cartilage from the Late Triassic Polzberg *Konservat-Lagerstätte* (Austria). PLoS ONE 17(4):e0264595. https://doi.org/10.1371/journal.pone.0264595

Lukeneder A, Lukeneder P (2023) New data on the marine Upper Triassic palaeobiota from the Polzberg Konservat-Lagerstätte in Austria. Swiss J Palaeontol 142:9. https://doi.org/10.1186/s13358-023-00269-3

Lukeneder A, Lukeneder P (2024a) Ammonoid taxonomy of the Carnian Polzberg Konservat-Lagerstätte in Austria. Jahrb Geol BA 162(1–4):27–55. https://opac.geologie.ac.at/ais312/dokumente/JB1621_027_A.pdf

Lukeneder P, Lukeneder A (2024b) Correction: Mineralized belemnoid cephalic cartilage from the late Triassic Polzberg *Konservat-Lagerstätte* (Austria). PLoS ONE 19(12):e0315418. https://doi.org/10.1371/journal.pone.0264595

Lukeneder A, Surmik D, Gorzelak P, Niedźwiedzki R, Brachaniec T, Salamon MA (2020) Bromalites from the Upper Triassic Polzberg section (Austria); insights into trophic interactions and food chains of the Polzberg palaeobiota. Sci Rep 10:20545. https://doi.org/10.1038/s41598-020-77017-x

Lukeneder P, Lukeneder A, Hiden TE, Reishofer G, Ritschl H, Sihorsch P (2023) Fenster in die Vergangenheit—Micro-CT bei prähistorischen Tintenfischknorpeln. Radiopraxis 16(01):37–39. https://doi.org/10.1055/a-1938-3926

Lukeneder P, Fuchs D, Lukeneder A (2024a) Morphology, taxonomy and trophic interactions of rostrum-less coleoids from the Late Triassic Polzberg *Konservat-Lagerstätte* (Lower Austria). Swiss J Palaeontol 143:21. https://doi.org/10.1186/s13358-024-00319-4

Lukeneder A, Lukeneder P, Sachsenhofer RF, Roghi G, Rigo M (2024b) Multi-proxy record of the Austrian Upper Triassic Polzberg *Konservat-Lagerstätte* in light of the Carnian Pluvial Episode. Sci Rep 14:11194. https://doi.org/10.1038/s41598-024-60591-9

Lukeneder S, Lukeneder A, Harzhauser M, İslamoğlu Y, Krystyn L, Lein R (2012) A delayed carbonate factory breakdown during the Tethyan-wide Carnian Pluvial Episode along the Cimmerian terranes (Taurus, Turkey). Facies 58:279–296.https://doi.org/10.1007/s10347-011-0279-8

Mueller S, Krystyn L, Kürschner WM (2016) Climate variability during the Carnian Pluvial Phase—A quantitative palynological study of the Carnian sedimentary succession at Lunz am See, Northern Calcareous Alps, Austria. Palaeogeogr Palaeoclimatol Palaeoecol 441:198–211. https://doi.org/10.1016/j.palaeo.2015.06.008

Oberhauser R (1980) Der Geologische Aufbau Österreichs: Geol B-A Wien, Springer, 699 p

Ogg JG (2015) The mysterious Mid-Carnian "Wet Intermezzo" global event. J Earth Sci 26:181–191. https://doi.org/10.1007/s12583-015-0527-x

Moser M, Pavlik W (2013) 74 Hohenberg, Geofast 1: 50 000, Geologische Karte der Republik Österreich, Geologische Bundesanstalt Wien. https://doi.org/10.24341/tethys.143

Pieroni V, Furrer H (2020) Middle Triassic gastropods from the Besano Formation of Monte San Giorgio. Switzerland. Swiss J Palaeontol 139:2. https://doi.org/10.1186/s13358-019-00201-8

Piller WE, Egger H., Erhart CW, Gross M, Harzhauser M, Hubmann B, van Husen D, Krenmayr H-G, Krystyn L, Lein R, Lukeneder A, Mandl G, Rögl F, Roetzel R, Rupp C, Schnabel W, Schönlaub HP, Summesberger H, Wagreich M, Wessely G (2004) Die Stratigraphische Tabelle von Österreich 2004 (sedimentäre Schichtfolgen). Kommission für die Paläontologische und stratigraphische Erforschung Österreichs. Österr Aka der Wiss Österr Stratigraph Komm, Wien. https://opac.geologie.ac.at/ais312/dokumente/Stratigraph_Tabelle_Oesterreich_2004.pdf

Rigo M, Preto N, Roghi G, Tateo F, Mietto P (2007) A rise in the Carbonate Compensation Depth of western Tethys in the Carnian (Late Triassic): Deep-water evidence for the Carnian Pluvial Event. Palaeogeogr Palaeoclimatol Palaeoecol 246(2–4):188–205. https://doi.org/10.1016/j.palaeo.2006.09.013

Rigo M, Mazza M, Karádi V, Nicora (2018) A New Upper Triassic conodont biozonation of the Tethyan Realm. In: Tanner LH (ed) The Late Triassic World: Topics in Geobiology 46:189–235. https://doi.org/10.1007/978-3-319-68009-5_6
Ruffell A, Simms MJ, Wignall PB (2015) The Carnian Humid Episode of the late Triassic: a review. Geol Mag 153:271–284. https://doi.org/10.1017/S0016756815000424
Ruttner A, Schnabel W (1988) 71 Ybbsitz, 1:50.000, Geologische Karte der Republik Österreich, Geologische Bundesanstalt Wien.
Schädel M, Eldijk T, Winkelhorst H, Reumer JWF, Haug J (2020) Triassic Isopoda—three new species from Central Europe shed light on the early diversity of the group. Bull Geosci 95(2):145–166. https://doi.org/10.3140/bull.geosci.1773
Schlager W, Schöllnberger W (1974) Das Prinzip stratigraphischer Wenden in der Schichtfolge der Nördlichen Kalkalpen. Mitt Geol Ges Wien 66:165–193
Schnabel W (1997) 58 Baden, 1:50.000, Geologische Karte der Republik Österreich, Geologische Bundesanstalt Wien.
Seilacher A (1970) Begriff und Bedeutung der Fossil-Lagerstätten. N Jahrb Geol Paläontol Mh 1970:34–39
Simms MJ, Ruffell AH (2018) The Carnian Pluvial Episode: from discovery, through obscurity, to acceptance. J Geol Soc 175(6):989–992. https://doi.org/10.1144/jgs2018-020
Simms MJ, Ruffell AH (1990) Climatic and biotic change in the late Triassic. J Geol Soc London 147:321–327. https://doi.org/10.1144/gsjgs.147.2.0321
Spengler E (1931) Erläuterungen zur Geologischen Spezialkarte. Blatt Schneeberg-St. Ägyd. Geologische Bundesanstalt, Wien, 108 p
Spengler E (1959) Versuch einer Rekonstruktion des Ablagerungsraumcs der Decken der Nördlichen Kalkalpen. 3. Teil: Der Ostabschnitt der Kalkalpen. Jb Geol BA 102:193–312
Spengler E (1963) Les zones de facies du trias des Alpes Calcaircs Septentrionales et leurs rapports avec la structure des nappes. Livre mem P Fallot, Paris 2:465–475
Stockar R (2010) Facies, depositional environment, and palaeoecology of the Middle Triassic Cassina beds (Meride Limestone, Monte San Giorgio, Switzerland). Swiss J Geosci 103:101–119. https://doi.org/10.1007/s00015-010-0008-2
Stockar R, Baumgartner PO, Condon D (2012) Integrated Ladinian bio-chronostratigraphy and geochrononology of Monte San Giorgio (Southern Alps, Switzerland). Swiss J Geosci 105:85–108. https://doi.org/10.1007/s00015-012-0093-5
Stur D (1874) Neue Aufschlüsse im Lunzer Sandsteine bei Lunz und ein neuer Fundort von Wengerschiefer im Pölzberg zwischen Lunzersee und Gaming. Verh K-K Geol RA 1:271–273
Stur D (1886) Vorlage des ersten fossilen Schädels von *Ceratodus* aus den ober triadischen Reingrabner Schiefern von Pölzberg nördlich bei Lunz. Verh K-K Geol RA 15:381–383
Teller F (1891) Über den Schädel eines fossilen Dipnoers, *Ceratodus Sturii* nov. spec. aus den Schichten der oberen Trias der Nordalpen. Abh K-K Geol RA 15:1–38
Thenius E (1974) Niederösterreich: Bundesländerserie, Heft Niederösterreich. Verh Geol BA Wien, 1–280
Tollmann A (1966) Geologie der Kalkvoralpen im Ötscherland als Beispiel alpiner Deckentektonik. Mitt Geol Ges Wien 58(1965):103–207
Tollmann A (1976) Analyse des klassischen nordalpinen Mesozoikums. Stratigraphie, Fauna und Fazies der Nördlichen Kalkalpen. Franz Deuticke, Wien, 580 p
Trauth F (1935) Die Aptychen der Trias. Sitzungs Ber Akad Wissen, Math-Natur Kl, Abt 1(144):455–483
Trauth F (1948) Geologie des Kalkalpenbereiches der Zweiten Wiener Hochquellenleitung. Abh Geol BAWien 26(1):1–99
Wagner L (1970) Die Entwicklung der Mitteltrias in den östlichen Kalkvoralpen im Raum zwischen Enns und Wiener Becken. unpupl PhD Thesis, Phil Fac Univ Wien, 202 p
Weitz J, Sachsenhofer R, Misch D, Lukeneder A (2026) Organic petrology and geochemistry of the Scheiblinggraben setion: detailed insights into the Carnian Pluvial Episode and comparison to time equivalent profiles. In: Lukeneder A (ed) Late triassic konservat-Lagerstätten within the

Carnian Pluvial Episode: 150 years of scientific excavations in Austria. Top Geobiol 55, Chap 23: 521–560. https://doi.org/10.1007/978-3-032-19777_23

Wessely G (2006) Die Geologie von Niederösterreich. Geol Österr Bundesl, Geologische Bundesanstalt, Wien, 416 p

Wöhrmann SFr. v. (1889) Die Fauna der sogenannten Cardita- und Raibler Schichten in den Nordtiroler und Bayrischen Alpen Jb K-K Geol RA 39:181–258

Wöhrmann SFr.v. (1893) Die Raibler Schichten. Jb K-K Geol RA 43/4:617–768

Zhang P, Yang M, Lu J, Jiang Z, Zhou K, Xu X, Wang Y, Wu L, Chen H, Zhu X, Guo Y, Ye H, Shao L, Hilton J (2023) Floral response to the Late Triassic Carnian Pluvial Episode. Front Ecol Evol 11:1199121. https://doi.org/10.3389/fevo.2023.1199121

Zhang P, Yang M, Lu J, Jiang Z, Vervoort P, Zhou K, Xu X, Chen H, Wang Y, He Z, Bian X, Shao L, Hilton J (2024) Four volcanically driven climatic perturbations led to enhanced continental weathering during the Late Triassic Carnian Pluvial Episode: EPSL:626:118517. https://doi.org/10.1016/j.epsl.2023.118517

Chapter 5
Clam Shrimp of the Upper Triassic Polzberg *Konservat-Lagerstätte*

Chavdar Rusev and Alexander Lukeneder

Abstract Today, clam shrimp inhabit ephemeral freshwater environments, but this was not always the case. Fossil members (*Euestheria*) of that group were found at the Polzberg *Konservat-Lagerstätte*, which shows all characteristics of a normal marine environment. Euestheriids have so far been known as inhabitants limited to freshwater and brackish areas. Hence, the paleobiology of this species and the paleoecology of this site characterized by marine influences is of interest. Seventy-eight clam shrimp fossils were examined and analyzed using light photography, scanning electron microscopy, and chemical analysis using energy dispersive X-ray spectroscopy. The average length (anterior to posterior) of the shells was 3.48–4.85 mm and 2.69–3.81 mm for the height (dorsal to ventral). The number of carapace growth lines ranged from 14 to 19. Based on the morphology and the ornamentation visible on the growth bands, these lower Carnian specimens were identified as *Euestheria minuta*. The main components of the shells of these fossils were calcium, phosphate, and silicium. The Polzberg region is known as a marine system, but Euestheriidae have primarily been found in freshwater areas. Two possible scenarios as to how these animals came to the Polzberg region can be forwarded. First, they may have been washed into the sea via freshwater rivers and lived there in brackish water, because extant species occur in brackish water. Second, it is also possible that the eggs were transferred. The latter could involve river currents, wind, or transport by predators. Further studies of *Euestheria* specimens and the Polzberg region are necessary to determine how this species came to occur there as fossils.

Keywords *Konservat-Lagerstätte* · Spinicaudata · Paleoenvironment · Late Triassic · Carnian Pluvial Episode

---

C. Rusev
iC Consulenten Zivitechniker GesmbH, Vienna, Austria
e-mail: c.rusev@ic-group.org

A. Lukeneder (✉)
Department of Geology and Paleontology, Natural History Museum Vienna, Vienna, Austria
e-mail: alexander.lukeneder@nhm.at

A. Lukeneder (ed.), *Late Triassic Konservat-Lagerstätten within the Carnian Pluvial Episode*, Topics in Geobiology 55, https://doi.org/10.1007/978-3-032-19777-1_5

5.1 Introduction

The family Euestheriidae was initially assigned to the suborder 'Conchostraca', but increasing doubts were later expressed. Euestheriidae are a clade of clam shrimp that are likely either a stem or crown group spinicaudatans (formerly known as conchostracans), hence being assigned to the Spinicaudata (Sell 2018). The fossil record of this family dates back to the Late Devonian, approximately 360 to 380 Ma ago (Olive et al. 2019). Spinicaudata living today have a bivalve carapace into which the soft body can be completely retracted, sometimes leaving the first or second antennae exposed. As the carapace does not moult with the rest of the body, typical concentric lines, growth lines, are formed on the surface of the shell. The unique shape of the carapace and the patterns of its adornments serve as key elements to distinguish between various genera (Jones 1862). The same structure on the left and right valve makes classification easier (Morton et al. 2017). Furthermore, the morphology of the carapace depends on environmental factors such as salinity, temperature and oxygen level (Horne 1971). A worldwide checklist for the Devonian to Cretaceous clam shrimp genus *Euestheria* was recently presented by Fang and Liao (2024).

This paper reports on the discovery of euestheriid clam shrimp fossils at the lower Carnian Polzberg *Konservat-Lagerstätte* (Late Triassic). At this site, the Carnian Pluvial Episode (CPE) led to the formation of argillaceous deposits within a stagnate basinal paleoenvironment (Lukeneder et al. 2024; Lukeneder 2026a, this volume). Frequent well-preserved and entire fossils (e.g., plants, ammonoids, bristle worms, fish) characterize the deposits of the Reingraben Formation (Lukeneder and Lukeneder 2021, 2023). The ecology of the lower Carnian euestheriids at that time is highlighted and the paleobiology of the fossil Euestheriidae is linked to the CPE.

5.2 Geologic Setting

The specimens were collected from the Polzberg locality in Lower Austria, south of Gaming and north of Lunz am See, a notable paleontological site dating back to the Upper Triassic (around 233 Ma) in the Northern Calcareous Alps (NCA). This site, located in the Reifling Basin of the Bajuvaric Lunz Nappe System (Fig. 5.1), is well-known for exposures of the Carnian Reingraben Formation, which are fossiliferous in the lowermost parts. Due to the exceptional fossil preservation, the site is classified as a *Konservat-Lagerstätte* (Seilacher 1970; Kimmig and Schiffbauer 2024). A *Konservat-Lagerstätte* is a sedimentary deposit that showcases entire and extraordinarily preserved fossil contents, capturing even fragile structures and soft tissues typically lost in fossil records. The Polzberg site is highly recognized for its Upper Triassic (lower Carnian) fossil collections (Lukeneder et al. 2020; Lukeneder and Lukeneder 2021, 2022a, b, 2023). The location is 4.5 km northeast of Lunz am See on the western slope of Mount Schindelberg (Lukeneder and Lukeneder 2021).

Historically, the Polzberg locality has been known and examined since the late nineteenth century and was first mentioned by Stur (1874). It was initially a black coal mining area. Subsequent fossil recovery efforts were made by geologists in 1885 and 1909 (Lukeneder and Lukeneder 2021, 2022a). The samples originate from marly limestones and marls deposited during the time interval known as the Carnian Pluvial Episode (CPE), a humid period noted for significant global climate changes. Terrestrial climates transitioned to a warmer and more humid state with increased precipitation and runoff. Concurrently, oceans experienced a decrease in the deposition of carbonate minerals, potentially caused by the demise of carbonate platforms and the decline of carbonate-precipitating organisms. The lithological changes and facies adaptations at Polzberg were caused by a global climate shift driven by volcanic activity related to the early rifting of Pangea, potentially including the formation of the Wrangellia Large Igneous Province (LIP; Dal Corso et al. 2020).

The main tectonic elements of the NCA in Lower Austria are the Frankenfels Nappe and, to the south, the Lunz Nappe. The Reifling Basin, within the Lunz Nappe is located Lunz Nappe between Polzberg and Großreifling (Krystyn 1991; Lukeneder and Lukeneder 2021, 2023; Lukeneder et al. 2024). The Reifling Basin was a marine

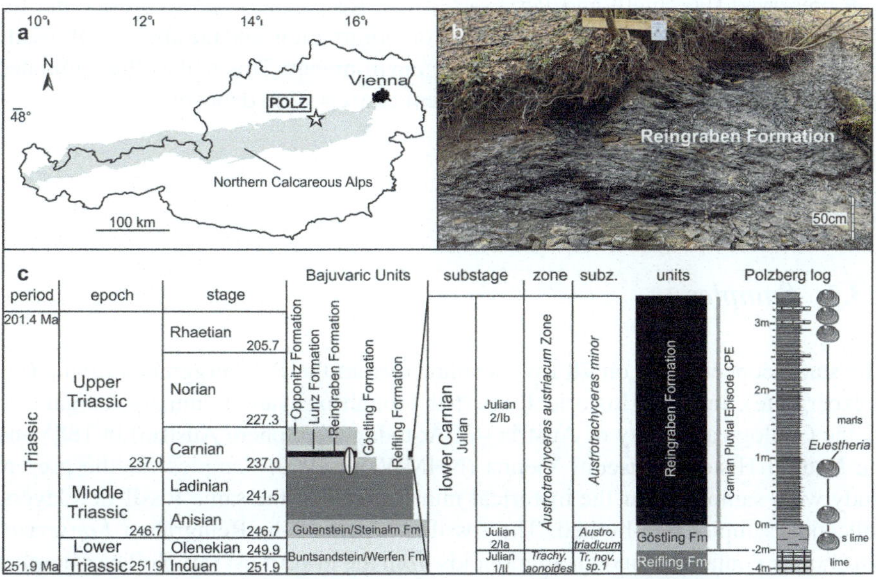

Fig. 5.1 **a** Locality map of the Polzberg area (Sch, Großreifling) in Lower Austria and the Austrian Northern Calcareous Alps (NCA, gray). **b** Outcrop situation at Polzberg (2023) with argillaceous deposits of the Reingraben Formation. Note marking spikes every 20 cm. **c** Stratigraphic position of the Scheiblingbach deposits during the lower Carnian. White ellipse: position of the Polzberg *Konservat-Lagerstätte*. Detailed Polzberg section with indicated sampling layers (*Euestheria* drawings) within the lower Carnian Reingraben Formation. Austro *Autrotrachyceras*, Trachy *Trachyceras*, chrono chronostratigraphy, subz subzone, Re Fm Reifling Formation, G Fm Göstling Formation, R Fm Reingraben Formation, lime limestone, s lime silicious limestone. Adapted after Lukeneder and Lukeneder (2021)

intraplatform basin during the Upper Triassic (Lukeneder and Lukeneder 2021). Festa et al. (2018) note that the occurrence of an intraplatform basin is a rare feature, especially when it involves a carbonate platform, and is interpreted as a signal of sin-depositional tectonics. Intraplatform basins such as the Reifling Basin were filled completely. The black deposits of the Reingraben Formation initially lack silt and are subsequently dominated by shales with more silt and sand intercalations. Finally, freshwater influx increased and epicontinental sediments were washed in from rivers; siliciclastic deposits then formed the Lunz Formation (Hornung et al. 2007a).

The Reingraben Formation is approximately 50 m thick and includes dark grey to black claystones, marlstones and occasional sandstone layers (Lukeneder and Lukeneder 2021). The shales are rich in pyrite and vary in calcium carbonate content from 86.9% in marly limestone to 2.9% in claystone/mudstone. The total organic carbon (TOC) amounts within the *Austrotrachyceras* abundance zone oscillate between 1.4 and 0.3% (Lukeneder and Lukeneder 2021). The shales are characterized by laminations of dark organic material and light-colored laminae with abundant posidoniid bivalves. The fauna found in the Reingraben Formation are mainly nektic organisms such as fish (actinopterygians) and cephalopods (ammonoids, coleoids). Fossil remnants obtained from the Polzberg *Konservat-Lagerstätte* are well preserved (Forchielli and Pervesler 2013). Researchers further noted that the high organic content in the sediment, a lack of bioturbation, and the absence of sessile organisms all point to a dysoxic or anoxic environment. This makes the laminated layers of the Reingraben Formation excellent conservation deposits.

5.3 Material and Methods

5.3.1 Samples

The study concentrated on 78 fossils sourced from the Polzberggraben ravine (i.e. Polzberg site). Sampling historically started through excavation campaigns organized by the Geological Survey of Austria (GBA; today GeoSphere Austria) in 1885 and the Natural History Museum Vienna (NHMW) in 1909. Samples for the present study were sampled near the historical mine tunnels in the same fossil-rich layers, collecting samples bed-by-bed. The fossil-rich part of the Reingraben *Konservat-Lagerstätte* contains abundant ammonoids from the lowermost sample/layer number 1 (Po –50 cm) up to the topmost layer 5 (Po 320 cm) in the section. The exact location of the fossil-rich locality is at N47°53′4.98″ and E15°4′28.15″.

5.3.2 Analytical Methods

The fossils were analyzed using various analytical tools and electronic instruments to gain insights into their structure and composition. In the Natural History Museum Vienna (NHMW), digital high-quality photographs of the samples were obtained using a Discovery.V20 Stereo Zeiss microscope (Zeiss, Oberkochen, Germany) with magnifications of 4.7× and 20×. Photographs were taken with the AxioCam MRc5 Zeiss (Zeiss, Germany). The photographs were documented and processed using the AxioVision SE64 Rel. 4.9 imaging system (Zeiss, Germany). Five selected samples (NHMW 2021/0123/0507, NHMW 2021/0123/0464, NHMW 2021/0123/0465, NHMW 2021/0123/0466 and NHMW 2021/0123/0471) were further analyzed using the scanning electron microscope (SEM) JSM-6610LV (Jeol, Japan) in a low vacuum setting. Additionally, energy dispersive X-ray spectroscopy (EDX) was performed at 15 kV voltages to determine the chemical composition of the preserved carapaces (i.e. 'shells'). These samples were uncoated and unpolished.

5.4 Results

5.4.1 Carapace Size and Sampling Layers

Mean sizes ranged from 3.48 to 4.85 mm for the length (anterior to posterior, x-axis) and 2.69 to 3.81 mm for the height (dorsal to ventral, y-axis, Fig. 5.2). Accordingly, the animals can be classified as "large" (Scholze and Schneider 2015; Sell 2018). Measurement points are the maximum bending points of the carapaces (see Sell 2018). The largest carapace measured had a max. length of 5.18 mm and a max. height of 4.33 mm. The maximum shell size was detected in a single specimen in layer 2 Po (Po 80–100 cm). Each layer is 20 cm thick, except layer 1 (Po −50 to 0 cm), which measures 50 cm. Most specimens were detected in layer 1 (Po −50 to 0 cm, n = 29) and in layer 4 (Po 300–320 cm, n = 43). Layers 1 and 4 are the most significant regarding the mean shell size distribution. Layer 2 (Po 80–100 cm), layer 3 (Po 280–300 cm) and layer 5 (Po 320–340 cm) yielded one, two and three specimens, respectively. Six well-preserved specimens of *Euestheria* show brownish (i.e. phosphatic) carapace preservation (Fig. 5.3). Specimen NHMW 2021/0123/0466 (layer 4, Po 300–320 cm Fig. 5.3a), specimen NHMW 2021/0123/0493 (layer 4, Po 300–320 cm, Fig. 5.3b), specimen NHMW 2021/0123/0468 (layer 4, Po 300–320 cm, Fig. 5.3c), four specimens are grouped on NHMW 2021/0123/0442 (layer 1, Po −50 to 0 cm, Fig. 5.3d), specimen NHMW 2021/0123/0494 (layer 4, Po 300–320 cm, Fig. 5.3e), and specimen **f** NHMW 2021/0123/0471 (layer 4; Po 300–320 cm, Fig. 5.3f).

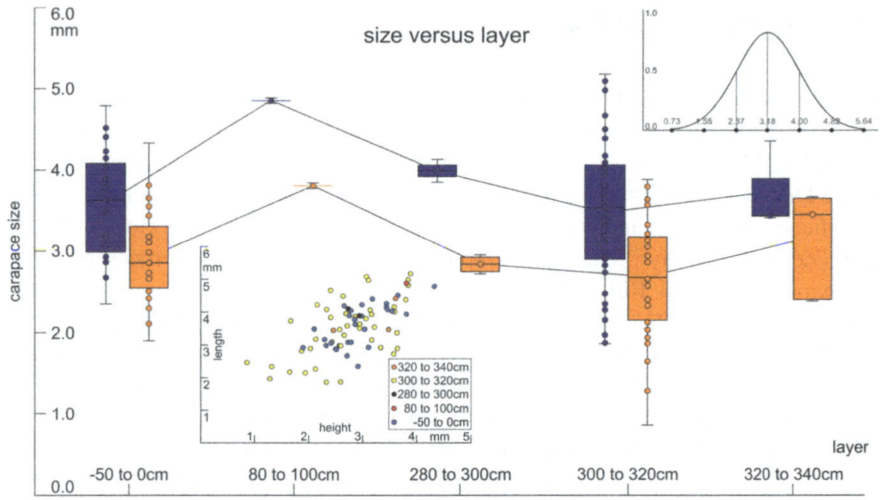

Fig. 5.2 Carapace size (mm) distribution within different sampling intervals. Box plots of measured length (blue) and height (orange) of carapaces in sampling intervals Po: −50 to 0 cm, 80 to 100 cm, 280 to 300 cm, 300 to 320 cm, 320 to 340 cm in mm (mean) at the different layers. Normal distribution (Gaussian) with standard deviation in upper right corner. Maximal length corresponds to anterior/posterior and height to dorsal/ventral size (Scholze and Schneider 2015)

5.4.2 Scanning Electron Microscope (SEM)

The micro-tuberculated patterns on the growth bands between growth lines (for terms see Sell 2018) are visible on the SEM images (but not on the digital photographs). Frequent, small tubercles are visible on the growth bands, with fewer on the growth lines (Figs. 5.3 and 5.4). Similar microgranular structures were described for *E. minuta* by Sell (2018; Text-fig. 4B) for the neotype from the Ladinian Erfurt Formation (Seebronn, Germany). They occur only on the uplifted growth bands of the carapace and are poorly or not at all visible on darker growth lines. These patterns on the growth bands are described as pitted (Scholze and Schneider 2015). The number of growth lines counted ranged between 14 and 19. Based on the number of growth lines, the size, shape and ornamentation of the carapace, and on a comparison with the available material shown in the recent literature, the specimens from Polzberg belong to the genus *Euestheria* (Fang and Liao 2024), as opposed to assumptions made by Sell (2018) most probably *E. minuta* and not *E. franconica.*

5.4.3 Energy Dispersive X-Ray Spectroscopy (EDX)

Carapace areas selected for analyses involved intact entire shells or shell fragments. No analyses were performed on the imprints in the sediment. Multiple measurements

Fig. 5.3 Euestheriids with carapace preservation. **a** *Euestheria minuta*, positive, lateral view, layer 4 (Po 300–320 cm), NHMW 2021/0123/0466. **b** *E. minuta*, positive, lateral view, layer 4 (Po 300–320 cm), NHMW 2021/0123/0493. **c** *E. minuta*, positive, lateral view, layer 4 (Po 300–320 cm), NHMW 2021/0123/0468. **d** Four specimens of *E. minuta*, positive, lateral view, layer 1 (Po −50 to 0 cm), NHMW 2021/0123/0442. **e** *E. minuta*, positive, lateral view, layer 4 (Po 300–320 cm), NHMW 2021/0123/0494. **f** *E. minuta*, positive, lateral view, layer 4 (Po 300–320 cm), NHMW 2021/0123/0471. Scale bars 500 μm

on the same points and different areas were done to improve precision (Fig. 5.5). Three different specimens were analyzed by elemental EDX mapping. Specimen 4308 (layer 4, Po 300–320 cm, NHMW 2021/0123/0467, Fig. 5.5a), specimen 4318 (layer 4, Po 300–320 cm, NHMW 2021/0123/0471, Fig. 5.5b), and specimen 4309 (layer 4, Po 300–320 cm, NHMW 2021/0123/0465, Fig. 5.5c). Note the indicated areas of elemental mapping. The main composition of the clam shrimp fossils is phosphate (P) and calcium (Ca). Most of the other element spikes visible on the spectrometry graphs reflect the argillaceous rock matrix composition with silicium

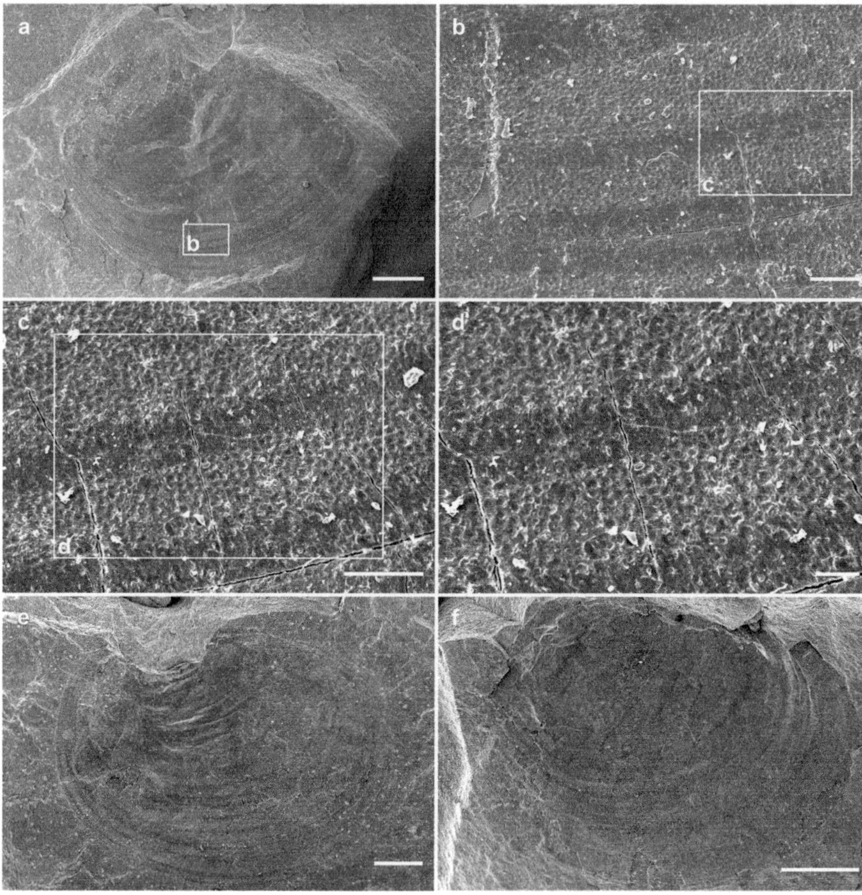

Fig. 5.4 SEM images of euestheriids with carapace growth bands, growth lines and micro-ornamentation. **a** *Euestheria minuta*, lateral view, layer 4 (Po 300–320 cm), NHMW 2021/0123/0465. **b** Magnification of **a** with growth lines and growth bands. **c** Magnification of **b** with tuberculation on growth bands, weaker on darker growth lines, anterior part of the carapace of the specimen. **d** Magnification of **c** with detailed tuberculation. **e** Entire specimen with visible growth bands and growth lines, layer 4 (Po 300–320 cm), NHMW 2021/0123/0466. **f** Entire specimen with visible growth bands and growth lines, layer 4 (Po 300–320 cm), NHMW 2021/0123/0471. Specimens coated with platinum. Scale bars: **d** 20 μm, **b, c** 50 μm, **a, e** 500 μm, **e** 1 mm

(Si), aluminium (Al), magnesium (Mg), potassium (K), and oxygen (O), characteristic for clay minerals (i.e. phyllosilicates). The chemical composition differs slightly depending on the position of the analyzed points on carapaces.

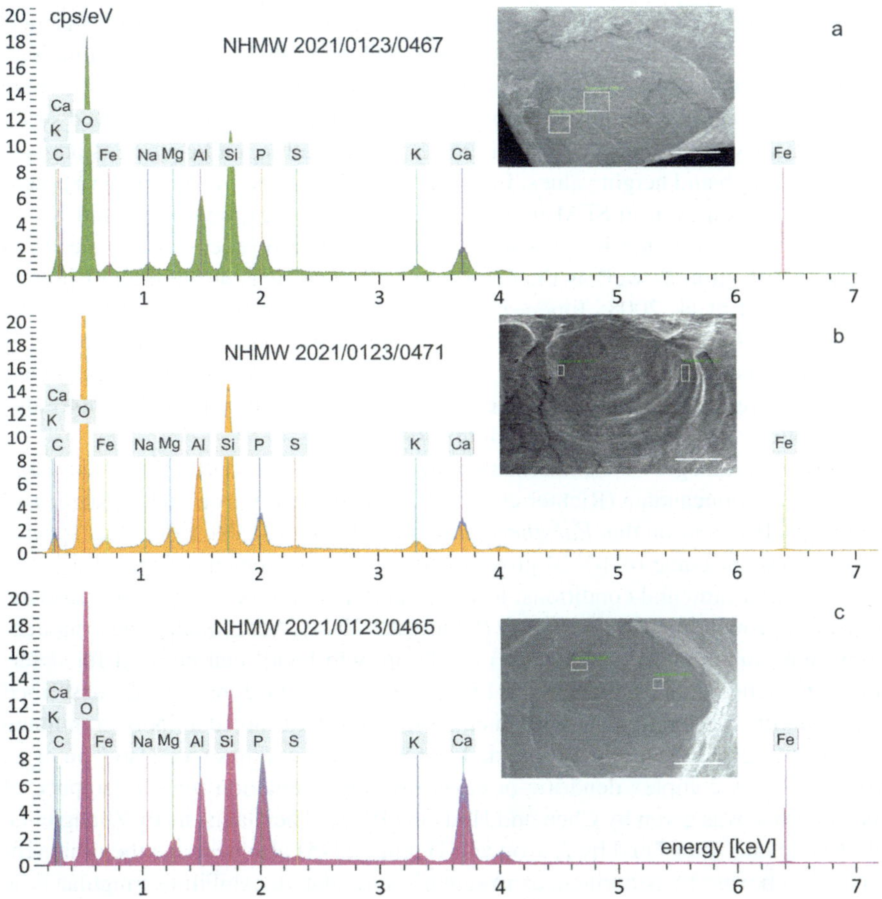

Fig. 5.5 Elemental spectra of specimens analyzed by energy dispersive X-ray spectroscopy (EDX). **a** Specimen 4308, lateral view, layer 4 (Po 300–320 cm), NHMW 2021/0123/0467. **b** Specimen 4318, layer 4 (Po 300–320 cm), NHMW 2021/0123/0471. **c** Specimen 4309, layer 4 (Po 300–320 cm), NHMW 2021/0123/0465. Note indicated areas of elemental mapping. Specimens prior to coating with carbon. Scale bars: **a** 800 μm, **b** 1 mm, **c** 800 μm

5.5 Discussion

5.5.1 Species Identification and Material Composition

As noted above, the suborder "Conchostraca" is no longer considered as a phylogenetic group, and the monophyletic groups Cladoceromorpha, Laevicaudata and Spinicaudata were generated. The family Euestheriidae was assigned to the Spinicaudata and is known to be extinct (Sell 2018). For size measurements, the fossils must be preserved in an appropriate orientation where the carapace is fully visible or where every individual must lie in the same orientation to enable comparison. Furthermore, Scholze and Schneider (2015) recommend using their carapace terminology when measuring the fossils for identification of genera and species. Fossilization may involve deformation, yielding diverging carapace outlines or poorly preserved structures such as growth lines. This complicates classification, and caution is advised regarding length and height values. In some cases, there is no indication of (a potential) compression even in SEM images at a high magnification (Holub and Kozur 1981). EDX can provide insights into the mineralization processes that occurred during fossilization as well as the environmental conditions at the time of preservation (Stigall et al. 2008; Jiménez et al. 2021). One problem with EDX is the interaction volume, which is due to the voltage of the electron beam: the higher the voltage, the larger the interaction volume. A high interaction volume can cause damage and reduce the resolution. However, as in this study, Orr et al. (2008) also used a voltage of 15 kV when examining fossils.

Growth line and growth band ornamentation are good features for systematics and occur in the Spinicaudata (Richter et al. 2007; Hegna and Astrop 2020). Kozur and Weems (2010) mention that *Euestheria minuta* is distinguishable from other species by its consistent range of 10–20 growth lines. The development of growth lines is linked to environmental conditions: low temperature and food scarcity decrease the number of growth lines in *Euestheria* (Olempska 2004). Additionally, the ornamentation (e.g., tuberculated, reticulated) on the growth bands can be used for determination (Scholze and Schneider 2015; see Hegna 2021). Jones (1862) described that *E. minuta* has a reticulated ornament, and Sell (2018) noted a finely granulated surface with grains or pits with rounded to round-oval outlines (see also Chu et al. 2013, 2016). A complex dendritic or cavernous ornamentation for the carapace of Euestheriidae was given by Chen and Hudson (1991). The diagnosis by Zhang et al. (1976) and later translated by Astrop and Hegna (2015) also describe the ornamentation in subovate Euestheriidae as irregularly reticulated, exhibiting irregular lirae with a high variation. This led the authors to conclude that this family is an extremely diverse group.

The main composition of the clam shrimp fossils exhibits elements with P and Ca within the carapaces. Additional element spikes reflect characteristic the clay mineralogy (i.e. phyllosilicates) of the attached rock matrix composition with Si, Al, Mg, K and O. Hegna et al. (2020) used Raman spectroscopy on extant Spinicaudata and detected a highly variable carapace composition with varying amounts of calcium phosphate and calcium carbonate, sometimes solely calcium phosphate. Carapaces of extant Spinicaudata examined by Stigall et al. (2008) contained calcium phosphate with silica, iron, aluminium, zinc and sulfur, not ruling out the additional presence of calcite (see also Hegna et al. 2020). Astrop et al. (2015) also examined extant Spinicaudata and found phosphate, calcium, carbon, oxygen and sodium. All these elements were detected in the fossils of the present study, except zinc. For Spinicaudata, a main preservation in calcium phosphate has been reported. Note, however, that modern clam shrimp have different compositions, highly variable due to external conditions as well (pers. comm. Hegna T 2025). Stigall et al. (2008) suppose that the calcium phosphate comes from the shells of the fossils, and where the calcium phosphate layers have exfoliated, they have been replaced by silica that has been deposited on the carapace in layers. According to them, this replacement takes place early in diagenesis, before the calcium phosphate recrystallizes, and results in fine details of the microstructure of the carapaces being better preserved (Stigall et al. 2008).

5.5.2 Paleoecology and Paleobiology

In the Polzberg subbasin, from where the samples were collected, an excellent preservation of fossils occurred due to anoxic/dysoxic conditions near seafloor and a lack of bioturbation in the sediment. Euxinic bottom water conditions limited and restricted the predator presence and allowed the fossils to be unaffected (Lukeneder and Lukeneder 2021). This led to the formation of the fossiliferous Reingraben Formation, comprising the Polzberg *Konservat-Lagerstätte*, which are rich in organic material (Savrda and Bottjer 1991; Lukeneder and Lukeneder 2021, 2023). Laminated shales discovered in a remote section of the Spiti Basin also indicate the beginning of anoxic conditions in the deeper shelf, occurring simultaneously with the 'Reingraben Event' in Austria (Hornung et al. 2007b). The Reingraben Formation deposits are relatively dark and laminated deposits that formed in a normal marine environment of the Reifling Basin (Forchielli and Pervesler 2013; Lukeneder et al. 2020). During deposition of the shales, the bottom conditions were inhospitable with little oxygen content or even lack of oxygen (Tintori 1992; Forchielli and Pervesler 2013; Lukeneder and Lukeneder 2023). The Reingraben Formation together with the siliciclastic-dominated deposits from the Lunz Formation interrupted the carbonate factory in the Tethys Ocean, and presumably the climate was wetter and warmer, leading to the deposition of the argillaceous Reingraben Formation (Mueller et al. 2015). Equivalent argillaceous shales were also detected in the Carnian from the Indian Himalaya of Spiti and also indicate a carbonate production collapse coupled with an increase in siliciclastic output (Hornung et al. 2007b). The silicious values found in the present study support this theory. Roghi et al. (2010) reported the demise of the carbonate platforms during the Carnian Pluvial Episode (CPE). A decline in carbonate production led to new environmental conditions in marine shelf and basin environments. Reduced biodiversity, changes in the sediment deposition and type, as well as altered ecosystem dynamics are some examples for a declining carbonate factory (Weidlich and Bernecker 2007).

Forchielli and Pervesler (2013) studied the Polzberg fauna and reported the predominance of nektic organisms. Marine nektic taxa such as cephalopods and fish dominate the soft-bodied fauna in Polzberg (Lukeneder and Lukeneder 2021, 2022a, 2023; Lukeneder et al. 2024). Extant 'Conchostraca" live in bodies of freshwater (Alarcón et al 2024), which sometimes have a relatively high amount of salt (Frank 1988). Fossil conchostracans were normally found in deposits formed in isolated ponds, freshwater lakes and brackish areas (Kozur and Weems 2010). Scholze et al. (2020), however, found samples in brackish, deltaic to shallow-marine deposits. Stur

(1874), Teller (1891) and Lukeneder and Lukeneder (2022a, b) described remains of the lungfish *Tellerodus*, which is assumed to be a freshwater taxon, and other 'conchostracan' shells indicating at least a sporadic and temporary influx of brackish water or freshwater into the otherwise isolated marine Reifling Basin (Polzberg subbasin). Recent studies (Lukeneder and Lukeneder 2022a, b, 2023) found additional evidence confirming freshwater or terrestrial influx in the area of investigation—plant remains of a member of the family Coniferophyta, with the genus *Voltzia*, was found alongside the 'Conchostraca' and various other fossil remains.

Fossil *Euestheria* have been found all over the world. Scholze et al. (2020) published a study from the Permian–Triassic transition in southwest China and noted that *Euestheria* sp. were recovered from yellow- to green-grey clay and siltstones, which they interpreted as freshwater coastal plain deposits. Another study from southwest China determined *Euestheria gutta* as markers of the Permian–Triassic transitional beds, which were in a marine siliciclastic setting (Chu et al. 2016). They found *Euestheria* in brownish-yellow siltstone and mudstone, which they thought hints to a marine environment. From the central United Kingdom, another study reports Middle Triassic (Ladinian) deposits of the Cotham Member in which *Euestheria minuta* was found (Morton et al. 2017). There they investigated Cromhall Quarry and described the deposits as grey-green calcareous mudstones, shales and laminated siltstones. Morton et al. (2017) assume that the 'Conchostraca' lived in brackish to freshwater intervals and that adult *Euestheria* died in shallow ponds, leaving their eggs in the dried sediment to hatch once new rains arrived. Kozur and Weems (2010) report that 'Conchostraca' are commonly found in sediments that formed in terrestrial and brackish environments. A study from Madagascar found "Conchostraca" in sandy marls with concretions rich in iron oxide (Shen et al. 2002). These data also hint to a freshwater to brackish habitat for *Euestheria. Limnestheria* from Ireland was found in very dark Carboniferous shales (Shen et al. 2002). The authors mention that *Euestheria mangaliensis* found in Hungary is common in brackish intercalations of marine beds. Tassi et al. (2015) studied a section in Argentina comprising *Euestheria* specimens consisting of fine-grained sandstones and interbedded mudstones, which the authors interpret as a playa lake or ephemeral rivers or streams. The lacustrine environment (see Alarcón et al. 2024) has been characterized by interbedded siltstones. The authors confirm a drastic increase in volcanic activity during the Middle Triassic time period based on the high number of pyroclastic deposits. Geyer and Kelber (2017) studied 'Conchostraca' from the Middle Keuper (Carnian, Upper Triassic) and conducted their study in the southern Germanic Basin, where one of the formations, the Coburg Sandstone Member, included *Euestheria kozuri.* The lithology of the Coburg Sandstone is predominantly sandstone, which is interstratified with minor grey very fine-grained sandstone, mudstone and siltstone, and interpreted as a brackish environment.

Olempska (2004) investigated Late Triassic 'Conchostraca' in southwestern Poland. The author describes the assemblage of *Euestheria* as being from lacustrine late Carnian claystones. The author notes that these animals occasionally occur in more saline environments such as playa lakes and coastal salt flats, but they die if the salinity of the water exceeds 5‰. *Euestheria* has been found in the claystones of the Krížna Unit in Slovakia. Sýkora et al. (2011) studied the Tržinovo Formation in the Fatric Basin and found the little 'Conchostraca' there. Even more interesting, the occurrence of freshwater to brackish *Euestheria* in the Tržinovo Formation reflects the carbonate productivity crisis during the Carnian in the northern Tethyan Realm, which in turn corresponds to the 'Reingraben Event' during the early Carnian. Chen et al. (2006) published a study on the geological ages of rock formations in China and on findings of *E. minuta* among other fossils, in the Xujiahe Formation, which is dated to the Upper Triassic (Carnian). This formation consisted of thick-bedded sandstone and black shale. Furthermore, they state that the Xujiahe Formation is not marine and appears with dark-colored mud- and siltstone, which is again evidence for freshwater to brackish water habitats for *Euestheria*. Scholze et al. (2016) mention that 'Conchostraca' have adapted to small inland (freshwater) water bodies during the Early Triassic in Central Germany and that they represent good indicators for such environments. That same study, however, also suggests that 'Conchostraca' could have occupied brackish water, such as the species *Magniestheria mangaliensis*, which had adapted to increased salinity. The researchers interpret the depositional environment as a low-energy, shallow inland body of water. Based on the above studies, it is possible that *E. minuta* found in the present studies were introduced into the marine Reinberg Basin by an inflow of freshwater, resulting in a brackish habitat. Jenisch et al. (2017) investigated the stratigraphic distribution of Spinicaudata during the Triassic and Jurassic in the Paraná Basin in central-eastern South America. The lithofacies there yielded, among other fossils, *E. minuta* in laminated mudstone beds which were deposited in floodplains. That study found 'Conchostraca' from the Lower to Upper Triassic period. Jenisch et al. (2017) note the importance of 'Conchostraca' as biostratigraphic markers due to their geographical distribution and high reproduction rates.

The presence of 'Conchostraca' within the basinal Polzberg area could be due to the transport and redeposition by aeolian mechanisms (Scholze et al. 2016). These organisms produce drought-resistant and freeze-resistant eggs (Kozur and Weems 2010) that can survive for a long time without water. In modern species, the 'conchostracan' life cycle from hatching to sexual maturation lasts only 5–23 days (Webb 1979), enabling them to occupy small (isolated) lakes and ponds that may not be present year around (Kozur and Weems 2010). Therefore, *Euestheria* eggs could be transported by freshwater currents and be brought to an otherwise marine environment, or winds can raise dust containing such eggs and relocate them (Tasch 1987; Havel et al. 2000; Graham and Wirth 2008; Vanschoenwinkel et al. 2009; Kozur and Weems 2010; Scholze et al. 2020). Both scenarios could help explain how the specimens ultimately ended up in a marine environment.

Finally, a biological transport by animals would also be possible. Lukeneder and Lukeneder (2023) outlined a trophic network for the organisms found in Polzberg (Fig. 5.6). Arrows within that scheme indicate the trophic direction from prey to predator. *Euestheria* could have been the prey of various trophic marine inhabitants such as cephalopods with the abundant *Austrotrachyceras* and the frequent coleoid *Phragmoteuthis*. Other predators at the Polzberg locality, such as actinopterygian fish, coelacanths and lungfish, most likely also fed on *Euestheria* (Lukeneder and

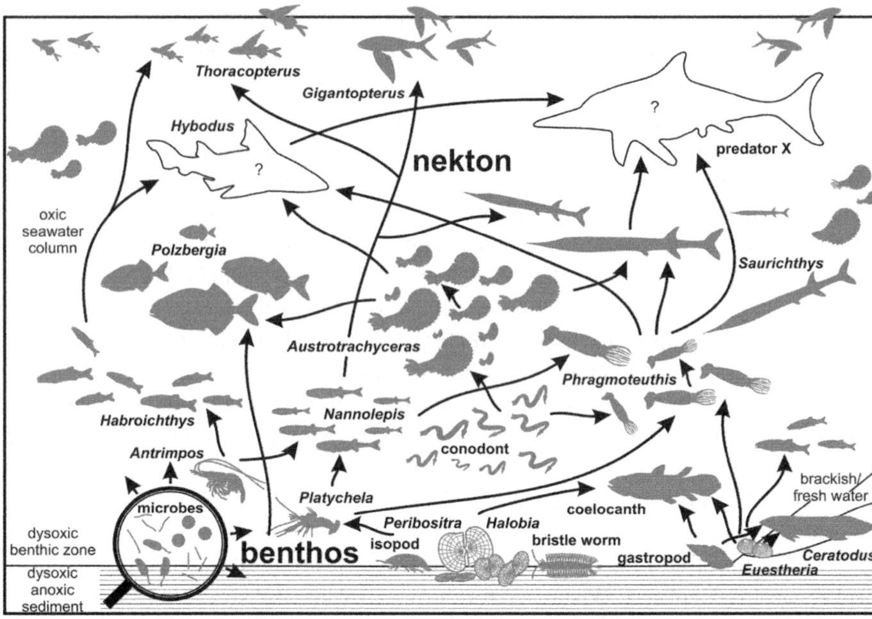

Fig. 5.6 Reconstructed trophic food web of the lower Carnian Polzberg paleobiota (gray) in the Reifling Basin based on direct and indirect evidence from the fossil record of the Polzberg *Konservat-Lagerstätte*. Ichthyopterygia not detected yet, most likely present. Figured members of paleobiota not to scale. Adapted after Lukeneder and Lukeneder (2021)

Lukeneder 2021). The living *Euestheria* would have been digested, but the eggs would have survived passage through the digestive tract and been excreted with the feces. As soon as the necessary conditions were established, those eggs would have hatched. These considerations also support the theory that *Euestheria* were mainly freshwater organisms and were moved to the marine Polzberg deposits by predators.

5.6 Conclusions

This study reports the 'conchostracan' taxon *Euestheria minuta* from the lower Carnian (Late Triassic) Polzberg *Konservat-Lagerstätte.* The marine and well-oxygenated environment of the Reifling Basin in the Carnian was a favorable habitat for numerous planktic and nektic taxa (e.g., ammonoids, coleoids, actinopterygian fish), however *Euestheria* is described in the literature as an inhabitant of freshwater to brackish water environments. Specimens detected in the argillaceous Reingraben Formation were probably redeposited or transported during periods of freshwater influx end enhanced rainfall from surrounding land masses. Frequent plant fossils along with the freshwater lungfish *Tellerodus* and allodapic layers exhibiting shallow-water inhabitants strengthen this interpretation. More evidence from future work is required to determine definitively whether *E. minuta* from Polzberg lived in the freshwater environments surrounding the Polzberg subbasin or inhabited the marine basin, or potentially both. Hence, neither a transport or redeposition into basinal marine environments nor an adaptation to full marine environments within the Reifling Basin during the deposition of the lower Carnian Reingraben Formation can be ruled out with certainty.

Acknowledgments The authors are particularly grateful to the owners of the outcrop area, Franziska and Hermann Hofreiter, Eva and Karl Jagersberger (all Gaming), for sampling and digging permission. I thank Petra Lukeneder, Lisa Marie Lukeneder and Felix Maximilian Lukeneder (all Gablitz) for their help with cleaning outcrops and sampling fossils. The work was done within the framework of and financially supported by projects of the Federal Government of Lower Austria (K3-F-964/001-2020 and K3-F-964/002-2023, Department Science and Research; headquarters in St. Pölten); the Austrian Academy of Sciences (headquarters in Vienna, project Polzberg Lukeneder 2021) represented by the National Committee for Geo/Hydro Sciences (Earth System Sciences Programme); and the Triassic project "Fenster in die Karnische Krise" of the Society of Friends of the Natural History Museum Vienna (Freunde des Naturhistorischen Museums Wien, CPE Lukeneder 2023). Wencke Wegner (SEM, microprobe) and Anton Englert (thin sections; both Vienna) are acknowledged for technical support. The manuscript greatly benefited from valuable comments by the reviewers Thomas Hegna (Fredonia) and Oscar F. Gallego (Corrientes), the production editor Ashok Arumairaj (Taramani), the in-house series editor Kathleen A. Ritterbush (Utah), the publishing editor Simon Shah-Rohlfs and the production editor Antje Endemann (both Heidelberg).

Author Contribution AL collected material and had the initial idea for the present study. CR provided documentation and first draft of the paper. CR, AL interpretation and writing the paper.

Funding Open access funding was enabled and the projects financially supported by the Federal Government of Lower Austria (K3-F-964/001-2020 and K3-F-964/002-2023, Department Science and Research; headquarters in St. Pölten); the Austrian Academy of Sciences (headquarters in Vienna, project Polzberg Lukeneder 2021) represented by the National Committee for Geo/Hydro Sciences (Earth System Sciences Programme); the Natural History Museum Vienna; and the Triassic project "Fenster in die Karnische Krise" of the Society of Friends of the Natural History Museum Vienna (Freunde des Naturhistorischen Museums Wien, CPE Lukeneder 2023) (to AL). The authors are responsible for the contents of this publication. The funders had no impact on conceptualization, design, data collection, analysis, decision to publish, or preparation of the manuscript.

Competing interests All authors stated that they have no competing interests.

Authors' contributions All authors stated that they have equally contributed to their chapters, read, edited and agreed to submit the final versions of their contributions.

Availability of data and materials All sources of information are provided in the text. Raw data related to the fossil material from Austria are available from the corresponding authors and Alexander Lukeneder upon request.

Additional information This article is part of the Special Volume: *Late Triassic Konservat-Lagerstätten within the Carnian Pluvial Episode. 150 Years of Scientific Excavations in Austria.*

Artificial Intelligence (AI) The chapter was produced without the use of AI.

Language The entire volume uses American English reviewed by a native scientist.

Received 2025.

Published online 2026.

References

Alarcón CM, Colombi CE, Gallego OF, Drovandi JM, Monferran MD (2024) Dispersal of Late Triassic clam shrimps across Pangea linking northwestern Gondwana and central Pangea rift basins. Sci Rep 14:15025. https://doi.org/10.1038/s41598-024-66015-y

Astrop TI, Hegna TA (2015) Phylogenetic relationships between living and fossil spinicaudatan taxa (Branchiopoda Spinicaudata): reconsidering the evidence. J Crustac Biol 35(3):339–354. https://doi.org/10.1163/1937240X-00002317

Astrop TI, Sahni V, Blackledge TA, Stark AY (2015) Mechanical properties of the chitin-calcium-phosphate "clam shrimp" carapace (Branchiopoda: Spinicaudata): implications for taphonomy and fossilization. J Crustac Biol 35(2):123–131. https://doi.org/10.1163/1937240X-00002332

Chen PJ, Hudson JD (1991) The Conchostracan fauna of the great estuarine group, Middle Jurassic, Scotland. Palaeontology 34(3):515–545. https://www.palass.org/sites/default/files/media/publications/palaeontology/volume_34/vol34_part3_pp515-545.pdf

Chen P, Li J, Matsukawa M, Zhang H, Wang Q, Lockley MG (2006) Geological ages of dinosaur-track-bearing formations in China. Cretac Res 27(1):22–32. https://doi.org/10.1016/j.cretres.2005.10.008

Chu DL, Tong JN, Yu JX, Song HJ, Tian L (2013) The conchostracan fauna from the Kayitou Formation of western Guizhou, China. Acta Palaeontol Sinica 52(3):265–280 (in Chinese with English abstract)

Chu D, Yu J, Tong J, Benton MJ, Song H, Huang Y, Song T, Tian L (2016) Biostratigraphic correlation and mass extinction during the Permian-Triassic transition in terrestrial-marine siliciclastic settings of South China. Glob Planet Change 146:67–88. https://doi.org/10.1016/j.gloplacha.2016.09.009

Dal Corso J, Bernardi M, Sun Y, Song H, Seyfulla LJ, Preto N, Gianolla P, Ruffell A, Kustatscher E, Roghi G, Merico A, Hohn S, Schmidt AR, Marzoli A, Newton RJ, Wignall PB, Benton MJ (2020) Extinction and dawn of the modern world in the Carnian (Late Triassic). Sci Adv 6(38):eaba0099. https://doi.org/10.1126/sciadv.aba0099

Fang S, Liao H (2024) A worldwide annotated checklist of Fossil (Devonian–Cretaceous) species of the Clam Shrimp Genus *Euestheria* (Branchiopoda: Diplostraca: Spinicaudata). Life 14(11):1438. https://doi.org/10.3390/life14111438

Festa V, Sabato L, Tropeano M (2018) 1:5,000 geological map of the upper Cretaceous intraplatform-basin succession in the "Gravina di Matera" canyon (Apulia Carbonate Platform, Basilicata, southern Italy). Ital J Geosci 137(1):3–15. https://doi.org/10.3301/IJG.2017.12

Forchielli A, Pervesler P (2013) Phosphatic cuticle in thylacocephalans: a taphonomic case study of *Austriocaris* (Arhropoda, Thylacocephala) from the Fossil-Lagerstätte Polzberg (Reingraben shales, Carnian, Upper Triassic, Lower Austria). Austrian J Earth Sci 106(1):46–61. https://ajes.at/images/AJES/archive/Band%20106_1/forchielli_pervesler_ajes_106_1.pdf

Frank PW (1988) Conchostraca. Palaeogeogr Palaeoclimatol Palaeoecol 62(1–4):399–403. https://doi.org/10.1016/0031-0182(88)90064-8

Geyer G, Kelber KP (2017) Spinicaudata ("Conchostraca", Crustacea) from the Middle Keuper (Upper Triassic) of the southern Germanic Basin, with a review of Carnian-Norian taxa and suggested biozones. Paläontol Z 92:1–34. https://doi.org/10.1007/s12542-017-0363-7

Graham TB, Wirth D (2008) Dispersal of large branchiopod cysts: potential movement by wind from potholes on the Colorado Plateau. Hydrobiologia 600:17–27. https://doi.org/10.1007/s10750-007-9171-5

Havel JE, Eisenbacher EM, Black AA (2000) Diversity of crustacean zooplankton in riparian wetlands: colonization and egg banks. Aquat Ecol 34:63–76. https://doi.org/10.1023/A:1009918703131

Hegna TA (2021) A redescription of the clam shrimp, *Orthestheria shupei* (Stephenson, in Stephenson & Stenzel, 1952) comb. nov., from the Cenomanian (Cretaceous) of Texas. Cretac Res 125(3):104864. https://doi.org/10.1016/j.cretres.2021.104864

Hegna TA, Astrop TI (2020) The Fossil record of the Clam Shrimp (Crustacea; Branchiopoda). Zool Stud 59:43. https://doi.org/10.6620/ZS.2020.59-43

Hegna TA, Czaja AD, Rogers DC (2020) Raman spectroscopic analysis of the composition of the clam-shrimp carapace (Branchiopoda: Laevicaudata, Spinicaudata, Cyclestherida): a dual calcium phosphate-calcium carbonate composition. J Crustac Biol 40(6):756–760. https://doi.org/10.1093/jcbiol/ruaa078

Holub V, Kozur H (1981) Revision einiger Conchostracen-Faunen des Rotliegenden und biostratigraphische Auswertung der Conchostracen des Rotliegenden. Geo Paläo Mitt Innsbruck 11(2):39–94

Horne FR (1971) Some effects of temperature and oxygen concentration on phyllopod ecology. Ecology 52(2):343–347. https://doi.org/10.2307/1934594

Hornung T, Brandner R, Krystyn L, Joachimski MM, Keim L (2007a) Multistratigraphic constraints on the NW Tethyan "Carnian crisis." Glob Triassic 41:59–67

Hornung T, Krystyn L, Brandner R (2007b) A Tethys-wide mid-Carnian (Upper Triassic) carbonate productivity crisis: evidence for the Alpine Reingraben Event from Spiti (Indian Himalaya)? J Asian Earth Sci 30(2):285–302. https://doi.org/10.1016/j.jseaes.2006.10.001

Jenisch AG, Lehn I, Gallego OF, Monferran MD, Horodyski RS, Faccini UF (2017) Stratigraphic distribution, taphonomy and paleoenvironments of Spinicaudata in the Triassic and Jurassic of the Paraná Basin. J S Am Earth Sci 80:569–588. https://doi.org/10.1016/j.jsames.2017.09.022

Jiménez VC, Monferran MD, Sperança MA, Castro JP, Catelani TA, Pellerano RG, Perino E, Pereira-Filho ER, Cabaleri NG, Gallego OF (2021) Combination of analytic techniques to chemical characterization and preservation of Jurassic clam shrimp carapaces from La Matilde formation, Patagonia. J S Am Earth Sci 109:103269. https://doi.org/10.1016/j.jsames.2021.103269

Jones TR (1862) A monograph of the fossil Estheriae. Monogr Palaeontogr Soc 14(62):vi–134. https://doi.org/10.1080/02693445.1863.12027934

Kimmig J, Schiffbauer JD (2024) A modern definition of Fossil-Lagerstätten. Trends Ecol Evol 39:621–624. https://doi.org/10.1016/j.tree.2024.04.004

Kozur HW, Weems RE (2010) The biostratigraphic importance of conchostracans in the continental Triassic of the northern hemisphere. Geol Soc Lond Spec Publ 334(1):315–417. https://doi.org/10.1144/SP334.13

Krystyn L (1991) Die Fossillagerstätten der alpinen Trias. In: Nagel D, Rabeder G (eds) Exkursionen im Jungpaläozoikum und Mesozoikum Österreichs. Österr Paläontol Ges Wien 1991:23–78

Lukeneder A (ed) (2026) Late Triassic Konservat-Lagerstätten within the Carnian Pluvial Episode: 150 Years of Scientific Excavations in Austria. Top Geobiol 55:600 p. https://doi.org/10.1007/978-3-032-19777-1

Lukeneder A, Lukeneder P (2021) The Upper Triassic Polzberg palaeobiota from a marine *Konservat-Lagerstätte* deposited during the Carnian Pluvial Episode in Austria. Sci Rep 11(1):16644. https://doi.org/10.1038/s41598-021-96052-w

Lukeneder A, Lukeneder P (2022a) Taphonomic history and trophic interactions of an ammonoid fauna from the Upper Triassic Polzberg palaeobiota. Sci Rep 12(1):7455. https://doi.org/10.1038/s41598-022-11496-y

Lukeneder P, Lukeneder A (2022b) Mineralized belemnoid cephalic cartilage from the late Triassic Polzberg *Konservat-Lagerstätte* (Austria). PLoS One 17(4):e0264595. https://doi.org/10.1371/journal.pone.0315418

Lukeneder A, Lukeneder P (2023) New data on the marine Upper Triassic palaeobiota from the Polzberg *Konservat-Lagerstätte* in Austria. Swiss J Plaeontol 142(9):1–18. https://doi.org/10.1186/s13358-023-00269-3

Lukeneder A, Surmik D, Gorzelak P, Niedźwiedzki R, Brachaniec T, Salamon MA (2020) Bromalites from the Upper Triassic Polzberg section (Austria). Insights into trophic interactions and food chains of the Polzberg palaeobiota. Sci Rep 10(1):20545. https://doi.org/10.1038/s41598-020-77017-x

Lukeneder A, Lukeneder P, Sachsenhofer RF, Roghi G, Rigo M (2024) Multi-proxy record of the Austrian Upper Triassic Polzberg *Konservat-Lagerstätte* in light of the Carnian Pluvial Episode. Sci Rep 14:11194. https://doi.org/10.1038/s41598-024-60591-9

Morton JD, Whiteside DI, Hethke M, Benton MJ (2017) Biostratigraphy and geometric morphometrics of conchostracans (Crustacea, Branchiopoda) from the Late Triassic fissure deposits of Cromhall Quarry, UK. Palaeontology 60(3):349–374. https://doi.org/10.5061/dryad.m17c8

Mueller S, Krystyn L, Kürschner WM (2015) Climate variability during the Carnian Pluvial phase—a quantitative palynological study of the Carnian sedimentary succession at Lunz am See, Northern Calcareous Alps, Austria. Palaeogeogr Palaeoclimatol Palaeoecol 441:198–211. https://doi.org/10.1016/j.palaeo.2015.06.008

Olempska E (2004) Late Triassic spinicaudatan crustaceans from southwestern Poland. Acta Palaeontol Pol 49(3):429–442. https://www.app.pan.pl/archive/published/app49/app49-429.pdf

Olive S, Pradel A, Martinez-Pérez C, Janvier P, Lamsdell JC, Gueriau P, Rabet N, Duranleau-Gagnon P, Cardenas-Rozo AL, Zapata Ramirez PA, Botella H (2019) New insights into Late Devonian vertebrates and associated fauna from the Cuche Formation (Floresta Massif, Colombia). J Vertebr Paleoontol 39(3):e1620247. https://doi.org/10.1080/02724634.2019.1620247

Orr PJ, Briggs DE, Kearns SL (2008) Taphonomy of exceptionally preserved crustaceans from the Upper Carboniferous of Southeastern Ireland. Palaios 23(5):298–312. https://doi.org/10.2110/palo.2007.p07-015r

Richter S, Olesen J, Wheeler WC (2007) Phylogeny of Branchiopoda (Crustacea) based on a combined analysis of morphological data and six molecular loci. Cladistics 23(4):301–336. https://doi.org/10.1111/j.1096-0031.2007.00148.x

Roghi G, Gianolla P, Minarelli L, Pilati C, Preto N (2010) Palynological correlation of Carnian humid pulses throughout western. Tethys 290(1–4):89–106. https://doi.org/10.1016/j.palaeo.2009.11.006

Savrda CE, Bottjer DJ (1991) Oxygen-related biofacies in marine strata: an overview and update. Geol Soc Lond Spec Publ 58:201–219. https://doi.org/10.1144/GSL.SP.1991.058.01.14

Scholze F, Schneider JW (2015) Improved methodology of 'conchostracan' (Crustacea: Branchiopoda) classification for biostratigraphy. Newsl Stratigr 48(3):287–298. https://doi.org/10.1127/nos/2015/0065

Scholze F, Schneider JW, Werneburg R (2016) Conchostracans in continental deposits of the Zechstein-Buntsandstein transition in central Germany: taxonomy and biostratigraphic implications for the position of the Permian-Triassic boundary within the Zechstein Group. Palaeogeogr Palaeoclimatol Palaeoecol 449:174–193. https://doi.org/10.1016/j.palaeo.2016.02.021

Scholze F, Shen S, Backer M, Wei H, Hübner M, Cui Y, Feng Z, Schneider JW (2020) Reinvestigation of conchostracans (Crustacea: Branchiopoda) from the Permian-Triassic transition in Southwest China. Palaeoworld 29(2):368–390. https://doi.org/10.1016/j.palwor.2019.04.007

Seilacher A (1970) Begriff und Bedeutung der Fossil-Lagerstätten. Neues Jahrb Geo Paläoontol Mh 1970:34–39

Sell J (2018) Die Euestheriidae (Conchostraca) des Oberen Muschelkalks und Unteren Keupers von Unterfranken. Semana 33:55–90

Shen Y, Garassino A, Teruzzi G (2002) Studies on Permo-Trias of Madagascar. 4. Early Triassic conchostracans from Madagascar. Atti Soc It Sci Nat Mus Civ Sto Nat Milano 143:3–11

Stigall AL, Babcock LE, Briggs DE, Leslie SA (2008) Taphonomy of Lacustrine interbeds in the Kirkpatrick Basalt (Jurassic), Antarctica. Palaios 23(6):344–355. https://doi.org/10.2110/palo.2007.p07-029r

Stur D (1874) Neue Aufschlüsse im Lunzer Sandsteine bei Lunz und ein neuer Fundort von Wengerschiefer im Pölzberg zwischen Lunzersee und gaming. Verh K-K Geol RA 1:271–273

Sýkora M, Siblík M, Soták J (2011) Siliciclastics in the Upper Triassic dolomite formations of the Krížna Unit (Malá Fatra Mountains, Western Carpathians): constraints for the Carnian Pluvial Event in the Fatric Basin. Geol Carpathica 62(2):121–138. https://doi.org/10.2478/v10096-011-0011-1

Tasch P (1987) Fossil Conchostraca of the Southern Hemisphere and continental drift. Geol Soc Am Mem 165:1–290. https://doi.org/10.1130/MEM165-p1

Tassi LV, Zavattieri AM, Gallego OF (2015) Triassic Spinicaudatan Fauna from the Cerro de Las Cabras Formation (Cuyo Basin), Mendoza Province (Argentina): description of new species and revision of previous records. Ameghiniana 52(2):241–264. https://doi.org/10.5710/AMGH.29.11.2014.2823

Teller F (1891) Über den Schädel eines fossilen Dipnoers, *Ceratodus Sturii* nov. spec. aus den Schichten der oberen Trias der Nordalpen. Abh K-K Geol RA 15:1–38. https://www.zobodat.at/pdf/AbhGeolBA_15_0001-0038.pdf

Tintori A (1992) Fish taphonomy and Triassic anoxic basins from the Alps: a case history. Riv Ital Paleontol Strati 97(3–4):393–408. https://doi.org/10.13130/2039-4942/8956

Vanschoenwinkel B, Gielen S, Seaman M, Brendonck L (2009) Wind mediated dispersal of freshwater invertebrates in a rock pool metacommunity: differences in dispersal capacities and modes. Hydrobiologia 635:363–372. https://doi.org/10.1007/s10750-009-9929-z

Webb JA (1979) A reappraisal of the palaeoecology of conchostracans (Crustacea: Branchiopoda). N Jahrb Geol Paläontol Abh 158:259–275. https://doi.org/10.1127/njgpa/158/1979/259

Weidlich O, Bernecker M (2007) Differential severity of Permian-Triassic environmental changes on Tethyan shallow-water carbonate platforms. Glob Planet Change 55(1–3):209–235. https://doi.org/10.1016/j.gloplacha.2006.06.014

Zhang WT, Chen PJ, Shen, YB (1976) Fossil Conchostracans of China. Science Press, Beijing, China, 325 p. (in Chinese)

Chapter 6
Decapod Crustaceans from the Carnian Deposits of Polzberg and Cave del Predil

Matúš Hyžný, Davide Di Franco, and Alexander Lukeneder

Abstract A synopsis of decapod crustaceans from the Late Triassic (Carnian) *Konservat-Lagerstätten* of Polzberg (Austria) and Cave del Predil (Italy) is presented. It is based on a re-examination of historical collections from GeoSphere Austria (Bundesanstalt für Geologie, Geophysik, Klimatologie und Meteorologie, ex Geological Survey of Austria), Vienna, and the Natural History Museum Vienna (Naturhistorisches Museum Wien). Among the more than 130 specimens studied, five species in four families are recognized: *Aeger straeleni* (Dendrobranchiata: Aegeridae), *Distaeger crassipes* (Dendrobranchiata: Aegeridae), *Clytiella spinifera* (Glypheidea: Clytiopsida), *Platychela trauthi* (Glypheidea: Platychelidae), and *Tetrachela raiblana* (Polychelida: Coleiidae). *Distaeger crassipes* was originally described as *Aeger crassipes* and later reassigned to the genera *Antrimpos* and *Anisaeger* by various authors. The current attribution to *Distaeger* is based on the recognition of the rostrum, a character of topmost taxonomic importance. *Bombur aonis* is regarded as a juvenile of *Distaeger crassipes*. Because of uneven sampling, comparisons between the decapod associations from Cave del Predil (=Raibl) and Polzberg should be made with caution. Nevertheless, relevant differences include different modes of preservation and dominance. At Cave del Predil, *Tetrachela raiblana* and *Distaeger crassipes* dominate the association, which is preserved as nearly completely flattened, dark brown to black specimens. At Polzberg, natantian shrimps and reptantian lobsters are present in similar numbers but are light brown to reddish and exhibit a higher relief of preserved cuticular surfaces.

M. Hyžný (✉)
Earth Science Institute, Slovak Academy of Sciences, Bratislava, Slovakia
e-mail: hyzny.matus@gmail.com

Department of Geology and Paleontology, Faculty of Natural Sciences, Comenius University, Bratislava, Slovakia

D. Di Franco
GeoSphere Austria, Vienna, Austria
e-mail: Davide.DiFranco@geosphere.at

A. Lukeneder
Department of Geology and Paleontology, Natural History Museum Vienna, Vienna, Austria
e-mail: alexander.lukeneder@nhm.at

A. Lukeneder (ed.), *Late Triassic Konservat-Lagerstätten within the Carnian Pluvial Episode*, Topics in Geobiology 55, https://doi.org/10.1007/978-3-032-19777-1_6

Keywords Natantian shrimps · Reptantian lobsters · Decapoda · Preservation · Carnian · Triassic

6.1 Introduction

Decapods are among the most diverse crustaceans, with almost 18,000 species, including more than 3,600 fossil species (Schweitzer et al. 2010; Schweitzer and Feldmann 2015). Decapods have been significant components of marine habitats since the Mesozoic, with ever increasing importance throughout the Cenozoic (Klompmaker et al. 2013; Schweitzer and Feldmann 2015). Interestingly, decapod crustaceans were apparently already well diversified by the Early Triassic (Smith et al. 2022), with most currently recognized infraorders experiencing major diversification events during the Jurassic and Cretaceous (Schweitzer and Feldmann 2015). In this respect, the diversification of decapods is a significant component of the Mesozoic Marine Revolution (Vermeij 1977; Kelley and Hansen 2001) during which the Paleozoic Evolutionary Fauna was gradually replaced by the Modern one, with decapods being a major constituent of the latter.

A major source of information on Triassic decapod faunas comes from localities with exceptional preservation, i.e., *Lagerstätten*, where (near-)complete individuals are preserved. Such faunas have been reported from the Lower Triassic of the western USA basin (Brayard et al. 2017; Smith et al. 2022), Middle Triassic of China (Feldmann et al. 2012; Schweitzer et al. 2014), and the Upper Triassic of Lombardy, Italy (Pinna 1974; Garassino and Teruzzi 1993) and Slovenia (Gašparič et al. 2020), among others. Upper Triassic *Lagerstätten* of Cave del Predil (Raibl), Italy, and Polzberg, Austria, have been known since the nineteenth (Bronn 1858; Reuss 1858; Lukeneder 2026, this volume) and the first half of the twentieth century (Glaessner 1931), respectively, and are among the first documented Triassic decapod assemblages. The present study provides a taxonomic revision of the decapod associations from Polzberg and Cave del Predil (Raibl), based on a re-examination of historical collections and in the view of the current classification framework.

6.2 Geologic Settings

The Polzberg specimens were collected from the Polzberg locality in Lower Austria, south of Gaming and north of Lunz am See, a notable paleontological site dating back to the Upper Triassic (around 233 Ma) in the Northern Calcareous Alps (NCA). This site is located in the Reifling Basin of the Bajuvaric Lunz Nappe System (Fig. 6.1) and is notable for exposures of the Carnian argillaceous Reingraben Formation, which are fossiliferous in their lowermost parts. Due to the exceptional fossil preservation, the site is classified as a *Konservat-Lagerstätte* (Seilacher 1970; Lukeneder et al. 2020; Lukeneder and Lukeneder 2021, 2022a, b, 2023; Lukeneder 2026, this volume). The location is situated 4.5 km northeast of Lunz am See on the western slope of Mount Schindelberg (Lukeneder and Lukeneder 2021).

The historical Cave del Predil (= Raibl) specimens derive from contemporaneous deposits of the Raibl Beds near the small village Cave del Predil, which is located in the Julian Alps of Friuli-Venezia Giulia (northern Italy (Fig. 6.1). Most of the fossil decapod material originates from historical excavations of limestone, marly limestone and black claystone deposits at the ravine Rinngraben (also Rinnengraben) near Cave del Predil. The location is situated 2 km north of the lake Lago del Predil and close to the Slovenian border, 1.5 km to the east.

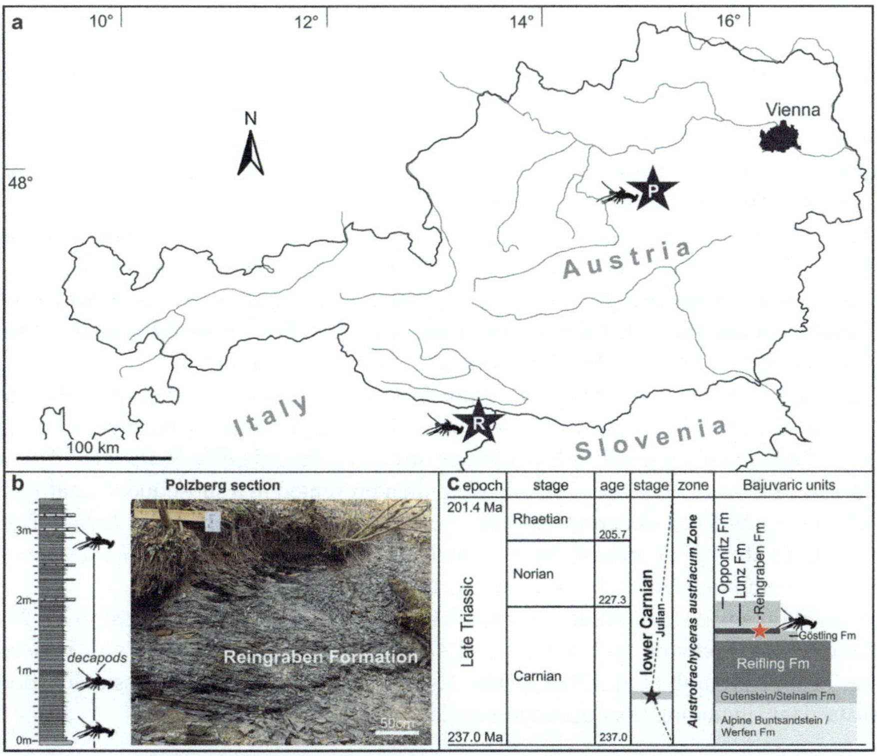

Fig. 6.1 **a** Locality map of the Upper Triassic *Lagerstätten* of Polzberg (P, Austria) and Cave del Predil (R, former = Raibl, Italy) areas, marked by black stars with indicated decapod portraits. **b** Outcrop situation at Polzberg with the detected argillaceous Reingraben Formation. Detailed Polzberg section with indicated sampling layers (decapod drawings) within the lower Carnian Reingraben Formation. **c** Stratigraphic position of the Polzberg deposits during the lower Carnian. Black star: the exact stratigraphic position within the Polzberg *Konservat-Lagerstätte* during the *Austrotrachyceras austriacum* Zone. Adapted after Lukeneder and Lukeneder (2021)

6.3 Previous Research

Decapod crustaceans from the Raibl Beds exposed at Cave del Predil (=Raibl), Italy, and the Reingraben Formation at Polzberg, Austria, have been studied since the nineteenth century. The first note is that by Bronn (1858), who described several species from the Raibl Beds, including *Bolina Raiblana*, *Aeger crassipes* and *Bombur* (?) *Aonis*. In his account on decapods from the Raibl Beds, Reuss (1858) reassigned *Bolina raiblana* to the genus *Tetrachela* and added one more species, *Stenochelus triasicus*, which was later recognized to be the same species as *T. raiblana* by Oppel (1862).

Seven decades later, Glaessner (1930) revised the decapod fauna of the Raibl Beds. Later, he described additional specimens from Cave del Predil and added decapod material from another co-eval location, the Reingraben Formation of Polzberg (Glaessner 1931). For the strata at Polzberg, he used the term "Lunzer Schichten", causing confusion in the scientific literature. Later on, he continued using the term "Lunz" for the Polzberg fauna, when comparing Polzberg and Cave del Predil (Glaessner 1965, p. 112).

Förster (1967a) presented a systematic overview of reptantian decapods of the Triassic. He included taxa from Cave del Predil and Polzberg, including *Clytiella spinifera* and *Platychela trauthi*; he considered *Platychela kahleri* a synonym of *P. trauthi*. In his work on Jurassic natantian decapods, Förster (1967b) made some relevant notes on the identity of *Bombur aonis* from the Raibl Beds.

In their catalogue of type and figured decapod specimens deposited in the Geological Survey of Austria (GBA, today GeoSphere Austria), Hyžný and Zorn (2016) included photographs of specimens originally described by Reuss (1858) and Glaessner (1930, 1931). More details were then presented in a subsequent catalogue dedicated to all fossil decapods deposited at the institution (Hyžný and Zorn 2020).

Audo et al. (2018) revised the polychelidan lobster *Tetrachela raiblana*, one of the most common elements in the decapod fauna of Cave del Predil.

Based on a re-examination of museum collections, Hyžný and Garassino (2022) attempted a taxonomic re-evaluation of *Aeger crassipes*. The species was revised once again by Smith et al. (2022), who also provided a list of all Triassic decapod crustaceans, including taxa occurring at Cave del Predil and Polzberg.

6.4 Material and Methods

The studied specimens represent historical acquisitions deposited in the GeoSphere Austria (Bundesanstalt für Geologie, Geophysik, Klimatologie und Meteorologie) in Vienna, and the Natural History Museum Vienna (Naturhistorisches Museum Wien). Specimens from Cave del Predil are typically dark, almost black, sometimes with patches of white, on a dark grey substrate. In contrast, specimens from Polzberg are usually brown. At both locations, decapod crustaceans are preserved as flattened individuals, although those from Polzberg have more relief.

In total, 132 specimens were studied, 111 from Cave del Predil and 20 from Polzberg. Only the material personally examined by the authors was included. The systematic section includes only material classified to genus and species level. Indeterminate material, i.e., specimens of shrimps not attributable to the genus level, was not included but is briefly mentioned in the Discussion section of this contribution.

No further preparation was performed on the studied specimens. Several methods were used when photographing specimens. For dry specimens, angled sidelight was used. In some cases, specimens were coated with ammonium chloride prior to photography to enhance contrast. Alternatively, uncoated specimens were immersed in ethanol and photographed using natural light from above.

Over the last decade, several works illustrating decapod specimens from Cave del Predil and Polzberg were published (Hyžný and Zorn 2016, 2020; Audo et al. 2018; Hyžný and Garassino 2022). Therefore, the present contribution focuses on previously undocumented specimens.

The investigated fossil material is housed in the collections of the GeoSphere Austria, Vienna, Austria (GBA), and the Natural History Museum Vienna, Austria (NHMW).

6.5 Systematic Paleontology

Malacostraca Latreille, 1802.

Decapoda Latreille, 1802.

Dendrobranchiata Bate, 1888.

Superfamily Penaeoidea Rafinesque, 1815.

Family Aegeridae Burkenroad, 1963.

Genus *Aeger* Münster, 1839.

Type species: *Macrourites tipularius* von Schlotheim, 1822, type by subsequent designation.

***Aeger straeleni* Glaessner,** 1930.

Figure 6.2.

1858 ?*Aeger crassipes* Bronn, p. 29, pl. 4, Fig. 5 (*pars*).

1929 *Aeger Straeleni* Glaessner, p. 53. (*nomen nudum*).

1930 *Aeger Straeleni* Glaessner—Glaessner, p. 140, pl. 6, Fig. 2, pl. 7, Fig. 1, pl. 9, Fig. 3 [erroneously indicated as *Aeger crassipes*].

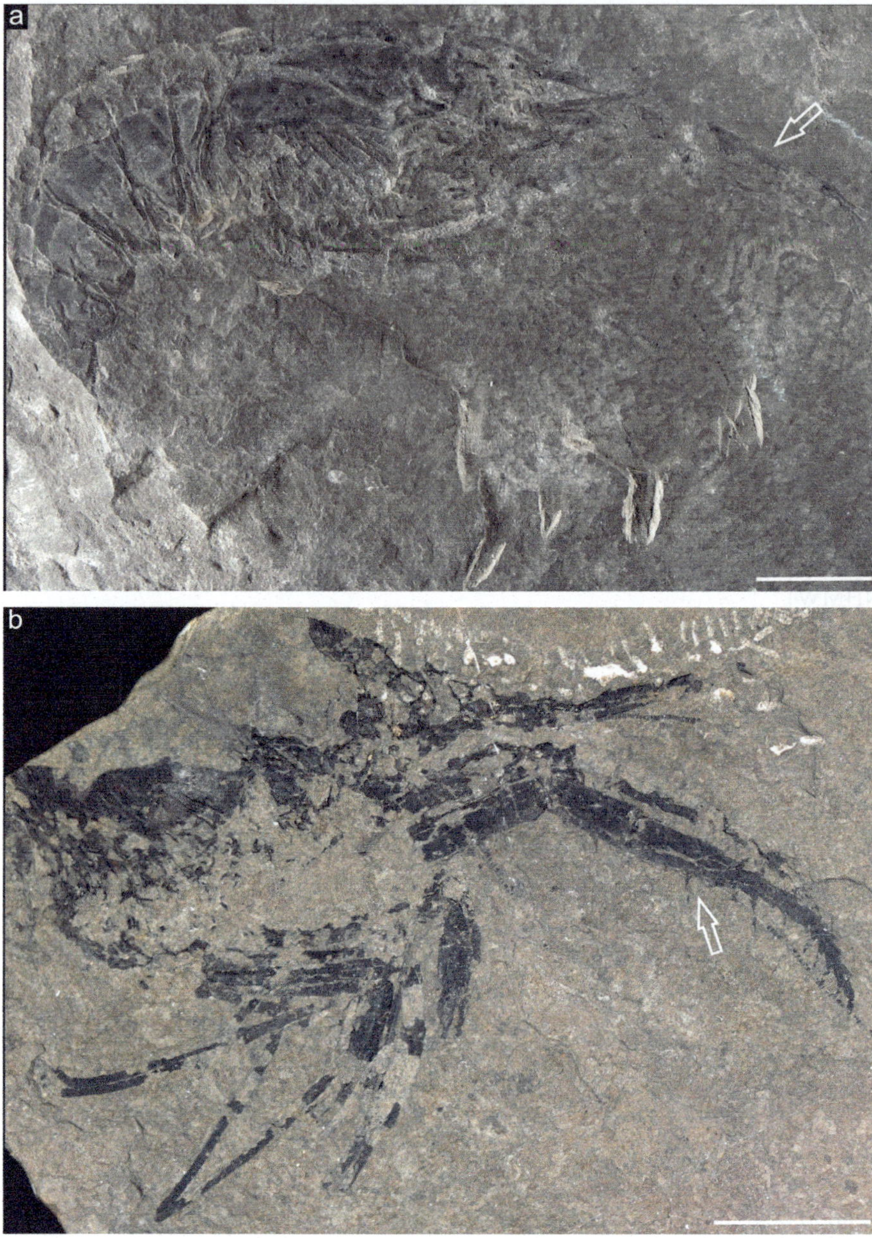

Fig. 6.2 *Aeger straeleni* Glaessner, 1930, from the Upper Triassic (Carnian) of Cave del Predil (= Raibl), Italy. **a** Holotype NHMW 1866/0040/0452. **b** GBA 2018/298/0142. White arrows: Mxp3 with movable spines. Scale bars = 10 mm

1931 *Aeger straeleni* Glaessner—Glaessner, p. 473.

1965 *Aeger straeleni* Glaessner—Glaessner, p. 112.

1967a *Aeger straeleni* Glaessner—Förster, p. 142.

2010 *Aeger straeleni* Glaessner—Schweitzer et al., p. 9. (list).

Type material: Holotype NHMW 1866/0040/0452, near-complete individual preserved in lateral view from Cave del Predil.

Other material: Three specimens from Cave del Predil: GBA 2018/298/0141, GBA 2018/298/0142; NHMW 1887/0009/0108; one specimen from Polzberg: NHMW 1910/0015/0022.

Description: See Glaessner (1930, p. 140).

Remarks: The species was described based on six near-complete specimens from Cave del Predil (Glaessner 1930), while an additional individual had already been illustrated by Bronn (1858, pl. 4, Fig. 5) as questionably belonging to *A. crassipes*. Glaessner (1930, p. 141) considered the species to be morphologically closest to *Aeger brodiei* Woodward, 1888, from the Lower Jurassic of the United Kingdom. Glaessner (1931) assigned one specimen from Polzberg to *A. straeleni*, but he did not figure it and noted only its smaller size in comparison to individuals from Raibl.

Aeger straeleni differs from the co-occurring and much more common *Distaeger crassipes* (Bronn, 1858) n. comb. especially by the robust maxilliped 3 (Mxp3) with large spines (Fig. 6.2); in *D. crassipes* the Mxp3 is comparatively more slender, and spines are less prominent (see below).

Occurrence: Upper Triassic (Carnian) of Cave del Predil, Italy, and Polzberg, Austria.

Genus *Distaeger* Schweitzer et al., 2014.

Type species: *Distaeger prodigiosus* Schweitzer et al., 2014, by original designation and monotypy.

***Distaeger crassipes*(Bronn, 1858) comb. nov.**

Figures 6.3, 6.4, 6.5, and 6.6.

1858 *Aeger crassipes* Bronn, p. 26, pl. 5, Figs. 1, 2 (*non* pl. 4, Fig. 5).

1858 *Bombur* (?) *aonis* Bronn, p. 40, pl. 4, Fig. 4.

1922 *Aeger crassipes* Bronn—Balss (1922), p. 128.

1929 *Bombur aonis* Bronn—Glaessner, p. 67.

1929 *Antrimpos crassipes* (Bronn)—Glaessner, p. 54.

1930 *Antrimpos crassipes* (Bronn)—Glaessner, p. 139, pl. 6, Fig. 1, pl. 10, Fig. 4 [erroneously indicated as *Aeger Straeleni*].

Fig. 6.3 *Distaeger crassipes* (Bronn, 1858) from the Upper Triassic (Carnian) of Cave del Predil, Italy. **a** GBA 2018/298/0196. **b** Detail of the rostrum. **c** Detail of rostral spines (white arrows). Scale bar in **a** = 10 mm, scale bar in **b** = 5 mm

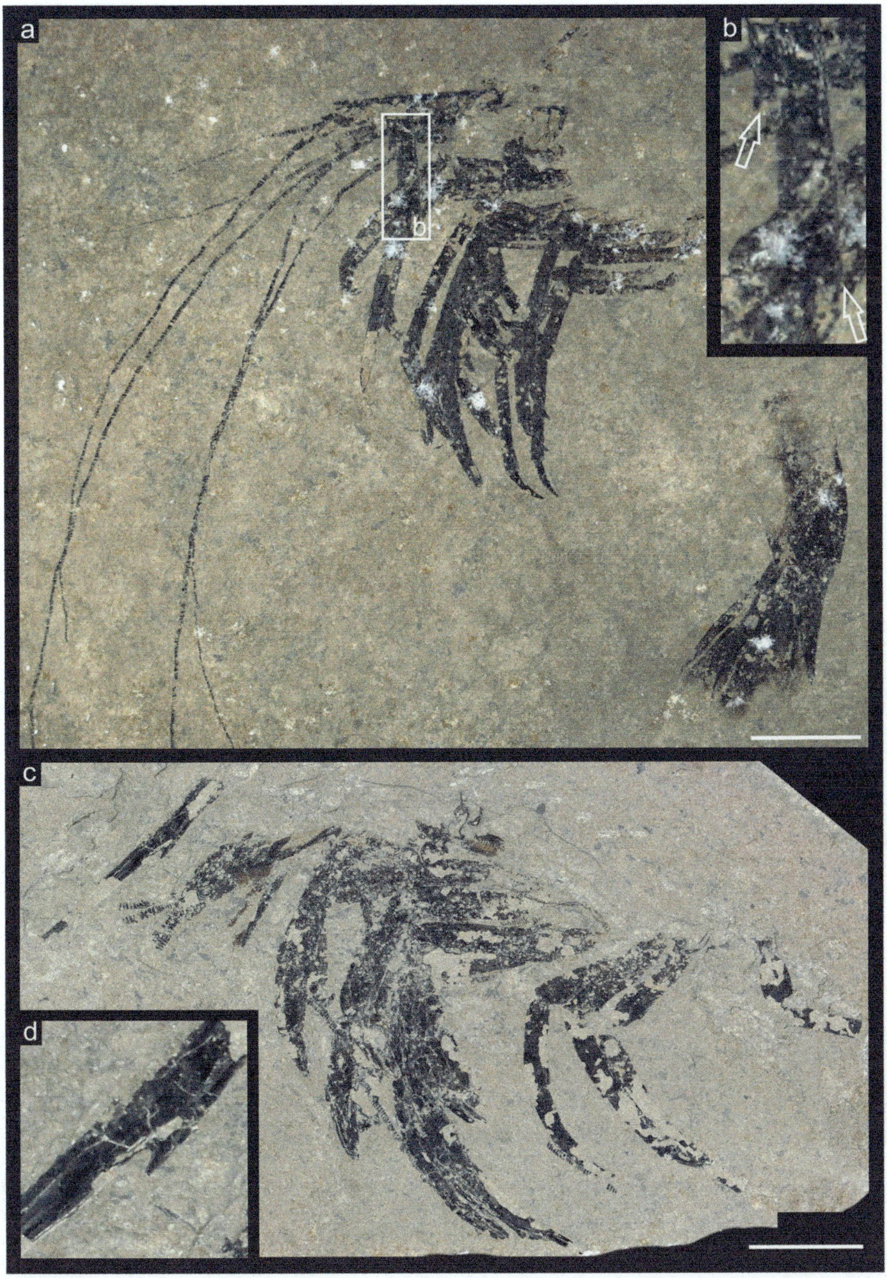

Fig. 6.4 *Distaeger crassipes* (Bronn, 1858) from the Upper Triassic (Carnian) of Cave del Predil (Raibl), Italy. **a** GBA 2018/298/0152. **b** Detail of the rostrum with rostral spines (white arrows). **c** GBA 2018/298/0143. **d** Detail of the rostrum. Scale bars = 10 mm

Fig. 6.5 *Distaeger crassipes* (Bronn, 1858) from the Upper Triassic (Carnian) of Cave del Predil (Raibl), Italy. **a** GBA 2018/298/0149. **b** Detail of Mxp3 with spines/setae (white arrows). **c** GBA 1930/002/0001, appendages. **d** The same specimen as in **c**, arrows show Mxp3 with setal pores. Specimen in **d** was whitened with ammonium chloride prior to photography. Scale bars = 10 mm

Fig. 6.6 *Distaeger crassipes* (Bronn 1858) from the Upper Triassic (Carnian) of Cave del Predil (Raibl), Italy. **a** GBA 2018/298/0144, adult individual in lateral view. **b** NHMW 1887/0009/0113, juvenile ("*Bombur aonis*") in lateral view. Both specimens are shown at the same scale. Scale bar = 10 mm

1930 *Bombur aonis* Bronn—Glaessner, p. 142.

1931 *Antrimpos crassipes* (Bronn)—Glaessner, p. 473.

1965 *Antrimpos crassipes* (Bronn)—Glaessner, p. 112, Fig. 1.

1965 *Bombur aonis* Bronn—Glaessner, p. 112.

1967a *Antrimpos crassipes* (Bronn)—Förster, p. 142.

1967b *Antrimpos crassipes* (Bronn)—Förster, p. 172.

1967b *Bombur aonis* Bronn—Förster, p. 172.

1998 *Antrimpos crassipes* (Bronn)—Müller (1998), p. 7.

2010 *Antrimpos crassipes* (Bronn)—Schweitzer et al., p. 9.

2014 *Antrimpos crassipes* (Bronn)—Garassino et al. (2014), Schweigert and Muscio, p. 17.

2016 *Antrimpos crassipes* (Bronn)—Hyžný and Zorn, p. 137, pl. 14, Figs. 1a–c.

2018 *Antrimpos crassipes* (Bronn)—Audo, Hyžný and Charbonnier, p. 42.

2020 *Antrimpos crassipes* (Bronn)—Hyžný and Zorn, p. 15. pl. 2, Fig. 1.

2022 Genus? *crassipes* Bronn, 1858—Hyžný and Garassino, p. 73, Figs. 1–4.

2022 *Anisaeger crassipes* (Bronn)—Smith et al., p. 49.

Type material: Hyžný and Garassino (2022, Fig. 1A) selected and figured the lectotype as the specimen without collection number from the "Originaliensammlung" collection, Institute of Geosciences, University of Heidelberg, Germany.

Other material: 25 specimens from Cave del Predil (Raibl): GBA 1930/002/0001 (Glaessner, 1930: pl. 6, Fig. 1), GBA 2018/298/0143, GBA 2018/298/0144, GBA 2018/298/0145, GBA 2018/298/0147, GBA 2018/298/0149, GBA 2018/298/0150, GBA 2018/298/0151, GBA 2018/298/0152, GBA 2018/298/0153, GBA 2018/298/0154/02, GBA 2018/298/0154/03, GBA 2018/298/0155, GBA 2018/298/0156, GBA 2018/298/0194, GBA 2018/298/0196, NHMW 1864/0052/0035, NHMW 1866/0040/0453, NHMW 1887/0009/0106, NHMW 1887/0009/0108, NHMW 1887/0009/0110, NHMW 1887/0009/0113, NHMW 1887/0009/0114, NHMW 1887/0009/0115, NHMW 1887/0009/0116. Four specimens from Polzberg: NHMW 1910/0015/0019, NHMW 1910/0015/0020, NHMW 1910/0015/0021, NHMW 1910/0015/0030.

Description: A description was provided by Hyžný and Garassino (2022, p. 73). New observations on the morphology of the rostrum and Mxp3 are presented below.

Remarks: Bronn (1858) described the species as a representative of *Aeger* Münster, 1839. Later, Glaessner (1930) re-assigned the species into *Antrimpos* Münster, 1839, but simultaneously noted that it was not possible to discuss the relationship between *A. crassipes* and its (supposed) congeners, apparently because of the poor preservation of *A. crassipes*. The original description by Bronn (1858) was based on four incomplete specimens without the cephalothorax. Glaessner (1930, 1931) mentioned further 23 specimens from Raibl (six of them with carapace) and four specimens from Polzberg (two of them with carapace). Hyžný and Garassino (2022) considered the attribution to *Antrimpos* as questionable, claiming that *A. crassipes* possesses elongate Mxp3, quite unlike any other species of *Antrimpos* described thus far (Schweitzer et al. 2010). Moreover, the genus has been considered a collective taxon (Glaessner 1969, p. R477; Hyžný and Garassino 2022, p. 72), without a reliable diagnosis. Accordingly, and based on the absence of a preserved rostrum, Hyžný and Garassino (2022) removed the species from *Antrimpos* and placed it in open nomenclature under "Genus?", waiting for better-preserved material before any reliable taxonomic assignment on the genus level could be made. Smith et al. (2022) reassigned the species to the genus *Anisaeger* Schweitzer et al. 2014. Interestingly, this was done without examining the specimens but based solely on the illustrations in Bronn (1858) and Glaessner (1930). Smith et al. (2022) noted the presence of a spineless rostrum, although the cited works do not illustrate the rostrum of *A. crassipes*.

Recently, additional specimens of *A. crassipes* were identified in the historical collections of GeoSphere Austria, likely representing the same specimens studied by Glaessner (1930). He mentioned (Glaessner 1930, p. 139) that in the studied material he was unable to recognize a preserved rostrum, a character of major taxonomic importance (Pérez Farfante and Kensley 1997; Poore and Ahyong 2023). Closer re-examination by us, however, revealed that the rostrum is indeed present in some specimens. Interestingly, in all cases the long rostrum is broken while the carapace is not preserved (Figs. 6.3 and 6.4). The element is sometimes mixed together with cephalothoracic appendages (Fig. 6.4a), partly obscuring its details. Nevertheless, two spines are apparent (Figs. 6.3c and 6.4b), one basal suprarostral spine and one subrostral spine at midlength. Such a rostrum is diagnostic of *Distaeger* Schweitzer et al., 2014, to which the discussed species is therefore assigned. Further examination of previously studied specimens revealed spines (Fig. 6.5a, b) and spine bases or setal pits (Fig. 6.5c, d) along the ventral margins of the Mxp3 propodus and dactylus, which further supports the attribution to *Distaeger* as diagnosed by Schweitzer et al. (2014). Interestingly, Glaessner (1930, p. 139) had mentioned that no spines or setae are observable.

Förster (1967b, p. 172) suggested that *Bombur aonis* (Fig. 6.6b), exhibiting a short rostrum and long antennules, represents a juvenile stage of *A. crassipes* (Fig. 6.6a), the most common shrimp in the studied strata. Specimens attributed to *B. aonis* by M.F. Glaessner (as documented by his handwritten labels) in the collections of GeoSphere Austria and Natural History Museum Vienna are distinctly smaller than those assigned to *A. crassipes*, while no *A. crassipes* specimens are comparable in size to those of *B. aonis*. Moreover, the preserved morphological features of *B. aonis* are consistent with juvenile *A. crassipes*. Thus, we share the opinion of Förster (1967b) and formally propose synonymizing *B. aonis* with *A. crassipes*.

Occurrence: Upper Triassic (Carnian) of Cave del Predil, Italy, and Polzberg, Austria.

Infraorder Glypheidea Van Straelen, 1925.

Superfamily Glypheoidea Winkler, 1881.

Family Clytiopsidae Beurlen, 1927.

Genus *Clytiella* Glaessner, 1931.

Type species: *Clytiella spinifera* Glaessner, 1931, by monotypy.

***Clytiella spinifera* Glaessner,** 1931.

Figure 6.7.

1931 *Clytiella spinifera* Glaessner, p. 474, text-Fig. 2, pl. 15, Fig. 1.

1967a *Clytiella spinifera* Glaessner—Förster, p. 144.

Fig. 6.7 *Clytiella spinifera* Glaessner, 1931 from the Upper Triassic (Carnian) of Polzberg, Austria; holotype GBA 1931/003/0001. Specimen was whitened with ammonium chloride prior to photography. Scale bar = 10 mm

1969 *Clytiella spinifera* Glaessner—Glaessner, p. R457, Fig. 261.1.

1998 *Clytiella spinifera* Glaessner—Müller, p. 8.

2010 *Clytiella spinifera* Glaessner—Schweitzer et al., p. 22.

2016 *Clytiella spinifera* Glaessner—Hyžný and Zorn, p. 137, pl. 16, Fig. 1a–b.

2020 *Clytiella spinifera* Glaessner—Hyžný and Zorn, p. 24, pl. 7, Fig. 7.

Type material: Holotype GBA 1931/003/0001, near-complete individual preserved in dorso-lateral view (Fig. 6.7). Glaessner (1931) mentioned the presence of two specimens of *Clytiella spinifera*, one of them explicitly designated as the holotype, the other one presumably the paratype. However, in the collections of GeoSphere Austria, only the holotype was located and refigured by Hyžný and Zorn (2016, pl. 16, Fig. 1) and Hyžný and Zorn (2020, pl. 7, Fig. 7). No other specimens fitting the description of Glaessner (1931, p. 474) attributable to *C. spinifera* were found in the collections.

Description: See Förster (1967a, p. 144).

Remarks: *Clytiella* is an enigmatic genus with obscure affinities, as noted already by Förster (1967a, p. 145). Poor preservation of the type material, particularly the lack of discernible characters on the carapace, makes the higher classification of the genus difficult.

Förster (1966, 1967a) placed *Clytiella* within the erymid subfamily Eryminae, whereas Glaessner (1969) suggested its attribution to the erymid subfamily Clytiopsinae. Schweitzer et al. (2010) treated the genus within Erymidae without subfamily attribution. Feldmann et al. (2012) and Karasawa et al. (2013) recognized a distinct family Clytiopsidae, with *Clytiella* being a member.

Hyžný and Zorn (2016, p. 137) and Hyžný and Zorn (2020, p. 24) expressed doubts about the assignment to erymoid lobsters and suggested affinities with polychelid lobsters. The re-examination of the type material revealed that the pleon of *Clytiella spinifera* is flattened, axially keeled and distinctly shorter than the carapace. Thus, the reconstruction of Glaessner (1931, text Fig. 2) is highly misleading: what was interpreted as a spiny dorsal crest on the carapace is a mere preservation artefact, and the length of pleonal somites in the reconstruction does not match the real specimen. Hyžný and Zorn (2020) noted that the holotype was preserved in a dorso-lateral orientation, not lateral as suggested by Glaessner (1931). Based on the following characters that are present in the holotype of *C. spinifera*, the genus was reassigned to the polychelidan family Coleiidae: (1) carapace flat and longer than wide, (2) pleon broad, shorter than carapace, and with axial keel, (3) chelipeds relatively stout and short.

Clytiella spinifera is somewhat similar to the co-occurring *Tetrachela raiblana*. Poor preservation of the holotype of *C. spinifera* raises the possibility that apparent differences may not be mere preservational artefacts. Both taxa are treated as being distinct here because of the morphology of P1 chelipeds. These are stout and tuberculate in *C. spinifera*, while in *T. raiblana* they are more slender and smooth. Also, the outline of P1 carpus is not identical in these two species. P1 chelipeds and the relatively short, broad pleon of *C. spinifera* are somewhat similar to those of the clytiopsid *Koryncheiros luopingensis* Feldmann et al., 2012, from the Anisian of China. Accordingly, we retain *Clytiella* in Clytiopsidae, although this attribution is considered provisional until more complete material becomes available.

Occurrence: Upper Triassic (lower Carnian) of Polzberg, Austria.

Family Platychelidae Glaessner, 1969.

Genus *Platychela* Glaessner, 1931.

Type species: *Platychela trauthi* Glaessner, 1931, by subsequent designation of Förster (1967a).

***Platychela trauthi* Glaessner,** 1931.

Figures 6.8 and 6.9.

1931 *Platychela trauthi* Glaessner, p. 478, text-Fig. 3, pl. 16.

1931 *Platychela kahleri* Glaessner, p. 479, text-Fig. 4, pl. 15, Fig. 2, pl. 17.

1967a *Platychela trauthi* Glaessner—Förster, p. 178, text-Fig. 15.

1969 *Platychela trauthi* Glaessner—Glaessner, p. R458, Fig. 263.

1993 *Platychela* Glaessner—Garassino and Teruzzi, Fig. 47.

1998 *Platychela trauthi* Glaessner—Müller, p. 8.

Fig. 6.8 *Platychela trauthi* Glaessner, 1931 from the Upper Triassic (Carnian) of Polzberg, Austria. **a** Holotype NHMW 1910/0015/0018. **b** Detail of large P1 claw and portion of antenna. **c** Detail of distal elements of P2. Scale bar in **a** = 10 mm, scale bars in **b** and **c** = 5 mm

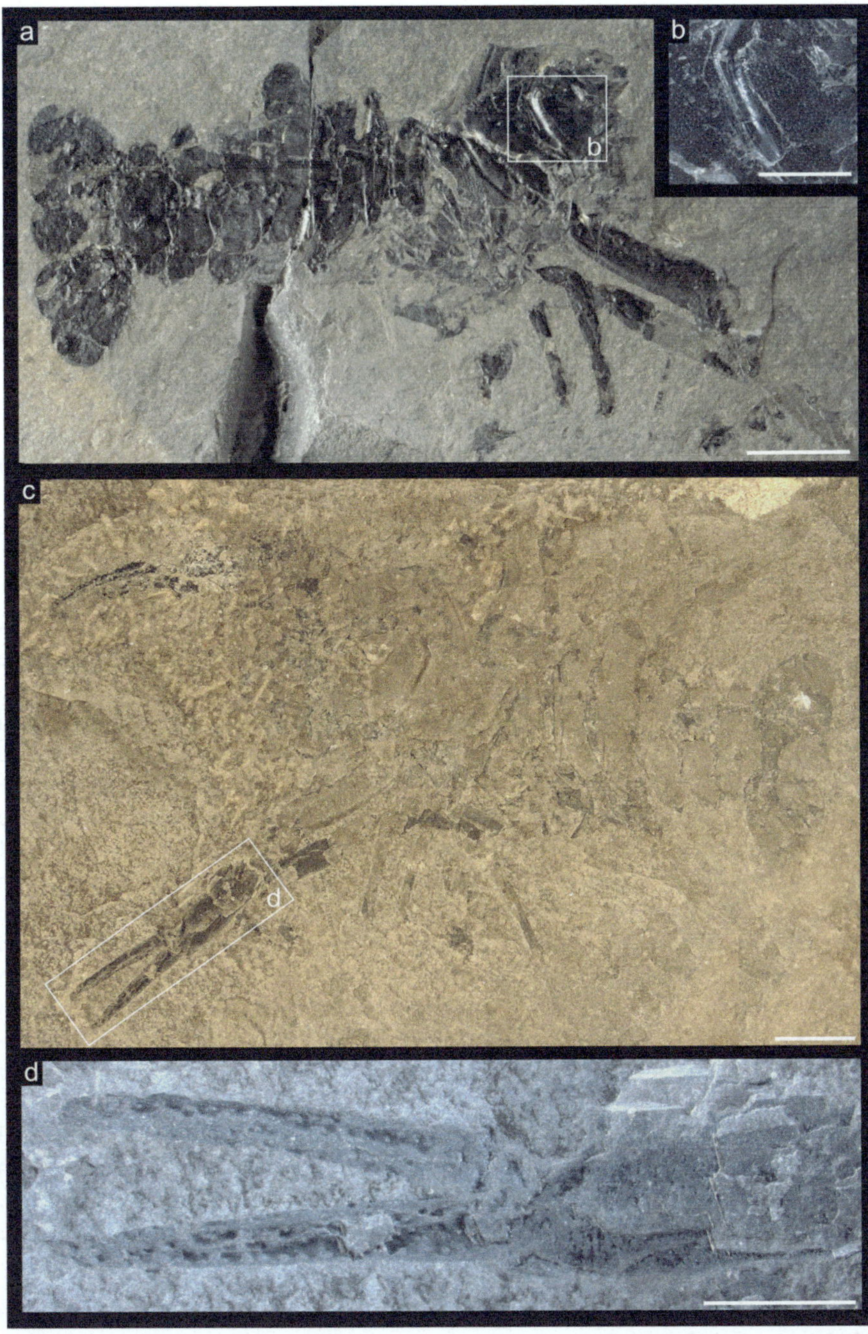

Fig. 6.9 *Platychela trauthi* Glaessner, 1931 from the Upper Triassic (Carnian) of Cave del Predil (Raibl), Italy. **a** NHMW 1887/0009/0105, part of the holotype of *Plachychela kahleri* Glaessner, 1931. **b** Detail of carapace ornamentation. **c** NHMW 1887/0009/0105, counterpart of the holotype of *P. kahleri*. **d** Detail of P1 claw (compare with Fig. 7b). Scale bars in **a**, **c**, and **d** = 10 mm, scale bar in **b** = 5 mm

2010 *Platychela trauthi* Glaessner—Schweitzer et al., p. 22.

2013 *Platychela trauthi* Glaessner—Karasawa et al., Fig. 12C.

2016 *Platychela trauthi* Glaessner—Hyžný and Zorn, p. 138, pl. 16, Figs. 2–3.

2020 *Platychela trauthi* Glaessner—Hyžný and Zorn, p. 20, pl. 4, Fig. 5.

Type material: Holotype of *Platychela trauthi* NHMW 1910/0015/0018 (part and counterpart; Fig. 6.8), paratype of *P. trauthi* NHMW 1910/0015/0024, paratype of *P. trauthi* GBA 1931/003/0002 (counterpart GBA 1931/003/0003/02), paratype of *P. trauthi* GBA 1931/003/0003/01, holotype of *P. kahleri* NHMW 1887/0009/0105 (part and counterpart; Fig. 6.9). All the material originally referred to *P. trauthi* comes from Polzberg, while that of *P. kahleri* originates from Cave del Predil.

Description: See Förster (1967a, p. 177).

Remarks: Glaessner (1931) described two species of *Platychela*, namely *P. trauthi* and *P. kahleri*. According to Glaessner (1931, 1965), *P. trauthi* occurs in the Reingraben Formation of Polzberg, whereas *P. kahleri* is known from the Raibl Beds of Cave del Predil. Förster (1967a) considered both species synonymous, a view followed by subsequent authors (e.g., Schweitzer et al. 2010; Karasawa et al. 2013). However, because Förster (1967a) did not provide a detailed discussion for this synonymy, Hyžný and Zorn (2020) considered both species distinct pending a more detailed revision.

The two species of *Platychela* presented by Glaessner (1931) differ from each other in rather minor aspects. Glaessner (1931) noted variations in the relative size of P1 chelae, and minor differences in the development of carapace grooves and the size and/or presence of spines/tubercles on the carapace and pleonal somites. Some of these differences can be explained as artefacts due to slightly different modes of preservation in Raibl and Polzberg *Lagerstätten*. The carapace and pleon of both species are tuberculate in a similar pattern, while more pronounced tuberculation in *P. trauthi* likely reflects ontogenetic variation, as these individuals were larger than the type material of *P. kahleri*. The comparatively larger P1 claws of *P. trauthi* can be attributed to intraspecific variation, allometric growth, or sexual dimorphism. All other observable characters, despite imperfect preservation, are identical or near-identical in both species. Thus, we concur with Förster (1967a) in treating *P. kahleri* as a subjective synonym of *P. trauthi*.

Poor preservation of *Platychela* type material has been emphasized in several studies. Karasawa et al. (2013) presented a morphological matrix of *Platychela* in their phylogenetic analysis, noting that over one fourth of all characters was unknown, while Charbonnier et al. (2015) excluded *Platychela* from their analysis due to the incomplete type material.

Occurrence: Upper Triassic (Carnian) of Cave del Predil, Italy, and lower Carnian of Polzberg, Austria.

Infraorder Polychelida Scholtz and Richter, 1995.

Superfamily Eryonoidea De Haan, 1841.

Family Coleiidae Van Straelen, 1925.

Genus *Tetrachela* Reuss, 1858.

Type species: *Bolina raiblana* Bronn, 1858, by monotypy.

Tetrachela raiblana (Bronn, 1858).

Figure 6.10.

1858 *Bolina Raiblana* Bronn, p. 22, pl. 4, Figs. 1–3.

1858 *Stenochelus triasicus* Reuss, p. 2, pl. 1, Fig. 1.

1858 *Tetrachela Raiblana* (Bronn)—Reuss, p. 5, pl. 1, Figs. 2–7.

1929 *Tetrachela raiblana* (Bronn)—Glaessner, p. 382.

1930 *Tetrachela raiblana* (Bronn)—Glaessner, p. 143, pl. 9, Fig. 1.

1931 *Tetrachela raiblana* (Bronn)—Glaessner, p. 480.

1965 *Tetrachela raiblana* (Bronn)—Glaessner, p. 112.

1969 *Tetrachela raiblana* (Bronn)—Glaessner, p. R470, Fig. 272.

1998 *Tetrachela raibliana* [sic!] (Bronn)—Müller, p. 9.

2009 *Tetrachela raiblana* (Bronn)—Ahyong, p. 370, Fig. 1A.

2010 *Tetrachela raiblana* (Bronn)—Schweitzer et al., p. 45. (list).

2013 *Tetrachela raiblana* (Bronn)—Feldmann et al., (2013) p. 8, Fig. 5.3.

2016 *Tetrachela raiblana* (Bronn)—Hyžný and Zorn, pp. 131, 138, pl. 3, Figs. 1–2, pl. 15, Fig. 2.

2018 *Tetrachela raiblana* (Bronn)—Audo et al., p. 46, Figs. 1–3, 4R.

2020 *Tetrachela raiblana* (Bronn)—Hyžný and Zorn, p. 24, pl. 7, Fig. 8.

Type material: The description of *Bolina raiblana* Bronn, 1858, is based upon seven syntypes. Only one of these syntypes (Bronn 1858, pl. 4, Fig. 3) was rediscovered in the "Originaliensammlung" collection housed in the Institute of Geosciences, University of Heidelberg. This specimen was figured and chosen as the lectotype by Audo et al. (2018, Fig. 1).

Other material: 77 specimens from Cave del Predil: GBA 1858/004/0002/01, GBA 1858/004/0002/02, GBA 1930/002/0002/01, GBA 1930/002/0002/02, GBA 2018/298/0146, GBA 2018/298/0154/01, GBA 2018/298/0160/01, GBA 2018/298/0160/02, GBA 2018/298/0161/01, GBA 2018/298/0161/02, GBA 2018/298/0162, GBA 2018/298/0163, GBA 2018/298/0164, GBA 2018/298/0165/01, GBA 2018/

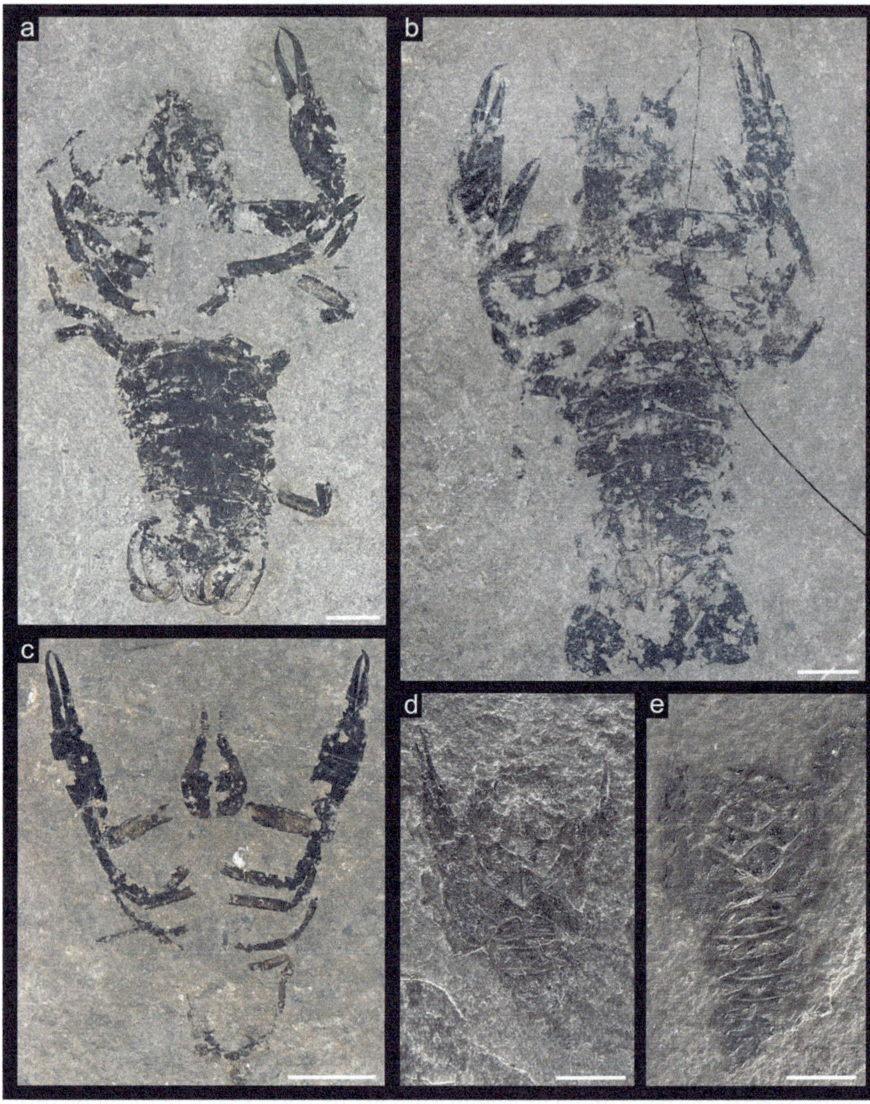

Fig. 6.10 *Tetrachela raiblana* (Bronn, 1858) from the Upper Triassic (Carnian) of Cave del Predil (Raibl), Italy. **a** GBA 2018/298/199. **b** GBA 2018/298/162. **c** GBA 2018/298/194/04. **d** NHMW 1922/0012/0001. **e** NHMW 1870/0054/0019. Specimens in **a**–**c** are preserved in ventral view and without the carapace; specimens in **d**–**e** are preserved in dorsal view and with the carapace. Scale bars = 5 mm

298/0165/02, GBA 2018/298/0166/03, GBA 2018/298/0166/04, GBA 2018/298/0167, GBA 2018/298/0168/01, GBA 2018/298/0168/02, GBA 2018/298/0169, GBA 2018/298/0170/02, GBA 2018/298/0170/03, GBA 2018/298/0171, GBA 2018/298/0172, GBA 2018/298/0173, GBA 2018/298/0174, GBA 2018/298/0175, GBA 2018/298/0176/01, GBA 2018/298/0176/02, GBA 2018/298/0176/03, GBA 2018/298/0176/04, GBA 2018/298/0177, GBA 2018/298/0178, GBA 2018/298/0179, GBA 2018/298/0180, GBA 2018/298/0181, GBA 2018/298/0182, GBA 2018/298/0183, GBA 2018/298/0184, GBA 2018/298/0185, GBA 2018/298/0186, GBA 2018/298/0187, GBA 2018/298/0188/01, GBA 2018/298/0188/02, GBA 2018/298/0189, GBA 2018/298/0190, GBA 2018/298/0191/01, GBA 2018/298/0191/02, GBA 2018/298/0192, GBA 2018/298/0193/01, GBA 2018/298/0193/02, GBA 2018/298/0193/03, GBA 2018/298/0193/04, GBA 2018/298/0199, NHMW 1864/0050/0019/01, NHMW 1864/0050/0019/02, NHMW 1866/0001/0041, NHMW 1866/0001/0042, NHMW 1866/0001/0043, NHMW 1866/0001/0044, NHMW 1866/0011/0455, NHMW 1866/0040/0454, NHMW 1870/0054/0019, NHMW 1870/0054/0020, NHMW 1887/0009/0110/01, NHMW 1887/0009/0110/02, NHMW 1887/0009/0110/03, NHMW 1887/0009/0110/04, NHMW 1887/0009/0110/05, NHMW 1887/0009/0110/06, NHMW 1887/0009/0112/01, NHMW 1887/0009/0112/02, NHMW 1887/0009/0112/03, NHMW 1887/0009/0112/04, NHMW 1922/0012/0001, NHMW 1922/0012/0002. Two specimens from Polzberg: NHMW 1910/0015/0025, NHMW 1910/0015/0027.

Glaessner (1930) reported 90 specimens of *Tetrachela raiblana*, deposited in several institutions. In the historical collections of NHMW and GBA we were able to identify 79 specimens, suggesting that these represent the material personally studied by Glaessner (1930). While Audo et al. (2018) revised *Tetrachela raiblana* based on a total of 43 specimens, only four of those belonged to the collection studied by Glaessner (1930).

Description: A detailed description was already presented by Audo et al. (2018) and is not repeated here.

Remarks: The species was originally described under the genus *Bolina* Münster, 1839. Reuss (1858) reassigned the species to the newly erected genus *Tetrachela* Reuss, 1858, and described a new taxon, *Stenochelus triassicus*. Notably, Bronn (1858) described *Bolina raiblana* based on preserved appendages only, while Reuss (1858) described *Stenochelus triassicus* based on the same animal preserved in dorsal view with the carapace. The identity of these two taxa was recognized already by Oppel (1862) and followed by all subsequent authors (see Audo et al. 2018 for more details).

Tetrachela raiblana was the sole species of the family Tetrachelidae Beurlen, 1930. Although Ahyong (2009) considered the family as the most primitive among polychelidan lobsters, its unique status has been questioned by Audo et al. (2018). Following their study, *Tetrachela* is here considered a representative of the family Coleiidae.

Occurrence: Upper Triassic (Carnian) of Cave del Predil, Italy, and lower Carnian of Polzberg, Austria.

6.6 Discussion

Taxonomically, the decapod faunas of both localities are largely similar, with *Clytiella* being the only taxon known from Polzberg but missing at Cave del Predil; all other decapod taxa occur at both locations. Nevertheless, there are two major differences between the two, as already noted by Gleassner (1931). The shrimps *Aeger straeleni* and *Distaeger crassipes*, as well as the polychelidan lobster *Tetrachela raiblana*, are distinctly larger at Cave del Predil than at Polzberg. The most striking difference between the two localities, however, is not the maximum size of the studied specimens, but the abundance and the mode of preservation of respective faunal elements. Glaessner (1931) observed that at Cave del Predil, the most common species is *T. raiblana*, which is preserved in most cases without the dorsal carapace (Fig. 6.10; see also Audo et al. 2018). At Polzberg, *T. raiblana* is known from only a few individuals, half of which are preserved with dorsal carapace intact. At Cave del Predil, the preservation of *T. raiblana* without a carapace suggests that the specimens represent moults. At Polzberg, the specimens of the lobsters *Clytiella*, *Platychela* and *Tetrachela*, supposedly living as part of the benthic epifauna, are often preserved near-complete. There are comparatively fewer natantian (actively swimming) shrimps at Polzberg than at Cave del Predil. Nevertheless, all the differences need to be interpreted with caution due to uneven sampling: 111 specimens studied here originate from Cave del Predil, while only 20 come from Polzberg. The number of indeterminate shrimps is three at Cave del Predil and eight at Polzberg.

6.7 Conclusions

The Late Triassic (Carnian) *Konservat-Lagerstätten* of Cave del Predil (Raibl) in northern Italy and Polzberg in Lower Austria have been known since the 19th and the first half of the twentieth century, respectively, and represent some of the first documented Triassic decapod assemblages. The present re-study is based on five identified taxa, including five species in four families: *Aeger straeleni* (Dendrobranchiata: Aegeridae), *Distaeger crassipes* (Dendrobranchiata: Ageridae), *Clytiella spinifera* (Glypheidea: Clytiopsida), *Platychela trauthi* (Glypheidea: Platychelidae), and *Tetrachela raiblana* (Polychelida: Coleiidae). This re-study is based on the examination of approximately 130 specimens from historical collections of the Natural History Museum Vienna (Naturhistorisches Museum Wien) and GeoSphere Austria (former Bundesanstalt für Geologie, Geophysik, Klimatologie und Meteorologie), Vienna. Current attribution of "*Aeger*" *crassipes* to *Distaeger* is based on the recognition of the diagnostic rostrum. *Bombur aonis* is regarded as a juvenile of *D. crassipes*. The higher classification of *Clytiella spinifera* is uncertain pending unearthing more complete material. Uneven sampling prevents relevant (semi)quantitative comparison between the two coeval decapod associations. Documented differences include mode of preservation and dominant species present. Nevertheless, both associations provide valuable data in understanding the Triassic diversification of decapod crustaceans, being a part of the Mesozoic Marine Revolution.

Acknowledgments The work was done within the framework of and financially supported by projects of the Federal Government of Lower Austria (K3-F-964/001-2020 and K3-F-964/002-2023, Department Science and Research; headquarters in St. Pölten); the Austrian Academy of Sciences (headquarters in Vienna, project Polzberg Lukeneder 2021) represented by the National Committee for Geo/Hydro Sciences (Earth System Sciences Programme); and the Triassic project "Fenster in die Karnische Krise" of the Society of Friends of the Natural History Museum Vienna (Freunde des Naturhistorischen Museums Wien, CPE Lukeneder 2023). The research of MH was supported by the Slovak Agency for Research and Development (APVV-22-0523) and the Slovak Scientific Grant Agency (VEGA 2/0106/23). The authors are particularly grateful to the owners of the outcrop area Franziska and Hermann Hofreiter as well as Eva and Karl Jagersberger (all Gaming) for sampling and digging permission at Polzberg. The manuscript greatly benefited from valuable comments by the reviewers Rok Gašparič (Kamnik) and Carrie E. Schweitzer (Ohio), the production editor Ashok Arumairaj (Taramani), the in-house series editor Kathleen A. Ritterbush (Utah), the publishing editor Simon Shah-Rohlfs and the production editor Antje Endemann (both Heidelberg).

Author Contribution AL, MH, DDF provided documentation on fossil material. MH conceptualization and research design, MH writing and documentation on fossil material.

Funding Open access funding was enabled and the projects financially supported by the Federal Government of Lower Austria (K3-F-964/001-2020 and K3-F-964/002-2023, Department Science and Research; headquarters in St. Pölten); the Austrian Academy of Sciences (headquarters in Vienna, project Polzberg Lukeneder 2021) represented by the National Committee for Geo/Hydro Sciences (Earth System Sciences Programme); the Natural History Museum Vienna; and the Triassic project "Fenster in die Karnische Krise" of the Society of Friends of the Natural History Museum Vienna (Freunde des Naturhistorischen Museums Wien, CPE Lukeneder 2023). The research of MH was supported by the Slovak Agency for Research and Development (APVV-22-0523) and the Slovak Scientific Grant Agency (VEGA 2/0106/23). The authors are responsible for the contents of this publication. The funders had no impact on conceptualization, design, data collection, analysis, decision to publish, or preparation of the manuscript.

Competing interests All authors stated that they have no competing interests.

Authors' contributions All authors stated that they have equally contributed to their chapters, read, edited and agreed to submit the final versions of their contributions.

Availability of data and materials All sources of information are provided in the text. Raw data related to the fossil material from Austria are available from the corresponding authors and Alexander Lukeneder upon request.

Additional information This article is part of the Special Volume: *Late Triassic Konservat-Lagerstätten within the Carnian Pluvial Episode. 150 Years of Scientific Excavations in Austria.*

Artificial Intelligence (AI) The chapter was produced without the use of AI.

Language The entire volume uses American English reviewed by a native scientist.

Received 2025.

Published online 2026.

References

Ahyong ST (2009) The polychelidan lobsters: phylogeny and systematics (Polychelida: Polychelidae). In: Martin JW, Crandall KA, Felder DL (eds) Decapod Crustacean phylogenetics. Crustacean Issues, vol 18. CRC Press, Boca Raton, London, New York, pp 369–396. https://doi.org/10.1201/9781420092592

Audo D, Hyžný M, Charbonnier S (2018) The early polychelidan lobster *Tetrachela raiblana* and its impact on the homology of carapace grooves in decapod crustaceans. Contrib Zool 87:41–57. https://doi.org/10.1163/18759866-08701004

Balss H (1922) Studies of the decapod fossils. Paläontol Z 5:123–147. https://doi.org/10.1007/BF03041549

Bate CS (1888) Report on the Crustacea Macrura collected by the Challenger during the years 1873–76. In: Report on the scientific results of the voyage of H.M.S. Challenger. Zoology 24(52):i–xc, 1–942. https://doi.org/10.5962/bhl.title.6513

Beurlen K (1927) Zur Stammesgeschichte der jurassischen Macruren. Paläontol Z 8:98–105. https://doi.org/10.1007/BF03160405

Beurlen K (1930) Vergleichende Stammesgeschichte: Grundlagen, Methoden und Probleme unter besonderer Berücksichtigung der höheren Krebse. Fortschritte der Geologie und Paläontologie 8:317–586

Brayard A et al (2017) Unexpected Early Triassic marine ecosystem and the rise of the modern evolutionary fauna. Sci Adv 3:e1602159. https://doi.org/10.1126/sciadv.1602159

Bronn HG (1858) Beiträge zur triasischen Fauna und Flora der bituminösen Schiefer von Raibl. Jahrbuch für Mineralogie, Geognosie, Geologie und Petrefakten-Kunde 1858:1–32

Burkenroad MD (1963) The evolution of the Eucarida (Crustacea, Eumalacostraca) in relation to the fossil record. Tulane Stud Geol 2:3–17

Charbonnier S, Audo D, Barriel V, Garassino A, Schweigert G, Simpson M (2015) Phylogeny of fossil and extant glypheid and litogastrid lobsters (Crustacea, Decapoda) from morphological characters. Cladistics 31:231–249. https://doi.org/10.1111/cla.12088

Feldmann RM, Schweitzer CE, Hu S, Zhang Q, Zhou C, Xie T, Huang J, Wen W (2012) Macrurous Decapoda from the Luoping Biota (Middle Triassic) of China. J Paleontol 86:425–441

Feldmann RM, Schweitzer CE, Karasawa H, Schweigert G, Garassino A (2013) Systematic descriptions: infraorder Polychelida. Treatise Online 60:1–10. https://doi.org/10.17161/to.v0i0.4593

Förster R (1966) Über die Erymiden: eine alte konservative Familie der mesozoischen Dekapoden. Palaeontographica A 125:61–175

Förster R (1967a) Die reptanten Dekapoden der Trias. Neues Jb Geol Paläontol Abh 128:136–194

Förster R (1967b) Zur Kenntnis natanter Jura-Dekapoden. Mitteilungen Bayerische Staatssammlung Paläontologie 7:157–174

Garassino A, Teruzzi G (1993) A new decapod crustacean assemblage from the Upper Triassic of Lombardy (Italy). Paleontologia Lombarda 1:3–27

Garassino A, Schweigert G, Muscio G (2014) *Acanthochirana triasica* n. sp. and *Antrimpos colettoi* n. sp. from the Upper Triassic of Italy. Gortania 35:11–18

Gašparič R, Audo D, Hitij T, Jurkovšek B, Kolar-Jurkovšek T (2020) One of the oldest polychelidan lobsters from the Upper Triassic of Slovenia. Neues Jb Geol Paläontol Abh 296:107–117

Glaessner MF (1929) Crustacea Decapoda. In: Pompeckj FJ (ed) Fossilium Catalogus, vol 41. W. Junk, Berlin

Glaessner MF (1930) Dekapodenstudien. Neues Jahrbuch Für Mineralogie, Geologie und Paläontologie 63:137–173

Glaessner MF (1931) Eine Crustaceenfauna aus den Lunzer Schichten Niederösterreichs. Jahrbuch der geologischen Bundesanstalt Wien 81:467–486

Glaessner MF (1965) Vorkommen fossiler Dekapoden (Crustacea) in Fisch-Schiefern. Senckenb Lethaea 46A:111–122

Glaessner MF (1969) Decapoda. In: Moore RC (ed) Treatise on invertebrate paleontology, Part R. Geological Society of America and University of Kansas Press
De Haan W (1833–1850) Crustacea. In: von Siebold PF (ed) Fauna Japonica. Lugduni-Batavorum
Hyžný M, Garassino A (2022) *Aeger crassipes* Bronn, 1858 revisited: implications for taxonomy. Ann Naturhist Mus Wien 122A:69–85
Hyžný M, Zorn I (2016) Catalogue of type and figured fossil decapods in Vienna. Jahrb Geol Bundesanst 156:127–177
Hyžný M, Zorn I (2020) Catalogue of fossil decapods in the collections of the Geological Survey of Austria. Abhandlungen der Geologischen Bundesanstalt 74:1–111
Karasawa H, Schweitzer CE, Feldmann RM (2013) Phylogeny and systematics of extant and extinct lobsters. J Crustac Biol 33:78–123. https://doi.org/10.1163/1937240X-00002111
Kelley PH, Hansen TA (2001) Mesozoic marine revolution. In: Briggs DEG, Crowther PR (eds) Palaeobiology II. Blackwell Science, pp 94–97
Klompmaker AA, Schweitzer CE, Feldmann RM, Kowalewski M (2013) The influence of reefs on the rise of Mesozoic marine crustaceans. Geology 41:1179–1182. https://doi.org/10.1130/G34768.1
Latreille PA (1802–1805) Histoire naturelle des crustacés et des insectes. F. Dufart, Paris
Lukeneder A (ed) (2026) Late Triassic Konservat-Lagerstätten within the Carnian Pluvial Episode: 150 Years of Scientific Excavations in Austria. Top Geobiol 55:600 p. https://doi.org/10.1007/978-3-032-19777-1
Lukeneder A, Lukeneder P (2021) The Upper Triassic Polzberg palaeobiota from Austria. Sci Rep 11:16644. https://doi.org/10.1038/s41598-021-96052-w
Lukeneder A, Lukeneder P (2022a) Taphonomic history and trophic interactions of an ammonoid fauna from the Upper Triassic Polzberg palaeobiota. Sci Rep 12:7455. https://doi.org/10.1038/s41598-022-11496-y
Lukeneder P, Lukeneder A (2022b) Mineralized belemnoid cephalic cartilage from the late Triassic Polzberg *Konservat-Lagerstätte* (Austria). PLoS ONE 17(4):e0264595. https://doi.org/10.1371/journal.pone.0264595
Lukeneder A, Lukeneder P (2023) New data on the marine Upper Triassic palaeobiota from the Polzberg *Konservat-Lagerstätte* in Austria. Swiss J Palaeontol 142:9. https://doi.org/10.1186/s13358-023-00269-3
Lukeneder A, Surmik D, Gorzelak P, Niedźwiedzki R, Brachaniec T, Salamon MA (2020) Bromalites from the Upper Triassic Polzberg section (Austria); insights into trophic interactions and food chains of the Polzberg palaeobiota. Sci Rep 10:20545. https://doi.org/10.1038/s41598-020-77017-x
Müller P (1998) Crustacea Decapoda. In: Flügel HW (ed) Catalogus Fossilium Austriae. Austrian Academy of Sciences, Vienna
Münster G (1839) Fossil decapods from Bavarian limestones. Beiträge zur Petrefakten-Kunde 2:1–88
Oppel A (1862) Ueber jurassische Crustaceen. Paläontologische Mitteilungen 1:1–120
Pérez Farfante I, Kensley BF (1997) Penaeoid and sergestoid shrimps and prawns of the world. Mémoires du Muséum national d'Histoire naturelle 175:1–233
Pinna G (1974) I Crostacei della fauna triassica di Cene. Memorie della Società Italiana di Scienze Naturali 21(1):1–34
Poore GCB, Ahyong ST (2023) Marine Decapod Crustacea: a guide to families and genera of the world. CSIRO Publishing/CRC Press
Rafinesque CS (1815) Analyse de la nature ou Tableau de lunivers et des corps organisés. Publishes by the author, Palermo. 224 pp. https://doi.org/10.5962/bhl.title.106607
Reuss AE (1858) Fossile Krebse aus den Raibler Schichten in Kärnten. Beiträge zur Paläontographie Österreichs 1:1–6
Schlotheim EF von (1822) Nachträge zur Petrefactenkunde. Becker, Gotha
Scholtz G, Richter S (1995) Phylogenetic systematics of reptant decapods. Zool J Linn Soc 113:289–328. https://doi.org/10.1006/zjls.1995.0011

Schweitzer CE, Feldmann RM (2015) Faunal turnover and niche stability in marine decapods. J Crustac Biol 35:633–649. https://doi.org/10.1163/1937240X-00002359
Schweitzer CE, Feldmann RM, Garassino A, Karasawa H, Schweigert G (2010) Systematic list of fossil decapods. Crustaceana Monographs 10:1–222
Schweitzer CE, Feldmann RM, Hu S, Huang J, Zhou C, Zhang Q, Wen W (2014) Penaeoid Decapoda from the Luoping Biota. J Paleontol 88:457–474. https://doi.org/10.1666/13-039
Seilacher A (1970) Begriff und Bedeutung der Fossil-Lagerstätten. N Jahrb Geol Paläontol Mh 1970:34–39
Smith CPA, Charbonnier S, Jenks JF, Bylund KG, Escarguel G, Olivier N, Fara E, Brayard A (2022) The Paris Biota decapod fauna and Triassic diversity. J Paleontol 96:1235–1263. https://doi.org/10.1017/jpa.2022.34
Van Straelen V (1925) Contribution à l'étude des Crustacés décapodes jurassiques. Mémoires Classe des Sciences Académie Royale de Belgique 7(1):1–462
Vermeij GJ (1977) The Mesozoic marine revolution. Paleobiology 3:245–258
Winkler TC (1881) Étude carcinologique sur *Pemphix, Glyphea* et *Araeosternus*. Archives du Musée Teyler 2:73–124
Woodward H (1888) A new species of *Aeger* from the Lower Lias. Geol Mag 5:385–387. https://doi.org/10.1017/S0016756800182433

Chapter 7
Bivalve Mass Occurrences in the Upper Triassic *Konservat-Lagerstätten* of Austria

Alexander Lukeneder

Abstract This contribution reports preliminary data on some Upper Triassic bivalve mass occurrences in the lower Carnian *Konservat-Lagerstätten* of the Northern Calcarous Alps of Lower Austria and Styria. New fossil material was obtained and examined from bed-by-bed collections in the well-laminated and argillaceous Reingraben Formation deposited during the Carnian Pluvial Episode. Fossils of various marine taxa were collected within the fossiliferous layers from the early Carnian (Julian 2, *Austrotrachyceras austriacum* Zone). The assemblages comprise ammonoids, coleoids, gastropods, thylacocephalan arthropods, branchiopods (*Euestheria*), rare coelacanth remains and acytinopterygian fishes, and bromalites along with frequent plant remains. All the detected lower Carnian paleobiotas from the Carnian Pluvial Episode are dominated by the pioneer taxa, the posidoniid bivalve *Peribositra* and the halobiid bivalve *Halobia rugosa*. The paleobiota of the basal Reingraben Formation is characterized by a fluctuating basinal environment with intermittent influx of freshwater, accompanied by the enhanced presence of land plants and colonization by abundant benthic posidoniid and halobiid bivalves. Under the prevailing low-oxygen conditions during the deposition of the Reingraben Formation, thin-shelled bivalves such as *Peribositra* and *Halobia* were the dominant epifaunal elements, adapted to life at or near the sea floor and/or within the carbon-rich and laminated sediment.

Keywords *Konservat-Lagerstätte* · Bivalves · *Peribositra* · *Halobia* · Early Carnian · Late Triassic · Austria

7.1 Introduction

The paleobiotas of fossiliferous sites known as *Konservat-Lagerstätte* (Seilacher 1970; Allison 1988; see also Krystyn 1991; Forchielli and Pervesler 2013; Lukeneder 2026, this volume, and references therein) are characterized by a special mode of

A. Lukeneder (✉)
Department of Geology and Paleontology, Natural History Museum Vienna, Vienna, Austria
e-mail: alexander.lukeneder@nhm.at

A. Lukeneder (ed.), *Late Triassic Konservat-Lagerstätten within the Carnian Pluvial Episode*, Topics in Geobiology 55, https://doi.org/10.1007/978-3-032-19777-1_7

fossil preservation, making them precious sources of paleobiological information (Briggs 2014; marine conservation deposit in Klug et al. 2024a, b). Such conservation *Lagerstätten* provide unique insights into paleocommunities, food chains and dietary habits, as well as into trophic interactions between the inhabitants of marine ecosystems (Lukeneder et al. 2020, and references therein). For the present contribution, excavations with bed-by-bed collecting were done during several field campaigns (60 days sampling) during four consecutive years between September 2020 and August 2025 (and still ongoing). The resulting details on the fauna and flora yielded new insights into the lower Carnian paleocommunities of the Northern Calcareous Alps (NCA) in northern Styria and Lower Austria (Lukeneder and Lukeneder 2021, 2023; Lukeneder et al. 2024a; Lukeneder 2026 this volume; Fig. 7.1). The presented Special Volume (Lukeneder 2026) provides more details on distinct taxonomic groups (thylacocephalans, bristle worms, conodontophorids, gastropods, bivalves, fish, belemnoids, bromalites), the geochemistry, geophysics, lithological aspects, as well as trophic interactions (food chains and dietary habits) between the inhabitants from the Upper Triassic marine ecosystems. These palebiotas were deposited during the Carnian Julian 2 Ib (*Austrotrachyceras austriacum* Zone, *A. minor* biohorizon), within a period known as the worldwide Upper Triassic Carnian Pluvial Episode (CPE, Rigo et al. 2007; Ruffell et al. 2015; Dal Corso et al. 2018, 2020; Simms and Ruffell 1990, 2018; Lukeneder et al. 2024a; Alverà et al. 2026; Lukeneder 2026, this volume; Wet Intermezzo, Kozur and Bachmann 2010; Ogg 2015; 'Carnian Crisis', Hornung et al. 2007; Hornung 2008). The CPE features an enhanced humidification of the Carnian climate, resulting in the deposition of the fossiliferous Reingraben Formation at Scheiblingbach and Polzberg. As noted by Lukeneder and Lukeneder (2021) as well as Lukeneder et al. (2020, 2024a), such laminated deposits provide excellent conditions for the preservation of entire and soft body fossils, a prerequisite for *Konservat-Lagerstätten*.

This preliminary study presents insights into the bivalve mass occurrences in the lower Carnian *Konservat-Lagerstätten* of Austria, formed within the frame of an intraplatform basinal environment, the Reifling Basin, during the CPE.

7.2 Material and Methods

Historical excavation campaigns, conducted by Joseph Haberfelner, to obtain the Polzberg fossils were organized by the Geological Survey of Austria (GBA; since 2023 GeoSphere Austria) in 1885 and the Natural History Museum of Vienna (NHMW) in 1909 (Stur 1886; Teller 1891; Griffith 1977). They yielded 6,397 fossil specimens. More recently, additional material (n 5,290) was collected by the author (and team) and private collectors (Birgitt and Karl Aschauer; n 282 donated to GBA collections; Lukeneder 2026 this volume). The fossil fauna from early collecting at Polzberg is deposited in various Austrian institutions. The enormous amount of historical and recent material in collections from the comparable Scheiblingbachgraben (n 4,602), Steinbachgraben (n 2,650), Rehberg (n 437), Polzberg (n 11,687),

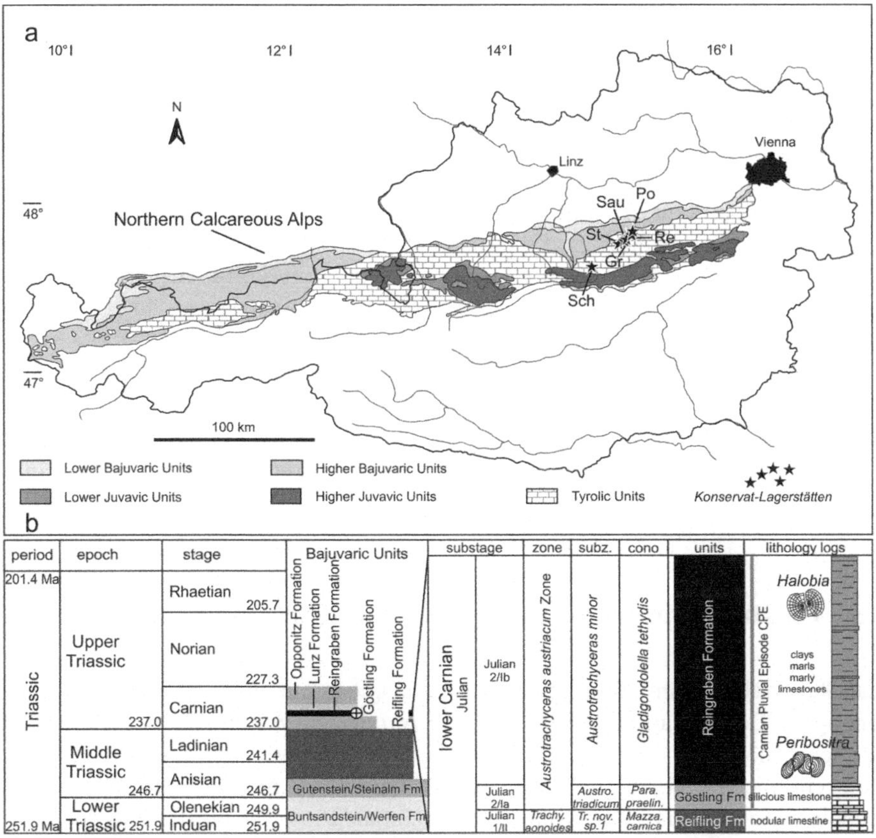

Fig. 7.1 **a** Map of the main Carnian *Konservat-Lagerstätten* localities (black asterisks) with bivalve mass occurrences in the Northern Calcareous Alps (NCA) of Austria. **b** Stratigraphic position of the Scheiblingbach deposits during the lower Carnian. White ellipse: position of the *Konservat-Lagerstätte* Scheiblingbach. Adapted after Lukeneder and Lukeneder (2021). Detailed Scheiblingbach section within the lower Carnian Reingraben Formation. Gr Großau, Po Polzberg, Re Rehberg, Sau Saugraben, Sch Scheiblingbach, St Steinbachgraben. Austro *Autrotrachyceras*, Trachy *Trachyceras*, chrono chronostratigraphy, subz subzone, Fm Formation. Adapted after Lukeneder and Lukeneder (2021)

Rad (n 124), and Hinterbrühl (n 47) localities encompasses approx. 30,000 fossil remains. They are held in distinct collections of the NHMW, the GeoSphere Austria, and the Departments of Geology (IGUW) and the Institute of Paleontology (IPUW), University of Vienna (DES; n 13), with smaller amounts in the Lower Austria State Collections (LASC; n 21). The by far highest numbers of specimens including holotypes and rare taxa are held at the NHMW, followed by the collection of the GeoSphere Austria.

In contrast to historical material from the same localities, the fossil material collected over the last decades involved detailed and bed-by-bed campaigns in numerous Late Triassic *Konservat-Lagerstätten* (Lower Austria: Polzberg, Rehgraben, Lehen, Saugraben, Großau, Steinbachgraben; Styria: Scheiblingbach) within deposits of the black, finely laminated Reingraben Formation. The new findings demonstrate that most detected fossils derive from the lowermost two calcareous meters of the Reingraben Formation (Fig. 7.1). The soft argillaceous deposits were also sampled and bear a restricted fauna, whereby rock-forming masses of bivalves are present in distinct levels (i.e. *Peribositra* and *Halobia* coquinas). The above numbers are exclusive of the uncountable masses of rock-forming bivalves (mainly *Peribositra*). *Peribositra* forms meter-thick coquinas in specific levels, hence representing the dominant benthic element at the various lower Carnian *Konservat-Lagerstätten* in the NCA (Fig. 7.1).

The Reingraben Formation is subdivided into the lower, more calcareous part exhibiting the *Konservat-Lagerstätten* (Figs. 7.2 and 7.3) and the upper, more argillaceous part with distinct levels of posidoniid coquinas and halobiid layers. Both parts are clearly dominated by posidoniids with *Periboitra* sp. (not countable, rock-forming numbers).

7.3 Biostratigraphy of the *Konservat-Lagerstätten*

The Reingraben Formation spans over a single ammonoid assemblage, representing the Lower Carnian *Austrotrachyceras austriacum* Zone (Julian 2, Fig. 7.1). The lowermost, fossiliferous part of the Reingraben Formation is dated based on the occurrence of *Austrotrachyceras minor*, *Paratrachyceras haberfellneri* and the accompanying *Carnites floridus* as Julian 2 (Julian 2/Ib). The biostratigraphic situation is comparable at all localities embracing the *Konservat-Lagerstätten* (Lukeneder et al. 2024a; Lukeneder 2026 this volume).

7.4 Geologic Setting and Lithology

The Reifling Basin is an elongated, approximately 300–500 km long, NE to SW striking intraplatform basin. It is characterized by the occurrence of specific lithological successions, here mainly the Lunzer Fazies (= Lunz Facies Zone LFZ) interdigitating with the Rohr Facies (= Rohrer Fazies Zone RFZ) to the east, and the Traunalpen Facies (= Traunalpen Fazies Zone TFZ) to the west (Spengler 1931; Thenius 1974; Tollmann 1976; Wessely 2006). The Lunz Facies Zone is the most important for the occurrence of the lower Carnian Reingraben Formation with the incorporated *Konservat-Lagerstätten* from Rad, Polzberg, Rehberg, Großau, Göstling an der Ybbs, and Scheiblingbachgraben (Fig. 7.1). Bajuvaric nappes (Lunz Nappe and Sulzbach Nappe) and the northern Tyrolic nappes (e.g., Reisalpen Nappe and Göller Nappe) are important in this respect.

Fig. 7.2 Outcrop situation at the most prominent localities with exposed deposits comprising the *Konservat-Lagerstätten* with abundant bivalves. **a** Detailed view on the basal Reingraben Formation at Polzberg near Gaming. **b** Basal Reingraben Formation of the Rehgraben locality, near Lunz am See. **c** Upper part of the Reingraben Formation in the Saugraben ravine, near Lunz am See. **d** Transition from the silicious Göstling Formation (right) into the overlying argillaceous Reingraben Formation (left) at the Großau drilling site, near Lunz am See. **e** Fossiliferous layers of the basal Reingraben Formation in the water of the Ybbs River, near Göstling (all Lower Austria). **f** Upper part of the Reingraben Formation with halobiid layers at the Scheiblingbach locality, near Großreifling (Styria)

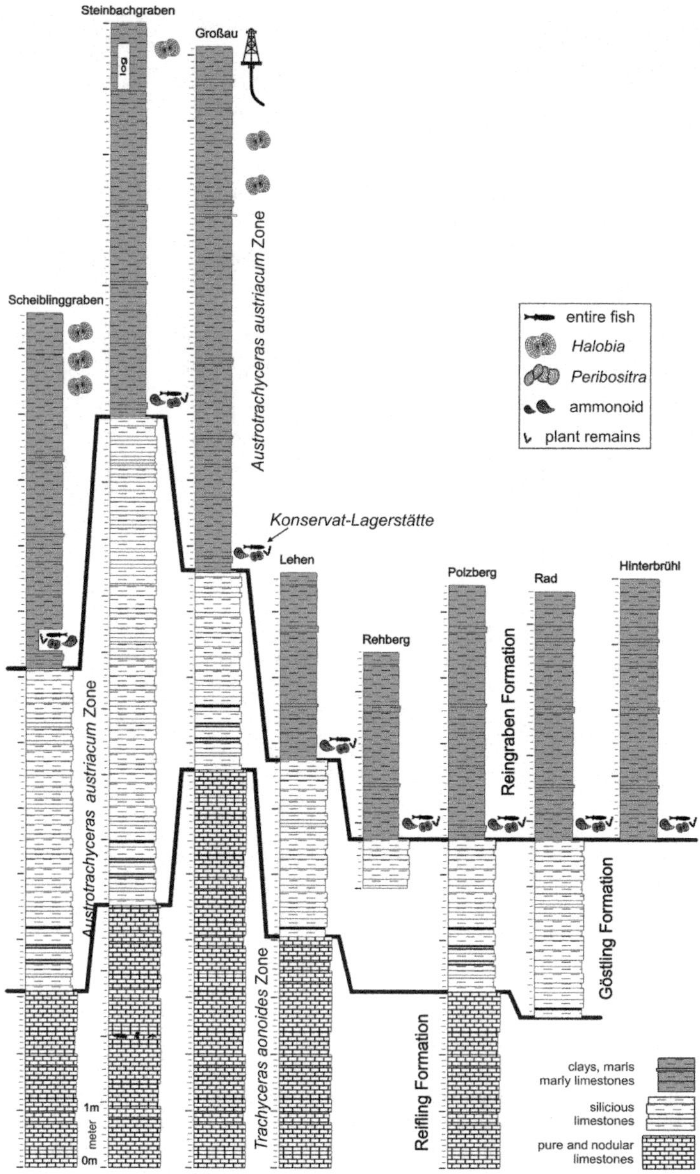

Fig. 7.3 Main localities of the marine Carnian *Konservat-Lagerstätten* Belt (CKLB), Northern Calcareous Alps of Styria and Lower Austria. Geographically arranged from west (Scheiblinggraben in Styria) to east (Hinterbrühl in Lower Austria). Position of the CKLB marked by fossil symbols at the basal argillaceous Reingraben Formation. Note the *Peribositra* dominance in the basal Reingraben Formation, whereas *Halobia* is abundant in the upper, more argillaceous parts of the Reingraben Formation. Note the drilling rig symbol at the top of the Großau log

After a strongly bioturbated top of the Göstling Formation, the overlying terrigeneous siliciclastic deposits of the Reingraben Formation (*A. austriacum* Zone, *A. minor* Subzone, Fig. 7.1) consist of a pronounced alternation of dark gray to black argillaceous marls ($CaCO_3$ 5–20%; Lukeneder et al. 2024a; Lukeneder 2026, this volume; Weitz et al. 2026, this volume) intercalated by rare limestone/dolomite layers. No macro-bioturbation is visible. This lithological change marks the 'Reingrabener Wende' (Schlager and Schöllnberger 1974; = Reingraben Turnover, Mueller et al. 2016) within the humid phase of the Carnian Pluvial Episode. The bivalve *Peribositra* appears in all ontogenetic stages and, in layer-/rock-forming quantities, mostly in the articulated butterfly position lumachelles (Figs. 7.3 and 7.4). Most of the fossils derive from the lowermost three calcareous meters of the Reingraben Formation, but macrofossils occur throughout the sections.

Polzberg (Po) Located on the western slope of Mount Schindelberg (1066 m), north of the Ois River, 4 km northeast of Lunz am See in Lower Austria (Stur 1874, 1886; Lukeneder and Lukeneder 2021; 1:50,000, geological map, sheet 71 Ybbsitz; Ruttner and Schnabel 1988 and sheet 72 Mariazell; Bauer and Schnabel 1997; Figs. 7.1, 7.2 and 7.3).

GPS (global positioning system): N 47°53′4.98″ and E 15°4′28.15″.

Taxa: bivalves occur with numerous indeterminate genera, rare adult *Halobia rugosa* (Fig. 7.4a) and mass occurrences of juvenile to adult *Peribositra* (Fig. 7.4a).

Geologic unit: lower part of the Reingraben Formation, NCA.

Facies: laminated marls with allodapic limestone intercalations, layers of bivalve lumachelles, thickness of bivalve level: approx. 5 m.

Sediment: laminated, no bioturbation.

Rehberg section (Re) Located near Lunz am See (Lower Austria) on the northwestern slope of Mount Maißzinken (1057 m), situated 2.5 km south to the Polzberg site, 700 m west of the lake Lunzer See on the Sulzbach Nappe within the Lunz Nappe System (High-Bajuvaric System; Geo map 71 Ybbsitz; Ruttner and Schnabel 1988; Figs. 7.1, 7.2 and 7.3). The section exhibits approx. 2.5 m of the Göstling Formation and, at the top, the argillaceous Reingraben Formation with characteristic entire fishes and ammonite abundance.

GPS: N 47°51′37.60″ and E 15°3′59.30″.

Taxa: rare *Halobia rugosa* and mass occurrences of *Peribositra*.

Geologic unit: lower part of the Reingraben Formation, NCA.

Facies: laminated marls with allodapic limestone intercalations, layers of bivalve lumachelles.

Thickness of bivalve level: approx. 3 m.

Sediment: laminated, no bioturbation.

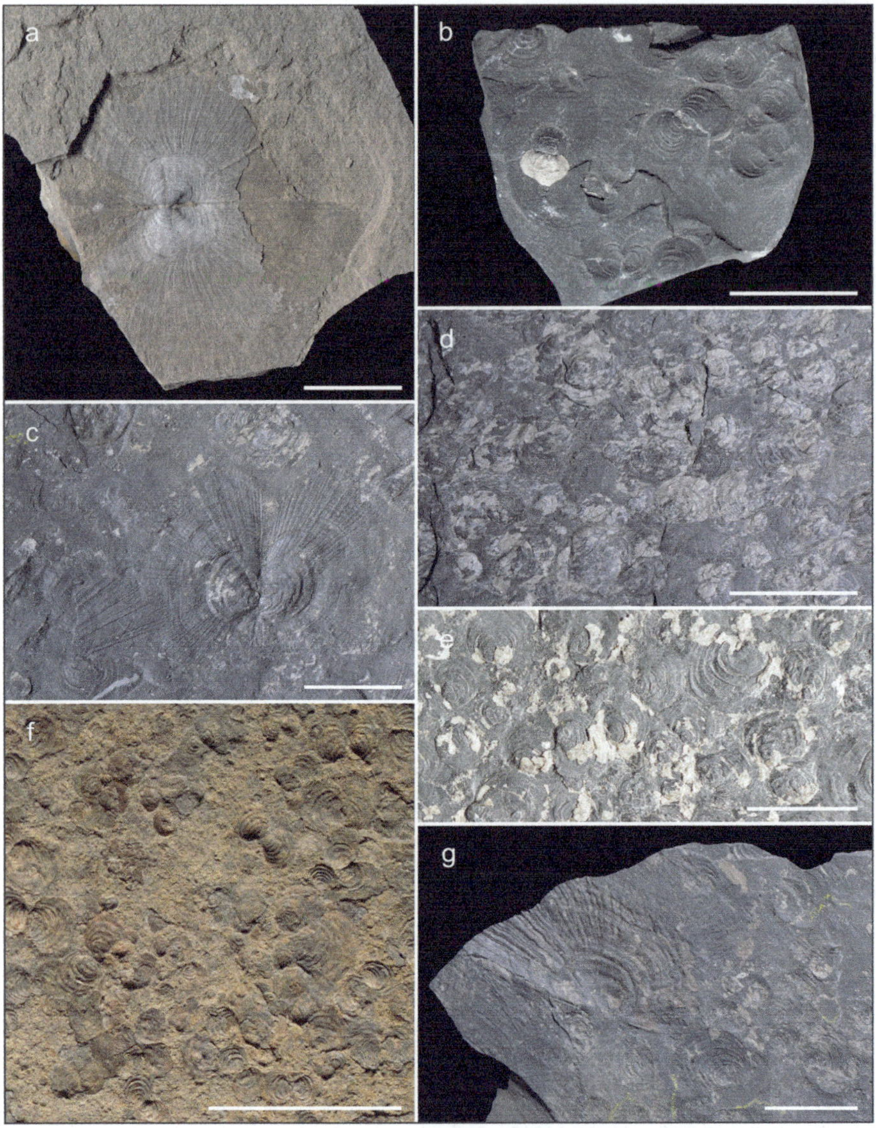

Fig. 7.4 Examples of the abundant and characteristic bivalve members of the lower Carnian *Konservat-Lagerstätten* paleobiota within the argillaceous Reingraben Formation. **a** *Halobia rugosa*, adult specimen, butterfly preservation, Polzberg, layer Po −50 to 0 cm, NHMW 2021/0123/0198. **b** *Peribositra* sp., juvenile specimens, accumulated in butterfly preservation, layer Gro 10.0 m, NHMW 2025/0127/0005., adult specimens, butterfly preservation, Polzberg, layer Po 120 to 140 cm, NHMW 2021/0016/0346. **c** *H. rugosa*, adult specimen, butterfly preservation, Saugraben, layer Sau 1, NHMW 2021/0123/0198. **d** *H. rugosa*, juvenile specimens, butterfly preservation, Saugraben, layer Sau 1, NHMW 2021/0127/0024. **e** *Peribositra* sp., coquina, single valves, Scheiblingbach, layer Sch 33, NHMW 2025/0127/0022. **f** *Peribositra* sp., juvenile specimens, accumulated in butterfly preservation, Großau drilling site, layer Gro 10.0 m, NHMW 2025/0127/0005. **g** *H. rugosa*, adult specimen, with mid-aged halobiids, Scheiblingbach, layer Sch 31 m, NHMW 2025/0127/0023. Scale bar: 10 mm

Lehen section (Le) Located near Lunz am See (Lower Austria) on the northern slope of Mount Großkopf (1147 m), situated 4.7 km southwest of the Rehberg site (Geo map 71 Ybbsitz; Ruttner und Schnabel 1988; Figs. 7.1, 7.2 and 7.3). The section exhibits the topmost Reifling Formation, approx. 4 m of the Göstling Formation and, at the top, the argillaceous Reingraben Formation.

GPS: N 47°50′1.26″ and E 15°1′6.83″.

Taxa: rare *Halobia rugosa* and mass occurrences of *Peribositra*.

Geologic unit: lower part of the Reingraben Formation, Sulzbach Nappe within the Lunz Nappe System, NCA.

Facies: laminated marls with allodapic limestone intercalations, layers of bivalve lumachelles.

Thickness of bivalve level: approx. 3.5 m.

Sediment: laminated, no bioturbation.

Saugraben section (Sau) Located in a small ravine in the Sulzbach valley on the Ahornberg, 2.7 km northwest of the Ybbs River (Figs. 7.1, 7.2 and 7.3). The fossiliferous site is located 2.4 km west of the Lunz am See center in the southwest of Lower Austria (Trauth 1948; geological map, 71 Ybbsitz, Ruttner and Schnabel 1988). Bivalves occur with frequent *Halobia rugosa* (Fig. 7.4c, d).

GPS: N 47°51′14.90″ and E 14°59′30.20″.

Taxa: rare *Halobia rugosa* and mass occurrences of *Peribositra*.

Geologic unit: upper part of the Reingraben Formation, Sulzbach Nappe within the Lunz Nappe System, NCA.

Facies: laminated marls with allodapic limestone intercalations, layers of bivalve lumachelles.

Thickness of bivalve level: approx. 6 m.

Sediment: laminated, no bioturbation.

Großau section (Gro) Located on the northeastern slope of Mount Großkopf (1147 m), 300 m southeast of the Ybbs River, in the Ypps valley connecting Göstling and Lunz am See (Figs. 7.1, 7.2 and 7.3). The drilling site is located 5 km southwest of the Lunz am See center and 5 km northeast of Göstling an der Ybbs in the southwest of Lower Austria (Trauth 1948; geological map, 71 Ybbsitz; Ruttner and Schnabel 1988).

GPS: N 47°49′24.40″ and E 15°00′0.80″.

Taxa: rare *Halobia rugosa* and mass occurrences of *Peribositra* (Fig. 7.4f).

Geologic unit: upper part of the Reingraben Formation, Sulzbach Nappe within the Lunz Nappe System, NCA.

Facies: laminated marls with allodapic limestone intercalations, layers of bivalve lumachelles.

Thickness of bivalve level: approx. 6 m.

Sediment: laminated, no bioturbation.

Steinbach section (St) Located near Göstling an der Ybbs (Lower Austria), on the northern slope of Mount Großschöntaler Berg (881 m), situated 4 km southwest of the Großau drilling site (Geo map 71 Ybbsitz; Ruttner and Schnabel 1988; Figs. 7.1 and 7.3). The outcop is located within and under the lower end of the Steinbach River and at the transition into the bigger Ybbs River, and on both banks of the Ybbs River, accessible only at low water of the river.

GPS: N 47°48′22.20″ and E 14°57′0.60″.

Taxa: mass occurrences of *Peribositra.*

Geologic unit: upper part of the Reingraben Formation, Sulzbach Nappe within the Lunz Nappe System, NCA.

Facies: laminated marls with allodapic limestone intercalations, layers of bivalve lumachelles.

Thickness of bivalve level: approx. 5 m.

Sediment: laminated, no bioturbation.

The **Rad section (Rad)** Located 2.5 km south of Kleinzell (Lower Austria) along a forest road on the northwestern slope of Mount Hirschkogel (990 m; Geofast map 74 Hohenberg; Moser and Pavlik 2013; Figs. 7.1 and 7.3).

GPS: N 47°57′7.90″ and E 15°43′37.50″.

Taxa: mass occurrences of *Peribositra.*

Geologic unit: lower part of the Reingraben Formation, Reisalpen Nappe of the Tyrolic System, NCA.

Facies: laminated marls with allodapic limestone intercalations, layers of bivalve lumachelles.

Thickness of bivalve level: approx. 4 m.

Sediment: laminated, no bioturbation.

Hinterbrühl locality (Hint) Located 4 km west of Mödling (Lower Austria) on the northern slope of Mount Schwarzkogel (406 m) along abandoned historical quarries (Tyrolic System, Geo map 58 Baden, Schnabel 1997; Figs. 7.1 and 7.3). Historical fossil specimens mirror the presence of the argillaceous Reingraben Formation.

GPS: N 48°4′53.45″ and E 16°14′20.05″.

Taxa: rare juvenile to adult *Peribositra.*

Geologic unit: upper part of the Reingraben Formation, Göller Nappe of the Ötscher Nappe System, NCA.

Facies: laminated marls with allodapic limestone intercalations, layers of bivalve lumachelles.

Thickness of bivalve level: approx. 2 m.

Sediment: laminated, no bioturbation.

Scheiblingbachgraben section (Sch) Located on the northern slope of Mount Lärchkogel (1087 m), south of Mount Reiflingberg (712 m), 600 m southwest of the Enns River, 300 m west of the Großreifling center in northern Styria (geological map, 100 Hieflau, Kreuss 2014; Figs. 7.1, 7.2 and 7.3).

GPS: N 47°39′52.50″, E 14°42′27.75″.

Taxa: mass occurrences of juvenile to adult *Peribositra* (Fig. 7.4e) in the basal part and abundant adult *Halobia rugosa* (Fig. 7.4g) in the upper part of the Reingraben Formation.

Geologic unit: lower and upper part of the Reingraben Formation, Reifling Unit part of Sulzbach Nappe, NCA.

Facies: laminated marls with allodapic limestone intercalations, layers of bivalve lumachelles.

Thickness of bivalve level: approx. 2 m.

Sediment: laminated, no bioturbation.

7.5 Bivalves of the Carnian *Konservat-Lagerstätten*

The dominating faunal element in the basal part of the Reingraben Formation, comprising the *Konservat-Lagerstätten*, is the posidoniid bivalve *Peribositra* (1–32 mm length). It is present in all ontogenetic stages and in layer-/rock-forming quantities (Fig. 7.4b, e, f), mostly in double-valved butterfly position (Lukeneder and Lukeneder 2021). In the uppermost parts from about 10 m upwards in the Reingraben Formation, the more clayey deposits appear with dominant halobiid bivalves of *Halobia rugosa* (5–65 mm length; Fig. 7.4a, c, g). Frequent halobiids in the upper part of the Reingraben Formation led to the misleading designation "Halobienschiefer" (Tollmann 1976) for the entire basal Reingraben deposits, though the dominant marine bivalve is *Peribositra*, often misidentified as juvenile *Halobia*. Thin sections of sediments from the lower part of the Reingraben Formation include microfossils such as frequent ostracods, microgastropods and rare radiolarians.

7.6 The *Konservat-Lagerstätten* Paleobiota

Special conditions such as stagnant environments resulted in the formation of *Konservat-Lagerstätten*, which are characterized by their excellent preservation of invertebrate, vertebrate and plant fossils (Seilacher 1970; Bernardi et al. 2011; Briggs 2014; Lukeneder and Lukeneder 2021, 2022a, 2022b, 2023, 2024; Lukeneder et al. 2024b). Recent data and fossil collections from Scheiblingbach and coeval Polzberg localities were gained by bed-by-bed sampling within the last decade (Lukeneder and Lukeneder 2023; Lukeneder et al. 2024a, 2026; Lukeneder 2026, this volume). The fossil material derives directly from above the top of the Göstling Formation (Fig. 7.3). Fossil communities from the lower Carnian *Konservat-Lagerstätte* show a characteristic preservation of benthic (epifaunal and infaunal) species along with nektic taxa such as fishes, ammonoids and belemnoids. The completeness of the bivalves and their preservation in butterfly position (still articulated, little fragmentation), ammonoids along with in situ anyptchi (= lower jaws) in or near body chambers of trachyceratid ammonoids, and entire belemnoid specimens with articulated arm crowns point to an autochthonous deposition of the animals or their remains with minimal or no subsequent post-mortem drift or transport. Within the *Konservat-Lagerstätten* paleobiotas (counted only from bed-by-bed collecting), filtering bivalves with *Peribositra* (uncountable, rock-forming numbers) and *Halobia rugosa* clearly dominate the benthic paleocommunities in the Reifling Basin, accompanied by frequent other inseterminate bivalve and gastropod taxa. Thin-shelled bivalves such as *Peribositra* and *Halobia* (similar to *Daonella*) were sessile benthos. They lived on the sea-floor but were not byssally attached (see also Waller and Stanley 2005; McRoberts 2010). These bivalves are considered to be opportunists, mainly living in oxygen-depleted (but not anoxic) settings, typically in deeper water below the wave base (pers comm. C. McRoberts 2022), where microorganisms served as prey (Fig. 7.5).

Periobositra dominates in the basal Reingraben Formation, the dominant benthic element in the upper part of the Reingraben deposits is *H. rugosa* (Fig. 7.3). In the prevailing low-oxygen conditions during the deposition of the Reingraben Formation, the thin-shelled *Peribositra* and *Halobia* were the dominant epifaunal elements, adapted to life at or near the sea floor and/or within the carbon-rich and laminated sediment. Their accumulations in layers and lumachelles represent episodic census assemblages (dense and almost monospecific) in an epibenthic environment subject to transient bottom oxygen conditions (anoxic-dysoxic; McRoberts 2010; pers comm. C. McRoberts 2022). The thin bivalve layers (i.e. few mm) dominated by juvenile individuals are attributed to a fairly rapid (episodic) change of oxygen concentrations in the bottom water. Shell beds with the larger halobiids may have persisted in the dysoxic or periodically oxic zones for longer time frames.

Triassic thin-shelled flat clams likely comprise opportunistic populations that have the ability to rapidly colonize and dominate stressed open marine or basinal environments such as the Reiflin Basin. Once colonization occurs, the posidoniid and/or halobiid population dominates the substrate, excluding almost all other benthic

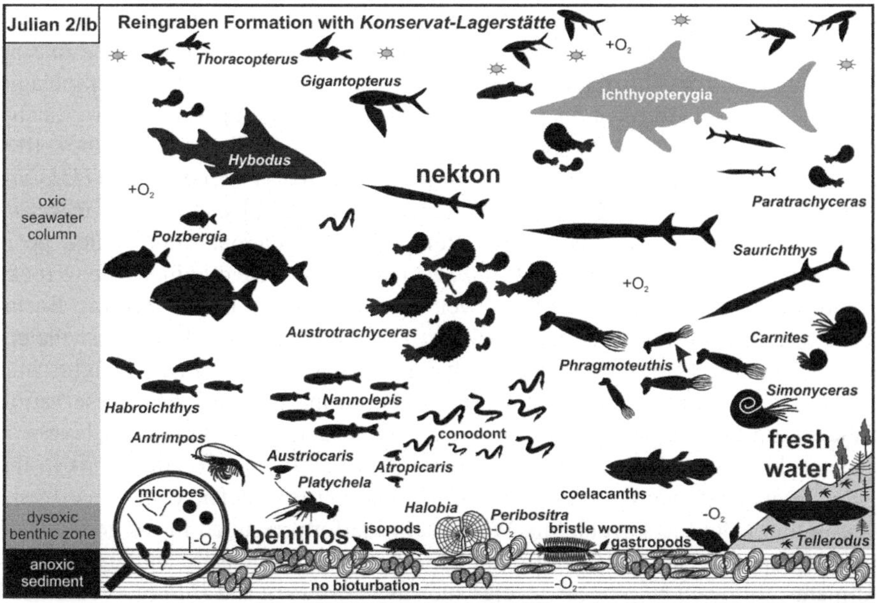

Fig. 7.5 Lower Carnian macrofaunal paleobiota (black) in the Reifling Basin based on direct evidence from the fossil record of all localities in the Carnian *Konservat-Lagerstätten* Belt. Direct evidence from fossils and bromalites. Ichthyopterygia not detected yet, most likely present. Figured members of paleobiota in the Julian2/Ib (Reingraben Formation), not to scale. Adapted after Lukeneder and Lukeneder (2021)

organisms. This dominance may continue even after detrimental conditions become favorable (McRoberts 2010). Subsequent mass mortality events occur, triggered by oxygen depletion or toxic environmental circumstances. This extinguishes all benthic life, followed by renewed colonization of the sea floor by the next generation of the same or differet thin-shelled bivalve taxa. The result is the episodic replacement of bivalve (i.e. thin-shelled, flat clams) populations.

7.7 Conclusions

New excavations involving bed-by-bed collecting from numerous Carnian *Konservat-Lagerstätten* (Hinterbrühl, Rad, Polzberg, Rehberg, Saugraben, Lehen, Großau, Steinbach, Scheiblingbach) were done during field campaigns over the last decade. Data on thousands of new fossils from the Late Triassic (early Carnian) yielded new details and insights into a previously unrecorded bivalve paleocommunity, introducing new bivalve material from the Austrian *Konservat-Lagerstätten.* Accordingly, the distinct *Konservat-Lagerstätten* localities encompass similar paleocommunities in the Reifling intraplatform basin from the Carnian Pluvial Episode

(CPE). All *Konservat-Lagerstätten* comprising bivalve mass occurrences (i.e. *Peribositra* and *Halobia*) are equivalent in age and paleobiota within the same marine basin. The basal deposits of the Reingraben Formation were formed at an epipelagic to upper mesopelagic position at the seafloor of the Reifling Basin, periodically disconnected from oxygenated bottom currents. These conditions affected the settlement by benthic organisms such as the thin-shelled bivalves *Peribositra* and *Halobia*. Low-oxygen ecosystems with limited bottom currents prevailed at all the *Konservat-Lagerstätten* sites at the basal of the Reingraben Formation. Thin-shelled posidoniid (*Peribositra*) and halobiid (*Halobia*) bivalves were the dominant (up to rock-forming) epifaunal elements in the lower Carnian deposits of the Reifling Basin, which contains the fossiliferous *Konservat-Lagerstätten* in the lowermost meters of the Reingraben Formation. The bivalve assemblage provides new insights into the lower Carnian marine ecosystem, deposited during the CPE. The preservation of these taxa in rock-forming masses (i.e., mass occurrences) in distinct layers or in meter-thick lumachelles points to unique paleoenvironmental conditions in the lower Carnian dysoxic bottom waters of the Reifling Basin. The occasional freshwater influx from the surrounding Wetterstein Platform resulted in oxygenation of bottom water currents, enhanced input of terrigenous siliciclastic and argillaceous sediments, followed by pioneer species such as *Peribositra* and *Halobia*. A detailed revision and description of the lower Carnian bivalve material from Austria will be performed by Christopher McRoberts in the near future.

Acknowledgments The author is particularly grateful to the owners of the important outcrop areas, the Austrian Federal Forests Company (Österreichische Bundesforste), for sampling and digging permission. Special thanks go to Lukas Nöst (forest district Großreifling) and Dominic Asztaller (forest dristrict Großreifling) for their advice and help during the field campaigns. I thank Petra, Lisa M. and Felix M. Lukeneder (all Gablitz), Veronika N.I. Koukal (Vienna), as well as Justine and Umberto Uprimny (Pichlern) for their help with cleaning outcrops and sampling fossils. The work was done within the framework of and financially supported by projects of the Federal Government of Lower Austria (K3-F-964/001-2020 and K3-F-964/002-2023, Department Science and Research; headquarters in St. Pölten); the Austrian Academy of Sciences (headquarters in Vienna, project Polzberg Lukeneder 2021) represented by the National Committee for Geo/Hydro Sciences (Earth System Sciences Programme); the Triassic project "Fenster in die Karnische Krise" of the Society of Friends of the Natural History Museum Vienna (Freunde des Naturhistorischen Museums Wien, CPE Lukeneder 2023); and the Natur- und UNESCO Geopark Steirische Esienwurzen (St Gallen, project Karnische Krise Lukeneder 2024). The manuscript greatly benefited from valuable comments by the reviewers Petra Lukeneder (Vienna) and Matthias Kranner (Hof), adapt to, the production editor Ashok Arumairaj (Taramani), the in-house series editor Kathleen A. Ritterbush (Utah), the publishing editor Simon Shah-Rohlfs and the production editor Antje Endemann (both Heidelberg).

Author Contribution AL collected material and designed research, provided documentation on fossil material and wrote the paper.

Funding Open access funding was enabled and the projects financially supported by the Federal Government of Lower Austria (K3-F-964/001-2020 and K3-F-964/002-2023, Department Science and Research; headquarters in St. Pölten); the Triassic project "Fenster in die Karnische Krise" of the Society of Friends of the Natural History Museum Vienna (Freunde des Naturhistorischen Museums Wien, CPE Lukeneder 2023); the Natural History Museum Vienna; and the Natur- und UNESCO Geopark Steirische Eisenwurzen (St Gallen, project Karnische Krise Lukeneder 2024) (to AL). The authors are responsible for the contents of this publication. The funders had no impact on conceptualization, design, data collection, analysis, decision to publish, or preparation of the manuscript.

Competing interests All authors stated that they have no competing interests.

Authors' contributions The author states that he has read, edited and agreed to submit the final version of the contribution.

Availability of data and materials All sources of information are provided in the text. Raw data related to the fossil material from Austria are available from the corresponding authors and Alexander Lukeneder upon request.

Additional information This article is part of the Special Volume: *Late Triassic Konservat-Lagerstätten within the Carnian Pluvial Episode. 150 Years of Scientific Excavations in Austria.*

Artificial Intelligence (AI) The chapter was produced without the use of AI.

Language The entire volume uses American English reviewed by a native scientist.

Received 2025.

Published online 2026.

References

Allison PA (1988) Konservat-Lagerstätten: Cause and classification. Paleobiology 14:331–344

Alverà G, Dal Corso J, Chu D, Cruciani G, Roghi G, Caggiati M, Song H, Song H, Tian L, Du Y, Haserbek T, Gianolla P (2026) Episodic aridification at the onset of the Carnian Pluvial Episode in Western Tethys. Glob Planet Change 256:105192. https://doi.org/10.1016/j.gloplacha.2025.105192

Bauer FK, Schnabel W (1997) Geologische Karte der Republik Österreich. Sheet Mariazell 72, 1:50.000. Geologische Bundesanstalt Wien

Bernardi M, Avanzini M, Bizzarini F (2011) Vertebrate fauna from the San Cassiano Formation (early Carnian) of the Dolomites region. Geo Alp 8:122–127

Briggs DEG (2014) Konservat-Lagerstätten 40 years on: the exceptional becomes *mainstream*. In: Laflamme M, Schiffbauer JD, Darroch SAF (eds) Reading and writing of the Fossil record: preservational pathways to exceptional fossilization. Paleontol Soc Papers 20:1–14. https://doi.org/10.1017/S1089332600002771

Dal Corso J, Bernardi M, Sun Y, Song H, Seyfullah LJ, Preto N, Gianolla P, Ruffell A, Kustatscher E, Roghi G, Merico A, Hohn S, Schmid, AR, Marzoli A, Newton RJ, Wignall PB, Benton MJ (2020) Extinction and dawn of the modern world in the Carnian (Late Triassic). Sci Adv 6(38):eaba0099. https://doi.org/10.1126/sciadv.aba0099

Dal Corso J, Ruffell A, Preto N (2018) The Carnian Pluvial Episode (Late Triassic): new insights into this important time of global environmental and biological change. J Geol Soc Lond 175:986–988. https://doi.org/10.1144/jgs2018-185

Forchielli A, Pervesler P (2013) Phosphatic cuticle in thylacocephalans: a taphonomic case study of (Arthropoda, Thylacocephala) from the *Fossil-Lagerstätte* Polzberg (Reingraben shales, Carnian, Upper Triassic, Lower Austria). AJES 106:46–61

Griffith J (1977) The Upper Triassic fishes from Polzberg bei Lunz. Zool J Linn Soc 60:1–93

Hornung T, Brandner R, Krystyn L, Joachimski MM, Keim L (2007) Multistratigraphic constraints on the NW Tethyan 'Carnin Crisis'. In: Lucas GS, Spielmann AJ (eds) The global triassic, vol 41. New Mex Mus Nat Hist Sci Bull, pp 59–67.

Hornung T (2008) The Carnian crisis in the Tethyan Realm. Dr. Müller, Saarbrücken, 1–235

Klug C, Scheyer TM, Klein N, Liu J, Stockar R (eds) (2024a) Special issue: 100 years of scientific excavations at UNESCO World Heritage Site Monte San Giorgio and Global Research on Triassic Marine Lagerstätten. Swiss J Palaeontol 143(Spec Issue):1–498, Springer, Basel

Klug C, Scheyer TM, Klein N, Liu J, Albisetti D, Furrer H, Stockar R (2024b) 100 years of scientific excavations at UNESCO World Heritage Site Monte San Giorgio and global research on Triassic marine Lagerstätten. Swiss J Palaeontol 143:37. https://doi.org/10.1186/s13358-024-00328-3

Kozur HW, Bachmann GH (2010) The middle Carnian Wet Intermezzo of the Stuttgart Formation (Schilfsandstein), Germanic Basin. Palaeogeogr Palaeoclimatol Palaeoecol 290:107–119. https://doi.org/10.1016/j.palaeo.2009.11.004

Kreuss O (2014) 100 Hieflau. Geologische Karte, 1: 50 000. Geofast, Geosphere Austria. https://doi.org/10.24341/tethys.101

Krystyn L (1991) Die Fossillagerstätten der alpinen Trias. In: Nagel D, Rabeder G (eds) Exkursionen im Jungpaläozoikum und Mesozoikum Österreichs. Österr Paläontol Ges Wien 1991:23–78

Lukeneder A (ed) (2026) Late Triassic Konservat-Lagerstätten within the Carnian Pluvial Episode: 150 years of scientific excavations in Austria. Top Geobiol 55:600 p. https://doi.org/10.1007/978-3-032-19777-1

Lukeneder A, Lukeneder P (2021) The Upper Triassic Polzberg palaeobiota from a marine *Konservat-Lagerstätte* deposited during the Carnian Pluvial Episode in Austria. Sci Rep 11:16644. https://doi.org/10.1038/s41598-021-96052-w

Lukeneder A, Lukeneder P (2022a) Taphonomic history and trophic interactions of an ammonoid fauna from the Upper Triassic Polzberg palaeobiota. Sci Rep 12:7455. https://doi.org/10.1038/s41598-022-11496-y

Lukeneder P, Lukeneder A (2022b) Mineralized belemnoid cephalic cartilage from the Late Triassic Polzberg *Konservat-Lagerstätte* (Austria). PLoS One 17(4):e0264595. https://doi.org/10.1371/journal.pone.0264595

Lukeneder A, Lukeneder P (2023) New data on the marine Upper Triassic palaeobiota from the Polzberg *Konservat-Lagerstätte* in Austria. Swiss J Palaeontol 142:9. https://doi.org/10.1186/s13358-023-00269-3

Lukeneder A, Lukeneder P (2024) Ammonoid taxonomy of the Carnian Polzberg *Konservat-Lagerstätte* in Austria. Jahrb Geol BA 162(1–4):27–55

Lukeneder A, Surmik D, Gorzelak P, Niedźwiedzki R, Brachaniec T, Salamon MA (2020) Bromalites from the Upper Triassic Polzberg section (Austria); insights into trophic interactions and food chains of the Polzberg palaeobiota. Sci Rep 10:20545. https://doi.org/10.1038/s41598-020-77017-x

Lukeneder A, Lukeneder P, Sachsenhofer RF, Roghi G, Rigo M (2024a) Multi-proxy record of the Austrian Upper Triassic Polzberg *Konservat-Lagerstätte* in light of the Carnian Pluvial Episode. Sci Rep 14:11194. https://doi.org/10.1038/s41598-024-60591-9

Lukeneder P, Fuchs D, Lukeneder P (2024b) Morphology, taxonomy and trophic interactions of rostrum-less coleoids from the Late Triassic Polzberg *Konservat-Lagerstätte* (Lower Austria). Swiss J Palaeontol 143:21. https://doi.org/10.1186/s13358-024-00319-4

Lukeneder A, Weitz J, Sachsenhofer RF, Misch D (2026) Drilling into the Carnian Pluvial Episode. In: Lukeneder A (ed) Late Triassic Konservat-Lagerstätten within the Carnian Pluvial Episode: 150 years of scientific excavations in Austria. Top Geobiol 55, Chap 15:315–352. https://doi.org/10.1007/978-3-032-19777-1_15

McRoberts C (2010) Paleoecological controls on Triassic flat clam biochronology: In: Di Stefano P, Balini M (eds) New developments on triassic integrated stratigraphy. Museo Geologico "GG Gemmerllaro", Palermo, Italy, September 12–16. Program and abstracts, p 38

Moser M, Pavlik W (2013) 74 Hohenberg, Geofast 1: 50 000, Geologische Karte der Republik Österreich, Geologische Bundesanstalt Wien. https://doi.tethys.at/10.24341/tethys.143

Mueller S, Krystyn L, Kürschner WM (2016) Climate variability during the Carnian Pluvial Phase—a quantitative palynological study of the Carnian sedimentary succession at Lunz am See, Northern Calcareous Alps, Austria. Palaeogeogr Palaeoclimatol Palaeoecol 441:198–211. https://doi.org/10.1016/j.palaeo.2015.06.008

Ogg JG (2015) The mysterious Mid-Carnian "Wet Intermezzo" global event. J Earth Sci 26:181–191. https://doi.org/10.1007/s12583-015-0527-x

Rigo M, Preto N, Roghi G, Tateo F, Mietto P (2007) A rise in the carbonate compensation depth of western Tethys in the Carnian (Late Triassic): deep-water evidence for the Carnian Pluvial Event. Palaeogeogr Palaeoclimatol Palaeoecol 246(2–4):188–205. https://doi.org/10.1016/j.palaeo.2006.09.013

Ruffell A, Simms MJ, Wignall PB (2015) The Carnian Humid Episode of the late Triassic: a review. Geol Mag 153:271–284. https://doi.org/10.1017/S0016756815000424

Ruttner A, Schnabel W (1988) 71 Ybbsitz, 1:50.000, Geologische Karte der Republik Österreich, Geologische Bundesanstalt Wien

Schlager W, Schöllnberger W (1974) Das Prinzip stratigraphischer Wenden in der Schichtfolge der Nördlichen Kalkalpen. Mitt Geol Ges Wien 66:165–193

Schnabel W (1997) 58 Baden, 1:50.000, Geologische Karte der Republik Österreich, Geologische Bundesanstalt Wien

Seilacher A (1970) Begriff und Bedeutung der Fossil-Lagerstätten. N Jahrb Geol Paläontol Mh 1970:34–39

Simms MJ, Ruffell AH (1990) Climatic and biotic change in the late Triassic. J Geol Soc Lond 147:321–327. https://doi.org/10.1144/gsjgs.147.2.0321

Simms MJ, Ruffell AH (2018) The Carnian Pluvial Episode: from discovery, through obscurity, to acceptance. J Geol Soc 175(6):989–992. https://doi.org/10.1144/jgs2018-020

Spengler E (1931) Erläuterungen zur Geologischen Spezialkarte. Blatt Schneeberg-St. Ägyd. Geologische Bundesanstalt, Wien, pp 1–108

Stur D (1874) Neue Aufschlüsse im Lunzer Sandsteine bei Lunz und ein neuer Fundort von Wengerschiefer im Pölzberg zwischen Lunzersee und Gaming. Verh K-K Geol RA 1:271–273

Stur D (1886) Vorlage des ersten fossilen Schädels von *Ceratodus* aus den ober triadischen Reingrabner Schiefern von Pölzberg nördlich bei Lunz. Verh K-K Geol RA 15:381–383

Teller F (1891) Über den Schädel eines fossilen Dipnoers, *Ceratodus Sturii* nov. spec. aus den Schichten der oberen Trias der Nordalpen. Abh K-K Geol RA 15:1–38

Thenius E (1974) Niederösterreich: Bundesländerserie, Heft Niederösterreich. Verh Geol BA Wien, pp 1–280

Tollmann A (1976) Analyse des klassischen nordalpinen Mesozoikums. Franz Deuticke, Wien, pp 1–580

Trauth F (1948) Geologie des Kalkalpenbereiches der Zweiten Wiener Hochquellenleitung. Abh Geol BAWien 26(1):1–99

Waller TR, Stanley GD (2005) Middle Triassic pteriomorphian bivalvia (Mollusca) from the New Pass Range, west-central Nevada: systematics, biostratigraphy, paleoecology, and paleobiogeography. J Paleontol 79:1–58. https://doi.org/10.1666/0022-3360(2005)79[1:MTPBMF]2.0.CO;2

Weitz J, Stojanovic KA, Sachsenhofer RF, Misch D, Lukeneder A (2026) Organic petrology and geochemistry of the Scheiblinggraben section: detailed insights into the Carnian Pluvial Episode and comparison to time equivalent profiles. In: Lukeneder A (ed) Late Triassic Konservat-Lagerstättenwithin the Carnian Pluvial Episode: 150 years of scientific excavations in Austria. Top Geobiol 55, Chap 23:527–566. https://doi.org/10.1007/978-3-032-19777-1_23

Wessely G (2006) Die Geologie von Niederösterreich. Geol Österr Bundesl, Geologische Bundesanstalt, Wien, 416 p

Chapter 8
3D Reconstruction and Microarchitecture of a Reticular Belemnoid Cephalic Cartilage from the Late Triassic Polzberg *Konservat-Lagerstätte*

Petra Lukeneder, Valentin Blüml, Christina Kaurin, and Alexander Lukeneder

Abstract The internal microarchitecture can be crucial for the interpretation of fossil structures. High-resolution micro-computer tomography (micro-CT) and 3D reconstructions of 14 fossil coleoid cephalic cartilages, embedded in dark, clayey limestones, provided insights into the internal structure of these unique Carnian (Upper Triassic) fossils. A well-developed branched, calcite-filled channel system was localized within the fossils. Measurements on these structures and how the channels are interlinked, along with actualistic comparisons, support the theory that these fossils were coleoid reticular cephalic cartilages, mineralized 234 million years ago. The findings provide new knowledge on the evolution of Triassic coleoids and in particular on the morphology and function of these fossil cephalopod cephalic cartilages. The present study combines classical morphometry from earlier work with modern 3D-visualization techniques of these unique fossils and discusses evolutionary relationships as well as the ecology of early coleoids. The structures are compared with today's reticular cephalic cartilages of coleoids, which inhabit the

P. Lukeneder (✉) · A. Lukeneder
Department of Geology and Paleontology, Natural History Museum Vienna, Vienna, Austria
e-mail: petra.lukeneder@nhm.at

A. Lukeneder
e-mail: alexander.lukeneder@nhm.at

P. Lukeneder
Institute of Applied Geology, Department of Landscape, Water and Infrastructure, BOKU University, Vienna, Austria

V. Blüml · C. Kaurin
Department of Evolutionary Biology, University of Vienna, Vienna, Austria
e-mail: valentin.blueml@univie.ac.at

C. Kaurin
e-mail: christina.kaurin@univie.ac.at

A. Lukeneder (ed.), *Late Triassic Konservat-Lagerstätten within the Carnian Pluvial Episode*, Topics in Geobiology 55, https://doi.org/10.1007/978-3-032-19777-1_8

pelagic realm and arc mainly represented by the neutrally-buoyant order Oegopsida, which already provides clues to the lifestyle of basal coleoids.

Keywords Coleoids · Cephalic cartilage · *Phragmoteuthis* · *Konservat-Lagerstätte* · Paleobiology · Triassic

8.1 The *Konservat-Lagerstätte* Polzberg

The Late Triassic Polzberg *Konservat-Lagerstätte*, the preservation of belemnoid soft tissues and probable mandibles of *Phragmoteuthis* specimens, as well as of trachyceratid soft tissues, have been previously addressed (Doguzhaeva et al. 2007a, b; Lukeneder and Lukeneder 2022b, 2022c; Lukeneder et al. 2023, 2024b; Lukeneder 2026, this volume). This study examines and presents for the first time the 3D reconstruction of the channel system of the Carnian cephalic cartilage of belemnoids.

The long-known *Konservat-Lagerstätte* Polzberg (GPS: N 47°53′5.90″ and E 15°4′27.70″; Lower Austria, ÖK 1:50.000, sheet 71 Ybbsitz and sheet 72 Mariazell; Austromap Online 2023; 712 m above sea level; Fig. 8.1) bears several (historical) names, among them: Unter Polzberg (Stur 1874), Pölzberg (Stur 1886, 1888), Polzberg-Graben (Teller 1891; Krystyn 1991), Schindelberggraben (Steindachner 1910, 1911; Glaessner 1931) (= Mount Schindlberg, 1066 m) and Polzberg (Forchielli and Pervesler 2013) (see also Lukeneder and Lukeneder 2021). In more recent years, the Polzberg fauna was extensively studied by Lukeneder and Lukeneder (2021, 2022a, b, c, 2023a, b), Lukeneder et al. (2024a, b). In the present study, the long-known enigmatic fossil structures from the Polzberg *Konservat-Lagerstätte* and the Rinngraben (also: Rinnengraben) locality near Cave del Predil (northern Italy) were studied morphometrically with a focus on the preserved internal channel system. Although discussed several times (Suess 1865; Doguzhaeva and Summesberger 2012; Lukeneder et al. 2020; Lukeneder and Lukeneder 2021, 2022b), these carbonized structures, which are often found in association with coleoid body parts (such as arm hooks, proostraca and phragmocones) were never studied by micro computed tomography (micro-CT). First insights were already reported by Lukeneder and Lukeneder (2022b), who described the peculiarities of these fossils in detail and compared the fossil morphology compared to recent coleoid cephalic cartilage. The here treated fossil cephalic cartilages were already described (Lukeneder and Lukeneder 2022b, c; Lukeneder et al. 2024b) and subdivided into three morphotypes, each of them assigned to one taxon (Lukeneder et al. 2024b): The type A (furcate) cephalic cartilage was assigned to *Phragmoteuthis polzbergensis*, type C cartilage to *Phragmoteuthis bisinuata* and type B cephalic cartilage (open ring) was suggested to belong to *Mojsisovicsteuthis* sp.. Additional fossil finds of 3D-preserved coleoid cephalic cartilage were made at the Rehberg locality (also known as Rechberg, at GPS 47°51′37.60″ N 15°3′59.30″ E), Steinbachgraben near Göstling (Lower Austria) (GPS 47°48′21.91″ N 14°56′59.13″ E), Scheiblingbachgraben near Großreifling (Styria) (GPS 47°39′52.5′″ N 14°42′28.4″ E) and Hinterbrühl (also Brühl) near

Vienna (Lower Austria) (near the former tavern "Zum goldenen Ochsen") (see also Fig. 8.1, Table 8.1). One questionable specimen is even known from the Middle Jurassic Christian Malford *Konservat-Lagerstätte* (UK; Donovan 1983). Rieber (1970) presented a similar Middle Triassic (Anisian) specimen from Monte San Giorgio (Switzerland), determined as "?*Phragmoteuthis ticinensis*", located within an arm crown.

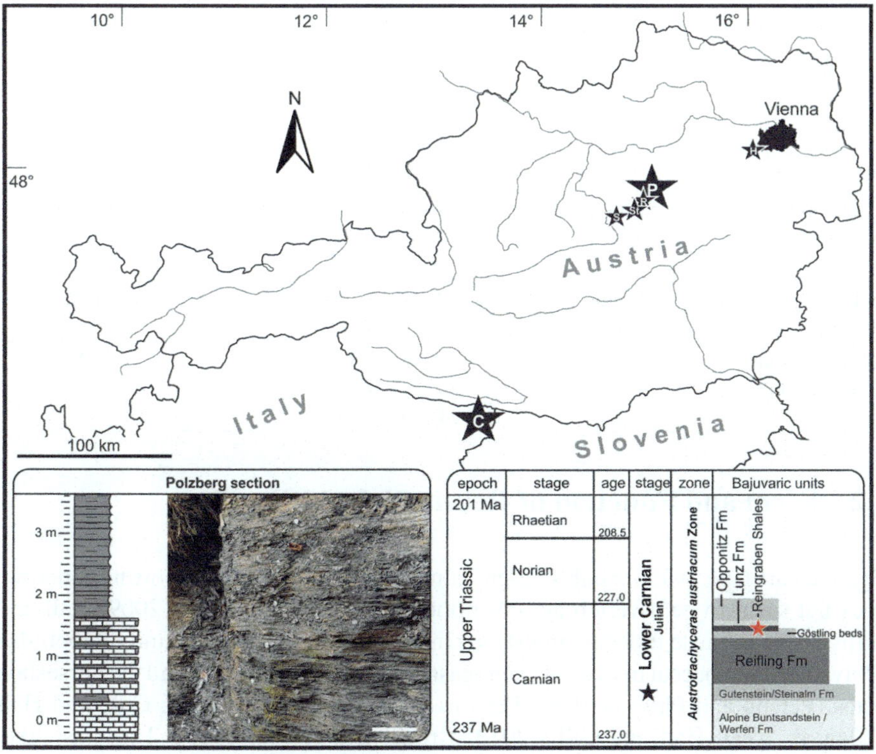

Fig. 8.1 Austrian map with star P marking the Polzberg *Konservat-Lagerstätte* (= Schindelberggraben ravine, Polzberg locality) and star C marking Cave del Predil (northern Italy; formerly: Raibl in Carinthia, Austria). Recent additional findings of coleoid cephalic cartilages in Scheiblingbachgraben in Styria (Austria) (star S), Steinbachgraben near Göstling (Lower Austria) (star St), Rehberg (star R) and Hinterbrühl (Lower Austria) (star H). Stratigraphic chart with the corresponding Bajuvaric Formations. The Reingraben Formation is marked with a red star. Lithological overview of the Polzberg main section with marking spikes every 20 cm. Adapted after Lukeneder et al. (2024b). The middle part of the section with clays and marls of the Reingraben Formation at Polzberg. Scale bar 20 cm

Table 8.1 Numbers of fossil coleoid cephalic cartilage investigated in the present study, specimens including their finding localities, GPS data, numbers of specimens and housing institutions

Locality nearby city/village	Federal district country	GPS data	Specimens n	Housing institution
Cave del Predil near Tarvis	Friuli Venezia Giulia Italy	Approx. N 46°26′38.73″ E 13°34′4.52″	9	NHMW GBA
Polzberg near Lunz am See	Lower Austria Austria	N 47°53′5.90″ E 15°4′27.70″	73	NHMW GBA
Rehberg near Lunz am See	Lower Austria Austria	N 47°51′37.60″N E 15°3′59.30″	7	NHMW
Steinbachgraben near Göstling	Lower Austria Austria	N 47°48′21.91” E 14°56′59.13”	3	NHMW
Hinterbrühl near Vienna	Lower Austria Austria	Approx. N 48°05′05.1″ E 16°14′56.0″	1	NHMW
Scheiblingbachgraben near Großreifling	Styria Austria	N 47°39′52.5″ E 14°42′28.4″	3	NHMW
Christian Malford near Swindon	Wiltshire UK	Approx. N 51°29′38.6″ E 2°03′59.4″	1?	NHMUK

8.2 Form and Function of Recent Coleoid Cartilage

Hyaline cartilage, with its light-weight connective and skeletal supporting function, does not seem to be a vertebrate invention (Cole and Hall 2004b, 2009; Hall and Gillis 2013) because it occurs in various invertebrates such as the family Sabellidae (Polychaeta), in buccal masses of caenogastropods, in cephalopods and the horseshoe crabs (Person and Philpott 1969; Person 1983; Wright et al. 2001; Cole and Hall 2004a, b, 2009; Hall and Gillis 2013; Brunet and Arendt 2016). The glass-like, translucent cephalic cartilage of living cephalopods has been described in detail and changes its color rapidly after death (Nixon 1998). The connective tissue cartilage can be chemically characterized by the presence of collagen and water. The branching processes of the chondrocytes (cartilage cells) form an interconnecting system of canaliculi (channels; Nixon 1998). The surrounding matrix consists of amorphous substance and fibers. In superficial zones of the cephalic cartilage, the fibrils (10–26 nm, Bairati et al. 1987) are arranged parallel to the surface. Nixon (1998) and Nixon and Young (2003) further describe histologically different types of cephalic cartilages in cephalopod groups. (1) A type, which resembles hyaline cartilage (Fig. 1 in Nixon, 1998), in species such as *Alloteuthis subulata*, *Chtenopteyx sicula*, *Gonatus fabricii* and *Teuthowenia megalops*. (2) The second type is described as a cephalic cartilage with a reticular (net-like, mesh-like, network) internal structure, which is known from the neutrally buoyant oegopsid squids such as *Abraliopsis* sp., the *Architeuthidae*, the *Mastigoteuthidae* and in the family *Onychoteuthidae*, (the latter being non-buoyant coleoids and strong, active swimmers (Nixon 1998; Nixon and Young 2003). The histological structure of the highly fenestrated reticular cephalic cartilage significantly differs from the typical hyaline type. This reticular network can be loose, exhibiting adaptations near the cartilage surface (see also Nixon and Young 2003). (3) A third type of knob-shaped, elastic cartilage is known from species such as *Histioteuthis meleagroteuthis* (Roper and Lu 1990).

8.3 Material and Methods

8.3.1 Paleontological Material

The examined fossilized cephalic cartilages are assigned to specimens of *Phragmoteuthis polzbergensis, P. bisinuata* and *Mojsisovicsteuthis* sp. (see also Lukeneder et al. 2024b). Until now, these fossils were known only from Cave del Predil (northern Italy) and Polzberg. Recent excavations proved the occurrence of these taxa also at Rehberg near Vienna. One additional single element ("wing-element") was found next to a specimen of a *Mastigophora brevipinnis* from the Middle Jurassic Christian Malford *Lagerstätte* (compare specimen at NHMUK 88607; Donovan 1983: Fig. 4). Overall, 97 well-preserved coleoid cephalic cartilages from recent bed-by-bed collections and historical collections were obtained for this study from the Natural History Museum Vienna (NHMW), the GeoSphere Austria (formerly Geological Survey of Austria), and the Museo Friulano di Storia Naturale Udine (Italy) (MFSN). Inventory numbers were already given in Lukeneder and Lukeneder (2022b) and Lukeneder et al. (2024b). In these earlier works, a digital caliper was used to measure all specimens. For EDX-Scanning Electron Microscopy (SEM) and microprobe analyzes see also Lukeneder and Lukeneder (2022b).

8.3.2 Scanning Electron Microscopy (SEM)

SEM images of the surface, as well as of the internal structures, of specimen NHMW 2012/0117/0024 and NHMW 2012/0117/0017 were taken using the Quanta™ 250 FEG from FEI (with a Schottky field emission source FEG-ESEM) from the Department of Material Sciences and Process Engineering (MAP) at the BOKU University, Vienna. For details on laboratory facilities and measurement data see Lukeneder and Lukeneder (2022b) and the supplementary material therein. The scanning electron microscope was equipped with an Everhardt Thornley SE-Detector, in low-vacuum settings with 15 kV accelerating voltage. The specimens were therefore not

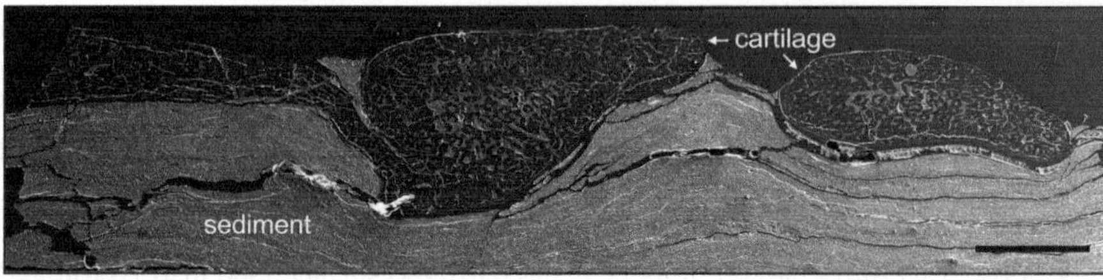

Fig. 8.2 Carbonized cephalic cartilage (black) with visible channel system (gray) of specimen NHMW 2012/0117/0024. Scanning electron microscopy (SEM) with additional geochemical composition analyzed by Microprobe and Energy Dispersive X-ray Spectroscopy (SEM-EDX). Stitched SEM images. Scale bar 1 mm

gold-sputtered (Fig. 8.2). Overview images were taken with a JEOL "Hyperprobe" JXA 8530-F field-emission electron microprobe (FE-EPMA) in combination with an online JEOL quantitative ZAF-correction program at the Central Research Laboratories of the NHMW (Lukeneder and Lukeneder 2022b, c). The samples were coated for EDX analyzes with an 8 nm carbon film. An accelerating voltage of 15 and 5 keV, a beam current of 5 nA, and fully focused electron beam (beam diameter of approx. 70–80 nm) were used. The Count Rate was 1055.00 cps/Ev (counts per second/electron Volt). EDX studies of carbonized structures and sediment were performed with a FEI Inspect-S scanning electron microscope with an EDAX Apollo XV SDD EDX detector at 15, 10 and 5 keV acceleration voltage. Spectra were acquired for 30–90 s to obtain a good signal to noise ratio, and intensities were corrected with the ZAF algorithm.

8.3.3 *Micro-computer Tomography (Micro-CT)*

Thirteen specimens of fossil cephalic cartilages were serially sectioned at the Core Facility for Micro-Computer Tomography (Vienna Micro-CT Lab), University of Vienna, Austria (Lukeneder and Lukeneder 2022b). A custom-built μ-CT scanner VISCOM X8060 NDT (Germany) scanned with varying scan parameters (scanning voltage with 150 kilovolts-kV; exposure time 1400 to 2000 ms-ms; pre-filter 0.75 mm-mm copper) and finally provided image-stacks with isometric voxel sizes (15.0, 16.5, 33.5, 17.5, 23.0, 25.5, 27.0, 33.5, 44.0 μm). Voxel size was 15 μm in NHMW 2012/0117/0028, 16 μm in NHMW 2021/0123/0165 and 16.5 μm in NHMW 2012/0117/0014. The stacks were combined, yielding a 3D volume, which was segmented afterwards. Specimen NHMW 2012/0117/0028 had a detected volume of 73.33 mm^3, specimen NHMW 2012/0117/0014 60.09 mm^3. For detailed measurements and additional data sets see the supplementary material in Lukeneder and Lukeneder (2022b). The specimens were segmented using the software Thermo Fisher Scientific Amira 2020 and finally virtually 3D-reconstructed in Drishti 2.7. The specimens NHMW 2012/0117/0014, NHMW 2012/0117/0028 and NHMW

2021/0123/0165 were selected for a segmentation and reconstruction of the channel system. After the segmentation, the volume of the channel systems was obtained. The differences in the cartilage volumes are mainly caused by the presence of only one element in specimen NHMW 2012/0117/0028 and both elements in NHMW 2021/0123/0165.

8.3.4 Measurements and 3D-Visualization

Diameters of representative channels were measured on the original Micro-CT slides of specimen NHMW 2012/0117/0028. The measurements were carried out in the software Thermo Fisher Avizo Amira 6.0. 3D-visualization was conducted following the segmentation of cartilage and lithic elements. Surface files generated during segmentation were applied as masks in Amira 2020 (Thermo Fisher Scientific), enabling the export of image stacks that isolated specific structures: (1) sedimentary matrix, (2) cartilage, and (3) hook elements (Fig. 8.3). The processed image stacks were subsequently imported into Drishti 2.7 for volume rendering. Visualizations were generated from either two or three overlaid volume stacks, each corresponding to a distinct anatomical or lithological component. Each image stack was linked to a specific image transfer function, which enabled differentiation of the respective materials. This approach enhanced contrast and allowed the clear visual separation of cartilage from stone within the three-dimensional renderings (Fig. 8.3). Additionally, surface renderings produced directly in Amira were utilized (Fig. 8.4).

8.3.5 Raman Spectroscopy

The composition of one specimen was analyzed by Raman spectroscopy at the Department of Mineralogy, University of Vienna. Raman spectroscopy was used for the in-depth characterization of the carbon phase of the fossil remains. For detailed measurements and additional data sets see the supplementary material in Lukeneder and Lukeneder (2022b). The Raman spectra were obtained using a Horiba LabRAM HR Evolution system equipped with an Olympus BX-series optical microscope and Si-based charge-coupled, Peltier-cooled, device detector. Spectra were excited with 532 nm emission of a frequency-doubled Nd-YAG laser (calcite, 12 mW at the sample; graphitic carbon, 0.012 mW). A 50 × lens (numerical aperture 0.55) was used to focus the light onto the surface of the sample. The light to be analyzed was dispersed with 1800 grooves per mm diffraction grating.

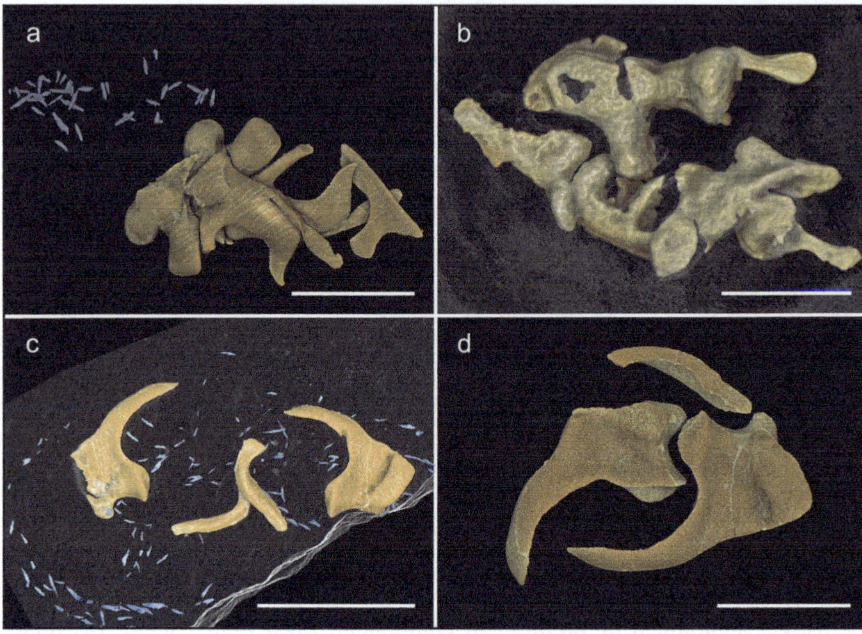

Fig. 8.3 Micro-CT reconstructions of morphotypes of carbonized cephalic cartilage. **a** NHMW 2021/0001/0002 Type A from Polzberg, attributed to *Phragmoteuthis polzbergensis*. **b** GBA 2006-011-0012, Type B cephalic cartilage from Cave del Predil, assigned to *Phragmoteuthis bisinuata*. Note the poorly defined morphology, which may also reflect taphonomic processes. **c** NHMW 2021/0124/0003, lateral view of Type C cartilage from Polzberg, attributed to *Mojsisovicsteuthis* sp., exhibiting ocular cartilage. **d** NHMW 2021/0124/0004 Type C from Polzberg, attributed to *Mojsisovicsteuthis* sp. Scale bars **a**, **c** 1 cm; **b**, **d** 0.5 cm

8.4 Results

8.4.1 The Channel System

Due to the small size of the Micro-CT scanned sample, resulting in a high scanning resolution, the specimen NHMW 2012/0117/0028 was selected as the best suited for further detailed examinations (Fig. 8.4). An overview on the distinct work steps is given (Fig. 8.4). Light microscopy revealed calcite-filled tube endings on the specimen (Fig. 8.4a). Micro-CT scanning images revealed the presence of these channels (Fig. 8.4b). Reconstruction of the image stack made it possible to capture the entire preserved structure (Fig. 8.4c). The additional segmentation of the channels provided, for the first time, detailed insights into the course of the canal system inside the fossil (Fig. 8.4d–f). First examinations of the channel system confirm a concentration of larger channels in the center of the structure, while channels with smaller diameters are found in the distal parts (Fig. 8.4f).

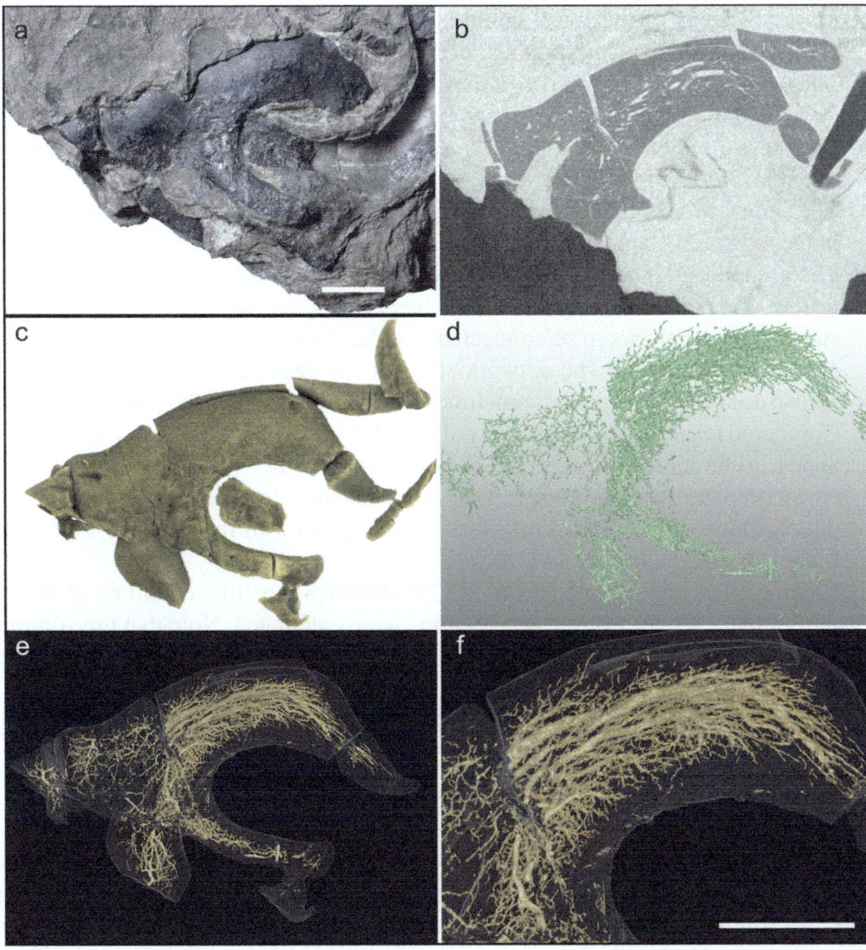

Fig. 8.4 Specimen NHMW 2012/0117/0028 was scanned with a scanning resolution of 15 μm (voxel size). **a** Photograph of the carbonized specimen in the host rock still partly covered with sediment. **b** Serial sectioning revealed a widely branched, calcite-filled channel system with thicker channels in the center of the fossil. **c** Surface view of the reconstructed fossil. STL files permit rotating the fossil to all sides. **d** The reconstruction of the channel system was essential because it is both an indicator for the origin of the fossil and provides information about its taphonomic history. **e** The translucent reconstruction of the preserved channel system gives first hints into the distribution of the channels. **f** Detail of **e** clarifies the presence of thick channels in the center of the fossil, whereas small branches are limited to the distal ends. Scale bars: **a**–**f** 2.5 mm

The distribution of the channel system (Fig. 8.4e, f) was already mentioned in Lukeneder and Lukeneder (2022a). The individual channels have repeating features, which are summarized below.

(1) The channel system follows the geometry of the fossil.
(2) Cracks in the fossil result in disruptions of channels.

(3) Marginal areas of the fossil could have been affected by, e.g. dissolution of connections by disintegration and decomposition of the organic structure.
(4) The geometry of the fossil was not affected, as confirmed by the degree of completeness of the channels in specific areas. No disruptions are visible in such areas, indicating that the original organic structure has not disintegrated much prior to its mineralization.
(5) The central area of the fossil is dominated by thicker channels, while the widely branched small channels are present distally (Fig. 8.4e, f)
(6) The branching of the channels deserves special attention. The structures do not follow each other linearly. The branches form local networks, whereby each starting point appears to be connected to at least three other starting points ("Y-geometry"; Fig. 8.5). The network starts at the large canals and the filaments spread out into lateral areas (sometimes nearly orthogonal to the large channels).
(7) Compared to cartilages from Cave del Predil (Fig. 8.3b), the Polzberg specimens appear to be more complete and better preserved. The former often appear to be incomplete near the distal ends. Additionally, the overall morphology of the cartilage differs, indicating that it belongs to a different taxon (see also Lukeneder et al. 2024b).
(8) Volumes: For specimen NHMW 2012/0117/0028 a cartilage volume of 66.75 mm^3 and a channel volume of 1.39 mm^3 were calculated. Note that the volume of the channels has possibly changed due to diagenesis or shrinkage of the cartilage substance (dehydration).

8.4.2 Composition of the Fossilized Cartilage

The geochemical composition analyzed by Microprobe and Energy Dispersive X-ray Spectroscopy (SEM-EDX) revealed the carbonized nature of the amorphous fossil. For detailed elemental data sets see the supplementary material in Lukeneder and Lukeneder (2022b). A thin calcareous fossil-surrounding seam/layer (approx. 20 μm thick) of amorphous structure with calcitic channel fillings is visible (Fig. 8.2). Microprobe elemental mapping (Fig. 6; S1 Data) of a thin-section at the fossil–sediment boundary indicates a high silicon (Si), aluminum (Al), potassium (K) and magnesium (Mg) content within the sediment. Aluminum, silicon and oxygen are the main sediment constituents. Carbon is the dominant element in the black fossil cartilage, with calcite-filled channels and the fossil-surrounding calcitic seam/layer. The outermost carbon layer of the fossil lacks sulphur (S), while the inner carbonized structure is rich in S, as is the sediment. Ti is generally distributed the same as S. The carbonized fossil is surrounded by an external layer with high amounts of calcium (Ca) and oxygen (O). A canalicular system, also rich in Ca and O, extends through the fossil structure. Raman spectroscopy of one specimen (NHMW 2021/0016/0397) reveals a composition of strongly disordered (i.e., sp2 hybrid bonded) carbon for the fossil specimens.

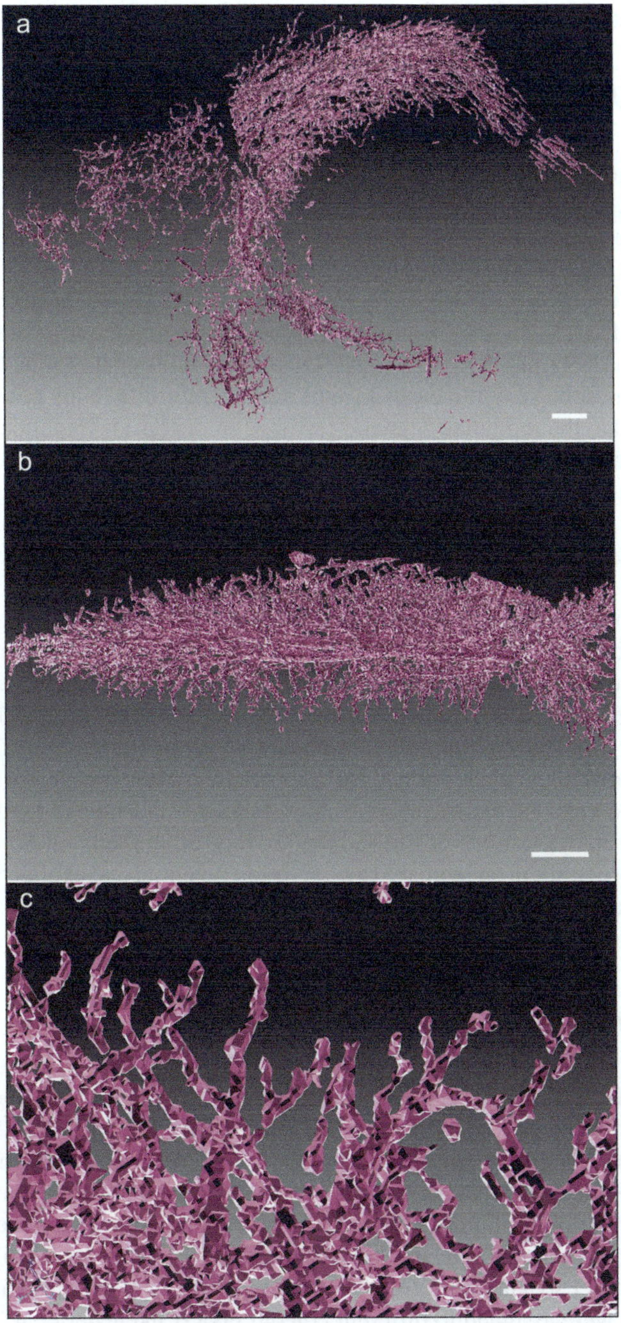

Fig. 8.5 Reconstruction of the preserved channel system of specimen NHMW 2012/0117/0028. **a** Lateral view indicating that the geometry of the fossil determines the course of the channels. **b** Apical view showing widely branched channel system spreading from thick central main channels. **c** Dichotomous branching of the channel system is visible at large magnifications. Scale bars: **a**, **b** 1 mm, **c** 100 μm

8.5 Discussion

8.5.1 Y-Shape

The dichotomous branching (Fig. 8.5a–c) is an important feature of the internal architecture of these fossils. One channel splits into two channels of approximately the same diameter. The branching type causes the reticular appearance observed in these fossils. One characteristic in the groups of modern oegopsid coleoids is the reticular structure of their cephalic cartilage (compare also Nixon and Young 2003) which can be interpreted as a light-weight construction. The loose arrangement of this cartilage type allows support and flexibility at the same time (Betts et al. 2022). Compared to the rather firm, spongy, hyaline cartilage of more benthically living coleoids it could be assumed that the reticulate cephalic cartilage in oegopsids might be an adaption to the pelagic mode of life (see also Clarke et al. 1979), enabling these coleoids to execute fast movements through the oceans without floating or sinking. Due to the presence of a semi-reticular head cartilage in the *Phragmoteuthis* group, a neutrally-buoyant mode of life could be proposed. Clements et al. (2017) report experimental evidence of a relationship between buoyancy mechanisms and preservation potential of coleoids in fossil deposits. Accordingly, neutrally buoyant decabrachians such as *Architeuthis* control buoyancy by ammonia (Clarke et al. 1979; Pratt and Anderson 2025). In the case of rapid burial, the ammonia in the decabrachian carcass decreases its ability to acquire the necessary pH value for authigenic calcium phosphate replacement. This significantly reduces the decabrachian preservation potential compared to vampyropod coleoids. Notably also non-buoyant coleoids, possessing a reticulate cephalic cartilage, exist, such as *Onychoteuthis* (Clarke et al. 1979; Nixon 1998). A non-buoyant mode of life and the absence of ammonia in *Phragmoteuthis* could therefore result in an increased preservation potential and explain the excellent preservation and abundance of coleoid fossils in Polzberg.

8.5.2 Significance for Coleoid Distribution

3D-preserved, fossil coleoid cephalic cartilages were first found at the localities Cave del Predil and Polzberg, later also from Monte San Giorgio (Anisian, Switzerland; Rieber 1970). Additional material stems from the Austrian Northern Calcareous Alps at the Rehberg Mountain (Lower Austria), Steinbachgraben Ravine (Lower Austria), Scheiblingbachgraben Ravine (Styria) and Hinterbrühl near Mödling (Lower Austria). Morphologically, the specimens from Cave del Predil differ from the finds in other localities by their morphology and taphonomic features. These morphotypes were assigned to different coleoid taxa (Lukeneder et al. 2024b). The findings of coleoid cephalic cartilages at these localities indicate a much wider coleoid distribution during the Carnian in these regions.

8.5.3 Possible Fossilization Settings

The role of microbial metabolic activity during fossilization processes appears to be complex and depends on many other factors (Janssen et al. 2022). Mineral precipitation, for example, requires special environmental conditions such as ion supersaturation (Janssen et al. 2022). The phosphate for the phosphatization of soft tissues can be gained from the decay of organic matter such as animal carcasses (Allison and Briggs 1991; Briggs and Kear 1993; Briggs 2003; Muscente et al. 2015; Schiffbauer et al. 2014; Janssen et al. 2022). The carbonate-phosphate-switch is regarded as one of the primary control factors on the precipitation of phosphate (Briggs and Wilby 1996; Schiffbauer et al. 2014). Under higher pH (>pH 6.38) and free Ca^{2+} and carbonate ions (CO^{3-}), calcium carbonate precipitates, whereas under low-pH conditions (<pH 6.38) and in the presence of free calcium ions (Ca^{2+}) and phosphate ions ($PO_4{}^{3-}$), calcium phosphate is formed (Burton and Walter 1990; Briggs and Kear 1993; Briggs and Wilby 1996; Janssen et al. 2022). Some studies indicate that organic matter itself acts as a substrate that promotes calcium-phosphate precipitation (Habraken et al. 2013).

8.6 Conclusions

Lukeneder and Lukeneder (2022b) presented the first detailed analyzis of the enigmatic carbonized fossil structures from Polzberg and homologized them with recent coleoid cephalic cartilage. The present report on the internal architecture of the reticular fossils supports this interpretation. Our study shows that coleoids were much more widespread in the Late Triassic Reifling Basin than previously assumed. The excellent preservation of the cephalic cartilage system is a prerequisite for a better understanding of the fossilization potential of fossil tissues in cephalopods in general, and of coleoids including decabrachians in particular. Clements et al. (2017) reported a reduced fossilization potential for decabrachians. We can confirm this based on the very different preservation states in coleoid hard tissues and soft tissues we observed. Although the exact preservation mechanisms in all the mentioned localities remain unknown, the details about the microarchitecture of the examined fossil cephalic cartilage takes us a step forward. Early on, Libbin et al. (1977) documented the in-vitro accumulation of mineral components by invertebrate cartilage. In particular, they showed that cranial cartilage of the coleoid *Loligo* mineralizes in vitro in hydroxyapatite-metastable media, but not in carbonate-metastable media. Their laboratory experiments showed rapid mineralization processes, faster than the expected decay of soft tissue. We consider different settings of taphonomic experiments to be helpful to clarify the fossilization processes and settings.

Acknowledgments We are very grateful to the owners of the Polzberg section, Franziska and Hermann Hofreiter Eva as well as Karl Jagersberger (all Gaming), for the digging permission during the whole project. We also want to thank Birgitt and Karl Aschauer (Waidhofen an der

Ybbs) for providing many fossil specimens for our studies. We are grateful to Dan Topa (SEM, microprobe), Leon Ploszczanski (SEM), Wencke Wagner (EDS-SEM), Goran Batic (mineralogical thin-sections) and Anton Englert (thin-sections) (all Vienna) for their technical support. Matthias Kranner (Vienna) helped with the SEM images. The present work was done within the framework of and financially supported by projects of the Federal Government of Lower Austria (K3-F-964/001-2020 and K3-F-964/002-2023, Department Science and Research; headquarters in St. Pölten); the Austrian Academy of Sciences (headquarters in Vienna, project Polzberg Lukeneder 2021) represented by the National Committee for Geo/Hydro Sciences (Earth System Sciences Programme); the Natural History Museum Vienna; and the Triassic project "Fenster in die Karnische Krise" of the Society of Friends of the Natural History Museum Vienna (Freunde des Naturhistorischen Museums Wien, CPE Lukeneder 2023). The manuscript greatly benefited from valuable comments by the reviewers René Hoffmann (Bochum) and Christian Klug (Zürich), the production editor Ashok Arumairaj (Taramani), the in-house series editor Kathleen A. Ritterbush (Utah), the publishing editor Simon Shah-Rohlfs and the production editor Antje Endemann (both Heidelberg).

Author Contribution AL initial idea for the present study. PL designed research. PL provided documentation on fossil material. CK, VB reconstructed the image stacks. PL provided first draft of the paper. PL, AL, CK, VB contributed to writing the paper.

Funding Open access funding was enabled and the projects financially supported by the Federal Government of Lower Austria (K3-F-964/001-2020 and K3-F-964/002-2023, Department Science and Research; headquarters in St. Pölten); the Austrian Academy of Sciences (headquarters in Vienna, project Polzberg Lukeneder 2021) represented by the National Committee for Geo/Hydro Sciences (Earth System Sciences Programme); the Natural History Museum Vienna; and the Triassic project "Fenster in die Karnische Krise" of the Society of Friends of the Natural History Museum Vienna (Freunde des Naturhistorischen Museums Wien, CPE Lukeneder 2023) (to AL). The authors are responsible for the contents of this publication. The funders had no impact on conceptualization, design, data collection, analysis, decision to publish, or preparation of the manuscript.

Competing interests All authors stated that they have no competing interests.

Authors' contributions All authors stated that they have equally contributed to their chapters, read, edited and agreed to submit the final versions of their contributions.

Availability of data and materials All sources of information are provided in the text. Raw data related to the fossil material from Austria are available from the corresponding authors and Alexander Lukeneder upon request.

Additional information The article is part of the Special Volume: *Late Triassic Konservat-Lagerstätten within the Carnian Pluvial Episode. 150 Years of Scientific Excavations in Austria.*

Artificial Intelligence (AI) The chapter was produced without the use of AI.

Language The entire volume uses American English reviewed by a native scientist.

Received 2025.

Published online 2026.

References

Allison PA, Briggs DEG (1991) Taphonomy: releasing the data locked in the fossil record. Plenum Press, 574 p

Bairati A, de Biasi S, Cheli F, Oggioni A (1987) The head cartilage of cephalopods. I. Architecture and ultrastructure of the extracellular matrix. Tissue Cell 19:673–685. https://doi.org/10.1016/0040-8166(87)90074-7

Betts G, Young KA, Wise JA, Johnson E, Poe B, Kruse DH, Korol O, Johnson JE, Womble M, DeSaix P (2022) Anatomy and physiology, 2nd edn. OpenStax, Rice University, pp 1–1418. ISBN 13: 978-1-711494-06-7

Briggs DEG (2003) The role of decay and mineralization in the preservation of soft-bodied fossils. Ann Rev Earth Planet Sci 31:275–301. https://doi.org/10.1146/annurev.earth.31.100901.144746

Briggs DEG, Kear AJ (1993) Fossilization of soft tissue in the laboratory. Science 259:1439–1442

Briggs DEG, Wilby PR (1996) The role of calcium carbonate-calcium phosphate switch in the mineralization of soft-bodied fossils. J Geol Soc Lond 153:665–668. https://doi.org/10.1144/gsjgs.153.5.0665

Brunet T, Arendt D (2016) Animal evolution: the hard problem of cartilage origins. Curr Biol 26:R685–R688. https://doi.org/10.1016/j.cub.2016.05.062

Burton EA, Walter LM (1990) The role of pH in phosphate inhibition of calcite and aragonite precipitation in seawater. Geochim Cosmochim Acta 54:797–808

Clarke MR, Denton EJ, Gilpin-Brown JB (1979) On the use of ammonium for buoyancy in squids. J Mar Biol Assoc UK 59:259–276

Clements T, Colleary C, De Baets K, Vinther J (2017) Buoyancy mechanisms limit preservation of coleoid cephalopod soft tissues in Mesozoic Lagerstätten. Palaeontology 60(1):1–14. https://doi.org/10.1111/pala.12267

Cole AG, Hall BK (2004a) Cartilage is a metazoan tissue; integrating data from nonvertebrate sources. Acta Zool 85:69–80. https://doi.org/10.1111/j.0001-7272.2004.00159.x

Cole AG, Hall BK (2004b) The nature and significance of invertebrate cartilages revisited: distribution and histology of cartilage and cartilage-like tissues within the Metazoa. Zoology 107:261–273. https://doi.org/10.1016/j.zool.2004.05.001

Cole AG, Hall BK (2009) Cartilage differentiation in cephalopod molluscs. Zoology 112:2–15. https://doi.org/10.1016/j.zool.2008.01.003

Doguzhaeva LA, Summesberger H (2012) Pro-ostraca of Triassic belemnoids (Cephalopoda) from Northern Calcareous Alps, with observations on their mode of preservation in an environment of northern Tethys which allowed for carbonization of non-biomineralized structures. N Jahrb Geol Palaontol Abh 266:31–38

Doguzhaeva LA, Summesberger H, Mutvei H, Brandstätter F (2007a) The mantle, ink sac, ink, arm hooks and soft body debris associated with the shells in Late Triassic coleoid cephalopod *Phragmoteuthis* from the Austrian Alps. Palaeoworld 16(4):272–284. https://doi.org/10.1016/j.palwor.2007.08.002

Doguzhaeva LA, Mapes RH, Summesberger H, Mutvei H (2007b) The preservation of body tissues, shell, and mandibles in the ceratitid ammonoid *Austrotrachyceras* (Late Triassic), Austria. In: Landman HN, Davis RA, Mapes RH (eds) Cephalopods present and past: new insights and fresh perspective. Springer, Dordrecht, pp 221–237. https://doi.org/10.1007/978-1-4020-6806-5_11

Donovan DT (1983) Mastrigophora OWEN 1856: a little-known genus of Jurassic coleoids. Neues Jahrbuch Geol Paläont Abh 165(3):484–495

Forchielli A, Pervesler P (2013) Phosphatic cuticle in thylacocephalans: a taphonomic case study of (Arthropoda, Thylacocephala) from the Fossil-Lagerstätte Polzberg (Reingraben shales, Carnian, Upper Triassic, Lower Austria). Austrian J Earth Sci 106:46–61

Glaessner MF (1931) Eine Crustaceenfauna aus den Lunzer Schichten Niederösterreichs. Jb Geol BA Wien 81:467–486

Habraken W, Tao J, Brylka L et al (2013) Ion-association complexes unite classical and non-classical theories for the biomimetic nucleation of calcium phosphate. Nat Commun 4:1507. https://doi.org/10.1038/ncomms2490

Hall BK, Gillis JA (2013) Incremental evolution of the neural crest, neural crest cells and neural crest-derived skeletal tissues. J Anat 222:19–31. https://doi.org/10.1111/j.1469-7580.2012.01495.x

Janssen K, Mähler B, Rust J, Bierbaum G, McCoy VE (2022) The complex role of microbial metabolic activity in fossilization. Biol Rev 97:449–465. https://doi.org/10.1111/brv.12806

Krystyn L (1991) Die Fossillagerstätten der alpinen Trias. In: Nagel D, Rabeder G (eds) Exkursionen im Jungpalaozoikum und Mesozoikum Österreichs. Wien (Österreichische Paläontologische Gesellschaft), pp 23–78

Libbin RM, Ozer R, Person P (1977) In vitro accumulation of mineral components by invertebrate cartilage. Calcif Tiss Res 22:67–75. https://doi.org/10.1007/BF02010347

Lukeneder A (ed) (2026) Late Triassic Konservat-Lagerstätten within the Carnian Pluvial Episode: 150 years of scientific excavations in Austria. Top Geobiol 55:600 p. https://doi.org/10.1007/978-3-032-19777-1

Lukeneder A, Lukeneder P (2021) The Upper Triassic Polzberg palaeobiota from a marine *Konservat-Lagerstätte* deposited during the Carnian Pluvial Episode in Austria. Sci Rep 11:16644. https://doi.org/10.1038/s41598-021-96052-w

Lukeneder A, Lukeneder P (2022a) Taphonomic history and trophic interactions of an ammonoid fauna from the Upper Triassic Polzberg palaeobiota. Sci Rep 12:7455. https://doi.org/10.1038/s41598-022-11496-y

Lukeneder P, Lukeneder A (2022b) Mineralized belemnoid cephalic cartilage from the late Triassic Polzberg *Konservat-Lagerstätte* (Austria). PLoS One 17(4):e0264595. https://doi.org/10.1371/journal.pone.0264595

Lukeneder A, Lukeneder P (2023a) Ammonoid taxonomy of the Carnian Polzberg *Konservat-Lagerstätte* in Austria. Jb Geol BA Wien 162(1–4):27–55

Lukeneder A, Lukeneder P (2023b) New data on the marine Upper Triassic palaeobiota from the Polzberg *Konservat-Lagerstätte* in Austria. Swiss J Palaeontol 142(9):1–18. https://doi.org/10.1186/s13358-023-00269-3

Lukeneder A, Surmik D, Gorzelak P, Niedźwiedzki TB, Salamon MA (2020) Bromalites from the Upper Triassic Polzberg section (Austria); insights into trophic interactions and food chains of the Polzberg palaeobiota. Sci Rep 10:20545. https://doi.org/10.1038/s41598-020-77017-x

Lukeneder A, Lukeneder P, Sachsenhofer R, Roghi G, Rigo M (2024a) Multi-proxy record of the Austrian Upper Triassic Polzberg *Konservat-Lagerstätte* in light of the Carnian Pluvial Episode. Sci Rep 14(1):11194. https://doi.org/10.1038/s41598-024-60591-9

Lukeneder P, Fuchs D, Lukeneder A (2024b) Morphology, taxonomy and trophic interactions of rostrum-less coleoids from the Late Triassic Polzberg *Konservat-Lagerstätte* (Lower Austria). Swiss J Palaeontol 143(21):1–23. https://doi.org/10.1186/s13358-024-00319-4

Lukeneder P, Lukeneder A (2022c) Comment on "Triassic coleoid beaks and other structures from the Calcareous Alps revisited." by Doguzhaeva et al. (2022). Acta Palaeontol Pol 67(4):963–965

Lukeneder P, Lukeneder A, Hiden TE, Reishofer G, Ritschl H, Sihorsch P (2023) Fenster in die Vergangenheit – Micro-CT bei prähistorischen Tintenfischknorpeln. Radiopraxis 16(01):37–39. https://www.thieme-connect.de/products/ejournals/abstract/10.1055/a-1938-3926

Muscente AD, Hawkins AD, Xiao S (2015) Fossil preservation through phosphatization and silicification in the Ediacaran Doushantuo Formation (South China): a comparative synthesis. Palaeogeogr Palaeoclimatol Palaeoecol 434:46–62

Nixon M (1998) Cephalic cartilage of Cephalopoda. Smithson Contrib Zool 586:31–38

Nixon M, Young JZ (2003) The brains and lives of Cephalopods. In: Nixon M, Young JZ (eds) Oxford University Press, Oxford, UK. ISBN 0-19-852761-6

Person P (1983) Invertebrate cartilages. In: Hall BK (ed) Cartilage, volume 1, Structure, function, and biochemistry. Academic Press, 31–57

Person P, Philpott DE (1969) The nature and significance of invertebrate cartilages. Biol 44:1–16. https://doi.org/10.1111/j.1469-185X.1969.tb00819.x
Pratt AM, Anderson FE (2025) Evolution of ammoniacal buoyancy within oegopsid squids and its relationship with depth. J Molluscan Stud 91(3):eyaf013. https://doi.org/10.1093/mollus/eyaf013
Rieber H (1970) *Phragmoteuthis*? *Ticinsensis* n. Sp. Ein Coleoidea Rest aus der Grenzbitumenzone (Mittlere Trias) des Monte San Giorgio (Kl. Tessin, Schweiz). Paläontol Z 44(H 1/2):32–40
Roper CFE, Lu CC (1990) Comparative morphology and function of dermal structures in oceanic squids (Cephalopoda). Smithson Contrib Zool 493:1–40. https://doi.org/10.5479/si.00810282.493
Schiffbauer JD, Wallace AF, Broce J, Xiao S (2014) Exceptional fossil conservation through phosphatization. Paleontol Soc Pap 20:59–82. https://doi.org/10.1017/S1089332600002801
Steindachner F (1910) Notizen. Jahresbericht für 1909. Ann K-K Naturhist Hofmus 24:1–70. https://www.jstor.org/stable/41768087
Steindachner F (1911) Notizen. Jahresbericht für 1910. Ann K-K Naturhist Hofmus 25:1–54
Stur D (1874) Neue Aufschlüsse im Lunzer Sandsteine bei Lunz und ein neuer Fundort von Wengerschiefer im Pölzberg zwischen Lunzersee und Gaming. Verh K-K. Geol RA, pp 271–273
Stur D (1886) Vorträge: Vorlage des ersten fossilen Schädels von *Ceratodus* aus den obertriadischen Reingrabner Schiefern von Pölzberg nördlich bei Lunz. Verh K-K Geol RA, Sitzung am 30. November:381–383
Stur D (1888) Geologische Aufnahmen und Untersuchungen im Felde: Exkursion zum Pölzberg bei Lunz [fingierter Titel]. Verh K-K Geol RA. Sitzung am 17. Jänner 1888
Suess E (1865) Über die Cephalopoden-Sippe *Acanthoteuthis* R Wagn. Sitz-Ber K Akad Wiss, Math-Naturwiss Cl 51:225–244
Teller F (1891) Über den Schädel eines fossilen Dipnoers, *Ceratodus Sturii* nov. spec. aus den Schichten der oberen Trias der Nordalpen. Abh K-K Geol RA 15:1–38
Wright GM, Keeley FW, Robson P (2001) The unusual cartilaginous tissues of jawless craniates, cephalochordates and invertebrates. Cell Tissue Res 304:165–174. https://doi.org/10.1007/s004410100374

Chapter 9
Polychaete Annelids from the Carnian Polzberg *Konservat-Lagerstätte*

Luke A. Parry, Alexander Lukeneder, and Petra Lukeneder

Abstract Polychaete annelid worms are common components of most modern marine ecosystems, encompassing ecological roles as diverse as sessile filter feeders and ambush predators. While several groups have evolved hard parts such as jaws and tubes that are common as fossils, most annelids are soft bodied and typically decay soon after death. *Konservat-Lagerstätten* where soft tissues are preserved provide our only window into the soft parts of ancient annelids, showcasing their early evolution and diversity, but such sites are distributed unevenly through time. Most well-known soft bodied annelid fossils are from the Palaeozoic, including several diverse biotas, but most Mesozoic species are poorly known, often from only known from single or few specimens. Here we describe two new polychaete species from the Late Triassic (early Carnian) Polzberg *Konservat-Lagerstätte* that include preservation of recalcitrant and biomineralized hard parts (jaws and chaetae) as well as remains of the body wall and gut. The first, *Machairascolex hofreiteri* gen. et sp. nov. is a member of the order Eunicida and preserves a complex, partially biomineralized, labidognath type jaw apparatus that identifies it as a member of the superfamily Eunicoidea, a diverse group of predatory worms that are well known from multiple species in Mesozoic *Lagerstätten*. The second is a fireworm (order Amphinomida) with evidence of biomineralized bristles that is maintained in open nomenclature. Amphinomids are rare in the fossil record, with only a handful of described species, two of which are from the Mesozoic.

Keywords Polychaetes · Annelids · *Konservat-Lagerstätte* · Early Carnian · Late Triassic · Austria

L. A. Parry (✉)
Department of Earth Sciences, University of Oxford, Oxford, UK
e-mail: luke.parry@earth.ox.ac.uk

A. Lukeneder · P. Lukeneder
Department of Geology and Paleontology, Natural History Museum Vienna, Vienna, Austria
e-mail: alexander.lukeneder@nhm.at

P. Lukeneder
Institute of Applied Geology, BOKU University, Vienna, Austria
e-mail: petra.lukeneder@nhm.at

A. Lukeneder (ed.), *Late Triassic Konservat-Lagerstätten within the Carnian Pluvial Episode*, Topics in Geobiology 55, https://doi.org/10.1007/978-3-032-19777-1_9

9.1 Introduction

Polychaete annelids are ubiquitous components of modern ecosystems, dominating marine sedimentary environments and forming key trophic links in marine food webs (Jumars et al. 2015). Their highly plastic, segmented body plan has allowed them to fill niches as disparate as epibenthic predators, sessile filter feeders and parasites and consequently annelids occupy most surface environments from the continents to the deep sea (Glasby et al. 2021). Today, Annelida encompass a range of body plans from terrestrial earthworms and leeches, gutless siboglinid tube worms and non-segmented sipunculans and echiurans (McHugh 1997; Struck et al. 2007; Weigert and Bleidorn 2016) but it is the polychaete body plan that is ancestral for annelids (Eibye-Jacobsen and Vinther 2012; Struck 2011), with these other groups nested within a paraphyletic assemblage of polychaete worms. As the soft bodies of polychaetes typically decay rapidly (Briggs and Kear 1993) most of our insights into the early diversification of annelids is from their record of recalcitrant jaws, tubes and other hard parts that have evolved multiple times independently in different subclades (Hints and Eriksson 2007; Ippolitov et al. 2014; Parry et al. 2019b; Vinn and Mutvei 2009). However, whole body fossils of polychaetes with soft parts are preserved in a variety of geological settings and taphonomic modes (Parry et al. 2018), providing much more complete windows into the morphology of ancient annelids. Unambiguous polychaetes first appear in the fossil record in exceptionally preserved faunas from the Cambrian (Chen 2020; Parry et al. 2016, 2019b) in which they are morphologically and ecologically diverse, encompassing sessile/infauna (Chen 2020), epibenthic (Eibye-Jacobsen 2004) and pelagic/nektonic (Yang et al. 2024) forms, showcasing an early diversification of lifestyles and forms in annelids. Some Ediacaran tube fossils have been interpreted as potential annelids (Schiffbauer et al. 2020; Yang et al. 2020) but this interpretation is not universally accepted (Dunn et al. 2022), and these tubes may have been produced by medusozoan cnidarians (Zhang et al. 2022).

Several diverse annelid faunas are known from post-Cambrian Paleozoic *Lagerstätten*, such as the Devonian Hunsrück Slate (Briggs and Bartels 2010; Gueriau et al. 2023; Högström et al. 2009) and the Carboniferous Mazon Creek (Hay 2002; Thompson 1979; Thompson and Johnson 1977). However, even within *Lagerstätten*, soft bodied annelid fossils are rare, and in some cases are known from single taxa or species even in deposits with very high-quality preservation (Farrell and Briggs 2007; Sutton et al. 2001). Nevertheless, Paleozoic polychaetes are often very well-preserved, retaining details of parapodia, head appendages, internal tissues that provide us some confidence in understanding their place in the annelid tree of life (Parry et al. 2019b). In contrast, the Mesozoic record of polychaetes is particularly well-studied for tube-dwelling taxa, which documents the assembly of the diversity of Serpulidae (Ippolitov et al. 2014), but soft bodied taxa from *Lagerstätten* are poorly understood. Unlike many Paleozoic examples, soft bodied polychaetes from the Mesozoic do not typically preserve details of the body outline and appendages, which are crucial taxonomic characters for diagnosing many of the extant families (Fauchald 1977). This is despite preservation in a range of lithologies (shales and lithographic limestones) and taphonomic modes (pyritization, entombment in nodules and phosphatization) that might each capture different features (Luque et al. 2015; Wilson et al. 2016). Exceptions are putative elytrae in *Palaeoaphrodite raetica*, a pair of anterior (presumably prostomial) appendages in *Homaphrodite speciosa* (Gall 1971) and the proboscis and caruncle in *Ankitokazoa bezanozanoi* (Alessandrello and Bracchi 2005). In the Cenomanian deposits of Lebannon, musculature is preserved in both eunicids and amphinomids, although more sparsely in the eunicids (Wilson et al. 2016). In the case of the latter, this musculature was used to identify *Rollinschaeta myoplena* as a member of the family Amphinomidae (Parry et al. 2015b). Notwithstanding these rare exceptions, most soft-bodied polychaetes from the Mesozoic are identified to family level based on the morphology and arrangement of hard parts (e.g., jaws and chaetae) as well as overall body shape, and the majority belong to the orders Eunicida and Phyllodocida (Errantia) and Amphinomida. The members of these orders are typically vagile and epibenthic, suggesting strong ecological and taxonomic filtering on the record of Mesozoic soft-bodied annelid fossils. Non-polychaete soft-bodied annelid fossils are also known from the Mesozoic, including a potential clitellate from continental rocks from the Early Triassic (Shcherbakov et al. 2020) and a Jurassic sipunculan (Alessandrello et al. 2004).

Triassic marine deposits include several records of soft-bodied errant/epibenthic taxa including a fireworm, *Ankitokazoa bezanozanoi* (Alessandrello and Bracchi 2005) and aphroditiform *Palaeoaphrodite anaboranoensis* (Alessandrello 1990) from the Lower Triassic of Madagascar and *Palaeoaphrodite raetica* from the Rhaetian of Lombardy (Alessandrello and Teruzzi 1986). The Anisian Grès à *Voltzia* contains two errant annelid taxa, *Eunicites triasicus* and *Homaphrodite speciosa* (Gall and Grauvogel 1967), both of which are found in brackish paleoenvironments (Gall and Grauvogel-Stamm 2005). A putative member of Arenicolidae (an infaunal family) *Archarenicola rhaetica* is described from black shales in Leicestershire (Horwood 1912) but the identity of this taxon is considered dubious (Rouse and Pleijel 2003) and warrants reinvestigation.

Jurassic soft-bodied annelid fossils are known from the Solnhofen Lagerstätte, which is dominated by errant worms. These include the genus *Eunicites* described by Ehlers (1868) which is represented by multiple described species in this deposit (Bracchi and Alessandrello 2005). Other taxa include the putatitive nereidid *Ctenoscolex* (Ehlers 1869), and *Meringosoma*, the latter of which is either an amphinomid (Ehlers 1869) or aphroditiform (Bracchi and Alessandrello 2005). A single eunicidan polychaete, *Melanoraphia maculata* (Arduini et al. 1982), is known from the Sinemurian of Osteno. This taxon was originally described as a lumbrinerid based on the (likely taphonomic) absence of anterior appendages, but most likely belongs to Euncidae or Onuphidae based on the morphology of the jaws (Bracchi and Alessandrello 2005). Four species of aphroditiform (*Palaeoaphrodite gallica*, *Palaeoaphrodite briggsiana*, *Palaeoaphrodite adeliae* and *Protopholoe rhodanitidis*) are described form the Jurassic La Voulte-sur-Rhône Lagerstätte (Alessandrello et al.

2004). A potential arenicolid, *Iubarenicola fischeri,* is also described from La Voulte-sur-Rhône (Alessandrello et al. 2004), which if correctly identified may be the only reliable record of exceptional preservation of an infaunal polychaete body fossil from the Mesozoic.

A diverse assemblage of annelids is known from the Late Cretaceous (Cenomanian) Lagerstätten of Lebanon including representatives of the extant families Amphinomidae, Aphroditidae, Eunicidae, Goniadidae, Lumbrinderidae and Oenonidae (Bracchi and Alessandrello 2005; Parry et al. 2015b). The new family Didonidae was erected to accommodate the genus *Didone* by Alessandrello and Bracchi (2005) which was suggested as morphologically intermediate between taxa with labidognath jaws (Eunicidae and Onuphidae) and Dorvilleidae, a proposal which is at odds with our current hypotheses of the interrelationships of these taxa based on jaw morphology in which dorvilleids are inferred as the sister group of all other eunicidans (Parry et al. 2019b; Paxton 2009). A single soft-bodied polychaete is known from the Cretaceous of Colombia, *Protopholoe colombiana* (Luque et al. 2015).

The relatively limited study of these fossils has resulted in issues with the systematics of Mesozoic soft-bodied polychaetes which has developed somewhat independently of the names applied to extant genera and species for their respective families. Consequently, it is uncertain (and perhaps unlikely) that all fossil genera represent natural groups with respect to extant genera (i.e. the fossil genera are parataxa defined by the character of being fossils), an issue that may be difficult to resolve given the level of detail preserved. This has resulted in some genera with long stratigraphic ranges. For example, six species of the genus *Palaeoaphrodite* are described, and the genus ranges from the Triassic to the Cretaceous. It is placed in the family Aphroditidae (Alessandrello and Teruzzi 1986; Bracchi and Alessandrello 2005). The fossil species vary in terms of characters preserved such as jaws, which are variably present and absent in different genera in this family (Rouse et al. 2022), and the arrangement of chaetae (which are typically a mix of robust more slender chaetae in different rami) but share a similar fusiform body shape. Aphroditidae contains 7 valid extant genera that encompass the variation exhibited by the fossil taxa with respect to the body shape, presence/absence of jaws, and chaetal distribution, such as if stout/acicular chaetae are present in the noto- and neuropodia or just the notopodia (Hutchings and McRae 1993). Mesozoic aphroditiforms are clearly in need of taxonomic revision (Bracchi and Alessandrello 2005).

Issues with the systematics of Mesozoic soft bodied annelids are further exemplified by the long ranging genus *Eunicites,* whole body fossils of which are known from the Jurassic to Eocene. Taxa described in this genus range into the Paleozoic however, and the genus is a mixture of species described from whole body specimens and isolated scolecodonts (individual jaw apparatus elements) that are unlikely to be congeneric (Bracchi and Alessandrello 2005; Ehlers 1868; Hinde 1879). Bracchi and Alessandrello (2005) recommended separating the scolecodont-based *Eunicites* species from those of whole-body impressions, proposing that a parataxonomic scheme should be established for the former. Such an approach has been discouraged for the study of eunicidan jaw apparatuses more generally (Eriksson et al. 2000). Further complicating this is that some of these taxa may belong to Onuphidae rather

Eunicidae, as both groups have similar labidognath type jaw apparatuses that may not be readily distinguishable in fossil material (Paxton 1986, 2009). For example, the jaw apparatus of *Eunicites diopatroides* resembles that of the onuphid *Diopatra,* but a relationship with Onuphidae was rejected based on the absence of preserved tubes that are characteristic of extant onuphids (Bracchi and Alessandrello 2005). However, the tubes of some Onuphidae are temporary and fragile (Paxton 1986) and even if tubes are genuinely absent in these taxa, they may nevertheless represent members of the onuphid stem group and not members of Eunicidae. These taxa would likewise benefit from revision with an explicit comparison with the jaws of the diversity of extant members of Eunicidae and Onuphidae.

Here we describe two new annelid body fossils from the Carnian Polzberg *Konservat-Lagerstätte* (Stur 1886; Teller 1891; Lukeneder and Lukeneder 2021; Lukeneder 2026 this volume) that are preserved as shale-hosted compression fossils (Figs. 9.1 and 9.2) that belong to two species. One was previously figured and assigned to the order Eunicida (Lukeneder and Lukeneder 2021) and preserves the characteristic jaw apparatus (Figs. 9.1 and 9.3, 9.4 and 9.5). These specimens are comparable to species placed in the genus *Eunicites* in terms of preservation, relative size and morphology. However, we have been able to describe the jaw apparatus in detail and provide a diagnosis for the species based primarily based on the morphology of the mandibles, and have erected a new genus *Machairascolex hofreiteri* gen. et sp. nov.

The second species (Figs. 9.2, 9.6 and 9.7) possesses a spindle-shaped body and elongate fascicles of chaetae and was previously tentatively assigned to the genus *Palaeoaphrodite* (Lukeneder and Lukeneder 2021). However, based on the chaetal morphology, likely composition and arrangement, it is here assigned to the order Amphinomida, but the family is uncertain. We have refrained from erecting a new taxon for this genus and have instead chosen to keep this material in open nomenclature (Bengtson 1988; Sigovini et al. 2016).

9.2 Material and Preservation

The body outlines of both species are defined by serially repeated chaetae (Figs. 9.1 and 9.2). In the case of the eunicids, these are the aciculae (Figs. 9.1a–c), with the external chaetae either difficult to identify or not preserved. In the amphinomid, only the external chaetae are visible (Fig. 9.2). The eunicid specimens preserve anterior jaw elements that include both the mandibles and maxillae (Figs. 9.1a, c, d and 9.3, 9.4 and 9.5). The jaws display a spectrum of preservation from in situ jaws within the anterior of the body (Fig. 9.1a) to limited amounts of disarticulation/disturbance of the aciculae and jaw elements (Fig. 9.1b) to isolated (but nevertheless articulated) maxillae with associations of dispersed aciculae (Figs. 9.1d and 9.5g). The jaw apparatus retaining articulation following some period of decay and dissociation from the body is consistent with postmortem observations from extant species, where

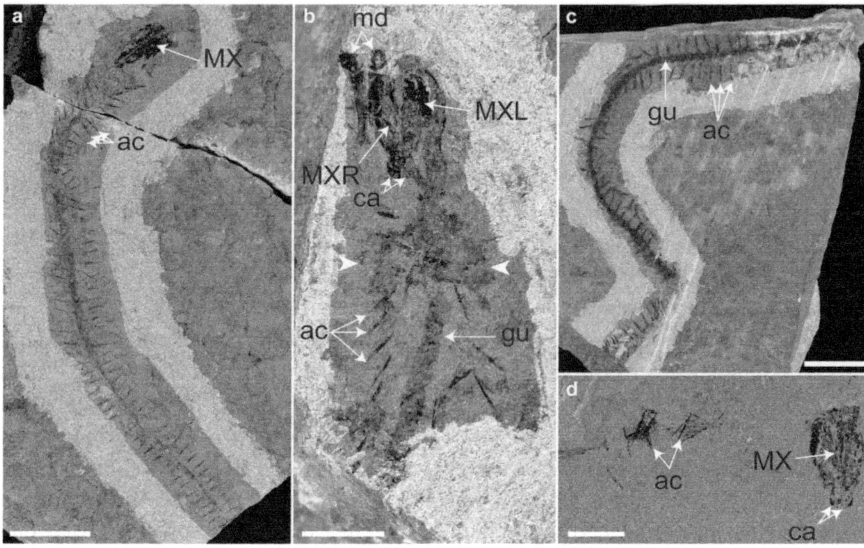

Fig. 9.1 Taphonomic variation in eunicidan specimens. **a** Posteriorly incomplete specimen of *Machairascolex hofreiteri* (holotype, NHMW 2025/0265/0001) with aciculae and jaw apparatus in life position. **b** Eunicoidea sp. B, posteriorly incomplete specimen in ventral view showing partial disarticulation of the aciculae and jaw elements, NHMW 2025/0265/0013. White arrowheads indicate the position of the margins of the body outline. **c** Eunicoidea indet. anteriorly and posteriorly incomplete specimen with aciculae in life position, NHMW 2025/0265/0004. **d** Eunicoidea sp. B, disarticulated aciculae and associated jaw apparatus, NHMW 2025/0265/0006. Scale bars: **a** 10 mm, **b**, **d** 5 mm. Abbreviations: ac—aciculae, ca—carriers, gu—gut, md—mandibles, MX—maxillae

the jaws can be extracted articulated through the mouth of partially decayed specimens (Paxton 2009). Several specimens are exposed in ventral view and so when the jaws are in situ, the mandibles are often obscured by the overlying maxillae, unless they have been displaced by disarticulation (Fig. 9.4b–g). Compaction of the jaw apparatus has resulted in near 2D preservation of the jaw apparatus, such that the outlines of jaw elements are visible through one another, revealing details that would otherwise be obscured by superposition (Fig. 9.5).

Almost all eunicid specimens preserve a dark axial trace that is continuous along the body (Figs. 9.1a–c and 9.3a–e). While this could represent the shrunken remains of the body wall and cuticle, it has relatively sharp, clearly defined margins and is considerably narrower (~10–30%) than the distance between the distal ends of the aciculae, which are within the body in life, while this dark axial trace is usually confined within the boundary defined by the proximal ends of the aciculae. In one specimen (Fig. 9.1b), this longitudinal structure is well preserved despite partial disarticulation of the aciculae which are embedded in a dark material that likely represents the remains of the body (see arrowed region in Fig. 9.1b). In one specimen, this feature projects outside of the body outline that is defined by the aciculae, which do show some signs of disarticulation, indicating that this structure has protruded out

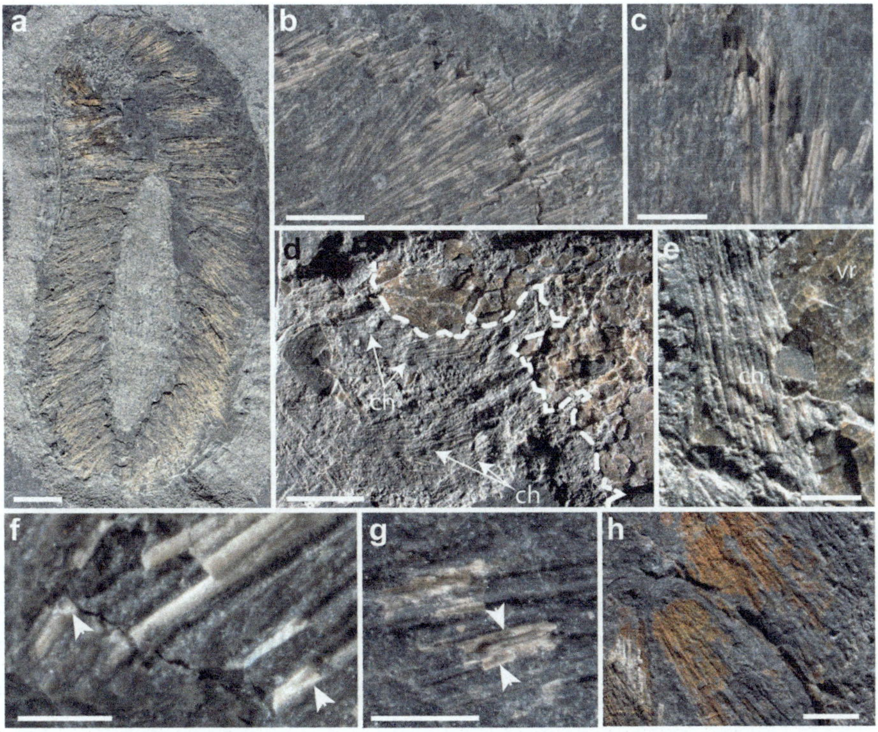

Fig. 9.2 Preservation of specimens assigned to the order Amphinomida. **a** Articulated specimen exposed in ventral view NHMW 2018/0103/0003. **b** Close up of chaetal fascicle. **c** Close up of chaetae showing 3D preservation of chaetae within the matrix. **d** Polychaete worm superimposed on fish specimen. Dashed line delineates the polychaete (left) and fish (right, *Pholidophoretes*) specimens, NHMW 2025/0265/0015. **e** Close up showing juxtaposition of vertebrate bone and chaetae. **f** Close up of broken chaetae showing full relief preservation and internal cavities (arrowheads), NHMW 2018/0103/0002. **g** Close up of chaetae showing internal cavities NHMW 2018/0103/0018. **h** Specimen where chaetae are preserved as iron mineral replacements, possibly during weathering, NHMW 2018/0103/0002. Scalebars: **a** 5 mm, **b** 2 mm; **c, e, g, h** 1 mm; **d** 4 mm; **f** 500 μm

of the body, following rupture of the outer body wall (Fig. 9.3e). In other eunicidan fossils where the body outline is preserved, it extends beyond the aciculae (e.g., *Eunicites orobicus*; Alessandrello and Bracchi 2004). Based on the features identified above, this structure therefore most likely represents the remains of the gut rather than the remains of the body cuticle.

In the eunicidan, the jaws and chaetae are preserved primarily as dark, presumably carbonaceous material that has undergone a substantial degree of flattening during compaction, such that individual elements of the maxillae of the jaws are difficult to differentiate from one another. The jaws are exposed in both dorsal and ventral views, with ventral views revealing the morphology of the cutting plates of the mandibles (Figs. 9.4b, c, f, j and 9.5a–d). The mandibles show more three-dimensional preservation, particularly in the region of the cutting plates, where some

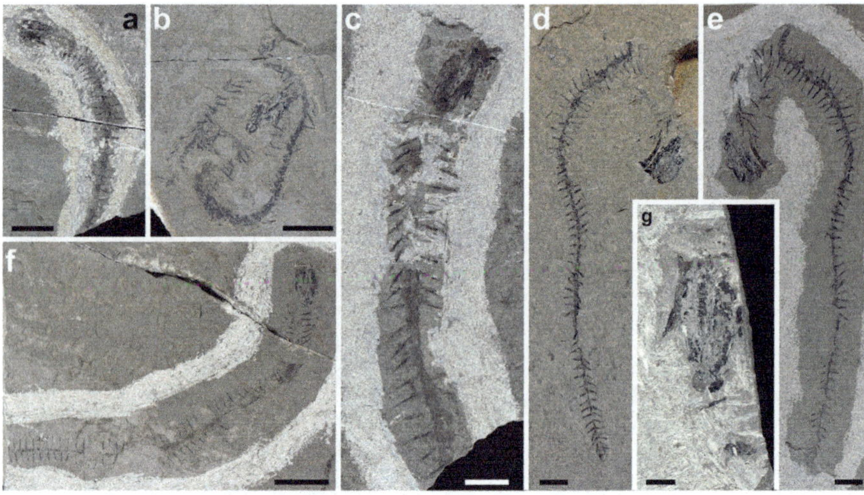

Fig. 9.3 Overview of specimens assigned to *Machairascolex hofreiteri*. **a** Anterior fragment in ventral view, NHMW 2025/0265/0002. **b** Partially disarticulated anterior fragment in ventral view, NHMW 2025/0265/0003. **c** Anterior fragment in dorsal view, NHMW 2025/0265/0005. **d**, **e** Anterior fragment in ventral view, NHMW 2025/0265/0004. **f** Anterior fragment in ventral view, NHMW 2025/0265/0011. **g** Isolated jaw apparatus, NHMW 2025/0265/0016. Scale bars: **a**, **b**, **f** 10 mm; **c–e**, **g** 5 mm

evidence of biomineralization is present, including complete series of growth lines (Fig. 9.4c). Biomineralization, if originally present, in the maxillae was likely quite weak. As many of the apparatuses are preserved in ventral views, some relief-based features are reversed in the fossil specimens. For example, the articulation mechanism between the MI elements of labidognath jaws is located on the dorsal surface of the element and consists of two furrows and a ridge on the left element and two ridges and a furrow on the right, but the opposite is observed in some fossil specimens (Fig. 9.4f, g) where this feature is viewed through the ventral surface.

In the amphinomid, rather than preserving as a dark and presumably organic material, the chaetae are an off-white three-dimensional material (Fig. 9.2a–g) or are preserved as impressions where the 3D material has become lost during splitting/preparation (e.g., Figure 9.2g) or as orange-red secondary iron oxide mineral replacement (Fig. 9.2h). In comparison to those of the eunicidans, they show substantial 3D preservation and relief, retaining an originally circular cross section (Fig. 9.2f), indicating that they have experienced limited compaction, unlike the organically preserved chaetae and jaw elements in the eunicidans. The chaetae are preserved throughout the matrix, rather than forming impressions on a single plane (Fig. 9.2c). These features suggest an original difference in composition to the chaetae of the eunicidans, indicating that they were biomineralized. Their light colour is like that of co-occurring scales in vertebrate specimens (including a polychaete and fish that are superimposed (Fig. 9.2d, e), indicating a likely originally partially phosphatic composition, consistent with that of modern fireworms (Righi et al. 2021). Chaetae

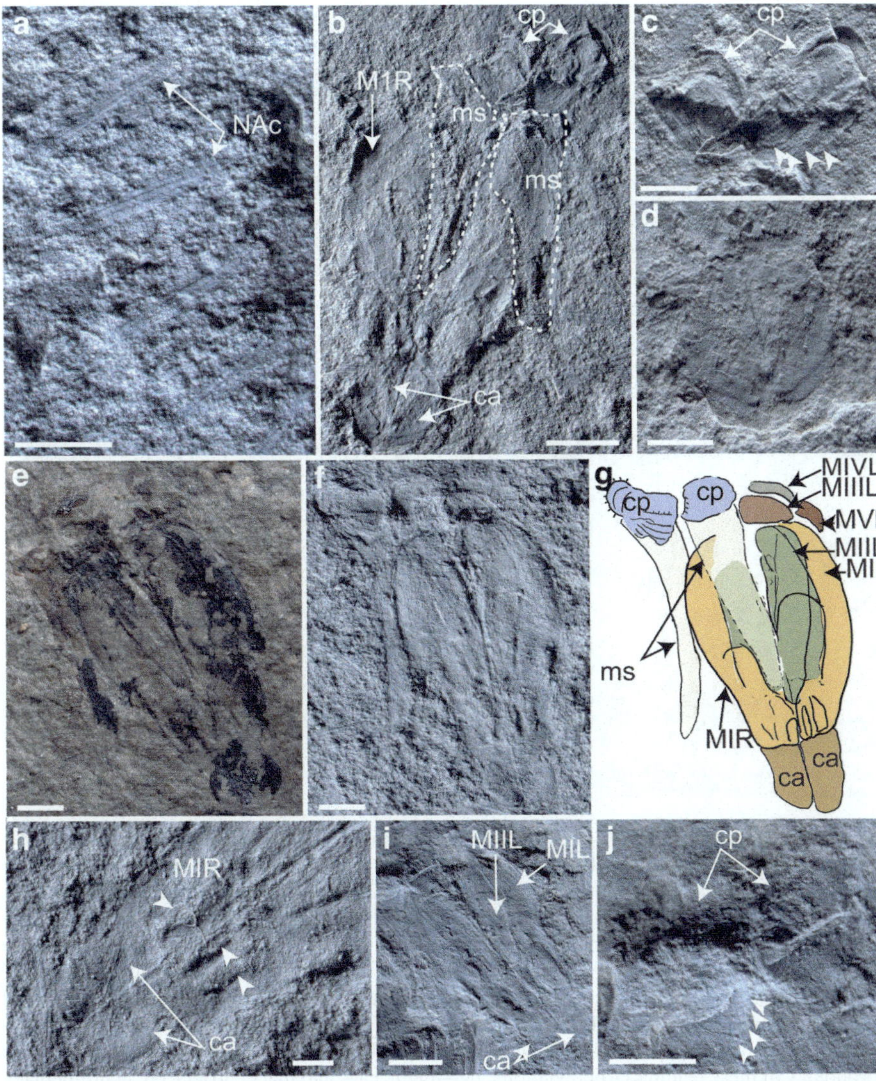

Fig. 9.4 Details of chaetae and ventral views jaw apparatuses in *Machairascolex hofreiteri*. **a** Close-up view of left notoaciculae from the anterior region of specimen, NHMW 2025/0265/0011. **b** Overview of jaw apparatus, NHMW 2025/0265/0003. **c** Detail of the cutting plates of the mandibles of specimen NHMW 2025/0265/0003. **d** Detail of the maxillary carriers of specimen, NHMW 2025/0265/0003. **e** Overview of jaw apparatus of specimen without coating to illustrate regions of carbonaceous preservation, NHMW 2025/0265/0011. **f** Overview of jaw apparatus of specimen NHMW 2025/0265/0011. **g** Interpretative drawing of specimen_to delineate individual jaw elements, NHMW 2025/0265/0011. Colour scheme for maxillary elements follows Kielan-Jaworowska (1966). **h** detail of carriers and articulation structure of the MI elements in specimen sp_1, NHMW 2025/0265/0002. Arrowheads indicate the positions of the furrows forming the articulation structure of the MI elements. **i** overview of the apparatus of specimen, NHMW 2025/0265/0002. **j** detail of the cutting plates of the mandibles of specimen, NHMW 2025/0265/0002. Images in panels **a**–**d**, **f**, **h**–**j** taken with specimen coated with ammonium chloride and low angle illumination. Abbreviations: ca—maxillary carriers, cp—cutting plate, ms—mandible shaft, MXL—left maxilla number X, MXR—right maxilla number X, NAc—notoaciculae. Scale bars: **a**, **b**, **i** 2 mm; **c**–**f**, **j** 1 mm; **h** 0.5 mm,

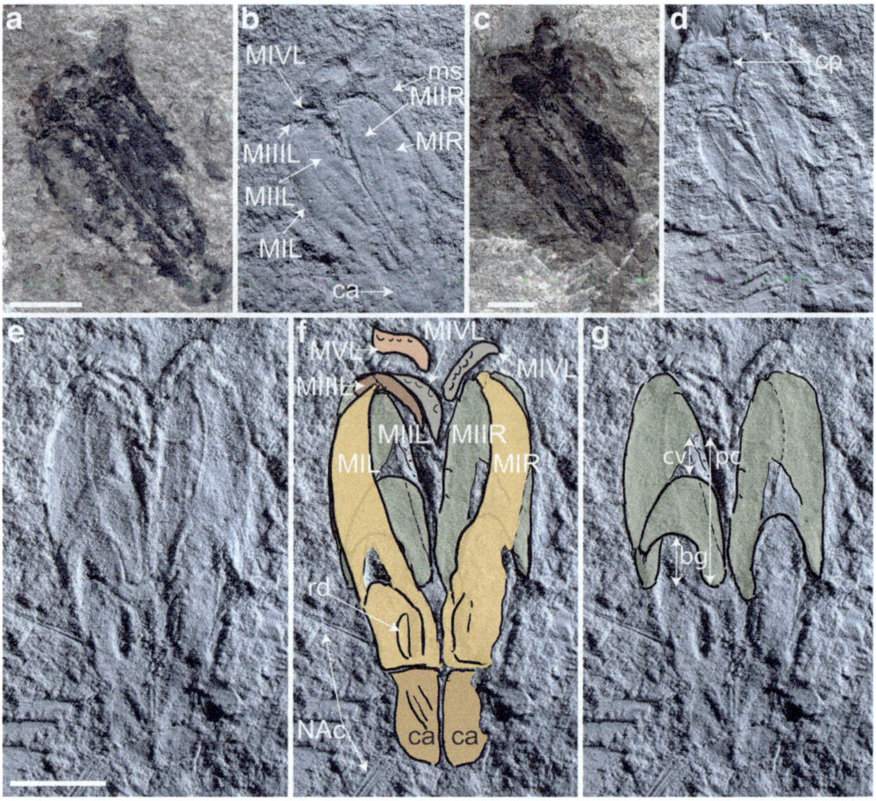

Fig. 9.5 Jaw apparatuses of *Machairascolex hofreiteri* in dorsal view. **a** Jaw apparatus of specimen with direct illumination, NHMW 2025/0265/0001. **b** Jaw apparatus of specimen under low angle light and coated with ammonium chloride, NHMW 2025/0265/0001. **c** Jaw apparatus of specimen with direct illumination NHMW 2025/0265/0005. **d** Jaw apparatus of specimen under low angle light and coated with ammonium chloride, NHMW 2025/0265/0005. **d** Close up of maxillae and carriers of specimen under low angle light and coated with ammonium chloride, NHMW 2025/0265/0005. **e** Same region as in **d** with superimposed interpretative drawing of maxillae and carriers. Colour scheme for maxillary elements follows Kielan-Jaworowska (1966). **g** Interpretative drawing as for panel **f** but with all elements except MIIIL and MIIIR removed to highlight the morphology of these features. Abbreviations: bg—bight, ca—maxillary carriers, cp—cutting plate, cv -cover, ms—mandible shaft, MXL—left maxilla number X, MXR—right maxilla number X, NAc—notoaciculae, pc—pulp cavity. Scale bars: **a**–**d** 2 mm; **e**–**g** 0.2 mm

preserved in cross section (Fig. 9.2f, g) show evidence of an open cavity within the chaeta, consistent with that present in Amphinomida (Müller et al. 2021; Righi et al. 2021), further consistent with biomineralization of these features.

Fireworms are rare in the fossil record and are currently only recognised from three species: *Rhaphidiophorus hystrix* from Mazon Creek (Thompson 1979), the Triassic *Ankitokazoa bezanozanoi* (Alessandrello and Bracchi 2005) from Madagascar and *Rollinschaeta myoplena* from the Cretaceous of Lebanon (Parry et al.

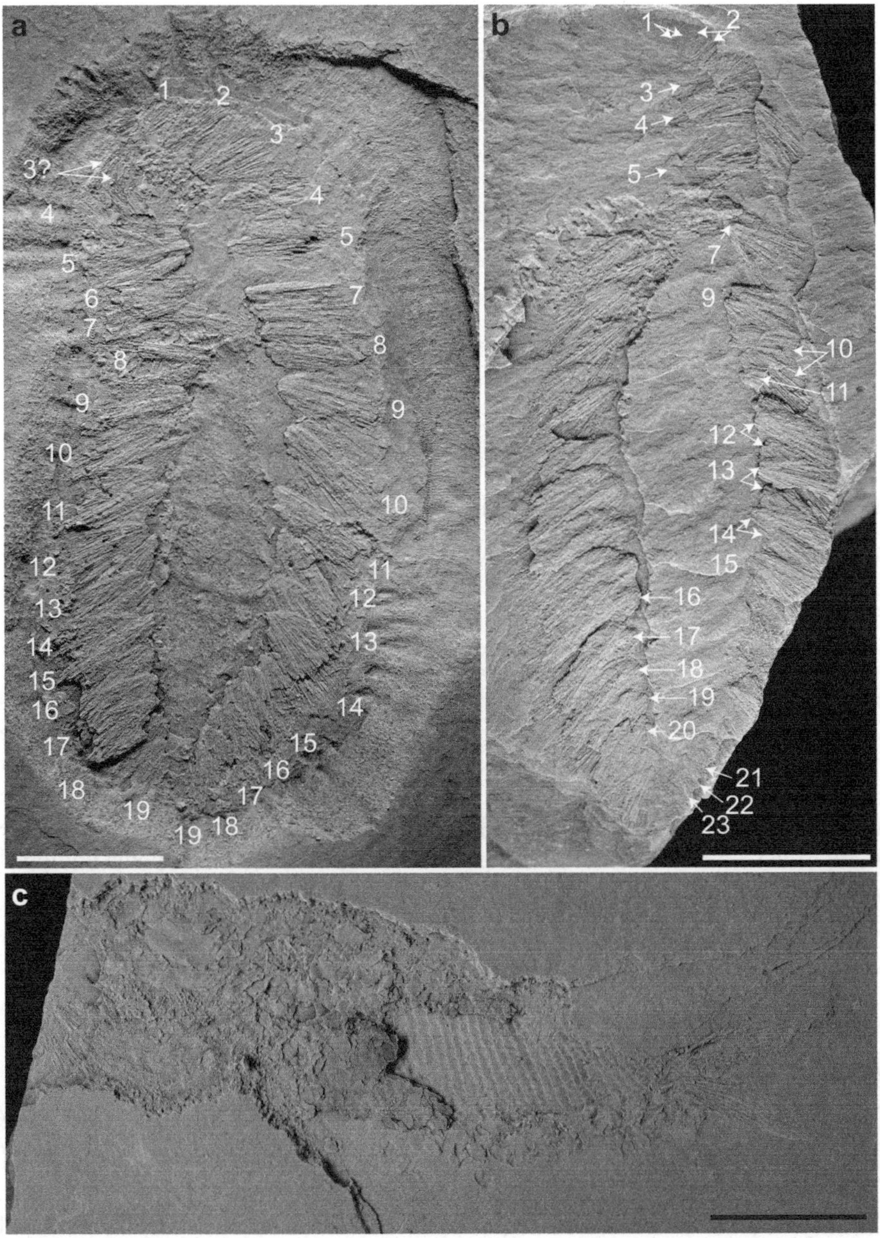

Fig. 9.6 Specimens of Amphinomida. A. **a** Articulated specimen exposed in ventral view, NHMW 2018/0103/0003. **b** Specimen with partial disarticulation of the chaetae of the anterior segments, NHMW 2018/0103/0002. **c** Specimen preserved superimposed preserved in association with an actinopterygian fish (*Pholidophoretes*), NHMW 2025/0265/0015. Scale bars all 10 mm

2015b). *Palaeocampa anthrax,* also from Mazon Creek, was previously identified as an amphinomid, but has recently been redescribed as a lobopodian panarthropod (Knecht et al. 2025).

9.3 Systematic Paleontology

Phylum Annelida Lamarck, 1802.

Subclass Errantia Audouin, 1832

Superfamily Eunicoidea Orensanz, 1990

Family uncertain.

Genus *Machairascolex* gen. nov.

Etymology: from the greek machaira, meaning curved sword, in reference to the curved blade like morphology of the first maxillae and scolex, worm.

Diagnosis: as for species by monotypy.

LSID:

Species *Machairascolex hofreiteri* sp. nov.

Diagnosis: Eunicidan polychaete with labidognath type jaw apparatus. Mandibles with ovoid cutting plates with prominent concentric growth rings that form complete ovals. Long axes of the cutting plates at a near right angle, such that the anterior margin of the mandibles forms a deep v-shape. Anterior margin of the cutting plate straight, without obvious cusps. Maxilla MIII with a bight approximately 20% of the length of the entire element. Mandible shafts gently curving. Neuropodia equipped with three aciculae in a parallel row.

Etymology: After Franziska and Hermann Hofreiter (both of Gaming), owners of the outcrop area.

LSID: urn:lsid:zoobank.org:act:7654BCD1-C0F6-4120-B6DE-3D5E9D925B9C

Holotype: NHMW 2025/0265/0001 (Fig. 9.1a, 9.5c–g and Fig. 9.5a, b),

Figured Material: (Fig. 9.1a), NHMW 2025/0265/0002 (Fig. 9.3a), NHMW 2025/0265/0003 (Fig. 9.3b), NHMW 2025/0265/0005 (Fig. 9.3C), NHMW 2025/0265/0011 (Fig. 9.3c), NHMW 2025/0265/0016 (Fig. 9.3g),

Locality: Polzberg *Konservat-Lagerstätte*, Lower Austria, N47°53′4.98″ and E15°4′28.15″

Additional Material: NHMW 2025/0265/0004, NHMW 2025/0265/0006, NHMW 2025/0265/0007, NHMW 2025/0265/0008, NHMW 2025/0265/0009, NHMW 2025/0265/0010, NHMW 2025/0265/0012, NHMW 2025/0265/0013, NHMW 2025/0265/0014, NHMW 2025/0265/0017.

Description: the available specimens are posteriorly incomplete, but range in length from 61 to 111 mm and maximally ~70 visible segments, based on the observed aciculae. Assuming that the distance between the distal tips of the aciculae are a reasonable proxy for body width, all specimens were of a similar body width, approximately 7 mm, with individual aciculae ~2.5 mm in length. The aciculae are present in groups of three and are robust, approximately 200 μm in width, and approximately straight (Fig. 9.4a). Only a single group of aciculae is observed per segment, identifying them as neuroaciculae, with chaetae of the notopodium either greatly reduced and not preserved, or potentially entirely absent. Additional incomplete material lacking the jaw apparatus consists of 7 specimens that are consistent in chaetal morphology and overall dimensions with *Machairascolex hofreiteri*, but would not be diagnosable to this species without the diagnostic characters of the jaws.

In *Machairascolex hofreiteri* the mandibles are not fused, though are closely associated in all specimens, even those that are partially disarticulated. Based on the super-positional relationships of the mandibles and maxillae, three specimens are preserved in ventral view (Fig. 9.4a–g) revealing the ventral surface of the cutting plates (Fig. 9.4c). The cutting plates of the mandibles are oval, arranged in a v-shape such that each cutting plate is inclined ~45° relative to the anterior posterior axis (Fig. 9.4c). Fine growth intervals are evident, converging on the adaxial side of the cutting plate (Fig. 9.4c, j). The anterior margin within the v-shape of each cutting plate appears relatively flat where it is complete and unbroken (Figs. 9.4c and 9.5d), without anterior cusps. The shafts of the mandibles diverge distally with a gentle curvature along their length (Fig. 9.4f, g).

In addition to the mandibles, the carriers and maxillae can be discerned. In all specimens, both the left and right MI and MII elements are present, but the smaller elements are only observed in three specimens, two in dorsal (Fig. 9.5a–g) and one in ventral view (Fig. 9.4e–g).

The MI elements are approximately 15% longer than the shafts of the mandibles and are falcate, without any evidence for denticles along their length (Fig. 9.5a–f), although they are typically partially obscured by the mandibles and MII. Furrows and ridges forming the articulation points between the proximal ends of the MI element are ~15% of the total length of the MI element. They are clear on the MIL element of one specimen in ventral view (Fig. 9.4h), where there one ridge and two furrows are evident, although visible with the opposite sense of relief as they are viewed through the ventral surface. This is corroborated by the specimen in dorsal view, where a prominent ridge is visible on MIL.

The MII elements are as long as the MI element excluding the region with the articulation structure. The MII elements have an irregular adaxial surface in specimens in ventral view, possibly due to the presence of denticles. The morphology of the MII elements is best seen in a specimen in dorsal view, where the outline of the pulp cavity, bight and surrounding margins can be discerned, despite the presence of the overlying MI element. The bight occupies ~20% of the length of the element posteriorly (Fig. 9.5g) while the pulp cavity is ~30% of the length of the element (based on MIIIL) approximately half of which is covered by the ventral cuticle of the jaw element. The carriers are short, approximately 30% of the total length of the MI element, ranging in absolute size from 1.5 to 2.5 mm.

On the left side of the specimen in ventral view, the anterior jaw elements are visible as crescent shaped impressions anterior of the MI and MII elements (Fig. 9.4f, g). The largest of these is immediately anterior to the MIL and MIIL and is therefore identified as MIIIL, with the more distal elements identified as MIVL and MVL. In the specimens exposed in dorsal view, three anterior elements are also observed and are consistent in location with those identified in the specimen in ventral view (Fig. 9.5b, e–f). In one dorsal view specimen, the MIVR element is present. Small denticles are evident on the smaller anterior elements, with five visible on MIVR (Fig. 9.5f). They cannot be discerned on MIIIL, but there are at least four on MIVL and MVL (Fig. 9.5f).

Remarks: The eunicidan described here can be diagnosed as a member of Eunicoidea (i.e. a clade of Onuphidae and Eunicidae) to the exclusion of other groups based on the presence of short carriers, forceps-like maxillae that articulate posteriorly with interlocking ridges and furrows and mandibles that are not fused and have distal calcified plates. The jaws of both Onuphidae and Eunicidae are highly similar and may be difficult to distinguish in fossil material (Paxton 1986). An exception is the presence of curved mandible cutting plates present in some members of Eunicidae, the presence of which would confidently place a taxon within this family, such as those in the genera *Palola* and *Lysidice* (Orensanz 1990; Zanol et al. 2007). Such curved mandibles may have a single origin within the family (Zanol et al. 2010) but have not thus far been identified in fossil material. In Onuphidae the MVI element is not present (Paxton 1986), unlike in Eunicidae, and this is also not evident in our specimens. While these characters may be sufficient to distinguish the two families from extant material, the anterior elements are relatively small and may be occluded by the larger MI and MII in compression fossils.

Although the specimens lack associated tubes, the absence of these cannot be used as diagnostic for Eunicidae, as members of the onuphid stem group would have lacked tubes based on their absence in outgroups. Consequently, we do not identify the present material below the superfamily Eunicoidea, which includes both Onuphidae and Eunicidae. This is distinct from the similar taxon Eulabidognatha, which encompasses these extant families and the fossil family Paulinitidae (Paxton 2009).

The rounded, ovoid mandible cutting plates with prominent concentric growth increments are distinctive, as in most onuphids and eunicids the anterior margin of the cutting plates is either relatively flat, with the long axis of the cutting plate perpendicular to the axis of the body (e.g., see illustration of *Marphysa* sp. in Parry et al. 2019a) and/or is ornamented with various cusps and notches (e.g., *Anchinothria, Nothria, Epidiopatra, Longibrachium, Rhamphobrachium* in Onuphidae, see illustrations in Paxton 1986), or are curved dorsally (e.g., *Marphysa* and *Palola,* see illustrations in Zanol et al. 2007). In *Machairascolex hofreiteri*, the long axis of cutting plates are angled such that the anterior margin of the mandibles forms a deep V shape, reminiscent of several members of Onuphidae, e.g. *Hartmanonuphis pectinata* and *Hirsutonuphis mariahirsuta* (Paxton 1986) and Eunicidae, e.g. *Eunice rubra* (Zanol et al. 2007). The prominence of the growth increments on the ventral surface of the cutting plates is also variable among members of the two families. In some Eunicidae, they are relatively smooth, and not readily discerned using SEM or microCT (Parry et al. 2019a; Paxton 2009), whereas concentric growth rings are illustrated in line drawings for many genera of Onuphidae in Paxton (1986). The curved mandible shafts are also distinctive, often being straight in Onuphidae (e.g. illustrations in Paxton (1986)) and Eunicidae, but they are strongly curved in *Lysidice ninetta* (e.g., Fig. 6D, E in Zanol et al. 2007)) but the cutting plates are curved dorsally and do not resemble those in *Machairascolex hofreiteri*. The species from Polzberg therefore appears morphologically distinctive based on the morphology of the jaws, justifying the creation of a new genus. Mandible morphology in *Machairascolex hofreiteri* therefore appears to be distinctive and mandibles may be a key taxonomic character for revising the taxonomic status of Mesozoic eunicidans, but have been largely overlooked and are frequently not figured in taxonomic works (Orensanz 1990).

Order Amphinomida Lamarck, 1818.

Family?Amphinomidae Lamarck, 1818 Genus uncertain.

Amphinomida sp. A

Figured Material: NHMW 2018/0103/0002 (Figs. 9.1b–c and 9.2a), NHMW 2018/0103/0001a (Fig. 9.1d–e and 9.2f), NHMW 2018/0103/0002 (Figs. 9.1f and 9.2b), NHMW 2025/0265/0018, (Fig. 9.1g), NHMW 2018/0103/0002 (Fig. 9.1h).

Locality: Polzberg *Konservat-Lagerstätte*, Lower Austria, N47°53′4.98″ and E15°4′28.15″

Additional Material: NHMW 2018/0103/0001a, b

Description: of the available material, two specimens are sufficiently complete to estimate the total number of segments, which range from ~19–23, with the specimens ranging in length from 43–50 mm exclusive of the chaetae. The chaetae are arrayed in flat bundles, with two bundles having a similar point of origin per segment, indicating that the parapodia are biramous. One set of chaetal bundles projects laterally, while the other is oriented more posteriorly, identifying them as the neurochaetae and notochaetae respectively. Both bundles have chaetae that are similar in morphology, containing a range of widths. In one specimen (Figs. 9.6a and 9.7a), the chaetae of segment 10 range from 40–150 μm. The total number of chaetae per bundle is difficult to assess as they are preserved through multiple levels in the matrix and so some are not visible. They must exceed 20 in both rami. The tips of the chaetae were either effaced during splitting or disappear into the matrix and so their morphology cannot be discerned. The chaetae were likely biomineralized in life and contained cavities (see preservation section and Fig. 9.2f, g).

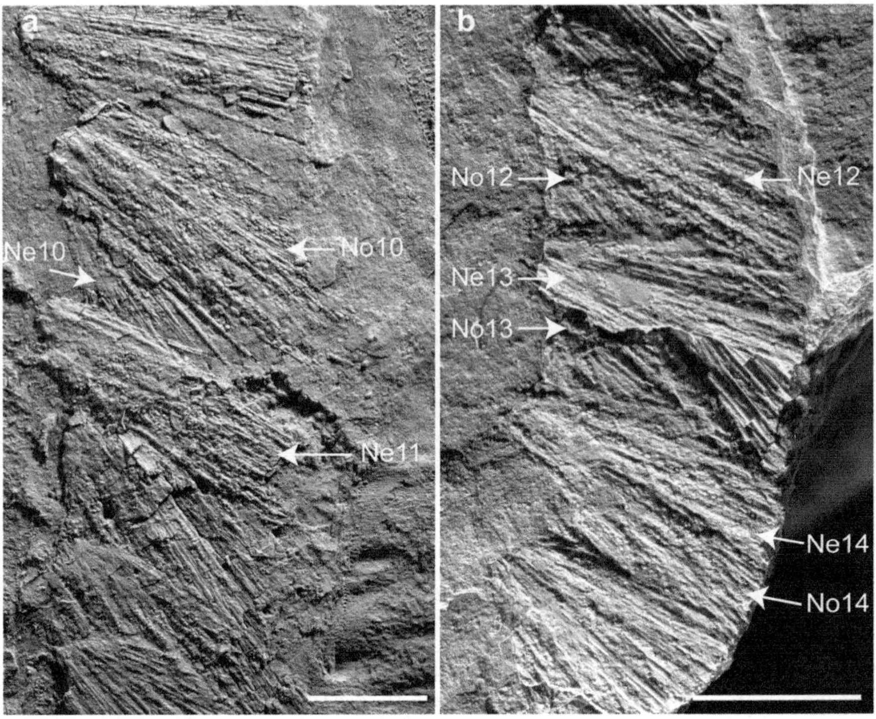

Fig. 9.7 Details of the parapodia of Amphinomida sp. A. **a** Detail of chaetae in the mid-body NHMW 2018/0103/0003. **b** Detail of the parapodia of the mid to posterior of the body, NHMW 2018/0103/0002. Abbreviations: NoX—notochaetae of segment X, Ne X—neurochaetae of segment X. Scale bars all 3 mm

Remarks: Polychaetes with spindle shaped bodies are present in several epibenthic families, including members of Aphroditiformia and Amphinomida (Borda et al. 2012, 2015; Rouse et al. 2022). Only Amphinomida have biomineralized bristles however, constraining Amphinomida sp. A as a member of this group (Müller et al. 2021; Righi et al. 2021). Assigning this species to one of the two extant families is more difficult however, due to the limited amount of anatomical information available. In Euphrosinidae, the chaetae are in dorsal transversal rows, unlike that of Amphinomida sp. A., whereas in Amphinomidae, they are in bundles/fascicles (Batts et al. 2022; Müller et al. 2021), which is likely the plesiomorphic state for Annelida (Parry et al. 2015a) and probably Amphinomida as well. Consequently, the new species could belong to the stem lineage of Amphinomida or Amphinomidae. Bifurcate chaetae are found in both families (Müller et al. 2021) but unfortunately the tips of the chaetae can't be resolved in the available material. The material also lacks evidence for informative soft tissues, such as the caruncle, which is preserved in other fossil fireworms (Alessandrello and Bracchi 2005; Parry et al. 2015b; Thompson 1979).

9.4 Conclusions

The polychaetes from the Polzberg *Konservat-Lagerstätte* show similarities with other sites of exceptional preservation preserving annelid body fossils. The most diverse sites include records of relatively large >10 cm eunicoids, most of which have been placed in the family Eunicidae (Bracchi and Alessandrello 2005; Ehlers 1868), with many in the genus *Eunicites*, a potential wastebasket taxon that is in need of revision (Bracchi and Alessandrello 2005). Here we erect a new genus based on mandible morphology, characters of which are frequently preserved in eunicid soft bodied fossils, the morphology of which is underutilised in the systematics of the group and further study of which may help to resolve the taxonomic status and interrelationships of Mesozoic eunicoids. The discovery of a new fireworm adds a key new record for this group, which despite their biomineralised bristles (Righi et al. 2021) are rare as fossils. They may be readily misidentified as members of other polychaete groups with spindle shaped bodies, such as aphroditiforms, for which many Mesozoic records are known, and so a comparative analysis of the preservation and composition of the chaetae of these fossils may further refine the record of Mesozoic fireworms.

Acknowledgments The authors are particularly grateful to the owners of the outcrop area, Franziska and Hermann Hofreiter, Eva and Karl Jagersberger (all Gaming), for sampling and digging permission. We are particularly grateful to Birgitt and Karl Aschauer (Waidhofen an der Ybbs), who made available fossil material for scientific investigations. We thank Lisa Marie Lukeneder and Felix Maximilian Lukeneder (both Gablitz) for their help with cleaning outcrops and sampling fossils. We thank Anna Alessandrello (Milan) for discussion on polychaete worms. The work was done within the framework of and financially supported by projects of the Federal Government of Lower Austria (K3-F-964/001-2020 and K3-F-964/002-2023, Department Science and Research; headquarters in St. Pölten); the Austrian Academy of Sciences (headquarters in Vienna, project Polzberg Lukeneder 2021) represented by the National Committee for Geo/Hydro Sciences (Earth System Sciences Programme); and the Triassic project "Fenster in die Karnische Krise" of the Society of Friends of the Natural History Museum Vienna (Freunde des Naturhistorischen Museums Wien, CPE Lukeneder 2023). Wencke Wegner (SEM, microprobe), Anton Fürst and Anton Englert (thin sections, Vienna) are acknowledged for technical support. We thank Peter Pormann (Manchester) for advice regarding the name *Machairascolex*. The manuscript greatly benefited from valuable comments by the reviewers Olev Vinn (Tartu) and Olle Hints (Tallinn), the production editor Ashok Arumairaj (Taramani), the in-house series editor Kathleen A. Ritterbush (Utah), the publishing editor Simon Shah-Rohlfs and the production editor Antje Endemann (both Heidelberg).

Author Contribution LP, AL, PL designed research. LP, AL provided documentation of fossil material. LP, AL contributed to writing the paper. AL made images. AL, PL collected the new material.

Funding Open access funding was enabled and the projects financially supported by the Federal Government of Lower Austria (K3-F-964/001-2020 and K3-F-964/002-2023, Department Science and Research; headquarters in St. Pölten); the Austrian Academy of Sciences (headquarters in Vienna, project Polzberg Lukeneder 2021) represented by the National Committee for Geo/Hydro Sciences (Earth System Sciences Programme); the Natural History Museum Vienna; and the Triassic project "Fenster in die Karnische Krise" of the Society of Friends of the Natural History Museum Vienna (Freunde des Naturhistorischen Museums Wien, CPE Lukeneder 2023) (to AL). LAP is funded by an Natural Environment Research Council Independent Research Fellowship (grant number NE/W007878/2). The authors are responsible for the contents of this publication. The funders had no impact on conceptualization, design, data collection, analysis, decision to publish, or preparation of the manuscript.

Competing interests All authors stated that they have no competing interests.

Authors' contributions All authors stated that they have equally contributed to their chapters, read, edited and agreed to submit the final versions of their contributions.

Availability of data and materials All sources of information are provided in the text. Raw data related to the fossil material from Austria are available from the corresponding authors and Alexander Lukeneder upon request.

Additional information The article is part of the Special Volume: *Late Triassic Konservat-Lagerstätten within the Carnian Pluvial Episode. 150 Years of Scientific Excavations in Austria.*

Artificial Intelligence (AI) The chapter was produced without the use of AI.

Language The entire volume uses American English reviewed by a native scientist.

Received 2025.

Published online 2026.

References

Alessandrello A (1990) *Palaeoaphrodite anaboranoensis* n. sp., a new species of polychaete annelid from the Lower Trias of Madagascar. Studies on Permo-Trias of Madagascar 2. Atti Soc Ital Sci Nat Mus Civ Stor Nat Milano 131(10):205–208

Alessandrello A, Bracchi G (2004) *Eunicites orobicus* n. sp. (Annelida, Polychaeta): a new eunicid polychaete from the Lower Rhaetian (Upper Triassic) of Italian Prealps (Ponte Giurino, Lombardy). Atti Soc ital sci nat Mus civ stor nat Milano 145(1):217–224

Alessandrello A, Bracchi G (2005) *Ankitokazoa bezanozanoi* n. gen. n. sp. (Annelida, Amphinomidae): a new fossil polychaete from the Olenekian (Lower Triassic) of the Ankitokazo Basin. Studies on the Permo-Trias of Madagascar 8. Atti Soc ital sci nat Mus civ stor nat Milano 146(2):155–163

Alessandrello A, Bracchi G, Riou B (2004) Polychaete, sipunculan and enteropneust worms from the lower Callovian (Middle Jurassic) of La Voulte-sur-Rhône (Ardèche, France). Mem Soc Ital Sci Natur Museo Civico Storia Natur Milano (FascI) 32:1–16

Alessandrello A, Teruzzi G (1986) *Palaeoaphrodite raetica* n. gen. n. sp., a new fossil polychaete annelid of the Rhaetic of Lombardy. Atti Soc Ital Sci Nat Mus Civ Stor Nat Milano 127(3–4):297–300

Arduini P, Pinna G, Terruzzi G (1982) *Melanoraphia maculata* ngn sp., a new fossil polychaete of the Sinemurian of Osteno in Lombardy. Atti Soc Ital Sci Nat Mus Civ Stor Nat Milano 123(4):462–468

Audouin JV (1832) Classification des Annélides, et description de celles qui habitent les côtes de la France. Proc Annal des Sci Nat 27:337–447

Batts RA, Whitman KN, Meißner K, Kocot KM (2022) Biodiversity and phylogeny of North Atlantic Euphrosinidae (Annelida). Diversity 14(11):996. https://doi.org/10.3390/d14110996
Bengtson P (1988) Open nomenclature. Palaeontol 31:223–227
Borda E, Kudenov JD, Bienhold C, Rouse GW (2012) Towards a revised Amphinomidae (Annelida, Amphinomida): description and affinities of a new genus and species from the Nile Deep-sea Fan, Mediterranean Sea. Zool Scri 41(3):307–325. https://doi.org/10.1111/j.1463-6409.2012.00529.x
Borda E, Yáñez-Rivera B, Ochoa GM, Kudenov JD, Sanchez-Ortiz C, Schulze A, Rouse GW (2015) Revamping Amphinomidae (Annelida: Amphinomida), with the inclusion of *Notopygos*. Zool Scri 44(3):324–333. https://doi.org/10.1111/zsc.12099
Bracchi G, Alessandrello A (2005) Paleodiversity of the free-living polychaetes (Annelida, Polychaeta) and description of new taxa from the Upper Cretaceous Lagerstätten of Haqel, Hadjula and Al-Namoura (Lebanon). Soc Ital Sci Nat Museo Civ Storia Nat Milano 32(3):1–64
Briggs DEG, Bartels C (2010) Annelids from the Lower Devonian Hunsrück Slate (lower Emsian, Rhenish Massif, Germany). Palaeontology 53(1):215–232. https://doi.org/10.1111/j.1475-4983.2009.00927.x
Briggs DEG, Kear AJ (1993) Decay and preservation of polychaetes: taphonomic thresholds in soft-bodied organisms. Paleobiology 19:107–135. https://doi.org/10.1017/S0094837300012343
Chen H, Parry LA, Vinther J, Zhai D, Xianguang H, Ma XA (2020) A Cambrian crown annelid reconciles phylogenomics and the fossil record. Nature 583:249–252. https://doi.org/10.1038/s41586-020-2384-8
Dunn F, Kenchington C, Parry L, Clark J, Kendall R, Wilby P (2022) A crown-group cnidarian from the Ediacaran of Charnwood Forest, UK. Nat Ecol Evol 6(8):1095–1104. https://doi.org/10.1038/s41559-022-01807-x
Ehlers E (1868) Ueber eine fossile Eunice aus Solenhofen (*Eunicites avitus*) nebst Bemerkungen ueber fossile Würmer überhaupt. Z Wiss Zoo 18:421–443
Ehlers, W (1869) Ueber fossile Würmer aus dem lithographischen Schiefer in Bayern: Palaeontographica (1846–1933):145–175
Eibye-Jacobsen D (2004) A reevaluation of *Wiwaxia* and the polychaetes of the Burgess Shale. Lethaia 37(3):317–335. https://doi.org/10.1080/00241160410002027
Eibye-Jacobsen D, Vinther J (2012) Reconstructing the ancestral annelid. J Zool Syst Evol Res 50(1):85–87. https://doi.org/10.1111/J.1439-0469.2011.00651.X
Eriksson M, Jeppsson L, Bergman CF, Hints O (2000) Paranomenclature and the rules of zoological nomenclature; with examples from fossil polychaete jaws (scolecodonts). Micropaleontol 46(2):186–188. https://doi.org/10.2113/46.2.186
Farrell U, Briggs DEG (2007) A pyritized polychaete from the Devonian of Ontario. Proc R Soc B Biol Sci 274(1609):499–504. https://doi.org/10.1098/rspb.2006.0063
Fauchald K (1977) The polychaete worm. Definitions and keys to the orders, families and genera. Nat Hist Mus Los Angeles County, Sci Ser 28:1–188. http://hdl.handle.net/10088/3435
Gall J-C, Grauvogel L (1967) Faunes du Buntsandstein III. Quelques annélides du Grès a Voltzia des Vosges. Ann Paleontol Paris 53(2):105–110
Gall J-C (1971) Faunes et paysages du Grès à Voltzia du Nord des Vosges. Essai Paléoécologique Sur Le Buntsandstein Supérieur: Mem Serv Carte Géol Als Lorr 314:1–318
Gall J-C, Grauvogel-Stamm L (2005) The early Middle Triassic 'Grès à Voltzia' Formation of eastern France: a model of environmental refugium. Compt Rend Palevol 4(6–7):637–652. https://doi.org/10.1016/j.crpv.2005.04.007
Glasby CJ, Erséus C, Martin P (2021) Annelids in extreme aquatic environments: diversity. adaptations and evolution. Diversity 13(2):98. https://doi.org/10.3390/d13020098
Gueriau P, Parry LA, Rabet N (2023) *Gilsonicaris* from the Lower Devonian Hunsrück slate is a eunicidan annelid and not the oldest crown anostracan crustacean. Biol Lett 19(8):20230312. https://doi.org/10.1098/rsbl.2023.0312
Hay AA (2002) Flabelligerida from the Francis Creek shale of Illinois. J Paleontol 76(4):764–766. https://doi.org/10.1666/0022-3360(2002)076%3c0764:FFTFCS%3e2.0.CO;2

Hinde GJ (1879) On Annelid jaws from the Cambro-Silurian, Silurian, and Devonian formations in Canada and from the Lower Carboniferous in Scotland: Quart J Geol Soc 35(1-4):370–389. https://doi.org/10.1144/GSL.JGS.1879.035.01-04.2

Hints O, Eriksson M (2007) Diversification and biogeography of scolecodont-bearing polychaetes in the Ordovician. Palaeogeogr Palaeoclimatol Palaeoecol 245:95–114. https://doi.org/10.1016/j.palaeo.2006.02.029

Högström AE, Briggs DEG, Bartels C (2009) A pyritized lepidocoleid machaeridian (Annelida) from the Lower Devonian Hunsrück Slate, Germany. Proc R Soc b: Biol Sci 276(1664):1981–1986. https://doi.org/10.1098/rspb.2008.1875

Horwood A (1912) II. – On *Archarenicola Rhætica*, sp. nov. Geol Mag (Decade V) 9(9):395–399. https://doi.org/10.1017/S001675680011550X

Hutchings P, McRae J (1993) The Aphroditidae (Polychaeta) from Australia, together with a redescription of the Aphroditidae collected during the Siboga Expedition. Rec Aust Mus 45(3):279–363. https://doi.org/10.3853/j.0067-1975.45.1993.24

Ippolitov A, Vinn, O, Kupriyanova E, Jäger M. (2014) Written in stone: history of serpulid polychaetes through time. Mem Mus Victoria 71:123–159. https://doi.org/10.24199/j.mmv.2014.71.12

Jumars PA, Dorgan KM, Lindsay SM (2015) Diet of worms emended: an update of polychaete feeding guilds: Ann Rev Mar Sci 7:497–520. https://doi.org/10.1146/annurev-marine-010814-020007

Kielan-Joworowska Z (1966) Polychaete jaw apparatuses from the Ordovician and Silurian of Poland and comparison with modern forms. Palaeontol Pol 16:152 p. https://www.palaeontologia.pan.pl/Archive/1966-16.pdf

Knecht RJ, McCall CR, Tsai C-C, Rabideau Childers RA, Yu N (2025) *Palaeocampa anthrax*, an armored freshwater lobopodian with chemical defenses from the Carboniferous: Comm Biol 8(1):1080. https://doi.org/10.1038/s42003-025-08483-0

Lamarck J-B d (1802) La nouvelle classes des Annélides. Bull Mus Hist Nat Paris, An X, Disc d'ouverture 27:483–517.

Lamarck J-B d M (1818) Histoire naturelle des animaux sans vertebres, Verdiere, Libraire,vol 6, no (1–2), pp 1–462. https://doi.org/10.5962/bhl.title.12712

Lukeneder A (ed) (2026) Late Triassic Konservat-Lagerstätten within the Carnian Pluvial Episode: 150 Years of Scientific Excavations in Austria. Top Geobiol 55:600 p. https://doi.org/10.1007/978-3-032-19777-1

Lukeneder A, Lukeneder P (2021) The upper Triassic Polzberg palaeobiota from a marine *Konservat-Lagerstätte* deposited during the Carnian Pluvial Episode in Austria. Sci Rep 11:16644. https://doi.org/10.1038/s41598-021-96052-w

Luque J, Hourdez S, Vinn O (2015) A new fossil bristle worm (Annelida: Polychaeta: Aphroditiformia) from the late Cretaceous of tropical America. J Paleontol 89(2):257–261. https://doi.org/10.1017/jpa.2014.22

McHugh D (1997) Molecular evidence that echiurans and pogonophorans are derived annelids. Proc Nat Acad Sci USA 94(15):8006–8009. https://doi.org/10.1073/pnas.94.15.800

Müller J, Schumacher A, Borda E, Rouse GW, Bartolomaeus T, Tilic E (2021) "Brittleworms": Ultrastructure and arrangement of the calcified chaetae of *Euphrosine* (Amphinomida, Annelida). Invertebr Biol 140(4):e12353. https://doi.org/10.1111/ivb.12353

Orensanz JM (1990) The eunicemorph polychaete annelids from Antarctic and Subantarctic Seas: With Addenda to the Eunicemorpha of Argentina, Chile, New Zealand, Australia, and the Southern Indian Ocean: biology of the Antarctic Seas XXI. Antarct Res Ser 52:1–183. https://doi.org/10.1029/AR052p0001

Parry L, Vinther J, Edgecombe GD (2015a) Cambrian stem-group annelids and a metameric origin of the annelid head. Biol Lett 11(10):20150763. https://doi.org/10.1098/rsbl.2015.0763

Parry LA, Edgecombe GD, Eibye-Jacobsen D, Vinther J (2016) The impact of fossil data on annelid phylogeny inferred from discrete morphological characters. Proc R Soc B 283(1837):20161378. https://doi.org/10.1098/rspb.2016.1378

Parry LA, Edgecombe GD, Sykes D, Vinther J (2019a) Jaw elements in Plumulites bengtsoni confirm that machaeridians are extinct armoured scaleworms: Proc R Soc Lond B 286(1907):20191247. https://doi.org/10.1098/rspb.2019.1247

Parry LA, Eriksson, M, Vinther J (2019b) 3. The Annelid Fossil Record. In: Purschke G, Böggemann M, Westheide W (eds) Handbook of zoology: Annelida, vol 1: Annelida Basal Groups and Pleistoannelida, Sedentaria I, De Gruyter, pp 69–88. https://doi.org/10.1515/9783110291582-003

Parry LA, Smithwick F, Nordén KK, Saitta ET, Lozano-Fernandez J, Tanner AR, Caron JB, Edgecombe GD, Briggs DE, Vinther J (2018) Soft-bodied fossils are not simply rotten carcasses–toward a holistic understanding of exceptional fossil preservation: BioEssays 40(1):1700167. https://doi.org/10.1002/bies.201700167

Parry LA, Wilson P, Sykes D, Edgecombe GD, Vinther J (2015b) A new fireworm (Amphinomidae) from the Cretaceous of Lebanon identified from three-dimensionally preserved myoanatomy: BMC Evol Biol 15:256. https://doi.org/10.1186/s12862-015-0541-8

Paxton H (1986) Generic revision and relationships of the family Onuphidae (Annelida: Polychaeta). Rec Aust Mus 38(1):1–74. https://doi.org/10.3853/j.0067-1975.38.1986.175

Paxton H (2009) Phylogeny of Eunicida (Annelida) based on morphology of jaws. Zoosymposia 2:241–264. https://doi.org/10.11646/zoosymposia.2.1.18

Righi S, Savioli M, Prevedelli D, Simonini R, Malferrari D (2021) Unravelling the ultrastructure and mineralogical composition of fireworm stinging bristles. Zoology 144:125851. https://doi.org/10.1016/j.zool.2020.125851

Rouse G, Pleijel F, Tilic E (2022) Annelida. Oxford University Press, pp 1–418. https://doi.org/10.1093/oso/9780199692309.003.0009

Rouse GW, Pleijel F (2003) Problems in polychaete systematics: In: Sigvaldadóttir E, Mackie ASY, Helgason GV, Reish DJ, Svavarsson J, Steingrimsson SA, Guðmundssonet G (eds) Advances in polychaete research. Hydrobiologia 496:175–189. https://doi.org/10.1007/978-94-017-0655-1_16

Schiffbauer JD, Selly T, Jacquet SM, Merz RA, Nelson LL, Strange MA, Cai Y, Smith EF (2020) Discovery of bilaterian-type through-guts in cloudinomorphs from the terminal Ediacaran period. Nat Commun 11:205. https://doi.org/10.1038/s41467-019-13882-z

Shcherbakov DE, Timm T, Tzetlin AB, Vinn O, Zhuravlev AY (2020) A probable oligochaete from an Early Triassic Lagerstatte of the southern Cis-Urals and its evolutionary implications. Acta Palaeontol Pol 65(2):219–233. https://doi.org/10.4202/app.00704.2019

Sigovini M, Keppel E, Tagliapietra D (2016) Open Nomenclature in the biodiversity era. Meth Ecol Evol 7(10):1217–1225. https://doi.org/10.1111/2041-210X.12594

Struck T (2011) Direction of evolution within Annelida and the definition of Pleistoannelida. J Zool Syst Evol Res 49(4):340–345. https://doi.org/10.1111/j.1439-0469.2011.00640.x

Struck T, Schult N, Kusen T, Hickman E, Bleidorn C, McHugh D, Halanych K (2007) Annelid phylogeny and the status of Sipuncula and Echiura. BMC Evol Biol 7:57. https://doi.org/10.1186/1471-2148-7-57

Stur D (1886) Vorlage des ersten fossilen Schädels von *Ceratodus* aus den ober triadischen Reingrabner Schiefern von Pölzberg nördlich bei Lunz. Verh KK Geol Reichsanst 15:381–383

Sutton MD, Briggs DEG, Siveter DJ, Siveter DJ (2001) A three-dimensionally preserved fossil polychaete worm from the Silurian of Herefordshire, England. Proc R Soc B Biol Sci 268:2355–2363. https://doi.org/10.1098/rspb.2001.1788

Teller F (1891) Über den Schädel eines fossilen Dipnoërs *Ceratodus sturii* nov. spec. aus den Schichten der oberen Trias der Nordalpen, Hölder. Abh K.K. Geol Reichsanst 15:1–38

Thompson I (1979) Errant polychaetes (Annelida) from the Pennsylvanian Essex fauna of northern Illinois. Palaeontogr Abt A 163(4–6):169–199

Thompson I, Johnson RG (1977) New fossil polychaete from Essex, Illinois. Fieldiana Geol 33(25):471–487

Vinn O, Mutvei H (2009) Calcareous tubeworms of the Phanerozoic. Eston J Earth Sci 58(4):286–296. https://doi.org/10.3176/earth.2009.4.07

Weigert A, Bleidorn C (2016) Current status of annelid phylogeny. Org Divers Evol 16:345–362. https://doi.org/10.1007/s13127-016-0265-7

Wilson P, Parry LA, Vinther J, Edgecombe GD (2016) Unveiling biases in soft-tissue phosphatization: extensive preservation of musculature in the Cretaceous (Cenomanian) polychaete *Rollinschaeta myoplena* (Annelida: Amphinomidae). Palaeontology 59(3):463–479. https://doi.org/10.1111/pala.12237

Yang B, Steiner M, Schiffbauer JD, Selly T, Wu X, Zhang C, Liu P (2020) Ultrastructure of Ediacaran cloudinids suggests diverse taphonomic histories and affinities with non-biomineralized annelids. Sci Rep 10(1):1–12. https://doi.org/10.1038/s41598-019-56317-x

Yang X, Aguado MT, Helm C, Zhang Z, Bleidorn C (2024) New fossil of Gaoloufangchaeta advances the origin of Errantia (Annelida) to the early Cambrian. R Soc Open Sci 11(4):231580. https://doi.org/10.1098/rsos.231580

Zanol J, Fauchald K, Paiva PC (2007) A phylogenetic analysis of the genus Eunice (Eunicidae, polychaete, Annelida). Zool J Linn Soc 150(2):413–434. https://doi.org/10.1111/j.1096-3642.2007.00302.x

Zanol J, Halanych KM, Struck TH, Fauchald K (2010) Phylogeny of the bristle worm family Eunicidae (Eunicida, Annelida) and the phylogenetic utility of noncongruent 16S, COI and 18S in combined analyses. Mol Phylogenet Evol 55(2):660–676. https://doi.org/10.1016/j.ympev.2009.12.024

Zhang G, Parry LA, Vinther J, Ma X (2022) Exceptional soft tissue preservation reveals a cnidarian affinity for a Cambrian phosphatic tubicolous enigma. Proc R Soc B 289(1986):20221623. https://doi.org/10.1098/rspb.2022.1623

Chapter 10
New Insights into the Fish Assemblage from the Early Carnian of Polzberg in Austria

Andrea Tintori, Davide Conedera, and Alexander Lukeneder

Abstract The present reappraisal of the fish fauna from Polzberg comes 50 years after the original description. As a result, several taxonomic changes are proposed, especially regarding the 'subholostean' component of the fauna, mainly belonging to Peltopleuriformes. A new combination is proposed for *Habroichthys gregarius* (Griffith, 1977), now ascribed to *Peltopleurus*. A new species, *Thoracopterus polzbergensis*, is erected for the scaled Polzberg flying fish, previously considered as *Thoracopterus niederristi*, originally described from Raibl. This reflects a few differences, mainly in the scale covering. The ichthyofauna from Polzberg is now considered to differ at least at the species level from that from Raibl. The latter is somewhat older and pre-CPE, but it also comes from a different paleoenvironment: Raibl is more closely related to a carbonate platform than the Reingraben Formation. Nonetheless, the general composition of the two ichthyofaunas does not show any major differences.

Keywords Actinopterygii · New taxon · *Konservat-Lagerstätte* · Polzberg · Late Triassic · Carnian Pluvial Episode

A. Tintori (✉)
TRIASSICA Institute for Triassic Lagerstätten, Perledo, Italy
e-mail: paleo.tintori@outlook.it

D. Conedera
Dipartimento di Geoscienze, Università di Padova, Padua, Italy
e-mail: davide.conedera@phd.unipd.it

A. Lukeneder
Department of Geology and Paleontology, Natural History Museum Vienna, Vienna, Austria
e-mail: alexander.lukeneder@nhm.at

A. Lukeneder (ed.), *Late Triassic Konservat-Lagerstätten within the Carnian Pluvial Episode*, Topics in Geobiology 55, https://doi.org/10.1007/978-3-032-19777-1_10

10.1 Introduction

Fossil fishes from Polzberg/Lunz have been known since the 19th century (Teller 1891), but Abel (1906) was the first to provide a list that enabled considering this assemblage a true 'fish fauna'. It remains unclear, however, on what grounds Abel (1906, p. 9) compiled his list, but one major consequence is that the Polzberg fish assemblage was considered by Griffith (1977) and Lehman (1978) to be similar to that of Raibl. Following Abel, the two assemblages were thought to share a number of species (*Pholidophorus bronni* Kner, 1866; *Peltopleurus splendens* Kner, 1866; *Pholidopleurus typus* Bronn, 1858; *Thoracopterus niederristi* Bronn, 1858). Unfortunately, the Abel (1906) 'names' were not related to actual specimens—no old labels seem to be present—so that Griffith (1977) described the Polzberg material without checking or discussing Abel's list of species/specimens. Moreover, he did not refer to the specimens he studied with a catalogue number. Those numbers were actually added only several years later, whereby only a few labels were created for the most important specimens. This sometimes makes it unclear which specimen he was writing about. Note that catalogue numbers are not written on the specimens themselves, so that that over the years misplacements could also have occurred.

To the best of our knowledge, *Pholidopleurus typus* is not present in the available fish material from Polzberg, as it is also clear that no semionotiforms are present in the 'original' fossil fish assemblages of either Polzberg or the 'Rossi Level' of Raibl. That species is present solely in the 'Rossi Level', which is around the Ladinian/Carnian Boundary or even late Ladinian (Tintori and Castelnuovo 2023).

Pholidophorus bronni was originally described from Raibl and it is now referred to *Knerichthys bronni* in Arratia's (2013) revision. Although Arratia stated that the species also occurred at Polzberg/Lunz (Arratia 2013, p. 106), Griffith (1977) erected *Pholidohoretes salvus* as the unique pholidophorid from Polzberg, without discussing previous attributions. Furthermore, it is not clear what Abel (1906) referred to as *Heterolepidotus cephalus* because Kner (1866) erected a *Pholidophorus cephalus* but did not figure it (although he cited a non-existent Pl. 4, Fig. 2). The same applies for *Pholidophorus latiusculus* Agassiz (1833), also described from the Norian fauna of Seefeld. In any case, it is likely that all these specimens are now labeled as *Pholidophoretes salvus*.

Regarding the Peltopleuriformes, Griffith (1977) did not consider the Raibl species of *Peltopleurus* and erected *Peltopleurus dirumptus* and *Habroichthys gregarius*, as well as the new genus and species *Nannolepis elegans*. He described the scaled flying fishes from Polzberg as *Thoracopterus niederristi* (originally from Raibl). Griffith (1977) regarded *Habroichthys*, *Nannolepis* and *Thoracopterus* as Luganoiformes, but this attribution is no longer supported by recent studies (Bürgin 1992; Tintori and Sassi 1992; Conedera et al. 2025; this paper). Furthermore, *H. gregarius* has to be moved to *Peltopleurus* as previously suggested by Bürgin (1992), Tintori et al. (2016), and Conedera et al. (2025). Moreover, a new species is herein proposed for the Polzberg *Thoracopterus* scaled flying fishes.

Accordingly, the two early Carnian ichthyofaunas from the Alps appear to be quite different, as it will be further demonstrated below. Abel's (1906) list lacks easily identifiable peculiar fishes such as *Saurichthys* and *Polzbergia*, the latter erected by Griffith (1977). This, however, can be explained by the relative rarity of these genera. A considerable number of palaeonisciform specimens still require a proper description, pending their scanty and fragmentary preservation (see also Griffith 1977). At present it is not possible to establish whether they all belong to a single species or to different taxa, since in most cases only body fragments have been collected.

Finally, note that the average preservation of Polzberg fossil fishes is very poor, as also emphasized by Arratia (2013, p. 107), who used the same words when describing *Knerichthys bronni* from Raibl (p. 51). In fact, fish specimens from both Raibl and Polzberg are usually split in part and counterpart, preventing a clear view of the single bone or scales from the outside. Furthermore, the limestone-marly limestone yielding the fossils rarely splits along very smooth surfaces; this is especially true for Polzberg, where the rock surfaces are often covered with bivalve shells. As a result, the bones and scales are frequently badly fragmented, and pieces can easily detach during collecting. In some cases, the moulds left on the rock after the bones are lost show a clearer outline of the individual elements than when the bone itself is still present.

10.2 The Polzberg *Fossil-Lagerstätte*

Fossils from the area around Polzberg, sometimes also known as coming from Lunz am See, a nearby small community (Stur 1874; Griffith 1977), have been known for more than a century (Stur 1874). They were recovered in the framework of mining activity, as was quite common in those times. Around 1885 and 1909, thousands of fossils were collected from the Polzberg locality during paleontological excavation campaigns by the Austrian Geological Survey (GBA; now GeoSphere Austria) and the Vienna Museum of Natural History (NHMW) (Lukeneder and Lukeneder 2021). Most of the material remained undescribed for a long time, with the exception of Abel's (1906) paper on flying fishes. Griffith (1977) then described all the available material. As reported by Lukeneder and Lukeneder (2021), "A detailed report on the paleobiota from the Polzberg area in the Northern Calcareous Alps is lacking so far. Though known for over 140 years, the Polzberg *Konservat-Lagerstätte*, its formation processes, and fossil assemblages are poorly understood".

Lukeneder and Lukeneder (2021) provided a first short review of this *Lagerstätte*, starting the revival of interest in this locality. This was also promoted by the relationships to the Carnian Pluvial Episode (CPE), which is now considered a major event in the Late Triassic history of life (Dal Corso et al. 2020). Most of the fossils have been recovered in the basal part of the Reingraben Formation, which is a much more siliciclastic paleoenvironment (Lukeneder and Lukeneder 2021; Lukeneder 2026, this volume) than the more carbonate-rich beds of the Predil Limestone in Raibl.

The dating for this assemblage has always been quite precise due to the common presence of the ammonoid *Austrotrachyceras minor* and due to the limited thickness (< 5 m) of the level yielding the Polzberg Fauna (Lukeneder and Lukeneder 2021). This contrasts with the Raibl Fauna, which was discovered more than 150 years ago in a unit about 200 m thick, where fishes are mainly known in the lower part (De Marco et al. 2000). Thus, the exact stratigraphic position of the Raibl 19th-century historical collection there remains uncertain. In the last decades, scattered finds have been reported from an area much wider than the classical one around the Raibl/Cave del Predil mines (N Italy), extending mainly westward to the Saisera Valley. This wide distribution suggests that more than one fish assemblage may be present within the entire sequence, which covers the lowest part of the early Carnian (Julian 1: *Daxatina canadensis* + *Trachyceras aon* + *T. aonoides* Zones). The sequence starts from its very base (De Marco et al. 2000; Tintori and Castelnuovo 2023), but still well before the deposition of the Polzberg Reingraben Formation, assigned to the *Austrotrachyceras austriacum* Zone (Julian 2).

As in most other marine Triassic *Lagerstätten* such as the Middle Triassic Monte San Giorgio (Italy and Switzerland), the Norian Zorzino Limestone from Lombardy and Forni Dolostone from Friuli (Italy; Tintori and Lombardo 2017) as well as the Carnian Raibl Fauna (pers. obs.; Italy), the Polzberg assemblage is very poor in shallow-water benthic organisms (Lukeneder and Lukeneder 2021). The only exception in the Alpine sites is the recently discovered site of the late Ladinian Pelsa/Vazzoler (Belluno Dolomites, Italy), where a fish-rich bed (Tintori 2018; Conedera et al. 2025) is interlayered with beds very rich in silicified benthic organisms transported from the rim of the intraplatform lagoon (Dominici et al. 2024). Regarding the Carnian benthic invertebrates, the San Cassiano Fauna appears to be the most diverse fauna in the Mesozoic up to the Pliocene (Roden et al. 2020). Here as well, the age is older than that of the Polzberg assemblage and the CPE, suggesting that some major bio-events in the marine environment took place well before the CPE itself.

10.3 Paleontological Description

Superclass Actinopterygii Cope, 1887.

Order Saurichthyformes Aldinger, 1937.

Family Saurichthyidae Owen, 1860 (*sensu* Stensiö 1925).

Genus *Saurichthys* Agassiz, 1833.

Saurichthys calcaratus Griffith, 1977.

1925 *Saurichthys*, Stensiö, p. 5.

1977 *Saurichthys calcaratus* Griffith, p. 4, Figs. 1–3, pl. 1.

1981 *Saurichthys calcaratus*, Beltan and Tintori, p. 53.

2020 *Saurichthys calcaratus*, Kogan et al., p. 485, Figs. 6–7.

Material. In addition to the type series of Griffith (1977)—also used by Kogan et al. (2020) for their recent revision of this species (NHMW 2007z0170/0001–0017)—three new specimens have been recently collected and added to the collection (NHMW2021/0123/0208a/b, ST 008, 0002 PT1). The maximum length of this species is confirmed as around 400 mm. This indicates a small-sized *Saurichthys* (*sensu* Tintori 2013), although still much larger than the Raibl *S. striolatus,* which usually did not reach 200 mm (Kogan et al. 2020). The new material consists of three isolated skulls, most probably preserved without the body. From a taphonomic point of view, this is quite curious because almost all other fish specimens from Polzberg are complete (although often slightly disarticulated). Kogan et al. (2020) also dealt with the post-cranial skeleton because the skull preservation is poor and no complete skull was found among the few specimens (max 10 individuals, probably 6 to 7). It is therefore impossible to state the ratio between the pre- and post-orbital regions. The only remark on the skull is the high variability in the outline of the operculum, which sometimes appears more squarish than in the holotype as already shown in Griffith (1977, Fig. 1a, b). However, the general preservation of the available material does not allow a thorough morphological analysis to support the presence of a second species. At present, *S. calcaratus* is considered to be the only *Saurichthys* species present in the Polzberg fish assemblage. This is not the case in Raibl, where *S. striolatus* is quite common, at least in the historical collections (Kogan et al. 2020), but *S. calcaratus* is also present, although rare.

Order Perleidiformes Berg, 1940.

Family Polzbergiidae Griffith, 1977.

Remarks. Lombardo and Tintori (2004), in erecting the new genus *Felberia*, discussed the order attribution of the Polzbergiidae, to which *Felberia* was assigned. Based on a different interpretation of certain dermal elements of the *Polzbergia* skull (see below), those authors rejected the order Polzbergiiformes as not being supported by sufficient diagnostic characters and ascribed the Family Polzbergiidae to the Perleidiformes. They also pointed out that no characters support the idea of Griffith (1977) that *Polzbergia* belonged to Chondrostei *Auctorum*, near Platisomiformes and Bobasatraniformes. Griffith's interpretation was also rejected because *Polzbergia* lacks a full heterocercal tail but has a hemiheterocercal one provided with epaxial rays.

Lombardo and Tintori (2004) and Lombardo et al. (2008) considered in this family the deep-bodied genera *Felberia* and *Stoppania*, although both papers underlined that the preservation state of *Polzbergia* specimens is somewhat problematic. Accordingly, the present revision supports the inclusion of these genera in the Polzbergiidae and provides evidence that, in the Triassic Middle Fish Fauna (TMFF) (Tintori et al. 2014) from western Tethys, the deep-bodied fishes belong to 'subholosteans' (mostly

Perleidiformes). In contrast, in the eastern Tethys (now southern China) this functional niche is occupied by neopterygians related to gynglymodians (Wen et al. 2012; Xu and Wu 2012).

Genus *Polzbergia* Griffith, 1977.

Polzbergia brochatus Griffith, 1977.

Material. Only three specimens are known and were used by Griffith (1977) when erecting the new genus and species. The holotype (NMW 2007z0170/0015, figured in pl. 2) (Fig. 10.1a) and the two referred specimens (NHMW 2007z0170/0016, the posterior body fragment, and NHMW 2007z0170/0017, a skull impression) are the only known specimens, making *Polzbergia* among the rarest actinopterygian taxa from Polzberg, together with *Phaidrosoma* (two known specimens, see below).

Remarks. The holotype slab shows, in addition to most of the body, the distal part of a tail and a fragment of a vertebral column showing neural and haemal spines. These two fragments do not belong to the main specimen, as already suggested by Griffith (1977). A few major errors in Griffith's (1977) description are related to the subopercle and preopercle, which he considered to be fused. In the holotype the left opercle and preopercle are disarticulated and clearly visible above the skull. The shape of the preopercle is clear: it is upside down and with the rounded posterior margin pointing forward, and the large infraorbital process is well visible encompassing the short margin that must have been in contact with the posterior margin of the maxilla. Thus, the preopercle cannot have been fused to the subopercle because the outline of the isolated bone is regular. Jaw teeth are thinner and longer than figured by Griffith (1977, Fig. 5), and the crown part is gently flattened and concave as in *Felberia* (Lombardo and Tintori 2004). The palatal teeth are also thinner, not adapted to a strictly durophagous (crushing) diet as proposed by Griffith (1977), although *Polzbergia* was most likely a bottom-dweller feeding on small benthic preys, as also indicated by the deep body, the heavy ornamentation on the skull bones (Fig. 10.1b), the scales, and the extended insertion of the dorsal and anal fins (Garbelli and Tintori 2015). A shallow-water necto-benthic lifestyle could explain the rarity of such fishes in the fossil assemblage of Polzberg because the preservation environment was probably quite distant from the habitat where they lived, as seen for *Felberia carnica* Lombardo and Tintori, 2004 in the Raibl ichthyofauna.

Order Peltopleuriformes Gardiner, 1967.

Family Peltopleuridae Brough, 1939.

Genus *Peltopleurus* Kner, 1866.

Peltopleurus dirumptus Griffith, 1977.

1977 *Peltopleurus dirumptus* Griffith, p. 20, Fig. 8, Pl.3A.

Diagnosis. (emended after Griffith 1977). Large *Peltopleurus* species up to 80 mm in s. l., about 50 transversal scale rows, longitudinal flank row with deep scales up to

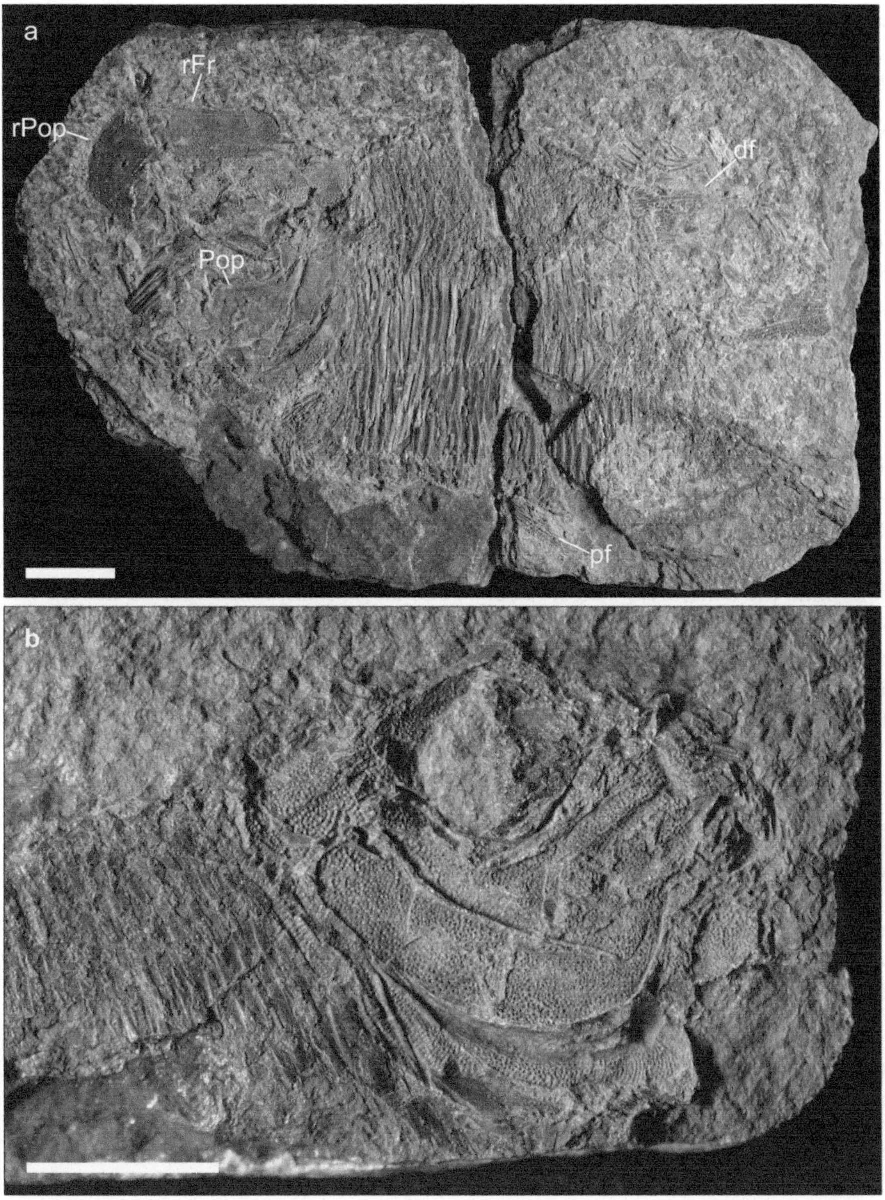

Fig. 10.1 *Polzbergia brochatus* Griffith, 1977. **a** the holotype NMW 2007z0170/0015. **b** NHMW 2007z0170/0017. df, dorsal fin; pf, pelvic fins; Pop, left preopercle; rFr, right frontal; rPop, right preopercle. Scale bars: 10 mm

the caudal region. Three dorsal and one ventral rows above the deep one. Otoliths. Deep flank scales with vertical anterior ridge on external surface, in situ covered by the preceding scale.

Material. Griffith (1977, Pl. 3A) figured the holotype, which is now NHMW 2007z0170/0018a, b (part and counterpart, Fig. 10.2a). This species is not common, counting only about 10 specimens.

Remarks. This species is characterized by a large size, although most Carnian *Peltopleurus* species are much larger than the Middle Triassic ones (Bürgin 1992), particularly compared to the late Ladinian species (Lombardo 1999; Tintori and Castelnuovo 2023; pers. obs.). Pending a very detailed revision of all the Polzberg specimens ascribed to *Peltopleurus*, including the newly collected material, it should be emphasized that Griffith (1977) considered as *P. dirumptus* the large specimens that are preserved in dorso-ventral view with the body totally open along the mid-ventral line so that both sides of the body are visible (Fig. 10.2b). In contrast, he ascribed to *P. gregarius* (see below) smaller specimens mostly preserved in lateral view. As the number of dorsal and ventral scale rows apparently differ in the two groups, the two species are probably valid, even if, as usual, the preservation makes it difficult to determine the conclusive states. A very detailed redescription of both Polzberg and Raibl peltopleurids is necessary to settle the Carnian *Peltopleurus* taxonomy. This is because the Raibl material also appears to show a high variability and because of the wider stratigraphical distribution of the *Peltopleurus* specimens (Santoro 2000; Tintori and Castelnuovo 2023).

Peltopleurus gregarius (Griffith, 1977) comb. nov.

1977 *Habroichthys gregarius* Griffith, p. 26, Figs. 10–13, pl. 3B/4A.

2021 *Habroichthys gregarius* Lukeneder and Lukeneder, Fig. 4D.

Diagnosis. (emended after Griffith 1977). A medium-large sized species of *Peltopleurus* reaching up to 60 mm in s.l., triangular vertical preopercle. Flank scales very deep up to the tail, a single dorsal and a single ventral flank row.

Holotype. The specimen indicated by Griffith (1977) as holotype was figured in Pl. 3B, now NHMW 2007z0170/0026, a nearly complete specimen, partially as a mould, with no traces of paired fins (Fig. 10.3a).

Paratypes. Griffith (1977) figured one of the three paratypes in Pl. 4A, now NHMW 2007z0170/0027 (Fig. 10.3b); the other two are now NHMW 2007z0170/0028/9.

Remarks. Griffith (1977) ascribed his *Habroichthys gregarius* to Habroichthyidae (Gardiner, 1967), considering this family to belong to Luganoiformes Lehman, 1978, although both Brough (1939) and Gardiner (1967) believed *Habroichthys* to be more closely related to Peltopleuriformes. We suggest that Griffith was misled in the suprafamiliar attribution because he used this Polzberg material that, as we show below, is quite different from the basic anatomy of *Habroichthys* as known from the original material from Monte San Giorgio. Since Bürgin (1992), Habroichthyidae

Fig. 10.2 *Peltopleurus dirumptus* Griffith, 1977. **a** the holotype NHMW 2007z0170/0018a. **b** NHMW 2007z0170/0025, mostly an imprint. Mx, maxilla; Op, opercle; ot, otoliths; Pop, preopercle; Sop, subopercle. Scale bars: 10 mm

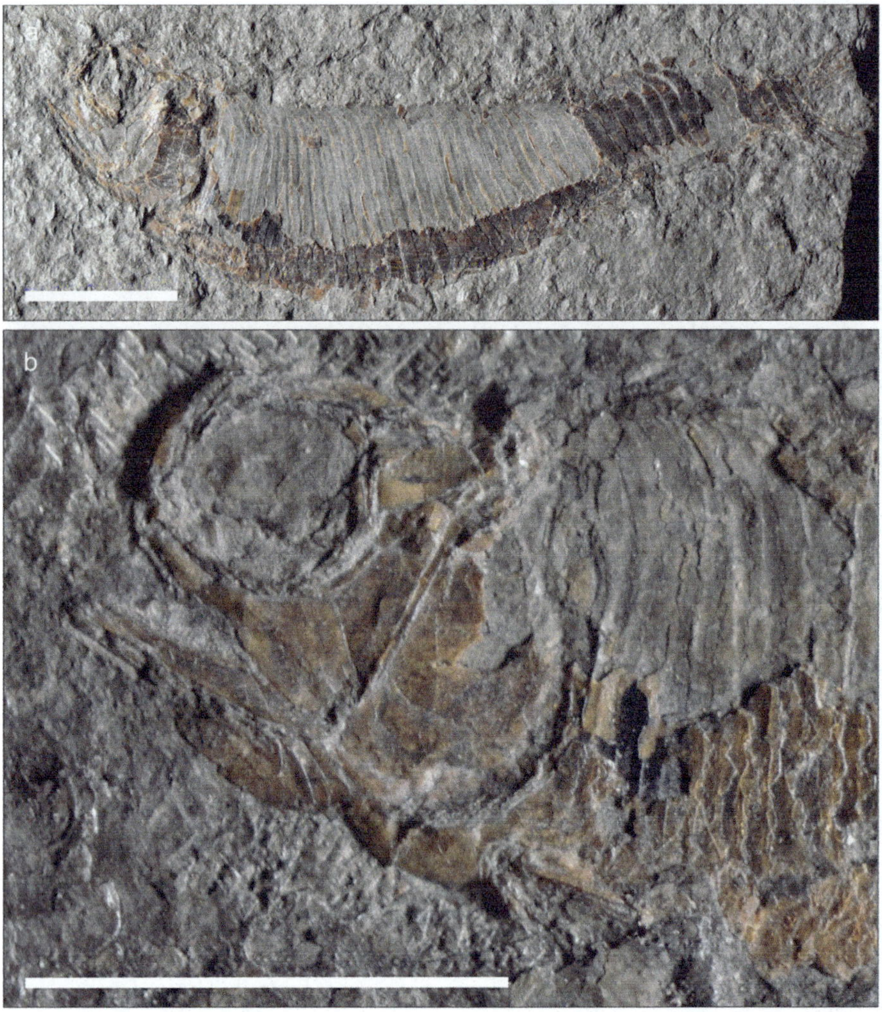

Fig. 10.3 *Peltopleurus gregarius* Griffith, 1977 comb. n. **a** the holotype (NHMW 2007z0170/0026). **b** specimen NHMW 2007z0170/0027, paratype, close-up of skull. Scale bars: 10 mm

is decisively considered to belong to Peltopleuriformes. Be that as it may, as *H. gregarius* is not a *Habroichthys* species and not related to any Luganoiformes, we move this species to Peltopleuriformes based mainly on the scale pattern, the preopercular bone outline, and the caudal and anal fin structures.

Bürgin (1992, p. 146) already noted that this species should not be ascribed to *Habroichthys*, especially because it lacks the major derived character of *Habroichthys*, namely the last large semicircular scale of the deep flank row, as stated in the original description of the genus by Brough (1939). Bürgin (1992), who added one more *Habroichthys* species, also based mostly on the rather poorly preserved specimens from Monte San Giorgio, noted other differences in *H. gregarius*. These regarded the shape of the flank scales and the skull anatomy, suggesting that this species might be ascribed to *Peltopleurus* or even to a new genus. However, the cited characters, such as the fusion of the frontal, parietal and dermopterotic bones, are now known to be quite variable also in *Habroichthys* species (Lin et al. 2011; Conedera et al. 2025), beyond in many other Peltopleuriformes, and therefore cannot be considered so important. Interestingly, Bürgin (1992) failed to underline another major character, the preopercular shape of *P. gregarius*, which clearly differs completely from that of all *Habroichthys* species (Bürgin 1992; Lin et al. 2011; Conedera et al. 2025). The latter exhibit a very peculiar and unique shape among Peltopleuriformes. Similarly, note the presence of teeth on both jaws as well as on palatal bones: true *Habroichthys* are always edentulous (Bürgin 1992; Lin et al. 2011; Conedera et al. 2025). Bürgin (1992), Lin et al. (2011), and Tintori et al. (2016) did not include *P. gregarius* in their comparisons for erecting new species of *Habroichthys*. Conedera et al. (2025) also noted that this species does not belong at all to *Habroichthys* itself. Be that as it may, none of the former authors formalized the attribution of *H. gregarius* to a different genus, as is done herein. *Peltopleurus* has become a genus with a very high interspecific variability, as also demonstrated by preliminary observations on Carnian specimens from Raibl (Santoro 2000; pers. obs.) and on late Ladinian specimens (Boeddu 2007; pers. obs.) based on very precise (bed-by-bed collecting) stratigraphic positions. In the light of this variability, we assume that *Habroichthys gregarius* should be moved to *Peltopleurus*.

Peltopleurus gregarius is quite common in the Polzberg fish assemblage, and the total number of available specimens is now around 100, which makes this species by far the most common fish taxon.

Genus *Nannolepis* Griffith, 1977.

Nannolepis elegans Griffith, 1977.

1977 *Nannolepis elegans* Griffith, p. 33, Figs. 14–15, pl. 3C/4B.

2021 *Nannolepis elegans* Lukeneder and Lukeneder, Figs. 4B, C.

Diagnosis. (emended after Griffith 1977). Medium-sized fish reaching about 60 mm in s.l., skull-roof bone with various kinds of fusion, opercle much larger than subopercle. Very high number of scale rows (about 75) with flank scales quite deep. Semiheterocercal tail with 1–2 epural rays. Large otoliths.

Material. The holotype is now NHMW 2007z0170/0117 (Fig. 10.4); only its skull was figured by Griffith (1977, pl. 4B). One paratype has been chosen (now NHMW 2007z0170/0148 + 0118 skull counterpart), but Griffith (1977, pl. 3C) figured another specimen, now NHMW 2007z0170/0147. In total, over 60 specimens are ascribed to this species, a few of them being part and counterpart so that the number of individuals should be at least 40–45. This makes it the third most common fish taxon of the assemblage.

Fig. 10.4 *Nannolepis elegans* Griffith, 1977. The holotype NHMW 2007z0170/0117. Scale bar: 10 mm

Remarks. Griffith (1977) assigned his new genus *Nannolepis* to Habroichthyidae. However, the scale pattern of *Nannolepis* differs completely from that of *Habroichthys*, this character being one of the most important for separating *Habroichthys* from other peltopleuriform genera. In our opinion, *Habroichthys* is still the only genus in this family, following Bürgin (1992) and Conedera et al. (2025). Nonetheless, the skull-roof bone pattern of *Nannolepis* is very consistent with that of several Peltopleuriformes such as a few *Peltopleurus* spp., most *Habroichthys* spp. and *Thoracopterus* spp., with the large convex rostral separating the nasals, very short supraorbital sensory canals stopping in the median part of the frontals, and dermopterotics usually fused to the parietals. Again, as already discussed for *P. gregarius*, Griffith's (1977) attribution to Luganoiformes is to be dismissed.

The dorsal view restoration by Griffith (1977, Fig. 14) is incorrect because the rostral is larger than figured and the supraorbital sensory canals are much shorter, ending at around the middle of the orbit. The suture lines between the frontals and with the dermopterotics are not so clear and, at least in NHMW 2007z0170/0118 (counterpart of NHMW 2007z0170/0148), there is no trace of the median suture. In this same specimen, large otoliths are clearly visible. Moreover, Griffith's (1977, Fig. 15) interpretation of the opercular series is very doubtful, and his branchiostegal ray 1 can be considered to actually be the very small subopercle, compared to the opercle, because there is a small displacement in this region and his gular is most probably the right mandible. Regardless, the most striking character of the genus is the very high number of longitudinal scale rows, which is quite unusual in Peltopleuriformes.

Family Thoracopteridae Griffith, 1977.

Genus *Thoracopterus* Bronn, 1858.

Type species. *Thoracopterus niederristi* Bronn, 1858 from the Lower Carnian of Raibl/Cave del Predil (Udine Province, Italy).

Other species. *T. wushaensis* Tintori et al., 2012, *T. telleri* (Abel, 1906), *T. polzbergensis* (this paper), *T. martinisi* Tintori and Sassi, 1992, and *T. magnificus* Tintori and Sassi, 1987.

Remarks. Tintori and Sassi (1992) moved *Gigantopterus telleri* to *Thoracopterus* and considered the restoration made by Griffith (1977) incorrect in a few important points. These included the relationships among rostral, nasals and frontals. Actually, for both *G. telleri* and *T. niederristi*, Griffith (1977) considered that the rostral does not contact the frontals and that the nasals meet each other behind the rostral. However, this is not the case (see below). Recently, Shen and Arratia (2022), without checking the material, accepted Griffith's (1977) interpretation without realizing that in no other Peltopleuriformes do the nasals meet along the midline. Rather, they are always separated by the large rostral, which in all Thoracopteridae is strongly convex, producing a conical tip of the snout. Furthermore, Shen and Arratia (2022) erected new genera for *T. wushaensis* and for the Norian species *T. martinisi* and *T. magnificus*, also keeping *Gigantopterus* as a valid genus. The conclusions of the latter authors Shen and Arratia (2022) are not accepted here due to numerous errors in the coding of their phylogenetic analysis (Tintori 2025) and because of the many characters that cannot be stated because of preservation issues.

Thoracopterus polzbergensis sp. nov.

1906 *Thoracopterus niederristi* Bronn (pro parte), Abel, p. 19.

1977 *Thoracopterus niederristi*, Griffith, p. 37, figs, 16–19, pls. 5–6.

1978 *Thoracopterus niederristi*, Lehman, p. 55, Fig. 1–4, pls. 1–6.

2021 *Thoracopterus niederristi*, Lukeneder and Lukeneder, Fig. 4E-F.

2022 *Thoracopterus niederristi*, Shen and Arratia, Fig. 6A. **LSID**: urn:lsid:zoobank.org:act:E06F1D87-40AB-40F3-B3F8-2EE1BA7E4322.

Ethimology. From Polzberg bei Lunz (Austria), where the specimens were collected.

Holotype. NHMW 2025/0260/0001, the almost complete specimen (Fig. 10.5a–c), figured by Griffith (1977, pls. 5-6A).

Paratypes. Complete specimens: NHMW 2007z0170/0183, the skull figured by Griffith (1977, pl. 6B), and the complete specimen figured by Shen and Arratia (2022, Fig. 6a), NHMW 1887-II.2, NHMW 2007z0170/0191, NHMW 2007z0170/0170, NHMW 2007z0170/0171, NHMW 2007z0170/0363.

Geographical and stratigraphical distribution. So far known only from the fossiliferous site of Polzberg near Lunz am See (southern Austria) in the basal part (3.5 m) of the Reingraben Shales (Julian 2, Lower Carnian, *Austrotrachyceras austriacum* Zone).

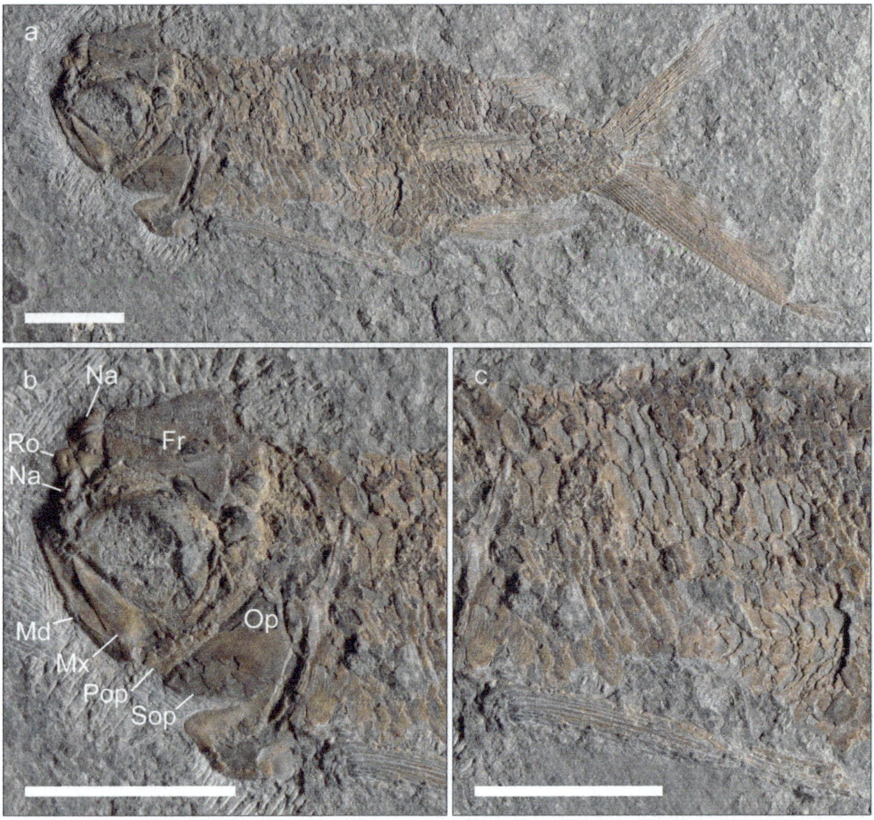

Fig. 10.5 *Thoracopterus polzbergensis* sp. nov. **a** the holotype (NHMW 2025/0260/0001). **b** close-up of the holotype skull. c) close-up of the anterior region of the body. Fr, frontal; Md, mandible; Mx maxilla; Na, nasal; Op, opercle; Pop, preopercle, Ro, rostral; Sop, subopercle. Scale bars: 10 mm

Diagnosis. Small *Thoracopterus*, up to about 70 mm in s.l., usually 60 mm or less. Skull-roof bones fused in variable ways (as in most Peltopleuriformes). Large conical rostral separating nasals. Strong dentition at least on the lower jaw. Body totally covered by scales, with about 45 transversal rows. Mid-flank row quite ventrally placed, with deep scales up to the pelvic fin insertion; at least eight dorsal and four ventral longitudinal rows other than the mid-flank one; scales of the first ventral row also deeper than longer. Scale ornamentation consists of vertical ridges in the anterior part of the scales. Pectoral fins over 60% of s. l., pelvic fins much shorter, about 20% of the s.l. (1/3 of the pectorals). Length of dorsal lobe of caudal fin around 60% of the ventral one.

Remarks and comparison. Among *Thoracopterus* species, the two from the Norian (*T. martinisi* and *T. magnificus*) show a completely naked body. *T. wushaensis* has a few rows forming a ventral half circle at the base of the tail (Tintori et al. 2012). *T. telleri*, in turn, was likely totally naked or with one ‘caudal’ row (see below) similar to *T. wushaensis*. Accordingly the comparison here is only with *T. niederristi*, the fully scaled species from Raibl/Cave del Predil.

Abel (1906), in his paper on the flying fishes, redescribed the original material of *T. niederristi* from Raibl and ascribed to this species also specimens collected in Polzberg/Lunz without providing any justification. Griffith (1977) and Lehman (1978) followed Abel and described the Polzberg specimens as *T. niederristi* without any comparison with the topotypical material from Raibl, although Griffith (1977, p. 37) wrote that he could check only few specimens.

Shen and Arratia (2022) provided no description of the Polzberg material and did not discuss the possibility that Raibl material could not be conspecific. In fact, they flatly followed Griffith (1977) even though Tintori and Sassi (1992) had already written that his skull restoration was incorrect. Shen and Arratia (2022) cast no doubt about *T. niederristi* from Polzberg. Furthermore, they published a picture of one of the Polzberg specimens (NHMW 2007z0170/0183, with no anatomical note but indicating the fins, where their dorsal fin is actually the distal part of the right pectoral (AT pers. obs. 2025). Those authors (p. 9) also wrote "It is important to note that the anatomical information on thoracopterids is incomplete due to poor preservation of the specimens. […] Better preserved specimens that provide more anatomical information, especially from the cranium and vertebral column and associated elements, which are almost unknown, are needed to clarify the taxonomy and phylogenetic position of *T. niederristi*". Unfortunately, as noted above, they used for the skull anatomy the descriptions and restorations by Griffith (1977) and Lehman (1978), which had already been shown to be inaccurate in several respects (especially regarding the Ro-Na-Fr relationships) by Tintori and Sassi (1992). Shen and Arratia (2022) did not examine any of the 'naked' Ladinian and Norian *Thoracopterus* species, where the vertebral column is quite well known, although they made a mistake in interpreting the *T. telleri* vertebral column (see below). All these issues casts serious doubt on their interpretation (phylogenetic analysis) of Thoracopteridae, including the erection of two new genera and keeping *Gigantopterus* as a valid one. (Following Shen and Arratia (2022), there should be five species of Thoracopteridae with four different genera.) A more detailed discussion is presented in Tintori (2025).

T. niederristi from Raibl typically shows 14–15 longitudinal scale rows: 3–4 ventral and 10 dorsal to the deep scale flank row. This stands in contrast to 3–4 and 8, respectively, in *T. polzbergensis* sp. nov. (7 in Griffith's restoration although he wrote that there are 8–9 rows in total and 5 in Lehman's restoration) for a total of 12–13. Note also in the latter species that the scales dorsal to the flank row are somewhat deeper than long as opposed to *T. niederristi*. Furthermore, the scales of *T. niederristi* are ornamented with elongate patches of ganoine especially in the trunk region (best seen in the holotype and in specimen GBM 1866/4/4/2), while those of *T. polzbergensis* sp. nov. show very evident vertical ridges in the anterior region as already noted by Griffith (1977, p. 45) and an almost smooth posterior region. Furthermore, in *T. niederristi* the posterior margin is pectinated, as in the holotype and in GBM 1866/4/4/2 (see also Abel (1906, p. 29)), while in *T. polzbergensis* sp. nov. the margin is completely smooth. Raibl specimens usually exceed 100 mm in length, while Polzberg individuals are much shorter, not exceeding 70 mm. *T.*

polzbergensis sp. nov. shows a few large teeth on the anterior part of the mandible, while such strong dentition is so far unknown in *T. niederristi*. Another difference concerns the presence of an area covered by small ray segments (the sickle sail = 'sensenförmiges Segel' in Abel (1906) or 'voile falciforme' in Lehman (1978)): it is clearly present in the *T. niederristi* holotype as well as in NHMW 1866-XI-451 from Raibl, but has not so far been detected in the Polzberg specimens (see Lehman (1978, p. 59)).

The flank deep scale row, where the sensory canal may have been lodged, although usually no trace of it is visible, is quite ventral. This is in agreement with a ventral position of the lateral line in modern flying fishes.

Thoracopterus from Raibl/Cave del Predil is very rare in the type locality. Only the holotype (split between London and Heidelberg), a few specimens in the Geological Survey of Austria (now GeoSphere), one at NHMW, one at the University of Vienna, one in Graz, one in Udine (Museo Friulano di Storia Naturale), one in Monfalcone (Gorizia), and one in Padova University Museo della Natura e dell'Uomo are known. Only the Udine and Monfalcone specimens have been recently collected (the latter by AT, who also prepared it). In Polzberg about 50 specimens are known, most of them belonging to the historical collections used by Abel (1906), Griffith (1977) and Lehman (1978). Since the descriptions by Griffith (1977) and Lehman (1978) now clearly refer to *T. polzbergensis* sp. nov., a detailed redescription of the *Thoracopterus* type species, *T. niederristi*, will be undertaken in the near future.

Thoracopterus telleri (Abel, 1906).

1906 *Gigantopterus telleri* Abel, p. 39, Figs. 6–8, pls. I/6, II, III/2.

1977 *Gigantopterus telleri* Griffith, p. 45, Figs. 20–21, pl. 7°, B, C.

1992 *Thoracopterus telleri* Tintori and Sassi, p. 266.

2012 *Thoracopterus telleri* Tintori et al., p. 40.

2013 *Gigantopterus*, Xu et al., Fig. 6.

2021 *Gigantopterus telleri* Lukeneder and Lukeneder, Fig. 4G, Table 1.

2022 *Gigantopterus telleri* Shen and Arratia, Figs. 3, 6B, 7C.

Holotype. The only specimen described and figured by Abel (1906) is to be considered as the holotype, and it is stored at the GeoSphere Austria collection (former k. k. Geologische Reichsanstalt, after that the Geologische Bundsanstalt, the Austrian Geological Survey) as 1906/2/2/1–3 (Fig. 10.6a). That specimen, however, was not used by Griffith (1977) in his redescription of this species.

Referred material. Specimens stored at the NHMuseum Wien used and figured by Griffith (1977, pl. 7 A-B) as NHMW 2007z0170/0205 (Fig. 10.6b) and NHMW 2007z0170/0366 (detail figured in pl. 7C). This latter specimen figured also by (Shen and Arratia 2022, Fig. 6b). Other specimens are under NHMW 2007z0170/0204–0206/0221. Specimen 2007z0170/0206 is the counterpart of NHMW 2007z0170/

Fig. 10.6 *Thoracopterus telleri* (Abel, 1906). **a** the holotype (GBA 1906/2/2/3). **b** specimen NHMW 2007z0170/0205. Fr, frontal; Md, mandible; Na, nasal; Ro, rostral. Scale bars: 20 mm

0366, where the pelvic fins and the v.c. elements are better visible, and NHMW 2007z0170/0205 is the counterpart of NHMW 2007z0170/0217. Possibly NHMW 2007z0170/0172, labeled as *T. niederristi*, is a juvenile of *T. telleri* because it lacks a scale covering.

Remarks. This species is the least known among the genus *Thoracopterus* due to the very poor preservation of the few available specimens. Nonetheless, the known skull anatomy is sufficient to assign it to *Thoracopterus* based on the highly consistent pattern shown by all the species especially in the skull-roof. Griffith (1977) and Lehman (1978) both interpreted the nasals as meeting each other behind the rostral (as also for their *T. niederristi*), but it is clear (specimen NHMW 2007z0170/0205) that the rostral, preserved almost completely in 3D, contacted the frontals separating the nasals. This latter character is synapomorphic for the Peltopleuriformes and its absence whould mean that *T. telleri* as well as *T. niederristi sensu* Griffith (1977) and Lehman (1978) could not be related to Peltopleuriformes. That incorrect attribution, however, was confirmed also by Shen and Arratia (2022), who apparently did not recognize this issue. As already noted in Tintori and Sassi (1992), the cranial anatomy supports the attribution to *Thoracopterus*, whereby the only differences compared to the type-species (*T. niederristi* from Raibl) are related to the scale covering. Shen and Arratia (2022) kept *Gigantopterus* as a valid name, although they did not provide justification beyond relying on the incorrect restoration by Griffith (1977) and Lehman (1978): without examining the actual specimens, they relied instead on photographs (as for all the other Triassic flying fishes). Finally, other characters (see below) are very similar in all the species. Acccordingly, and because Shen and Arratia's (2022) matrix is imprecise (Tintori 2025), their taxonomical statements are not accepted regarding maintaining the validity of *Gigantopterus* or changing the genus name of the other three species of *Thoracopterus*.

The dentition in the lower jaw consists of a row of very stout teeth at least in the anteriormost region (nothing is known for the posterior part and the maxilla), with smaller tooth rows arranged along the lingual side. In other *Thoracopterus* species, such as *T. magnificus* and *T. martinisi*, the teeth are also arranged in at least three rows, with those in the lower jaw being much larger anteriorly and with an external main row (Tintori and Sassi 1992), thus again in a very similar pattern. The parasphenoid, observed in specimen NHMW2007z0170/0217, is very similar to that of *T. wushaensis* (Tintori et al. 2012, Fig. 5), showing a well-developed dentition covering most of the palatal bones.

The estimated standard length of the holotype (possibly the largest specimen) is about 120–130 mm, but most specimens are less than 100 mm, thus well within the size range of other *Thoracopterus* species. The exception is *T. polzbergensis* sp. n., which is much smaller, reaching at maximum 75 mm in s. l.. *T. magnificus*, although most specimens are below 80 mm, can reach 120 mm, similar to *T. wushaensis* (85–120 mm). *T. martinisi* was probably the largest species, reaching 200 mm, with several specimens above 100 mm (Tintori and Sassi 1992; Tintori et al. 2012; pers. obs.).

Shen and Arratia (2022, fig 6B, C) figured specimen NHMW 2007z0170/0366, an incomplete fossil without skull. They misidentified some fins: (their pelvic fin at the bottom of the picture is actually the dorsal lobe of the caudal fin, while the left pelvic is not labeled, positioned directly above the right one). Traces of the dorsal fin and fragments of the vertebral column are present, and most of the specimens appear well articulated. Fig. 6C depicts the detail of the base of the caudal fin with a row of scales (they are covered with ganoine!) that Shen and Arratia interpreted as chordacentra. Note, however, that in their matrix—character 95—*T. telleri* is coded as having no centra!). Importantly, regarding this scale row, in no other specimen of *T. telleri* showing the caudal region are similar scales visible, whereas in *T. wushaensis* the group of scales surrounding the ventral part (not all around the pedicle as in Xu et al. 2013) of the caudal pedicle are known in most specimens (Tintori et al.

2012). On a few specimens, mainly on NHMW 2007z0170/0366, black marks along the pectoral fin rays suggest the presence of some colored spots in live fishes. In front of the dorsal fin in NHMW 2007z0170/0366 there is an elongate (mid-dorsal?) scale with small tubercles, followed by a smaller one and at least two basal fulcra. A black film outlining the body shape is present in NHMW 2007z0170/0205 (mainly in the anterodorsal region), and in NHMW 2007z0170/0218, a specimen preserved in dorso-ventral view, the mark seems to extend up to the pelvic fins insertion. *T. wushaensis* exhibits an antero-dorsal field of scales, whereby *T. telleri* possibly features something similar, although much thinner and without ganoine covering so that the preservation is very poor. Griffith (1977, pl. 7B) figured the inner side of the pectoral fins—'sensenförmiges Segel' in Abel (1906) or '*voile falciforme*' in Lehman (1978)—that is clearly visible in a few specimens; here we figured the same specimen used by Griffith (1977, pl. 7B), NHMW 2007z0170/0205. This same structure has been recorded so far also in *T. wushaensis* (Tintori et al. 2012, Fig. 10) and *T. magnificus* (Tintori and Sassi 1992).

Class Neopterygii Regan, 1923.

Clade Halecomorphi Cope, 1872.

Order ?Panxianichthyiformes Sun et al., 2016.

Remarks. Griffith (1977) ascribed his new genus to the Parasemionotidae (Parasemionotiformes), although he emphasized the confused state of the group. Since then, Grande and Bemis (1998) published a comprehensive study of advanced halecomorphs, but they did not examine the Parasemionotiformes in detail due to several problems related to the very high number of genera erected for the Early Triassic specimens, apparently unsupported by the anatomical characters (Grande and Bemis 1998, p. 597). For *Phaidrosoma*, the preopercle, being not ovoidal, prevents this genus from being ascribed to Parasemionotiformes following Grande and Bemis (1998, p. 578). A number of Middle-Late Triassic genera has been considered to represent Halecomorphi, and Sun et al. (2016) attempted to clarify the position of at least some by establishing the order Panxianichthyformes. This is because such Middle Triassic genera are clearly different from the Late Permian/Early Triassic Parasemionotiformes (typical of the TEFF, see Tintori et al. 2014), as well as from the more advanced Ionoscopiformes-Amiformes (Section B of Halecomorphi *sensu* Grande and Bemis 1998), which are mostly developed starting from the Late Jurassic. Since the skull anatomy of *Phaidrosoma* remains poorly known, its attribution to Panxianichthyformes should be considered as being tentative.

Genus Phaidrosoma Griffith, 1977.

Phaidrosoma lunzensis Griffith, 1977.

Holotype. A complete specimen in part and counterpart (Griffith 1977, pl. 8, 9A), now NHMW 2007z0170/0222a, b (Fig. 10.7a).

Fig. 10.7 *Phaidrosoma lunzensis* Griffith, 1977. **a** the holotype, NHMW 2007z0170/0222a; **b-c** NHMW 2007z0170/0223a, **b** close-up of skull; **c** complete specimen. Ang, angular; D, dental; Na, nasal; Q, quadrate; Qj, quadratojugal, Ro, rostral; Sy, symplectic. Scale bars: **a** 50 mm; **b** 10 mm; **c** 40 mm

Referred material. Specimen NHMW 2007z0170/0223a, b (Fig. 10.7b, c), part and counterpart, mostly complete specimen but with scanty preservation of the skull.

Remarks. The main character missing to definitively assign *Phaidrosoma* to Panxianichthyformes is the absence of a detectable supramaxilla in the two known specimens. Furthermore, the rostral appears to be relatively large (NHMW 2007z0170/0223a) compared to the narrow bone in other genera, although it shows a very thick ornamented anterior margin, similar to that seen in *Panxianichthys,* for instance. In the same specimen the left lower jaw and its articulation are quite well preserved, and a symplectic is apparently present ventrally to the quadrate and quadratojugal. The postero-ventral infraorbital is probably much larger, being broken in the holotype, and the scales are posteriorly pectinated as in most Panxianichthyformes. Thus, although still somewhat doubtful, the suprageneric position of *Phaidrosoma* among the Panxianichthyformes is highly probable.

Clade Teleostei *sensu* Arratia, 1999.

Family Ichthyokentemidae Griffith and Patterson, 1963.

Genus *Elpistoichthys* Griffith, 1977.

Diagnosis (after Griffith 1977). Dermal bones of head moderately to strongly ornamented; nasal contact in mid-line short; opercle only slightly larger than subopercle, with oblique ventral edge; preopercular slightly curved, ventral part considerably expanded, sensory canal closer to anterior than posterior edge, many and long tubules; one large and one very small suborbital; dermosphenotic not more than twice as high as wide; infraorbital 3 considerably enlarged; a single, small supraorbital, frontal contributing to orbital margin; jaw articulation level with rear border of the orbit; maxilla long, extending beyond the orbit, toothless; coronoid process of mandible of moderate size, tapering gradually anteriorly and not sharply marked off from dentigerous portion of mandible. Cleithrum relatively large, with moderate curvature and distinct posterior notch. Scales rhomboid, at most only slightly deeper than long, external surface variously ornamented, rear edge pectinated or smooth.

Remarks. Arratia (2013, p. 2) in her revision of Triassic pholidophorids, still considered *Elpistoichthys* in the Ichthyokentemidae although she wrote that "its taxonomic position remains uncertain".

'Elpistoichthys' pectinatus Griffith, 1977.

Holotype. The specimen now NHMW 2007z0170/0225a, b (Fig. 10.8a), figured also by Griffith (1977, pl. 9B).

Paratypes. Specimens NHMW 2007z0170/0226/7/8, the one figured by Griffith (1977, pl. 10A) is now NHMW2007z0170/0228.

Other material. Specimens NHMW 2007z0170/0229–0274 are now registered, some of them in part and counterpart. The specimen figured by Griffith (1977, pl. 10B) as being associated with supposed ?*Habroichthys* has not been identified, but it is clear that the two specimens near the body of the fish are thylacocephalans, probably *Microcaris striata* (Glaessner, 1931).

Diagnosis. (after Griffith 1977). A species of *Elpistoichthys* reaching 155 mm in s. l.. Dermal bones of skull strongly ornamented with irregular, short ridges of ganoine. Fin-ray numbers: pectoral 16–17; pelvic 12; dorsal 15; anal 14; caudal 21/22–23. External surface of scales with large central depression, posterior edges pectinated (Fig. 10.8b, c).

'Elpistoichthys' striolatus Griffith, 1977.

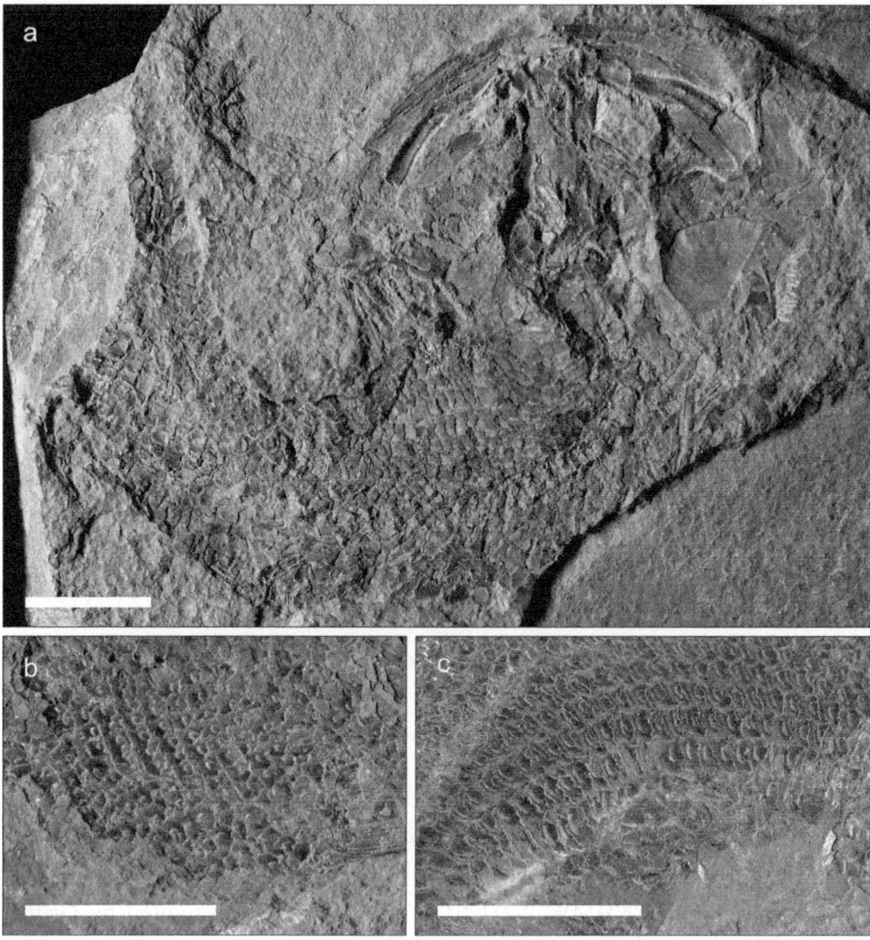

Fig. 10.8 *Elpistoichthys pectinatus* Griffith, 1977. **a** the holotype, NHMW 2007z0170/0225a; **b** close-up of scales in NHMW 2007z0170/0229; **c** close-up of scales in paratype NHMW 2007z0170/0228. Scale bars: 10 mm

Remarks. Griffith (1977) apparently erected this species solely based on differences in the scales compared to the type species *E. pectinatus*, a somewhat smaller size and minor differences in a few skull dermal bones. However, as already pointed out, the general preservation of skulls is quite poor, making detailed description of several bones nearly impossible. In our opinion, the difference in the scale 'ornamentation' is quite important, as discussed below, so that the attribution to the genus *Elpistoichthys* may be questioned, although this would imply a complete revision of all the material, which is beyond the goal of this paper. We will tentatively use the generic name in quotation marks.

When erecting this species, Griffith (1977, p. 70) stated that the holotype was "an individual in counterpart. It is poorly preserved but does show some details of the head region". Unfortunately, no picture of the 'holotype' was published and no further indications were provided; the only picture for this species is a close-up of a few scales from an unknown specimen (Griffith 1977, pl 11B). Catalogue numbers were given in 2007, well after Griffith's publication, based solely on a few labels left by Griffith himself together with some of the specimens (AT pers. obs. 1988). During the recent examination of the material in 2025, one of us (AT) realized that the supposed holotype actually consists of two different individuals, rather than a single specimen in part and counterpart as stated by Griffith (1977). Specimen NHMW 2007z0170/0275a is an almost complete specimen with the head split open and seen in dorso-ventral view, although no bones are preserved and only their impressions are visible. The body is somewhat twisted, and the vertebral column can be traced beneath the scales up to the caudal region, where it appears near one of the edges of the specimen. No caudal fin is visible. Specimen NHMW 2007z0170/0275b shows very little of the skull, and traces of the vertebral column are along the middle part of the body, not along the edge. Furthermore, there are remains of a small crustacean near the tail, but nothing similar is visible on NHMW 2007z0170/0275a, the supposed counterpart. We must therefore choose a lectotype between the two specimens bearing the same catalogue number (2007z0170/0275a, b). Since the number is not inscribed on the specimens themselves, we must rely on the labels in their boxes, as the specimens have not been figured. Following what Griffith (1977) wrote about 'showing some details of the head region' we consider as the lectotype the specimen that shows the better preserved skull region, although the other, considered as counterpart at that time, has a much better preserved caudal region.

Lectotype. An almost complete specimen, NHMW 2007z0170/0275a (Fig. 10.9a).

Syntype. An almost complete specimen, NHMW 2007z0170/0275b (Fig. 10.9b).

Other material. A few specimens from recent collecting in Polzberg have been added to the original material of Griffith, although, as usual, the preservation is generally poor.

Geographic and stratigraphic distribution. The type locality is Polzberg near Lunz am See (Austria), lower Carnian Reingraben Formation. Its presence in the Lower Carnian of the area around Raibl/Cave del Predil (Italy) remains to be proven.

Diagnosis. (after Griffith 1977). A species of *Elpistoichthys* reaching 70 mm in standard length. Fin-ray count: pectoral 18; pelvic 9+; dorsal 12+; anal 9+; caudal 16–17/16–17. External surfaces of scales with fine striations anteriorly but otherwise smooth; scales noticeably thinner than those of *E. pectinatus*; bone-cell spaces either absent or very infrequent.

Remarks. As noted above, a major difference between the two *Elpistoichthys* species erected by Griffith (1977) is related to the scales, in addition to the body size, which is approximately half in '*E.*' *striolatus*. The thin vertical ridges on the anterior part of the scales (Fig. 10.9c) are a quite rare feature and have so far been considered peculiar to

Fig. 10.9 *'Elpistoichthys' striolatus* Griffith, 1977. **a** the lectotype NHMW 2007z0170/0275a, former part of the 'holotype' in Griffith (1977); **b** the syntype NHMW 2007z0170/0275b (former counterpart of the 'holotype' in Griffith (1977); **c** specimen NHMW 2007z0170/0292, close-up of the scales in the abdominal region, external surface in the square on the left, imprint of the external surface in the square on the right. Scale bars: 10 mm

the genus *Placopleurus* (Brough 1939, Conedera et al. accepted). As reported above, the similar structures in *T. polzbergensis* sp. n. are somewhat different in being a kind of terraces rather than true ridges. *Placopleurus* is so far considered to be a Middle Triassic genus because the youngest species is from the late Ladinian Pelsa/Vazzoler Lagerstätte (Conedera et al. accepted) and it is not known in any of the different fish assemblages of Raibl (latest Ladinian-early Carnian, Tintori and Castelnuovo 2023).

A thorough revision of the *Elpistoichthys* material is needed in order to clarify the true generic status of '*E.' striolatus;* however, we believe it is likely that a new genus will need to be erected for this species.

Order Pholidophoriformes, *sensu* Arratia, 2013.

Family Pholidophoridae *sensu* Arratia, 2013.

Genus *Pholidophoretes* Griffith, 1977.

Diagnosis. (after Arratia 2013). Autapomorphies are identified with an asterisk [*]. Pholidophorids of small to medium size, with small head, about 20–21% of standard length*. Ganoine ornamentation on dermal bones almost nonexistent. Nasals well developed, with extensive median contact. Opercular roughly triangular, with strongly inclined lower edge. Preopercle with sensory canal closer to posterior than to anterior margin. A single suborbital; five infraorbitals; infraorbital 3 considerably enlarged and extending posteriorly, ventral to the suborbital. Maxilla of moderate size, not markedly deep; posterior margin characteristically notched. Two supramaxillae; supramaxilla 1 very small; supramaxilla 2 large, slightly more than 50% maxillary length. Dorsal fin positioned opposite to anal fin*; fringing fulcra present along anterior edges of all fins. Ganoid scales moderately thick; rhomboid with smooth posterior margins. Anterior scales of lateral line series twice as deep as broad.

Pholidophoretes salvus Griffith, 1977.

1977 *Pholidophoretes salvus* Griffith, p. 72, Figs. 27–30, Pl. 11C.

2013 *Pholidophoretes salvus*, Arratia, p. 107, Fig. 90–94.

Holotype. NHMW 2007z170/0293a, b.

Paratype. NHMW 2007z0170/0094.

Diagnosis. Same as for genus.

Remarks. As the species has been recently revised by Arratia (2013), nothing more needs to be added.

10.4 Conclusions

The first consideration is that the Polzberg fish assemblage preserved in the deposits of the Reingraben Formation is now well dated (Lukeneder and Lukeneder 2021) and it is more recent than the Raibl Limestone (De Marco et al. 2000) from where the Raibl fossil fishes stem. Accordingly, all three different Raibl fish assemblages (Tintori and Castelnuovo 2023) should be considered as pre-CPE. Ever since Abel (1906), the Polzberg ichthyofauna was considered to be very similar to that of the Raibl/Cave del Predil. Griffith (1977), however, in the only comprehensive description of the actinopterygian taxa, recognized only a single species in common, namely *T. niederristi*. This view has now been proven to be incorrect because the scaled flying fish from Polzberg differs from the type species of *Thoracopterus* from Raibl. The two assemblages differ not only at species level. At the genus level, there are only a few taxa, such as *Peltopleurus* and *Thoracopterus*, in common. They also show a somewhat different quantitative composition, at least for neopterygians, although it should be noted that most of the collecting at both sites took place over a century ago, without the modern bed-by-bed sampling. Only very recently has such work been carried out in Polzberg and in Raibl, but so far on very limited surfaces or stratigraphical intervals. These include the so-called ‘Rossi level’ in Raibl, representing the lowermost assemblage, probably around the Ladinian/Carnian boundary or late Ladinian (Tintori and Castelnuovo 2023). Furthermore, in Polzberg the fishes occur within a short, about 3-m-thick interval at the base of the Reingraben Formation (Lukeneder and Lukeneder 2021). The Raibl ichthyofauna is widespread over the whole Predil Limestone, over 200 m thick (De Marco et al. 2000; Tintori and Castelnuovo 2023), currently hindering a detailed comparison pending a review of all the available Raibl fish material. Importantly, the ‘subholosteans’ *Auctorum* (Peltopleuriformes, Perleidiformes, Pholidopleuriformes) are by far the most common in both sites. They are dominated by Peltopleuriformes, with *Peltopleurus* being very common in both sites (in Raibl in all three assemblages, AT pers. obs.) and by *Thoracopterus*, the latter especially in Polzberg, whereas *Nannolepis* has yet to be recorded in Raibl. *Pholidopleurus* is very common in what we now consider to be the ‘original’ assemblage of Raibl, but is absent in Polzberg (*Pholidopleurus* is present also in the Norian Zorzino Limestone Fauna from Lombardy following Tintori and Lombardo (2017), suggesting that its absence in Polzberg may be related to paleoenvironmental conditions). Regarding neopterygians, they are poorly known from Raibl, although pholidophorids, halecomorphs, *Marcopoloichthys* and *Eosemionotus* are recorded (Arratia 2013; Tintori and Castelnuovo 2023). In contrast, in Polzberg especially pholidophorids such as *Pholidophoretes* are quite common (Griffith 1977), but halecomorphs are very rare. *Elpistoichthys* is common in Polzberg but probably rare in Raibl (AT pers. obs.). *Marcopoloichthys* is present in Raibl but absent in Polzberg, although this genus is widespread from Anisian to Norian in several Alpine and Chinese fish assemblages (Tintori et al. 2007; Arratia 2022; Tintori and Castelnuovo 2023; pers. obs. by the authors). Quite intriguing is the almost complete absence of ginglymodians in both Carnian ichthyofaunas (apart from *Eosemionotus*, a typical Middle Triassic genus, from the ‘Rossi Level’), whereas they are quite common in the Middle Triassic (Tintori et al. 2007; López-Arbarello et al. 2011, 2019) and Norian (Tintori 1983, 1998; Tintori et al. 2014).

Overall, only a few differences are evident between pre-CPE (Raibl assemblages) and CPE (Polzberg) fish faunas. This is probably due mostly to the different paleoenvironments, rather than to a true bioevent related to a strong climate change, as no substantially new taxa appear among the Polzberg fish. Regarding actinopterygians, it is now evident that the late Ladinian (Middle Triassic) was a major turning point for fishes, as well as for marine reptiles and benthic invertebrates, as shown especially by the Pelsa/Vazzoler Lagerstätte (Belluno Dolomites, Italy), the Kalkschieferzone (Monte San Giorgio, Italy/Switzerland), and the Xingyi Fauna (Guizhou, China) fish assemblages (Tintori et al. 2012, 2014, 2015; Tintori 2018; Lu et al. 2018; Dominici et al. 2024; Conedera et al. 2025, accepted).

Acknowledgments All the photographic material regarding types and referred specimens has been provided by A. Lukeneder and we acknowledge this time-consuming work owing also to the somewhat problematic preservation of the specimens. E. Bernard and A. McArdle provided detailed photographs of *T. niederristi* holotype (counterpart) stored in NHMUK, London. D. Di Franco provided images of *T. niederristi* from Raibl stored at Geosphere (formerly Austrian Geological Survey), M. Maslo of the specimen in the Institute of Paleontology of the University of Vienna and L. Giusberti of the specimen in the Padua Museum collection. The authors are particularly grateful to the owners of the outcrop area, Franziska and Hermann Hofreiter as well as Eva and Karl Jagersberger (all Gaming), for sampling and digging permission. The work was done within the framework of and financially supported by projects of the Federal Government of Lower Austria (K3-F-964/001-2020 and K3-F-964/002-2023, Department Science and Research; headquarters in St. Pölten); the Austrian Academy of Sciences (headquarters in Vienna, project Polzberg Lukeneder 2021) represented by the National Committee for Geo/Hydro Sciences (Earth System Sciences Programme); and the Triassic project "Fenster in die Karnische Krise" of the Society of Friends of the Natural History Museum Vienna (Freunde des Naturhistorischen Museums Wien, CPE Lukeneder 2023). The manuscript greatly benefited from valuable comments by the reviewers Tomaž Hitij (Ljubljana) and Cristina Lombardo (Milan), the production editor Ashok Arumairaj (Taramani), the in-house series editor Kathleen A. Ritterbush (Utah), the publishing editor Simon Shah-Rohlfs and the production editor Antje Endemann (both Heidelberg).

Author Contribution AT, DC designed research, provided documentation on fossil material. AL documented the fossil material.

Funding Open access funding was enabled and the projects financially supported by the Federal Government of Lower Austria (K3-F-964/001–2020 and K3-F-964/002–2023, Department Science and Research; headquarters in St. Pölten); the Austrian Academy of Sciences (headquarters in Vienna, project Polzberg Lukeneder 2021) represented by the National Committee for Geo/Hydro Sciences (Earth System Sciences Programme); the Natural History Museum Vienna; and the Triassic project "Fenster in die Karnische Krise" of the Society of Friends of the Natural History Museum Vienna (Freunde des Naturhistorischen Museums Wien, CPE Lukeneder 2023). The authors are responsible for the contents of this publication. The funders had no impact on conceptualization, design, data collection, analysis, decision to publish, or preparation of the manuscript.

Competing interests All authors stated that they have no competing interests.

Authors' contributions All authors stated that they have equally contributed to their chapters, read, edited and agreed to submit the final versions of their contributions.

Availability of data and materials All sources of information are provided in the text. Raw data related to the fossil material from Austria are available from the corresponding authors and Alexander Lukeneder upon request.

Additional information The article is part of the Special Volume: *Late Triassic Konservat-Lagerstätten within the Carnian Pluvial Episode. 150 Years of Scientific Excavations in Austria.*

Artificial Intelligence (AI) The chapter was produced without the use of AI.

Language The entire volume uses American English reviewed by a native scientist.

Received 2025.

Published online 2026.

References

Abel O (1906) Fossile Flugfische. Jahrbuch der Kaiserlich Königlichen Geologischen Reichsanstalt. Im selbstverlage des Autors, Wien 56:1–88

Agassiz L (1833) Recherches sur les poissons fossiles. Imprimerie de Petitpierre

Aldinger H (1937) Permische ganoidfische aus Ostgrönland. Meddelelser om Grønland, Reitzels Forlag

Arratia G (1999) The monophyly of Teleostei and stem-group teleosts. Mesoz Fishes 2:265–334

Arratia G (2013) Morphology, taxonomy, and phylogeny of Triassic pholidophorid fishes (Actinopterygii, Teleostei). JVP 33:1–138. https://doi.org/10.1080/02724634.2013.835642

Arratia G (2022) The outstanding suction-feeder Marcopoloichthys furreri new species (Actinopterygii) from the middle Triassic Tethys realm of Europe and its implications for early evolution of neopterygian fishes. Foss Rec 25:231–261

Beltan L, Tintori A (1981) The genus Saurichthys (Pisces, Actinopterigii) during the Gondwana period. In: Proceedings of the Fifth International Gondwana Symposium, Rotterdam, pp 53–59

Berg LS (1940) Classification of fishes both recent and fossil. Trav Inst Acad Sc URSS 52:87–517

Boeddu I (2007) Il genere Peltopleurus Kner, 1866 (Osteichthyes, Actinopterygii) nel Ladinico superiore (Triassico Medio) di Meride (TI, Svizzera). In: Università degli Studi di Milano, Master thesis

Bronn HG (1858) Beiträge zur triasischen Fauna und Flora der bituminösen Schiefer von Raibl, nebst Anhang über die Kurr'sche Sippe Chiropteris aus dem Lettenkohlen-Sandsteine: Mit 10 Tafeln Abbildungen. Schweizerbart, Stuttgart, pp 129–142

Brough J (1939) The Triassic fishes of Besano, Lombardy. Bull Br Mus (Nat Hist) 117

Bürgin T (1992) Basal ray-finned fishes (Osteichthyes; Actinopterygii) from the middle triassic of Monte san Giorgio (Canton Tessin, Switzerland): systematic palaeontology with notes on functional morphology and palaeoecology. Abh Schweiz Palaeontol Ges 114

Conedera D, Hitij T, Monari S et al (2025) The miniature fish *Habroichthys* (Actinopterygii; Peltopleuriformes): seven new species from the middle Triassic of Italy and Slovenia reveal insights on its palaeobiology, palaeobiogeography, and palaeoecology. Geobios. https://doi.org/10.1016/j.geobios.2025.06.001

Conedera D, Hitij T, Lombardo C, Andreetti M, Gatto R, and Tintori A (accepted) New species of the miniature fish Placopleurus (Neopterygii) from the Middle Triassic of Italy and Slovenia reveal the oldest near-homocercal tail. CJZ

Cope ED (1872) Observations on the systematic relations of the fishes. Am Nat 5:579–593. https://doi.org/10.1086/270831

Cope ED (1887) Zittel's manual of palaeontology. Am Nat 21:1014–1019

Dal Corso J, Bernardi M, Sun Y et al (2020) Extinction and dawn of the modern world in the Carnian (late Triassic). Sci Adv 6:eaba0099. https://doi.org/10.1126/sciadv.aba0099

De Marco S, Fantoni R, Ponton M, Scotti P (2000) Cave del Predil: la successione di Raibl. In: 80a Riunione Estiva della Società Geologica Italiana, Guida alle escursioni, Trieste, pp 102–115

Dominici S, Danise S, Tintori A (2024) A middle Triassic Cassian-type fauna (Pelsa-Vazzoler Lagerstätte) and the adaptive radiation of the modern evolutionary fauna. Pap Palaeontol 10:e1579. https://doi.org/10.1002/spp2.1579

Garbelli C, Tintori A (2015) A preliminary study on the ornamentation patterns of ganoid scales in some Mesozoic actinopterygian fishes. Boll Soc Paleontol Ital 54:219–228

Gardiner BG (1967) Further notes on palaeoniscoid fishes with a classification of the Chondrostei. Bull Br Mus (Nat Hist), Geol 14:146–206

Glaessner MF (1931) Eine crustaceenfauna aus den lunzer schichten Niederösterreichs. Jb Geol B-A 81:467–486

Grande L, Bemis WE (1998) A comprehensive phylogenetic study of Amiid fishes (Amiidae) based on comparative skeletal anatomy. An empirical search for interconnected patterns of natural history. JVP 18:1–696. https://doi.org/10.1080/02724634.1998.10011114

Griffith J (1977) The upper Triassic fishes from Polzberg bei Lunz. Austria Zool J Linn Soc 60:1–93. https://doi.org/10.1111/j.1096-3642.1977.tb00834.x

Griffith J, Patterson C (1963) The structure and relationships of the Jurassic fish Ichthyokentema purbeckensis. Bull Br Mus (Nat Hist)

Kner R (1866) Die Fische der bituminösen Schiefer von Raibl in Kärnthen. Sitzungsber Akad Wiss Wien 53:152–197

Kogan I, Tintori A, Licht M (2020) Locomotor function of scales and axial skeleton in middle-late Triassic species of Saurichthys (Actinopterygii). Riv Ital Paleontol Stratigr 126:475–498

Lehman J-P (1978) Note sur les poissons du Trias de Lunz. In: I. Thoracopterus Bronn. Ann. Naturhist. Mus. Wien, pp 53–66

Lin H, Sun Z, Tintori A et al (2011) A new species of Habroichthys Brough, 1939 (Actinopterygii; Peltopleuriformes) from the Pelsonian (Anisian, middle Triassic) of Yunnan Province, South China. N Jb Geol Paläontol Abh 262:79–89. https://doi.org/10.1127/0077-7749/2011/0186

Lombardo C (1999) Sexual dimorphism in a new species of the actinopterygian Peltopleurus from the Triassic of northern Italy. Palaeontology 42:741–760. https://doi.org/10.1111/1475-4983.00095

Lombardo C, Rusconi M, Tintori A (2008) New perleidiform from the lower Ladinian (middle Triassic) of the northern Grigna (northern Italy). Riv Ital Paleontol Stratigr 114:263–272

Lombardo C, Tintori A (2004) New perleidiforms from the Triassic of the southern Alps and the revision of Serrolepis from the Triassic of Württemberg (Germany). Meso Fishes 3:179–196

López-Arbarello A, Buergin T, Furrer H, Stockar R (2019) Taxonomy and phylogeny of Eosemionotus Stolley, 1920 (Neopterygii: Ginglymodi) from the middle Triassic of Europe. Palaeontol Electron:1–64. https://doi.org/10.26879/904

López-Arbarello A, Sun Z-Y, Sferco ME et al (2011) New species of Sangiorgioichthys Tintori and Lombardo, 2007 (Neopterygii, Semionotiformes) from the Anisian of Luoping (Yunnan Province, South China). Zootaxa 2749:25–39. https://doi.org/10.11646/zootaxa.2749.1.2

Lu H, Jiang D-Y, Motani R et al (2018) Middle Triassic Xingyi Fauna: showing turnover of marine reptiles from coastal to oceanic environments. Palaeoworld 27:107–116. https://doi.org/10.1016/j.palwor.2017.05.005

Lukeneder A (ed) (2026) Late Triassic Konservat-Lagerstätten within the Carnian Pluvial Episode: 150 Years of Scientific Excavations in Austria. Top Geobiol 55:600 p. https://doi.org/10.1007/978-3-032-19777-1

Lukeneder A, Lukeneder P (2021) The upper Triassic Polzberg palaeobiota from a marine *Konservat-Lagerstätte* deposited during the Carnian pluvial episode in Austria. Sci Rep 11:16644. https://doi.org/10.1038/s41598-021-96052-w

Owen R (1860) Palaeontology, or a systematic summary of extinct animals and their geological relations. Adam and Charles Black

Regan CT (1923) The skeleton of *Lepidosteus*, with remarks on the origin and evolution of the lower Neopterygian fishes. Proc Zool Soc Lond 93:445–461. https://doi.org/10.1111/j.1096-3642.1923.tb02191.x

Roden VJ, Hausmann IM, Nützel A et al (2020) Fossil liberation: a model to explain high biodiversity in the Triassic Cassian formation. Palaeontology 63:85–102. https://doi.org/10.1111/pala.12441

Santoro E (2000) Il genere Peltopleurus Kner, 1886 (Osteichthyes, Actinopterygii) del Triassico Medio Superiore. In: Università degli Studi di Milano, Master thesis

Shen C, Arratia G (2022) Re-description of the sexually dimorphic peltopleuriform fish Wushaichthys exquisitus (middle Triassic, China): taxonomic implications and phylogenetic relationships. J Syst Palaeontol 19:1317–1342

Stensiö E (1925) Triassic fishes from Spitzbergen, part 2. K Sven Vetensk Akad Handl 2:1–261

Stur D (1874) Neue Aufschlüsse im Lunzer Sandsteine bei Lunz und ein neuer Fundort von Wengerschiefer im Pölzberg zwischen Lunzersee und Gaming. Verh K-K Geol Reichsanst 1:271–273

Sun Z, Tintori A, Yaozhong X et al (2016) A new non-parasemionotiform order of the Halecomorphi (Neopterygii, Actinopterygii) from the middle Triassic of Tethys. J Syst Palaeontol 15:223–240. https://doi.org/10.1080/14772019.2016.1181679

Teller F (1891) Über den Schädel eines fossilen Dipnoers, Ceratodus Sturii nov. spec. aus den Schichten der oberen Trias der Nordalpen. Abh K-K Geol Reichsanst 15:1–38

Tintori A (1983) Hypsisomatic semionotidae (Pisces, Actinopterygii) from the upper Triassic of Lombardy (N. Italy). Riv Ital Paleontol Stratigr 88:417–442

Tintori A (1998) Fish biodiversity in the marine Norian (late Triassic) of northern Italy: the first neopterygian radiation. Ital J Zool 65:193–198. https://doi.org/10.1080/11250009809386812

Tintori A (2013) A new species of Saurichthys (Actinopterygii) from the middle Triassic (early Ladinian) of the northern Grigna Mountain (Lombardy, Italy). Riv Ital Paleontol Stratigr 119:287–302

Tintori A (2018) Il Ladinico superiore della Laguna Vazzoler-Pelsa (Civetta, Dolomiti Agordine). Frammenti 8:43–47

Tintori A, Castelnuovo N (2023) The new fish fauna from "Rossi-level" in cave del Predil/Raibl. Geol Paleontol Friuli Adjac Areas:61–64

Tintori A, Hitij T, Jiang D et al (2014) Triassic actinopterygian fishes: the recovery after the end-Permian crisis. Integr Zool 9:394–411. https://doi.org/10.1111/1749-4877.12077

Tintori A, Lombardo C (2017) The Zorzino limestone Actinopterygian Fauna from the late Triassic (Norian) of the southern Alps. In: Tanner LH (ed) The late Triassic world: earth in a time of transition. Springer International Publishing, Cham, pp 315–350

Tintori A, Lombardo C, Kustatscher E (2016) The Pelsonian (Anisian, middle Triassic) fish assemblage from Monte Prà della Vacca/Kühwiesenkopf (Braies Dolomites, Italy). N Jb Geol Paläontol Abh 282:181–200. https://doi.org/10.1127/njgpa/2016/0612

Tintori A, Sassi D (1987) Pesci volanti del genere Thoracopterus nel Norico Lombardo. Nota preliminare Riv Ital Paleontol Stratigr 93:337–346

Tintori A, Sassi D (1992) *Thoracopterus* bronn (Osteichthyes: Actinopterygii): a gliding fish from the upper Triassic of Europe. JVP 12:265–283. https://doi.org/10.1080/02724634.1992.10011459

Tintori A, Sun Z, Ni P et al (2015) Oldest stem Teleostei from the late Ladinian (middle Triassic) of southern China. Riv Ital Di Paleontol E Strat 121:285–296

Tintori A, Sun Z-Y, Lombardo C et al (2007) New specialized basal neopterygians (Actinopterygii) from Triassic of the Tethys realm. Geol Insubr 10:13–20

Tintori A, Sun Z-Y, Lombardo C et al (2012) A new "flying" fish from the upper Ladinian (middle Triassic) of Wusha /Guizhou Province, southern China. Gortania 33:39–50

Tintori A (2025) A Flying Fish Across the Tethys: Thoracopterus wushaensis Tintori et al., 2012, from the Pelsa/Vazzoler Fossil-Lagerstätte (Late Ladinian, Middle Triassic) in the Dolomites (Italy). Diversity 2025, 17, 847. https://doi.org/10.3390/d17120847

Wen W, Zhang Q-Y, Hu S-X et al (2012) A new basal actinopterygian fish from the Anisian (middle Triassic) of Luoping, Yunnan Province, Southwest China. Acta Palaeontol Pol 57:149–160

Xu G, Wu F (2012) A deep-bodied ginglymodian fish from the middle Triassic of eastern Yunnan Province, China, and the phylogeny of lower neopterygians. Chin Sci Bull 57:111–118. https://doi.org/10.1007/s11434-011-4719-1

Xu G-H, Zhao L-J, Gao K-Q, Wu F-X (2013) A new stem-neopterygian fish from the middle Triassic of China shows the earliest over-water gliding strategy of the vertebrates. Proc R Soc B 280:1–7. https://doi.org/10.1098/rspb.2012.2261

Chapter 11
Scanning and Computed Tomography of Upper Triassic *Tellerodus* and *Coelacanthus* from the Polzberg *Konservat-Lagerstätte*

Alexander Lukeneder, Viola Winkler, and Michael Klein

Abstract This study portrays non-destructive methods of 3D visualization applied on paleontological marine vertebrates, starting with 3D scanning and computed tomography yielding high-quality 3D animations, reconstructions, and video clips. We present the first such visualization of the lungfish *Tellerodus sturii* and the coelacanth *Coelacanthus lunzensis*, both from the Upper Triassic Polzberg *Konservat-Lagerstätte* in the Northern Calcareous Alps (Lower Austria). 3D scanning, computed microtomography, paleontological data, and 3D reconstructions were included in the production of animated presentations on Sketchfab and 3D life reconstructions of the extinct organisms. These approaches provide state-of-the-art knowledge on the shape and mode of life of these enigmatic taxa as well as their habitats. The results are based on the historical holotypes, a complete skull of *Tellerodus*, and an almost entire skeleton of *Coelacanthus*. Object-based combined analyses from computed tomography and various computed 3D facility programs improve our understanding of morphological details. An additional goal was to share 3D models with scientists from various scientific fields from around the world and to connect this community with paleontological methods and data. The use of a 3D viewer (implemented on Sketchfab) opens new perspectives and enables discussion of fossil taxa without being present at the institution storing the original material. In the present case, the models and tools can serve as starting points for discussions on the morphology and

Supplementary Information The online version contains supplementary material available at https://doi.org/10.1007/978-3-032-19777-1_11.

A. Lukeneder (✉)
Department of Geology and Paleontology, Natural History Museum Vienna, Vienna, Austria
e-mail: alexander.lukeneder@nhm.at

V. Winkler
Department of Central Research Laboratories, Natural History Museum Vienna, Vienna, Austria
e-mail: viola.winkler@nhm.at

M. Klein
7Reasons Medien GmbH, Vienna, Austria
e-mail: mk@7reasons.net

A. Lukeneder (ed.), *Late Triassic Konservat-Lagerstätten within the Carnian Pluvial Episode*, Topics in Geobiology 55, https://doi.org/10.1007/978-3-032-19777-1_11

taxonomy of extinct marine inhabitants of the Reifling Basin and within the lower Carnian Polzberg paleobiota. We demonstrate the wide range of potential applications and analytical techniques as well as outline possible limitations of 3D models in paleontology and earth sciences. Data, image details, and reconstructions can be discussed online, and the morphological details and motion-style can be adapted as necessary to better represent the Upper Triassic marine paleobiota from Austria.

Keywords 3D scanning · Computed tomography · *Tellerodus* · *Coelacanthus* · 3D viewer · Late Triassic

11.1 Introduction

The creation of 3D models of fossil surfaces (e.g., of vertebrates) helps to reconstruct, visualize and interpret finds and may be even used as basis for constructing mechanical models. Such models of fossil specimens have become increasingly popular, improving the reconstructions of morphologies, volumes, and sizes of fossils in a sample, as well as more accurately depicting biostratinomic and diagenetic processes. The fossil assemblages from Polzberg reflect a marine environment in the lower Carnian, one with intermittent shallow-water influence as indicated by the findings of freshwater inhabitants such as the lungfish *Tellerodus* (Stur 1886; Teller 1891; Lehman 1975; Martin 1982; Lukeneder and Lukeneder 2021), which is of great interest for the scientific community.

The main faunal elements of the Polzberg *Konservat-Lagerstätte* are benthic bivalves, ceratitid ammonoids, anaptychi and upper jaws of ammonoids, coleoids, gastropods, arthropods, thylacocephala, euestheriids, polychaetes, conodontophorids, actinopterygiid fishes, chondrichtyes, lungfish, coelacanths, bromalites, and plants (for detailed summary and description see Lukeneder and Lukeneder 2021, 2022a, b, 2023, 2024; Lukeneder et al. 2024a, b; Lukeneder 2026, this volume). No macrofaunal elements from this site have previously been 3D scanned or analyzed by microcomputed tomography (microCT), except upper Triassic bromalites (Lukeneder et al. 2020), cephalopod cartilage (microCT; Lukeneder and Lukeneder 2022a, b; Lukeneder et al. 2026, this volume), conodontophorids (Lukeneder et al. 2026), and recently hybodontid chondrichthyes (Stumpf et al. 2026, this volume).

Laser scanning (airborne, terrestrial, or desktop), point cloud data generated from digital images, or computed tomography (CT) are increasingly employed for geoscientific research. They offer spatial resolutions down to microns. While CT (including macroCT, microCT, and nanoCT) is a common method, it faces challenges with poorly contrasting material densities, e.g. between sediment and the actual specimen. To address this, several complementary techniques have emerged, providing 3D datasets on paleontological specimens through both surface and volume scanning methods. Numerous papers (Marschallinger 2001; Mees et al. 2003; Sutton 2008; Maloof et al. 2010; Mayrhofer and Lukeneder 2010; Kruta et al. 2011; Marschallinger

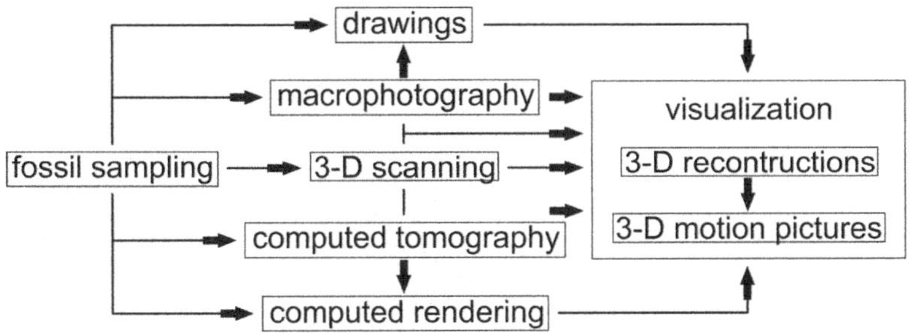

Fig. 11.1 Main processing steps from sampling of fossil material, over drawings and scanning to visualization (3D: reconstruction, animation, motion pictures)

et al. 2011; Lukeneder 2012; Saupe et al. 2012) show the multitude of applications of 3D geometrical models in geological and paleontological studies for better visualization. The great variability, the wide range of applications, and the analytical techniques in the fossil record have been demonstrated for dinosaurs (Rayfield et al. 2001, 2007; Balanoff et al. 2008; Witmer and Ridgely 2009; Fortuny et al. 2011; Tsuihiji et al. 2011; Knoll et al. 2012; Das et al. 2017), lizards (Polcyn et al. 2002), birds (Rowe et al. 2001; Degrange et al. 2010; Zelenitsky et al. 2011), fish (Gai et al. 2011; Argyriou et al. 2018), mammals (Luo et al. 2002), molluscs (Hoffmann and Zachow 2011; Lukeneder et al. 2014; Cherns et al. 2021 Lukeneder and Lukeneder 2022a, b; Lukeneder et al. 2024a, b), brachiopods (Gaspard et al. 2011), bromalites (Lukeneder et al. 2020), plants (Friis et al. 2007; Scott et al. 2009), algal or acritarch fossils (Cunningham et al. 2012), and protists (Donoghue et al. 2006; Dong et al. 2010; Huldtgren et al. 2011). The latter papers also outline possible limitations of 3D models in earth sciences and in paleontology.

The main goal of this paper is to present methods and possibilities for visualizing paleontological material based on 3D scanning, microCT, and paleontological features (e.g., size, morphology; Fig. 11.1). We utilized the 3D laboratory infrastructure at the Natural History Museum in Vienna to show these possibilities with the fossil species *Tellerodus sturii* and *Coelacanthus lunzensis*. High-resolution reconstructions of these fishes demonstrate the increasing importance of 3D visualization techniques in paleontology.

11.2 Material and Methods

The shape data derive from historically described species, namely *Tellerodus sturii* (Stur 1886; Teller 1891; Lehman 1975; Martin 1982; Schultz 2013; Lukeneder and Lukeneder 2021; Fig. 11.2) and *Coelacanthus lunzensis* (Reis 1900; Schultz 2013; Lukeneder and Lukeneder 2021). The fossil lungfish *Tellerodus sturii* (holotype, GBA 1891/001/0001 1-10) and the coelacanth fish *Coelacanthus lunzensis* (holotype, GBA 1891/001/0002) originate from the basal Reingraben Formation at the Polzberg

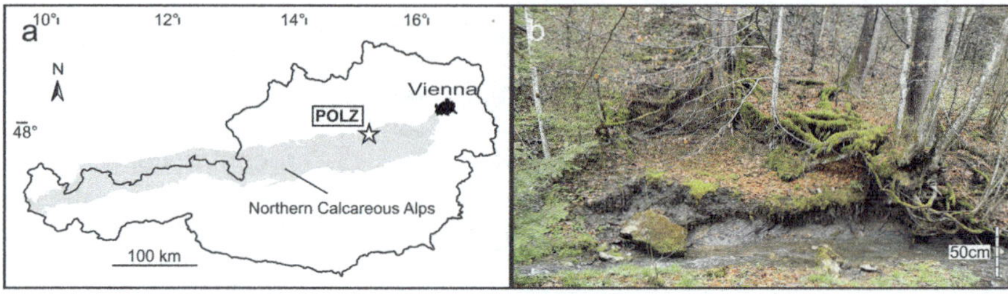

Fig. 11.2 **a** Locality map of the Polzberg area (POLZ, Lunz am See) in Lower Austria with marked position of the Austrian Northern Calcareous Alps (gray). **b** Detailed view of the historical, collapsed tunnel within basal boundary layers of the argillaceous Reingraben Formation

Konservat-Lagerstätte (Northern Calcareous Alps, Lower Austria; Fig. 11.2). The material was collected during an historical excavation campaign organized by the GeoSphere Austria (formerly Geological Survey of Austria, GBA) in 1885 and the Natural History Museum Vienna (NHMW) in 1909 (Stur 1886; Teller 1891). Both historical findings derive from the lowermost three calcareous meters of the Reingraben Formation, collected from excavated tunnels that have since collapsed.

Non-invasive techniques (macrophotography, microcomputed tomogrophy, 3D scanning) enable processing great volumes of information on fossils without alteing the rare and fragile material, an important consideration especially regarding holotypes.

11.2.1 Macrophotography

Macro-photographs of the fossil *Tellerodus* and *Coelacanthus* specimens (Figs. 11.3a and 11.4a) and the drawings in the original literature, figured as *Ceratodus Sturii* in Teller (1891, Taf IV = pl. 4, Figs. 1–4; Fig. 11.3b) and figured as *Coelacanthus Lunzensis* in Reis (1900, Taf IX = pl. 9 Fig. 11.4b), were taken with a Nikon Digital Camera, D 5200 SLR, lens Micro SX SWM MICRO 1:1 Ø52 Nikon AF-S, processed by the free graphic software tool digiCamControl version V.2.1.2.0 at the NHMW. Digital high-quality photomicrographs were taken using a Discovery.V20 Stereo Zeiss microscope. The magnifications were ×10 ×20 and ×40 in incident light mode. Data from the AxioCam MRc5 Zeiss were processed and documented using the AxioVision SE64 Rel. 4.9 imaging system at the NHMW.

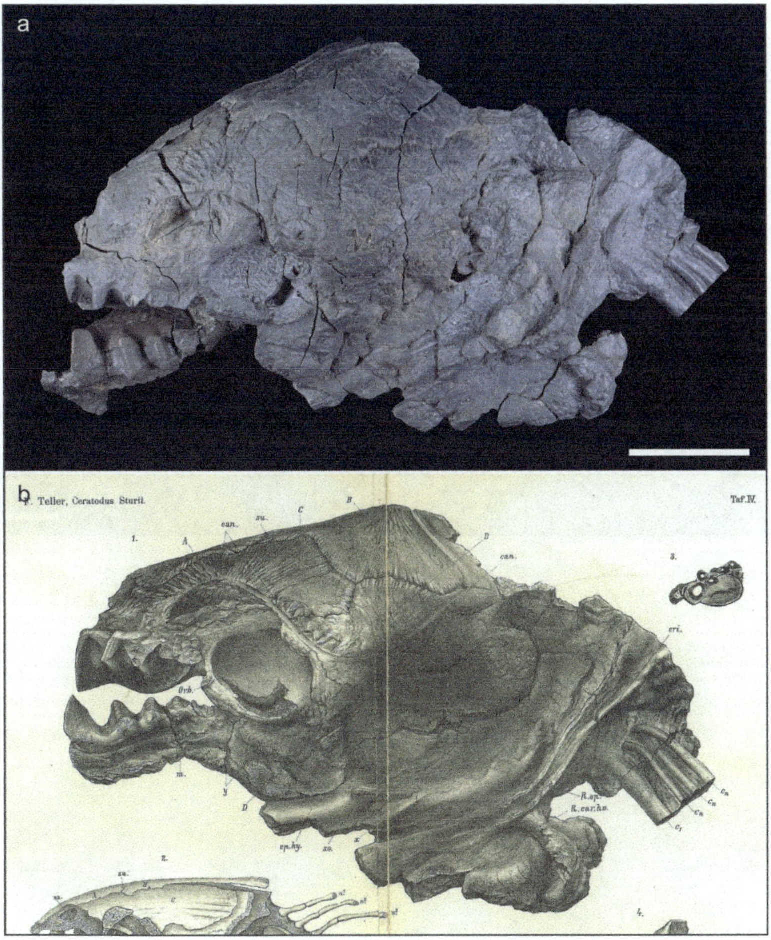

Fig. 11.3 **a** Entire skull of *Tellerodus sturii*, historical material from collapsed tunnel, normal light, cast of holotype, digitally reconstructed and 3D printed, holotype GBA 1891/001/0001 1-10. **b** Original historical drawing of the holotype, figured as *Ceratodus Sturii* in Teller (1891, Taf IV = pl. 4, Figs. 1–4). Scale bar: 5 cm

11.2.2 Microcomputed Tomogrophy (MicroCT)

The tomography scans on some of the fragments of the holotypes of *Tellerodus sturii* and *Coelacanthus lunzensis* were made at the NHMW. The microCT facility (YXLON FF35 CT, equipped with FXE Direct Beam and Elmer Y. Panel 4343 CT detector; Fig. 11.5) at the NHMW was used. For each scan, 3600 projection images were captured and then reconstructed with the software Siemens CERA. The pre-filters, voltage, and current were chosen based on the density of the individual fragments. This approach captured the basal information on morphological features.

Fig. 11.4 **a** Entire skeleton of *Coelacanthus lunzensis*, historical material from collapsed tunnel, normal light, holotype GBA 1891/001/0002. **b** Original historical drawing of the holotype, figured as *Coelacanthus Lunzensis* in Reis (1900, Taf IX = pl. 9)

The resulting datasets were processed and visualized using the software Dragonfly 3D World. The datasets including the scanning and reconstruction parameters are available under a CC-BY license in the data repository of the NHMW (*Coelacanthus lunzenis*: https://doi.org/10.57756/3h9knh; *Tellerodus sturii*: https://doi.org/10.57756/6nhtk4).

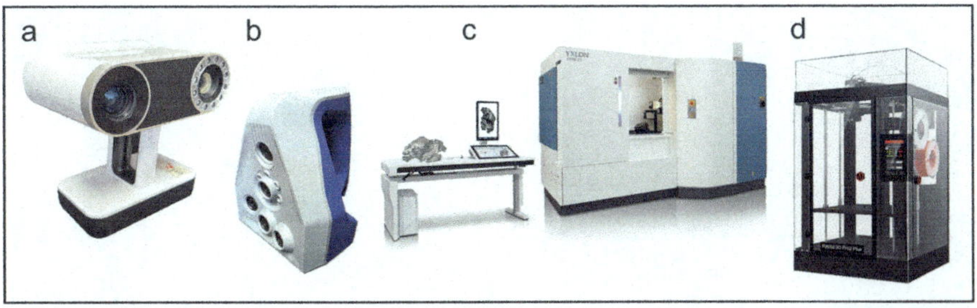

Fig. 11.5. 3D scanning facilities, microCT, and printing devices at the Natural History Museum Vienna. **a** Artec Leo scanner. **b** Artec Space Spider scanner. **c** Micro-computed tomography device YXLON FF35 CT. **d** Raise3D Pro 2 Plus 3D printer

11.2.3 Structured Light Scanning

To create the true-to-scale 3D surface models of the fossils, a handheld structured light scanner (Artec Space Spider; Fig. 11.5) was used. The scanner's 3D resolution is up to 0.1 mm with a 3D point accuracy of up to 0.05 mm and a texure resolution of 1.3 megapixels (mp). After scanning, the collected data were aligned and processed in the software Artec Studio Professional 15 and 17. The 3D scanner works on the basis of structured light technology and can capture even complex surfaces without damaging the objects' surfaces. The scanner was used for both the *Tellereodus sturii* skull and the *Coelacanthus lunzensis* fragments.

Low-resolution versions of the 3D models created at the NHMW are presented in the 3D museum hosted on the platform Sketchfab (http://sketchfab.com/NHM Wien) and are mostly available for download under a CC-BY-NC licence. The fragments of *T. sturii* (https://skfb.ly/o6XNC and https://doi.org/10.57756/huq2cx) and *C. lunzensis* https://skfb.ly/pAqWq and https://doi.org/10.57756/ehw7mb) were digitally assembled in Blender to illustrate how the pieces fit together (Fig. 11.6).

The 3D models can subsequently be used as a basis for 3D prints. For the 3D print of the assembled *Tellerodus* fragments the fused deposition modeling (FDM) process was utilized. The fossil was printed using the material PLA and a Raise 3D Pro2 Plus printer, which offers a build volume up to 305 × 305 × 605 mm (Fig. 11.7).

11.2.4 3D Visualization and Multimedia Formats

The modeling procedure for the fossil fish material is based on line drawings on which successive cross-sections were fitted. Additionally, the series of macrophotographs, 3D scans, and microCT images turned out to be an essential source to understand the morphology and shape of the two fossil fish specimens. This knowledge was accompanied by the morphological study of extant species of lungfish (i.e. genus *Neoceratodus*) and coelacanth fish (i.e. genus *Latimeria*). Both were used to draw

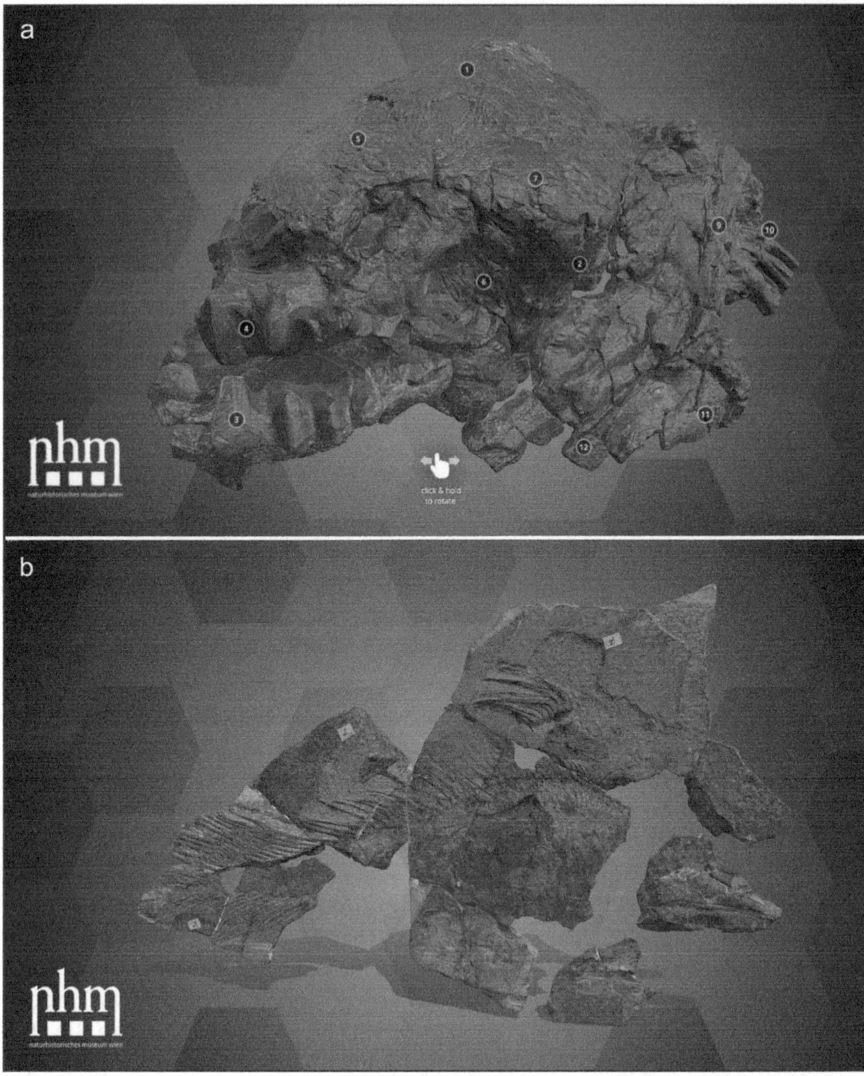

Fig. 11.6 Animated 3D model of **a** the fossil lungfish *Tellerodus sturii* and **b** the coelacanth *Coelacanthus lunzensis* in the 3D viewer Sketchfab allows a new perspective and discussion of original material fossil material. https://sketchfab.com/NHMWien; **a** https://skfb.ly/o6XNC; **b** https://skfb.ly/pJxKV

the cross-sections used in the later-processed 3D models. RGB spectrum was used for lighting of bumps and dents. This helped to refine the visual appearance and understand various delicate structures. The process involved projecting this normal map onto a surface inside a 3D Realtime Editor and then introducing a virtual light

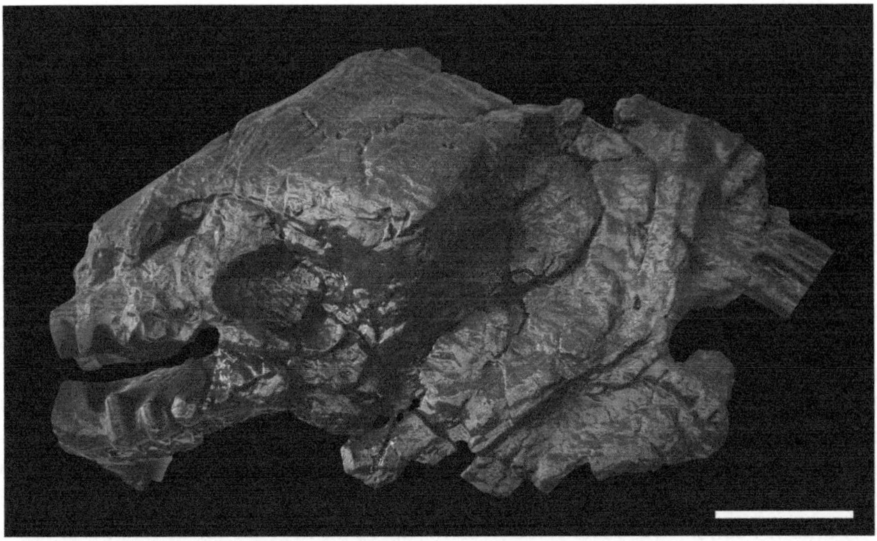

Fig. 11.7. 3D print in natural size of *Tellerodus sturii* can subsequently be used as a basis for 3D prints. For the 3D print of the assembled *Tellereodus* fragments, the fused deposition modeling (FDM) process was utilized. The fossil was printed using the material PLA and a Raise 3D Pro2 Plus printer, the volume is built up to 305 × 305 × 605 mm. Scale bar: 5 cm

source, which was tilted in different directions to gain plasticity from highlights and shadows.

The 3D modeling procedure was carried out in different stages using different 3D modeling and animation software facilities and programs (e.g., Nurbs and polygon editor like 3ds Studio Max® from Autodesk®; Ebert et al. 2003), each specialized for its defined task. The first step was to achieve a precise volume model of the fossil fish, followed by refining its geometric shape by adjusting scale and patterns to place the discrete spines. The final texture was created by coloration and image information directly on its surface in preparation for the final rendering and animation process.

The outlines of the *Tellerodus* skull and the fragments of *Coelacanthus* were vectorized and imported into a CAD Application (Rhinoceros® 4.0), which is orientated to create 3D geometry from so-called NURBS (Non Unified Rational B-Splines, Foley et al. 1996; Bishop et al. 2021), and hereafter used as a guideline to extrude the prepared cross-sections along its edges. Various other techniques had been tested before, including the manual extrusion of the sections or their replication, but with unsatisfactory results.

To refine surface features, the NURBS Model was brought into 3D Studio Max® from Autodesk®. The detailed morphological features visible on the macrophotographs and CT images of the source material were used to determine morphological criteria. These were then modeled and placed by hand. The soft tissue of the fishes was processed by 3D Studio Max®, starting with a basic shape, which was subsequently refined by precision of morphological characters (Fig. 11.8). The resulting geometries were exported to an additional program specialized in 3D texturing and painting (3D Coat©). UV mapping refers to the process of manually assigning or creating custom texture coordinates for the vertices of a 3D model in order to present the surface appearance including details such as RBG Color, specularity, and bump mapping, which are basics for directly painting on the 3D surface of the generated models. UVW describes the coordinates of the underlying mesh in its x-, y-, and z-axis and translates a given texture or coloration to distinct areas or morphological features. For the actual process of painting, various maps were gathered from existing images and placed in the background of the application; the 3D model was placed in front, and parts of these images were projected onto a special marked area.

The final geometries were then re-imported to 3D Studio Max© for the animation and rendering setup. For the animation of the animal's body, virtual "elements" were created by the inverse kinematic bones system and linked to a hierarchy with inverse kinematic constraints. This allows the animator to transform several parts of the geometrical model into desired positions, but can also be aided through a scripted (GUI/graphical user interface) to simulate movements parametrically. The surrounding scenery was assembled with a simple plane for the sea floor, a parametric geometry serving as a water plane, and the setting of virtual lights and ambient influences (e.g., lights through waves; Fig. 11.8; Supplementary Material 11.1).

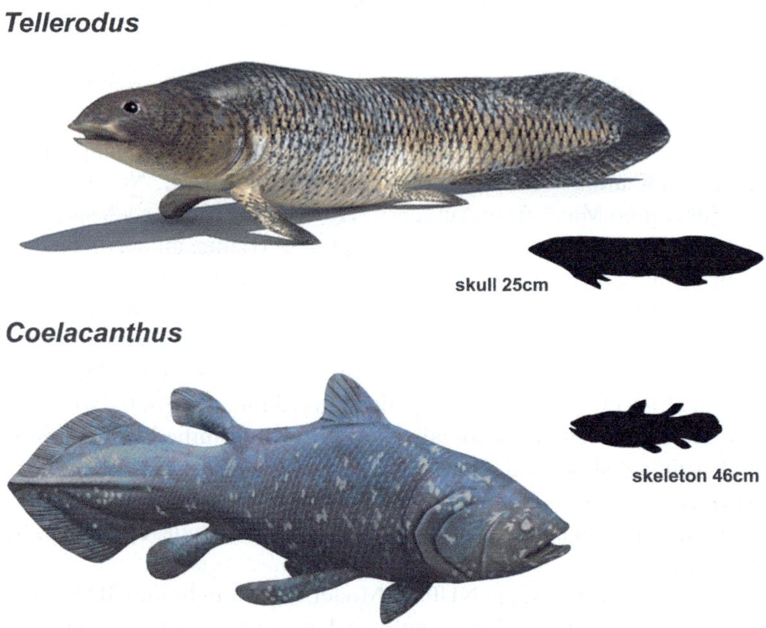

Fig. 11.8 3D reconstructions of Upper Triassic *Tellerodus sturii* and *Coelacanthus lunzensis* from the Carnian *Konservat-Lagerstätte* of Polzberg. Morphology based on detailed photography, 3D scans, and 3D imaging. In black, size values for skull in *Tellerodus* and entire skeleton in *Coelacanthus*

11.3 Visualization and Motion Pictures

11.3.1 3D Model on Sketchfab

The online platform Sketchfab facilitates the sharing of 3D surface models, providing an intuitive, browser-based viewer (Fig. 11.6). This embedded viewer allows users to explore the 3D models interactively without the need to download additional software. Users can zoom, rotate, and move the model freely. Moreover, the viewer offers the option to disable the texture, which is particularly useful for studying fossils because it eliminates the distraction of color, enabling a clearer examination of the surface details. To activate this feature, users can either press the "3" key or enable the Matcap option within the Model Inspector (accessible via the "i" key).

Fossil lungfish *Tellerodus sturii*

3D surface scan of the fossil lungfish *Tellerodus sturii.*
The fossil was found in the Polzberg mining tunnels, Lower Austria, in 1886.
The type material is stored at the GeoSphere Austria.
A cast of this unique fossil lungfish is exhibited in Hall 8 of the NHMW.
Specimen: *Tellerodus sturii* (Teller 1891)
Inventory number: GBA 1891/001/0001
Repository: GeoSphere Austria, Department of Paleontology and Collections (contact: Davide Di Franco), Neulinggasse 38, A-1030 Vienna
Scanned and edited by Anna Haider and Viola Winkler (NHMW)
Link: https://skfb.ly/o6XNC and https://doi.org/10.57756/huq2cx.

Coelacanth *Coelacanthus lunzensis*

3D surface scan of the fossil coelacanth fish *Coelacanthus lunzensis.*
The fossil was found in the Polzberg mining tunnels, Lower Austria, in 1886.
The type material is stored at the GeoSphere Austria.
A cast of this unique fossil coelacanth is stored in the repository of the NHMW.
Specimen: *Coelacanthus lunzensis* (Teller 1891)
Inventory number: GBA 1891/001/0002 1-10
Repository: GeoSphere Austria, Department of Paleontology and Collections (contact: Davide Di Franco), Neulinggasse 38, A-1030 Vienna
Scanned and edited by Viola Winkler (NHMW)
Link: https://skfb.ly/pJxKV and https://doi.org/10.57756/ehw7mb.

11.4 Discussion

In recent years, the digitization of materials using 3D scanners has gained popularity due to the increased accessibility of the necessary infrastructure (Sketchfab: http://sketchfab.com/NHMWien). One such method, structured light scanning, was used here to capture the surface of two fossil specimens. Other methods, such as photogrammetry, can also produce detailed 3D models, including color information, but they require manual scaling afterward. This manual scaling can introduce discrepancies between the original specimen and the digital replica, potentially leading to inaccurate measurements. However, photogrammetry typically provides superior color detail, which can be advantageous.

Capturing 3D models at such high resolution increases accessibility, allowing scientists and even the broader public to study the fossils without the need to physically send or loan the original specimens. This is especially valuable for rare and valuable objects such as holotypes, where 3D digitization can enhance the safety and preservation of the original material. Additionally, 3D printing based on the surface models can create both accurate replicas, which can-with proper manual coloring–even mimic the original's color, and hands-on objects for study. This enables researchers to examine the structures in detail without the risk of damaging the originals.

MicroCT imaging offers a non-invasive method for studying the internal structures of fossils. This can reveal both the internal anatomy of specimens and areas hidden by sediment, without the need for physical preparation (Lukeneder et al. 2020; Lukeneder and Lukeneder 2022a, b; Lukeneder et al. 2026, this volume; Stumpf et al. 2026, this volume). This approach not only reduces the risk of damage to valuable, irreplaceable fossils, but also enables detailed morphometric analyses. The resulting data can be shared with other researchers, much like 3D surface models, further reducing the handling of the fossils.

Compared to clinical CT, microCT for paleontological material provides higher resolution and utilizes higher voltage because it is not limited by the need to image live tissue. The data can be interactively explored, allowing for virtual slicing from any angle and providing 3D visualizations of hidden structures, which may yield new insights.

For microCT to be effective, the X-rays must be able to pass through the scanned objects. In cases of very dense or large samples, this may not be possible with institutional microCT systems.

11.5 Conclusions

The integration of various non-invasive 3D imaging techniques enables a detailed analysis of paleontological fossils, as shown by the two specimens presented here. These methods also provide the opportunity to share the resulting data on online platforms, such as Sketchfab, thereby engaging both the scientific community and the broader public interested in fossils. This approach promotes the easy sharing of materials stored in repositories, making them accessible for studies and collaboration.

The production of high-quality 3D data further reduces the need for handling the original specimens, minimizing the risk of damage or degradation. This is particularly crucial for rare and valuable specimens such as holotypes. Importantly, the data generated through these techniques are true to scale, enabling precise measurements without physical interaction with the original material. Finally, 3D printing technologies enable creating physical replicas based on these digital models.

In summary, microCT imaging enhances the study of fossil material by providing high-resolution, non-invasive, and three-dimensional imaging capabilities, without the need for physical extraction. It facilitates detailed analysis of both external and internal features, helps preserve delicate specimens, and provides a wealth of data for future research. The technique opens new doors for understanding the anatomy and possibly even behavior and evolution of extinct organisms in detail. As this technique advances, even well-studied specimens may yield new findings, offering novel insights into previously inaccessible details.

When combined with traditional methods, these non-invasive imaging techniques serve as valuable supplements, furthering our ability to achieve a more comprehensive understanding of paleontological specimens. The study presents results of an object-based combined analysis from 3D scanning, computed tomography, and various computed 3D facility programs performed on one of the most complete lungfish skulls worldwide, the holotype of *Tellerodus sturii*, and on the most complete specimen of an Upper Triassic coelacanth with the holotype of *Coelacanthus lunzensis*. This is the first 3D visualization and animation of these holotypes from the lower Carnian Polzberg *Konservat-Lagerstätte* (Northern Calcareous Alps, Lower Austria). Based on the morphological characteristics, the extinct species were reconstructed. The size and body shape reconstructions are also supported by actuopaleontological studies on modern and extant members of the lungfish and coelacanthids. 3D scanning, computed tomography, and paleontological data are combined with 3D reconstructions, yielding animated 3D models of the type material of the two species online in Sketchfab and museum online, accessible for a broad scientific community. These case studies once again demonstrate the possibilities of non-destructive methods for 3D visualization of paleontological material. The resulting images are realistic, and the reconstructed animals are put into context with their paleoenvironment. High-quality 3D models can simply be shared amongst paleontologists. Data, images, and short-clips of fossil type material can be discussed online and subsequently adapted morphologically or published in diverse online or print media, and ultimately printed out as physical models for analyses.

Acknowledgments The author is particularly grateful to the owners of the outcrop area, Eva and Karl Jagersberger as well as Franziska and Hermann Hofreiter (all Gaming), for sampling and digging permission. Anna Haider (Malta) helped in scanning the type material at the GeoSphere Austria (former Geologische Bundesanstalt, Geological Survey of Austria, Vienna). We thank the GeoSphere Austria for the opportunity to scan the fossil type material of *Tellerodus sturii* and *Coelacanthus lunzensis* (Repository GeoSphere Austria, Davide Di Franco). The work was done within the framework of and financially supported by projects of the Federal Government of Lower Austria (K3-F-964/001-2020 and K3-F-964/002-2023, Department Science and Research; headquarters in St. Pölten); the Austrian Academy of Sciences (headquarters in Vienna, project Polzberg Lukeneder 2021) represented by the National Committee for Geo/Hydro Sciences (Earth System Sciences Programme); and the Triassic project "Fenster in die Karnische Krise" of the Society of Friends of the Natural History Museum Vienna (Freunde des Naturhistorischen Museums Wien, CPE Lukeneder 2023). The manuscript greatly benefited from valuable comments of the reviewers Veronika N. I.

Koukal (Vienna) and Petra Lukeneder (Vienna), the production editor Ashok Arumairaj (Taramani), the in-house series editor Kathleen A. Ritterbush (Utah), the publishing editor Simon Shah-Rohlfs and the production editor Antje Endemann (both Heidelberg).

Author Contribution AL designed research, provided documentation on fossil material. VW 3D scanning, computed tomography, and printing of fossil material. AL, VW interpretation of data. MK visualization and animation.

Funding Open access funding was enabled and the projects financially supported by the Federal Government of Lower Austria (K3-F-964/001-2020 and K3-F-964/002-2023, Department Science and Research; headquarters in St. Pölten), and the Triassic project "Fenster in die Karnische Krise" of the Society of Friends of the Natural History Museum Vienna (Freunde des Naturhistorischen Museums Wien, CPE Lukeneder 2023); and the Natural History Museum Vienna (to AL). The authors are responsible for the contents of this publication. The funders had no impact on conceptualization, design, data collection, analysis, decision to publish, or preparation of the manuscript.

Competing interests All authors stated that they have no competing interests.

Authors' contributions All authors stated that they have equally contributed to their chapters, read, edited and agreed to submit the final versions of their contributions.

Availability of data and materials All sources of information are provided in the text. Raw data related to the fossil material from Austria are available from the corresponding authors and Alexander Lukeneder upon request.

Additional information The article is part of the Special Volume: *Late Triassic Konservat-Lagerstätten within the Carnian Pluvial Episode. 150 Years of Scientific Excavations in Austria.*

Artificial Intelligence (AI) The chapter was produced without the use of AI.

Language The entire volume uses American English reviewed by a native scientist.

Received 2025.

Published online 2026.

References

Argyriou T, Giles S, Friedman M, Romano C, Kogan I, Sánchez-Villagra MR (2018) Internal cranial anatomy of Early Triassic species of *Saurichthys* (Actinopterygii: Saurichthyiformes): implications for the phylogenetic placement of saurichthyiforms. BMC Evol Biol 18(161). https://doi.org/10.1186/s12862-018-1264-4

Balanoff AM, Norell MA, Grellet-Tinner G, Lewin MR (2008) Digital preparation of a probable neoceratopsian preserved within an egg, with comments on microstructural anatomy of ornithischian eggshells. Naturwissenschaften 95:493–500. https://doi.org/10.1007/s00114-008-0347-2

Bishop PJ, Falisse A, De Groote F, Hutchinson JR (2021) Predictive simulations of running gait reveal a critical dynamic role for the tail in bipedal dinosaur locomotion. Sci Adv 7(39):article abi7348. https://doi.org/10.1126/sciadv.abi734

Cherns L, Spencer ART, Rahman IA, Garwood RJ, Reedman C, Burca G, Turner MJ, Hollingworth NTJ, Hilton J (2021) Correlative tomography of an exceptionally preserved Jurassic ammonite implies hyponome-propelled swimming. Geology 50(4):397–401. https://doi.org/10.1130/G49551.1

Cunningham JA, Thomas CW, Bengtson S, Kearns SL, Xiao S, Marone F, Stampanoni M, Donoghue PCJ (2012) Distinguishing geology from biology in the Ediacaran Doushantuo biota relaxes constraints on the timing of the origin of bilaterians. Proc R Soc B, Biol Sci. https://doi.org/10.1098/rspb.2011.2280

Das AJ, Murmann DC, Cohrn K, Raskar R (2017) A method for rapid 3D scanning and replication of large paleontological specimens. PLoS ONE 12(7):e0179264. https://doi.org/10.1371/journal.pone.0179264

Degrange FJ, Tambussi CP, Moreno K, Witmer LM, Wroe S (2010) Mechanical Analysis of Feeding Behavior in the Extinct “Terror Bird” *Andalgalornis steulleti* (Gruiformes: Phorusrhacidae). PLoS ONE 5(8):e11856. https://doi.org/10.1371/journal.pone.0011856

Dong XP, Bengtson S, Gostling NJ, Cunningham JA, Harvey THP, Kouchinsky A, Val’kov AK, Repetski JE, Stampanoni M, Marone F, Donoghue PCJ (2010) The anatomy, taphonomy, taxonomy and systematic affinity of *Markuelia*: Early Cambrian to Early Ordovician scalidophorans. Palaeontology 53(6):1291–1314. https://doi.org/10.1111/j.1475-4983.2010.01006.x

Donoghue PCJ, Bengtson S, Dong XP, Gostling NJ, Huldtgren T, Cunningham JA, Yin C, Yue Z, Peng F, Stampanoni M (2006) Synchrotron X-ray tomographic microscopy of fossil embryos. Nature 442:680–683. https://doi.org/10.1038/nature04890

Ebert DS, Musgrave FK, Peachey D, Perlin K, Worley S (2003) Texturing and modeling: a procedural approach, 3rd edn. Morgan Kaufmann Publishers, Elsevier, 687 p

Foley JD, van Dam A, Feiner SK, Hughes John F (1996) Computer graphics—principles and practice, 2nd edn. Addison-Wesley, 1175 p

Fortuny J, Galobart À, De Santisteban C (2011) A new capitosaur from the Middle Triassic of Spain and the relationships within the Capitosauria. Acta Palaeontol Pol 56(3):553–566. https://doi.org/10.4202/app.2010.0025

Friis EM, Crane PR, Pederse KR, Bengtson S, Donoghue PCJ, Grimm GW, Stampanoni M (2007) Phase-contrast X-ray microtomography links Cretaceous seeds with Gnetales and Bennettitales. Nature 450:549–552. https://doi.org/10.1038/nature06278

Gai Z, Donoghue PCJ, Zhu M, Janvier P, Stampanoni M (2011) Fossil jawless fish from China foreshadows early jawed vertebrate anatomy. Nature 476:324–327. https://doi.org/10.1038/nature10276

Gaspard D, Putlitz B, Baumgartner L (2011) X-ray computed tomography—a promising tool for brachiopod shell investigations. In: Proceedings IAMG 2011, Salzburg 2011, pp 518–533

Hoffmann R, Zachow S (2011) Non-invasive approach to shed new light on the buoyancy business of chambered cephalopods (Mollusca). In: Proceedings IAMG 2011, Salzburg 2011, pp 1–11. https://doi.org/10.5242/iamg.2011.01638

Huldtgren T, Cunningham JA, Yin C, Stampanoni M, Marone F, Donoghue PCJ, Bengtson S (2011) Fossilized nuclei and germination structures identify Ediacaran “Animal Embryos” as encysting protists. Science 334(6063):1696–1699. https://doi.org/10.1126/science.1209537

Knoll F, Witmer LM, Ortega F, Ridgely RC, Schwarz-Wings D (2012) The braincase of the basal sauropod dinosaur Spinophorosaurus and 3D. Reconstructions of the cranial endocast and inner ear. PLoS ONE 7(1):e30060. https://doi.org/10.1371/journal.pone.0030060

Kruta I, Landman N, Rouget I, Cecca F, Tafforeau P (2011) The role of ammonites in the Mesozoic marine food web revealed by jaw preservation. Science 331(6013):70–72. https://doi.org/10.1126/science.1198793

Lehman JP (1975) A propos de *Ceratodus sturii* Teller, 1891. Bull Mus Nati Hist Nat Paris 375:241–246. https://www.zobodat.at/pdf/VerhGeolBundesanstalt_1982_0021-0031.pdf

Lukeneder A (2012) Computed 3D visualisation of an extinct cephalopod using computer tomographs. Comput Geosci 45:68–74. https://doi.org/10.1016/j.cageo.2012.04.003

Lukeneder A (ed) (2026) Late Triassic Konservat-Lagerstätten within the Carnian Pluvial Episode: 150 Years of Scientific Excavations in Austria. Top Geobiol 55:600 p. https://doi.org/10.1007/978-3-032-19777-1

Lukeneder A, Lukeneder P (2021) The Upper Triassic Polzberg palaeobiota from a marine *Konservat-Lagerstätte* deposited during the Carnian Pluvial Episode in Austria. Sci Rep 11:16644. https://doi.org/10.1038/s41598-021-96052-w

Lukeneder A, Lukeneder P (2022a) Taphonomic history and trophic interactions of an ammonoid fauna from the Upper Triassic Polzberg palaeobiota. Sci Rep 12:7455. https://doi.org/10.1038/s41598-022-11496-y

Lukeneder P, Lukeneder A (2022b) Mineralized belemnoid cephalic cartilage from the Late Triassic Polzberg *Konservat-Lagerstätte* (Austria). PLoS ONE 17(4):e0264595. https://doi.org/10.1371/journal.pone.0264595

Lukeneder A, Lukeneder P (2023) New data on the marine Upper Triassic palaeobiota from the Polzberg *Konservat-Lagerstätte* in Austria. Swiss J Palaeontol 142:9. https://doi.org/10.1186/s13358-023-00269-3

Lukeneder A, Lukeneder P (2024) Ammonoid taxonomy of the Carnian Polzberg Konservat-Lagerstätte in Austria. Jahrb Geol BA 162(1–4):27–55. https://opac.geologie.ac.at/ais312/dokumente/JB1621_027_A.pdf

Lukeneder A, Surmik D, Gorzelak P, Niedźwiedzki R, Brachaniec T, Salamon MA (2020) Bromalites from the Upper Triassic Polzberg section (Austria); insights into trophic interactions and food chains of the Polzberg palaeobiota. Sci Rep 10:20545. https://doi.org/10.1038/s41598-020-77017-x

Lukeneder A, Lukeneder P, Sachsenhofer RF, Roghi G, Rigo M (2024a) Multi-proxy record of the Austrian Upper Triassic Polzberg *Konservat-Lagerstätte* in light of the Carnian pluvial episode. Sci Rep 14:11194. https://doi.org/10.1038/s41598-024-60591-9

Lukeneder P, Fuchs D, Lukeneder A (2024b) Morphology, taxonomy and trophic interactions of rostrum-less coleoids from the Late Triassic Polzberg *Konservat-Lagerstätte* (Lower Austria). Swiss J Palaeontol 143:21. https://doi.org/10.1186/s13358-024-00319-4

Lukeneder P, Valentin B, Kaurin C, Lukeneder A (2026) 3D reconstruction and microarchitecture of a reticular belemnoid cephalic cartilage from the Late Triassic Polzberg *Konservat-Lagerstätte*. In: Lukeneder A (ed) Late Triassic Konservat-Lagerstätten within the Carnian Pluvial Episode: 150 years of scientific excavations in Austria. Top Geobiol, Chap 8:131–147. https://doi.org/10.1007/978-3-032-19777-1_8

Lukeneder S, Lukeneder A, Gusenbauer C (2014) Computed tomography and laser scanning of fossil cephalopods (Triassic and Cretaceous). Denisia 32:81–92. https://www.zobodat.at/pdf/DENISIA_0032_0081-0092.pdf

Luo ZX, Kielan-Jaworowska Z, Cifelli RL (2002) In quest for a phylogeny of Mesozoic mammals. Acta Palaeontol Pol 47:1–78. https://www.paleo.pan.pl/acta/acta47/app47-001.pdf

Maloof AC, Rose CV, Beach R, Samuels BM, Calmet CC, Erwin DH, Poirier GR, Yao N, Simons FJ (2010) Possible animal-body fossils in pre-Marinoan limestones from South Australia. Nat Geosci 3:653–659. https://doi.org/10.1038/ngeo934

Marschallinger R (2001) Three-dimensional reconstruction and visualization of geological materials with IDL—examples and source code. Comput Geosci 27(4):419–426. https://doi.org/10.1016/S0098-3004(00)00126-6

Marschallinger R, Hofmann P, Daxner-Höck G, Ketcham RA (2011) Solid modeling of fossil small mammal teeth. Comput Geosci 37(9):1364–1371. https://doi.org/10.1016/j.cageo.2010.07.011

Martin M (1982) Revision von *Tellerodus sturii* (Teller) 1891. Verh Geol BA 1982(2):21–31. https://www.zobodat.at/pdf/VerhGeolBundesanstalt_1982_0021-0031.pdf

Mayrhofer S, Lukeneder A (2010) 3D Modellierung eines Karnischen Ammoniten Massenvorkommens (Taurus, Türkei; FWF P22109 B17). In: Marschallinger R, Wanker W, Zobl F (eds) Beiträge zur COGeo 2010, p 12. https://doi.org/10.5242/cogeo.2010.0000
Mees F, Swennen R, Van Geet M, Jacobs P eds (2003) Applications of X-ray computed tomography in the geosciences. Geol Soc Lond Spec Publ 215:1–243. https://doi.org/10.1144/gsl.sp.2003.215.01.01
Polcyn MJ, Rogers JV, Kobayashi Y, Jacobs LL (2002) Computed tomography of an Anolis lizard in Dominican amber: systematic, taphonomy, biogeography, and evolutionary implications. Palaeontol Electron 5(1):1−13. http://palaeo-electronica.org/paleo/2002_1/amber/issue1_02.htm
Rayfield EJ, Milner AC, Xuan VB, Yound PG (2007) Functional morphology of spinosaur 'crocodile-mimic' dinosaurs. J Vert Paleontology 27:892–901. https://doi.org/10.1671/0272-4634(2007)27[892:FMOSCD]2.0.CO;2
Rayfield EJ, Norman DB, Horner CC, Horner JR, Smith PM, Thomason JJ, Upchurch P (2001) Cranial design and function in a large theropod dinosaur. Nature 409:1033–1037. https://doi.org/10.1038/35059070
Reis OM (1900) *Coelacanthus lunzensis* Teller. Jb K-K Geol RA 50(2):187–192. https://www.zobodat.at/pdf/JbGeolReichsanst_050_0187-0192.pdf
Rowe T, Ketcham RA, Denison C, Colbert M, Xu X, Currie PJ (2001) Forensic palaeontology—the *Archaeoraptor* forgery. Nature 410:539–540. https://doi.org/10.1038/35069145
Saupe EE, Fuente RP, Selden PA, Delclos X, Tafforeau P, Soriano C (2012) New *Orchestina* Simon, 1882 (Araneae: Oonopidae) from Cretaceous ambers of Spain and France: first spiders described using phase-contrast X-ray synchrotron microtomography. Palaeontology 55(1):127–143. https://doi.org/10.1111/j.1475-4983.2011.01123.x
Schultz O (2013) Pisces. In: Piller WE (ed) Austrian Academy of Sciences. Catal Foss Austr 3:1–576
Scott AC, Galtier J, Gostling NJ, Smith SY, Collinson ME, Stampanoni M, Marone F, Donoghue PCJ, Bengtson S (2009) Scanning electron microscopy and synchrotron radiation X-ray tomographic microscopy of 330 million year old charcoalified seed fern fertile organs. Microsc Microanal 15(2):166–173. https://doi.org/10.1017/S1431927609090126
Stumpf S, Lukeneder A, Winkler V (2026) First hybodontiform record reveals chondrichthyan role in Carnian food webs of the Upper Triassic Reifling Basin, Austria. In: Lukeneder A (ed) Late Triassic Konservat-Lagerstätten within the Carnian Pluvial Episode: 150 years of scientific excavations in Austria. Top Geobiol, Chap 21:481–498. https://doi.org/10.1007/978-3-032-19777-1_21
Stur D (1886) Vorlage des ersten fossilen Schädels von *Ceratodus* aus den ober triadischen Reingrabner Schiefern von Pölzberg nördlich bei Lunz. Verh K-K Geol RA 15:381–383. https://www.zobodat.at/pdf/VerhGeolBundesanstalt_1886_0381-0383.pdf
Sutton MD (2008) Tomographic techniques for the study of exceptionally preserved fossils. Proc R Soc B 275(1643):1587–1593. https://doi.org/10.1098/rspb.2008.0263
Teller F (1891) Über den Schädel eines fossilen Dipnoers, *Ceratodus Sturii* nov. spec. aus den Schichten der oberen Trias der Nordalpen. Abh K-K Geol RA 15:1–38. https://www.zobodat.at/pdf/AbhGeolBA_15_0001-0038.pdf
Tsuihiji T, Watabe M, Tsogtbaatar K, Tsubamoto T, Barsold R, Suzuki S, Lee AH, Ridgely R, Kawahara Y, Witmer LM (2011) Cranial osteology of a juvenile specimen of Tarbosaurus bataar (Theropoda, Tyrannosauridae) from the Nemegt Formation (Upper Cretaceous) of Bugin Tsav, Mongolia. J Vertebr Paleontol 31(3):497–517. https://doi.org/10.1080/02724634.2011.557116
Witmer LM, Ridgely RC (2009) New insights into the brain, braincase and ear region of tyrannosaurs (Dinosauria, Theropoda), with implications for sensory organization and behaviour. Anat Rec 292:1266–1296. https://doi.org/10.1002/ar.20983
Zelenitsky DK, Therrien F, Ridgely RC, McGee AR, Witmer LM (2011) Evolution of olfaction in non-avian dinosaurs and birds. Proc R Soc Lond B 278:3625–3634. https://doi.org/10.1098/rspb.2011.0238

Chapter 12
The Plant Fossils of the Carnian (Late Triassic) Polzberg *Konservat-Lagerstätte* in Austria

Evelyn Kustatscher, Roberta Branz, Miriam Slodownik, Guido Roghi, and Alexander Lukeneder

Abstract The Upper Triassic Polzberg *Konservat-Lagerstätte* (Lower Austria) yields a diverse fossil assemblage from the Reingraben Formation, deposited during the Carnian Pluvial Episode (CPE). While the site is renowned for its diverse marine fauna, bed-by-bed collections and a revision of historical specimens also reveal a small but informative macroflora. Ninety-two plant fossils were examined, of which 77 could be assigned to major taxonomic groups. The assemblage is dominated by conifers, mainly *Brachyphyllum*, *Pagiophyllum*, and *Pelourdea*, with rare *Elatocladus* and several indeterminate reproductive organs. The preservation is generally good, yielding cuticles in some of the conifer remains, despite likely long-distance transport from emergent landmasses into the intraplatform basin. The Polzberg

Supplementary Information The online version contains supplementary material available at https://doi.org/10.1007/978-3-032-19777-1_12.

E. Kustatscher (✉) · R. Branz
Museum of Nature South Tyrol, Bozen/Bolzano, Italy
e-mail: e.kustatscher@tiroler-landesmuseen.at

E. Kustatscher
Department of Natural History, Collection and Research Centre (SFZ), Tirolean State Museums, Hall in Tirol, Austria

Department für Geo- und Umweltwissenschaften, Paläontologie und Geobiologie, Ludwig-Maximilians-Universität, München, Germany

Bayerische Staatssammlung für Paläontologie und Geobiologie, München, Germany

M. Slodownik
Botany Department, Trinity College Dublin, Dublin, Ireland
e-mail: miriam.slodownik@tcd.ie

G. Roghi
Institute of Geosciences and Earth Resources - CNR, Padova, Italy
e-mail: guido.roghi@igg.cnr.it

A. Lukeneder
Department of Geology and Paleontology, Natural History Museum Vienna, Vienna, Austria
e-mail: alexander.lukeneder@nhm.at

A. Lukeneder (ed.), *Late Triassic Konservat-Lagerstätten within the Carnian Pluvial Episode*, Topics in Geobiology 55, https://doi.org/10.1007/978-3-032-19777-1_12

macroflora contrasts sharply with the coeval, swamp-dominated Lunz flora and with the rich, diverse local palynoflora, both of which indicate a wider spectrum of hygrophytic and mesophytic taxa. This disparity reflects the combined effects of taphonomic selection and bias of robust elements. Our results highlight the need to integrate macrofossil, palynological, and sedimentological data when reconstructing terrestrial vegetation from marine settings.

Keywords Reingraben Formation · Carnian · CPE · Lower Austria · Terrestrial ecosystem

12.1 Introduction

The Polzberg site in the Northern Calcareous Alps (NCA) of Lower Austria is a *Konservat-Lagerstätte* of Carnian age, known for over 150 years (Stur 1874, 1886; Fig. 12.1a, b). In the past 60 years it has undergone comprehensive examination from sedimentological and paleontological perspectives (Allison 1988; Krystyn 1991; Forchielli and Pervesler 2013; Lukeneder 2026, this volume). Since 2021, the Museum of Natural History Vienna (Naturhistorisches Museum Wien) has conducted extensive excavations at the site. A meticulous bed-by-bed collecting approach has yielded a substantial fossil collection (>5200 specimens) providing exceptional insights into the composition of the marine paleocommunity during the Carnian including food chains, dietary habits, trophic interactions, and the taphonomic processes affecting marine organisms (Lukeneder et al. 2020; Lukeneder and Lukeneder 2021, 2022b). A substantial portion of this fossil assemblage is still under study, among them bivalves, gastropods, arthropods, crustaceans, polychaetes, fishes, and bromalites (coprolites and regurgitalites).

This fossil assemblage was deposited during the Carnian Pluvial Episode (CPE; Lukeneder and Lukeneder 2021; Lukeneder 2026, this volume). The increased precipitation during the CPE led to low oxygenation of bottom sediments, resulting in the deposition of laminated sediments of the fossiliferous Reingraben Formation at Polzberg. These deposits provide optimal conditions for the preservation of both entire and soft body fossils, which are essential for *Konservat-Lagerstätten.* The in situ findings of benthic and nectic taxa underscore the significant value of the unique Polzberg paleobiota and the autochthonous deposition of the inhabitants within the paleohabitat. The biota reflects a marine environment with intermittent shallow-water influence, evidenced by discoveries of freshwater species such as the lungfish *Tellerodus* (Stur 1886; Lehman 1975; Teller 1891; Martin 1982; Lukeneder and Lukeneder 2021), the branchiopod *Euestheria* (Stur 1874; Lukeneder and Lukeneder 2021), and terrestrial plants.

This paper provides a first overview of the plant fossils of the Polzberg *Konservat-Lagerstätte*, which were transported from nearby islands to the basin where they were ultimately deposited. This provides an unique glimpse into the terrestrial ecosystems of the Northern Calcareous Alps during the Carnian.

12.2 Material and Methods

The Reingraben Formation at Polzberg yielded a distinctive ammonoid assemblage belonging to the Julian (lower Carnian) *Austrotrachyceras austriacum* Zone (Julian 2; Fig. 12.1c). The lowermost, fossiliferous portion of the Reingraben Shales is dated to the Julian 2 (Julian 2/Ib, *A. minor* biohorizon) based on the occurrence of *Austrotrachyceras minor*, *Paratrachyceras haberfellneri* along with the associated species *Carnites floridus* and *Simonyceras simonyi*. The co-occurrence of these ammonoids, combined with the mass occurrence of the benthic bivalves *Peribositra and Halobia rugosa* in the fossiliferous layers, supports the biostratigraphical assignment to the Julian (Mueller et al. 2016; Lukeneder and Lukeneder 2021, 2022a, b). According to the ammonoid zonation and the specific duration of ammonoid zones in the late Carnian (Mueller et al. 2016; Dal Corso et al. 2020), the analyzed section represents approximately 500,000 years of deposition.

All plant remains discussed in this paper come from historical collections or recent excavations at the Polzberggraben (Northern Calcareous Alps, NCA) near Polzberg (also referred to as Schindelberggraben or Polzberg in numerous collections and publications; Lukeneder and Lukeneder 2021). Thus, the assignment of fossils and samples to the locality Schindelberg is synonymous with the locality Polzberg (historically spelled also as Pölzberg; Stur 1874, 1886; Lukeneder and Lukeneder 2021; 1:50 000, geological map, sheet 71 Ybbsitz; Ruttner and Schnabel 1988, and sheet 72 Mariazell; Bauer and Schnabel 1997; Fig. 12.1). The excavation area is situated

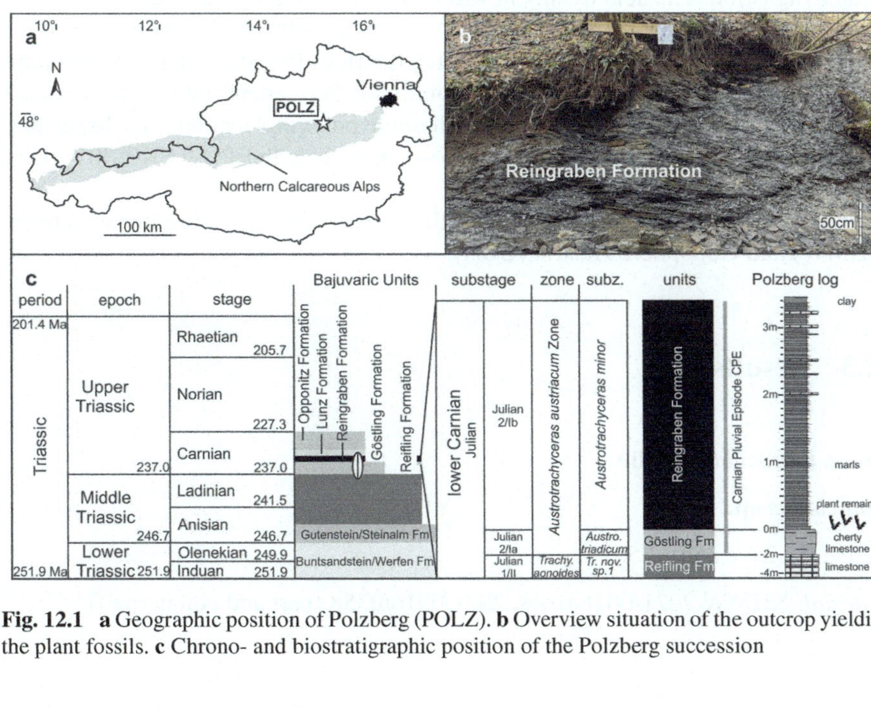

Fig. 12.1 **a** Geographic position of Polzberg (POLZ). **b** Overview situation of the outcrop yielding the plant fossils. **c** Chrono- and biostratigraphic position of the Polzberg succession

on the western slope of Mount Schindelberg (elevation 1066 m; N 47°53′4.98″, E 15°4′28.15″), located to the north of the Ois River, and approximately 4 km northeast of Lunz am See (Niederösterreich, Fig. 12.1a). Bed-by-bed sampling was feasible in the lower calcareous unit of the section, whereas in the upper argillaceous portion of the section the sampling interval was set at 20 cm due to the soft nature of the deposits.

Plant remains were photographed with a digital reflex camera (NIKON D5200). The macrofossils were measured from digital photographs using ImageJ1.53a (Schneider et al. 2012). For the Schulze method (see Nowak et al. 2022), the organic matter was placed in a glass tube and a small amount of crystalline potassium chlorate (KCl) was added (an amount slightly exceeding the sample's mass). Then, under a fume hood, 3 ml of 30% nitric acid (HNO_3) was added. After each day for the following three days, 1 ml of 67.5% nitric acid was added. On the fifth day, the acid was decanted and the residue washed with deionized water. Then, a few drops of 5% potassium hydroxide (KOH) were added to the residue, and after a few seconds diluted with deionized water to stop the reaction; in case of an instantaneous reaction, water was added immediately. For light microscopy, residues were mounted on glass slides in glycerin jelly and photographed with a Leica DM 2500 LED microscope and a Leica DMC4500 camera (for details see Nowak et al. 2022). Measurements were taken from the images with the Fiji distribution of ImageJ version 1.53c (Schindelin et al. 2012).

Samples were collected along the section also for palynological analyses. All samples were crushed to grains of several mm in size to increase sample surface area for acid digestion. The acid treatment was carried out by Rae Jones laboratory, Paly Parlour (UK) with 28% hydrochloric acid (HCl) and 60% hydrofluoric acid (HF) and treatment with additional HCl to dissolve fluorides formed by HF. The resultant residue was sieved at 10 μm, without ultrasonic bath. Permanent slides were produced by mounting the residue on cover slips with dilute polyvinyl alcohol as a dispersant, dried and subsequently mounted on a microscope slide with EPO-TEK® 301 epoxy resin.

The fossils and slides are housed at the Natural History Museum of Vienna (NHMW) and GeoSphere Austria (GBA).

12.3 Results

Genus *Elatocladus* Halle

***Elatocladus* sp.**

Figures 12.2a–c and 12.5a

Material: NHMW 2021/0016/0348, 2021/0016/0358 (part and counterpart)

Preservation: A single articulated leafy shoot is preserved as a partial compression and partial impression, with cuticles preserved.

Description: The apical shoot fragment is 50 mm long and 28 mm wide with a 4 mm wide axis. The leaves are densely arranged in a helical pattern. The leaves are broad lanceolate in shape with rounded apices and broad bases. Leaf dimensions vary, with lengths ranging from 13–15 mm and widths from 3.5–4.5 mm.

The cuticle was extracted from a single sample (NHMN 2021/0016/0358). The cuticle fragments are small, very thin, and exhibit few recognizable features. The epidermal cells are approximately isodiametric, with an average width of 26 μm and length of 37 μm. No papillae are present on the periclinal walls. No stomata were observed in the small cuticle fragments.

Discussion: The genus *Elatocladus* was established by Halle (1913) to encompass "sterile coniferous branches of the radial or the dorsiventral type, which do not show any characters that permit them to be included in one of the genera instituted for more peculiar forms." Although our specimen presents some unique features, we choose to retain it within *Elatocladus* because it is the so far only specimen of its kind and the cuticle is insufficiently indicative. This classification remains tentative and may be revised with the discovery of additional specimens with diagnostic characters, including better preserved cuticle fragments.

Genus *Pelourdea* Seward

Species *Pelourdea vogesiaca* (Schimper et Mougeot) Seward

Figure 12.2d–g

Material: NHMW 2021/0016/0327, 2021/0016/0331, 2021/0016/0334, 2021/0016/0341–0343, 2021/0016/0345, 2021/0016/0349, 2021/0016/0351, 2021/0016/0353, 2021/0016/0356, 2021/0016/0360, 2021/0016/0364, 2021/0016/0370, 2021/0016/0372, 2021/0016/0375, 2021/0016/0377, 2021/0016/0379, 2021/0016/0381–0383, 2021/0016/0390, 2021/0016/0394, 2024/0272/0001–0003, 2024/0272/0012

Preservation: In total, 27 isolated leaf compressions of *Pelourdea vogesiaca* were identified, preserved primarily on the surface of small hand pieces that retain only portions of the leaves. Nonetheless, some key morphological features such as the leaf base, apex, margin, and surface structure are discernible.

Description: The leaf fragments are lanceolate with an acute apex. The most complete fragments are up to 220 mm long, and 18–20 mm wide, tapering apically (NHMW 2024/0272/0001, NHMW2021/0016/0394). In some cases (NHMW2021/0016/0349; NHMW2021/0016/0364), longitudinal striations are spaced at intervals of <1 mm. The base is crescent-shaped to truncate and up to 20 mm wide (NHMW 2024/0272/0002). One specimen (NHMW2021/0016/0349; Fig. 12.2e) exhibits a distinct area of missing leaf compression bordered by a thickened edge.

Discussion: The leaves exhibit distinct longitudinal striations, indicative of parallel veins that extend from the base and terminate distally at the leaf margin and apex. The crescent-shaped bases correspond to the *Pelourdea* leaves described by Ash (1987) as lanceolate and helically arranged around an unbranched stem.

Yuccites vogesiacus was initially described by Schimper et Mougeot (1844). However, the name *Yuccites* Schimper et Mougeot is considered illegitimate under International Code of Botanical Nomenclature (ICBN) (Turland et al. 2018) because *Yuccites* Martius holds nomenclatural priority. Seward (1917) established the genus *Pelourdea* Seward, designating *Pelourdea vogesiaca* (Schimper et Mougeot) Seward as the type species, for *Yuccites* type leaves sharing morphological similarities with *Cordaites* but that cannot be assigned to the order Cordaitales.

Desmiophyllum imhoffii (Heer) Florin from the Carnian of Neuewelt (near Basel, Switzerland) and Germany (e.g., Kelber and Hansch 1995) resembles this species in its lanceolate leaf shape and acute apex. However, *D. imhoffii* lacks reported details on the leaf bases and features a pattern of interconnected, interstitial veins (see Kelber and Hansch 1995, Fig. 158), in contrast to the crescent shaped leaf base and parallel veins of *P. vogesiaca*. Kelber and Hansch (1995, p. 73), however, interpreted the interconnected, interstitial veins of *D. imhoffii* as a possible taphonomic artefact, suggesting that *D. imhoffii* may represent a junior synonym for *P. vogesiaca*.

As described above, specimen NHMW2021/0016/0349 (Fig. 12.2e) shows a distinct area of missing lamina bordered by a thickened edge, here interpreted as a reaction rim indicative of arthropod herbivory.

Distribution: *Pelourdea vogesiaca* is a common fossil in the Early and Middle Triassic plant fossil assemblages across Europe (e.g., Kustatscher et al. 2019; Forte et al. 2021). Confirmed records from the Carnian come from Monte Pora (Bergamasc Alps), Dogna (Julian Alps), and Raibl/Cave del Predil (Carnic Alps) in Italy (Passoni and Van Konijnenburg-van Cittert 2003; Kustatscher et al. 2019, 2023; Dobruskina et al. 2001). Recently they have been described also from the Norian of Preone, Rio Seazza (Carnic Pre-Alps; Kustatscher et al. 2023).

Pelourdea sp.

Material: NHMW 2021/0016/0338, 2021/0016/0347, 2021/0016/0354, 2021/0016/0363, 2021/0016/0365, 2021/0016/0373, 2024/0272/0008, 2024/0272/0012, 2021/0016/0392, 2024/0272/0014

Preservation: Eight isolated, partly preserved and heavily charcoalified leaf compressions of *Pelourdea* sp. were identified. They are typically three-dimensionally deposited and thus only partially exposed with only few observable morphological features such as surface structure and leaf shape.

Description: The leaf fragments are lanceolate with a pointed apex and are generally preserved curved. The most complete fragment is 90 mm long and 2–7 mm wide (NHMW 2024/0272/0012). In some cases (NHMW 2024/0272/0012) longitudinal striations are visible spaced at intervals of 0.5–0.8 mm. In one case a putative plant-animal interaction is visible (MHMW 2021/0016/0347).

Fig. 12.2 a-i *Elatocladus* sp. and *Pelourdea* sp. **a** Leafy shoot of *Elatocladus* sp. (NHMW 2021/0016/0348). **b** Counterpart to **a** (NHMW 2021/0016/0358). **c** Close-up of **a**. **d** Leaf of *Pelourdea vogesiaca* (NHMW 2024/0272/0001). **e** Leaf of *P. vogesiaca* with arthropod feeding damage (NHMW 2024/0016/0349). **f** Leaf tip of *P. vogesiaca* with (NHMW 2024/0272/0003). **g** Putative leaf base of *P. vogesiaca* (NHMW 2024/0272/0002). **h** *Pelourdea* sp. with typical curled leaf shape and non-planar deposition (NHMW 2024/0272/0008). **i** *Pelourdea* sp. (NHMW 2024/0272/0014). Scales = 10 mm

Fig. 12.3 Conifer shoots and their preservation modes.**a** Leafy branch of *Brachyphyllum* sp. 1 (GBA 2021/0170/0001). **b** Leafy branch of *Brachyphyllum* sp 1. (NHMW 2021/0170/0147). **c** Tip of leafy branch of *Brachyphyllum* sp. 1 (NHMW 2024/0272/0015). **d** Leafy branch of *Brachyphyllum* sp. 2 with preserved silhouette (NHMW 2021/0123/0033). **e** Leafy branch of *Brachyphyllum* sp. 2 with anatomically preserved leaves (NHMW 2024/0272/0004). **f** Close-up of **i**. **g** *Brachiophyllum* sp. 3 (NHMW2001/0123/0001). **h** Close-up of **g**. **i** Leafy branches of *Pagiophyllum* sp. 1 (NHMW 2021/0016/0380). **j** Close-up of **i**. **k** Amorph articulated conifer shoot (NHMW 2021/0016/0393). No leaf structures are preserved, and branches often have an organic halo in the surrounding sediment. Scales of **a–j** = 10 mm, scale of **k** = 50 mm

Fig. 12.4 Reproductive organs. **a** Putative ovuliferous complex (NHMW 2024/0272/0009). **b** Counterpart to **a** (NHMW 2024/0272/0009). **c** Seed morphotype 1 (NHMW 2024/0272/0006). **d** Putative orvuliferous complex 2 (2024/0272/0016). **e** Reverse view of seed morphotype 2 (2024/0272/0016). **f** *Carpolithes* sp. (NHMW 2024/0272/0007). **g** Putative reproductive organ (NHMW 2021/0016/0393). **h** Putative reproductive organ, potential counterpart of **g** (NHMW 2021/0016/0355). Scale of **a–f** = 5 mm. Scale for **g** and **h** = 10 mm

Discussion: The leaves are distinguishable from the typical *Pelourdea vogesiaca*-type of leaves by the narrower leaf widths, strongly curved leaf arrangement in the sediment, smaller apex angles, and often three-dimensional and folded deposition in the sediment. In contrast, *P. vogesiaca* leaves are lanceolate with larger apex angles and are typically deposited flat on the sediment surface. Similar leaf fragments from the Norian of Preone were also informally ascribed to the genus *Pelourdea* (Kustatscher et al. 2023: Fig. 8b). A detailed study of all lanceolate leaves from the Late Triassic would be necessary to understand whether the lanceolate leaf type belongs to *P. vogesiaca* or represents a distinct taxon.

Distribution : *Pelourdea* sp. has been described also from the Norian of Preone, Rio Seazza (Carnic Pre-Alps; Kustatscher et al. 2023).

Genus *Brachyphyllum* Brongniart

***Brachyphyllum* sp. 1**

Figure 12.3a–c

Material: GBA 2021/0170/0001, NHMW 2021/0170/0147, 2024/0272/0015

Preservation: Three specimens were assigned to *Brachyphyllum* sp. 1. These fossils are typically preserved as leafy, bifurcating branches preserved as compressions.

Description: This conifer shoot type is composed of penultimate shoot fragments up to 160 mm long. The axes of the shoots are 5–9 mm wide and arise at an angle of ~20–30°. The shoots are covered by helically arranged, imbricate scale-like leaves. The leaf attachments are 2.5–6 mm wide. The individual leaves reach 7 mm in length. The leaves are slightly curled towards the adaxial sides. The leaf apexes are typically obtuse, sometimes acute and are apically curved towards the axis.

Discussion: The specimens figured by Dobruskina et al. (2001) from the Carnian of the Julian Alps (Dobruskina et al. 2001: pl. 4, Fig. 4; pl. 8, Figs. 1–3) and the Karawanken mountains (Slovenia; Dobruskina et al. 2001: pl. 17, Fig. 5) as *Voltzia foetterlei* resemble our specimens closely. The organic material is better preserved, and the shoot fragments bear terminal female cones. Also, the specimen described by Passoni (1999) as *Brachyphyllum* sp. 2 (Passoni 1999: pl. 8, Fig. 2) resembles our specimens in stem organization and leaf arrangement.

Distribution: Conifer shoots resembling this taxon were already described from the Carnian of Monte Pora and the Julian Alps (Italy) as well as the Karavanke Mountains (Slovenia; Dobruskina et al. 2001; Passoni 1999).

***Brachyphyllum* sp. 2**

Figures 12.3d–f and 12.5d–e

Material: NHMW 2021/0123/0004, 2021/0123/0033, 2021/0016/0392

Preservation: Four charcoalified compressions of conifer shoots were assigned to *Brachyphyllum* sp. 2, all yielding cuticle fragments. One specimen (NHMW 2021/

0272/0004; Fig. 3I) is well-preserved, exhibiting clear leaf details, whereas the other two are largely amorphous, with mainly their silhouettes remaining intact.

Description: The longest shoot fragment reaches 86 mm in length and 11 mm in width. The leaves are arranged in a close helix. The base is strongly keeled and spindle-shaped in front view (7.5 × 5 mm). In lateral view the apex slightly protrudes (7.5 × 4.8 mm) at an angle of ~43–51°. The leaf apex is typically bent slightly towards the adaxial side of the leaf, and occasionally towards the apical side.

The cuticle is moderately thick, with randomly arranged, dicyclic or incomplete dicyclic stomatal complexes. These complexes consist of 6–8 trapezoidal subsidiary cells, devoid of papillae (the first cycle presents cells with an average width of 61 μm and average length of 95 μm, while the second cycle has cells with an average length of 154 μm and average width of 82 μm). The epidermal cells are isodiametric and lack papillae (average length 154 μm, average width 123 μm).

Discussion: The specimens figured by Dobruskina et al. (2001: pl. 6, Figs. 2–8) from the Carnian of the Julian Alps as *Voltzia haueri* Stur resemble our specimens closely, but the organic material is better preserved in the Julian Alps and the shoot fragments bear terminal female cones.

Distribution: This taxon may also be present in the Carnian of the Julian Alps (Italy; Dobruskina et al. 2001).

Brachyphyllum sp. 3

Figures 12.3g, h and 12.5f, g

Material: NHMW 2021/0123/0001

Preservation: A single specimen, preserved as a compression of a leafy branch fragment with cuticle preserved, is assigned to *Brachyphyllum* sp. 3. The cuticular features were poorly visible when observed under transmitted light microscopy but became clearer under epifluorescence.

Description: The fragment is 22 mm long and 3.6 mm wide. The leaves are arranged in a helix, with an imbricated leaf arrangement, clasping the axis. The leaf shape is roughly triangular; they are up to 2.3 mm wide and 3.1 mm long. The leaf base is broad, and the apex is rounded with an obtuse apex angle and curled towards the adaxial leaf side.

The cuticle is thick, with randomly arranged monocyclic stomatal complexes consisting of 6–8 trapezoidal subsidiary cells. The stomatal complexes are very distinct because the subsidiary cells are larger and more cutinized than the epidermal cells. Each subsidiary cell has a small papilla extending towards the stomatal aperture. The epidermal cells are polygonal/isodiametric and lack papillae.

Discussion: *Brachyphyllum* sp. 3 shares similarities with *Brachyphyllum* sp. 2 in terms of branch and leaf sizes, and some leaf shapes in lateral view. Unlike *Pagiophyllum* sp. 1, where leaves arise at a 90° angle from the branch, the leaves of

Brachyphyllum sp. 3 are closely imbricated. The cuticles of *Brachyphyllum* sp. 3 differs from those of *Brachyphyllum* sp. 1 and *Pagiophyllum* sp. 1 by being much thicker, and from *Pagiophyllum* sp. 1 because the latter has peculiar monocyclic stomatal complexes with polar and lateral subsidiary cells. *Brachyphyllum* sp. 2 differs from *Brachyphyllum* sp. 3 by the dicyclic stomatal complexes of the former.

Genus *Pagiophyllum* Heer

***Pagiophyllum* sp. 1**

Figures 12.3i, j and 12.5b, c

Material: NHMW 2021/0016/0361, 2021/0013306/0367, 2021/0016/0378, 2021/0016/0380, 2021/0016/0388

Preservation: In total, five specimens of *Pagiophyllum* sp. 1 were identified. They are preserved as compressions of articulated, frequently bifurcating, or single leafy shoots. Cuticles were extracted from two samples (NHMW 2021/0016/0380, 2021/0016/0388; Fig. 12.5b, c).

Description: The conifer shoot fragments are up to 70 mm long and 100 mm wide. Ultimate shoots arise alternately at an angle of 30–60° and are up to 60 mm long and 7 mm wide, with a 1.5–2 mm wide axis and leaves arranged in an open helix. The leaves are strongly keeled, appearing subtriangular in lateral view. The rounded apex is curved towards the adaxial side of the leaf. The leaves are 3–4 mm long with a 1.5–2.5 mm wide base, extending at a ~90° angle from the axis.

The cuticle is thin, with monocyclic stomatal complexes arranged in loosely spaced rows. The complexes consist of 4–6 subsidiary cells, with the polar cells being less cutinized than the 2–4 lateral cells. The subsidiary cells are trapezoidal, sometimes slightly rounded. Occasionally they feature small papillae, but these never completely cover the stomatal aperture. The epidermal cells are generally rectangular, often elongated (average length 85 μm, average width 23 μm), and do not exhibit papillae.

Discussion: This conifer is characterized by widely spaced, keeled leaves, slightly resembling in attachment angle and shape the leaves of *Voltzia recubariensis* (Massalongo ex De Zigno) Schenk. The cuticle is, however, peculiar because it is not composed of a typical monocyclic stomata complex: there seems to be a slight differentiation into polar and lateral cells. It is possible that these shoots do not represent typical *Voltzia* shoots but belong to some other conifer group. The shoots resemble those of *Geinitzia*, which are, however, mainly known from the Cretaceous onwards. This is why, for now, we keep the shoot fragments in the fossil taxon *Pagiophyllum* sp. 1, typical for conifers with leaves that are much longer than wide. Future finds of more and better-preserved material will permit us to understand whether conifers of the family Pinaceae could indeed have been present already during the CPE in the Southern Alps.

Indeterminate woody branches of conifers

Figures 12.3k, 12.5e and 12.6

Material: NHMW 2021/0016/0329, 2021/0016/0330, 2021/0016/0336, 2021/0016/0385, 2021/0016/0386, 2021/0016/0387, 2021/0016/0339, 2021/0016/0393, 2024/0272/0004, 2024/0272/0010, 2024/0272/0015

Preservation: In contrast to the *Brachyphyllum* and *Pagiophyllum* specimens, which retain a visible leaf morphology, most of the preserved branches lack any foliar detail and are typically macromorphologically not well preserved. Notably, these shoots frequently exhibit what appears to be an organic halo in the surrounding sediment. Strangely enough, three fragments yielded, although badly preserved, cuticles (NHMW 2021/0016/0330; 2021/0016/0329; 2021/0016/0386).

Description: Ultimate shoots emerge from the main shoot at an angle of 30°, while penultimate shoots extend at an angle of 70°. The angle of shoot emergence decreases from proximal to distal within a branch. The shoot fragments vary in length, reaching at least up to 285 mm, with widths of up to 9.5 mm.

The cuticle is relatively thick, with randomly arranged monocyclic stomatal complexes. These complexes consist of 5–7 trapezoidal subsidiary cells, which are devoid of papillae (average length 35 μm, average width 23 μm). Stomatal complexes are difficult to distinguish from the epidermal cells due to the uniform cutinization and similar cell shape. In rare cases, however, subsidiary cells exhibit thicker anticlinal walls than the surrounding epidermal cells. The epidermal cells are polygonal/isodiametric and lack papillae (average length 35 μm, average width 24 μm).

Discussion: Although poorly preserved, the branches resemble either *Brachyphyllum* sp. 1 or *Brachyphyllum* sp. 2 in branch arrangement and overall width of the axes. If referable to those species, the intermediate specimens could provide information on branching patterns due to their comparatively larger preserved size.

The cuticle of the intermediate woody branches, although very badly preserved in this material, differs from *Brachyphyllum* sp. 2 in that the latter exhibits dicyclic stomatal complexes, with subsidiary cells being more cutinized and with a different shape from the epidermal cells. This suggests that these remains could rather belong to *Brachyphyllum* sp. 1. This issue can only be resolved once we recover cuticle samples of the latter species (Table 12.1).

Indeterminate reproductive organs

Putative ovuliferous complex 1

Figure 12.4a, b

Material: NHMW 2024/0272/0009

Preservation: A single specimen is preserved as a compression of an ovuliferous complex.

Description: The ovuliferous complex is spatulate, with a length of 19 mm and a width of 20 mm. The base constricts to a width of 7 mm and has a truncated form. The distal margin is crescent-shaped, featuring a ~1.5 mm wide ridge. A narrow furrow

Table 12.1 Characters of the four conifer shoot morphotypes described in this paper

Conifer shoot morphotypes	Leaf type	Leaf arrangement	Leaf length [mm]	Leaf width [mm]	Total branch width
Brachyphyllum sp. 1	Scale-like	Helical, imbricated, leaf clasping branch	≤ 7	2.5–6	5–9
Brachyphyllum sp. 2	Broad based needle-like	Helical, leaves extending at a 43–51° angle	≤ 4.5	~7.5 × 5	≤ 11
Brachyphyllum sp. 3	Scale-like	Helical, imbricated, leaves clasping branch	≤ 3.1	≤ 2.3	3.6
Pagiophyllum sp.	Scale-like	Helical, leaves extending at a 90° angle from the branch	3–4	1.5–2.5	≤ 7 (axis 1.5–2)
Conifer shoot morphotypes	Branching angle	Cuticle thickness	Stomatal apparatus	Stomatal arrangement	Epidermal cell shape
Brachyphyllum sp. 1	20–30	–	–	–	–
Brachyphyllum sp. 2	–	Moderately thick	Dicyclic or incomplete dicyclic, 6–8 subsidiary cells	Random	Isodiametric
Brachyphyllum sp. 3	–	Thick	Monocyclic, 6–8 subsidiary cells	Random	Polygonal/ isodiametric
Pagiophyllum sp.	30–60	Thin	Monocyclic, 4–6 subsidiary cells, occasionally small papillae	Loosely spaced rows	Rectangular

is present adjacent to this ridge. On the surface, the ovuliferous complex displays striations following the curvature of the distal margin as well as fine longitudinal striations.

Discussion: This ovuliferous complex seems to have heavily fused segments which would suggest that it might belong to the Cheirolepidiaceae rather than to the Voltziaceae. Unfortunately, so far only one, poorly preserved specimen has been found. More details would be needed to solve its botanical affinity.

Putative ovuliferous complex 2

Figure 12.4d, e

Material: NHMW 2024/0272/0016

Preservation: The fossil is a three-dimensionally preserved compression and completely extracted from the matrix.

Description: The putative ovuliferous complex (OC) is spatulate and is slightly three-dimensionally curved (convex towards the putative abaxial side). It is 12 mm long and has a maximum width of up to 7 mm. The base is rounded and constricted to a width of 2 mm. The apex is slightly pointed with an obtuse apex angle of 125°. On the putative adaxial side of the OC, a crescent-shaped ridge, approximately 2 mm from the apex, roughly follows the distal margin. The surface of the seed is, with exception of the distal, adaxial ridge, entirely papillate. The distal margin is slightly convex and irregular, whereas the proximal margin appears relatively straight and slightly rugged.

Discussion: This ovuliferous complex looks very different from the typical Voltzialean conifer scales of the Triassic. However, the Carnian is a key period for the radiation of conifer families (Dal Corso et al. 2020; Roghi et al. 2022).

Seed morphotype 1

Figure 12.4c

Material: NHMW 2024/0272/0006

Preservation: The fossil is preserved as part and counterpart of a charcoalified compression fossil.

Description: The seed is wedge-shaped (or cuneate) and in total 9.5 mm long, 4 mm wide distally, 1 mm wide proximally. The seed is flanged by a lateral wing that extends towards the distal margin and slightly increases in width (0.3–0.8 mm).

Discussion: Winged seeds are characteristic of Araucariaceae-type conifers as well as of Permian conifers (Majonicaceae; Looy and Stevenson 2014). Both families are apparently present during the Carnian in the paleotropics (e.g., Roghi et al. 2006, 2022).

***Carpolithes* sp.**

Figure 12.4f

Material: NHMW 2021/0016/0323, 2021/0016/0335, 2024/0272/0007

Preservation: Three specimens of similar shape and size are preserved as compression fossils.

Description: The seeds are approximately round with a slightly constricted apex and a rounded to slightly truncated base. The seeds measure 7–8 mm in length and 6 mm in width. The seeds have parallel, discontinuous striae along the longitudinal axis.

Discussion: The preservation is insufficiently good to discern any further details or to enable an assignment other than to "seeds".

Putative reproductive organs

Figure 12.4g, h

Material: NHMW 2021/0016/0355, 2021/0016/0389

Preservation: Two fossils are preserved as charcoalified compression, partly impression. The two specimens seem to represent part and counterpart, assuming that the surrounding sediment has been altered through preparation, so that the two parts no longer align.

Description: The woody putative reproductive organ is flat, more or less rounded, with a diameter of 39–44 mm and four slightly more angular corners. A relatively well-preserved, smooth ~2-mm-wide rim runs along the margin. The medial surface features distinct parallel striations that extent, though less distinctly, into the smooth margin; they likely represent foliation resulting from the charcoalification process, similar to other woody fossils from the assemblage (see Fig. 12.6e).

Discussion: The specimen is too badly preserved for details that could help to assign this structure to any plant group. Nonetheless, it is clearly not a wood fragment with rounded corners because it did yield some cuticles, even if they were poorly preserved.

12.4 Discussion

12.4.1 Quantitative Composition

A total of 92 plant remains were recovered from the Polzberg locality, of which 77 have been identified to some extent. The remaining 15 consist of indiscernible organic matter and are not considered here. The most abundant genus is *Pelourdea*, which potentially includes two species (*Pelourdea vogesiaca* and *Pelourdea* sp.) and accounts for 45% of all identified fossils (Fig. 12.7; Supplementary material 12.1). Conifer taxa including *Brachyphyllum* and *Pagiophyllum*, along with a substantial

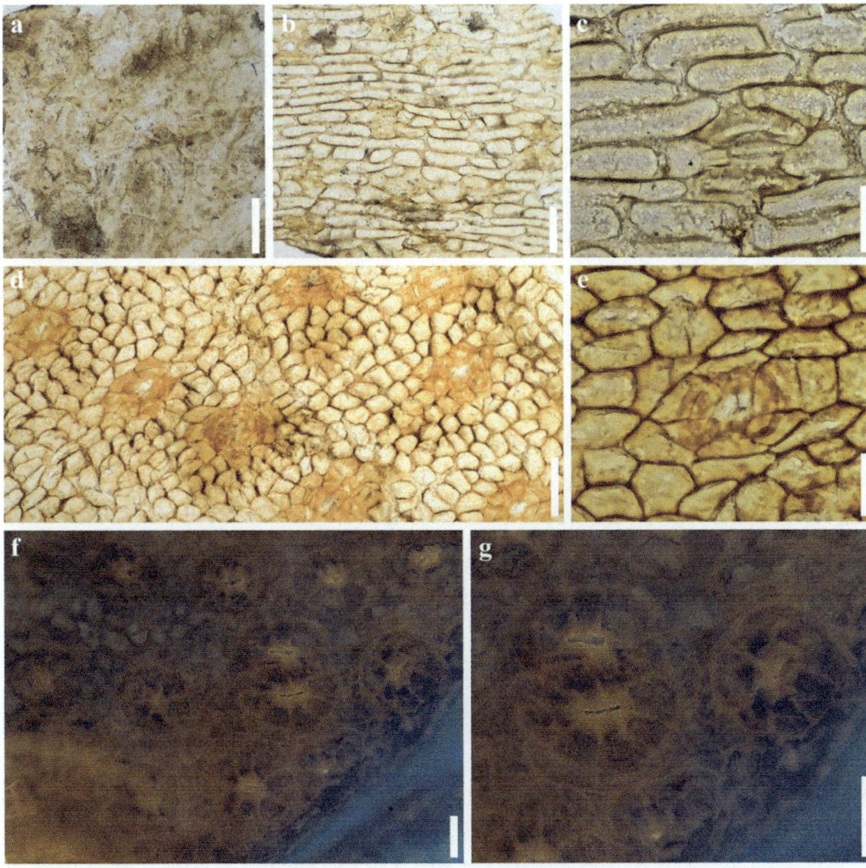

Fig. 12.5 Transmitted light micrographs (**a–e**) and epifluorescence micrographs (**f, g**) of cuticles. **a** *Elatocladus* sp. (NHMW 2021/0016/0348). **b** *Pagiophyllum* sp. 1 (NHMW 2021/0016/0388). **c** *Pagiophyllum* sp. 1 (NHMW 2021/0016/0388) close-up of stoma. **d** *Brachyphyllum* sp. 2 (NHMW 2021/0016/0392). **e** Indet. conifer shoot (NHMW 2021/0016/0339) close-up of stoma. **f** *Brachyphyllum* sp. 3 (NHMW 2021/0123/0001). **g** *Brachyphyllum* sp. 3 (NHMW 2021/0123/0001) close-up of stoma. Scale for **a**, **b** and **d** = 100 μm. Scale for **c** and **e–g** = 50 μm

number of indeterminate branches with likely conifer affinity, collectively constitute 32% of the determined plant fossils. Five types of reproductive organs of unknown affinity represent together 9% of the assemblage, while indeterminate wood accounts for 13%.

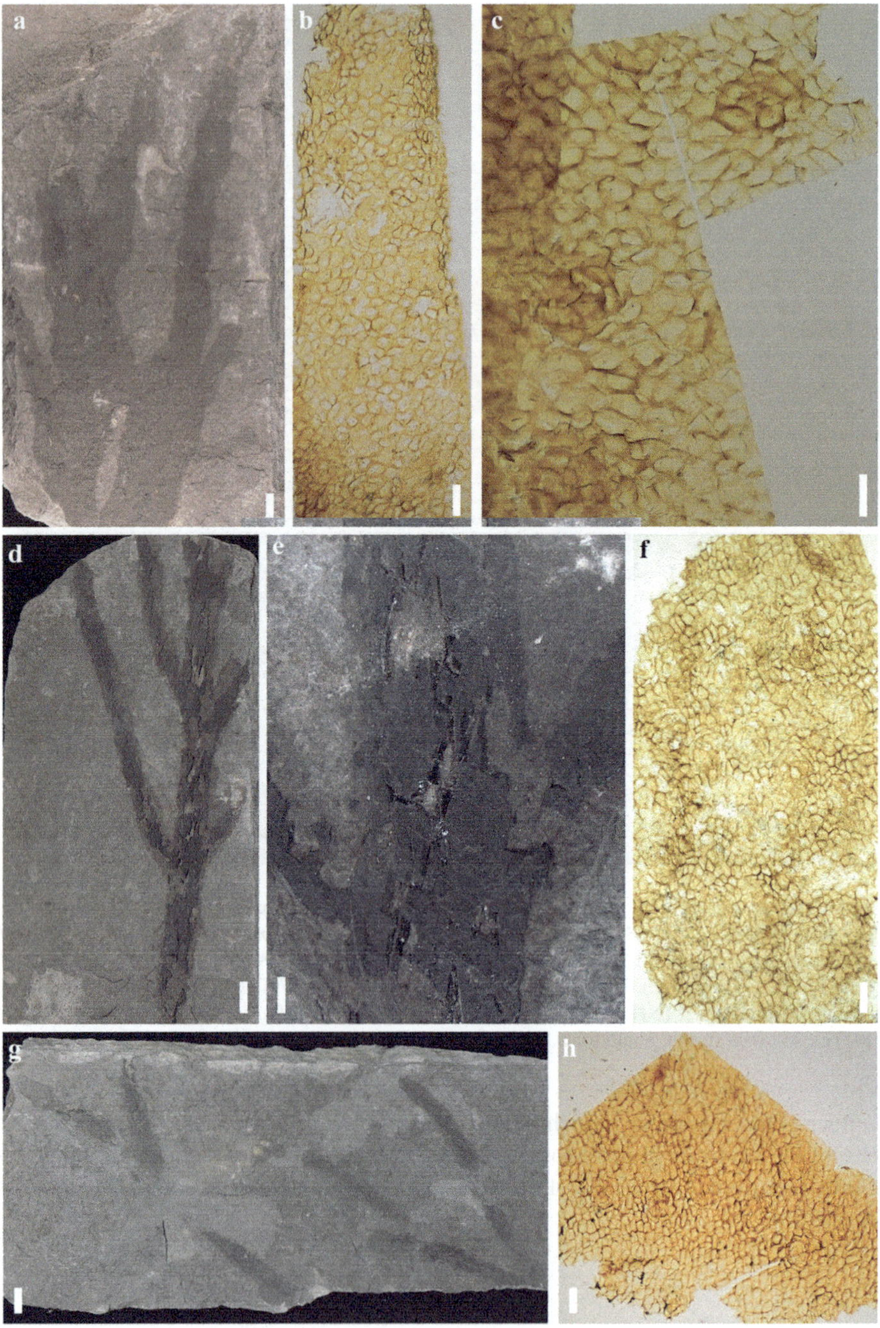

Fig. 12.6 Indeterminate branch fragments of putative conifer affinity. **a** Conifer shoot fragment (NHMW 2021/0016/329). **b** Cuticle fragment (NHMW 2021/0016/329). **c** Cuticle fragment showing stomatal details (NHMW 2021/0016/329). **d** Conifer shoot fragment (NHMW 2021/0016/330). **e** Conifer shoot fragment (NHMW 2021/0016/330) showing detail of leaves. **f** Cuticle fragment (NHMW 2021/0016/330). **g** Conifer shoot fragment (NHMW 2021/0016/386). **h** Cuticle fragment (NHMW 2021/0016/389). Scale for **a, d, g** = 10 mm, for **b, f, h** = 75 μm, for **c** = 50 μm and for **h** = 2 mm

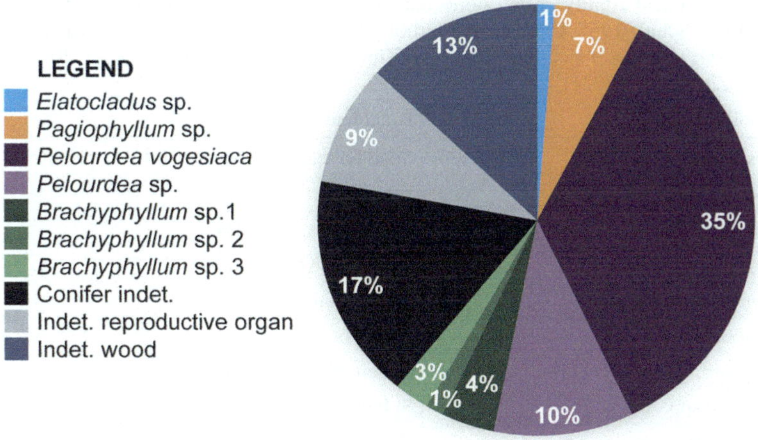

Fig. 12.7 Pie chart showing the percentage of fossil specimens representing different plant taxa, wood, and reproductive organs in the Polzberg assemblage

12.4.2 Comparison with the Palynoflora from Polzberg

The Reingraben Formation at the Polzberg succession has yielded a rich and diverse assemblage of spores and pollen grains (Table 12.2), along with marine palynomorphs such as acritarchs, dinoflagellate cysts, and chitinous remains, including scolecodonts and foraminiferal linings. Fern spores are particularly diverse and include representatives of Dicksoniaceae (*Converrucosisporites*), Dicksoniaceae or Matoniaceae (*Concavisporites*), Dipteridaceae (*Combaculatisporites*), and Osmundaceae (*Todisporites*, *Verrucosisporites*), as well as several taxa not yet assigned to specific families (*Anapiculatisporites*, *Lunzisporites*, *Punctatisporites*, *Trachysporites*).

Lycophyte spores are also diverse, with representatives of Isoetales (*Aratrisporites*), Selaginellales (*Kraeuselisporites*, *Uvaesporites*), Lycopodiales (*Lycopodiacidites*), and general lycophyte affinity (*Densosporites*). Sphenophytes (*Calamospora*) are present but less diverse.

Among gymnosperms, conifers are the most diverse group, including taxa assigned to Araucariaceae (*Araucariacites*), Cheirolepidiaceae (*Camerosporites*, *Classopollis*, *Duplicisporites*, *Partitisporites*), Patokaeaceae (*Enzonalasporites*, *Patinasporites*, *Pseudoenzonalasporites*, *Vallasporites*), Podocarpaceae (*Platysaccus*, *Samaropollenites*), and Voltziaceae (*Klausipollenites*, *Lunatisporites*, *Ovalipollis*, *Triadispora*). Additional conifer pollen grains with uncertain affinity include *Brachysaccus* and *Ellipsovelatisporites*. Other gymnosperm groups are less diverse, among them seed ferns (*Alisporites*), Bennettitales (*Aulisporites*, *Leschikisporis*), and Cycadales/Ginkgophytes/Bennettitales (*Cycadopites*).

The palynological assemblage reflects a complex terrestrial ecosystem, dominated by swamp or marsh (floodplain) vegetation, with additional contributions

Table 12.2 Botanical affinity of the spores and pollen of the Polzberg succession (Couper 1958; Townrow 1962; Schulz 1967; Clement-Westerhof 1974; Van Konijnenburg-van Cittert 1978, 1993; Looy et al. 2001; Barron et al. 2006; Bonis and Kürschner 2012; Costamagna et al. 2018; Baranyi et al. 2019; Zhang et al. 2021)

Spores/pollen	Botanical affinity
Alisporites sp.	Seed ferns, Corystospermales
Anapiculatisporites telephorus	Ferns
Aratrisporites spp.	Lycophytes, Isoetales
Araucariacites australis	Conifers, Araucariaceae
Aulisporites astigmosus	Bennettitales
Brachysaccus sp.	Conifers
Calamospora lunzensis	Sphenophytes, Equisetales
Calamospora spp.	Sphenophytes, Equisetales
Camerosporites secatus	Conifers, Cheirolepidiaceae
Classopollis cf. *meyeriana*	Conifers, Cheirolepidiaceae
Combaculatisporites mesozoicus	Ferns, Dipteridaceae
Concavisporites spp.	Ferns, Dicksoniaceae and Matoniaceae
Converrucosisporites tumulosus	Ferns, Dickoniaceae
Cycadopites spp.	Cycadales, Ginkgophytes, Bennettitales
Densosporites sp.	Lycophytes
Distalanulatisporites punctus	Indet
Duplicisporites continuus	Conifers, Cheirolepidiaceae
Duplicisporites granulatus	Conifers, Cheirolepidiaceae
Duplicisporites verrucosus	Conifers, Cheirolepidiaceae
Ellipsovelatisporites spp.	Conifers
Enzonalasporites vigens	Conifers, Patokaeaceae
Klausipollenites sp.	Conifers, Voltziaceae
Kraeuselisporites cooksonae	Lycophytes, Selaginellales
Kraeuselisporites spp.	Lycophytes, Selaginellales
Leschikisporis aduncus	Bennettitales
Lunatisporites acutus	Conifers, Voltziaceae
Lunzisporites spp	Ferns
Lycopodiacidites kuepperi	Lycophytes, Lycopodiales
Ovalipollis ovalis	Conifers, Voltziaceae
Ovalipollis pseudoalatus	Conifers, Voltziaceae
Partitisporites spp.	Conifers, Cheirolepidiaceae
Patinasporites densus	Conifers, Patokaeaceae
Platysaccus spp.	Conifers, Patokaeaceae
Pseudoenzonalasporites summus	Conifers, Patokaeaceae
Punctatisporites sp.	Ferns

(continued)

Table 12.2 (continued)

Spores/pollen	Botanical affinity
Retusotriletes hercinicus	Lycophytes, Selaginellales
Samaropollenites speciosus	Conifers, Podocarpaceae
Spiritisporites cf. *spirabilis*	Indet
Todisporites minor	Ferns, Osmundaceae
Todisporites spp.	Ferns, Osmundaceae
Trachysporites spp.	Ferns
Triadispora spp.	Conifers, Voltziaceae
Uvaesporites sp.	Lycophytes, Selaginellales
Vallasporites ignacii	Conifers, Patokaeaceae
Verrucosisporites morulae	Ferns, Osmundaceae

from coastal pioneer communities and upland hinterland floras (Lukeneder et al. 2024). This reconstruction is consistent with the diverse landscape inferred from the Lunz plant macrofossil assemblages (Pott and Lukeneder 2026, this volume), but contrasts with the comparatively low-diversity macrofossil flora recovered from Polzberg (this study). This discrepancy reflects differences in taphonomic and spatial resolution: plant macrofossils record local vegetation shaped by depositional and transport biases, whereas palynological assemblages integrate a broader regional signal of vegetation and paleoclimate (Costamagna et al. 2018). The long transport of the plant macrofossils therefore yields a very contradictory result.

12.4.3 Comparison with the Lunz Flora

The most important and best-studied *Fossil-Lagerstätte* of the Northern Calcareous Alps from a plant perspective is the Lunz Flora (e.g., Pott and Lukeneder 2026, this volume). It is also referred to the CPE and, thus, coeval with Polzberg *Konservat-Lagerstätte*. The plant fossil assemblage there is rich in sphenophytes, such as the 5–10 m high *Equisetites arenaceus,* as well as the smaller, herbaceous *Equisetites conicus* and *Neocalamites merianii* (Pott et al. 2008a, 2018). The ferns of the Lunz flora were very species-rich, including the eusporangiate Marattiales *Asterotheca merianii*, *Symopteris lunzensis* and *Danaeopsis lunzensis* as well as the leptosporangiate Osmundales *Todites linnaeifolius* (Pott et al. 2018). *Phlebopteris lunzensis* and *Cynepteris reticulata* belong to the Matoniaceae, whereas *Dictyophyllum serratum, Dictyophyllum lunzensis, Clathropteris reticulata, Camptopteris lunzensis,* and *Digitopteris repanda* are representatives of the Dipteridaceae (both leptosporangiate Gleicheniales). Some ferns such as *Mertensides bullatus* could not be assigned so far to any fern family, but potentially belong to the Marattiales (Pott et al. 2018). A peculiar feature in this context is the co-occurrence of both the typically more Paleozoic eusporangiate and the more modern leptosporangiate ferns (Pott and Lukeneder 2026, this volume). This mixture of "more Paleozoic" and "more Mesozoic" elements seems typical of the Carnian (CPE) floras (Dal Corso et al. 2020; Roghi et al. 2022).

The seed plants were also important elements of the Lunz flora. Ginkgophytes may be represented by *Arberophyllum florinii*, although it remains unclear whether that species represents a ginkgophyte or a fern-like plant (Pott and Lukeneder 2026, this volume).

The two most common plants in the entire Lunz plant fossil assemblage are the bennettites *Pterophyllum filicoides* and *Pterophyllum brevipenne* with pinnated leaves. Large, incompletely pinnate or entire leaves up to 70 cm long assignable to *Anomozamites lunzensis*, *Nilssoniopteris haidingeri* and *Nilssoniopteris angustior* are also comparatively common (Pott and Lukeneder 2026, this volume). The order Bennettitales is also represented by reproductive units such as *Bennetticarpus wettsteinii*, *Haitingeria krasseri*, *Westersheimia pramelreuthensis*, *Pramelreuthia haberfelneri*, *Sturiella langeri*, *Lunzia austriaca*, *Leguminanthus siliquosus* and *Cycadolepis wettsteinii* (Pott 2016, 2021; Pott et al. 2017; Pott and Lukeneder 2026, this volume). Among the cycads, the most important genus is *Nilssonia* with four species (Pott et al. 2007a). Other leaves belong to *Pseudoetenis cornelii* (Pott et al. 2007b). Reproductive organs of the cycadophytes are abundant and include megasporophylls of the *Dioonitocarpidium* type, such as *Dioonitocarpidium keuperianum* and *Dioonitocarpidium liliensternii* (Pott 2019).

Interestingly, conifers such as *Elatocladus lipoldii* and a possible *Voltzia* (if not an equisetalean strobilus) are rather underrepresented (Pott et al. 2008b; Pott and Krings 2010; Pott and Lukeneder 2026, this volume). In contrast, these are the dominant plant group in the Polzberg fossils assemblage. This may reflect the fact that the Lunz plant fossil assemblage was deposited in a deltaic swamp setting rather than in a marine basin like the Polzberg plant fossil assemblage. The plant fossils from Polzberg could, in addition, also have been transported from a general coastal setting rather than from the swampy area. This suggests that some of those conifers could have a potential mangrove nature. This would explain why so many conifer remains were transported into the sea, even though conifers are generally considered to inhabit the hinterland areas.

12.4.4 Polzberg, a Konservat-Lagerstätte *of the CPE*

Konservat-Lagerstätten, with their excellent preservation of invertebrate and vertebrate fossils, provide unique insights into paleocommunities, foodchains and dietary habits, as well as into trophic interactions between the inhabitants of marine ecosystems (e.g., Seilacher 1970; Bernardi et al. 2011; Briggs 2014; Lukeneder and Lukeneder 2021). Findings of otherwise rare fossil groups and the soft body preservation of numerous organisms help shed light on otherwise hidden organisms.

Polzberg is one of the few known Triassic *Konservat-Lagerstätten* and is of particular significance because its succession was deposited during the Carnian Pluvial Episode (CPE). The sedimentary succession here clearly records the transition from an arid to humid climate, as well as an increased sedimentary input into the basin (Lukeneder and Lukeneder 2021). The CPE affected not only marine ecosystems but also triggered substantial changes in terrestrial environments. The Carnian was marked by elevated extinction rates among several tetrapod groups, including rhynchosaurs and dicynodonts, alongside a major diversification of early dinosaurs (Benton et al. 1986, 2018; Bernardi et al. 2018). Plant communities experienced significant radiation and diversification during this interval, with a shift from xerophytic to hygrophytic taxa. These include Bennettitales, several modern conifer families (e.g., Cheirolepidiaceae), and ferns (Hymenophyllaceae, Matoniaceae, Dipteridaceae). Peltaspermales and Corystospermales also reached peak diversity in the Carnian (Dal Corso et al. 2020).

Data on palynomorphs of the Northern Calcareous Alps (Mueller et al. 2016) support a general trend from a dry climate in the Julian 1 to more humid and warmer conditions during the early Julian 2, corresponding to the deposition of the Göstling Formation and the Reingraben Formation (Lukeneder and Lukeneder 2021; Lukeneder et al. 2024). The increasing humidification is reflected by a change in lithology from the basal Julian sequence in the Polzberg area (Lukeneder and Lukeneder 2021). The Polzberg *Konservat-Lagerstätte* yields a low-diversity assemblage of plant fossils, dominated by conifers. Despite the long-distance transport from their original terrestrial habitats to the basinal setting, the preservation of plant remains is relatively good.

12.4.5 The Paleoenvironment of the Polzberg Konservat-Lagerstätte

During the Carnian, the Polzberg subbasin and the Reifling Basin were located at the north-western Tethyan margin in latitudes of 15°–30° N (e.g., Lukeneder et al. 2012; Lukeneder and Lukeneder 2021). The lowermost nodular limestones of the Reifling Formation were deposited on the paleoslope of the Reifling Basin. At the base of the Julian 2, the Reifling Formation (*Trachyceras aonoides* Zone, *T. aonoides* Subzone) is replaced by the limestone deposits with organic-rich mudstones of the Göstling Formation in the *Austrotrachyceras austriacum* Zone (*A. austriacum* Subzone, biohorizon of *A. triadicum*). These are overlain by deposits of the Reingraben Formation (*A. austriacum* Subzone, biohorizon of *A. minor*).

The Reingraben Formation, characterized by dark, laminated deposits, was deposited in an epipelagic to upper mesopelagic setting, which was periodically disconnected from deoxygenated bottom currents. In these dysoxic conditions, halobiid bivalves were the dominant, sometimes rock-forming, epifaunal elements. The overlying oxygenated water column was inhabited by nektic/nektobenthic ammonoids and actinopterygian fishes, occupying the top predator niche (Lukeneder and Lukeneder 2021).

This interpretation is supported by the dominance of benthic posidoniids and a diverse nektonic fauna, including abundant ammonoids and fishes (Griffith 1977; Forchielli and Pervesler 2013; Lukeneder et al. 2020; Lukeneder and Lukeneder 2021). Bottom-water conditions during deposition ranged from dysoxic to anoxic.

Geochemical data (Lukeneder et al. 2024; Chraiki et al. 2026, this volume) indicate significant variability in oxygen availability throughout the deposition of the Carnian successions in the basin. The plant fossils here described are the allochthonous components of the fossil assemblage, together with the freshwater lungfish and the branchiopod *Euestheria*. Most other fossils are, however, autochthonous, evident by the preservation of delicate anatomical structures and the exceptional in situ preservation of buccal (anaptychi) elements within or near the body chambers of the ammonoids. A long-term ecological shift over ~500,000 years is recorded in the transition from the dominant ceratitid *Austrotrachyceras minor* to the smaller and less ornamented *Paratrachyceras haberfellneri*.

12.5 Conclusions

The Polzberg *Konservat-Lagerstätte* provides a rare, well-preserved macrofloral record from a basinal marine setting during the Carnian Pluvial Episode (CPE). The assemblage is quantitatively dominated by conifers, primarily *Pelourdea* and waste basket conifer shoot morphotypes (*Brachyphyllum, Pagiophyllum*), supplemented by rare *Elatocladus* and several indeterminate reproductive organs. This composition differs markedly from coeval macrofloras such as the Lunz assemblage, which is rich in ferns, sphenophytes, and bennettites. The discrepancy reflects strong taphonomic and transport biases. The Polzberg material represents an allochthonous plant fossil assemblage, transported from nearby terrestrial ecosystems into the basinal environment, which selectively preserved robust elements over more delicate hygrophytic taxa.

Comparison with the rich and diverse local palynoflora, which records a broad range of plant groups and ecological niches, further underscores that the macroflora is an incomplete, environmentally filtered subset of the regional vegetation. These results demonstrate that macrofossil assemblages from basinal settings cannot be interpreted isolated. Their taxonomic composition must be evaluated alongside sedimentological and palynological evidence. The Polzberg flora thus offers a window into the more resistant components of Carnian vegetation and provides valuable context for understanding plant dispersal, preservation pathways, and the terrestrial-marine interface during the CPE.

Acknowledgments We are particularly grateful to the landowners of the outcrop area, Franziska and Hermann Hofreiter as well as Eva and Karl Jagersberger (all Gaming), for sampling and digging permission. Furthermore, we thank Chris Mays (Naturhistorisches Museum Wien) for access to the collection. The work was done within the framework of and financially supported by projects of the Federal Government of Lower Austria (K3-F-964/001-2020 and K3-F-964/002-2023, Department Science and Research; headquarters in St. Pölten), the Austrian Academy of Sciences (headquarters in Vienna, project Polzberg Lukeneder 2021) represented by the National Committee for Geo/Hydro Sciences (Earth System Sciences Programme), and the Triassic project "Fenster in die Karnische Krise" of the society of friends of the Natural History Museum Vienna (Freunde des Naturhistorischen Museums Wien, CPE Lukeneder 2023). The manuscript greatly benefited from valuable comments of Dragana Životić (Belgrade) and Alexander Zdravkov (Sofia), the production editor Ashok Arumairaj (Taramani), the in-house series editor Kathleen A. Ritterbush (Utah), the publishing editor Simon Shah-Rohlfs and the production editor Antje Endemann (both Heidelberg).

Author Contribution EK, AL designed the research. AL collected the new material. EK, AL and MS provided documentation on fossil material. RB carried out the cuticle analyzes and descriptions. GR studied palynologic data. EK, AL, RB and MS contributed to writing the paper. MS and AL produced the figures.

Funding This project (including open access publishing fees) was supported by the Austrian Academy of Sciences (projects K3-F-964/001-2020, and K3-F-964/002-2023 to AL); the Natural History Museum Vienna; the Society of Friends of the Natural History Museum Vienna (project "Fenster in die Karnische Krise" to AL), the Natural History Museum Vienna; and the Joint Research Project SNF-ST (project WETTEST, CUP H33C23001340003, Dekret 35/2024 to EK and Elke Schneebeli). MS received funding from the European Union's HORIZON EUROPE program under the Marie Skłodowska-Curie grant agreement No 101210648. The authors are responsible for the contents of this publication. The funder had no impact on conceptualization, design, data collection, analysis, decision to publish, or preparation of the manuscript.

Competing interests All authors stated that they have no competing interests.

Authors' contributions All authors stated that they have equally contributed to their chapters, read, edited and agreed to submit the final versions of their contributions.

Availability of data and materials All sources of information are provided in the text. Raw data related to the fossil material from Austria are available from the corresponding authors and Alexander Lukeneder upon request.

Additional information The article is part of the Special Volume: *Late Triassic Konservat-Lagerstätten within the Carnian Pluvial Episode. 150 Years of Scientific Excavations in Austria.*

Artificial Intelligence (AI) The chapter was produced without the use of AI.

Language The entire volume uses American English reviewed by a native scientist.

Received 2025.

Published online 2026.

References

Allison RA (1988) Konservat-Lagerstätten: cause and classification. Palaeobiology 14:331–344. https://doi.org/10.1017/S0094837300012082

Ash SR (1987) Growth habit and systematics of the Upper Triassic plant Pelourdea poleoensis, Southwestern U.S.A. Rev Palaeobot Palynol 51:37–49. https://doi.org/10.1016/0034-6667(87)90018-2

Baranyi V, Rostási A, Raucsik B, Kürschner WM (2019) Palynology and weathering proxies reveal climatic fluctuations during the Carnian Pluvial Episode (CPE, Late Triassic) from marine successions in the Transdanubian Range (western Hungary). Glob Planet Change 177:155–172. https://doi.org/10.1016/j.gloplacha.2019.03.013

Barrón E, Gómez JJ, Goy A, Pieren AP (2006) The Triassic-Jurassic boundary in Asturias (northern Spain): palynological characterization and facies. Rev Palaeobot Palynol 138:187–208. https://doi.org/10.1016/j.revpalbo.2006.01.005

Bauer FK, Schnabel W (1997) Geologische Karte der Republik Österreich. Blatt Mariazell 72, 1:50.000. Geologische Bundesanstalt, Wien.

Benton MJ (1986) More than one event in the Late Triassic mass extinction. Nature 321:857–861. https://doi.org/10.1038/321857a0

Benton MJ, Bernardi M, Kinsella C (2018) The Carnian Pluvial Episode and the origin of dinosaurs. JGS 175:1019–1026. https://doi.org/10.1144/jgs2018-049

Bernardi M, Avanzini M, Bizzarini F (2011) Vertebrate fauna from the San Cassiano Formation (early Carnian) of the Dolomites region. Geo. Alp 8:122–127

Bernardi M, Gianolla P, Petti FM, Mietto P, Benton MJ (2018) Dinosaur diversification linked with the Carnian Pluvial Episode. Nat Commun 9:1499. https://doi.org/10.1038/s41467-018-03996-1

Bonis NR, Kürschner WM (2012) Vegetation history, diversity patterns, and climate change across the Triassic/Jurassic boundary. Paleobiology 38(2):240–264. https://doi.org/10.1666/10026.1

Briggs DEG (2014) Konservat-Lagerstätten 40 years on: the exceptional becomes mainstream. In: Laflamme M, Schiffbauer JD, Darroch SAF (eds) Reading and writing of the fossil record: preservational pathways to exceptional fossilization. Paleontol Soc Pap 20:1–14

Chraiki I, Lukeneder A, Lukeneder P, Gier S, Ghnahallah M, Dondoglio E, Fontaine C, El Khouri A, Aupissy C, El Albani A (2026) Redox conditions and paleoenvironment of an Austrian Upper Triassic *Konservat-Lagerstätte* with the marine Polzberg paleobiota as evidence for the Carnian Pluvial Episode. In: Late Triassic *Konservat-Lagerstätten* within the Carnian Pluvial Episode: 150 years of scientific excavations in Austria. Top Geobiol 55, Chap 13:xx-xx. https://doi.org/10.1007/978-3-032-19777-1_13

Clement-Westerhof JA (1974) In situ pollen from gymnospermous cones from the Upper Permian of the Italian Alps—a preliminary account. Rev Palaeobot Palynol 17:63–73. https://doi.org/10.1016/0034-6667(74)90092-X

Couper RA (1958) British Mesozoic microspores and pollen grains: a systematic and stratigraphic study. Palaeontographica Abt B 103:75–179

Costamagna LG, Kustatscher E, Scanu GG, Del Rio M, Pittau P, van Konijnenburg-van Cittert JHA (2018) A palaeoenvironmental reconstruction of the Middle Jurassic of Sardinia (Italy) based on integrated palaeobotanical, palynological and lithofacies data assessment. Palaeobio Palaeoenv 98, 111–138. https://doi.org/10.1007/s12549-017-0306-z

Dal Corso J, Bernardi M, Sun Y, Song H, Seyfullah LJ, Preto N, Gianolla P, Ruffell A, Kustatscher E, Roghi G, Merico A, Hohn S, Schmidt AR, Marzoli A, Newton RJ, Wignall PB, Benton MJ (2020) Extinction and dawn of the modern world in the Carnian (Late Triassic). Sci Adv 6(38):eaba0099. https://doi.org/10.1126/sciadv.aba0099

Dobruskina IA (2001) Upper Triassic flora from "Raibl beds" of Julian Alps (Italy) in Karavanke Mts. (Slovenia). Geologija 44:263–290

Forchielli A, Pervesler P (2013) Phosphatic cuticle in thylacocephalans from the Fossil-Lagerstätte Polzberg (Reingraben shales, Carnian). Austrian J Earth Sci 106:46–61

Forte G, Kustatscher E, Van Konijnenburg-van Cittert JHA (2021) Conifer diversity in the Middle Triassic: new data from the Fossillagerstätte Kühwiesenkopf/Monte Prà della Vacca (Pelsonian, Anisian) in the Dolomites (NE Italy). Int J Plant Sci 182:445–467. https://doi.org/10.1086/714280

Griffith J (1977) The Upper Triassic fishes from Polzberg bei Lunz. Zool J Linn Soc 60:1–93

Halle TG (1913) The Mesozoic flora of Graham Land. Wiss Ergebn Schwed Südpolarexped 1901–1903 3(14):1–123

Kelber K-P, Hansch W (1995) Keuperpflanzen. Die Enträtselung einer über 200 Millionen Jahre alten Flora. Museo 11:1–157

Krystyn L (1991) Die Fossillagerstätten der alpinen Trias. In: Vasilček W, Krystyn L, Golebiowski R (eds) Exkursionen im Jungpaläozoikum und Mesozoikum Österreichs. Österreichische Paläontologische Gesellschaft, Wien, pp 23–78

Kustatscher E, Nowak H, Forte G, Roghi G (2019) Triassic macro- and microfloras of the Eastern Southern Alps. Geo. Alp 16:5–43

Kustatscher E, Simonetto L, Roghi G, Nowak H, Schneebeli-Hermann E, Muscio G (2023) Triassic macrofloras of the Udine Province (Eastern Southern Alps), Gortania. Geol Paleontol Paletnol 45:93–112. https://doi.org/10.57614/GORT20380410.2023.45.6

Lehman J-P (1975) A propos de Ceratodus sturii Teller, 1891. Bull Mus Natl Hist Nat Paris 345:241–246

Looy CV, Stevenson RA (2014) Earliest occurrence of autorotating seeds in conifers: the Permian (Kungurian–Roadian) *Manifera talaris* gen. et sp. nov. Int J Plant Sci 175(7):841–854. https://doi.org/10.1086/676973

Looy CV, Twitchett RJ, Dilcher DL, Van Konijnenburg-van Cittert JHA, Visscher H (2001) Life in the end-Permian dead zone. Proc Natl Acad Sci USA 98:7879–7883. https://doi.org/10.1073/pnas.131218098

Lukeneder A (ed) (2026a) Late Triassic Konservat-Lagerstätten within the Carnian Pluvial Episode: 150 Years of Scientific Excavations in Austria. Top Geobiol 55:600 p. https://doi.org/10.1007/978-3-032-19777-1

Lukeneder A (2026b) The marine Upper Triassic paleobiota from the Scheiblingbach *Konservat-Lagerstätte* in Austria. In: Lukeneder A (ed) Late Triassic Konservat-Lagerstätten within the Carnian Pluvial Episode: 150 years of scientific excavations in Austria. Top Geobiol, Chap 19:437–467. https://doi.org/10.1007/978-3-032-19777-1_19

Lukeneder A, Lukeneder P (2021) The Upper Triassic Polzberg palaeobiota from a marine *Konservat-Lagerstätte* deposited during the Carnian Pluvial Episode in Austria. Sci Rep 11:16644. https://doi.org/10.1038/s41598-021-96052-w

Lukeneder A, Lukeneder P (2022a) Taphonomic history and trophic interactions of an ammonoid fauna from the Upper Triassic Polzberg palaeobiota. Sci Rep 12:7455. https://doi.org/10.1038/s41598-022-11496-y

Lukeneder A, Lukeneder P (2024a) Ammonoid taxonomy of the Carnian Polzberg *Konservat-Lagerstätte* in Austria. Jahrb Geol BA 162(1–4):27–55.

Lukeneder A, Lukeneder P, Sachsenhofer RF, Roghi G, Rigo M (2024) Multi-proxy record of the Austrian Upper Triassic Polzberg *Konservat-Lagerstätte* in light of the Carnian Pluvial Episode. Sci Rep 14:11194. https://doi.org/10.1038/s41598-024-60591-9

Lukeneder A, Surmik D, Gorzelak P, Niedźwiedzki R, Brachaniec T, Salamon MA (2020) Bromalites from the Upper Triassic Polzberg section (Austria): insights into trophic interactions and food chains of the Polzberg palaeobiota. Sci Rep 10:20545. https://doi.org/10.1038/s41598-020-77017-x

Lukeneder P, Lukeneder A (2022b) Mineralized belemnoid cephalic cartilage from the Late Triassic Polzberg *Konservat-Lagerstätte* (Austria). PLoS ONE 17(4):e0264595. https://doi.org/10.1371/journal.pone.0264595

Lukeneder S, Lukeneder A, Harzhauser M, İslamoğlu Y, Krystyn L, Lein R (2012) A delayed carbonate factory breakdown during the Tethyan-wide Carnian Pluvial Episode along the Cimmerian terranes (Taurus, Turkey). Facies 58:279–296

Martin M (1982) Revision von *Tellerodus sturii* (Teller) 1891. Verh Geol Bundesanst 1982:21–31
Mueller S, Krystyn L, Kürschner WM (2016) Climate variability during the Carnian Pluvial Phase—a quantitative palynological study of the Carnian succession at Lunz am See, Northern Calcareous Alps, Austria. Palaeogeogr Palaeoclimatol Palaeoecol 441:198–211
Nowak H, Kustatscher E, Roghi G, Van Konijnenburg-van Cittert JHA (2022) Variability of in situ spores in some leptosporangiate ferns from the Triassic in Italy and Austria. Rev Palaeobot Palynol 296:104544. https://doi.org/10.1016/j.revpalbo.2021.104544
Passoni L (1999) Studio della macroflora e dei palinomorfi del Carnico di Colle di Verano (Monte Pora, Lombardia) (PhD–thesis). Università di Milano.
Passoni L, van Konijnenburg-van Cittert JHA (2003) New taxa of fossil Carnian plants from Mount Pora (Bergamasc Alps, Northern Italy). Rev Palaeobot Palynol 123:321–346. https://doi.org/10.1016/S0034-6667(02)00226-9
Pott C (2016) *Westersheimia pramelreuthensis* from the Carnian (Upper Triassic) of Lunz, Austria: more evidence for a unitegmic seed coat in early Bennettitales. Int J Plant Sci 177:771–791. https://doi.org/10.1086/688708
Pott C (2019) The cycadalean megasporophyll *Dioonitocarpidium* in the Carnian (Late Triassic) flora of Lunz am See, Austria. Paläontol Z 93:517–530. https://doi.org/10.1007/s12542-019-00459-w
Pott C (2021) Die fossile Flora der Lunz-Formation (Karnium) von Lunz am See (Niederösterreich). In: Hauschke N, Bachmann GH, Franz M (eds) TRIAS. Aufbruch in das Erdmittelalter. Dr. Pfeil, München, pp 447–454
Pott C, Bouchal JM, Choo TYS, Yousif R, Bomfleur B (2018) Ferns and fern allies from the Carnian (Upper Triassic) of Lunz am See, Lower Austria: A melting pot of Mesozoic fern vegetation. Palaeontographica Abteilung B 297, 1–101. https://doi.org/10.1127/palb/2018/0059
Pott C, Fischer T, Aschauer B (2017) *Lunzia austriaca* – a bennettitalean microsporangiate structure with Cycadopites-like in situ pollen from the Carnian (Upper Triassic) of Lunz, Austria. Grana 56:321–338. https://doi.org/10.1080/00173134.2017.1282010
Pott C, Kerp H, Krings M (2007a) Morphology and epidermal anatomy of *Nilssonia* (cycadalean foliage) from the Upper Triassic of Lunz (Lower Austria). Rev Palaeobot Palynol 143:197–217. https://doi.org/10.1016/j.revpalbo.2006.07.007
Pott C, Kerp H, Krings M (2008a) Sphenophytes from the Carnian (Upper Triassic) of Lunz am See (Lower Austria). Jahrb Geol Bundesanst 148:183–199
Pott C, Krings M (2010) Gymnosperm foliage from the Upper Triassic of Lunz, Lower Austria: an annotated checklist and identification key. Geo. Alp 7:19–38
Pott C, Krings M, Kerp H (2008b) The Carnian (Late Triassic) flora from Lunz in Lower Austria: palaeoecological considerations. Palaeoworld 17:172–182. https://doi.org/10.1016/j.palwor.2008.03.001
Pott C, Lukeneder A (2026) A deltaic swamp forest: Coastal wetland vegetation during the climax of the Carnian Pluvial Episode. In: Lukeneder A (ed) Late Triassic Konservat-Lagerstättenwithin the Carnian Pluvial Episode: 150 years of scientific excavations in Austria. Top Geobiol, Chap 18:405–436. https://doi.org/10.1007/978-3-032-19777-1_18
Pott C, Van Konijnenburg-van Cittert JHA, Kerp H, Krings M (2007b) Revision of the *Pterophyllum* species (Cycadophytina: Bennettitales) in the Carnian (Late Triassic) flora from Lunz, Lower Austria. Rev Palaeobot Palynol 147:3–27. https://doi.org/10.1016/j.revpalbo.2007.03.005
Roghi G, Gianolla P, Kustatscher E, Schmidt AR, Seyfullah LJ (2022) An exceptionally preserved terrestrial record of LIP effects on plants in the Carnian (Upper Triassic) amber-bearing section of the Dolomites, Italy. Front Earth Sci 10:900586. https://doi.org/10.3389/feart.2022.900586
Roghi G, Kustatscher E, Van Konijnenburg-van Cittert JHA (2006) Late Triassic Plant from Julian Alps (Italy). Boll Soc Pal Ital 45(I):133–140
Ruttner A, Schnabel W (1988) Geologische Karte der Republik Ös-terreich, Blatt 71 Ybbsitz, 1:50,000. Geologische Bundesanstalt, Wien
Schneider CA, Rasband WS, Eliceiri KW (2012) NIH Image to ImageJ: 25 years of image analysis. Nat Methods 9(7):671–675

Schimper WP, Mougeot A (1844) Monographie des plantes fossiles du Grès Bigarré de la chaîne des Vosges. Guillaume Engelmann, Leipzig

Schulz E (1967) Sporenpaläontologische Untersuchungen rätoliassischer Schichten im Zentralteil des Germanischen Beckens. Paläontol Abh B 2:427–633

Seilacher A (1970) Begriff und Bedeutung der Fossil-Lagerstätten. Neues Jahrb Geol Paläontol Abh 1970:34–39

Seward AC (1917) Fossil Plants. A Textbook for Students of Botany and Geology. Vol. III: Pteridospermae, Cycadofilices, Cordaitales, Cycadophyta. Cambridge University Press, 559 p

Stur D (1874) Neue Aufschlüsse im Lunzer Sandsteine bei Lunz und ein neuer Fundort von Wengerschiefer im Pölzberg. Verh K-K Geol Reichsanst 1:271–273

Stur D (1886) Vorlage des ersten fossilen Schädels von Ceratodus aus den obertriadischen Reingrabner Schiefern von Pölzberg nördlich bei Lunz. Verh K-K Geol Reichsanst 15:381–383

Teller F (1891) Über den Schädel eines fossilen Dipnoers, Ceratodus sturii nov. spec., aus den Schichten der oberen Trias der Nordalpen. Abh K-K Geol Reichsanst 15:1–38

Townrow JA (1962) On Pteruchus, a microsporophyll of the Corystospermaceae. Bull Br Mus Nat Hist (Geol) 6:287–320

Turland NJ, Wiersema JH, Barrie FR, Greuter W, Hawksworth DL, Herendeen PS, Knapp S, Kusber W-H, Li D-Z, Marhold K, May TW, McNeill J, Monro AM, Prado J, Price MJ, Smith GF (eds) (2018) International Code of Nomenclature for algae, fungi, and plants (Shenzhen Code). Koeltz Botanical Books, Glashütten, Regnum Vegetabile

Van Konijnenburg-van Cittert JHA (1978) Osmundaceous spores in situ from the Jurassic of Yorkshire, England. Rev Palaeobot Palynol 26:125–141. https://doi.org/10.1016/0034-6667(78)90009-X

Van Konijnenburg-van Cittert JHA (1993) A review of the Matoniaceae based on in situ spores. Rev Palaeobot Palynol 78:235–267. https://doi.org/10.1016/0034-6667(93)90012-Y

Zhang J, Lenz OK, Wang P, Gao Y, Hornung J (2021) The Eco-Plant model and its implication on Mesozoic dispersed sporomorphs for Bryophytes, Pteridophytes, and Gymnosperms. Rev Palaeobot Palynol 293:104503. https://doi.org/10.1016/j.revpalbo.2021.104503

Chapter 13
Redox Conditions and Paleoenvironment of an Austrian Upper Triassic *Konservat-Lagerstätte* with the Marine Polzberg Paleobiota as Evidence for the Carnian Pluvial Episode

Ibtissam Chraiki, Alexander Lukeneder, Petra Lukeneder, Susanne Gier, Mohamed Ghnahalla, Elea Dondoglio, Claude Fontaine, Anna El Koury, Christian Aupissy, and Abderrazak El Albani

Abstract Multi-proxy investigations were conducted on clay minerals and redox-sensitive trace elements of the lower Carnian basinal succession from the Polzberg *Konservat-Lagerstätte* in the Northern Calcareous Alps (Austria). The marine Polzberg succession including the *Konservat-Lagerstätte* was studied using whole rock geochemistry and mineralogy, which enabled detailed and robust calibration

Supplementary Information The online version contains supplementary material available at https://doi.org/10.1007/978-3-032-19777-1_13.

I. Chraiki (✉) · E. Dondoglio · C. Fontaine · A. El Koury · C. Aupissy · A. El Albani
UMR-CNRS IC2MP, Université de Poitiers, Poitiers, France
e-mail: ibtissamchraiki@gmail.com

E. Dondoglio
e-mail: elea.dondoglio@etu.univ-poitiers.fr

C. Fontaine
e-mail: claude.fontaine@univ-poitiers.fr

C. Aupissy
e-mail: christian.pierrick.s.aupissy@univ-poitiers.fr

A. El Albani
e-mail: abder.albani@univ-poitiers.fr

A. Lukeneder · P. Lukeneder
Department of Geology and Paleontology, Natural History Museum Vienna, Vienna, Austria
e-mail: alexander.lukeneder@nhm.at

P. Lukeneder
e-mail: petra.lukeneder@nhm.at

S. Gier
Department of Geology, University of Vienna, Vienna, Austria

A. Lukeneder (ed.), *Late Triassic Konservat-Lagerstätten within the Carnian Pluvial Episode*, Topics in Geobiology 55, https://doi.org/10.1007/978-3-032-19777-1_13

of the Polzberg succession. The Polzberg section revealed the paleoceanographic and paleoclimatological history of the Carnian Reifling Basin and traces the global climatic Carnian Pluvial Episode during that period. The age of the Upper Triassic Reingraben Shales at the Polzberg section is refined as early Carnian (Julian 2) within the *Austrotrachyceras austriacum* Zone. Ammonoids and conodonts provide a detailed biostratigraphic subdivision that forms the basis for analyses of the clay minerals and major, trace, and rare earth elements mirroring the paleoenvironmental and paleoclimate evolution of the Reifling Basin. The Polzberg succession includes lithological and facies changes, from carbonate-dominated to argillaceous deposits, similar to those of coeval units in the Mediterranean Tethys. The Carnian was characterized by a weak (~1‰) positive $\delta^{13}C$ trend. This was punctuated by a positive shift in the lower Carnian corresponding to the initiation of the Carnian Pluvial Episode, a period representing the onset of early/late Carnian transitional global greenhouse conditions. The type and composition of clay minerals and distribution of trace elements highlight a shift from carbonate-dominated and oxic (water column and sea floor) conditions in the Reifling Formation to argillaceous and anoxic/euxinic bottom water and pore water conditions in the Reingraben Shales with increasing terrestrial influx in the upper part of the Polzberg section.

Keywords Redox conditions · Paleoenvironment · *Konservat-Lagerstätte* · Carnian Pluvial Episode · Late Triassic · Austria

13.1 Introduction

During the Late Triassic an important global climatic episode took place, referred to as either the Carnian Pluvial Episode (CPE; Dal Corso et al. 2012, 2020; Miller et al. 2017; Lukeneder and Lukeneder 2021; Pecorari et al. 2023; Lukeneder 2026, this volume), Carnian Pluvial Event (CPE; Rigo et al. 2007; Roghi et al. 2010; Dal Corso et al. 2015), Carnian Pluvial Phase (CPP; Mueller et al. 2016) or Carnian Crisis (CC; Hornung et al. 2007a, b; Kohút et al. 2018). Depositional records and sections, especially from localities around the western Tethys, documented a climatic transition to warmer and more humid conditions (Dal Corso et al. 2012, 2020; Lukeneder and Lukeneder 2023). These climatic perturbations were mainly related to the increasing volcanic activity of the Wrangellia large igneous province, which triggered episodes of climate warming, enhanced rainfall and freshwater influx in the marine basins, increased burial of organic carbon, and humidification (Mueller et al. 2016; Miller et al. 2017; Dal Corso et al. 2012, 2020; Lukeneder and Lukeneder 2021; Pecorari et al. 2023). The evidence shows that during the early Carnian (Julian), global

e-mail: susanne.gier@univie.ac.at

M. Ghnahalla
Ecole Supérieure Polytechnique, Nouakchott, Mauritania
e-mail: mohamed.ghnahalla@esp.mr

temperatures began to rise, shifting the system towards warm and humid climatic conditions (Rigo and Joachimski 2010a, b; Rigo et al. 2012; Trotter et al. 2015). This was the initial phase of the CPE, lasting ca. 2 million years. The humid and warm conditions peaked at the Julian/Tuvalian boundary (lower Carnian /upper Carnian boundary (Dal Corso et al. 2012, 2015, 2020).

The Upper Triassic Polzberg section (for location see Fig. 13.1) comprises a unique *Konservat-Lagerstätte* (Seilacher 1970; Allison 1988), and is here presented and compared to several other reference sections that document the Carnian Pluvial Episode.

The Polzberg reference section is one of the best-preserved and -studied sections in Europe because of its well-studied biostratigraphy and taxonomy (Lukeneder and Lukeneder 2024), taphonomy and trophic relations (Lukeneder and Lukeneder 2021, 2022) of various marine inhabitants of the Reifling Basin. This section therefore represents the key section for the lower Carnian *Konservat-Lagerstätte* worldwide and, as such, serves as an Upper Triassic high-quality depositional archive of the marine ecosystem for this period. Known taxa from historical and recent collections of the Polzberg paleobiota range from invertebrates, vertebrate trace fossils, to plants. This reveals the entire biota and provides new details on the Carnian food webs from this unique *Konservat-Lagerstätte*.

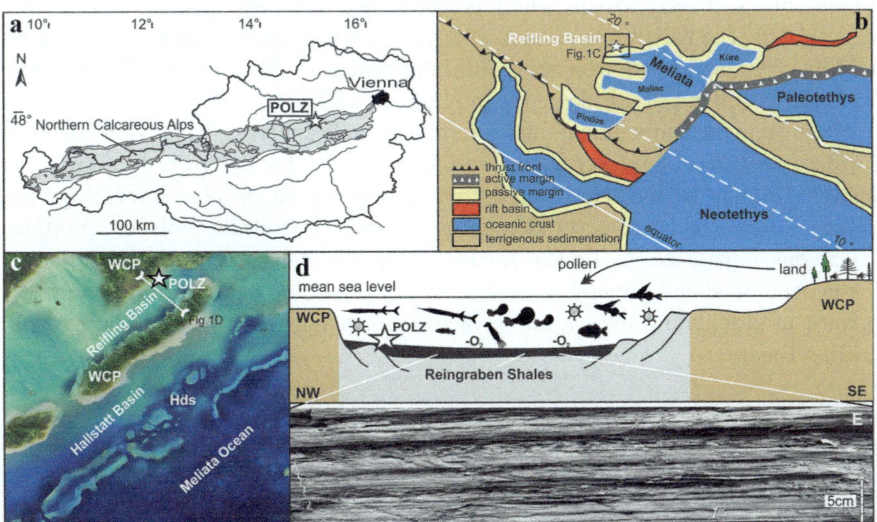

Fig. 13.1 **a** Locality map of the Polzberg area (POLZ) in Lower Austria, with marked position of the Austrian Northern Calcareous Alps (in gray). **b** Paleogeographic position of the Reifling Basin (white asterisk) at the northern rim of the Meliata Ocean. **c** Position of the Reifling Basin within the Wetterstein carbonate platforms (WCP). **d** Transect through the Reifling Basin with reconstruction of basinal environments, position of Polzberg (white asterisks), low oxygen conditions ($-O_2$) and surrounding land with plants. **e** Detailed view on argillaceous layers of the Reingraben Formation. NW northwest, SE southeast, WCP Wetterstein Carbonate Platform, Hds Hallstatt deep swell

The Carnian paleobiota from the Polzberg *Konservat-Lagerstätte* (Krystyn 1991; Forchielli and Pervesler 2013; Lukeneder and Lukeneder 2021) was historically and recently studied for several topics. During the last 6 years, results have been published on the paleoenvironment and on the trophic chain (Lukeneder et al. 2020; Lukeneder and Lukeneder 2021), taxonomy and taphonomy of ammonoids (Lukeneder and Lukeneder 2022a, 2024), including belemnoid cartilage investigations (Lukeneder and Lukeneder 2022b).

Although the lower Carnian Polzberg *Konservat-Lagerstätte* hae been known for over 140 years, the formation processes here are still not well understood. New excavations with bed-by-bed collecting of lithological samples and fossils were done during four field campaigns in 2021 and 2022. The presented details on the fauna and flora yielded new insights into the Upper Triassic Polzberg paleocommunity of the Northern Calcareous Alps (Lower Austria) (Lukeneder et al. 2020; Lukeneder and Lukeneder 2021, 2022a, b). Nonetheless, a detailed geochemical study remains to be reported. Deposited during the Carnian Julian 2 Ib (*Austrotrachyceras austriacum* Zone, *A. minor* biohorizon), the Polzberg *Konservat-Lagerstätte* (PK-L) depicts a short but precise fingerprint of the CPE (Ruffell et al. 2016; Lukeneder et al. 2012; Lukeneder and Lukeneder 2021). The deposition of fossiliferous Reingraben Shales at Polzberg, as a result of the severe environmental changes during the CPE, features enhanced humid conditions during the lower Carnian. This improves the preservation of entire and soft body fossils, which is a prerequisite for and typical of *Konservat-Lagerstätten.* The paleobiota reflects a marine environment deposited in a laminated and low oxygenation environment with intermittent shallow-water influence (Stur 1874, 1886; Teller 1891; Lukeneder and Lukeneder 2021).

The present paper documents and comprehensively analyzes the Polzberg deposits in order to establish a reference section (Polzberg, Po) for the lower Carnian in the western Tethyan region. Here we present a new record on clay mineralogy and on major, trace, and rare earth element geochemistry from the Polzberg Northern Calcareous Alps - NCA, Lower Austria, spanning the early Carnian (Julian 2). The geochemical and mineralogical framework presented here is an important step forward in elucidating the evolution of the paleoenvironmental conditions of the Reifling Basin.

13.2 Material and Methods

Samples were crushed and handpicked to avoid contamination by recent particles on the sample surface and secondary calcite veins or weathered material.

Lithology and facies. A total of 76 thin sections were made to study the lithology, microfacies and biostratigraphically important microfossils (radiolaria, ostracods, foraminifera).

Geochemistry. Samples for $CaCO_3$, TOC, S, and N content were collected with a spacing of 0.1 to 0.2 m and combined with grey-scale quantification. Calcium carbonate content was measured using a carbonate bomb technique. Total carbon

(TC) content was measured using a LECO WR-12 analyzer and TOC content was calculated as the difference between TC and $CaCO_3$. $CaCO_3$, S and TOC were measured at the Institute for Earth Sciences (Karl-Franzens-University, Graz, Austria).

Whole Rock geochemistry was conducted on carbonate samples collected from the Reifling (3 samples), the Göstling (2 samples) and the Reingraben Formations (27 samples). The samples were processed at the "Service d'Analyse des Roches et Minéraux" (SARM) of the "Centre de Recherches Pétrographiques et Géochimiques" (CRPG) in Nancy, France. Each sample underwent powdering in an agate mortar, and approximately 0.5 g was fused with lithium metaborate ($LiBO_2$) and dissolved in dilute nitric acid (HNO_3). The analysis included determining major elements, trace elements and rare earth elements using inductively coupled plasma atomic emission spectrometry (ICP-AES) and inductively coupled plasma mass spectrometry (ICP-MS), following the methods outlined by Carignan et al. (2007). Pearson's correlation coefficient was calculated using the "Past" software. Details regarding the normalization of REY data and the procedures for calculating Ce anomalies, Pr anomalies, Eu anomalies, enrichment factor and chemical index of alteration are available in the Supplementary material 13.7.

Mineralogy. The mineralogical composition was examined through X-ray diffraction (XRD) analysis, focusing on a subset of 32 representative agate mortar powdered samples. The XRD analysis was carried out using an Xpert Pro Bragg/Bentano diffractometer at the University of Poitiers, utilizing CuKα radiation with operating conditions set at 40 kV and 40 mA. The analysis covered a 2–65° 2θ angular range, with 0.017° 2θ step sizes per 0.7 s. After air drying and ethylene–glycol solvation, the clay fraction underwent analysis over an angular range of 2–35° 2θ at 0.017° 2θ with 0.7 s step sizes. Mineralogical identification was conducted using the Xpert High-score pro software, comparing the data to the International Centre for Diffraction Data (ICDD) database, as outlined by Moore and Reynolds (1989). To determine the relative percentages of clay minerals in the samples, a semi-quantitative analysis was employed using classical air-dried oriented preparations (Brindley and Brown 1980; Moore and Reynolds 1997). Following the suggestion of Hillier (2003), the Chung method (Chung 1974) of Reference Intensity Ratio was modified by using the peak area of the major reflection for each mineral. The net areas were determined using Fytyk software, and the RIR values were considered as 1 due to the crystalline structure and aluminosilicate composition of the identified clay minerals. On the other hand, we opted for air-dried preparations rather than glycolate preparations because the signal of the IS mixed layers disappear in glycolate patterns due to their low organization in the stack.

13.3 Geologic Setting

Upper Triassic outcrops at Polzberg (= Schindelberg locality) are located on the western slope of Mount Schindelberg (1066 m), 4 km northeast of Lunz am See in Lower Austria (Stur 1874, 1886; Ruttner and Schnabel 1988; Bauer and Schnabel 1997; Lukeneder and Lukeneder 2021; Fig. 13.1). Within the Lunz Nappe, a tectonic major unit of the Northern Calcareous Alps (NCA) in Lower Austria, deposits of the former intraplatform Reifling Basin (Krystyn 1991; Lukeneder and Lukeneder 2021) are located between Polzberg and Großreifling during the Upper Triassic. The exact position of the fossiliferous localities in the southern area of the Lunz Nappe within the lower fossiliferous part of the Reingraben Formation (Tollmann 1976; Piller et al. 2004) was determined south of the market town Gaming in the federal district Scheibbs by global positioning system (GPS): N 47°53′4.98" and E 15°4′28.15".

13.4 Deposits at the Polzberg *Konservat-Lagerstätte*

The lithologies at Polzberg can be subdivided into three main lithological units that are, in stratigraphic order, the Reifling Formation, the Göstling Formation and the Reingraben Formation (Fig. 13.2).

Lithology and microfacies. The lowermost nodular limestones with cherty nodules of the Reifling Formation were deposited on the paleoslope of the Reifling Basin (Lukeneder and Lukeneder 2021). At the base of the Julian 2, the Reifling Formation (*Trachyceras aonoides* Zone, *T. aonoides* Subzone) is replaced by the limestone deposits with organic-rich and cherty limestones of the Göstling Formation in the *Austrotrachyceras austriacum* Zone (*A. austriacum* Subzone, biohorizon of *A. triadicum*). These are overlain by terrigenous siliciclastic deposits of the Reingraben Formation (*A. austriacum* Subzone, biohorizon of *A. minor*; Fig. 13.3).

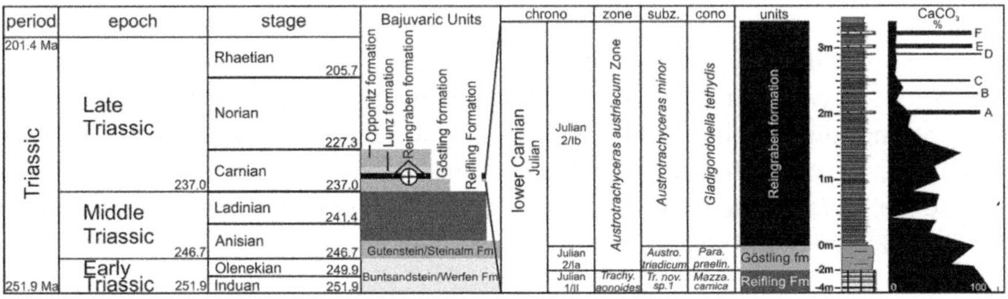

Fig. 13.2 Stratigraphic position of the Polzberg deposits during the lower Carnian. White circle: position of the Polzberg *Konservat-Lagerstätte.* Adapted after Lukeneder and Lukeneder (2021). Detailed Polzberg section within the lower Carnian Reingraben Formation, with $CaCO_3$ content. Austro *Autrotrachyceras*, Trachy *Trachyceras*, chrono chronostratigraphy, subz subzone, Re Fm Reifling Formation. A-F intercalated limestone layers. Adapted after Lukeneder and Lukeneder (2021)

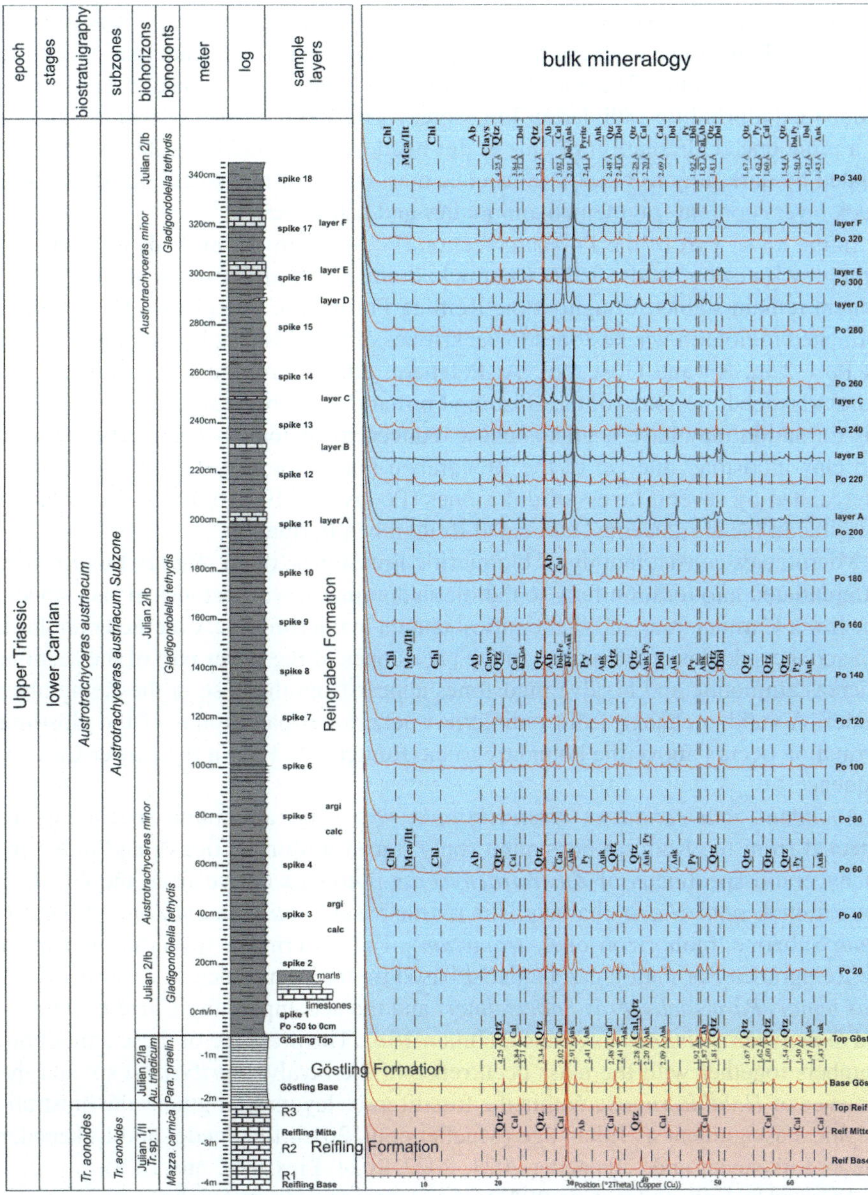

Fig. 13.3 X-ray diffraction patterns of the bulk mineralogy of the Polzberg section. Chl-chlorite, Mca/Ilt-mica/illite, Ab-albite, Qtz-quartz, Dol-dolomite, Cal-calcite, Ank-ankerite, Py-pyrite

The Reifling Formation consists of light grey and grey-green limestones as peloidal, filamentous wackestone and packstones. Rare fish and marine reptile remnants are visible. The Göstling Formation consists of black, mm- to cm-bedded calciturbiditic radiolarian limestones. The Reingraben Formation consists of a pronounced alternation of cm- to dm-bedded limestone/dolomite layers intercalated by dark grey to black argillaceous marl layers of mm to cm thickness. The black layers contain fragmented ammonites and fish scales.

In the lower part of the Polzberg section (Po −2 m to −4 m, beds R1 and R2), light grey limestone beds, 0.2–0.4 m thick, dominate. Bed-thickness and carbonate content remain stable throughout the Reifling Formation. The change to dark fine-laminated radiolaria wackestones to packstones of the Göstling Formation is abrupt (at Po −2 m, R3 and Göst top). At Polzberg, only approximately two meters of radiolaria-rich limestones of the Göstling Formation are observable; the thickness of this formation may have been somehow reduced by tectonics. The marl-limestone and clay mudstone interval of the Reingraben Formation (Po 0 cm to 340 cm) is intercalated by several layers of dolostones (Po layer A-F). No macro-bioturbation is visible in the argillaceous, laminated Reingraben Formation.

Microfacies consist mainly of biomicritic limestones, often slightly bioturbated, differentiated into foraminifera- to radiolaria dominated wackestones to packstones. The lower Carnian interval is generally abundant in microfossils. Calcified radiolaria and sponge spicules are abundant in the packstones of the lower part of the section, whereas ammonite and bivalve mudstones appear from the base of the Reingraben Shales. A sudden change from radiolaria wackestones/packstones of the Göstling Formation occurs above the boundary to the Reingraben Fm. in the Lower Carnian sequence.

Ammonite and conodont biostratigraphy. No solid ammonite specimens are detected from the Reifling Formation (only traced in thin sections), and only rare trachyceratid specimens of *Austrotrachyceras patroclus* derive from the Göstling Formation *A. patroclus* is characteristic for the basal *A. austriacum* Zone (*A. austriacum* Subzone, biohorizon of *A. triadicum*). The lowermost, fossiliferous part of the Reingraben Formation can be dated based on the occurrence of *Austrotrachyceras minor*, *Paratrachyceras haberfellneri* and the accompanying *Carnites floridus* and *Simonyceras simonyi* as Julian 2 (Julian 2/Ib). The presence of these ammonoid members together with the mass occurrence of the bivalve *Peribositra* sp. and the presence of *Halobia rugosa* within the fossiliferous layers strengthens the biostratigraphical assignment as Julian age (Mueller et al. 2016; Lukeneder and Lukeneder 2022b; Lukeneder and Lukeneder 2021, 2022, 2024; Figs. 13.2 and 13.3).

As noted by Lukeneder et al. (2024), conodonts are restricted to the lowermost part of the section, in deposits of the Reifling and Göstling formations (*Gladigondolella tethydis*; *Gladigondolella malayensis malayensis*, *Gladigondolella* sp.). The only exception in the deposits of the Reingraben Formation are ramiform conodonts, accumulated in microcoprolites. The conodont association collected at the top of the Reifling Formation is represented by typical Julian forms including *Paragondolella polygnathiformis, P. inclinata, P. foliata, P. tadpole, P. praelindae, Paragondolella* aff. *P. foliata* (sensu Rigo et al. 2007), *Gladigondolella tethydis, G. malayensis*

malayensis and *"Mosherella" longnanensis*. The presence of *P. polygnathiformis* constrains the age of the samples to Carnian, while *P. praelindae* together with *Paragondolella* aff. *P. foliata* date the top of the Reifling Formation to Julian 2 (Rigo et al. 2007, 2018). It is noteworthy that the genus *Gladigondolella* is a stenothermal species that prefers cool waters (Trotter et al. 2015): it became extinct at the Julian/Tuvalian boundary and has never been found along with Tuvalian conodont species (Rigo et al. 2018).

13.5 Results

13.5.1 Carbonate Content

$CaCO_3$ (calcite equivalent percentages) values of the Polzberg section (Po) fluctuate between low values in the Reingraben Formation with 2.9% (Po 40 T, clay) to maxium values in the Reifling Formation of 94.7% (Po R1; Fig. 13.2). This reflects the evolution of the Polzberg sequence (Lukeneder et al. 2024). Basal limestones of the Reifling Formation exhibit $CaCO_3$ values between 87.4% and 94.7%, limestones of the Göstling Formation have lower values (58.4 to 78.5%), and the Reingraben Formation features 80.1% basally, decreasing towards the clayey topmost layers with 6.0–7.0%.

13.5.2 Total Organic Carbon and Sulfur

The Reifling Formation contains very low amounts of TOC (0.09–0.15%) and sulfur (0.00–0.02%). The corresponding values in the laminated Göstling Formation are higher: 0.67–1.97% and 0.06–0.33%, respectively. The TOC values in the Reingraben Formation are 0.59–1.44%, the sulfur contents 0.44–1.78%.

13.5.3 Mineralogy

The bulk samples of the Reifling Formation consist of almost pure calcite, only a small amount of quartz is present. Sample Göstling base shows a mixture of calcite and quartz in almost equal proportions, Göstling top additionally contains high amounts of dolomite/ankerite and minor amounts of clay minerals such as chlorite, mica or illite. The Reingraben Formation of the Polzberg profile (0–340 cm) has a rather uniform composition, being composed of quartz, albite, pyrite, chlorite, muscovite and varying amounts of Mg-calcite, ankerite and dolomite. The carbonates are more abundant in the lower part of the profile (0–140 cm) than in the higher part (160–340 cm), where the amount of carbonates decreases. In layers A-F, ankerite, dolomite and calcite are the dominant minerals; the samples contain only small amounts of quartz (Fig. 13.3). Clay mineralogy shows that illite and / or mica and chlorite are predominant in the Polzberg section, with a high enrichment of clay minerals in the Reingraben Formation compared to the Reifling and Göstling Formations (Supplementary material 13.1).

In the carbonate strata at the base of the sequence, the clay fraction of the Reifling Formation consists of chlorite and illite, primarily Well Crystallized Illite (WCI) associated with some Low Crystallized Illite (LCI) (Fig. 13.4a). Moving upward, the Göstling Formation starts with a similar signature but exhibits abundant kaolinite at the top (Fig. 13.4b). The presence of a bulge at 10.35 Å likely indicates a very small quantity of Illite/Smectite (I/S) mixed layers.

The abrupt sedimentary transition, marked by the onset of marl deposits of the Reingraben Formation, is accompanied by a significant contribution of I/S mixed layers (Figs. 13.4c and 13.5, Supplementary material 13.1). The broad band between 11.46 Å and 10.67 Å reveals, rather than a single interstratified unit (Fig. 13.4c), a compositional range in which the proportion of illite remains predominant. While chlorite and illite are equally abundant, the presence of kaolinite drops drastically in relation to the top Göstling content. This assemblage is consistent in both the lower (Figs. 13.4c and 13.5) and upper (Figs. 13.4d and 13.4e) parts of Reingraben, although the intensity ratios vary slightly and the positions of the I/S components show moderate shifts. In the intercalated dolomitic layers of the upper part of Reingraben, the I/S mixed layers content decreases, while the kaolinite content rises sharply (Figs. 13.4f and 13.5).

Kaolinite is known to be formed as a residual weathering product under a long-lasting warm and wet climate at high rates of chemical weathering (Vogt et al. 2010). Mixed-layer I/S minerals can be derived from incomplete chemical weathering of illite (Chamley 1989). During weathering, interlayer cations are leached out of the clay minerals by percolating waters. The progressive extraction of K^+ ions from the interlayer sheet causes the illite to transform into smectite during chemical weathering; the first stages of the reaction form regularly ordered mixed-layer minerals (Meunier 2005). The occurrence of illite and chlorite resulting from physical weathering and mixed/layer I/S and kaolinite resulting from chemical weathering suggests a change in climate from cool and dry to warm and wet. The abrupt increase of kaolinite in the "Top Göst" is indicative of the start of the CPE.

13.5.4 Redox-Sensitive Elements

Concentrations of trace elements, including U, Th, Cu, Zn, Ni, Co, V, and Cr, as well as elemental ratios such as Ni/Co, V/V + Ni, V/Cr, Cu/Zn, and U/Th, are commonly employed to reveal sediment redox depositional conditions (Jones and Manning 1994; Armstrong-Altrin et al. 2015; Chraiki et al. 2023). Using these trace

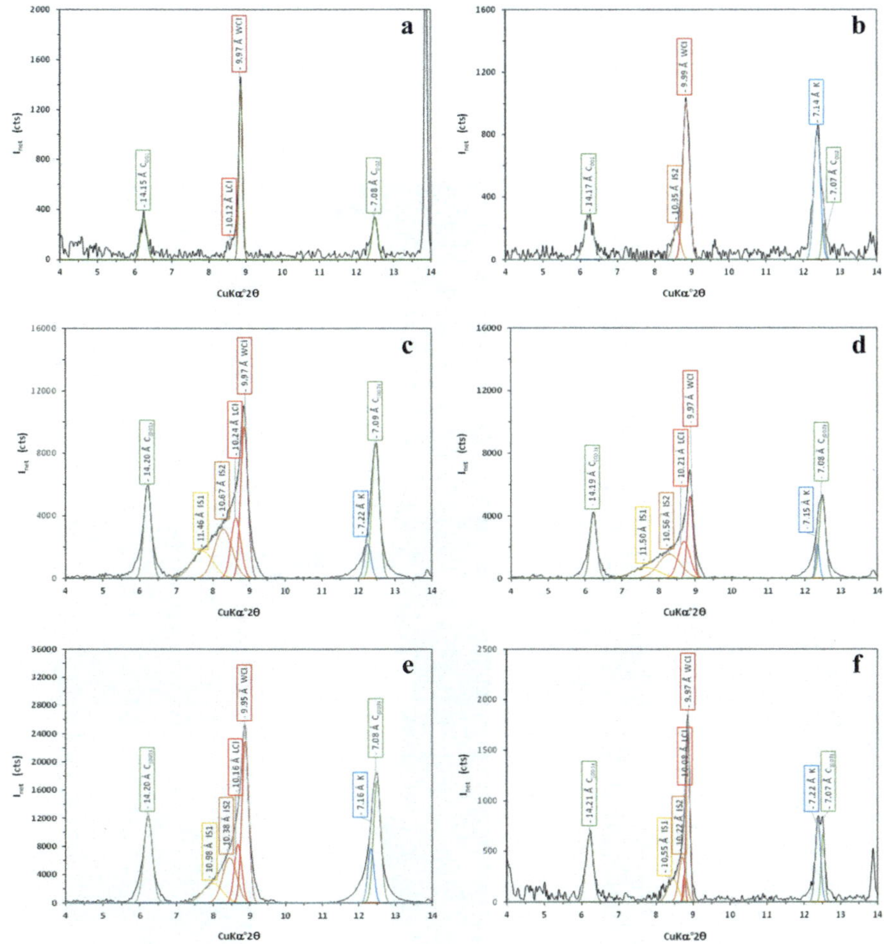

Fig. 13.4 Deconvolution of the X-ray diffractogram patterns (background subtracted) of Re-M (**a**), Go-T (**b**), Po-040 (**c**), Po-200 (**d**), Po-280 (**e**), and Po-D (**f**) samples in oriented preparation showing the evolution of proportions of phases in the clay mineral fraction: (C001, C002) chlorite, (IS1, IS2) mixed-layers of illite/smectite-type, (LCI, WCI) low and well crystallized illite, and (K) kaolinite

element concentrations and ratios as redox proxies involves correlating the trace metal concentration with immobile Al, assumed to be a detrital index (Abedini and Calagari 2017; Chraiki et al. 2023).· Pearson's correlation analysis for the studied section reveals a strong positive correlation among all redox-sensitive elements and Al. The exceptions are U, which shows no correlation with Al, and Mo, which shows a moderate positive correlation (Table 13.1). This suggests that trace element concentrations are more closely associated with the detrital fraction and that trace-metal proxies given by Jones and Manning (1994) cannot be applied in this case to assesss the redox conditions of the paleoenvironment.

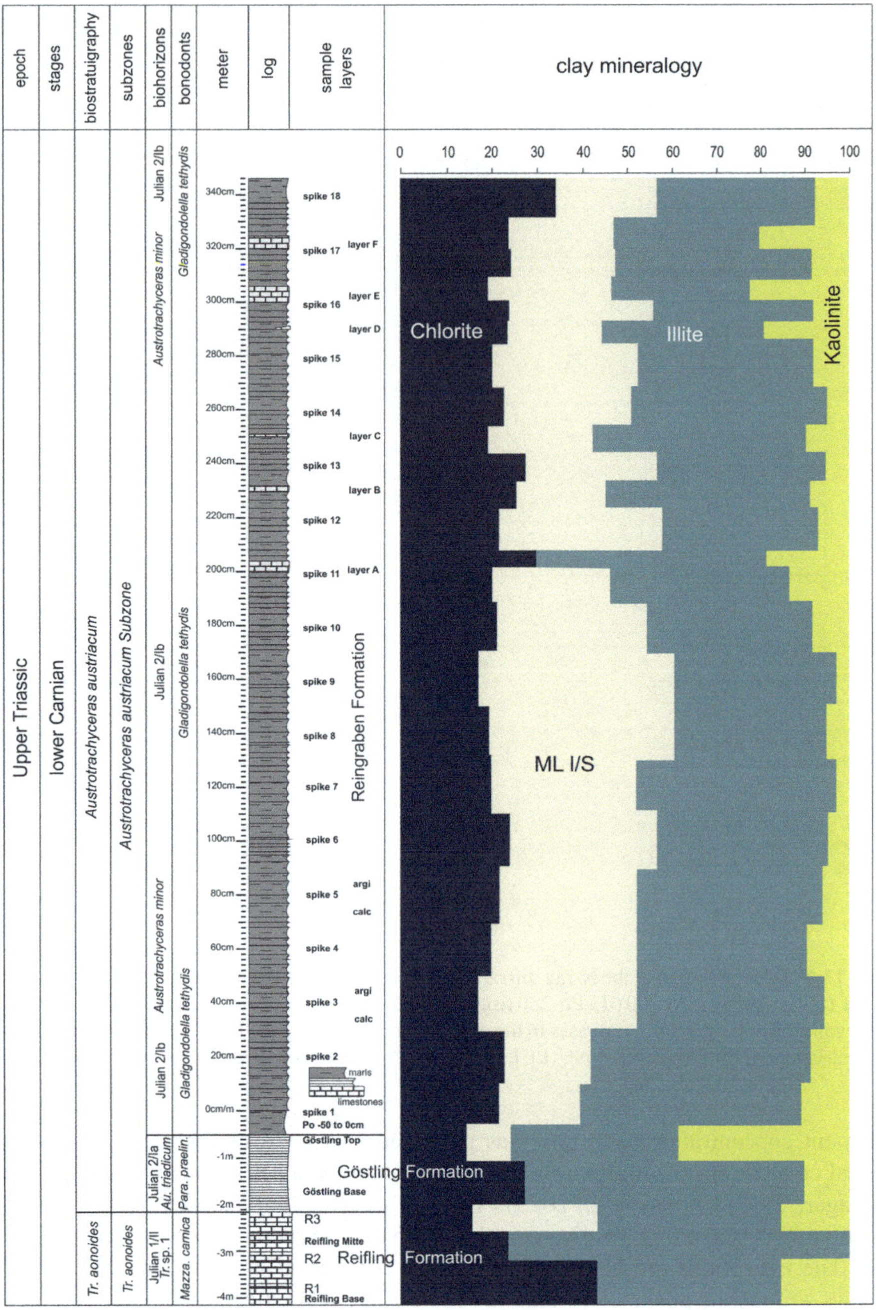

Fig. 13.5 Evolution of relative contents of chlorite, mixed-layers illite/smectite-type (IS1 + IS2), illite (LCI + WCI), and kaolinite during the sedimentation of Carnian deposits

Table 13.1 Pearson's correlation coefficient (r) of redox-sensitive trace elements in (ppm) and Al_2O_3%. Values in bold are different from 0 at significance level alpha = 0.05

	Co	Cu	Mo	Ni	Th	U	V	Zn	Al_2O_3
Co	**1**								
Cu	**0.96**	**1**							
Mo	**0.66**	**0.77**	**1**						
Ni	**0.97**	**0.98**	**0.72**	**1**					
Th	**0.99**	**0.94**	**0.67**	**0.94**	**1**				
U	0.07	0.26	**0.63**	0.22	0.04	**1**			
V	**0.97**	**0.97**	**0.77**	**0.98**	**0.96**	0.250	**1**		
Zn	**0.89**	**0.90**	**0.82**	**0.89**	**0.91**	0.365	**0.93**	**1**	
Al_2O_3	**0.99**	**0.95**	**0.67**	**0.95**	**1.00**	0.048	**0.97**	**0.90**	**1**

Trace metals have different sensitivities to changing redox conditions, and the lack of enrichments in V, U, Cr, Cu, Ni, Zn and Mo in fine-grained siliciclastic sediments and sedimentary rocks is generally considered to be an indicator of oxic deposition (Tribovillard et al. 2006). The enrichment factors for these elements indicate that all samples exhibit high enrichment in Mo and U (EF > 3), and moderate enrichment in Ni and Zn (Fig. 13.6). The remaining redox-sensitive elements (Co, V, Cr, Cu, Th) are not enriched and are predominantly associated with the detrital fraction. The uranium (U) enrichment factor (EF) ranges from 9.29 to 217.5 with a mean of 92.03 in the Reifling and Göstling Formations and decreases in the Reingraben Formation (range 1.17–6.65; mean 2.5). The (Mo) EF ranges from 46.59 to 357.55, with a mean of 194.34 in Reifling and Göstling, and from 10.55 to 32.14, with a mean of 18.6 in the Reingraben Formation (Fig. 13.6, Table 13.1). The enrichment factors for Zn and Ni exhibit a similar trend to Mo and U, displaying higher levels in the lower section, with a mean of 12.3 for Zn (EF) and 5.21 for Ni (EF). In contrast, the Reingraben Formation shows lower enrichment, with mean values of 1.39 for Zn EF and 1.45 for Ni EF (Table 13.1, Fig. 13.6).

The pronounced enrichment of redox-sensitive elements in Reifling and Göstling, which sharply decreases from the base of the Reingraben Formation (Fig. 13.6), should be interpreted with caution in view of their extremely low Al contents (<1.2%). To confirm the enrichment of these deposits with redox-sensitive trace elements, we examined the fluctuation of their concentrations throughout the section (Supplementary material 13.2). Trace element concentrations at Reifling generally fall below the PAAS (Post-Archean average Australian Shale) values. They increase at Göstling, peaking in the sample "Top Göst" and then generally decreasing in the Reingraben deposits, with concentrations often surpassing the PAAS values. This suggests that the Göstling deposits are most enriched in trace elements, followed by Reingraben samples. In contrast, the high enrichment factors in the Reifling samples are more likely associated with their low Al content (<1.2%). The absence of trace element enrichment in Reifling and the pronounced enrichment in the Göstling and Reingraben Formations imply oxic bottom water conditions for the Reifling Formation

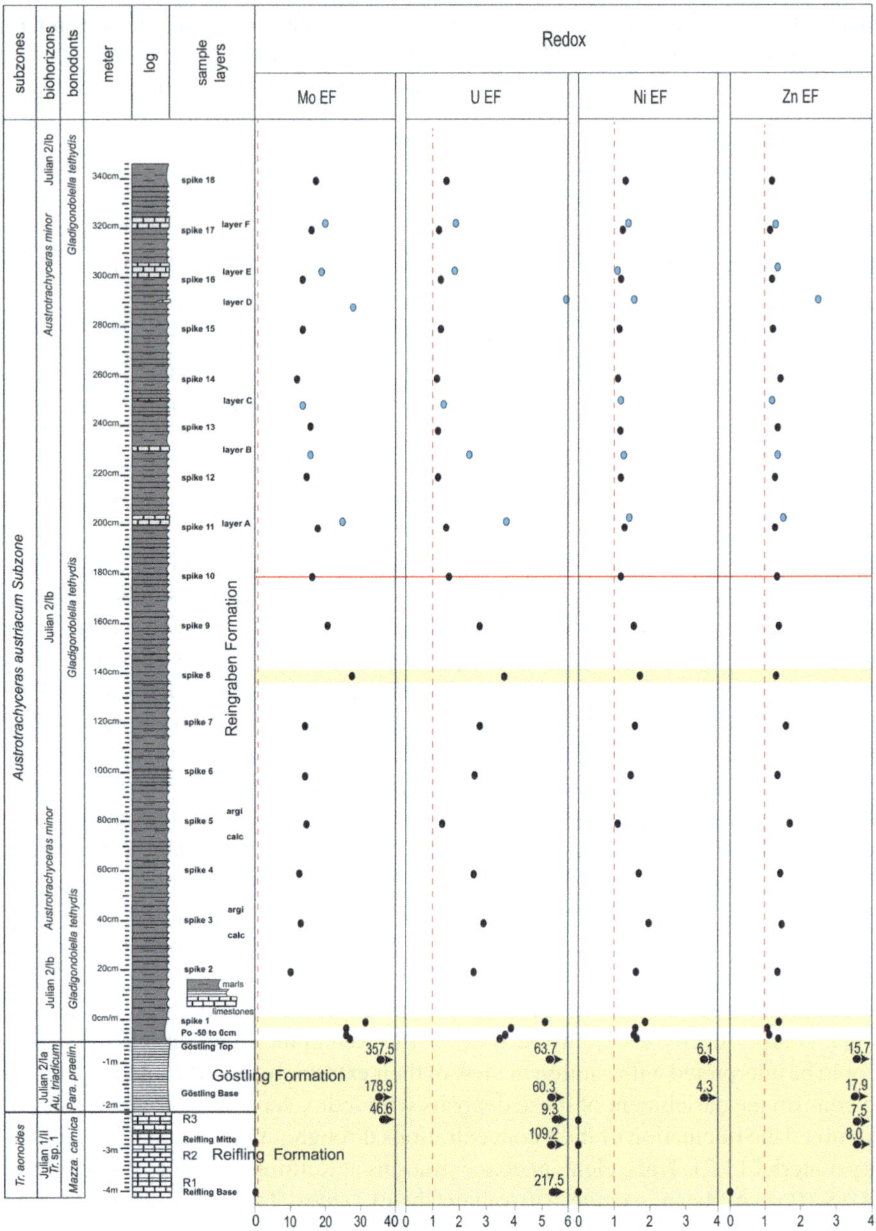

Fig. 13.6 Enrichment factors of redox-sensitive trace elements through the Upper Triassic Polzberg section. Blue spots: values for the dolomitic layers

and anoxic to euxinic bottom water conditions for the Göstling and Reingraben Formations.

13.5.5 Mo-EF vs U-EF Covariation

Authigenic U-Mo covariation patterns in marine sediments provide insights into bottom water redox conditions, the operation of metal-oxyhydroxide particulate shuttles, and the amount of deep-water restriction (Algeo and Tribovillard 2009; Tribovillard et al. 2012). Authigenic U is absorbed by marine sediments at the Fe(II)–Fe(III) redox boundary (i.e. suboxic conditions), whereas authigenic Mo enrichment requires the presence of H_2S (i.e. euxinic conditions). Additionally, particulate shuttles may enhance the absorption of aqueous Mo to the sediment, but aqueous U is not impacted by this process. U-EF vs Mo-EF plots of the Polzberg section show that one sample of the Reifling Formation exhibits high Mo enrichment compared to U, with $(Mo/U)_{EF}$ ratios superior to that of seawater (SW) (Fig. 13.7). In contrast, Mo is below the detection limit of 0.5 ppm for the rest of the Reifling samples. Samples from the Göstling Formation express the highest enrichment factors for U and Mo, with a $(Mo/U)_{EF}$ superior or equal to $(Mo/U)_{SW}$. Samples of the basal part of Reingraben (Po(−50) to Po-160) exhibit UEF >2 and Mo/U ratios ranging between (Mo/U)SW and (Mo/U)(3*SW). The upper part of Reingraben exhibits a low $U_{EF} < 2$ and high Mo/U ratios >10, located next to the particulate shuttle area (Fig. 13.7). These patterns collectively suggest euxinic bottom water conditions for the Göstling and Reingraben Formations. Particulate shuttles played a role in enhancing the Mo concentration at Reingraben, specifically in its upper part (Po180- Po-340).

Scott and Lyons (2012) offer a new paleoredox proxy based on Mo content to distinguish between conditions where sulfide was abundant in bottom waters and those where sulfide was restricted to pore fluids. Mo enrichment suggests that dissolved sulfide was present in the paleoenvironment, but only in pore fluids, where its concentration was higher than the crustal average but less than 25 ppm, which is the case for Göstling and Reingraben deposits (Supplementary material 13.2). In contrast, in euxinic environments where hydrogen sulfide is prevalent in the water column throughout the year and dissolved Mo is abundant due to efficient interaction with the open ocean, enrichments demonstrate a much larger range, with values frequently >60 ppm and reaching 100 ppm (Scott and Lyons 2012).

13.5.6 Ce Anomaly

The examination of the REE (Rare Earth Element) composition in carbonates offers insights into paleo-environmental conditions (Webb and Kamber 2000; Nothdurft et al. 2004). For instance, the Ce/Ce* values of carbonates can record water column

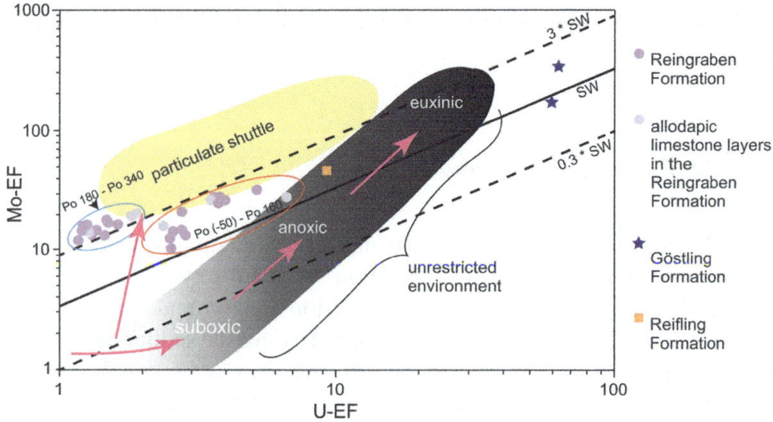

Fig. 13.7 U–EF versus Mo–EF for the Upper-Triassic Polzberg section. Gray field: the "unrestricted marine" (UM) trend, characteristic of the eastern tropical Pacific; green field: the "particulate shuttle" (PS) trend, characteristic of depositional systems such as the Cariaco Basin in which intense redox cycling of metal (especially Mn−) oxyhydroxides occurs within the water column (modified from Algeo and Tribovillard 2009)

redox conditions (Tostevin et al. 2016). Nevertheless, note that Ce anomalies in original aquatic carbonates may be influenced by diagenesis (Shields and Stille 2001; Nothdurft et al. 2004) and detritus (Nothdurft et al. 2004). According to Shields and Stille, diagenetic exchange post-deposition can lead to REE patterns becoming increasingly Ce-enriched and Eu-depleted, with decreased Dy_N/Sm_N ratios. This diagenetic exchange has the potential to obscure or diminish primary features associated with REE distribution patterns, and if this exchange may be identifiable through negative correlations between Ce/Ce* versus Eu/Eu* and Dy_N/Sm_N (Shields and Stille 2001). In our study, all samples exhibited slightly positive Eu anomalies ranging from 1.02 to 1.2, except for the sample "Reif Base", which displays a negative Eu/Eu* value of 0.97. The absence of correlations between Ce/Ce* versus Eu/Eu* and DyN/SmN, stemming from diagenetic alteration (Shields and Stille 2001), does not appear in our samples (Supplementary material 13.3). This suggests that the calculated Ce anomalies in our samples were not significantly altered during diagenesis.

Based on the significant enrichment of silicate-derived REE relative to chemically produced carbonate REE content, detrital contamination is one of the most likely sources of variations in carbonate REE distribution patterns from inherited primary features. Such effects could be determined by using concentrations and correlations among chemical elements such as bulk rock Al content and Ce/Ce* profiles of leachable carbonates (Webb and Kamber 2000; Nothdurft et al. 2004). The Pearson correlation of Ce versus Al demonstrates a strong positive correlation for the Polzberg section (R2 = 0.91) (Table 13.2). This means that the bulk Ce/Ce* data cannot be effectively used to infer the redox depositional conditions of the Polzberg section.

Table 13.2 Pearson's correlation coefficient (r) REE + Y elements and Al. Values in bold are different from 0 at significance level alpha = 0.05

	La	Ce	Pr	Nd	Sm	Eu	Gd	Tb	Dy	Y	Ho	Er	Tm	Yb	Lu	Al_2O_3
La	1															
Ce	1.00	1.00														
Pr	0.99	0.99	1.00													
Nd	0.99	0.99	1.00	1.00												
Sm	0.97	0.97	0.99	0.99	1.00											
Eu	0.95	0.95	0.97	0.98	1.00	1.00										
Gd	0.96	0.96	0.98	0.99	1.00	1.00	1.00									
Tb	0.97	0.97	0.99	0.99	1.00	0.99	1.00	1.00								
Dy	0.98	0.98	0.99	0.99	1.00	0.99	1.00	1.00	1.00							
Y	0.98	0.98	0.99	0.99	0.98	0.97	0.98	0.99	0.99	1.00						
Ho	0.99	0.98	0.99	0.99	0.99	0.98	0.99	0.99	1.00	0.99	1.00					
Er	0.99	0.99	0.99	0.99	0.98	0.96	0.97	0.98	0.99	0.99	1.00	1.00				
Tm	0.99	0.98	0.98	0.97	0.96	0.95	0.96	0.97	0.98	0.97	0.99	1.00	1.00			
Yb	0.98	0.98	0.97	0.97	0.95	0.93	0.94	0.95	0.97	0.96	0.98	0.99	1.00	1.00		
Lu	0.98	0.97	0.97	0.96	0.95	0.92	0.94	0.95	0.96	0.96	0.98	0.99	1.00	1.00	1.00	
$A_{12}O_3$	0.92	0.91	0.89	0.87	0.86	0.83	0.84	0.85	0.88	0.87	0.91	0.94	0.96	0.97	0.97	1.00

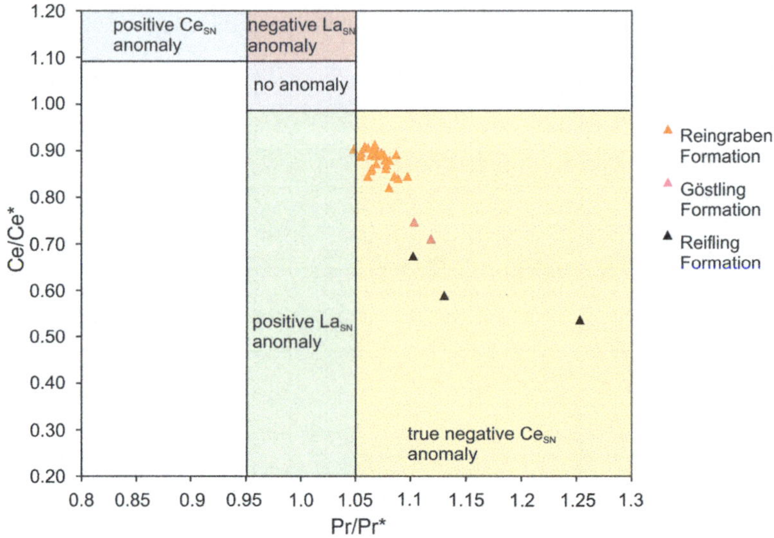

Fig. 13.8 PAAS normalized REE + Y spectra of the Polzberg section

13.5.7 PAAS Normalized REE + Y Patterns

Göst and Reif samples display fractionated PAAS-normalized REE patterns, whereas the remaining Polzberg samples show flat REE patterns (Fig. 13.8). The Pearson correlation of REE + Y and Al_2O_3% yields values >0.8 (Table 13.2), indicating that REE + Y are positively correlated with Al_2O_3%, reflecting terrestrial detrital inputs for REE + Y.

13.5.8 Salinity

The interaction between seawater and freshwater results in intricate patterns within marginal marine environments characterized by pronounced physicochemical gradients (Anadón et al. 2002). This refers primarily to salinity, which is an essential chemical feature of water masses that critically regulates the homeostatic balance in aquatic biological systems (Jewuła et al. 2022; Chraiki et al. 2023). Several trace element ratios were used as geochemical paleoredox proxies. This approach is based on the observation that certain elements that are not sensitive to redox conditions exhibit relatively higher concentrations in seawater. These include B, Y, S and Sr in seawater. In freshwater they include Ga, Ho, TOC and Ba, when normalized for salinity (Retallack et al. 2020; Wei and Algeo 2020; Xiao et al. 2020). The Pearson correlation of these trace elements with Al_2O_3 reveals that Ga, Y, Ho and B are positively correlated with Al_2O_3 (Table 13.3), indicating a detrital signature for these elements. In contrast, Ba and Sr do not exhibit such a positive correlation and may be effectively utilized as a paleosalinity proxy.

The Sr/Ba ratio corresponds to an important paleosalinity proxy that is based on the fact that seawater leans towards Sr enrichment as opposed to Ba in freshwater environments (Webb and Kamber 2000; Nagarajan et al. 2007). Thus, Sr/Ba sedimentary ratios of <0.2–0.5, and >0.5 are characteristic of freshwater, brackish-saline, and full marine facies, respectively. However, this proxy may be affected by a high carbonate content (Algeo and Maynard 2004) because carbonate has the potential to increase bulk Sr/Ba ratios relative to the clay mineral fraction. For this reason, we screened for the influence of carbonate-hosted Sr by plotting the Sr versus CaO%. The samples show no correlation between Sr and CaO% (Supplementary material 13.4), which implies that the Sr/Ba proxy is operational in this case. Our compiled dataset reveals Sr concentrations of 336.6–502.5 ppm, with an average of 419.3 ppm for the lower part of the sction and 93–697 ppm with a mean of 396 ppm for the Reingraben Formation. Ba exhibits values ranging from 8.4–46.9 ppm with an average of 22.3 ppm for the lower part of the section, and 77–1603 ppm with an average of 646 ppm for the Reingraben Formation (Supplementary material 13.6). Corresponding Sr/Ba ratios in the range of 9.2–42.2 and 0–9 were obtained for the lower and upper part, respectively (Fig. 13.9). The Sr/Ba ratios suggest, in general, that samples from the Reifling and Göstling Formations were deposited in fully marine water. The Reingraben Formation differs somewhat in theis respect: it exhibits Sr/Ba values characteristic of marine salinity in the lower part, while the upper part (from Po-200 to Po-340) shows values characteristic of deposition in brackish water. This suggests a mix of fresh- and marine water during their deposition (Fig. 13.9). However, the intercalated carbonate layers in the upper part of the Reingraben deposits exhibit high Sr/Ba values characteristic of marine salinity, indicating the influence of short normal marine periods.

Table 13.3 Pearson's correlation coefficient (r) between salinity-sensitive trace elements and Al. Values in bold are different from 0 at significance level alpha = 0.05

	Ba	Ga	Sr	Y	Ho	B	Al_2O_3
Ba	**1**						
Ga	0.48	**1**					
Sr	−0.03	−0.65	**1**				
Y	**0.57**	**0.87**	−0.34	**1**			
Ho	**0.57**	**0.90**	−0.39	**0.99**	**1**		
B	**0.51**	**1.00**	−0.61	**0.89**	**0.92**	**1**	
Al_2O_3	0.49	**1.00**	−0.64	**0.87**	**0.91**	**1.00**	**1**

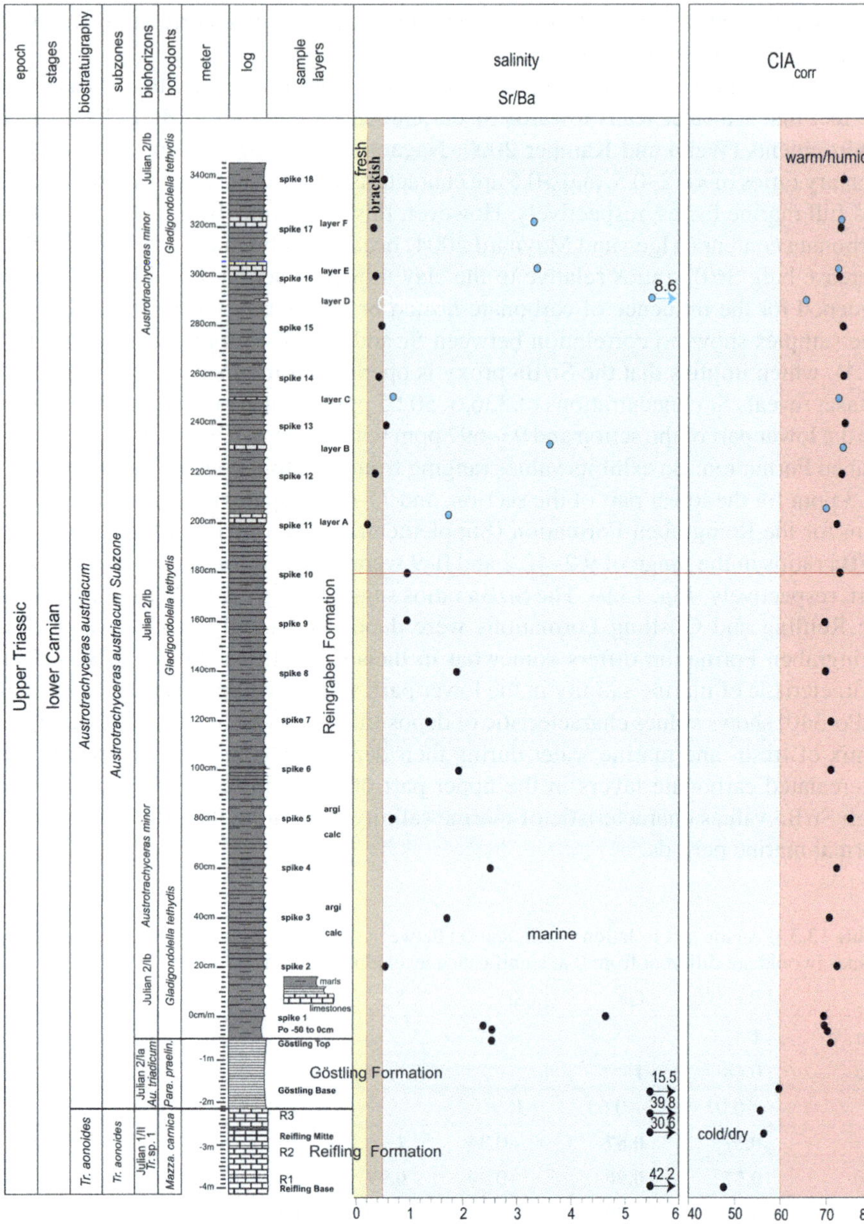

Fig. 13.9 Stratigraphic profile of Sr/Ba ratios as a salinity proxy of the Polzberg section. Blue area: samples deposited in a marine environment; yellow area: samples deposited in brackish water. Chemical Index of Alteration (CIA) shows a trend from cold/dry to warm/humid climate

13.5.9 Chemical Weathering and Sediment Provenance

Chemical weathering has a significant impact on the mineralogy and major elemental geochemistry of siliciclastic sediments (Nesbitt and Young 1982; McLennan 1993; Wang et al. 2020). The evaluation of the degree of chemical weathering can be theoretically facilitated by quantitative parameters such as the Chemical Index of Alteration (CIA; Nesbitt and Young, 1982). High CIA values, typically exceeding 60, indicate the removal of labile cations (e.g., Ca^{2+}, Na^{+}, K^{+}) in comparison to stable residual constituents such as Al^{3+} and Ti^{4+} during the weathering process (Nesbitt and Young 1982; Fedo et al. 1995; Wang et al. 2020). Conversely, low CIA values, generally below 60, suggest the virtual absence of chemical weathering and may signify cool and/or arid conditions.

The Polzberg samples, when plotted on the A-CN-K ternary diagram (Fig. 13.10a), exhibit CIA values ranging from 63.86 to 70.82 in the upper section and from 52.44 to 68.40 in the lower section (Fig. 13.9). Most samples deviate slightly from the ideal weathering trend on the A-CN-K diagram, leaning toward the A-K axis. This deviation implies that pervasive k-metasomatism might have impacted the overall sample composition (Nesbitt and Young 1982; Fedo et al. 1995), necessitating a correction of CIA values.

Utilizing discriminant diagrams, specifically Th/Sc versus La/Sc (Bankole et al. 2020; Bhatia and Crook 1986) (Supplementary material 13.5), helps to deduce the parent rock source. This approach places the samples in the transitional zone between mafic and felsic fields, closely aligned with TTG affinity, indicating an intermediate composition (Supplementary material 13.5). This affirms TTG as the probable dominant parent rock source for our samples. The K-corrected CIA values (CIAcorre) range from 65.65 to 75.21, averaging 71.81 in the upper section, and from 47.56 to 70.96, averaging 58.07 in the lower section. These values imply weak chemical weathering (<60), characteristic of a dry and cold climate in the lower part of the section, transitioning to a warm and humid climate moving upward (from the sample "Top Göst") (Figs. 13.9 and 13.10b). This result is consistent with the increase in the relative percentage of Kaolinite, reaching up to 38% in the "Top Göst" (Fig. 13.5). Such values are typically associated with prolonged warm and wet climates, signifying the onset of the warm and humid CPE.

13.6 Evolution of the Polzberg Succession

The Polzberg locality differs from all other Carnian localities deposited during the CPE by the presence of a worldwide unique feature, namely the lower Carnian Polzberg *Konservat-Lagerstätte* (Lukeneder et al. 2024). The formation of that special preservational stage was due to the particular position of an intraplatform basin, the Reifling Basin. That basin was separated from the western open Tethyan Ocean by the Wetterstein carbonate platforms (Wcp) and the Hallststatt deepwater

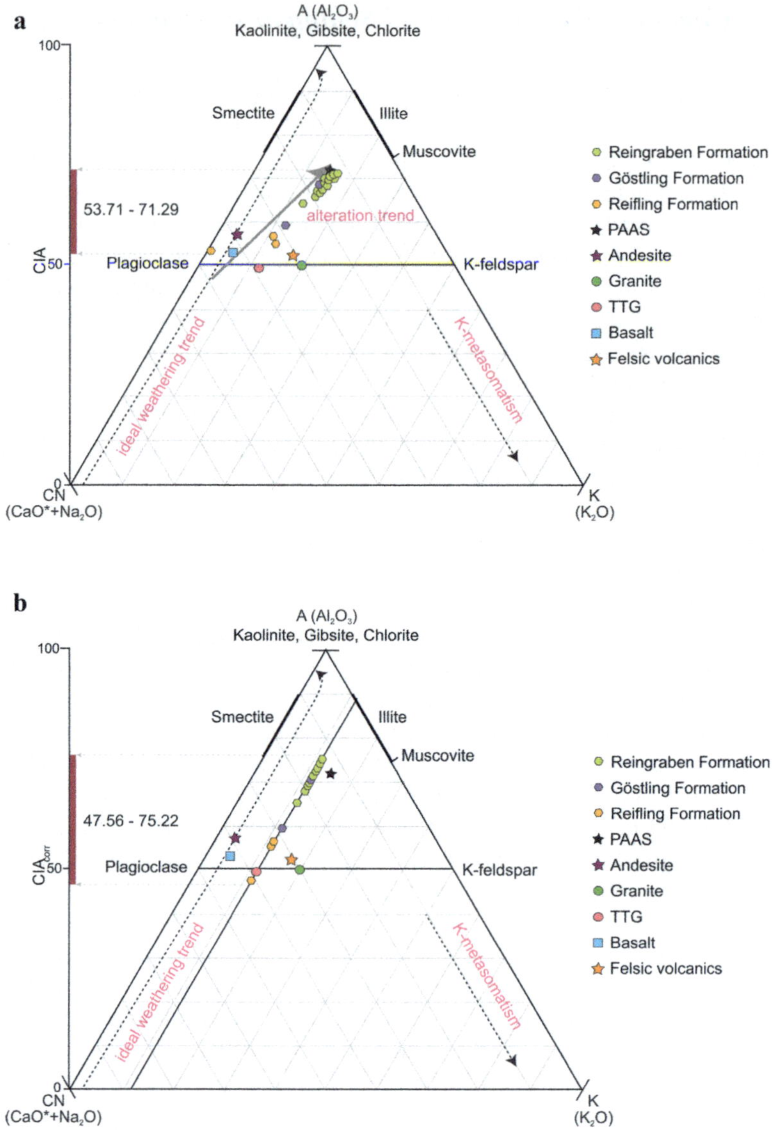

Fig. 13.10 **a** A-CN-K ternary diagram with uncorrected data. **b** A-CN-K ternary diagram with corrected data. CIA–Chemical Index of Alteration, CIAcorr–Corrected CIA, A–Al_2O_3, CN–CaO* + Na_2O, K–K_2O (all in molar proportions), CaO*–CaO incorporated in the silicate fraction of the sample. Note: Post-Archean average Australian Shale (PAAS: (Taylor and McLennan 1985); andesite; granite; TTG; felsic volcanics and basalt (Condie 1993)

swells (Hds) (Fig. 13.1). Due to its protected position, the basin reacted more sensitively to environmental changes such as regional tectonics, sea level changes and enhanced freshwater influx with increased terrigenous input. The effects of the CPE are detected worldwide and are both numerous and diverse. Environmental and geochemical changes are given from over 25 sites in Dal Corso et al. (2020). Humidification, enhanced siliciclastic, major freshwater input and the crisis in the microbial platforms are only a few of the important and notable adaptations on the ecosystems around the world (Rigo and Joachimski 2010a, b; Rigo et al. 2012; Trotter et al. 2015; Mueller et al. 2016; Miller et al. 2017; Dal Corso et al. 2012, 2020; A. Lukeneder A et al. 2024; Lukeneder and Lukeneder 2021; Pecorari et al. 2023; Lukeneder 2026, this volume).

The depositional environment was influenced by the evolution of climate and environmental conditions. The sequence shows three main lithologies and two main lithological transitions: the grey radiolaria limestones of the Reifling Formation, the transitional laminated black silicious limestones of the the black Göstling Formation, and the laminated argillaceous shaly Reingraben Formation. A transition is observed in Late Triassic successions from several other localities in the NCA of Austria (e.g., Göstling, Grossau, Mendlingbach, Saugraben, Scheiblinggraben, Steinbachgraben, Rehberg, Saugraben) (Krystyn 1991; Hornung et al. 2007a, b). Although minor differences occur in the lithotypes and in the thickness of formations and members, the lithological boundaries occur consistently at a regional scale. This is evidence for a change in basinal sedimentation throughout the NCA and the western Tethys in general during the Late Triassic (Hornung et al. 2007a, b; Lukeneder et al. 2024; Lukeneder 2026, this volume).

13.7 Conclusions

High-resolution studies on the Polzberg section including the Polzberg *Konservat-Lagerstätte* reveal the early Late Triassic paleoceanographic history of the Reifling Basin (Northern Calcareous Alps) in the framework of the worldwide Carnian Pluvial Episode (CPE). We investigated basinal paleoenvironments in the lower Carnian as recorded in the transitional sequence of Polzberg based on geochemical data and the analysis of clay mineralogy and trace elements. They show evidence for deposition during the onset of a humid and warmer episode. A cascade of lithological changes reflects the paleoenvironmental and climate change during the lower Carnian *Austrotrachyceras austriacum* Zone (Julian 2; Carnian 2). A lithological change from carbonate-dominated limestones (Reifling Formation), over black laminated limestones (Göstling Formation) and up to argillaceous to shaly deposits (Reingraben Formation) appear to be triggered by the Wrangellian Large Igneous Province (LIP). Increased emission of volcanic gases into the atmosphere resulted in positive shifts of temperatures and rainfall along with negative shifts of $\delta^{13}C$. Carbonate factories and carbonate platforms (Wetterstein Platform) were interrupted or even vanished. Freshwater influx and terrestrial elements increased; subsequently, anoxic areas formed in basinal environments (Reifling Basin). Carbonate platforms dominated by microbial ecosystems declined and shifted to less productive metazoan-dominated environments. Clay contents and composition along with the distribution of trace elements mark the onset of the CPE in the Tethys Ocean and at the Polzberg section at the base of the Göstling Formation at the transition from Julian 1 to Julian 2. The lithology (thin sections, SEM), mineralogy (bulk- and clay mineralogy), and geochemistry ($CaCO_3$, TOC, trace elements) show a transition from a carbonate to an argillaceous deposit during the formation of the Polzberg *Konservat-Lagerstätte*, which was deposited during the onset of the worldwide warmer and more humid CPE. This indicates that the Reifling Basin and the adjacent Wetterstein platforms were affected by peculiar oceanographic conditions during this humid episode. The result was a severe transformation of the sea water composition towards an oxygen-depleted setting near the sea floor and in the sediment. Redox-sensitive elements indicate that euxinic to anoxic conditions prevailed in the upper Julian (Julian 2/1b) Reifling Basin, coinciding with the peak of the Wrangellia LIP manifested in the entire Panthalassa-Tethys Ocean system. This study improves our understanding of the marine adaptations and paleoceanographic changes during the Carnian. High-resolution studies on the mineralogy and geochemistry of the Polzberg section and the formation of the Polzberg *Konservat-Lagerstätte* within the NCA provide a reliable reference for future analyses of other localities and sections.

Acknowledgments The work was done within the framework of and financially supported by projects of the Federal Government of Lower Austria (K3-F-964/001-2020 and K3-F-964/002-2023, Department Science and Research; headquarters in St. Pölten); the Austrian Academy of Sciences (headquarters in Vienna, project Polzberg Lukeneder 2021) represented by the National Committee for Geo/Hydro Sciences (Earth System Sciences Programme); the Triassic project "Fenster in die Karnische Krise" of the Society of Friends of the Natural History Museum Vienna (Freunde des Naturhistorischen Museums Wien, CPE Lukeneder 2023); and La Région Nouvelle Aquitaine and l'ANR-BIOGEN. We thank the owner of the outcrop area, Franziska and Hermann Hofreiter as well as Eva and Karl Jagersberger (all Gaming), for sampling and digging permission. The manuscript greatly benefited from valuable comments by the reviewers Matija Križnar (Ljubljana) and Franz Ottner (Vienna), the production editor Ashok Arumairaj (Taramani), the in-house series editor Kathleen A. Ritterbush (Utah), the publishing editor Simon Shah-Rohlfs and the production editor Antje Endemann (both Heidelberg).

Author Contribution AL designed the research. IC and AL wrote the manuscript. AL, PL, SG sampled, prepared and documented the fossil material. AL basic work, biostratigraphy, lithology. PL ecology, paleobiology. IC, AEA, ED, MG, CF bulk- and clay mineralogy. IC, AEA, MG, AEK whole rock geochemistry.

Funding Open access funding enabled and projects financially supported by the Federal Government of Lower Austria (K3-F-964/001-2020 and K3-F-964/002-2023, Department Science and Research; headquarters in St. Pölten); the Austrian Academy of Sciences (headquarters in Vienna, project Polzberg Lukeneder 2021) represented by the National Committee for Geo/Hydro Sciences (Earth System Sciences Programme); and Triassic project "Fenster in die Karnische Krise" of the Society of Friends of the Natural History Museum Vienna (Freunde des Naturhistorischen Museums Wien, CPE Lukeneder 2023) and the Natural History Museum Vienna (to AL). The authors are responsible for the contents of this publication. The funders had no impact on conceptualization, design, data collection, analysis, decision to publish, or preparation of the manuscript.

Competing interests All authors stated that they have no competing interests.

Authors' contributions All authors stated that they have equally contributed to their chapters, read, edited and agreed to submit the final versions of their contributions.

Availability of data and materials All sources of information are provided in the text. Raw data related to the fossil material from Austria are available from the corresponding authors and Alexander Lukeneder upon request.

Additional information The article is part of the Special Volume: *Late Triassic Konservat-Lagerstätten within the Carnian Pluvial Episode. 150 Years of Scientific Excavations in Austria.*

Artificial Intelligence (AI) The chapter was produced without the use of AI.

Language The entire volume uses American English reviewed by a native scientist.

Received 2025.

Published online 2026.

References

Abedini A, Calagari AA (2017) Geochemistry of claystones of the Ruteh formation, NW Iran: implications for provenance, source-area weathering, and paleo-redox conditions. N Jahrb Mineral Abh 194(2):107–123. https://doi.org/10.1127/njma/2017/0040

Algeo TJ, Maynard JB (2004) Trace-element behavior and redox facies in core shales of upper Pennsylvanian Kansas-type cyclothems. Chem Geol 206(3–4):289–318. https://doi.org/10.1016/J.CHEMGEO.2003.12.009

Algeo TJ, Tribovillard N (2009) Environmental analysis of paleoceanographic systems based on molybdenum-uranium covariation. Chem Geol 268(3–4):211–225. https://doi.org/10.1016/j.chemgeo.2009.09.001

Allison PA (1988) *Konservat-Lagerstätten*: cause and classification. Paleobiology 14(4):331–344. https://doi.org/10.1017/S0094837300012082

Anadón P, Gliozzi E, Mazzini I (2002) Paleoenvironmental reconstruction of marginal marine environments from combined paleoecological and geochemical analyses on ostracods. In: Geophysical monograph series. Blackwell Publishing Ltd, vol 131, pp 227–247. https://doi.org/10.1029/131GM12

Armstrong-Altrin JS, Nagarajan R, Balaram V, Natalhy-Pineda O (2015) Petrography and geochemistry of sands from the Chachalacas and Veracruz beach areas, western Gulf of Mexico, Mexico: constraints on provenance and tectonic setting. J S Am Earth Sci 64:199–216. https://doi.org/10.1016/J.JSAMES.2015.10.012

Bankole OM, El AA, Meunier A, Poujol M, Bekker A (2020) Elemental geochemistry and Nd isotope constraints on the provenance of the basal siliciclastic succession of the middle Paleoproterozoic Francevillian group, Gabon. Precambrian Res 348:105874. https://doi.org/10.1016/j.precamres.2020.105874

Bauer FK, Schnabel W (1997) Geologische Karte der Republik Österreich. Sheet Mariazell 72, 1:50.000. Geologische Bundesanstalt Wien

Bhatia MR, Crook KAW (1986) Trace element characteristics of graywackes and tectonic setting discrimination of sedimentary basins. Contrib Mineral Petrol 92(2):181–193. https://doi.org/10.1007/BF00375292

Brindley GW, Brown G (1980) Crystal structures of clay minerals and their X-ray identification. Mineralogical Society, London, vol 5, pp 1–495. https://doi.org/10.1180/mono-5

Carignan J, Hild P, Mevelle G, Morel J, Yeghicheyan D (2007) Routine analyses of trace elements in geological samples using flow injection and low pressure on-line liquid chromatography coupled to ICP-MS: a study of geochemical reference materials BR, DR-N, UB-N, An-G and GH. Geostand Geoanalytical Res 25(2–3):187–198. https://doi.org/10.1111/j.1751-908X.2001.tb00595.x

Chamley H (1989) Clay sedimentology. Springer-Verlag Berlin Heidelberg, pp 1–623. https://doi.org/10.1007/978-3-642-85916-8

Chraiki I, Fru EC, Somogyi A, Bouougri EH, Bankole O, Ghnahalla M, El AA (2023) Blooming of a microbial community in an Ediacaran extreme volcanic lake system. Sci Rep 13:9080. https://doi.org/10.1038/s41598-023-36031-5

Chung FR (1974) Quantitative interpretation of X-ray diffraction patterns of mixtures. I. Matrix-flushing method for quantitative multicomponent analysis. J App Crystallogr 7:519–525. https://doi.org/10.1107/S0021889874010375

Condie KC (1993) Chemical composition and evolution of the upper continental crust: contrasting results from surface samples and shales. Chem Geol 104:1–37. https://doi.org/10.1016/0009-2541(93)90140-E

Dal Corso J, Bernardi M, Sun Y, Song H, Seyfullah LJ, Preto N, Gianolla P, Ruffell A, Kustatscher E, Roghi G, Merico A, Hohn S, Schmid, AR, Marzoli A, Newton RJ, Wignall PB, Benton MJ (2020) Extinction and dawn of the modern world in the Carnian (Late Triassic). Sci Adv 6(38):eaba0099. https://doi.org/10.1126/sciadv.aba0099

Dal Corso J, Mietto P, Newton RJ, Pancost RD, Preto N, Roghi G, Wignall PB (2012) Discovery of a major negative $\delta^{13}C$ spike in the Carnian (Late Triassic) linked to the eruption of Wrangellia flood basalts. Geology 40(1):79–82. https://doi.org/10.1130/G32473.1

Dal Corso J, Gianolla P, Newton RJ, Franceschi M, Roghi G, Caggiati M, Raucsik B, Budai T, Haas J, Preto N (2015) Carbon isotope records reveal synchronicity between carbon cycle perturbation and the Carnian Pluvial Event in the Tethys realm (Late Triassic). Glob Planet Change 127:79–90. https://doi.org/10.1016/j.gloplacha.2015.01.013

Fedo CM, Nesbitt HW, Young GM (1995) Unraveling the effects of potassium metasomatism in sedimentary rocks and paleosols, with implications for paleoweathering conditions and provenence. Geology 23(10):921–924. https://doi.org/10.1130/0091-7613(1995)023%3c0921

Forchielli A, Pervesler P (2013) Phosphatic cuticle in thylacocephalans: a taphonomic case study of *Austriocaris* (Arthropoda, Thylacocephala) from the *Fossil-Lagerstätte* Polzberg (Reingraben Shales, Carnian, Upper Triassic; Lower Austria). Austrian J Earth Sci 106:46–61. https://www.ajes.at/images/AJES/archive/Band%20106_1/forchielli_pervesler_ajes_106_1.pdf

Hillier S (2003) Quantitative analysis of clay and other minerals in sandstones by X-ray powder diffraction (XRPD). Int Assoc Sedimentol Spec Publ 34:213–251. https://doi.org/10.1002/978 1444304336.ch11

Hornung T, Krystyn L, Brandner RA (2007b) A Tethys-wide mid-Carnian (Upper Triassic) carbonate productivity crisis: evidence for the Alpine Reingraben Event from Spiti (Indian Himalaya)? J Asian Earth Sci 30(2):285–302. https://doi.org/10.1016/j.jseaes.2006.10.001

Hornung T, Brandner R, Krystyn L, Joachimski MM, Keim L (2007) Multistratigraphic constraints on the NW Tethyan Carnian Crisis. In Lucas SG, Spielmann JA (eds) The Global Triassic. New Mexico Mus Nat Hist Scie Bull 41:59–67. https://www.researchgate.net/publication/291974578

Jewuła K, Środoń J, Kuligiewicz A, Mikołajczak M, Liivamägi S (2022) Critical evaluation of geochemical indices of palaeosalinity involving boron. Geochim Cosmochim Acta 322:1–23. https://doi.org/10.1016/j.gca.2022.01.027

Jones B, Manning DAC (1994) Comparison of geochemical indices used for the interpretation of palaeoredox conditions in ancient mudstones. Chem Geol 111(1–4):111–129. https://doi.org/10.1016/0009-2541(94)90085-X

Kohút M, Hofmann M, Havrila M, Linnemann U, Havrila J (2018) Tracking an upper limit of the Carnian Crisis and/or Carnian stage in the Western Carpathians (Slovakia). Int J Earth Sci 107(1):321–335. https://doi.org/10.1007/s00531-017-1491-8

Krystyn L (1991) Die Fossillagerstätten der alpinen Trias. In: Nagel D, Rabeder G (eds) Exkursionen im Jungpaläozoikum und Mesozoikum Österreichs. Österr Paläontol Ges Wien 1991:23–78

Lukeneder A (ed) (2026) Late Triassic Konservat-Lagerstätten within the Carnian Pluvial Episode: 150 Years of Scientific Excavations in Austria. Top Geobiol 55:600 p. https://doi.org/10.1007/978-3-032-19777-1

Lukeneder A, Lukeneder P (2021) The Upper Triassic Polzberg palaeobiota from a marine *Konservat-Lagerstätte* deposited during the Carnian Pluvial Episode in Austria. Sci Rep 11(1):16644. https://doi.org/10.1038/S41598-021-96052-w

Lukeneder A, Lukeneder P (2022a) Taphonomic history and trophic interactions of an ammonoid fauna from the Upper Triassic Polzberg palaeobiota. Sci Rep 12(1):7455. https://doi.org/10.1038/S41598-022-11496-y

Lukeneder A, Lukeneder P (2023) New data on the marine upper Triassic palaeobiota from the Polzberg *Konservat-Lagerstätte* in Austria. Swiss J Palaeontol 142:9. https://doi.org/10.1186/s13358-023-00269-3

Lukeneder A, Lukeneder P (2024) Ammonoid taxonomy of the Carnian Polzberg *Konservat-Lagerstätte* in Austria. Jahrb Geol BA (GeoSphere Austria) 162:27–55

Lukeneder A, Surmik D, Gorzelak P, Niedźwiedzki R, Brachaniec T, Salamon MA (2020) Bromalites from the Upper Triassic Polzberg section (Austria); insights into trophic interactions and food chains of the Polzberg palaeobiota. Sci Rep 10:20545. https://doi.org/10.1038/s41598-020-77017-x

Lukeneder A, Lukeneder P, Sachsenhofer RF, Roghi G, Rigo M (2024) Multi-proxy record of the Austrian Upper Triassic Polzberg Konservat-Lagerstätte in light of the Carnian Pluvial Episode. Sci Rep 14:11194. https://doi.org/10.1038/s41598-024-60591-9

Lukeneder P, Lukeneder A (2022a) Comment on Triassic coleoid beaks and other structures from the Calcareous Alps revisited by Doguzhaeva et al. (2022). Acta Palaeontol Pol 67(4):963–965. https://doi.org/10.4202/app.01016.2022

Lukeneder P, Lukeneder A (2022b) Mineralized belemnoid cephalic cartilage from the late Triassic Polzberg *Konservat-Lagerstätte* (Austria). PLoS ONE 17(4):e0264595. https://doi.org/10.1371/journal.pone.0315418

Lukeneder S, Lukeneder A, Harzhauser M, İslamoğlu Y, Krystyn L, Lein R (2012) A delayed carbonate factory breakdown during the Tethyan-wide Carnian Pluvial Episode along the Cimmerian terranes (Taurus, Turkey). Facies 58:279–296. https://doi.org/10.1007/s10347-011-0279-8

McLennan SM (1993) Weathering and Global Denudation. J. Geology 101(2):295–303. https://doi.org/10.1086/648222

Meunier A (2005) Clays. Springer-Verlag Berlin Heidelberg, pp 1–472. https://doi.org/10.1007/b138672

Miller CS, Peterse F, Da Silva AC, Baranyi V, Reichart GJ, Kürschner WM (2017) Astronomical age constraints and extinction mechanisms of the Late Triassic Carnian crisis. Sci Rep 7:2557. https://doi.org/10.1038/S41598-017-02817-7

Moore DM, Reynolds RC (1989) X-ray diffraction and the identification and analysis of clay minerals. Oxford University Press, New York, 332 p

Moore DM, Reynolds RC (1997) X-ray diffraction and the identification and analysis of clay minerals, 2nd edn. Oxford University Press, New York, 378 p

Mueller S, Krystyn L, Kürschner WM (2016) Climate variability during the Carnian Pluvial Phase—a quantitative palynological study of the Carnian sedimentary succession at Lunz am See, Northern Calcareous Alps, Austria. Palaeogeogr Palaeoclimatol Palaeoecol 441(1):198–211. https://doi.org/10.1016/J.PALAEO.2015.06.008

Nagarajan R, Madhavaraju J, Nagendra R, Selvamony Armstrong-Altrin J, Moutte J (2007) Geochemistry of Neoproterozoic shales of the Rabanpalli Formation, Bhima Basin, Northern

Karnataka, southern India: implications for provenance and paleoredox conditions. Rev Mex Cienc Geol 24(2):150–160

Nesbitt HW, Young GM (1982) Early proterozoic climates and plate motions inferred from major element chemistry of lutites. Nat 299(5885):715–717. https://doi.org/10.1038/299715a0

Nothdurft LD, Webb GE, Kamber BS (2004) Rare earth element geochemistry of Late Devonian reefal carbonates, Canning Basin, Western Australia: confirmation of a seawater REE proxy in ancient limestones. Geochim Cosmochim Acta 68(2):263–283. https://doi.org/10.1016/S0016-7037(03)00422-8

Panahi A, Young GM, Rainbird RH (2000) Behavior of major and trace elements (including REE) during Paleoproterozoic pedogenesis and diagenetic alteration of an Archean granite near Ville Marie, Quebec, Canada. Geochim Cosmochim Acta 64(13):2199–2220. https://doi.org/10.1016/S0016-7037(99)00420-2

Pecorari M, Caggiati M, Dal Corso J, Cruciani G, Tateo F, Chu D, Gianolla P (2023) Weathering and sea level control on siliciclastic deposition during the Carnian Pluvial Episode (Southern Alps, Italy). Palaeogeogr Palaeoclimatol Palaeoecol 617:111495. https://doi.org/10.1016/J.PALAEO.2023.111495

Piller WE, Egger H, Gross M, Harzhauser M, Hubmann B, Van Husen D, Krenmayr H-G, Krystyn L, Lein R, Mandl G, Rögl F, Roetzel R, Rupp C, Schnabel W, Schönlaub HP, Summesberger H, Wagreich M (2004) Die Stratigraphische Tabelle von Österreich 2004 (sedimentäre Schichtfolgen). Kommission für die Paläontologische und stratigraphische Erforschung Österreichs (Österreichische Akademie der Wissenschaften und Österreichische Stratigraphische Kommission, 2004)

Retallack G, Gates N, York N, Feldshuhorn M (2020) Boron paleosalinity proxy for deeply buried Paleozoic and Ediacaran fossils. Palaeogeogr Palaeoclimatol Palaeoecol 540:109536. https://doi.org/10.1016/j.palaeo.2019.109536

Rigo M, Joachimski MM (2010a) Palaeoecology of Late Triassic Conodonts: constraints from Oxygen Isotopes in Biogenic Apatite. Acta Palaeontol Pol 55(3):471–478. https://doi.org/10.4202/APP.2009.0100

Rigo M, Preto N, Roghi G, Tateo F, Mietto P (2007) A rise in the carbonate compensation depth of western Tethys in the Carnian (Late Triassic): deep-water evidence for the Carnian Pluvial event. Palaeogeogr Palaeoclimatol Palaeoecol 246(2–4):188–205. https://doi.org/10.1016/J.PALAEO.2006.09.013

Rigo M, Trotter JA, Preto N, Williams IS (2012) Oxygen isotopic evidence for Late Triassic monsoonal upwelling in the northwestern Tethys. Geology 40(6):515–518. https://doi.org/10.1130/G32792.1

Rigo M, Mazza M, Karádi V, Nicora A (2018) New Upper Triassic conodont biozonation of the Tethyan realm. In Tanner L (ed) The Late Triassic World. Topics in Geobiology, vol 46, pp 189–235. https://doi.org/10.1007/978-3-319-68009-5_6

Roghi G, Gianolla P, Minarelli L, Pilati C, Preto N (2010) Palynological correlation of Carnian humid pulses throughout western Tethys. Palaeogeogr Palaeoclimatol Palaeoecol 290(1–4):89–106. https://doi.org/10.1016/J.PALAEO.2009.11.006

Ruffell A, Simms M, Wignall PB (2016) The Carnian Humid Episode of the late Triassic: a review. Geol Mag 153(2):271–284. https://doi.org/10.1017/S0016756815000424

Ruttner A, Schnabel W (1988) Geologische Karte der Republik Österreich, Sheet Ybbsitz 71, 1:50.000. Geologische Bundesanstalt

Scott C, Lyons TW (2012) Contrasting molybdenum cycling and isotopic properties in euxinic versus non-euxinic sediments and sedimentary rocks: refining the paleoproxies. Chem Geol 324–325:19–27. https://doi.org/10.1016/j.chemgeo.2012.05.012

Seilacher A (1970) Begriff und Bedeutung der Fossil-Lagerstätten (Concept and meaning of fossil lagerstätten). Neues Jahrb Geol Paläontol Mh 34–39

Shields G, Stille P (2001) Diagenetic constraints on the use of cerium anomalies as palaeoseawater redox proxies: an isotopic and REE study of Cambrian phosphorites. Chem Geol 175(1–2):29–48. https://doi.org/10.1016/S0009-2541(00)00362-4

Stur D (1886) Vorlage des ersten fossilen Schädels von Ceratodus aus den ober triadischen Reingrabner Schiefern von Pölzberg nördlich bei Lunz. Verh k. k. Geol Reichsanstalt 15:381–383

Stur D (1874) Neue Aufschlüsse im Lunzer Sandsteine bei Lunz und ein neuer Fundort von Wengerschiefer im Pölzberg zwischen Lunzersee und Gaming. Verh kk Geol Reichsanstalt 271–273

Taylor SR; McLennan SM (1985) The continental crust: Its composition and evolution. Blackwell Scientific, Oxford, pp 1–312. https://doi.org/10.1017/S0016756800032167

Teller F (1891) Über den Schädel eines fossilen Dipnoers, Ceratodus Sturii nov. spec. aus den Schichten der oberen Trias der Nordalpen. Abh kk Geol Reichsanst 15(3):1–38

Tollmann A (1976) Analyse des klassischen nordalpinen Mesozoikums. Franz Deuticke, Wien, 580 p

Tostevin R, Shields GA, Tarbuck GM, He T, Clarkson MO, Wood RA (2016) Effective use of cerium anomalies as a redox proxy in carbonate-dominated marine settings. Chem Geol 438:146–162. https://doi.org/10.1016/j.chemgeo.2016.06.027

Tribovillard N, Algeo TJ, Lyons T, Riboulleau A (2006) Trace metals as paleoredox and paleoproductivity proxies: an update. Chem Geol 232(1–2):12–32. https://doi.org/10.1016/j.chemgeo.2006.02.012

Tribovillard N, Algeo TJ, Baudin F, Riboulleau A (2012) Analysis of marine environmental conditions based onmolybdenum-uranium covariation—applications to Mesozoic paleoceanography. Chem Geol 324–325:46–58. https://doi.org/10.1016/j.chemgeo.2011.09.009

Trotter JA, Williams IS, Nicora A, Mazza M, Rigo M (2015) Long-term cycles of Triassic climate change: a new δ18O record from conodont apatite. Earth Planet Sci Lett 415:165–174. https://doi.org/10.1016/J.EPSL.2015.01.038

Vogt T, Clauer N, Larqué P (2010) Impact of climate and related weathering processes on the authigenesis of clay minerals: examples from circum-Baikal region, Siberia. CATENA 80(1):53–64. https://doi.org/10.1016/j.catena.2009.08.008

Wang P, Du Y, Yu W, Algeo TJ, Zhou Q, Xu Y, Qi L, Yuan L, Pan W (2020) The chemical index of alteration (CIA) as a proxy for climate change during glacial-interglacial transitions in earth history. Earth-Sci Rev 201:103032. https://doi.org/10.1016/j.earscirev.2019.103032

Webb GE, Kamber BS (2000) Rare earth elements in Holocene reefal microbialites: a new shallow seawater proxy. Geochim Cosmochim Acta 64(9):1557–1565. https://doi.org/10.1016/S0016-7037(99)00400-7

Wei W, Algeo TJ (2020) Elemental proxies for paleosalinity analysis of ancient shales and mudrocks. Geochim Cosmochim Acta 287:341–366. https://doi.org/10.1016/j.gca.2019.06.034

Xiao D, Cao J, Luo B, Tan X, Xiao W, He Y, Li K (2020) Neoproterozoic postglacial paleoenvironment and hydrocarbon potential: a review and new insights from the Doushantuo Formation Sichuan Basin, China. Earth-Sci Rev 212:103453. https://doi.org/10.1016/j.earscirev.2020.103453

Chapter 14
Foraminifera from the Upper Triassic Carnian Pluvial Episode in Lower Austria

Luka Gale and Alexander Lukeneder

Abstract The lower Carnian marl-dominated Reingraben Formation in the Northern Calcareous Alps was deposited during the Carnian Pluvial Episode (CPE), a global environmental perturbation. Here we report on foraminiferal assemblages from the Reifling, Göstling and Reingraben Formations from three localities in Austria, including the Polzberg *Konservat-Lagerstätte*. The analysis is based on thin sections and residue samples. The total assemblage contains *Gsollbergella gsollbergensis* (Zaninetti), Ophthalmidiidae, *Cornuspira* sp.?, *Ammobaculites* sp.?, *Duostomina biconvexa* Kristan-Tollmann, *Endotriada kuepperi* (Oberhauser)?, *Lamelliconus*? *multispira* (Oberhauser), *Lamelliconus* ex gr. *ventroplanus* (Oberhauser), *Trocholina umbo* Frentzen, *Aulotortus sinuosus* Weynschenk?, *Aulotortus tenuis* (Kristan)?, *Parvalamella friedli* (Kristan-Tollmann)?, undetermined sessile foraminifera, and uniserial Lagenida. Except for *G. gsollbergensis*, foraminifera are not abundant in the samples. The total assemblage bears most similarities with assemblages from the Tor Formation and its equivalents in the Southern Alps.

Keywords Foraminifera · Carnian Pluvial Episode · Reingraben Formation · Late Triassic · Austria

L. Gale (✉)
University of Ljubljana, Faculty of Natural Sciences and Engineering, Department of Geology, Ljubljana, Slovenia
e-mail: luka.gale@ntf.uni-lj.si; luka.gale@geo-zs.si

Geological Survey of Slovenia, Ljubljana, Slovenia

A. Lukeneder
Department of Geology and Paleontology, Natural History Museum Vienna, Vienna, Austria
e-mail: alexander.lukeneder@nhm.at

A. Lukeneder (ed.), *Late Triassic Konservat-Lagerstätten within the Carnian Pluvial Episode*, Topics in Geobiology 55, https://doi.org/10.1007/978-3-032-19777-1_14

14.1 Introduction

The lower Carnian (Julian 2) Göstling and Reingraben Formations from the Reifling Basin of the Northern Calcareous Alps (Eastern Alps) were deposited during the Carnian Pluvial Event (CPE; sensu Simms and Ruffel 1989). They yield important insights into this environmental perturbation (Lukeneder and Lukeneder 2021; Lukeneder et al. 2024). The CPE was presumably caused by periods of volcanic activity of the Wrangellia Large Igneous Province, which led to global warming and intensification of the hydrological cycle. This is reflected in increased terrestrial influx into marginal marine areas and increased burial of organic carbon (Preto et al. 2010; Dal Corso et al. 2012; Ogg 2015). The CPE is also closely related to notable changes in marine and terrestrial faunas, including increased levels of extinction in several animal groups (Simms and Ruffel 1989; Dal Corso et al. 2020), a marked change in carbonate platforms (Hornung et al. 2007; Breda et al. 2009; Stefani et al. 2010; Gattolin et al. 2015; Caggiati et al. 2017), and a shift in land tetrapods (Brusatte et al. 2011; Bernardi et al. 2018). From the paleoecological perspective, the *Konservat-Lagerstätte* from the Reingraben Formation at Polzberg near Lunz am See (Northern Calcareous Alps, Lower Austria) presents an important window into the composition of marine biota during the CPE. Collecting and research (Stur 1874, 1886; Teller 1891; Griffith 1977) has intensified and become more systematic in recent years (Doguzhaeva et al. 2006, 2007; Doguzhaeva and Summersberger 2012; Forchielli and Pervesler 2013; Lukeneder et al. 2020, 2021, 2024; Lukeneder, 2026). These efforts have led to a long list of well-preserved animal taxa representing ammonites, coleoids, bivalves, gastropods, echinoids, thylacocephalan arthropods, crustaceans, polychaetes, actynopterygians, cartilaginous fishes and sarcopterygians, along with bromalites, some plant remains, and very rare conodont elements (Lukeneder et al. 2021).

In contrast to this extensive list of fossil groups, foraminifera were not yet studied from the Polzberg locality, even though relatively well-preserved and diverse foraminiferal faunas are known from other localities and equivalent lithostratigraphic units deposited during the CPE (e.g., Oberhauser 1957, 1960; Kristan-Tollmann 1960; Koehn-Zaninetti 1969; Rettori et al. 1998; Kolar-Jurkovšek et al. 2005; Carillat and Martini 2009; Forel et al. 2020; Li et al. 2025). This paper briefly presents the foraminiferal taxa determined from the Reingraben Formation and the underlying Reifling and Göstling Formations (sensu Lukeneder et al. 2024) at Polzberg, as well as from previously unreported Steinbach and Großau sites.

14.2 Geologic Setting of Outcrops

The stratigraphic and sedimentological description of the Polzberg section was previously reported in Lukeneder and Lukeneder (2021, 2023), and Lukeneder et al. (2024). The Upper Triassic outcrops at Polzberg (Po) are located on the western

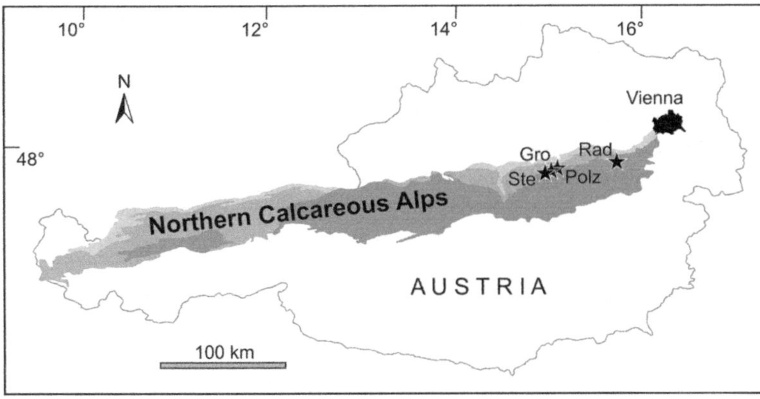

Fig. 14.1 Locality map of the Rad (Rad), Polzberg (Polz), Großau drilling (Gro) and Steinbachgraben (Ste) sites in Lower Austria, with marked position of the Austrian Northern Calcareous Alps (in grey)

slope of Mount Schindelberg (1066 m), north of the Ois River, 4 km northeast of Lunz am See in Lower Austria (Stur 1874, 1886; Lukeneder and Lukeneder 2021; 1:50 000, geological map, sheet 71 Ybbsitz; Ruttner and Schnabel 1988 and sheet 72 Mariazell; Fig. 14.1). Within the Lunz Nappe, the Reifling Basin—an Upper Triassic intraplatform basin—is located between Polzberg and Großreifling (Krystyn 1991; Lukenerder and Lukeneder 2021). The exact position of the fossiliferous locality in the southern area of the Lunz Nappe within the lower, fossiliferous part of the Reingraben Formation (see Tollmann 1976; Piller et al. 2004) was determined by GPS (global positioning system): N 47°53′4.98″ and E 15°4′28.15″.

The Großau section (Gro) is located on the northeastern slope of Mount Großkopf (1147 m), 300 m southeast of the Ybbs River, in the Ypps valley connecting Gösling and Lunz am See (Fig. 14.1). The drilling site (47°49′24.40″N, 15°00′0.80″E) is located 5 km southwest of the Lunz am See center and 5 km northeast of Göstling an der Ybbs in southwestern Lower Austria (Trauth 1948; geological map, 71 Ybbsitz, Ruttner and Schnabel 1988). The southernmost tectonic element of the Northern Calcareous Alps (NCA) in southeastern areas of Lower Austria around Großau is the Lunz Nappe, occasionally interpreted as a distinct Sulzbach Nappe.

The Steinbach section (St) is located near Göstling an der Ybbs (Lower Austria), 4 km southwest of the Großau drilling site on the Sulzbach Nappe within the Lunz Nappe System (High-Bajuvaric System; Geo map 71 Ybbsitz; Ruttner und Schnabel 1988). The outcrop is located at the lower end of the Steinbach River and at the transition into the larger Ybbs River, on both banks of the Ybbs River, and is only accessible at low water. The section exhibits the Reifling Formation, approx. 10 m of the Göstling Formation, and at the top the argillaceous Reingraben Formation with frequent intercalated calcareous layers. The calcareous layers are of allodapic character with transported fragments and fossils (e.g., foraminifera).

The Rad section (Rad) south of Kleinzell (Lower Austria) is located at a forest road on the tectonic Reisalpen Nappe of the Tyrolic System (Geofast map 74 Hohenberg; Moser and Pavlik 2013). Single oolitic layers are present above the Göstling Formation. They contain bivalves and foraminifera, which were transported into basin from the edge or slope of the basin or from nearby surrounding platforms within the basal argillaceous Reingraben Formation.

14.3 Stratigraphy

The Reifling Formation (*Trachyceras aonoides* Zone, *T. aonoides* Subzone; Julian 1/II) was deposited on the paleoslope of the Reifling Basin (Lukeneder and Lukeneder 2021; Lukeneder et al 2024). It consists of light grey and grey-green nodular limestone composed of peloidal, filamentous wackestone and packstone (Fig. 14.2). The conodonts *Paragondolella polygnathiformis*, *P. inclinata*, *P. foliata*, *P. praelindae*, *Paragondolella* aff. *foliata*, *Gladigondolella tethydis*, *G. malayensis*, and "*Mosherella*" *longnanensis* were determined (Lukeneder et al. 2024). The Göstling Formation (base of Julian 2/Ia, *Austrotrachyceras austriacum* Zone, *A. minor* Subzone) comprises black mm- to cm-bedded calciturbidites and radiolarian limestone with organic-rich mudstone. The conodonts *G. tethydis*, *G. malayensis malayensis*, and *Gladigondolella* sp. were recorded there (Lukeneder et al. 2024). The Reingraben Formation (*A. austriacum* Zone, *A. minor* Subzone; Julian 2/Ib) is represented by an alternation of dark grey to black argillaceous marl, intercalated by rare limestone/dolomite (foraminiferal- to radiolarian-dominated wackestone to packstone) beds. The Reingraben Formation is followed by the siliciclastic Lunz Formation (Julian 2/II), which is no longer part of the sampled succession.

14.4 Material and Methods

For 28 thin sections of standard sizes (28 mm × 47 mm), 4 samples were taken from the Reifling Formation, 4 from the Göstling Formation, and 15 from the Reingraben Formation. Thin sections were investigated with a polarizing microscope. Foraminifera were also found in two residue samples from the Reifling Formation, one residue sample from the Göstling Formation, and one residue sample from the Reingraben Formation. The position of the corresponding localities and sampled logs are all from the Sulzbach Nappe in Lower Austria (Figs. 14.1 and 14.2). Acetolysis was performed with acetic acid. The chemical composition of foraminiferal tests was analyzed using scanning electron microscopy coupled with energy-dispersive X-ray spectroscopy (SEM-EDS). Analytical work was carried out on a ThermoFisher Scientific Quattro S field-emission scanning electron microscope, equipped with an Oxford Instruments UltimMax 64 EDS detector. The instrument operated under high-vacuum conditions, with an accelerating voltage of 10 kV and a beam current of

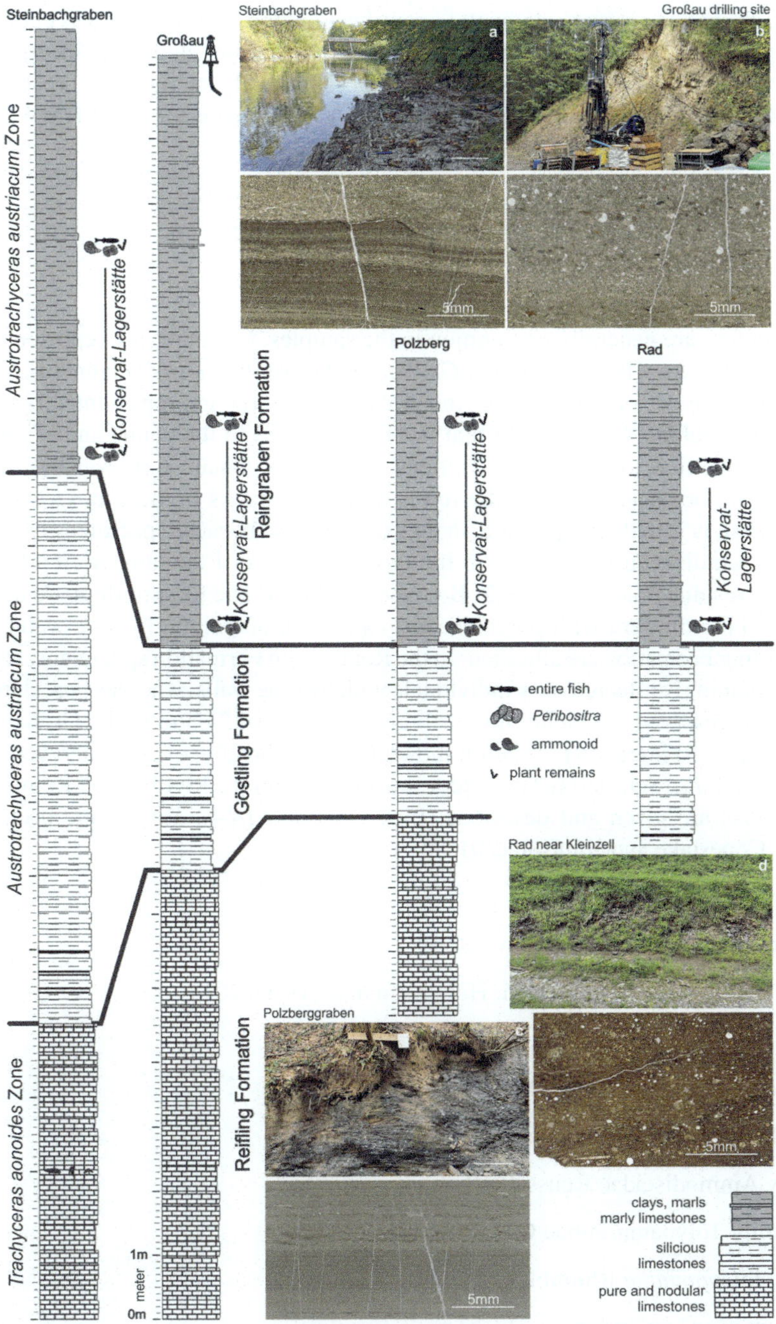

Fig. 14.2 Stratigraphic position of the foraminifera-bearing deposits. Adapted after Lukeneder et al. (2021). Detailed sections within the foraminifera-bearing layers in the lower Carnian Reingraben Formation

10 nA. To ensure accurate elemental quantification, the EDS system was calibrated using pure silicon for light elements and cobalt for heavier elements. Spectral data were collected over a 60-s acquisition interval. The fossil material is incorporated in the collections of the Natural History Museum Vienna (main number NHMW 2025/ 00261/).

14.5 Foraminifera Taxonomy

Foraminifera are generally uncommon in the samples. Most taxa are represented by a few specimens. The exception is *Gsollbergella gsollbergensis* (Zaninetti), which is present in most samples of the Reingraben Formation and represented by many specimens. Table 14.1 lists taxa that were determined to the genus and/or species level. In addition to these, thin sections commonly contain uniserial lageninids. These were not determined to the lower taxonomic ranks because observation of the aperture is usually not possible in thin sections and because specimens from thin sections are difficult to compare to the type specimens, if the latter were obtained from the residue (Gale et al. 2022; Barros et al. 2025). The foraminifera are unfragmented. Foraminifera with porcelaneous, calcitic or aragonitic hyaline walls from the Reifling Formation are silicified. In the Reingraben Formation specimens, silicification seemingly affected only individuals with hyaline walls, whereas porcelaneous walls are preserved as calcite. Interestingly, sample Gro 25.8 m in the Großau drill core from the Reingraben Formation shows numerous bundles of calcimicrobes of the type *Girvanella*. Selected foraminifera specimens from the Reifling and Reingraben Formations are given and described (Figs. 14.3 and 14.4). Taxonomic principles follow Copestake and Johnson (2014).

Systematic Paleontology

Phylum Foraminifera d'Orbigny 1826

Class Tubothalamea Pawlowski, Holzmann and Tyszka 2013

Order Spirillinida Hohenegger and Piller 1975

Suborder Ammodiscina Mikhalevich 1980

Superfamily Ammodiscoidea Reuss 1862

Family Ammodiscidae Reuss 1862

Subfamily Tolypammininae Cushman 1928

Genus *Tolypammina* Rhumbler 1895

***Tolypammina*? sp. indet.**

Figure 14.4i–j

Table 14.1 Foraminifera determined from the Reifling, Göstling and Reingraben Formations in this study

	Reifling Formation	Göstling Formation	Reingraben Formation
Tolypammina? sp. indet	x		
Neotolypammina discoidea	x		
Gsollbergella gsollbergensis	?		x
Ophthalmidiidae			x
Planiinvoluta? *deflexa*?	x		
Calcivertellidae? gen. et sp. indet			x
Cornuspira? sp. A			x
Lituotuba? sp. A	x		
Ammobaculites? sp.			x
Gaudryinopsis kelleri	x		
Duostomina biconvexa			x
Endotriada kuepperi?	x		x
Lamelliconus? *multispira*			x
Lamelliconus ex gr. *ventroplanus*			x
Trocholina cf. *umbo*			x
Aulotortus sinuosus?			x
Aulotortus tenuis?			x
Parvalamella friedli?			x
Foraminifera indet. sp. A			x
Uniserial Lagenida	x	x	x

Material: Isolated specimens from the Reifling Formation (sample NHMW 2025/0261/0003).

Description: Undivided tube is elliptical in cross-section, variously dense and loosely coiled, unattached. The wall is siliceous, agglutinated. The test measures 0.69–0.72 mm; the final lumen is up to 0.15 mm in diameter; the wall is 0.02 mm thick.

Genus *Neotolypammina* Urošević and Sudar (1992)

Neotolypammina discoidea (Trifonova 1962)

Figure 14.4h

1962 *Tolypammina discoidea* n. sp. – Trifonova 1962, p. 165, pl. 1, Figs. 10–13; pl. 2, Figs. 1–2.

1992 *Tolypammina discoidea* Trifonova 1962. – Trifonova, p. 11, pl. 6, Fig. 32; pl. 7, Figs. 5, 6, 9–11; pl. 8, Figs. 1–3.

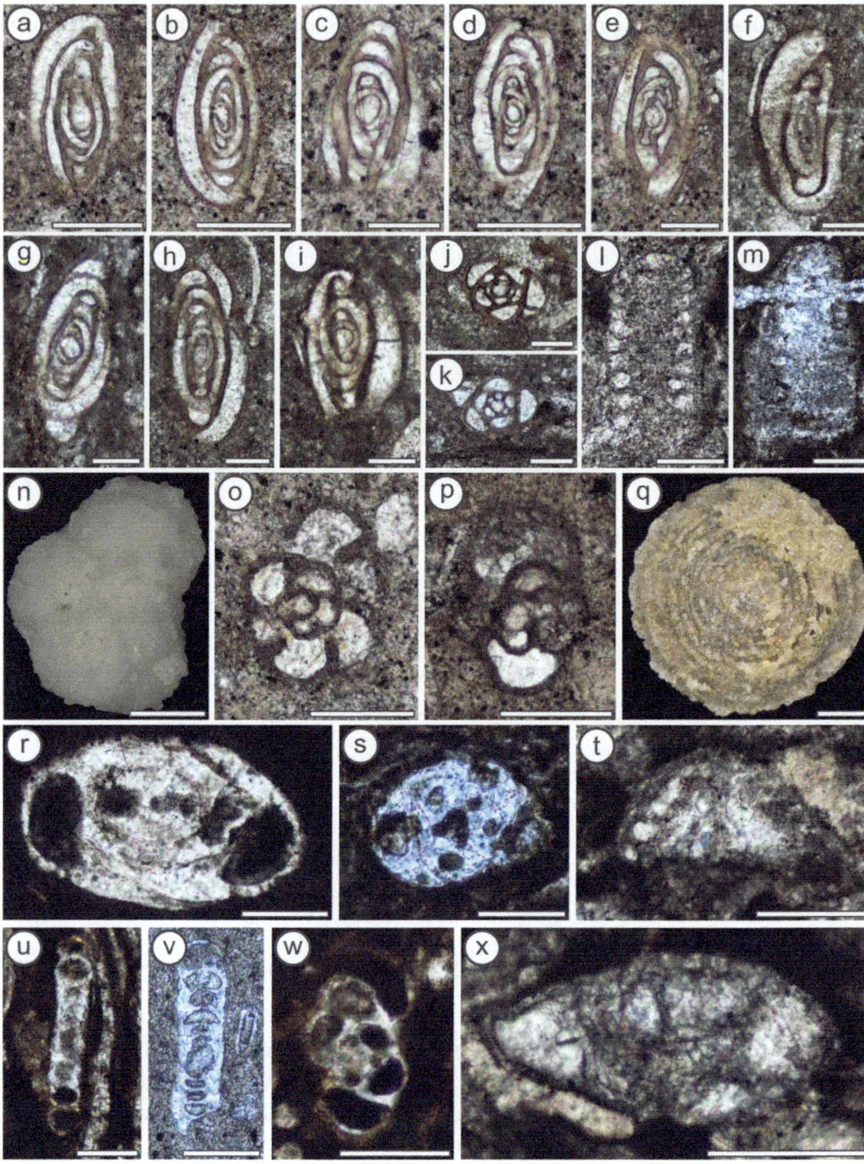
a
b
c
d
e
f
g
h
i
j
k
l
m
n
o
p
q
r
s
t
u
v
w
x

◂**Fig. 14.3** Triassic foraminifera from the Reifling and Reingraben Formations. **a–k** *Gsollbergella gsollbergensis* (Zaninetti). **a–e** Großau drill core, sample Gro 25.8 m, Reingraben Formation, NHMW 2025/0261/0022. **f, h–j** Steinbach, sample St 39.5 m, Reingraben Formation, NHMW 2025/0261/0014. **g, k** Stainbach, sample St 44 m, Reingraben Formation, NHMW 2025/0261/0019. **l–m** *Lamelliconus*? *multispira* (Oberhauser). ***l*** Großau drill core, sample Gro 23.6 m, Reingraben Formation, NHMW 2025/0261/0020. ***m*** Sample NHMW 2025/0261/0023, Reingraben Formation, NHMW 2025/0261/0023. **n–p** *Endotriada kuepperi* (Oberhauser)?. **n** Polzberg, sample Top Reif PO, Reifling Formation, NHMW 2025/0261/0004. **o–p** Großau drill core, sample Gro 25.8 m, Reingraben Formation, NHMW 2025/0261/0022. **q** *Lamelliconus* ex gr. *ventroplanus* (Oberhauser). Kleinzell, sample Rad, Reingraben Formation, NHMW 2025/0261/0023. **r** *Aulotortus sinuosus* Weynschenk?. Großau drill core, sample Gro 23.6 m, Reingraben Formation, NHMW 2025/0261/0020. **s** *Parvalamella friedli* (Kristan-Tollmann)?. Sample Rad, Reingraben Formation, NHMW 2025/0261/0023. **t** *Trocholina* cf. *umbo* Frentzen. Großau drill core, sample Gro 23.4 m, Reingraben Formation, NHMW 2025/0261/0020. **u –v** *Aulotortus tenuis* (Kristan)?. ***u*** Sample Rad, Reingraben Formation, NHMW 2025/0261/0023. ***v*** Sample NHMW 2025/0261/0023, Reingraben Formation, NHMW 2025/0261/0023. **w** Foraminifera indet. sp. A. Sample Rad, Reingraben Formation, NHMW 2025/0261/0023. **x** *Duostomina biconvexa* Kristan-Tollmann. Großau drill core, sample Gro 23.6 m, Reingraben Formation, NHMW 2025/0261/0020. Scale bars: for figures c, e–k, m, s, u–w scale is 0.05 mm; for figures a, b, d, l, n, o, p, r, t, x scale is 0.1 mm; for figure q scale is 0.2 mm

1992 *Neotolypammina discoidea* (Trifonova 1962). – Urošević and Sudar, p. 224, pl. 3, Figs. 5, 7.

Material: Isolated specimens from the Reifling Formation (samples NHMW 2025/0261/0003, NHMW 2025/0261/0004).

Description: The test is attached. Apparently undivided half-tube is initially coiled in four whorls. This is followed by a devolute last segment. The wall is agglutinated. The planispiral part measures 0.35–0.38 mm in diameter; the final width of the lumen in the uncoiled part is 0.04 mm; the wall thickness is 0.03 mm.

Stratigraphic distribution: Anisian to Norian (Trifonova 1992).

Order Miliolida Delage and Hérouard 1896
Suborder Nubeculariina Jones in Griffith and Henfrey 1875
Superfamily Nubecularioidea Jones in Griffith and Henfrey 1875
Family Ophthalmidiidae Wiesner 1920
Genus *Gsollbergella* Zaninetti 1979.

***Gsollbergella gsollbergensis* (Zaninetti** 1969)

Figure 14.3a–k

1969 *Agathammina austroalpina* Kristan-Tollmann et Tollmann. – Koehn-Zaninetti, p. 57, pl. 8, Fig. A–D, text-Fig. 11A–H.

1969 *Agathamminoides gsollbergensis* gen n., sp. n. – Zaninetti, p. 699, text-Fig. 1.

1971 *Agathammina spiroloculiformis* n. sp. – Oravecz-Scheffer, p. 91, pl. 2, Figs. 1–5.

Material: Numerous specimens in thin sections from samples of the Reingraben Formation (samples NHMW 2025/0261/0014—NHMW 2025/0261/0023).

a
b
c
d
e
f
g
h
i
j
k
l
m
n
o
p

◂**Fig. 14.4** Triassic foraminifera from the Reifling Formation and Reingraben Formation. **a** ?*Ammobaculites* sp. Großau drill core, sample Gro 25.8, Reingraben Formation, NHMW 2025/0261/0022. **b** *Gaudryinopsis kelleri* (Tappan). Polzberg, sample Top Reif Polz, Reifling Formation, NHMW 2025/0261/0003. **c–d** *Cornuspira*? sp. A. Großau drill core, sample Gro 25.8, Reingraben Formation, NHMW 2025/0261/0022. **e–f** *Planiinvoluta*? *deflexa* Leischner?. **e** Polzberg, sample Top Reif Polz, Reifling Formation, NHMW 2025/0261/0003. *f* Polzberg, sample Base Reif Polz, Reifling Formation, NHMW 2025/0261/0003. **g** *Ammodiscus* or *Cornuspira* sp. (the wall is silicified). Polzberg, sample Top Reif Polz, Reifling Formation, NHMW 2025/0261/0003. **h** *Neoptolypammina discoidea* (Trifonova). Polzberg, sample Top Reif Polz, Reifling Formation, NHMW 2025/0261/0003. **i–j** Polzberg, sample Top Reif Polz, Reifling Formation, NHMW 2025/0261/0003. **k** *Lituotuba*? sp. A. Polzberg, sample Top Reif Polz, Reifling Formation, NHMW 2025/0261/0003. **l** Undetermined foraminifera with porcelaneous wall. Großau drill core, sample Gro 25.8 m, Reingraben Formation, NHMW 2025/0261/0022. **m** Lagenida indet. Großau drill core, sample Gro 24.0 m, Reingraben Formation, NHMW 2025/0261/0021. **n** *Protodentalina* sp. Steinbach, sample St 44 m, Reingraben Formation, NHMW 2025/0261/0019. **o–p** Calcivertellinae? gen. et sp. indet. Großau drill core, sample Gro 25.8 m, Reingraben Formation, NHMW 2025/0261/0022. Scale bars: for figures d, m scale is 0.05 mm; for figures a, c, e–h, k, l, n–p scale is 0.1 mm; for figures b, i, j scale is 0.2 mm

Description: Small ovoid test, composed of elongated chambers arranged in quinqueloculine arrangement and separated by rudimental septa. Chambers extend along the entire length/height of the test and are widest near their proximal end and narrowest at the distal end. In transverse section, the test is roughly isometric ("circular"), smooth. Between 10 and 13 chambers, half-circular/reniform in shape, can be counted, not including a large, circular proloculus in the middle. Wall is homogeneous, slightly translucent, light or dark brown (porcelaneous). Length of the specimens is 0.14–0.24 mm, the diameter 0.1–0.13 mm.

Remarks: The genus *Gsollbergella* Zaninetti 1979 closely resembles *Agathammina* Neumayr 1887, from which it differs by the presence of rudimentary septa. We consider the genus to be monospecific, that is, that the two species currently being placed in this genus—*Gsollbergella gsollbergensis* (Zaninetti 1969) and *Gsollbergella spiroloculiformis* (Oravecz-Scheffer 1971)—are synonymous, with *G. gsollbergensis* having the priority. We note here that Zaninetti (1979) incorrectly named *G. spiroloculiformis* as the type species of *Gsollbergella*, having believed that it was published in 1968. In the case of the studied thin sections, many specimens clearly show short septa with a triangular transect, typical of *G. gsollbergensis*. In contrast, no constrictions are visible in other specimens. In the absence of the specimens with constrictions, the latter specimens would likely be proclaimed for *Agathammina austroalpina* Kristan-Tollmann and Tollmann 1964 (see also Zaninetti 1969). *Ophthalmidium*-like chambers can also be seen in the "*Agathammina austroalpina*" specimen on plate 1, Fig. 7 in Oravecz-Scheffer (1971).

Stratigraphic distribution: *Gsollbergella gsollbergensis* has so far been reported from the Carnian (Oravecz-Scheffer 1971; Zaninetti 1969).

Superfamily Calcivertelloidea Loeblich and Tappan 1964 nom. transl. Vachard (2016)

Family Calcivertellidae Loeblich and Tappan 1964 nom. transl. Reitlinger in Vdovenko et al. 1993
Genus *Planiinvoluta* Leischner 1961

***Planiinvoluta*? *deflexa* Leischner** 1961?

Figure 14.4e–f

1961 *Planiinvoluta deflexa* n. g. n. sp. – Leichner, p. 12, pl. 10, Figs. 15–22; pl. 12, Figs. 7b, 8b.

Material: Several isolated specimens from the Reifling Formation (samples NHMW 2025/0261/0003, NHMW 2025/0261/0004).

Description: Test is probably bilocular (proloculus is not visible). Most of the test consists of an undivided, half-tubular chamber that is attached on one side. The lumen is nearly constant in size. From the lower side, it appears as if the deuteroloculus meanders across the surface. On the upper side, though, the test is almost planispiral and is reminicscent of *Cornuspira*. The diameter of the test is 0.27–0.38 mm; the lumen is 0.041–0.056 mm wide, and the wall is 0.03 mm thick, white, and finely-crystalline and siliceous.

Remarks: The determination is uncertain due to the undetermined type of wall (agglutinated or silicified porcelaneous). Due to the visually homogenous wall, distinct from the clearly agglutinated wall of *N. discoidea* (Trifonova), we conclude a porcelaneous wall. According to Leischner (1961), *P. deflexa* differs from *Planiinvoluta carinata* Leischner 1961 based on the curvature of the attached surface, larger test (diameter 0.72 mm vs diameter 0.39 mm), and in a thicker wall (0.010–0.030 mm in *P. deflexa*, 0.010–0.015 mm in *P. carinata*). Koehn-Zaninetti (1969), Wernli (1971), Zaninetti (1976), Trifonova (1993), and Böhm et al. (1999) all doubted the existence of two separate species. In contrast, Blau et al. (1995) as well as Kristan-Tollmann (1990) considered both species valid. We consider *P. deflexa* distinct from *P. carinata* based on the less dense coiling (1 coil per 0.1 mm of length in *P. deflexa* vs. 2–4 coils per 0.1 mm length in *P. carinata*).

Stratigraphic distribution: Anisian (Blau et al. 1995; Martini et al. 1996) to Rhaetian (Leischner 1961; Salaj et al. 1983; Kristan-Tollmann 1990); according to Wernli (1971) as well as He and Norling (1991) also Jurassic.

Calcivertellidae? gen. et sp. indet.

Figure 14.4o–p

Material: Specimens from thin sections from the Reingraben Formation (samples NHMW 2025/0261/0003, NHMW 2025/0261/0022).

Description: The test is flat (originally probably being attached) on one side and convex on the other side. Globular proloculus is followed by what is probably an undivided tubular second chamber. The wall is homogeneous, light-brown and slightly transparent, interpreted as porcelaneous.

Suborder Miliolina Loeblich and Tappan 1988
Superfamily Cornuspiroidea Schultze 1854 nom. translat. Bogdanovich in Subbotina et al. (1981)
Family Cornuspiridae Schultze 1854
Subfamily Cornuspirinae Schultze 1854 nom. translat. Rhumbler (1904)
Genus *Cornuspira* Schultze 1854

***Cornuspira?* sp. A**

Figure 14.4c–d

Material: Several specimens from the Reigraben Formation (samples NHMW 2025/0261/0020—NHMW 2025/0261/0023); axial and equatorial transects in thin sections.

Description: Test is bilocular, discoidal, with almost parallel sides, tending to be slightly biconcave, reaching the largest thickness in the last coil. The deuteroloculus is wound in 5–6 (7?) coils. The wall is homogeneous, brown as in *Gsollbergella*, and thus interpreted to be porcelaneous. However, we do not exclude the possibility that it is aragonitic, and perhaps preserved differently from other involutinids. The test diameter is 0.08–0.17 mm. The size of the proloculus is 0.01–0.02 mm. The final height of the lumen is 0.01–0.02 mm.

Class Globothalamea Pawlowski, Holzmann and Tyszka 2013
Subclass Textulariana Mikhalevich 1980
Order Lituolida Lankester 1885
Suborder Lituolina Lankester 1885
Family Lituotubidae Loeblich and Tappan 1984
Genus *Lituotuba* Rhumbler 1895

***Lituotuba?* sp. A**

Figure 14.4k

Material: Isolated specimens from the Reifling Formation (sample NHMW 2025/0261/0003).

Description: The test is roughly planispiral. The wall is siliceous, finely agglutinated. Transverse growth-constrictions (suture-like deepenings) are visible on the last whorl, but the presence of septa could not be shown with the methodology used here.

Superfamily Lituoloidea Blainville 1827
Family Lituolidae Blainville 1827
Subfamily Ammomarginulininae Podobina 1978
Genus *Ammobaculites* Cushman 1910

***Ammobaculites* ? sp.**

Figure 14.4a

Material: One specimen in a longitudinal transect from a thin section from the Reingraben Formation (sample NHMW 2025/0261/0022).

Description: The initial part of the test is not clearly visible. Ten chambers are arranged uniserially in an uneven line. Chambers have a pyriform to circular lumen. They are connected through simple, centrally positioned foramina. The wall is agglutinated. The test height is 1.1 mm.

Suborder Verneuilinina Kaminski and Mikhalevich in Kaminski 2004
Superfamily Verneuilinoidea Cushman 1911
Family Verneuilinidae Cushman 1911
Subfamily Verneuilinoidinae Suleymanov 1973
Genus *Gaudryinopsis* Podobina 1975

***Gaudryinopsis kelleri* (Tappan 1955)**

Figure 14.4b

1955 *Gaudryina kelleri* Tappan, n. sp. – Tappan, p. 47, pl. 13, Fig. 19.

1992 *Gaudryina triassica* Trifonova, 1961. – Urošević and Sudar, p. 4, Figs. 10–12.

2022 *Gaudryinopsis kelleri* (Tappan 1955). – Gale et al., p. 274, Fig. 6. [cum syn.]

Material: Several specimens from the residue from the top of the Reifling Formation (sample NHMW 2025/0261/0003).

Description: Elongated test with a very short, triangular-shaped triserial initial part, followed by nine "pairs" of nearly globular chambers in biserial arrangement. The aperture is interiomarginal. The wall is agglutinated. The height of the test is 1.15–1.19 mm.

Stratigraphic distribution: Carnian—Lower Jurassic (Gale et al. 2022).

Order Robertinida Loeblich and Tappan 1984
Suborder Duostominina Mikhalevich 2013
Superfamily Duostominoidea Brotzen 1963
Family Duostominidae Brotzen 1963
Genus *Duostomina* Kristan-Tollmann 1960

Duostomina biconvexa **Kristan-Tollmann 1960**

Figure 14.3x

1960 *Duostomina biconvexa* nov. gen. nov. spec. – Kristan-Tollmann, p. 68, pl. 17, Figs. 1, 2; pl. 18, Fig. 2.

1998 *Duostomina biconvexa* Kristan-Tollmann. – di Bari and Baracca, p. 126, pl. 2, Fig. 2.

Material: Specimens from thin sections from the Reingraben Formation (samples NHMW 2025/0261/0020—NHMW 2025/0261/0022); axial and equatorial transect.

Description: A multilocular test with chambers in a low trochospire with a lenticular, biconvex axial transect. The spiral side is of the same height as the umbilical. The sides of the test are angular. The umbilicus is filled with a lamellar wall. The wall is recrystallized into spar, partly silicified. The largest diameter of the test is 0.1–0.23 mm. The test height is 0.05–0.11 mm.

Remarks: *Duostomina biconvexa* differs from *Duostomina rotundata* Kristan-Tollmann 1960 in having angular sides of the test. Kristan-Tollmann (1990) illustrated transitional forms towards *Duostomina alta* Kristan-Tollmann 1960 and *Duostomina turboidea* Kristan-Tollmann 1960.

Stratigraphic distribution: Anisian—Carnian (di Bari and Baracca 1998), according to Gale et al. (2011), also Norian and Rhaetian.

Class Fusulinata Gaillot and Vachard 2007 emend. Vachard, Krainer and Lucas 2013
Subclass Fusulinana Maslakova 1990 nom. translat. Vachard et al. (2010a, b)
Order Endothyrida Fursenko 1958
Superfamily Endoteboidea Vachard, Martini, Rettori and Zaninetti 1994 nom. translat. Vachard, Krainer and Lucas 2013.
Family Endotebidae Vachard, Martini, Rettori and Zaninetti 1994
Genus *Endotriada* Vachard, Martini, Rettori and Zaninetti 1994

Endotriada kuepperi (**Oberhauser** 1960)?

Figure 14.3n–p

1960 *Endothyra küpperi* nov. sp. – Oberhauser, p. 16, pl. 3, Figs. 7, 14, 22.

1969 *'Endothyra' küpperi* Oberhauser 1960. – Koehn-Zaninetti, p. 41, pl. 6, Fig. C; text-Fig. 7.

1995 *Endoteba* ex gr. *controversa* Vachard and Razgallah 1988. – Rettori, p. 73, pl. 8, Figs. 1–4.

? 1996 *Endoteba kuepperi* (Oberhauser). – Pronina and Vuks, pl. 1, Fig. 1.

Material: Few specimens in thin sections and residue from the top of the Reifling Formation and from the Reingraben Formation (samples NHMW 2025/0261/0003, NHMW 2025/0261/0020, NHMW 2025/0261/0022).

Description: Slightly elongated test consists of very low trochospirally, nearly planispirally enrolled chambers, separated by well-formed septa. The initial part of the test is not fully visible. It is followed by two coils, the first consisting of approximately four, and the second of six chambers. The aperture is at the base of the last chamber. The wall is dark brown and slightly recrystallised. The largest test diameter is 0.19–0.30 mm; the wall is approximately 0.013 mm thick.

Remarks: The specimens morphologically closely resemble *E. kuepperi*, described from Reingraben Formation by Oberhauser (1960), and *E. controversa*, described from Permian beds by Vachard and Razgallah (1988) and later extended into the Rhaetian (Vachard et al. 1994; Rettori 1995). The illustrated specimens are closer to *E. kuepperi* in size (although smaller) and in the number of chambers. It is also possible that *E. kuepperi* has a thinner wall than *E. controversa*, that is 0.02 mm in the former (as noted in pl. 3, Fig. 22 in Oberhauser 1960) compared to 0.04–0.07 mm in the latter (measured from Rettori 1995). Based on the relatively thin wall, we place '*Endothyra' kuepperi* into the genus *Endotriada* Vachard, Martini, Rettori and Zaninetti. The difference in wall thickness between *Endoteba* and *Endothyra* was notspecified in Vachard et al. (1994). However, every species placed by Vachard et al. (1994) in *Endoteba* possesses a wall thicker than 0.04 mm, whereas the wall in *Endotriada tyrrehnica* Vachard, Martini, Rettori and Zaninetti is only 0.017 mm thick. Pronina and Vuks (1996) instead placed '*E.*' *kuepperi* in the genus *Endoteba.*

Stratigraphic distribution: So far, this species was unambiguously reported from the Carnian (Oberhauser 1960; Koehn-Zaninetti 1969), but it is likely that some specimens from the Triassic were misidentified as other species of the genera *Endothyra* or *Endoteba.*

Class Foraminifera incertae sedis
Order Involutinida Hohenegger and Piller 1977
Family Trocholinidae Kristan-Tollmann 1963 emend. Rigaud et al. (2013)
Subfamily Lamelliconinae Zaninetti et al. 1987 emend. Rigaud et al. (2013)
Genus *Lamelliconus* Piller 1978

***Lamelliconus? multispira* (Oberhauser 1957)**

Figures 14.3l–14.3

1957 *Trocholina (Trocholina) multispira* sp. nov. – Oberhauser, p. 261, pl. 20, Figs. 1–14; text-Fig. 1.

1969 *Trocholina multispira* Oberhauser 1957. – Koehn-Zaninetti, p. 136, pl. 8, Fig. E, H; text-Fig. 43.

1995 *Lamelliconus multispirus* (Oberhauser 1957). – Rettori, p. 84, pl. 13, Figs. 1, 3, 4, 9, 10.

Material: Rare specimens in thin sections from the Reingraben Formation (samples NHMW 2025/0261/0020, NHMW 2025/0261/0023).

Description: High conical test with well-rounded apical end and parallel or slightly diverging sides. Deuteroloculus is wound in 13–15 coils. Lumen of deuteroloculus increases slowly, then remains constant in diameter. Proloculus is approximately 0.02 mm in diameter, as is the final height of the deuteroloculus lumen. Test height is 0.17–0.28 mm; largest test diameter is 0.10–0.14 mm.

Stratigraphic distribution: Anisian to Carnian (Rettori 1995).

***Lamelliconus* ex gr. *ventroplanus* (Oberhauser 1957)**

Figure 14.3q

1957 *Trocholina (Trocholina) ventroplana* sp. nov. – Oberhauser, p. 262, pl. 20, Figs. 15–23; text-Figs. 1, 2.

1957 *Trocholina (Trocholina) biconvexa* sp. nov. – Oberhauser, p. 263, pl. 21, Figs. 26–37; text-Figs. 3–5.

1969 Trochlina cf. ventroplana Oberhauser 1957. – Koehn-Zaninetti, p. 139, Fig. 44f–j.

1995 *Lamelliconus* ex gr. *ventroplanus* (Oberhauser 1957). – Rettori, p. 82, pl. 12, Figs. 1–11.

Material: Few specimens from the residue of one sample from the Reingraben Formation (sample NHMW 2025/0261/0023).

Description: Test is low-conical, biconvex, with widely diverging sides. Tubular deuteroloculus is wound in a low trochospire in 12 coils and is visible on the eroded spiral side of the test. The final diameter of the test is 0.84 mm.

Stratigraphic distribution: Ladinian to Carnian (Rettori 1995).

Subfamily Trocholininae Kristan-Tollmann 1963 emend. Rigaud et al. (2013)
Genus *Trocholina* Paalzow 1922

Trocholina* cf. *umbo Frentzen 1941

Figure 14.3t

1941 *Trocholina umbo* nov. spec. – Frentzen, p. 306, pl. 1, Fig. 12.

1978 *Trocholina umbo* Frentzen 1941. – Piller, p. 81, pl. 20, Figs. 9–11, 13, 14, 16, 17.

Material: Few specimens from the residue of one sample from the Reingraben Formation (samples NHMW 2025/0261/0020, NHMW 2025/0261/0021, NHMW 2025/0261/0023).

Description: Conical test consists of proloculus and deuteroloculus coiled in approximately six coils. Papillae are visible in the umbilical infill. The wall was recrystallized into calcite (?), and in some places it was silicified. The test height is 0.08–0.11 mm, the largest diameter 0.17–0.22 mm.

Remarks: The specimens best suit the description and size of the type material of *Trocholina umbo* Frentzen, which was described as having a height one-half of the test diameter. However, *Frentzenella crassa* (Kristan 1957) and *Trocholina intermedia* Frentzen 1941 are morphologically very similar. The former is larger (height of the type specimens is 0.59–1.26 mm). The latter was described as having a relatively higher test and smaller deuteroloculus compared to the final test diameter (Frentzen 1941).

Stratigraphic distribution: Upper Triassic and Lower Jurassic (Piller 1978), as well as Middle Jurassic (Zsiborás and Győrög 2020).

Family Involutinidae Bütschli 1880
Subfamily Aulotortinae Zaninetti 1984
Genus *Aulotortus* Weynschenk 1956

Aulotortus sinuosus Weynschenk 1956?

Figure 14.3r

1956 *Aulotortus sinuosus* Weynschenk, n. sp. – Weynschenk, p. 27, pl. 6, Figs. 1–3; text-Figs. 1, 2.

p.p. 1978 *Aulotortus sinuosus* Weynschenk 1956. – Piller, p. 45, pl. 2, Figs. 1–7; pl. 3; pl. 4, Figs. 1–3, 5–11, 15, 16; pl. 5, Figs. 8, 10–16; text-Figs. 4 [non pl. 4, Figs. 13, 14; ? non pl. 5, Figs. 1–7, 9? pl. 4, Figs. 4, 12].

Material: Rare in the Reingraben Formation (samples NHMW 2025/0261/0020).

Description: The test is oval, ellipsoidal in axial section. The deuteroloculus is reniform to half-circular in shape, wound in a planispire or in very low trochospire in three coils. The wall is silicified, but the lateral extensions of L2 laminae, reaching over the "umbilical" region and tapering, remain visible. The test diameter is 0.33 mm.

Remaks: The specimen is attributed to *A. sinuosus*, which essentially shows a planispiral coiling of the deuteroloculus. Deviations from a perfect planispire are common in this species and were the basis for division of this species into several distinct species or subspecies (e.g., Oberhauser 1964; Salaj et al. 1967, 1983; Jendrejáková 1972), all considered to be junior synonyms (di Bari and Laghi 1994).

Stratigraphic distribution: Anisian—Rhaetian (Piller 1978).

***Aulotortus tenuis* (Kristan** 1957)?

Figure 14.3u–v

1957 *Angulodiscus tenuis* nov. gen. nov. spec. – Kristan, p. 280, pl. 22, Fig. 18.

1969 *Involutina tenuis* (Kristan), 1957. – Koehn-Zaninetti, p. 116, Fig. 33 A–J.

1978 *Aulotortus tenuis* (Kristan 1957). – Piller, p. 62, pl. 12, Figs. 1–12.

Material: Few specimens from the thin section from the Reingraben Formation (sample NHMW 2025/0261/0023).

Description: Bilocular foraminifera with tubular second chamber enrolled planispirally. The test is flattened, with parallel sides and of a uniform thickness. Four coils are visible. The lumen is oval in transect. In the last coil, the deuteroloculus is evolute, and the wall of the last coil does not extend over the previous part of the test, giving a slightly detached look. The wall of the specimen is silicified. The test diameter is 0.13–0.16 mm.

Stratigraphic distribution: Ladinian?, Norian—Rhaetian (Piller 1978).

Subfamily Parvalamellinae Rigaud, Martini and Rettori 2012

Genus *Parvalamella* Rigaud, Martini and Rettori 2012

***Parvalamella friedli* (Kristan-Tollmann** 1962)?

Figure 14.3s

1969 *Involutina gaschei* (Koehn-Zaninetti et Brönnimann), 1968. – Koehn-Zaninetti, p. 127, pl. 11, Fig. E, F; text-Fig. 38.

1969 *Involutina gaschei* (Koehn-Zaninetti et Brönnimann) *praegaschei*, n. subsp. – Koehn-Zaninetti, p. 130, Fig. 39.

1978 *Aulotortus friedli* (Kristan-Tollmann 1962). – Piller, p. 55, pl. 8, Figs. 1–8; pl. 9, Figs. 1–16; pl. 10, Figs. 1–14.

1985 *Aulotortus friedli* (Kristan-Tollmann 1962). – Ciarapica and Zaninetti (1985), p. 71, pl. 1, Figs. 1–9; pl. 2, Figs. 1–8; pl. 3, Figs. 1–9; text-Fig. 1 A-F.

2010 *Aulotortus friedli* (Kristan-Tollmann 1962). – Senowbari-Daryan et al., p. 578, Figs. 12a-d, 13.

2010 *Aulotortus friedli* (Kristan-Tollmann 1962). – Chablais et al., (2010) p. 155, Figs. 4–5.

2012 *Parvalamella friedli* (Kristan-Tollmann 1962) n. comb., emend.—Rigaud et al., p. 249, text-Fig. 4A–4F.

Material: Few specimens from the thin sections from the Reingraben Formation (samples NHMW 2025/0261/0017, NHMW 2025/0261/0023).

Description: The test is small, globular, comprising a coiled deuteroloculus. The wall is silicified. The lumen is slightly constricted, probably due to silicification. The size of the specimens is 0.1–0.13 mm.

Remaks: Due to silicification of the wall, the genus cannot be determined with certainty. However, it appears that only aragonitic tests were silicified during diagenesis.

Stratigraphic distribution: Ladinian—Rhaetian (Rigaud et al. 2012).

Subfamily Involutinida *incertae sedis.*

Foraminifera indet. sp. A

Figure 14.3w

Material: A single specimen from the thin section from the Reingraben Formation (sample NHMW 2025/0261/0023).

Description: The specimen is visible in equatorial transect. The outline of the wall is S-shaped, with widely rounded margins. The lumen is elliptical, then half-circular to reniform in transect. Its size notably increases. The test is multilocular, and the transect resembles *Labalina* Azbel 1988, with chambers initially in a quinqueloculinid (?) arrangement, followed by three or four chambers in a sigmoidal arrangement on each side of the test. The wall is thin, and the sutures between chambers are visible in the outline of the test. The wall is silicified, as is the case for specimens from the same formation that originally had aragonitic walls (Duostominidae, Involutinina), whereas *Labalina* has a porcelaneous wall. The larger diameter of the test is 0.09 mm.

14.6 Discussion

The early Carnian shift towards a more humid climate is doubtlessly among the most interesting events of the Triassic. In shallow marine areas, the CPE is often reflected in an increased siliciclastic input (Hornung et al. 2007; Ogg 2015; Li et al. 2025). A major shift in platform carbonate producers has been observed in the western Neotethys (Flügel 2002; Russo 2005; Hornung et al. 2007; Breda et al. 2009; Stefani et al. 2010; Gattolin et al. 2015; Caggiati et al. 2017), as well as in the Panthalassa Ocean (Peybernes et al. 2016). More recently, the CPE is becoming recognized as one of the most important biotic turnover events, which, in the marine realm, was associated with possible declines in biodiversity in several marine animal groups (Benton 1986; Zhang et al. 2018; Dal Corso et al. 2020). This led to important new ecological opportunities and innovations (e.g., rise of calcareous nannoplankton—Falkowski et al. 2004; Erba 2004; Preto et al. 2013; Dal Corso et al. 2020), and to the origination of symbiosis among scleractinian corals and zooxanthellate dinoflagellates—Stanley and Swart 1995; Stanley 2003; Stanley and van de Schootbrugge 2009; Kiessling 2010; Frankowiak et al. 2021). The influence of the CPE on the evolution of foraminifera is still understudied at present. Although several authors have described foraminiferal assemblages from the time of the CPE, the stratigraphic resolution of those studies is often relatively low, and the focus was not on the changes in foraminiferal associations, but on taxonomy (e.g., Koehn-Zaninetti 1969; Trifonova 1977) and/or biostratigraphy (e.g., Salaj et al. 1983; Bérczi-Makk 1996). Lists of taxa thus mostly allow for a general comparison between localities but do not address the impact of the CPE on the evolution of foraminifera. The sole exception is the study by Li et al. (2025), performed at the Erguan section in Guizhou Province, South China, and the Quxia section in the Tibet Autonomous Region, China. While both sections represent deposition on the shallow marine shelf, the Erguan section was located in the Paleotethys Ocean, and the Quxia section in the southern Neotethys Ocean. Li et al. (2025) found no evidence for significant extinctions among foraminifera in these regions. While assemblages from the Erguan section did show a decrease in diversity and abundance, no change was seen in the Quxia section (Li et al. 2025). Those authors thus suggest that the impact of the CPE on foraminiferal assemblages was regional rather than global.

In the following text and in Table 14.2, we summarize foraminiferal reports from several Carnian formations of a more siliciclastic character and deposited in a somewhat deepened environment. Because different authors may have arrived at different interpretations of the species, Table 14.2 shows data simplified to the level of genus or even family.

Focusing on the taxonomic composition of foraminiferal assemblages from the Reingraben Formation (Oberhauser 1957, 1960; Kristan-Tollmann 1960; Koehn-Zaninetti 1969; this study), we find the largest similarities with the assemblages reported from the Tor and similar formations from the Southern Alps, which could be explained by the similarity in lithological composition and depositional environment.

Table 14.2 Foraminiferal assemblages from various clastics-bearing/dominated Carnian formations, simplified to the genus or family level. See text for localities and references

	Reingraben Fm	Tor Fm	Predil Lst	"Beli Potok" development	Lesno Brdo Lst	San Cassiano Fm	Veszprém Marl	Tržinovo Fm	Mufara Fm	Sanqiao and Zhamure Fm
Annulina	+									
Ammodiscus						+			+	+
Glomospira	+					+				+
Rectoglomospira								+		
Glomospirella/Pilamminella		+	+	+	+	+		+	+	+
Reophax						+				
Ammobaculites	+					+	+		+	+
Palaeolituonella						+			+	
Textularia						+				+
Gaudryinopsis/Gaudryina/ Gaudryinella							+			
Trochammina		+	+		+					
Piallina			+			+				+
Earlandia										+
Endotriada/Endoteba	+			+		+		+	+	+
Cornuspira	+									
Hoyenella		+							+	+
Meandrospira?							+			
Semimeandrospira						+				
Agathammina			+							+

(continued)

Table 14.2 (continued)

	Reingraben Fm	Tor Fm	Predil Lst	"Beli Potok" development	Lesno Brdo Lst	San Cassiano Fm	Veszprém Marl	Tržinovo Fm	Mufara Fm	Sanqiao and Zhamure Fm
Gsollbergella	+		+		+	+	+			
Ophthalmidiidae	+		+			+	+	+	+	+
Ophthalmipora							+			
Gheorghianina						+				
Turriglomina			+						+	+
Decapoalina										+
Miliolipora										+
Cucurbita									+	
Hydrania									+	
Tignumparina									+	
Orthotrinacria									+	
Duostominidae	+		+			+	+	+	+	+
Involutinidae	+	+	+	+	+	+	+	+	+	+
Trocholinidae	+		+		+	+		+		+
Vaginulinidae	+					+	+			+
Nodosariidae	+	+	+		+	+	+	+	+	+
Protonodosariidae										+
Abriolina										+
Oberhauserellidae						+	+	+		
Sessile forms	+									+

Combining results from previous studies (Oberhauser 1957, 1960; Kristan-Tollmann 1960; Koehn-Zaninetti 1969) and this study, the Julian Reingraben Formation of the *A. austriacum* Zone contains a foraminiferal assemblage of *Annulina metensis* Terquem (taxon inquirendum), *Ammobaculites* sp.?, *Glomospira* spp., *Endotriada kuepperi* (Oberhauser)?, *Gsollbergella gsollbergensis* (Oravecz-Scheffer), *Ophthalmidium exiguum* Koehn-Zaninetti, Calcivertellidae? gen. et sp. indet., *Cornuspira* sp.?, Duostominidae, *Lamelliconus* spp., *Trocholina* cf. *umbo* Frentzen, *Aulotortus sinuosus* Weynschenk?, *Aulotortus tenuis* (Kristan)?, *Parvalamella friedli* (Kristan-Tollmann)?, and several lageninids (e.g., *Lenticulina, Nodosaria, Dentalina, Lingulina, Frondicularia, Pseudonodosaria, Austrocolomia*).

The uppermost Julian Tor Formation in the eastern Southern Alps is composed of marly limestone and dolomite, bivalve coquinas (tempestites), siltstone, claystone and marlstone (De Zanche et al. 2000; Gale et al. 2015). Foraminiferal assemblages comprise *Trochammina almtalensis* Koehn-Zaninetti, *Parvalamella friedli* (Kristan-Tollmann), *Triadodiscus eomesozoicus* (Oberhauser)?, *Angulodiscus impressus* Kristan-Tollmann, *Glomospirella vulgaris* Ho?, *Pilamminella kuthani* (Salaj), *Hoyenella sinensis* Ho?, and '*Frondicularia woodwardii* Howchin' auct. (Ogorelec et al. 1984; Gale et al. 2015). The Predil Limestone ('Calcare del Predil') was stratigraphically placed in the Julian *Trachyceras aon* Zone (i.e. slightly older than the *A. austriacum* Zone of the Reingraben Formation). *Glomospirella* sp., *Trochammina* sp., *Piallina bronnimanni* Martini et al., *Lamelliconus artiskomorphos* di Bari and Laghi, *L. cordevolicus* (Oberhauser), *L. multispira* (Oberhauser), *L. procerus* (Liebus), *Prorakusia salaji* di Bari and Laghi, *Aulotortus sinuosus* Weynschenk, *Turriglomina carnica* (Dager), *T. conica* (He), *Agathammina* sp.?, *Gsollbergella spiroloculiformis* (Oravecz-Scheffer), *Ophthalmidium* sp., *Nodosaria* sp., *Duostomina* sp., and *Variostoma* sp. were determined (Rettori et al. 1998).

The "Beli Potok" development from the Karavanke Mountains, Southern Alps, is characterized by platy limestone with chert, overlying "massive" platform dolomite. Based on the conodont *Nicoraella*? *budaensis* Kozur and Mock, the Carnian age was determined (Kolar-Jurkovšek et al. 2005; Forel et al. 2020), but the precise position relative to the CPE remains unclear. Limestone with chert continues upwards into an alternation of dark grey limestone and marly limestone with marl intercalations. Limestone beds are often characterized by the shells of the bivalve *Posidonia* or sponge spicules (Ramovš 1993; Kolar-Jurkovšek et al. 2005). Foraminiferal assemblages of the "Beli Potok" development contain involutinid, ammodiscid and 'endothyrid' foraminifera, as well as a few duostominids. *Aulotortus sinuosus* Weynschenk, *A. tumidus* (Kristan-Tollmann), and less common *Prorakusia salaji* di Bari and Laghi (in our opinion *Parvalamella friedli*), *Pilamminella kuthani* (Salaj), and *Endotriada kuepperi* (Oberhauser) were determined.

Oblak (2001) studied black medium-bedded limestone intercalated with shale near Lesno Brdo in the northern Dinaric Mountains, External Dinarides, and illustrated *Gsollbergella spiroloculiformis* (Oravecz-Scheffer), *Glomospirella* sp. (determined as *Gandinella falsofriedli* (Salaj et al.)), *Trochammina* cf. *almtalensis* Koehn-Zaninetti, *Parvalamella friedli* (Kristan-Tollmann)? (determined as *Gandinella falsofriedli* (Salaj et al.)), *Triadodiscus eomesozoicus* (Oberhauser), *Aulotortus sinuosus* Weynschenk, *Lamelliconus procerus* (Liebus), and *Nodosaria* sp. The formally unnamed stratigraphic unit contains a rich bivalve fauna (Jelen 1990) that includes *Umbrostrea*? *montiscaprilis* (Klipstein). According to Ruvinetti (2004), this bivalve is late Julian in age.

The San Cassiano Formation from the central Southern Alps in the Dolomites spans the Julian from the *Trachyceras* to the *Austriacum* zones (De Zanche et al. 1993; Gianolla et al. 1998). From the *Trachyceras aonoides* Subzone part of the San Cassiano Formation, di Bari and Baracca (1998) determined *Reophax asperus* Cushman and Waters, *Ammodiscus annulinoides* Kristan-Tollmann, *Ammodiscus infimus* (Strickland), *Ophthalmidium lucidum* (Trifonova), *Ophthalmidium triadicum* (Kristan), *Ophthalmidium fusiforme* (Trifonova), *Pragsoconulus robustus* Oberhauser, several species of Duostominidae, *Lenticulina* spp., rare Aulotortidae (*Aulotortus sinuosus* Weynschenk, *A. tumidus* (Kristan-Tollmann)), Triadodiscidae (*Lamelliconus ventroplanus* (Oberhauser), *L. procerus* (Liebus), *L. ventroplanus* (Oberhauser)), and Nodosaridae. Lithologically, the section comprises marl, alternating with calciturbidites and micritic limestone (di Bari and Baracca 1998).

From the *Daxatina* cf. *canadensis* Subzone part of the San Cassiano Formation at Prati di Stuores, Broglio Loriga et al. (1999) determined *Ammobaculites hiberensis* Márquez and Trifonova, *Textularia* sp., *Reophax eominutus* Kristan-Tollmann, *R. asperus* Cushman and Waters, *R. horridus* Cushman, *Ammodiscus* spp., Duostominidae, *Lenticulina* spp., nodosariids, *Endotriada kuepperi* (Oberhauser), *Gsollbergella spiroloculiformis* (Oravecz-Scheffer), *Semimeandrospira* sp., *Piallina tethydis* Rettori and Zaninetti, *Palaeolituonella meridionalis* (Luperto), *Gheorghianina* sp., *Ophthalmidium* spp., *Glomospira perplexa* Franke, *Glomospirella hemigordiformis* Tollmann and Kristan-Tollmann, Oberhauserellidae, *Aulotortus* gr. *sinuosus* Weynschenk, and *Lamelliconus* spp.

The Veszprém Marl Formation from the Transdanubian Central Range, Hungary, spans the entire Carnian period. The middle part is dominated by claystones. It contains numerous Duostominidae and Nodosariidae. Additional species present are *Gaudryina triassica* Trifonova, *Gaudryinella clavuliniformis* Trifonova, *Ammobaculites* sp., *Meandrospira*? *karnica* Oravecz- Scheffer, *Ophthalmipora* sp., *Gsollbergella spiroloculiformis* (Oravecz- Scheffer), *Ophthalmidium triadicum* Kristan, *Lenticulina* sp., *Aulotortus* cf. *sinuosus* Weynschenk, and *Schmidita* cf. *inflata* Fuchs (Oravecz-Scheffer 1987).

The Tržinovo Formation on the Krížna Unit in Malá Fatra Mountains, Western Carpathians, contains *Glomospirella* sp., *Rectoglomospira senecta* Trifonova, *Pilamminella kuthani* Salaj, *P.* cf. *gemerica* Salaj, *Endotriada* sp., *Triadodiscus eomesozoicus* (Oberhauser), *Aulotortus sinuosus* Weynschenk, *Parvalamella friedli* (Kristan-Tollmann), *Semiinvoluta* sp., *Prorakusia* sp., *Frentzenella turris* (Frentzen), *Ophthalmidium exiguum* Koehn-Zaninetti, *Duostomina* cf. *alta* Kristan-Tollmann, *Oberhauserella mesotriassica* (Oberhauser), and *Nodosaria ordinata* Trifonova (Sykora et al. 2011).

The upper Ladinian/lower Carnian to upper Carnian Mufara Formation from Sicily is dominated by clay, marl, and micritic limestone, deposited along a distally steepened shelf or ramp with sponge patch reefs, ranging from a shallow water to deeper water offshore zones (Martini et al. 2007; Carrillat and Martini 2009). Foraminiferal assemblages consist of *Cucurbita* cf. *infundibuliformis* Jablonsky, *Hydrania dulloi* Senowbari-Daryan, *Ophthalmidium tori* Zaninetti and Brönnimann, *Ophthalmidium* sp., Nodosariidae, Duostominidae, *Tignumparina zeissi* Senowbari-Daryan, *Glomospirella parallela* Kristan-Tollmann, *Endotriada tyrrhenica* Vachard et al., *Glomospira* spp., *Ammodiscus* sp., *Orthotrinacria gracilis* Zaninetti et al., *Hoyenella* aff. *inconstans* (Michalik et al.), *Turriglomina mesotriasica* (Koehn-Zaninetti), *Aulotortus* sp., *Endotebanella* sp., and *Palaeolituonella meridionalis* (Luperto) (Carrillat and Martini 2009). With genera such as *Cucurbita*, *Hydrania*, *Orthotrinacria*, and *Palaeolituonella*, the assemblage bears a more 'reefal' character than the one from Reingraben Formation.

The Carnian Sanqiao Formation from South China and the Zhamure Formation from the Tibet Autonomous Region of China are both dominated by Miliolida and Nodosariida. The dominant species in the Sanqiao Formation are *Piallina bronnimanni* Martini et al., *Lenticulina* sp., *Agathammina austroalpina* Kristan-Tollmann and Tollmann, and *Austrocolomia marschalli* Oberhauser. Other species are *Tolypammina gregaria* Wendt, *Ammodiscus* spp., *Glomospirella* spp., *Glomospira* spp., '*Endothyra'* spp., *Earlandia tintinniformis* (Mišík), *Ophthalmidium tori* Zaninetti and Brönnimann, *Planiinvoluta* spp., *Hoyenella shengi* (He), *Turriglomina mesotriasica* (Koehn-Zaninetti), *Abriolina mediterranea* Luperto, *Aulotortus sinuosus* Weynschenk, *Parvalamella friedli* (Kristan-Tollmann), *Variostoma catilliforme* Kristan-Tollmann, *Variostoma* sp., *Diplotremina astrofimbriata* Kristan-Tollmann, and various uniserial lageninids.

In the Zhamure Formation, the main foraminiferal species are *Decapoalina triadica* (Langer), *Nodosaria* spp., *Lenticulina* spp., *Agathammina austroalpina* Kristan-Tollmann and Tollmann, and *Ophthalmidium leischneri* (Kristan-Tollmann). Other species are *Glomospira* sp., *Glomospirella* sp., *Ammodiscus* spp., *Textularia* sp., *Vaginulina* spp., *Vervilleina* sp., *Endothyra* sp., *Agathammina parafusiformis* Salaj et al., *Ophthalmidium carpathicum* Samuel and Borza, *Lamelliconus biconvexus* (Oberhauser), *Turriglomina mesotriasica* (Koehn-Zaninetti), *Planiinvoluta carinata* Leischner, *Miliolipora tamarae* Gale et al., *Astacolus* spp., *Lenticulina* spp., and uniserial lageninids (Li et al. 2025).

14.7 Conclusions

The lower Carnian Reingraben Formation from the Northern Calcareous Alps contains a modest foraminiferal assemblage, partly affected by silicification. The specimens are unfragmented. *Gsollbergella gsollbergensis* (Zaninetti) is locally present in large numbers. Other taxa include Ophthalmidiidae, *Cornuspira* sp.?, *Ammobaculites* sp.?, *Duostomina biconvexa* Kristan-Tollmann, *Endotriada kuepperi*

(Oberhauser)?, *Lamelliconus*? *multispira* (Oberhauser), *Lamelliconus* ex gr. *ventroplanus* (Oberhauser), *Trocholina* cf. *umbo* Frentzen, *Aulotortus sinuosus* Weynschenk?, *Aulotortus tenuis* (Kristan)?, *Parvalamella friedli* (Kristan-Tollmann)?, undetermined sessile foraminifera, and uniserial Lagenida.

Acknowledgments The work was done within the framework of and financially supported by projects of the Federal Government of Lower Austria (K3-F-964/001-2020 and K3-F-964/002-2023, Department Science and Research; headquarters in St. Pölten); the Austrian Academy of Sciences (headquarters in Vienna, project Polzberg Lukeneder 2021) represented by the National Committee for Geo/Hydro Sciences (Earth System Sciences Programme); and the Triassic project "Fenster in die Karnische Krise" of the Society of Friends of the Natural History Museum Vienna (Freunde des Naturhistorischen Museums Wien, CPE Lukeneder 2023). The research of MH was supported by the Slovak Agency for Research and Development (APVV-22-0523) and the Slovak Scientific Grant Agency (VEGA 2/0106/23). The authors are particularly grateful to the owners of the outcrop area at Polzberg, namely Franziska and Hermann Hofreiter as well as Eva and Karl Jagersberger (all Gaming), and Felix Pöchhacker (Lunz am See), owner of the Großau drilling site, for sampling and digging permission. We thank Petra Lukeneder (Gablitz) for her help with cleaning outcrops and sampling fossils, and Aleš Šoster (Ljubljana) for doing the EDS analysis of the foraminiferal walls. The manuscript greatly benefited from valuable comments by the reviewers Ágnes Görög (Budapest) and Roberto Rettori (Perugia), the production editor Ashok Arumairaj (Taramani), the in-house series editor Kathleen A. Ritterbush (Utah), the publishing editor Simon Shah-Rohlfs and the production editor Antje Endemann (both Heidelberg).

Author Contribution AL collected new material. LG, AL designed research and provided documentation on fossil material.

Funding Open access funding was enabled and the projects financially supported by the Federal Government of Lower Austria (K3-F-964/001-2020 and K3-F-964/002-2023, Department Science and Research; headquarters in St. Pölten); the Austrian Academy of Sciences (headquarters in Vienna, project Polzberg Lukeneder 2021) represented by the National Committee for Geo/Hydro Sciences (Earth System Sciences Programme); the Triassic project "Fenster in die Karnische Krise" of the Society of Friends of the Natural History Museum Vienna (Freunde des Naturhistorischen Museums Wien, CPE Lukeneder 2023); the Natural History Museum Vienna; the Natur- und UNESCO Geopark Steirische Eisenwurzen (St Gallen, project Karnische Krise Lukeneder 2024) (to AL); and by the Slovenian Research and Innovation Agency (research core funding No. P1-0011—Regional Geology). The authors are responsible for the contents of this publication. The funders had no impact on conceptualization, design, data collection, analysis, decision to publish, or preparation of the manuscript.

Competing interests All authors stated that they have no competing interests.

Authors' contributions All authors stated that they have equally contributed to their chapters, read, edited and agreed to submit the final versions of their contributions.

Availability of data and materials All sources of information are provided in the text. Raw data related to the fossil material from Austria are available from the corresponding authors and Alexander Lukeneder upon request.

Additional information The article is part of the Special Volume: *Late Triassic Konservat-Lagerstätten within the Carnian Pluvial Episode. 150 Years of Scientific Excavations in Austria.*

Artificial Intelligence (AI) The chapter was produced without the use of AI.

Language The entire volume uses American English reviewed by a native scientist.

Received 2025.

Published online 2026.

References

Azbel AY (1988) Upper Jurassic miliolids of the European part of the USSR and western Kazakhstan. In: Mikhalevich VI (ed) Classification, ecology and stratigraphy of the Miliolata (Foraminifera). Proceedings of the Zoological Institute, Acad Sci USSR 184:11–43

Barros IS, Haig DW, McCartain E (2025) Upper Triassic (Norian-Rhaetian) variostomatids (Foraminifera), Timor-Leste: Systematics, paleoenvironmental, and biostratigraphic implications. J Foram Res 55(2):188–217. https://doi.org/10.61551/gsjfr.55.2.188

Benton MJ (1986) More than one event in the late Triassic mass extinction. Nature 321:857–861. https://doi.org/10.1038/321857a0

Bérczi-Makk A (1996) Foraminifera of the Triassic formations of Alsó Hill (Northern Hungary). Part 3: Foraminifer assemblage of the basinal facies. Acta Geol Hung 39(4):413–459

Bernardi M, Gianolla P, Petti FM, Mietto P, Benton MJ (2018) Dinosaur diversification linked with the Carnian Pluvial Episode. Nat Commun 9:1499. https://doi.org/10.1038/s41467-018-03996-1

Blainville HMD de (1827) Manuel de malacologie et de conchyliologie. 649–664, Levrault, Paris

Blau J, Wenzel B, Senff M, Lukas V (1995) Die Foraminiferen des oberen Buntsandsteins (Röt) und des unteren Muschelkalks (germanische Trias: Skyth, Anis) in Nordhessen. Geol Paläontol Mitt 20(Festschrift zum 60. Geburtstag von Helfield Mostler):13–33.

Böhm F, Ebli O, Krystyn L, Lobitzer H, Rakús M, Siblík M (1999) Fauna, stratigraphy and depositional environment of the Hettangian-Sinemurian (Early Jurassic) of Adnet (Salzburg, Austria). Abh Geol B-A 56(2):143–271

Breda A, Preto N, Roghi G, Furin S, Meneguolo R, Ragazzi E, Fedele P, Gianolla P (2009) The Carnian Pluvial Event in the Tofane area (Cortina d'Ampezzo, Dolomites, Italy). GeoAlp 6:80–115

Broglio Loriga C, Cirilli S, De Zanche V, di Bari D, Gianolla P, Laghi GF, Lowrie W, Manfrin S, Mastandrea A, Mietto P, Muttoni G, Neri C, Posenato R, Rechichi M, Rettori R, Roghi G (1999) The Prati di Stuores/Stuores Wiesen section (Dolomites, Italy): a candidate Global Stratotype Section and Point for the base of the Carnian stage. Riv It Paleontol Strat 105(1):37–78

Brotzen F (1963) Evolutionary trends in certain calcareous Foraminifera on the Paleozoic-Mesozoic boundary. In: Koenigswald GHR, Emeis JD, Buning WL, Wagner CW (eds) Evolutionary trends in Foraminifera. Elsevier Publ Com, Amsterdam, pp 66–78

Brusatte SL, Benton MJ, Lloyd GT, Ruta M, Wang SC (2011) Macroevolutionary patterns in the evolutionary radiation of archosaurs (Tetrapoda: Diapsida). Earth Environ Sci Trans Royal Soc Edinburgh 101:367–382. https://doi.org/10.1017/S1755691011020056

Bütschli O (1880) Erster Band. Protozoa. In: Bronn HG (ed) Klassen und Ordnungen des Thier-Reichs, I Band, I Abteilungen. CF Winter Leipzig: 1–224

Caggiati M, Gianolla P, Breda A, Celarc B, Preto N (2017) The start-up of the Dolomia Principale/Hauptdolomit carbonate platform (Upper Triassic) in the eastern Southern Alps. Sedimentology 65(4):1097–1131. https://doi.org/10.1111/sed.12416

Carrillat A, Martini R (2009) Palaeoenvironmental reconstruction of the Murafa Formation (Upper Triassic, Sicily): High resolution sedimentology, biostratigraphy and sea-level changes. Palaeogeogr, Palaeoclimatol, Palaeoecol 283(1–2):60–76. https://doi.org/10.1016/j.palaeo.2009.09.004

Chablais J, Martini R, Onoue T (2010) *Aulotortus friedli* from the Upper Triassic gravitational flow deposits of the Kumagawa River (Kyushu, Southwest Japan). Paleontol Res 14(2):151–160. https://doi.org/10.2517/1342-8144-14.2.151

Ciarapica G, Zaninetti L (1985) Le cas de "*Glomospirella friedli-Angulodiscus*? *gaschei*" (= *Aulotortus friedli*, Aulotortinae, Involutinidae, Foraminifere, Trias): Analyse structurale et révision taxonomique. Arch Sci Geneve 38(1):71–86

Copestake P, Johnson B (2014) Lower Jurassic foraminifera from the Llanbedr (Mochras Farm) borehole, North Wales, UK. 1–403, Palaeontolographical Society, London

Cushman JA (1910) A monograph of the Foraminifera of the North Pacific Ocean. Part I. Astrorhizidae and Lituolidae. Bull US Nat Mus 71:1–134

Cushman JA (1911) A monograph of the Foraminifera of the North Pacific Ocean. Part II. Textulariidae. Bull US Nat Mus 71:1–108

Cushman JA (1928) Foraminifera; their classification and economic use. Spec Publ Cushman Lab Foram Res 1:1–401

Dal Corso J, Mietto P, Newton RJ, Pancost RD, Preto N, Roghi G, Wignall PB (2012) Discovery of a major negative δ13C spike in the Carnian (Late Triassic) linked to the eruption of Wrangellia flood basalts. Geology 40(1):79–82. https://doi.org/10.1130/G32473.1

Dal Corso J, Bernardi M, Sun Y, Song H, Seyfullah LJ, Preto N, Gianolla P, Ruffell A, Kustatscher E, Roghi G, Merico A, Hohn S, Schmid, AR, Marzoli A, Newton RJ, Wignall PB, Benton MJ (2020) Extinction and dawn of the modern world in the Carnian (Late Triassic). Sci Adv 6(38):eaba0099. https://doi.org/10.1126/sciadv.aba0099

Delage Y, Hérouard E (1896) Traité de Zoologie Concrète, vol. 1, La cellule et les protozoaires. 1–584, Schleicher Frères, Paris

De Zanche V, Gianolla P, Mietto P, Siorpaes C, Vail PR (1993) Triassic sequence stratigraphy in the Dolomites (Italy). Mem Sci Geol 45:1–27

De Zanche V, Gianolla P, Roghi G (2000) Carnian stratigraphy in the Raibl/Cave del Predil area (Julian Alps, Italy). Eclogae Geol Helv 93:331–347

di Bari D, Baracca A (1998) Late Triassic (Carnian) foraminifers of Northeastern Cortina d'Ampezzo (Tamarin, San Cassiano Fm., Dolomites, Italy). Ann Mus Civ Rovereto, Sez Archeol Stor Sci Nat 12(1996):117–146

di Bari D, Laghi GF (1994) Involutinidae Bütschli (Foraminiferida) in the Carnian of the northeastern Dolomites (Italy). Mem Sci Geol 46:93–118

Doguzhaeva LA, Summesberger H (2012) Pro-ostraca of Triassic belemnoids (Cephalopoda) from Northern Calcareous Alps, with observations on their mode of preservation in an environment of northern Tethys which allowed for carbonization of nonbiomineralized structures. Neues Jahrb Geol Palaontol Abh 266:31–38. https://doi.org/10.1127/0077-7749/2012/0266

Doguzhaeva L, Summesberger H, Mutvei H (2006) An unique upper Triassic coleoid from the Austrian Alps reveals pro-ostracum and mandibule ultrastructure. Acta Univ Carol Geol 49:69–82

Doguzhaeva LA, Summesberger H, Mutvei H, Brandstaetter F (2007) The mantle, ink sac, ink, arm hooks and soft body debris associated with the shells in Late Triassic coleoid cephalopod *Phragmoteuthis* from the Austrian Alps. Palaeoworld 16:272–284. https://doi.org/10.1016/j.palwor.2007.08.002

Erba E (2004) Calcareous nannofossils and Mesozoic oceanic anoxic events. Mar Micropal 52:85–106. https://doi.org/10.1016/j.marmicro.2004.04.007

Falkowski PG, Katz ME, Knoll AH, Quigg A, Raven JA, Schofield O, Taylor FJR (2004) The evolution of modern eukaryotic phytoplankton. Science 305:354–360. https://doi.org/10.1126/science.1095964

Flügel E (2002) Triassic reef patterns. In: Kiessling W, Flügel E, Golonka J (eds) Phanerozoic reef patterns. SEMP Spec Publ 72:391–463

Forchielli A, Pervesler P (2013) Phosphatic cuticle in thylacocephalans: a taphonomic case study of (Arthropoda, Thylacocephala) from the Fossil-Lagerstatte Polzberg (Reingraben shales,

Carnian, Upper Triassic, Lower Austria). Aust J Earth Sci 106:46–61. https://doi.org/10.3847/1538-4357/ab9819

Forel M-B, Kolar-Jurkovšek T, Jurkovšek B (2020) Ostracods from the 'Raibl Beds' (Carnian, Late Triassic) of Belca section in Karavanke Mountains, northwestern Slovenia. Geodiversitas 42(21):377–407. https://doi.org/10.5252/geodiversitas2020v42a21

Frankowiak K, Roniewitz E, Stolarski J (2021) Photosymbiosis in Late Triassic scleractinian corals from the Italian Dolomites. Peer J 9:e11062. https://doi.org/10.7717/peerj.11062

Frentzen K (1941) Die Foraminiferenfaunen des Lias, Doggers und unteren Malms der umgebung von Blumberg (Oberes Wutachgebiet). Beitr Naturkund Forsch Oberrheingeb 6:125–402

Fursenko AV (1958) Fundamental state of development of foraminiferal faunas in the geological past. Trudy Instituta Geolog Nauk, Akad Nauk Belorusskoi SSR 1:10–29

Gaillot J, Vachard D (2007) The Khuff Formation (Middle East) and time-equivalents in Turkey and South China: biostratigraphy from Capitanian to Changhsingian times (Permian), new foraminiferal taxa, and palaeogeographical implications. Coloquios Paleontol 57:37–223

Gale L, Rettori R, Marini R, Šmuc A, Kolar-Jurkovšek T, Rožič B (2011) Duostominidae (Foraminiferea, Robertinida) from the Upper Triassic beds of the Slovenian Basin (Southern Alps, Slovenia). Riv It Paleontol Strat 117(3):375–397

Gale L, Celarc B, Caggiati M, Kolar-Jurkovšek T, Jurkovšek B, Gianolla P (2015) Paleogeographic significance of Upper Triassic basinal succession of the Tamar Valley, northern Julian Alps (Slovenia). Geol Carpathica 66(4):269–283. https://doi.org/10.1515/geoca-2015-0025

Gale L, Korat L, Marinšek M, Grădinaru E (2022) Use of X-ray micro-computed tomography on selected Upper Triassic (Rhaetian) foraminifera from the western Black Sea shelf, offshore Romania. Riv It Paleontol Strat 128:267–282

Gattolin G, Preto N, Breda A, Franceschi M, Isotton M, Gianolla P (2015) Sequence stratigraphy after the demise of a high-relief carbonate platform (Carnian of the Dolomites): Sea-level and climate disentangled. Palaeogeogr, Palaeoclimatol, Palaeoecol 423:1–17. https://doi.org/10.1016/j.palaeo.2015.01.017

Gianolla P, De Zanche V, Mietto P (1998) Triassic sequence stratigraphy in the Southern Alps (Northern Italy): definition of sequences and basin evolution. In: De Graciansky PC, Hardenbol J, Jacquin T, Vail PR (eds) Mesozoic and Cenozoc sequence stratigraphy of European basins. SEPM Spec Publ 60:719–746

Griffith J (1977) The Upper Triassic fishes from Polzberg bei Lunz. Zool J Linn Soc 60:1–93

Griffith JW, Henfrey A (1875) The micrographic dictionary, vol. 1 (3rd ed.). 879 p, Van Voorst, London.

He Y, Norling E (1991) Upper Triassic Foraminifera and stratigraphy of Mianzhu, Sichuan province, China. Sveriges Geologiska Undersökning, Avhandlingar Och Uppsatser 76:1–47

Hohenegger J, Piller W (1975) Ökologie und systematische Stellung der Foraminiferen im gebankten Dachsteinkalk (Obertrias) des Nördlichen Toten Gebirges (Oberösterreich). Palaeogeogr, Palaeoclimatol, Palaeoecol 18:241–276. https://doi.org/10.1016/0031-0182(75)90035-8

Hohenegger J, Piller W (1977) Die Stellung der Involutinidae Bütschli und Spirillinidae Reuss im System der Foraminiferen. Neues Jahrb Geol Paläontol Monatsch 7:407–418

Hornung T, Spatzenegger A, Joachimski MM (2007) Multistratigraphy of condensed ammonoid beds of the Rappoltstein (Berchtesgaden, southern Germany): unravelling palaeoenvironmental conditions on 'Hallstatt deep swells' during the Reingraben Event (Late Lower Carnian). Facies 53:267–292. https://doi.org/10.1007/s10347-006-0101-1

Jelen B (1990) The Carnian bivalves (Mollusca) from Lesno brdo, Slovenia, NW Yugoslavia and their paleobiological significance. Geologija 31(1):11–127

Jendrejáková O (1972) *Involutina muranica* n. sp. in der oberen Trias der Westkarpaten. Geol Zb—Geol Carpathica 23(1):197–200.

Kaminski M (2004) The year 2000 classification of the agglutinated foraminifera. In: Bubik M, Kaminiski MA (eds) Proceedings of the Sixth International Workshop on agglutinated foraminifera. Gryzbowski Found Spec Publ 8:237–255.

Kiessling W (2010) Reef expansion during the Triassic: Spread of photosymbiosis balancing climatic cooling. Palaeogeogr, Palaeoclimatol, Palaeoecol 290:11–19. https://doi.org/10.1016/j.palaeo.2009.03.020

Koehn-Zaninetti L (1969) Les foraminifères du trias de la region de l'Almtal (Haute-Autriche). Jahrb Geol B-A Spec. 14:1–155

Kolar-Jurkovšek T, Gaździcki A, Jurkovšek B (2005) Conodonts and foraminifera from the "Raibl Beds" (Carnian) of the Karavanke Mountains, Slovenia: stratigraphical and palaeobiological implications. Geol Quart 49(4):429–438

Kristan E (1957) Ophthalmidiidae und Tetrataxinae (Foraminifera) aus dem Rhät der Hohen Wand in Nieder-Österreich. Jahrb Geol B-A 100:269–298

Kristan-Tollmann E (1960) Rotaliidea (Foraminifera) aus der Trias der Ostalpen. Jahrb Geol B-a, Sdb. 5:47–78

Kristan-Tollmann E (1962) Stratigraphisch wertvolle Foraminiferen aus Obertrias- und Liaskalken der voralpinen Fazies bei Wien. Erdöl Zeitschr 1962:228–233

Kristan-Tollmann E (1963) Entwicklungsreihen der Trias-Foraminiferen. Paläontol Zeitschr 37:147–154

Kristan-Tollmann E (1990) Rhät-Foraminiferen aus dem Kuta-Kalk des Gurumugl-Riffes in Zentral-Papua/Neuguinea. Mitt Österr Geol Ges 82:211–289

Kristan-Tollmann E, Tollmann A (1964) Das mittelostalpine Rhät-Standardprofil aus dem Stangalm-Mesozoikum (Kärnten). Mitt Geol Ges Wien 56(2):539–589

Krystyn L (1991) Die Fossillagerstätten der alpinen Trias. In: Nagel D, Rabeder G (eds) Exkursionen im Jungpaläozoikum und Mesozoikum Österreichs. Österr Paläontol Ges Wien 1991:23–78

Lankester ER (1885) Protozoa. In: Encyclopaedia Britannica 19:830–866

Leischner W (1961) Zur Kenntnis der Mikrofauna und –flora der Salzburger Kalkalpen. Neues Jahrb Geol Paläontol Abh 112(1):1–47

Li Z, Yang X, Dal Corso J, Wang F, Jia E, Dai X, Yuan Z, Chen X, Lai J, Li X, Liu X, Jiang S, Wang B, Wu K, Chu D, Song H, Tian L, Song H (2025) No extinction in foraminifera during the Carnian Pluvial Episode (Late Triassic). Global Planet Change 251:104817. https://doi.org/10.1016/j.gloplacha.2025.104817

Loeblich ARJr, Tappan H (1964) Treatise on Invertebrate Palenotology. Part C. Protista 2. Sarcodina, chiefly "thecamoebians" and Foraminiferida. The Geological Society of America & The University of Kansas Press, Lawrence, Vol 1:936 p

Loeblich AR, Tappan H (1984) Suprageneric Classification of the Foraminiferida (Protozoa). Micropaleontology 30(1):1–70

Loeblich ARJr, Tappan H (1988) Foraminiferal genera and their classification. Van Nostrand Reinhold Company, New York, 970 p

Lukeneder A (ed) (2026) Late Triassic Konservat-Lagerstätten within the Carnian Pluvial Episode: 150 years of scientific excavations in Austria. Top Geobiol 55:600 p. https://doi.org/10.1007/978-3-032-19777-1

Lukeneder A, Lukeneder P (2021) The Upper Triassic Polzberg palaeobiota from a marine *Konservat-Lagerstätte* deposited during the Carnian Pluvial Episode in Austria. Sci Rep 11:16644. https://doi.org/10.1038/s41598-021-96052-w

Lukeneder A, Lukeneder P (2023) New data on the marine Upper Triassic palaeobiota from the Polzberg *Konservat-Lagerstätte* in Austria. Swiss J Palaeontol 142:9. https://doi.org/10.1186/s13358-023-00269-3

Lukeneder A, Surmik D, Gorzelak P, Niedźwiedzki R, Brachaniec T, Salamon MA (2020) Bromalites from the Upper Triassic Polzberg section (Austria); insights into trophic interactions and food chains of the Polzberg palaeobiota. Sci Rep 10:20545. https://doi.org/10.1038/s41598-020-77017-x

Lukeneder A, Lukeneder P, Sachsenhofer RF, Roghi G, Rigo M (2024) Multi-proxy record of the Austrian Upper Triassic Polzberg Konservat-Lagerstätte in light of the Carnian Pluvial Episode. Sci Rep 14:11194. https://doi.org/10.1038/s41598-024-60591-9

Martini R, Peybernes B, Zaninetti L, Fréchengues M (1996) Découverte de foraminiferes dans les intervalles transgressifs de deux séquences de dépôt anisiennes (Muschelkalk) du bassin de la Weser (Hesse, Allemagne du Nord). Geobios 24(5):505–511. https://doi.org/10.1016/S0016-6995(96)80021-0

Martini R, Cirilli S, Saurer C, Abate B, Ferruzza G, Lo Cicero G (2007) Depositional environment and biofacies characterisation of the Triassic (Carnian to Rhaetian) carbonate succession of Punta Bassano (Marettimo Island, Sicily). Facies 53:389–400. https://doi.org/10.1007/s10347-007-0115-3

Maslakova NI (1990) Criteria of establishing of higher taxa in Foraminifera. Systematics and phylogeny of Invertebrata. Izdatelstvo Nauka, Moscow, pp 22–27

Mikhalevich VI (1980) Systematics and evolution of the Foraminifera in view of the new data on their cytology and ultrastructure. In: Principy postroeniya makrosistemy odnokletochnyih zhivotnyh. Trudy Zoologicheskogo Instit, AN SSSR 94:42–61

Mikhalevich VI (2013) New insight into the systematics and evolution of the foraminifera. Micropaleontology 59:493–527

Moser M, Pavlik W (2013) GEOFAST—Zusammenstellung ausgewählter Archivunterlagen der Geologischen Bundesanstalt 1:50.000—74 Hohenberg. Wien

Neumayr M (1887) Die natürlichen Verwandschaftsverhältnisse der schalentragenden Foraminiferen. Sitz-Ber K Akad Wiss, Math-Naturwiss Kl 95:156–186

Oberhauser R (1957) Ein Vorkommen von Trocholina und Paratrocholina in der ostalpinen Trias. Jahrb Geol B-A 100:257–267

Oberhauser R (1960) Foraminiferen und Mikrofossilien "incertae sedis" der ladinischen und karnischen Stufe der Trias aus den Ostalpen und Persien. Jahrb Geol B-A Sdb 5:5–46

Oberhauser R (1964) Zur Kenntnis der Foraminiferengattungen Permodiscus, Trocholina und Triasina in der alpinen Trias und ihre Einordnung zu den Archaedisciden. Verh Geol B-A 2:196–210

Oblak K (2001) Carnian Foraminifera from the Kucler quarry at Lesno Brdo (Slovenia). Geologija 44(2):305–316. https://doi.org/10.5474/geologija.2001.022

Ogg JG (2015) The mysterious mid-Carnian "Wet Intermezzo" global event. J Earth Sci 26:181–191. https://doi.org/10.1007/s12583-015-0527-x

Ogorelec B, Jurkovšek B, Šribar L, Jelen B, Stojanovič B, Mišič M (1984) Carnian beds at Tamar and at Log pod Mangartom. Geologija 27:107–158

Oravecz-Scheffer A (1971) The representatives of the superfamily Miliolacea (Foraminifera) in the Carnian deposits, borehole Bakonyszűcs-1, Transdanubia. Hungary. Inst Geol Publ Hung 1968:89–105

Oravecz-Scheffer A (1987) Triassic Foraminifers of the Transdanubian Central Range. Geol Hung 50:3–134

Orbigny AD d' (1826) Tableau méthodique de la classe des Céphalopodes. Ann Sci Nat 7:96–169, 245–314

Paalzow R (1922) Die Foraminiferen der Parkinsoni-Mergel von Heidenheim am Hahnenkamm. Abh Nat Ges Nürnberg 22:1–35

Pawlowski J, Holzmann M, Tyszka J (2013) New supraordinal classification of Foraminifera: Molecules meet morphology. Mar Micropaleontol 100:1–10. https://doi.org/10.1016/j.marmicro.2013.04.002

Peybernes C, Chablais J, Onoue T, Escarguel G, Martini R (2016) Paleoecology, biogeography, and evolution of reef ecosystems in the Panthalassa Ocean during the Late Triassic: Insights from reef limestone oft he Sambosan Accretionary Complex, Shikoku, Japan. Palaeogeogr, Palaeoclimatol, Palaeoecol 457:31–51. https://doi.org/10.1016/j.palaeo.2016.05.032

Piller W (1978) Involutinacea (Foraminifera) der Trias und des Lias. Beit Paläontol Österr 5:1–164

Piller WE, Egger H, Erhart CW, Gross M, Harzhauser M, Hubmann B, van Husen D, Krenmayr HG, Krystyn L, Lein R, Lukeneder A, Mandl GW, Rögl F, Roetzel R, Rupp C, Schnabel W, Schönlaub HP, Summsberger H, Wagreich M, Wessely G (2004) Die Stratigraphische Tabelle von Österreich 2004 (sedimentäre Schichtfolgen). Kommission für die Paläontologische und

stratigraphische Erforschung Österreichs, Österreichische Akademie der Wissenschaften und Österreichische Stratigraphische Kommission, Wien
Podobina VM (1975) Foraminifera of the Upper Cretaceous and Paleogene from the West Siberian depression, their importance for stratigraphy. 1–290, Tomsk University Press, Tomsk
Podobina VM (1978) Systematics and phylogeny of the Haplophragmiidae. 1–91, Tomsk Universitet, Tomsk
Preto N, Kustatscher E, Wignall PB (2010) Triassic climates—State of the art and perspectives. Palaeogeogr, Palaeoclimatol, Palaeoecol 290:1–10. https://doi.org/10.1016/j.palaeo.2010.03.015
Preto N, Agnini C, Rigo M, Sprovieri M, Westphal H (2013) The calcareous nannofossil *Prinsiosphaera* achieved rock-forming abundances in the latest Triassic of western Tethys: consequences for the $\delta^{13}C$ of bulk carbonate. Biogeosciences 10:6053–6068. https://doi.org/10.5194/bg-10-6053-2013
Pronina GP, Vuks VJa (1996) New data on the Triassic foraminifers of Crimea. Ann Mus Civ Rovereto 11:215–228
Ramovš A (1993) Stratigraphic development of Triassic in northern Julian Alps and in western Karavanke Mountains—correlation. Min Metall Quart 40:103–114
Rettori R (1995) Foraminiferi del Trias inferiore e medio della Tetide: revisione tassonomica, stratigrafia ed interpretazione filogenetica. Uni Geneve, Publ Dép Géol Paléontol 18(1995):1–107
Rettori R, Loriga C, Neri C (1998) Lower Carnian foraminifers from the type locality of the Calcare del Predil (Raibl Group, northeastern Italy). Riv It Paleontol Strat 104(3):369–380
Reuss AE (1862) Entwurf einer systematischen Zusammenstellung der Foraminiferen. Sitz-Ber K Akad Wiss, Math-Naturwiss Cl 44:355–396
Rhumbler L (1895) Entwurf eines naturlichen Systems der Thalamophoren. Nachr Ges Wiss Göttingen, Math-Phys Kl 1:51–98
Rhumbler L (1904) Systematische Zusammenstellung der recenten Reticulosa. Arch Protistenkunde 3:181–294
Rigaud S, Martini R, Rettori R (2012) Parvalamellinae, a new subfamily for Triassic glomospiroid Involutinidae. J Foram Res 42(3):245–256. https://doi.org/10.2113/gsjfr.42.3.245
Rigaud S, Blau J, Martini R, Rettori R (2013) Taxonomy and phylogeny of the Trocholinidae (Involutinina). J Foram Res 43:317–339. https://doi.org/10.2113/gsjfr.43.4.317
Russo F (2005) Biofacies evolution of the Triassic platforms of the Dolomites, Italy. Ann Univ Studi Ferrara Museol Scient Natural Vol. Spec. 2005:33–44. https://doi.org/10.15160/1824-2707/353
Ruttner A, Schnabel W (1988) Geologische Karte der Republik Österreich. Sheet Ybbsitz 71, 1:50.000. Wien: Geologische Bundesanstalt
Ruvinetti R (2004) Benthic molluscs and brachiopods: Paleoclimatic and paleoenvironmental evolution of the Middle Carnian of the eastern Southern Alps. [PhD thesis]. 1–190, Università degli Studi di Ferrara, Ferrara
Salaj J, Biely A, Bystrický J (1967) Trias-Foraminiferen in Den Westkarpaten. Geol Práce 42:119–136
Salaj J, Borza K, Samuel O (1983) Triassic foraminifers of the West Carpathians. Geologický ústav Dionýsa Štúra, Bratislava, 213 p
Schultze MJS (1854) Über den Organismus der Polythalamien (Foraminiferen), nebst Bemerkungen über die Rhizopoden im allgemeinen. Ingelmann, Leipzig, 68 p
Senowbari-Daryan B, Rashidi K, Torabi H (2010) Foraminifera and their associations of a possibly Rhaetian section of the Nayband Formation in central Iran, northeast of Esfahan. Facies 56:567–596. https://doi.org/10.1007/s10347-010-0221-5
Simms MJ, Ruffell AH (1989) Synchroneity of climatic change and extinctions in the Late Triassic. Geology 17:265–268. https://doi.org/10.1130/0091-7613(1989)017%3c0265:SOCCAE%3e2.3.CO;2

Stanley GD Jr (2003) The evolution of modern corals and their early history. Earth-Sci Rev 60:195–225. https://doi.org/10.1016/S0012-8252(02)00104-6

Stanley GD Jr, Swart PK (1995) Evolution of the coral-zooxanthenllae symbiosis during the Triassic: a geochemical approach. Paleobiology 21(2):179–199. https://doi.org/10.1017/S0094837300013191

Stanley GD Jr, van de Schootbrugge B (2009) The evolution of the coral-algal symbiosis. In: van Oppen MJH, Lough JM (eds) Coral bleaching. Ecological Studies 205. 7–19, Springer, Berlin, Heidelberg. https://doi.org/10.1007/978-3-540-69775-6_2

Stefani M, Furin S, Gianolla P (2010) The changing climate framework and depositional dynamics of Triassic carbonate platforms from the Dolomites. Palaeogeogr, Palaeoclimatol, Palaeoecol 290:43–57. https://doi.org/10.1016/j.palaeo.2010.02.018

Stur D (1874) Neue Aufschlüsse im Lunzer Sandsteine bei Lunz und ein neuer Fundort von Wengerschiefer im Pölzberg zwischen Lunzersee und Gaming. Verh K-K Geol RA 1:271–273

Stur D (1886) Vorlage des ersten fossilen Schädels von *Ceratodus* aus den ober triadischen Reingrabner Schiefern von Pölzberg nördlich bei Lunz. Verh K-K Geol RA 15:381–383

Subbotina NN, Voloshinova NA, Azbel' AYa (1981) Introduction to the study of foraminifera (classification of calcareous foraminifera of the Meso-Cenozoic). Vsesoyuznyy Neftyanoy Nauchno-issledovatel'skii Geologorazvedochnyy Institut (VNIGRI), Leningrad

Suleymanov IS (1973) Some questions on the systematics of the family Verneuilinidae Cushman 1927 in connection with conditions of the habitat. Dokladari Uzbekiston SSR, Fanlar Akademiyasining, Tashkent 1973(8):35–36

Sykora M, Siblík M, Soták J (2011) Siliciclastics in the Upper Triassic dolomite formations of the Krížna Unit (Malá Fatra Mountains, Western Carpathians): constraints for the Carnian Pluvial Event in the Fatric Basin. Geol Carpathica 62:121–138. https://doi.org/10.2478/v10096-011-0011-1

Tappan H (1955) Foraminifera from the Arctic slope of Alaska, part 2, Jurassic Foraminifera. Profess Papers US Geol Survey 236B:21–90

Teller F (1891) Über den Schädel eines fossilen Dipnoers, *Ceratodus sturii* nov. spec. aus den Schichten der oberen Trias der Nordalpen. Abh K-K Geol RA 15:1–38

Tollmann A (1976) Analyse des klassischen nordalpinen Mesozoikums. Franz Deuticke, Wien, 580 p

Trauth F (1948) Geologie des Kalkalpenbereiches der zweiten Wiener Hochquellenleitung. Abh Geol B-A 26:1–99

Trifonova E (1962) Upper Triassic foraminifera from the surroundings of Kotel—the Eastern Balkan. Ann Dir Gen Rech Geol 12:141–170

Trifonova E (1977) Foraminiferen aus der Trias des Ostbalkans. Palaeontol Stratigr Lithol 6:47–63

Trifonova E (1992) Taxonomy of Bulgarian Triassic foraminifera. I. Families Psammosphaeridae to Nodosinellidae. Geol Balc 22(1):3–50

Trifonova E (1993) Taxonomy of Bulgarian Triassic foraminifera. II. Families Endothyriidae to Ophthalmidiidae. Geol Balc 23(2):19–66

Urošević D, Sudar M (1992) Anisian foraminifera of Merica (Zlatar Mountain, Western Serbia). Ann Géol Péninsule Balk 56(1):211–235

Vachard D (2016) Macroevolution and Biostratigraphy of Paleozoic Foraminifers. Stratigr Timesc, 257–323. https://doi.org/10.1016/bs.sats.2016.10.005

Vachard D, Razgallah S (1988) Importance phylogénétique d'un nouveau endothyroide: *Endoteba controversa* n.gen. n.sp. (Permien du Jebel Tebaga, Tunisie). Geobios 21(6):805–811. https://doi.org/10.1016/S0016-6995(88)80096-2

Vachard D, Martini R, Rettori R, Zaninetti L (1994) New classification of Triassic Endothyroid Foraminifera. Geobios 27:543–557. https://doi.org/10.1016/S0016-6995(94)80249-1

Vachard D, Krainer K, Lukas SG (2010a) Pennsylvanian (Late Carboniferous) calcareous microfossils from Cedro Peak (New Mexico, USA). Part 2: Smaller foraminifers and fusulinids. Ann Paléontol 99:1–42

Vachard D, Pille L, Gaillot J (2010b) Palaeozoic foraminifera: systematics, palaeoecology and responses to global changes. Rev Micropaleontol 53:209–254

Vdovenko MV, Rauser-Chernousova DM, Reitlinger EA, Sabirov AA (1993). Guide for the systematics of Paleozoic smaller foraminifera. Rossiiskaya Akademiya Nauk, Komissiya po Mikropaleontologii, 'Nauka', Moscow, 128 p

Wernli R (1971) *Planiinvoluta carinata* Leischner, 1961 (Foraminifere) dans l'Aalénien supérieur du Jura Méridional (France). Arch Sci Geneve 24(2):219–226

Weynschenk R (1956) *Aulotortus*, a new genus of Foraminifera from the Jurassic of Tyrol. Austria. Contrib Cushman Found Foram Res 7(1):26–28

Wiesner H (1920) Zur Systematik der Miliolideen. Zool Anz 51(1/2):13–20

Zaninetti L (1969) "*Agathamminoides*" gen. n., un nouveau genre de foraminiferes du Trias alpin. Riv It Paleontol 75(4):697–704

Zaninetti L (1976) Les Foraminiferes du Trais. Essai de synthese et corrélation entre les domaines Mésogéens Européen et Asiatique. Riv It Paleontol 82(1):1–258

Zaninetti L (1984) Les Involutinidae (Foraminifères), Proposition pour une subdivision. Rev Paléob 3:205–207

Zaninetti L, Ciarapica G, Martini R, Salvini-Bonnard G, Rettori R (1987) *Turriglomina scandonei*, n. sp., dans le calcaires recifaux de Trias moyen (Ladinien) en Apennin Meridional. Rev Paléobiol 6(2):177–182

Zhang ZT, Sun YD, Wignall PB, Fu JL, Li HX, Wang MY, Lai XL (2018) Conodont size reduction and diversity losses during the Carnian Humid Episode in SW China. J Geol Soc 175:1027–1031. https://doi.org/10.1144/jgs2018-002

Zsiborás G, Gőrög A (2020) Aalenian–lower Bajocian foraminiferal fauna from the Ammonitico Rosso series of Bakonycsernye (Hungary). Part I: taxonomy and biostratigraphy. J Foram Res 50:41–72

Chapter 15
Drilling into the Carnian Pluvial Episode

Alexander Lukeneder, Johannes A. Weitz, Reinhard F. Sachsenhofer, and David Misch

Abstract This contribution reports the first data from a core drilling into the Carnian Pluvial Episode (CPE). The Großau drill site in the Northern Calcareous Alps of Lower Austria is located in a small private quarry between the famous Polzberg *Konservat-Lagerstätte* in Lower Austria and the similar Scheiblingbach locality in Styria. All three sites expose the Upper Triassic succession representing the CPE in the intraplatform Reifling Basin. The Upper Triassic sediments at the Großau drill site are overturned and include, in stratigraphic order, the calcareous Reifling Formation, the silicious Göstling Formation, and the well-laminated Reingraben Formation. Gamma and magnetic susceptibility measurements were carried out on the 30-m-long drill core and a 13-m-long profile in the Großau quarry. Bulk geochemical parameters and biomarker compositions were determined on fresh core material. In addition, core samples were sent for various analyses to scientists around the world. New fossils of various marine taxa were collected from the drill core and the nearby quarry. Fossiliferous layers are concentrated within the lowermost meters of the Reingraben deposits from the early Carnian (Julian 2, *Austrotrachyceras austriacum* Zone). The newly collected assemblage comprises ammonoids (*Austrotrachyceras*, *Paratrachyceras*), bivalves (*Peribositra, Halobia*, div. indet taxa), and acytinopterygian fishes (entire and scales, *palaeoniscids*). Bromalites (coprolites and regurgitalites) produced by piscivorous actinopterygians and durophagous fish

Supplementary Information The online version contains supplementary material available at https://doi.org/10.1007/978-3-032-19777-1_15.

A. Lukeneder (✉)
Department of Geology and Paleontology, Natural History Museum Vienna, Vienna, Austria
e-mail: alexander.lukeneder@nhm.at

J. A. Weitz · R. F. Sachsenhofer · D. Misch
Department Applied Geosciences and Geophysics, Technical University of Leoben, Leoben, Austria
e-mail: johannes.weitz@unileoben.ac.at

R. F. Sachsenhofer
e-mail: reinhard.sachsenhofer@unileoben.ac.at

D. Misch
e-mail: david.misch@unileoben.ac.at

A. Lukeneder (ed.), *Late Triassic Konservat-Lagerstätten within the Carnian Pluvial Episode*, Topics in Geobiology 55, https://doi.org/10.1007/978-3-032-19777-1_15

accompany the Großau paleobiota along with frequent plant remains on bedding planes of the Göstling Formation (*Voltzia,* div. indet plants). The lithology of the studied succession, together with the preliminary data on the Großau paleobiota and the geochemical proxies, characterize a fluctuating basinal environment with intermittent influx of freshwater. This was accompanied by the presence of land plants and colonization by abundant benthic posidoniid bivalves in the normal marine Reifling Basin. The basal Reingraben deposits contain a *Konservat-Lagerstätte,* an equivalent to the Polzberg *Konservat-Lagerstätte.* Within that special environment and those depositional conditions, the nektic community was dominated by ammonoids and acytinopterygian fishes. The prerequisites for high-quality preservation – calm and oxygen-reduced conditions – prevailed at the sea floor of the Großau *Konservat-Lagerstätte* during the deposition of black intermittent layers in the Göstling Formation and in the basal part of the Reingraben Formation. These paleoenvironmental conditions prevailed within the entire Reifling Basin during the early Carnian (Julian 2, *A. austriacum* Zone). New in situ findings of benthic and nektic taxa highlight the great value of the additional Großau paleobiota in comparison to the well-known contemporaneous Polzberg paleobiota. Overall, the drilled succession points to a carbonate platform decline following warmer and wetter conditions during the CPE.

Keywords Drilling · Carnian Pluvial Episode · Early Carnian · Late Triassic · Austria

15.1 Introduction

Numerous studies have been conducted on lower Carnian deposits of the Northern Calcareous Alps (Fig. 15.1a), mainly in Lower Austria. The main focus was placed on the paleobiota from the Polzberg *Konservat-Lagerstätte* and the Carnian Pluvial Episode (CPE), a time interval with enhanced worldwide humidification and carbonate platform demise (Wet Intermezzo Ogg 2015; CPE, Lukeneder et al. 2012; Ruffel et al. 2015; Dal Corso et al. 2018; Simms and Ruffel 2018; Lukeneder et al. 2024; Alverà et al. 2026; Lukeneder 2026, this volume).

A large number of papers was published on the Upper Triassic paleobiota from the well-known lower Carnian Polzberg *Konservat-Lagerstätte* in Lower Austria (Stur 1874, 1886; Teller 1891; Reis 1900; Abel 1906; Glaessner 1931, 1965; Griffith 1977; Martin 1982; Krystyn 1991; Forchielli and Pervesler 2013; Schädel et al. 2020). The contemporaneous Scheiblingbach (= Scheiblinggraben) *Konservat-Lagerstätte* in Styria near the border with Lower Austria exhibits similar lithologies and similar flora and fauna assemblages. The main difference between the Polzberg and Scheiblingbach localities is the duration of sampling: fossils have been collected at Polzberg since the 1870s and at Scheiblingbach since the 1970s. Accordingly, Polzberg can look back on a sampling history that is 100 years longer. Recently, additional material from the two sites has become available due to the first author's intensified collecting

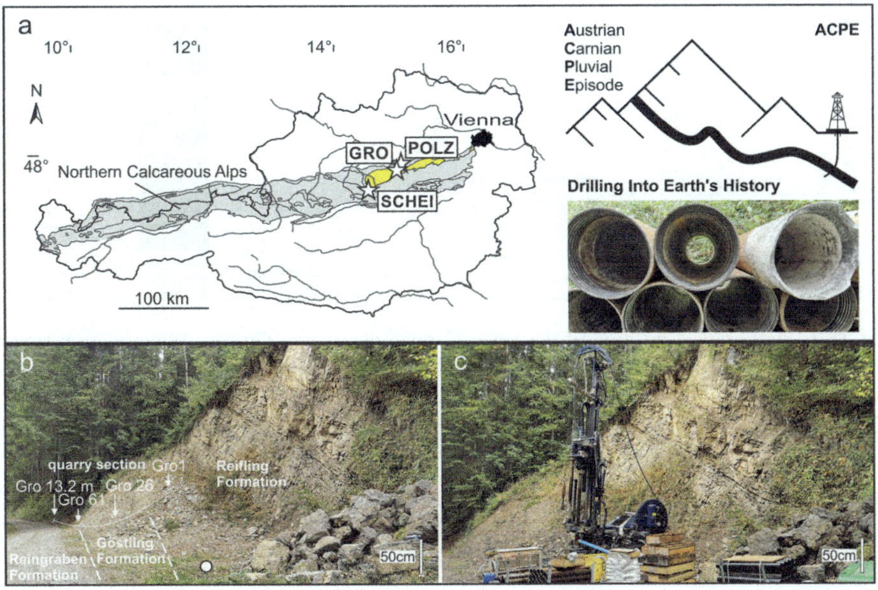

Fig. 15.1 **a** Location of the Northern Calcareous Alps in Austria (in gray) with position of the Großau drilling site (GRO) in Lower Austria and the emblem of the Austrian Carnian Pluvial Episode (ACPE) drilling project. The Lunz Nappe and Sulzbach Nappe system is highlighted in yellow. POLZ Polzberg, SCHEI Scheiblingbach. **b** Detailed view of the Großau quarry with the boundaries between the inverse Reifling Formation, Göstling Formation, and Reingraben Formation. The circle marks the position of the drill hole. **c** Mobile core-drilling facility (Rotomax XL B) of the PORR Bau GmbH company at the Großau drilling site

activities since 2021. Together with historical paleontological material from the official Austrian collections (Natural History Museum Vienna, GeoSphere Austria) and material from private collections (B and K Aschauer), this material has been used in studies on the paleoenvironment and the trophic food web (Lukeneder et al. 2020; Lukeneder and Lukeneder 2021), taphonomy, trophic history, and taxonomy of ammonoids (Lukeneder and Lukeneder 2022a, 2024). Another focus was on unique mineralized belemnoid cartilages (Lukeneder and Lukeneder 2022b; Lukeneder et al. 2024b).

The Polzberg and Scheiblingbach paleobiotas were deposited in an intraplatform basin setting (Reifling Basin) during the Carnian Julian 2 Ib (*Austrotrachyceras austriacum* Zone, *A. minor* biohorizon; Lukeneder and Lukeneder 2021; Lukeneder et al. 2024). In this setting, the CPE caused the change of the limestone of the Reifling Formation (formalized) to the siliceous laminated limestone of the Göstling Formation (= Göstlinger Kalke; not formalized; frequently noted as Göstling member of the uppermost part of the Reifling Formation) and finally to the laminated shales of the Reingraben Formation (= Reingraben Shales, Reingrabener Schiefer Fig. 15.2). As noted by Lukeneder and Lukeneder (2021) and Lukeneder et al. (2020, 2024a),

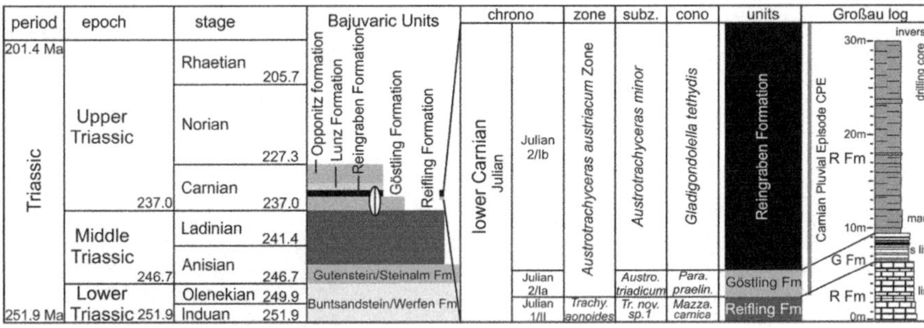

Fig. 15.2 Stratigraphic position of the lower Carnian deposits at Großau. White ellipse: stratigraphic range of the Großau drill core. Detailed Großau core section within the lower Carnian Reifling Formation, the Göstling Formation, and the Reingraben Formation. Note the marked lithological transition during the Carnian Pluvial Episode. Austro *Autrotrachyceras*, Trachy *Trachyceras*, chrono chronostratigraphy, subz subzone, Re Fm Reifling Formation, G Fm Göstling Formation, R fm Reingraben Formation, lime limestone, s lime silicious limestone. Adapted after Lukeneder et al. (2021). Numerical age data according to the ICS International Chronostratigraphic Chart ICC 2024/12 (Cohen et al. 2013)

laminated deposits provide excellent conditions for the preservation of entire and soft body fossils, a prerequisite for *Konservat-Lagerstätten.*

In 2023 a scientific borehole was drilled in the context of the 'Austrian Carnian Pluvial Episode (ACPE) drilling project' in a small quarry at Großau (Lower Austria; 9 km SW of Polzberg; Fig. 15.1). The drill core provides a unique opportunity to study the lithology along with the fauna and flora paleocommunities in the Reifling Basin of the Northern Calcareous Alps between the Polzberg and Scheiblingbach localities.

The present paper presents paleocommunities for the first time the lithological variations observed in the drill core, along with organic geochemical data and preliminary results on the fauna and flora paleocommunities. Furthermore, fossil taxa and the preliminary fossil assemblage are compared to the contemporaneous deposits and fossil remains of the Polzberg and Scheiblingbach *Konservat-Lagerstätten.* This will help to strengthen the correlation of the Polzberg and Scheiblingbach *Konservat-Lagerstätten* (35 km geographical distance; cf. Lukeneder and Lukeneder 2021, 2023 or Lukeneder et al. 2024a).

More details on specific taxonomic groups, lithological aspects, inorganic geochemical proxies, geophysical data, along with important food chains and dietary habits as well as trophic interactions amongst the inhabitants from the Upper Triassic marine ecosystems are presented within this Special Volume (Harzhauser et al. 2026, Kustatscher et al. 2026, Lukeneder 2026, Weitz et al. 2026, all this volume).

15.2 Material and Methods

The Großau borehole (Fig. 15.1a) was drilled in a small private quarry (owner Felix Pöchhacker) exposing limestone of the Reifling Formation, which dips with approximately 35° towards southeast. The quarry is located on the northeastern slope of Mount Großkopf (1147 m) at an altitude of 620 m in the Ypps valley between Lunz am See and Göstling an der Ybbs (Lower Austria; Fig. 15.1b). The drill site is located 300 m southeast of the Ybbs River, 5 km southwest of the Lunz am See center and 5 km northeast of Göstling an der Ybbs (47° 49′ 24.40″ N, 15° 00′ 0.80″ E).

The aim was to drill the upper and lower boundaries of the Göstling Formation and the fossiliferous lower part of the argillaceous Reingraben Formation. The drill site Großau was selected because drilling was not allowed by the private owners at the historical site Polzberg. The borehole was drilled in October of 2023 during a five-day drilling campaign by PORR Bau GmbH using a mobile core-drilling rig (Rotomax XL B; Fig. 15.1c). The borehole was cored over its entire length (30 m; 1 m core intervals). The core diameter was changed from 15 cm (Gro 0–2 m) to 12 cm (Gro 2–30 m). Two core fragments were stored in one wooden box, respectively (Fig. 15.3). Due to the inverted position of the Großau section, the older Reifling Formation was drilled at the surface, followed by the Göstling Formation and the Reingraben Formation in the deepest part. The dip of the bedding planes is below 10° in the shallow part of the drill core but increases in the deeper part (locally up to 30°). Thus, the core depth might slightly overestimate the true thickness of the Reingraben Formation.

The Upper Triassic succession was described lithologically, both in the drill core and in the newly excavated Großau quarry. Macrofossils (ammonoids, bivalves, fish remains, plants, trace fossils) were collected directly from the core and the corresponding layers in the outrop during field campaigns in 2023 and 2024. Microfossils (radiolaria, ostracods, planktonic foraminifera) were investigated by a total of 46 thin sections. Drill cores, macrosfossils, and thin sections are stored in the Natural History Museum in Vienna (NHMW, Austria).

Low-field magnetic susceptibility (χ, MS) was measured directly in the field with a hand-held, highly sensitive (1×10^{-7}, measured in SI units; results are indicated throughout as 'value $\times 10^{-6}$ SI'), low-level SM-30 MS meter (GF Instruments) bearing a 50-mm pick-up coil. Each bed was measured in well-bedded lithologies (e.g., limestone or dolomite layers). In lithologies without distinct bedding (e.g., clay deposits, Reingraben Formation), a measuring interval between 5 and 50 cm was used; 112 measurements of MS were performed in sediments of the Reifling Formation ($n = 12$), Göstling Formation ($n = 10$) and Reingraben Formation ($n = 90$). Each single bed was measured at least three times and the average calculated. To prevent measurement errors due to the sensitivity of the hand-held instruments (uneven terrain and therefore variable distance between coil and rock surface), a fresh smooth surface was prepared for each measured point.

The gamma response (gamma-ray logs) was measured in the field with 0.1–0.2 m spacing using a hand-held gamma-ray scintillometer ('Compact Gamma

Fig. 15.3 Photographs of the Großau drill core. Each drill box is 1 m long and contains 2 m of core. Since the succession is overturned, the core begins with the Reifling Formation (0.0–6.5 m core depth), continues with the Göstling Formation (6.5–9.6 m), and ends in the Reingraben Formation (9.6–30.0 m). The attached sketch shows the arrangement of drill boxes, core depth (in m), and the different formations in varying shades of gray

Surveyor', Gf Instruments). The gamma log of the Reifling Formation (n = 12), Göstling Formation (n = 10), and Reingraben Formation (n = 72) is based on 94 measurements.

Geochemical analyses and Rock–Eval pyrolysis were carried out at the Chair of Energy Geosciences (Technical University of Leoben; Austria). An Eltra Helios CS elemental analyzer was used to determine total carbon (TC; %), total sulfur (TS; %), and total organic carbon (TOC; %) contents. TOC contents were measured on samples pre-treated with concentrated phosphoric acid. The accuracy of C and S measurements is better than 1%. Calcite equivalent percentages were derived using the equation $calcite_{equiv.} = 8.334 \times (TC-TOC)$.

Rock–Eval pyrolysis was performed in duplicate using a Rock–Eval 6 instrument. The determined parameters include the amounts of free hydrocarbons (S1 [mg HC/g rock]), hydrocarbons generated during thermal cracking (S2 [mg HC/g rock]), and the temperature of maximum hydrocarbon generation (Tmax [°C]). S1 and S2 were used to calculate the hydrogen index ($HI = 100 \times S2/TOC$) and the production index ($PI = S1/[S1 + S2]$). The relative error for S2, HI, and Tmax is about 2.0, 1.0, and 0.3%, respectively. Based on their TOC content and stratigraphic position, six samples were selected for biomarker analyses. 5–10 g of the sample material was extracted for about 1 h at 75 °C and 100 bar using dichloromethane as solvent and a Dionex ASE 350 Accelerated Solvent Extractor. The extract was concentrated to ~0.5 ml using a Zymark TurboVap 500 closed cell concentrator. Afterwards, asphalthenes were precipitated from a hexane:dichloromethane solution (80:1 according to volume) and separated by centrifugation. The hexane-soluble fractions were further separated into heterocompounds (NSO) as well as saturated and aromatic hydrocarbons by medium-pressure liquid chromatography using a Köhnen-Willsch MPLC instrument. Proportionate amounts of internal standards (squalene for aliphatics and 1,1′-binaphthyl for aromatics) were added to each sample prior to measurement. For detailed GCMS measurement conditions see Weitz et al. (2026).

15.3 Geologic Setting and Lithology

The Großau drill site is located in the southern part of the Lunz Nappe of the Northern Calcareous Alps (NCA), which is occasionally interpreted as a separate nappe (Sulzbach Nappe; Trauth 1948; geological map, 71 Ybbsitz, Ruttner and Schnabel 1988). Within the Lunz Nappe, the NE-SW elongated Reifling Basin (Krystyn 1991; Lukeneder and Lukeneder 2021) extended between Polzberg (Lower Austria) and Großreifling (northernmost Styria). During the Late Triassic, the Reifling Basin was an intraplatform basin surrounded by Wetterstein platforms, which formed between Pangea in the northwest and the Meliata Ocean in the southeast. The Meliata Ocean was a small oceanic basin within the Tethyan Realm, separated from the Tethys Ocean by the island chain of the Cimmerian continent (Hornung and Brandner 2005; Lukeneder et al. 2012). The Reifling Formation was deposited on the paleoslope and basinal areas of the Reifling Basin (Krystyn et al. 2008; Lukeneder and Lukeneder 2021; Lukeneder et al. 2024).

15.3.1 Großau Drill Core

The (inverse) lithological column at Großau includes in stratigraphic order the older Reifling Formation (Figs. 15.3, 15.4 and 15.5a, b), the Göstling Formation (Figs. 15.3, 15.4 and 15.5c, d), and the younger Reingraben Formation (Figs. 15.3, 15.4 and 15.5e, f).

The drill core starts at the surface (Gro 0 m) with limestones of the Reifling Formation (0.0–6.5 m, layers Gro 0 m to Gro 6.5 m; Fig. 15.4). It consists of light grey and grey-green nodular limestones composed of peloidal, filamentous wackestone and packstones with strong bioturbation and a bed-thickness of 5–60 cm. The carbonate content is uniformly high (85–95 wt% $calcite_{equiv.}$) throughout the formation (Lukeneder et al. 2024). The Reifling Formation is overlain (deeper in the core) by the Göstling Formation (6.5–9.6 m, layer Gro 6.6 m to Gro 9.6 m; Fig. 15.4). The Göstling Formation consists of black, mm- to cm-bedded calciturbiditic radiolarian limestones (60–82 wt% $calcite_{equiv.}$) with organic-rich mudstones ($CaCO_3$ 28–53 wt% $calcite_{equiv.}$; Supplementary material 15.1, bulk geochemistry). The transition from the Reifling Formation to dark, fine-laminated radiolaria wackestones to packstones of the Göstling Formation is abrupt and occurs between layers Gro 6.48 m and Gro 6.5 m (Fig. 15.4). Layers with intense bioturbation alternate with laminated layers in which bioturbation is rare or absent. This reflects different oxygen availability during deposition. In the Großau drill core, the radiolarian-rich Göstling limestones are only 3.1 m thick. The top of the Göstling Formation is strongly bioturbated (surface of layer Gro 9.6 m).

The overlying terrigeneous siliciclastic deposits of the Reingraben Formation (9.6–30.0 m, layers Gro 9.6 m to Gro 30.0 m; Fig. 15.4) consist of a pronounced alternation of dark grey to black argillaceous shales (3–20 wt% $calcite_{equiv.}$; Supplementary material 15.1, bulk geochemistry) intercalated by rare allodapic limestone/dolomite layers (e.g., 79 wt% $calcite_{equiv.}$ at Gro 20.0 m; Gro 19.5 m to Gro 21.5 m); no macro-bioturbation is visible. The bivalve *Peribositra* is present in all ontogenetic stages and, in layer-/rock-forming quantities (lumachelle), mostly in articulated butterfly position.

15.3.2 Großau Quarry Section

The Großau quarry exposes an inverted 13-m-long section that can be easily correlated with distinct layers in the drill core (Figs. 15.4 and 15.5) and corresponding fossiliferous layers with dominant fossil taxa (Fig. 15.6). It comprises light grey limestones of the Reifling Formation (layers Gro 1 to Gro 25; drill core Gro 6.5 m

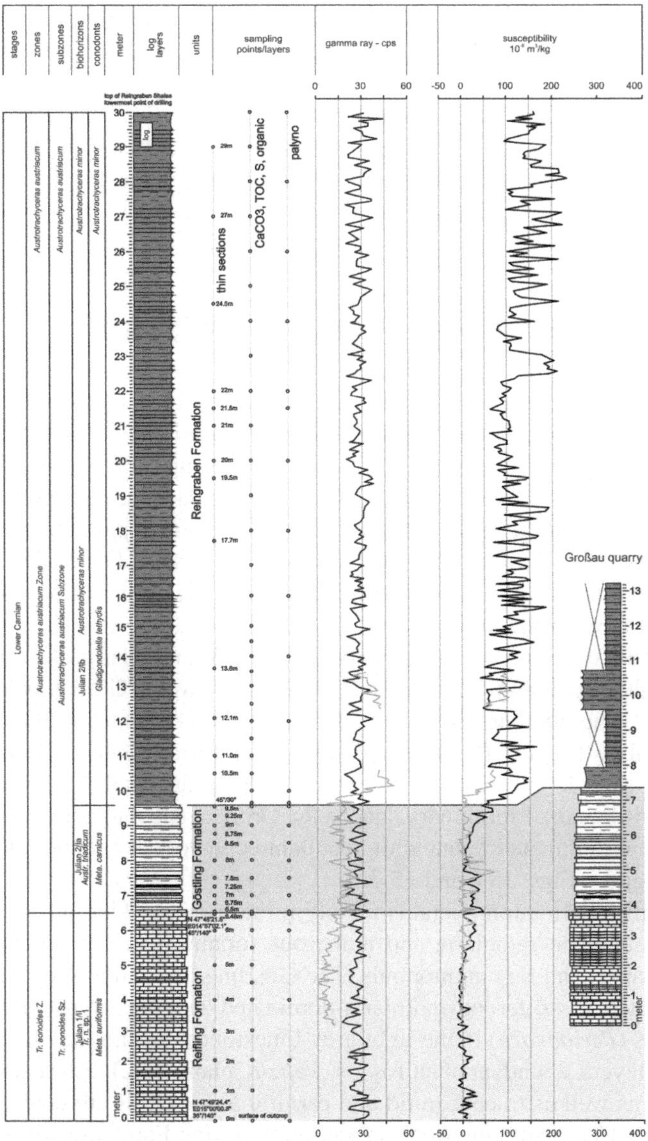

Fig. 15.4 Compiled logs of the Großau drill core (left) and the Großau quarry (right). From left to right: biostratigraphic information (stage, ammonoid and conodont zonal schemes); log of the Großau drill core (with layer numbers; drilling depth and lithological units); position of sampling layers/points (Gro) for (i) thin sections, (ii) carbonate ($CaCO_3$), total organic carbon (TOC) and sulfur (S) contents and biomarkers (organic), along with (iii) palynological (palyno) samples; gamma-ray and magnetic susceptibility logs (black: drill core logs; light gray: quarry logs); log of the Großau quarry section. Dark gray: Reifling Formation; light grey: Göstling Formation; white: Reingraben Formation. CPE, Carnian Pluvial Episode

to Gro 0.0 m (Figs. 15.7), the laminated and silicified limestones of the Göstling Formation (layers Gro 62 to Gro 61; drill core Gro 9.6 m to Gro 6.5 m), which are only slightly thicker (3.5 m) than in the drill core, and the lowermost 6 m of the Reingraben Formation (layers Gro 62 to quarry Gro 13.2 m; drill core roughly equivalent to Gro 15.5 m to Gro 9.6 m). Plant fossils indicating enhanced terrigenous input are abundant on the upper surface of numerous layers in the Göstling Formation (e.g., layers Gro 54, 60, 61).

15.4 Fossil Material from Drill Core and Quarry

Macrofossil material from the Upper Triassic Großau drill site was obtained from the drill core and the Großau quarry. Recent excavations (2023–2025) provided 237 macrofossil specimens of various taxonomic groups (Fig. 15.6). Most fossils are found in the lowermost two meters of the argillaceous Reingraben deposits (bivalves, ammonoids, entire fish) and the shaly layers separating the limestone beds of the Göstling Formation (plant fossils, trace fossils, fish scales; Fig. 15.7). Rock-forming masses of bivalves (posidoniid coquinas) occur within distinct levels in the Reingraben Formation.

The frequent occurrence of macrofossils (n = 273) and their preservation quality (entire, fragile morphologies) yield special insights into the Großau paleobiota and specific taxa. The number above is exclusive of the uncountable masses of rock-forming bivalves (mainly *Peribositra*). *Peribositra* forms coquinas and, in specific levels, meter-thick layers, hence representing the dominant benthic element at Großau (quarry and drill core). Preliminary data on the Großau paleobiota (excl. posidoniids) show a nekton-dominated fauna with abundant cephalopods (ammonoids) and fish (actinopterygians, Figs. 15.6 and 15.7).

The Reifling Formation contains mainly microfossils, which are visible in thin sections as abundant radiolaria and numerous foraminifera accompanied by filamentous bivalves and rare ammonoids. The Göstling Formation shows a change to finely laminated, fossiliferous (radiolaria-dominated) silicious limestones with abundant bivalves (*Peribositra*) in the light grey limestones. Within the black laminated more marly layers, abundant plant fossils (*Voltzia*, plants spp.), fish scales (no entire specimens), as well as trachyceratid and carnitid ammonoids are detected on layer surfaces (Fig. 15.6). Both the Reifling and the Göstling Formations feature distinct layers with intense bioturbation (e.g., quarry Gro 64, Gro 61). The lowermost Reingraben Formation exhibits the Großau *Konservat-Lagerstätte* with distinct levels of halobiid coquinas. Apart from this, ammonoids dominate in the basal part with austrotrachyceratids and paratrachyceratids, actinopterygian fishes, accompanied by frequent bromalites (Fig. 15.6).

The autochthonous macrobenthic community is dominated by countless posidoniid bivalves, in distinct layers and meter-thick coquina levels (Fig. 15.7). These small (0.1–3.2 cm), thin-shelled clams dominate throughout the section, both with sporadic presence and in rock-forming accumulations within distinct

Fig. 15.5 Thin sections of the prevailing lithofacies in the Reifling Formation, the Göstling Formation, and the Reingraben Formation from the Großau drill core. **a** Radiolarian-rich limestone, filamentous wackestone to packstone, Gro 0.0 m, basal Reifling Formation. **b** Radiolarian-rich limestone, filamentous wackestone to packstone, Gro 6.48 m, top Reifling Formation. **c** Laminated, silicious and radiolarian-rich limestone, wackestone to packstone, Gro 7.0 m, basal Göstling Formation. **d** Laminated, silicious and radiolarian-rich limestone, wackestone to packstone, Gro 9.0 m, top Göstling Formation. **e** Laminated, argillaceous marlstone with ammonoid shell (top), laminated wackestones to mudstones, Gro 10.5 m, basal Reingraben Formation. **f** Laminated, argillaceous marlstone with ammonoid shell (bottom), top is formed by an allodapic limestone layer, laminated wackestone to mudstone, Gro 27.0 m, basal Reingraben Formation. Scale bars: 1 mm

Fig. 15.6 Invertebrate, vertebrate, and plant members of the lower Carnian Großau paleobiota within the argillaceous Reingraben Formation (**a**–**g**) and the silicious Göstling Formation (**h**–**k**). **a** *Austrotrachyceras minor*, positive, lateral view, quarry layer Gro 10.0 m, NHMW 2025/0176/0001. **b** *Paratrachyceras haberfellneri*, positive, lateral view, quarry layer Gro 10.0 m, NHMW 2025/0176/0002. **c** *P. haberfellneri*, negative fragments, lateral view, quarry layer Gro 10.2 m, NHMW 2025/0127/0003. **d** *Anaptachus lunzensis*, lower jaw of *A. minor*, positive, lateral view, isolated specimen, with two *A. minor* fragments, quarry layer Gro 10.2 m, NHMW 2025/0127/0004. **e** *Peribositra* sp., juvenile specimens, accumulated in butterfly preservation, layer Gro 10.0 m, NHMW 2025/0127/0005. **f** *Peribositra* sp., juvenile and adult specimens, juveniles accumulated in butterfly preservation, layer Gro 10.2 m, NHMW 2025/0127/0006. **g** Palaeoniscid fish, almost complete, note paired pits of otoliths (dissolved), layer Gro 10.0 m, NHMW 2025/0127/0007. **h** *A. minor*, positive, lateral view, quarry layer Gro 52 (6.15 m), NHMW 2025/0127/0008. **i** *Peltopleurus* fish scale, lateral view, layer Gro 46 (5.55 m), NHMW 2025/0127/0009. **j** *Chondrites* on a trace fossil layer surface, layer Gro 54 (6.5 m), NHMW 2025/0127/0010. **k** Accumulated plant remains on layer surface, layer Gro 54 (6.5 m), NHMW 2025/0127/0011. Yellow line marks boundary of Göstling versus Reingraben findings. Scale bars: 1 cm

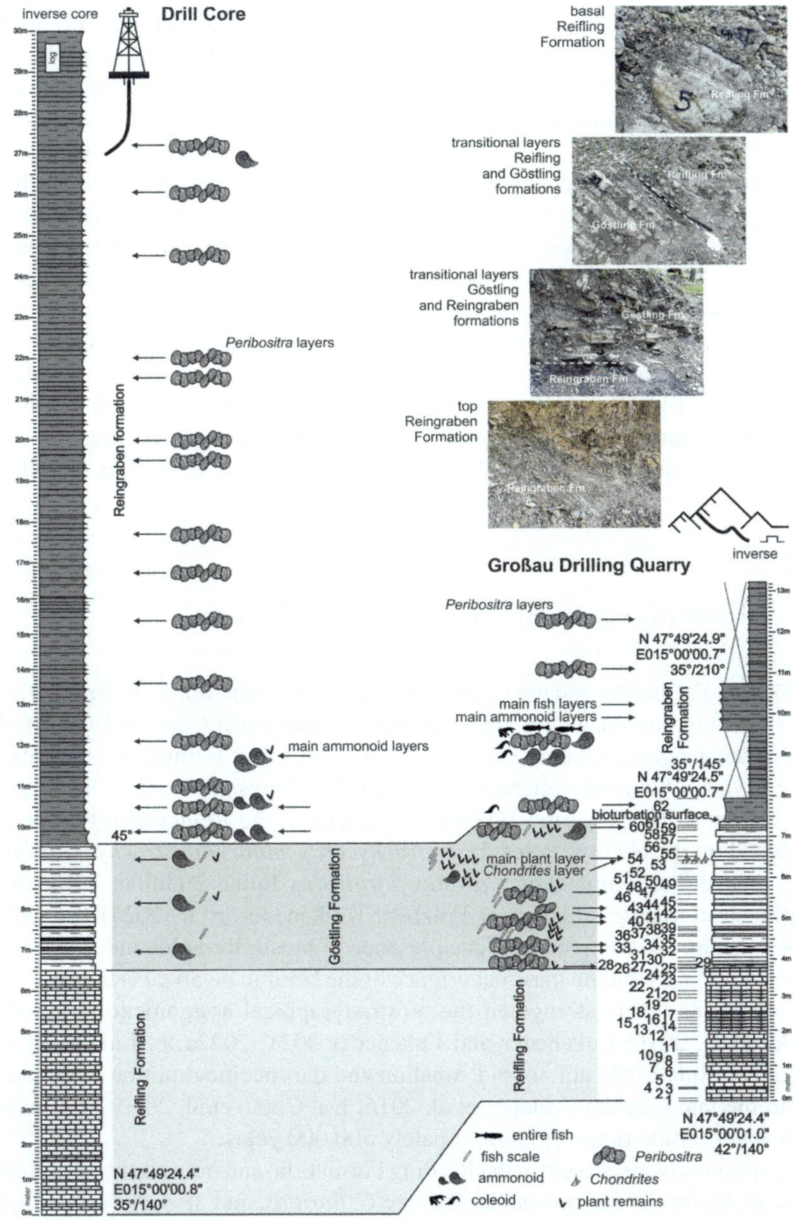

Fig. 15.7 Summary of the occurrence of macrofossils (fishes, ammonoids, bivalves, plant remains) in the Großau drill core (left) and the Großau quarry section (right). The distance between the Reifling/Göstling Formation boundary in the drill hole and in the quarry is 6 m. Images show the basal Reifling Formation, indicating the transitional intervals at the Reifling/Göstling Formations and Göstling/Reingraben Formations, and the topmost Reingraben Formation of the Großau quarry

layers. Cephalopods, especially ammonoids, dominate the nekton, followed by fish (actinopterygians), accompanied by frequent bromalites with coprolites and regurgitalites as well as allochthonous plant remains in the Göstling Formation.

The main faunal elements in the Reingraben deposits are ceratitid ammonoids *Austrotrachyceras minor* (Fig. 15.6a) and rare *Paratrachyceras haberfellneri* (Fig. 15.6b, c). These cephalopods are accompanied by anaptychi of trachyceratids (Fig. 15.6d), benthic bivalves *Peribositra* (Fig. 15.6e, f), abundant bromalites (coprolites and regurgitalites), and frequent actinopterygian fishes (entire and single scales, Fig. 15.6g). On layer surfaces of the Göstling Formation, rare ceratitid ammonoids with *Austrotrachyceras* (Fig. 15.6h), frequent actinopterygian fish scales (Fig. 15.6i), trace fossils with *Chondrites* (Fig. 15.6j), and, in distinct layers, accumulated plant remains (Coniferopsida) along with undetermined plant remains are visible (Fig. 15.6k). The dominating faunal element *Peribositra* (1–16 mm length) appears in all ontogenetic stages and, in layer/rock-forming quantities, mostly in double-valved butterfly position. The Großau collection is housed at the NHMW (Vienna).

15.5 Biostratigraphy of the Großau Drill Core

The Göstling Formation and the younger Reingraben Formation in the drill core span over a single ammonoid assemblage, representing the Lower Carnian *Austrotrachyceras austriacum* Zone (Julian 2, Fig. 15.2). The Göstling Formation is ascribed to the *A. triadicum* Subzone, whereas the Reingraben deposits belong to the higher *A. minor* Subzone. The lowermost, fossiliferous part of the Reingraben Formation is dated based on the occurrence of *Austrotrachyceras minor*, *Paratrachyceras haberfellneri*, and the accompanying *Carnites floridus* as Julian 2 (Julian 2/Ib), similar to the biostratigraphic situation at Polzberg (Lukeneder et al. 2024) and Scheiblingbach (Chap. 19, this volume). The presence of biostratigraphic index ammonoid members together with the mass occurrence of the benthic bivalve *Peribositra* within the fossiliferous layers strengthen the biostratigraphical assignment as Julian age (Mueller et al. 2016; Lukeneder and Lukeneder 2021, 2022a, b, Lukeneder et al. 2024). According to the ammonoid zonation and the specific duration of ammonoid zones in the late Carnian (Mueller et al. 2016; Dal Corso et al. 2020), the analysed section spans a time range of approximately 500,000 years.

No ammonoids are present in the Reifling Formation, and only trachyceratid specimens with *Austrotrachyceras patroclus*, rare *C. floridus*, and *Sirenites* sp. specimens are present in the Göstling Formation. *A. patroclus* is characteristic for the basal *Austrotrachyceras austriacum* Zone (*A. triadicum* Subzone, Julian 2/Ia). Conodonts are restricted to the lowermost part of both the Scheiblingbach and Polzberg sections, in deposits of the Reifling and Göstling Formations (*Gladigondolella tethydis*, *Gladigondolella malayensis malayensis*, *Gladigondolella* sp.; Fig. 15.2). The conodont association collected at the contemporaneous and corresponding top of the Reifling Formation at Polzberg and Scheiblingbach is represented by typical Julian forms, including *Paragondolella polygnathiformis*, *P. inclinata*, *P. foliata*, *P. tadpole*, *P. praelindae*, *Paragondolella* aff. *Par. foliata* (sensu Rigo et al. 2007), *Gladigondolella tethydis*, *G. malayensis malayensis*, and *'Mosherella' longnanensis*. The presence of *P. polygnathiformis* constrains the age of the samples to the Carnian, while *P. praelindae* together with *Paragondolella* aff. *Par. foliata* date the top of the Reifling Formation to the Julian 2 (Rigo et al. 2007, 2018).

15.6 Multi-proxy Data of the Großau Drill Core

Various geochemical and geophysical data have been recorded along the studied section. To avoid weathering effects, geochemical data are restricted to the drill core.

15.6.1 Gamma-Ray and Magnetic Susceptibility Logs

The gamma-ray (GR) and magnetic susceptibility (χ) logs of the studied drill core and the quarry section are displayed in Fig. 15.4. As expected, the GR log of the quarry section is characterized by low values in the Reifling limestone (<15 cps), intermediate values in the Göstling limestone with mudstone layers (~15 cps), and relatively high (generally 30–45 cps) in the Reingraben Formation (Fig. 15.4). In contrast, the GR log of the drill core shows minor differences between the drilled formations, as indicated by the mean values of the Reifling Formation (25.9 cps), Göstling Formation (26.1 cps), and the Reingraben Formation (29.8 cps).

The magnetic susceptibility values (SI) from the drill core are low in the Reifling Formation (2.3×10^{-6} m^3/kg), slightly higher in the Göstling Formation (22.2×10^{-6} m^3/kg), and significantly higher in the Reingraben Formation (113.8×10^{-6} m^3/kg) (Fig. 15.4). The highest values are recorded near the shale-rich top of the Reingraben Formation (max 222×10^{-6} m^3/kg at Gro 26.6 m).

15.6.2 Bulk Geochemistry – Carbonate Content, Total Organic Carbon, Sulfur, Rock–Eval Parameters

Bulk geochemical parameters are summarized in Supplementary material 15.1 (bulk geochemistry) and are plotted versus stratigraphic position in Fig. 15.8. The Reifling Formation is characterized by high carbonate contents (58–95 wt%, mean 81 wt% calcite$_{equiv.}$), except for a few shale-rich layers in its uppermost part (e.g., Gro 6.0 m, 34 wt% calcite$_{equiv.}$). TOC (<0.15 wt%) and sulfur contents (<0.01 wt%) are low. The Rock–Eval parameters S1 (<0.01 mg HC/g rock) and S2 (<0.05 mg HC/g rock) are very low. The HI varies between 16 and 50 mg HC/g TOC.

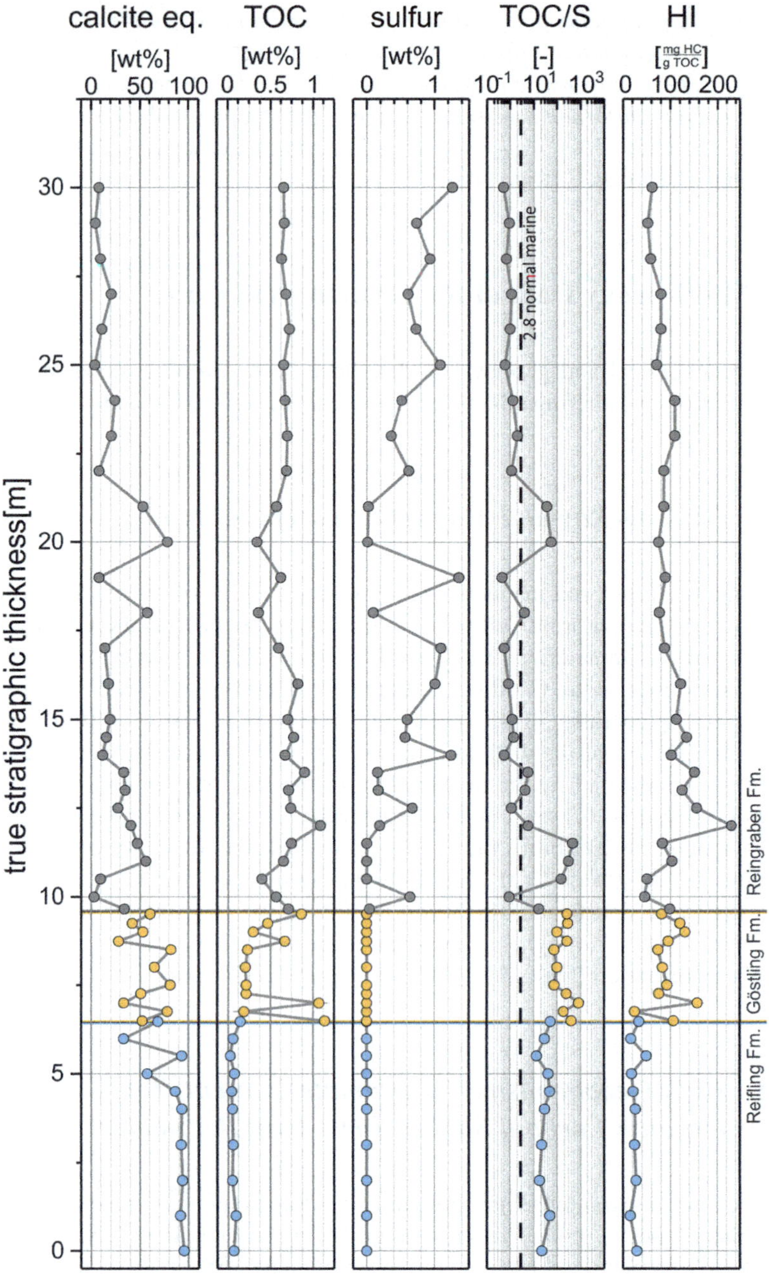

Fig. 15.8 Plot of bulk geochemical data (Supplementary material 15.1, bulk geochemistry). Columns from left to right: carbonate content (calcite eq.), total organic carbon (TOC) content, sulfur (S) content, TOC/S ratio, and hydrogen index (HI)

The overlying Göstling Formation is characterized by strongly varying carbonate contents (28–82 wt%, mean 57 wt% $calcite_{equiv.}$). The TOC content (0.19–1.14 wt%) is higher than in the Reifling Formation and shows a weak negative correlation with the carbonate content ($R^2 = 0.33$, $n = 11$). The sulfur content is as low as in the Reifling Formation, but S1 (<0.06 mg HC/g rock) and S2 (<1.69 mg HC/g rock) are considerably higher, which also results in higher HI values (24–157 mg HC/g TOC).

The Reingraben Formation is characterized by significantly lower carbonate contents (3–47 wt%, mean 25 wt% $calcite_{equiv.}$) but also includes some carbonate-rich intervals/layers (e.g., Gro 20.0 m, 79 wt% $calcite_{equiv.}$). The TOC content (0.34–1.01 wt%) is similar to that in the Göstling Formation, but the sulfur content is higher (max. 1.4 wt%). S1 (<0.18 mg HC/g rock) and S2 (<2.53 mg HC/g rock) are low, but higher than in the other formations. The HI ranges from 46 to 232 mg HC/g TOC. It peaks about 2.4 m above the base of the Reingraben Formation, from where it steadily decreases upwards.

Generally, low S2 values lead to biased T_{max} values in samples of all formations. The average T_{max} of samples with S2 > 0.2 mg HC/g rock is 437 °C. Similarly, production index (PI) values (0.00–0.17) are probably influenced by the very small S1 and S2 peaks. PI values for samples with S2 > 0.2 mg HC/g rock range from 0.02 to 0.12 (mean 0.06).

15.6.3 Organic Petrography—Vitrinite, Inertinite and Liptinite

Organic petrological investigations were restricted to six samples with relatively high TOC contents. While samples from the Reifling Formation were not investigated (<0.15 wt% TOC), two samples from the Göstling Formation (1.08 and 0.87 wt% TOC, respectively) and four samples from the Reingraben Formation (1.09–0.72 wt%) were selected for semi-quantitative maceral analysis. Maceral percentages are listed in the Supplementary material 15.1 (organic petrography) and are plotted in Fig. 15.9.

The organic matter of both samples (Gro 7.00 m, Gro 9.50 m) from the Göstling Formation is dominated by liptinite macerals. The relative amounts of macerals are given on a mineral matter free basis (relative to all measured macerals). Lamalginite (36–41 vol%) is the prevailing liptinite maceral. Liptodetrinite (12–16 vol%), telalginite (<9 vol%), and sporinite (<2 vol%) are present in smaller quantity. The lower sample (Gro 7.00 m) contains a very well-preserved and diverse assemblage of microfossils with liptinite characteristics (brown under reflected light and strong fluorecence under UV excitation), among them holothurian axial plates (Fig. 15.10, XII–XIV). The relative abundance of vitrinite (6–18 vol%) is lower than that of inertinite (19–26 vol%).

In two samples from the lower part of the Reingraben Formation (Gro 12.00 m, Gro 13.50 m), the liptinite content is elevated (≈80 vol%), which is mainly due to

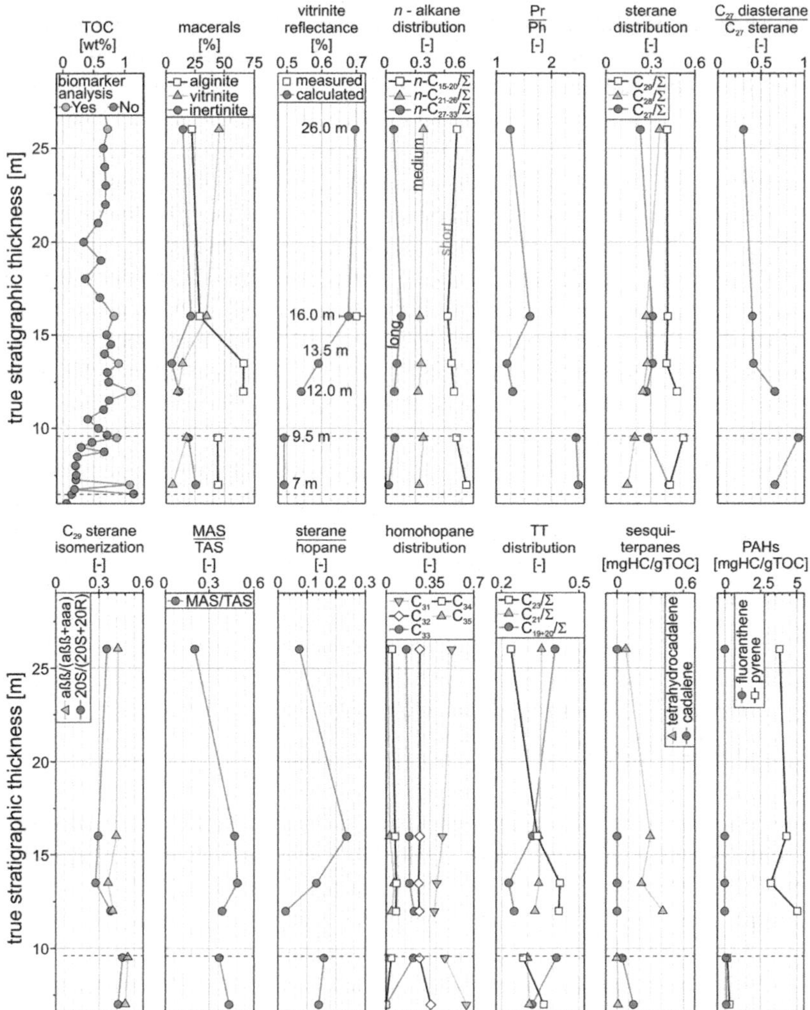

Fig. 15.9 Stratigraphic plots showing total organic carbon (TOC) content (Supplementary material 15.1, bulk geochemistry), maceral percentages (Supplementary material 15.1, organic petrography), and biomarker data (Supplementary material 15.1, hydrocarbon composition). Samples used for petrographic and biomarker analysis are highlighted in the TOC column. Maceral percentages are given as vol% (mmf, mineral matter free basis). C_{29} sterane isomerization (20S/(20S + 20R), ββ/(αα + ββ)), calculated and measured vitrinite reflectance, and monoaromatic/triaromatic steroid (MAS/TAS) ratio are indicative of thermal maturity. *n*-Alkane, sterane, and tricyclic terpane (TT) ratios provide information on OM source and depositional environment. The sterane/hopane ratio is indicative of eukaryotic to prokaryotic input. The abundance of sesquiterpanes (cadalene, tetrahydrocadalene) is indicative also for diterpenoid derivatives and the input of land plants. Polyaromatic hydrocarbons (PAHs, fluoranthene and pyrene) are shown as proxies for wildfires. Pristane/phytane (Pr/Ph), C_{27} diasterane/C_{27} sterane, and the distribution of $C_{31\text{-}35}$ αβ extended homohopanes are related to the depositional environment (e.g., redox conditions)

the high proportion of lamalginite (65 vol%). Higher up in the section (Gro 16.00 m, Gro 26.00 m), the proportions of liptinite (40 vol%) and lamalginite (≈ 25 vol%) are lower. The vitrinite (10–35 vol%) and inertinite (5–21 vol%) contents show the opposite trend compared to liptinite. Vitrinite reflectance of the sample Gro 16.00 m is 0.70 ± 0.05%Rr.

15.6.4 *Organic Geochemistry—*n*-Alkanes, Isoprenoids, Steroids, Hopanoids, and Aromatic Compounds*

The hydrocarbon composition was determined for the same samples used for organic petrographic analysis (Supplementary material 15.1, hydrocarbon composition). Characteristic biomarker ratios are plotted together with maceral percentages in Fig. 15.9. The extract yield is about 0.2 mg/g rock in samples from the Göstling Formation and the upper part of the Reingraben Formation, while it is significantly higher (0.68–0.97 mg HC/g rock) in the lower part of the Reingraben Formation. Extract yields correlate very well with the amount of S1 hydrocarbons ($R^2 = 0.95$).

The extracts of Göstling samples contain higher amounts of asphaltenes (avg. 28.2 wt%) and lower relative amounts of aliphatic (avg. 28 wt%) and aromatic hydrocarbons (avg. 4.5 wt%) than Reingraben samples (avg. 3.7 wt%; 41 wt%, 14.4 wt%, respectively). The relative proportion of NSO compounds is similar in both formations (~40 wt%; Supplementary material 15.1, hydrocarbon composition).

n-Alkanes and isoprenoids

In the Göstling Formation, *n*-alkanes (234, 1115 μg/g TOC) and isoprenoids (62, 103 μg/g TOC) are less abundant than in the Reingraben Formation (2370–7330 μg/g TOC, 272–1430 μg/g TOC, Fig. 15.11). Furthermore, the range of *n*-alkanes is smaller in the Göstling (n-$C_{15\text{-}28}$) than in the Reingraben Formation (n-$C_{13\text{-}33}$). Short chain *n*-alkanes ($\sum n$-$C_{15\text{-}20}/\sum n$-alkanes: 0.53–0.69) are more abundant than medium chain ($\sum n$-$C_{21\text{-}26}/\sum n$-alkanes: 0.28–0.32) and long chain *n*-alkanes ($\sum n$-$C_{27\text{-}33}/\sum n$-alkanes: 0.02–0.13). The carbon preference index (CPI; according to Bray and Evans 1961) is high in the lowermost sample (1.63) but significantly lower in the remaining samples (1.06–1.21). Generally, n-C_{17} is more abundant than n-C_{18}. The pristane/phytane (Pr/Ph) ratio is higher in the Göstling (≈2.5) than in the Reingraben Formation (≈1.25). The Pr/n-C_{17} ratio is high in the Göstling Formation (2.36 and 1.54, respectively) and significantly lower in the Reingraben Formation (1.18–0.86). The Ph/n-C_{18} is below 1 (0.96–0.58) in all samples.

Steroids

Steranes were identified in all samples. In contrast to the samples from the Reingraben Formation, which show widely varying concentrations (21–257 μg/g TOC), both Göstling samples (Gro 7.00 m, Gro 9.50 m) have similar concentrations (40 and 43 μg/g TOC). The distributions of regular steranes of all samples are dominated by

C_{27} steranes (C_{27} steranes/$C_{27\text{-}29}$ steranes: 0.41–0.52). The highest values occur in the uppermost Göstling and the lowermost Reingraben sample. The relative amount of C_{28} steranes increases continuously upwards (0.14–0.36). C_{29} steranes are abundant only in the lower part of the Göstling Formation (C_{29} steranes/ $C_{27\text{-}29}$ steranes: 0.42). In all other samples, the C_{29} steranes/$C_{27\text{-}29}$ steranes ratio varies between 0.23 and 0.31. Diasteranes are found in all samples. They are generally less abundant than their sterane counterparts. C_{27} Diasterane/C_{27} sterane ratios appear to increase in the Göstling Fm. (0.52–0.93). In the Reingraben Fm. they decrease upwards (0.66–0.30).

The 20S/(20S + 20R) isomerization of the C_{29} steranes is about 0.45 in the Göstling Formation and slightly lower in the Reingraben Formation (0.28–0.38). The average $\beta\beta/(\alpha\alpha + \beta\beta)$ isomerization ratio in the Göstling and Reingraben Formations are 0.48 and 0.40, respectively. Monoaromatic steroids (MAS) and triaromatic steroids (TAS) were detected in the aromatic fraction. The MAS/TAS ratio is generally between 0.37 and 0.49 and is lower (0.20) only in the uppermost sample (Gro 26.00 m).

Hopanoids

Hopanoids were identified in all samples in relatively high quantities (273–2952 μg/g TOC). Therefore, the sterane/hopane ratio is generally very low (0.03–0.24). The minimum values are observed in the Reingraben Formation (Gro 12.00 m and 26.00 m). The 22S/(22S + 22R) C_{31} hopane isomerization ratio has reached its equilibrium value (0.58 ± 0.01, Seifert and Moldowan 1980).

The C_{29} $\alpha\beta$ hopane is more abundant than C_{29} Ts hopane and C_{30} diahopane in the Göstling Formation. At the transition from the Göstling to the Reingraben Formation, C_{29} Ts and C_{30} diahopane abundances increase drastically and all three compounds occur in similar concentrations (samples Gro 12.00 m, Gro 13.50 m). Above, C_{29} Ts and C_{30} diahopane concentrations decrease at similar rates. The extended $\alpha\beta$ homohopanes of the Göstling Formation show an exponential decrease and $C_{34\text{-}35}$ is not present, whereas in the Reingraben Formation the decrease up to C_{35} is rather linear. The C_{29} $\beta\alpha$ hopane is present in the Reingraben Formation in small quantities but is below detection limit in the Göstling Formation. Similarly, C_{30} $\alpha\beta$ hopane and C_{30} $\beta\alpha$ moretane are more prominent in the Reingraben Formation. Gammacerane could not be identified in any sample.

Tricyclic terpanes (TT) are present in all samples (38.2–220 μg/g TOC). The ratio of $C_{19\text{-}20}$ TT over the sum of all extended tricyclic terpanes (ETT) is high in the Göstling Formation (0.41–0.45), low in the lower samples from the Reingraben Formation (0.25–0.23), and increases in the upper part of the Reingraben Formation (0.32–0.41). The C_{23} TT/ETT ratio shows the opposite trend with maximum values (0.42–0.43) in the lower part of the Reingraben Formation. The C_{21} TT/ETT ratio gradually increases upwards from 0.27 to 0.36. The concentration of C_{24} tetracyclic terpane is similar to that of C_{23} TT. Isomers of the tricyclic terpanes are generally not prominent and are absent in the Reingraben Formation.

Aromatic compounds

Trimethylnaphthalene and tetramethylnaphthalene were found in all samples (Fig. 15.12). Dimethylnaphthalene was identified in all samples, except Gro 9.50 m. Ethylnaphthalene and methylnaphthalene were not detected. In the Göstling Formation, phenanthrene (4.78 – 6.33 μg/g TOC) is more abundant than methylphenanthrenes (MP, 1.87–2.39 μg/g TOC), while dimethylphenanthrenes (DMP) are not present in quantifiable amounts. In the Reingraben Formation, MP (14.9–85.2 μg/g TOC) is more abundant than phenanthrene (4.40–52.0 μg/g TOC) and DMP (2.95–10.4 μg/g TOC).

The methylphenanthrene index (MPI-1; Radke and Welte 1983) increases upwards from 0.15 in the Göstling Formation to 0.49 in the uppermost sample in the Reingraben Formation. The calculated vitrinite reflectance (Rc; Radke and Welte 1983) increases from 0.49 to 0.70%Rc.

Dibenzothiophene could not be identified in any samples. However, methylated dibenzothiophenes were identified in all samples, albeit in lower concentrations than methylphenanthrenes (Fig. 15.12).

Diterpenoid derivatives were identified as well. Cadalene and simonellite are present in the Göstling Formation but absent in the Reingraben Formation. In contrast, tetrahydrocadalene, tetrahydroretene and retene are more abundant in the latter, where their concentration decreases upwards. The fernane 22,25,29,30-tetranorferna-5,7,9-triene could be identified only in the Göstling Formation and was more abundant in the upper sample (Gro 9.50 m, Hauke et al. 1992).

Fluoranthene was detected only in the Göstling Formation, while pyrene and its methylated homologues occur in all samples, but are particularly abundant in the Reingraben Formation. Alkylbenzenes (especially C_{21}-alkylbenzene) are abundant in the Reingraben Formation and rare in the Göstling Formation.

15.7 Discussion

15.7.1 The Großau Paleobiota Compared to the Scheiblingbach and Polzberg Paleobiota

The Großau drilling site is located between the Polzberg and Scheiblingbach sites. Paleogeographically these locations represent the Reifling Basin. Therefore the preliminary data presented here enable a comparison of the Großau paleocommunity with other contemporaneous paleobiotas in the Late Triassic Reifling Basin. Of special interest are the paleoenvironmental conditions during the Carnian Pluvial Episode, which caused the formation of lower Carnian *Konservat-Lagerstätten*. Special conditions such as stagnant environments are required for the formation of *Konservat-Lagerstätten* at the basal fossiliferous Reingraben Formation, characterized by their excellent preservation of invertebrate, vertebrate, and plant fossils (Seilacher 1970; Bernardi et al. 2011; Briggs 2014; Lukeneder and Lukeneder 2021, 2023).

Most of the fossil material from the Großau, Polzberg, and Scheiblingbach sections was found in the basal part of the argillaceous Reingraben deposits directly above the top of the Göstling Formation (Fig. 15.4). Hence, the recently collected fossil material from Großau can be directly correlated with the coeval (lower Carnian; Julian 2/Ib) fossil collections from the Polzberg and Scheiblingbach localities (Fig. 15.6). The basal Reingraben deposits were formed at an epipelagic to upper mesopelagic position at the sea floor of the intraplatform Reifling Basin (Lukeneder and Lukeneder 2021; Lukeneder et al. 2024a).

Within the three paleobiotas, benthic bivalves with *Peribositra* (uncountable, rock-forming numbers), accompanied by frequent other bivalve taxa (indet) and gastropods, clearly dominate throughout the basal intervals of the Reingraben Formation. Nektic members are dominated at all localities by ammonoids with the trachyceratids *Austrotrachyceras minor* and *Paratrachyceras haberfellneri* (Fig. 15.6). More sampling is needed at the Großau site to quantitatively compare the faunal elements on percentage level as calculated for the Scheiblingbach and Polzberg paleobiotas (see Lukeneder 2026, this volume). The bed-by-bed collection yielded 4,602 fossil remains from Scheiblingbach and 5,290 specimens from Polzberg (plus 6,397 in historical collections), whereas only 237 fossils have been detected in Großau to date.

Fossil communities from all three sites – Großau, Scheiblingbach, and the Polzberg *Konservat-Lagerstätte* – show a characteristic preservation of benthos (epifaunal and infaunal) along with nektic taxa such as fish, ammonoids and belemnoids (Fig. 15.6). The complete preservation of bivalves, frequently found in butterfly position (still articulated, little fragmentation), ammonoids together with in situ aptychi (= lower jaws) in or near body chambers of trachyceratid ammonoids, and entire belemnoid specimens with arm-hook crowns indicate autochthonous deposition of the animals or their remains, with minimal or no post-mortem drift or transport.

Recent findings of thousands of new fossils from the comparable lower Carnian localities (Großau, Scheiblingbach, Polzberg), including well-preserved vertebrate, invertebrate, and plant fossils of the *Konservat-Lagerstätten*, mirror the trophic conditions of the paleo-food web at the time of deposition (Seilacher 1970; Briggs 2014; Lukeneder and Lukeneder 2021, 2023). At the time of deposition of the basal Reingraben Formation, oxygen-depleted bottom water conditions prevailed. Such conditions at the sediment–water boundary and within the sediment reduce or prevent the activity of benthic scavengers on fragmenting or decaying shells or carcasses lying on the seafloor (Tintori 1992; Furrer 2003; Beardmore and Furrer 2016; Lukeneder and Lukeneder 2021, 2023, Lukeneder et al. 2024a, b). The presence of specific faunal elements as the lungfish *Tellerodus* and the frequent conchostracan shells (both believed to be fresh- to brackish water inhabitants) indicate a sporadic influx of freshwater or sedimentation from surrounding shallow-water or terrestrial areas into the marginal areas of the restricted Reifling Basin. Plant remains with abundant foliated trunks of the Coniferophyta *Voltzia* and frequent, partly accumulated plant fragments support the interpretation of sporadic freshwater influx into the otherwise fully marine basin.

In marginal areas of the Reifling Basin, represented by the Scheiblingbach-Großau-Polzberg deposits, calm and stagnant environments with oxygen-depleted conditions were hostile to scavengers at the sea floor (Furrer 2003). This created the prerequisites for excellent preservation of entire fish carcasses, in situ buccal masses, cephalic cartilages, muscle scars, and ink sacs in the lower Carnian (Julian 2/Ib).

15.7.2 The Paleoenvironment of the Großau Drill Site Deposits

The sedimentary succession exposed in the Großau drill core and the Großau quarry was formed in a relatively deep normal marine environment within the intraplatform Reifling Basin (Lukeneder and Lukeneder 2021, 2023). This interpretation is based on the dominance of macro- and microfossils with benthic posidoniids, a diverse nektic fauna with abundant ammonoids and actinopterygiid fishes, and a microfacies dominated by radiolarians and filamentous bivalves.

Within the lower Carnian succession, the facies changed from silicious bioturbated Reifling limestone, over silicious laminated Göstling deposits, to argillaceous shales and marls with frequent freshwater influx of the fossiliferous Reingraben Formation. The general trend from carbonate to clastic sediments is also reflected by the obtained gamma-ray (GR) and magnetic susceptibility (MS) logs, both from the drill core and the Großau quarry (Fig. 15.4).

The bottom water was well-oxygenated during deposition of the Reifling limestone and strongly restricted (dysoxic to anoxic) during deposition of the Göstling and Reingraben Formations in the entire Reifling Basin. Strongly oxygen-depleted conditions are based mainly on the presence of fully laminated sediments (without or with scarce bioturbation), the preservation of frequent entire fish carcasses with fragile skeletons, masses of double-valved posidoniid bivalves, accumulated plant debris and fragments in distinct layer surfaces, the presence of framboidal pyrite, and the absence of sessile benthic organisms (see Lukeneder et al. 2020; Lukeneder and Lukeneder 2021, 2022a, b).

The Pr/Ph ratio is a widely used redox parameter but is also influenced by the input of transported land plants (Didyk et al. 1978). In the case of the Großau drill core, a decrease in the Pr/Ph ratio from the Göstling Formation ($\approx$2.4) to the Reingraben Formation (1.2–1.6; Fig. 15.9) probably reflects greater oxygen restriction during deposition of the Reingraben Formation. Since the formation of diasteranes is favoured under oxic conditions and catalysed by clay minerals (Rubinstein et al. 1975; van Kaam-Peters et al. 1998; Peters et al. 2005; Killops and Abbott 2022; Wang et al. 2024), stronger oxygen restriction during deposition of the Reingraben Formation is also supported by lower C_{27} diasterane/C_{27} sterane ratios in the Reingraben (0.66–0.30) versus the Göstling Formation ($\approx$0.8; Fig. 15.9). Finally, stronger oxygen restriction is also indicated by a change in the distribution of C_{31} to C_{35} $\alpha\beta$ extended homohopanes (Rullkötter and Philip 1981). This is characterized by an exponential decline in concentrations with increasing carbon number in the Göstling Formation and by a linear decline in the Reingraben Formation.

The Polzberg succession within the Reifling Basin was marine with intermittent limited freshwater input (Lukeneder and Lukeneder 2021, 2022a, b). The freshwater input probably caused sulphate-limited conditions, which are responsible for the (very) high TOC/S ratios (Berner 1984) in the Göstling Formation and in some parts of the Reifling Formation (Fig. 15.8). Sulphate-limited conditions also agree with the absence of dibenzothiophene (Hughes et al. 1995). Episodical freshwater input is also indicated by the abundance of C_{21} tricyclic terpane (Zumberge 1987; Tao et al. 2015; Xiao et al. 2024).

Low energy in the bottom water system with limited or absent bottom water currents on the sea floor, along with anaerobic to dysaerobic conditions, led to the extraordinary preservation of the Großau paleobiota in the lowermost part of the Reingraben deposits (Großau *Konservat-Lagerstätte*), a situation well known from the lower Carnian Polzberg and Scheiblingbach equivalents. The geochemical results for comparable Polzberg and Scheiblingbach localities (Lukeneder et al. 2024a, see Weitz et al. 2026, this volume), along with the laminated fabric and facies, abundant organic matter, and high amounts of sulfur, indicate that the assemblage was deposited under conditions of intermittent oxygen depletion associated with stable water masses. These analyses suggest an environment with fluctuating carbonate productivity or oscillating siliciclastic and terrestrial input (Fig. 15.4).

15.7.3 Thermal Maturity

The average T_{max} (437 °C) and PI (0.07) of samples with S2 values > 0.2 mg HC/g TOC indicate early oil window maturity, which is also supported by the vitrinite reflectance of a sample from the Reingraben Formation (Gro 16.00 m; 0.70 ± 0.05% Rr; Peters and Cassa 1994). Early oil window maturity is further supported by the extent of isomerization reactions. The 22S/22(S + R) isomerization ratio of C_{31} 17α,21β-hopanes has reached its equilibrium value (0.58 ± 0.01, Seifert and Moldowan 1980), while the C_{29} sterane 20S/20(S + R) and ββ/(αα + ββ) isomerization ratios have not (avg. 0.37 and 0.43, respectively; Fig. 15.9).

Sterane isomerization values suggest slightly higher maturity in the Göstling Formation (20S/20(S + R): 0.43–0.46; ββ/(αα + ββ): 0.48 –0.50) than in the Reingraben Formation (20S/20(S + R): 0.33 ± 0.05; ββ/(αα + ββ): 0.40 ± 0.04). However, given the low thickness of the succession, the differences are more likely related to facies variations than to higher maturity (Seifert and Moldowan 1981; Mackenzie et al. 1983; Cornford et al. 1986; Peters et al. 2005; Weitz et al. 2026, this volume, Fig. 15.9). This fits with the observation that MPI-1 and the corresponding calculated vitrinite reflectance (Rc; Radke and Welte 1983) show an upward increase in maturity (0.49–0.69%Rr; Fig. 15.9). Overall, an early oil window maturity is very likely and also supported by a low MAS/TAS ratio (0.20–0.49; Mackenzie 1984; Hoffmann et al. 1984; Peters et al. 1990, 2005; Fig. 15.9).

15.7.4 Organic Matter Sources and Kerogen Type

Organic geochemical and organic petrographic data enable conclusions to be drawn on the source of organic matter. The *n*-alkane distribution is dominated by the short (*n*-C_{15-20}), over mid (*n*-C_{21-26}) and long (*n*-C_{27-33}) chain *n*-alkanes in all samples (Fig. 15.9). This distribution can be indicative of marine organic matter and/or elevated maturity (Peters et al. 2005). Although a minor maturity effect cannot be completely ruled out, the value of the the C_{31} $\alpha\beta$(R)/C_{30} $\alpha\beta$ ratio (≥ 0.25) also suggests the predominance of marine organic matter (Peters et al. 2005). High concentrations of *n*-C_{17} and the absence of methylhopanes (Gelpi et al. 1970; Neunlist and Rohmer 1985; Zundel and Rohmer 1985), in combination with prevalent lamalginite macerals, clearly indicate a major contribution of algae to the organic matter in the Göstling Formation. Their relative contribution increases in the lower five meters of the Reingraben Formation before dropping.

High amounts of algal material are also indicated by sterane distributions, which are dominated in all samples by C_{27} sterane (Fig. 15.9), indicative of marine organic matter (Huang and Meinschein 1979; Peters et al. 2005). The highest relative amounts were recorded in the uppermost part of the Göstling and the lower part of the Reingraben Formation. Similarly, the relative amount of C_{23} tricyclic terpane, which is derived mainly from marine algae (Moldowan et al. 1983; Zumberge 1987; Simoneit et al. 1990; Grande et al. 1993; Greenwood et al. 2000; Tao et al. 2015; Philp et al. 2021), peaks in the lower part of the Reingraben Formation. A positive correlation between the relative amount of C_{23} tricyclic terpane and lamalginite ($R^2 = 0.69$) supports the algal origin of C_{23} tricyclic terpane (Weitz et al. 2026, Chap. 23, this volume; Fig. 15.9).

A relatively high terrigenous contribution to the organic matter can be identified based on a high CPI (caused by abundant *n*-C_{27} and *n*-C_{31}, Peters et al. 2005) and a high relative abundance of C_{29} steranes for the lower sample from the Göstling Formation (Gro 7.00 m). Interestingly, in addition to abundant inertinite macerals (Fig. 15.9), this sample also contains a diverse phytoplankton assemblage (Fig. 15.10).

C_{19-20} tricyclic terpanes are usually associated with higher plants (Aquino Neto et al. 1982). Accordingly, the trend of the C_{19-20} TT/ETT ratio shows a relatively high terrigenous input for the Göstling Formation, a drastic decrease in the lower Reingraben Formation, and a gradual increase in the upper part of the Reingraben Formation (Fig. 15.9). This trend is also supported by HI values, which decrease gradually above a depth of 12 m, and by an upward increase in vitrinite macerals.

In addition, sesquiterpanes (tetrahydrocadalene, cadalene) and diterpenoid derivatives (simonellite, tetrahydroretene, retene) prove the contribution of higher plants and gymnosperms, respectively. Like fernanes, simonellite and cadalene are present only in the Göstling Formation. In comparison, tetrahydrocadalene, tetrahydroretene, and retene values are higher in the Reingraben Formation and continually decrease above sample Gro 12.00 m (Simoneit 1986; Alexander et al. 1988; Peters et al. 2005). Generally, sterane/hopane ratios are low (average 0.24) and point to intense

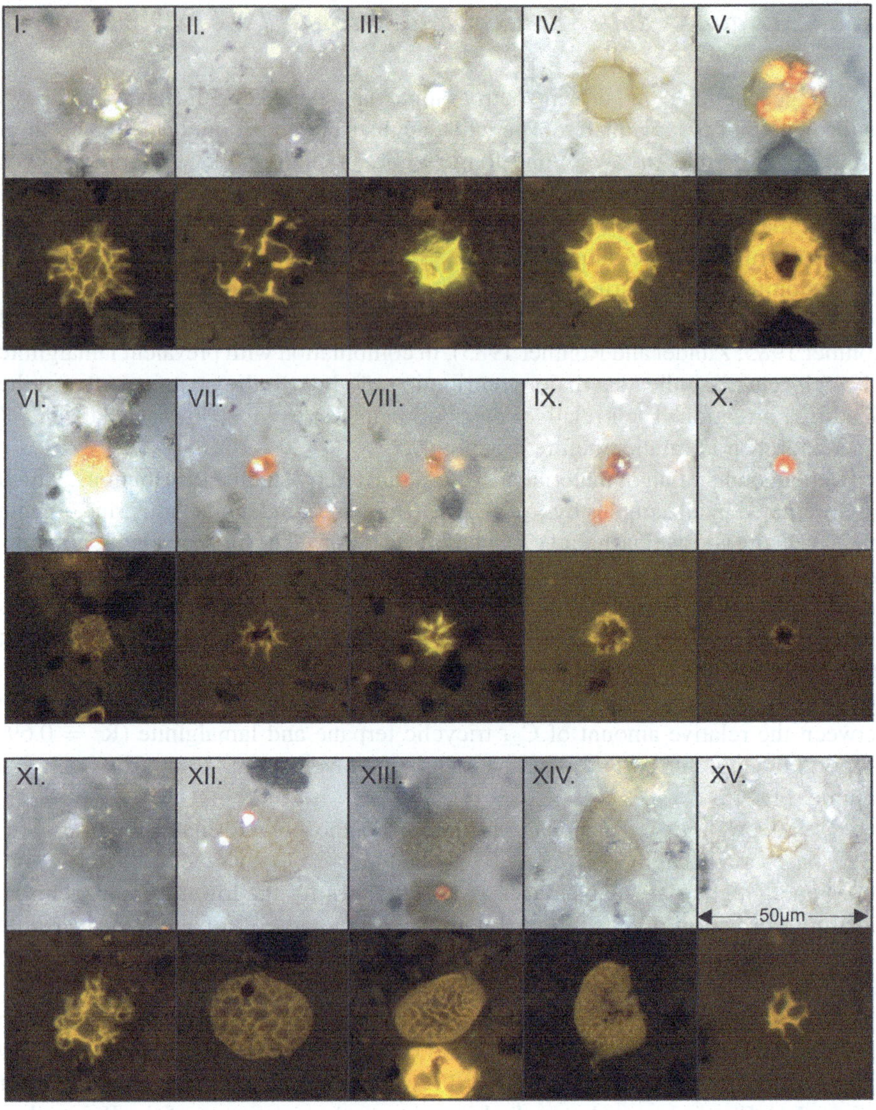

Fig. 15.10 Microphotographs of microfossils with preserved organic matter exhibiting liptinite characteristics (strong fluorecence under blue light irradiation); the scale is the same in all images. XII and XIII may be holothurian axial plates (e.g., Schlumberger 1890). The remaining photos show alginite macerals representing different types of planktonic algae

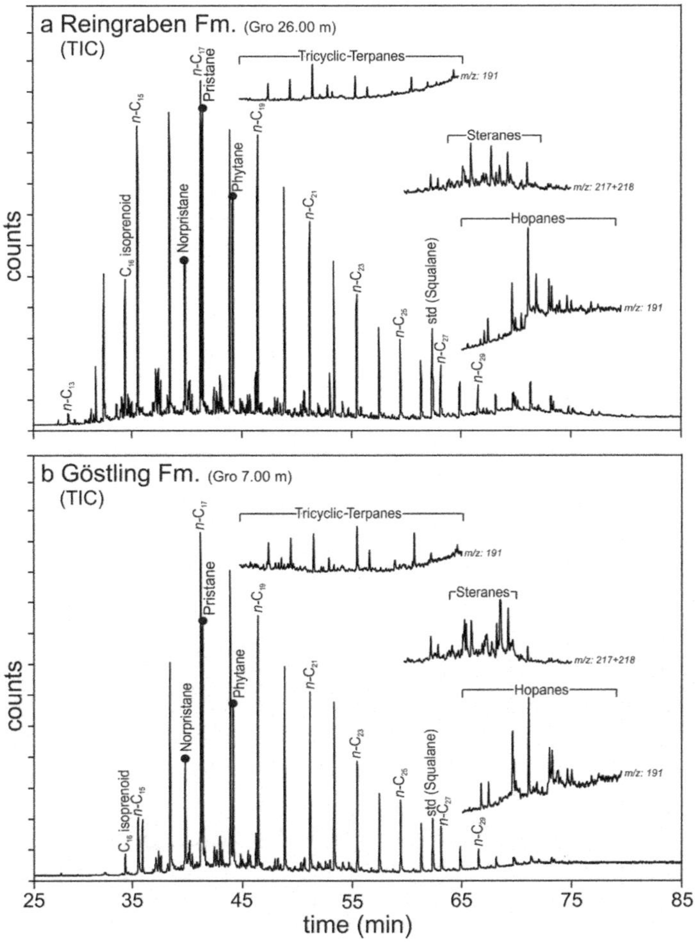

Fig. 15.11 Total ion chromatogram of the aliphatic fraction, as well as the typical mass-to-charge ratios for tricyclic terpanes, steranes, and hopanes, of sample Gro 7.00 m (**a** Göstling Formation) and Gro 26.00 m (**b** Reingraben Formation)

reworking of organic matter by prokaryotes (archaea and/or bacteria). This suggests that the HI of the primary organic matter might have been higher (Peters et al. 2005).

The HI of the Reifling Formation is ≤50 mg HC/g TOC and classifies the organic matter as Type IV kerogen. The average HI of the Göstling and Reingraben Formations is 97 mg HC/g TOC, implying the presence of Type III kerogen. Only the sample Gro 9.60 m (232 mg HC/g TOC) may be classified as Type II-III kerogen. Similar low HI values have been determined for samples from the Polzberg site (Lukeneder et al. 2024). Considering the excellent preservation of fossils, the abundance of liptinite macerals, and the early oil window maturity, the HI values in the Göstling and Reingraben Formations are surprisingly low. Probably, apart from microbial reworking,

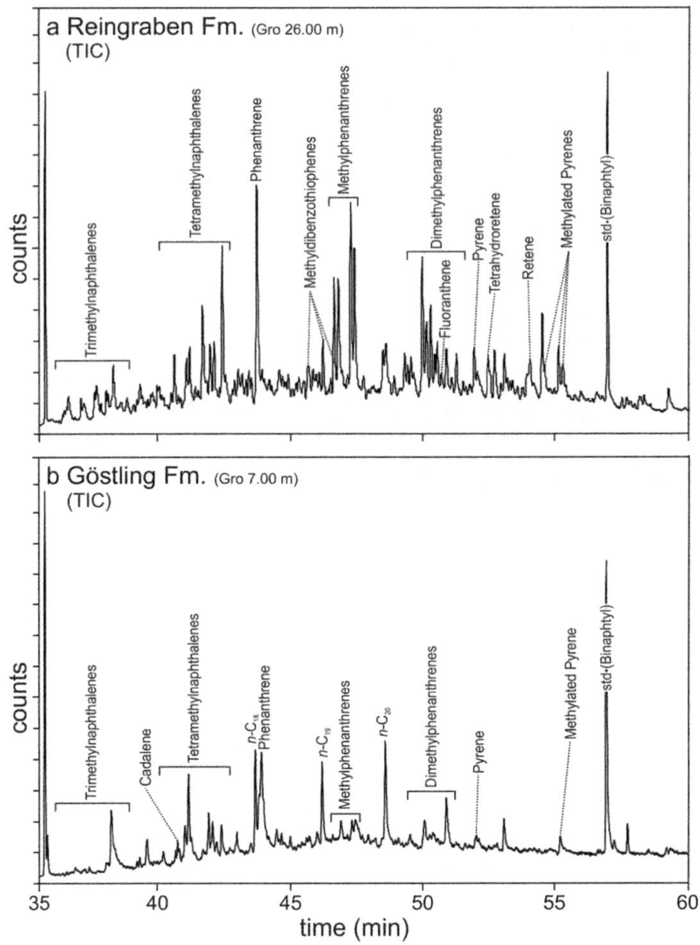

Fig. 15.12 Total ion chromatogram of the aromatic fraction. **a** sample Gro 7.00 m, Göstling Formation. **b** Gro 26.00 m, Reingraben Formation

the low HI values in both sites reflect partial organic matter oxidation within the water column.

15.8 Conclusions

A new drill site, recent excavations, and data on the geophysics and geochemistry from the Großau drill core obtained during fieldwork in 2023 and 2024 are presented. Preliminary data based on 237 new macrofossils and numerous thin sections for microfacies and microfossils from the Late Triassic (early Carnian) provide new details and insights into a new paleocommunity, introducing the Großau *Konservat-Lagerstätte*. We show that the Großau drill site yields information about paleoenvironmental conditions in an intermediate position between Polzberg and Scheiblingbach in the Reifling intraplatform basin from the Carnian Pluvial Episode. The Großau section with Reifling Formation, Göstling Formation, and Reingraben Formation (with *Konservat-Lagerstätte*) is equivalent in age and composition of detected paleobiota (verterbrates, invertebrates, plants) to the well-known Polzberg *Konservat-Lagerstätte* and the newly introduced Scheiblingbach *Konservat-Lagerstätte* (Lukeneder 2026, this volume) in the same marine basin. The basal sediments of the Reingraben Formation were deposited at an epipelagic to upper mesopelagic position on the sea floor of the NE-SW elongated Reifling Basin, periodically disconnected from oxygenated bottom currents. Geochemical proxies show that oxygen depletion reached a maximum during the depositon of the lower part of the Reingraben Formation. Lithologies, thickness of units, and the content of paleobiota are almost identical between the Großau drilling site and the Scheiblingbach and Polzberg localities. Although preliminary, the new Großau fossil biota provides new insights into the Late Triassic lower Carnian marine ecosystem, deposited during the Carnian Pluvial Episode. The excellent preservation of fragile skeletons and rare to missing bioturbation point to unique paleoenvironmental conditions in the lower Carnian dysoxic bottom water along the basinal borders of the Reifling Basin. The Großau drill site and the Polzberg site are assumed to be located on the northern slope of the Reifling Basin, whereas the Scheiblingbach site was situated on the southern slope. The periodical freshwater influx with enhanced transport of terrigenous sediment and organic matter from the continental realm and the surrounding Wetterstein platforms into basinal areas is supported by biomarker data and bulk geochemical data. This shows the evolution from normal marine conditions with good oxygen availability (Reifling Formation), over periodically sub- to dysoxic conditions (Göstling Formation), to a more siliciclastic-dominated system with an initially relatively increased contribution of marine organic matter (relative decrease of terrigenous organic matter) and more reducing conditions (Reingraben Formation). The thermal overprint of the sedimentary succession resulted in early oil window maturity. Future work is planned on various topics such as bed-by-bed fossil collecting along with additional and more detailed analyses on geochemistry or geophysics, palynology, and organic carbon content.

Acknowledgments The author is particularly grateful to the owners of the outcrop area, Felix Pöchhacker and his family, for drilling, digging, and sampling permission. The first author thanks Petra Lukeneder (Gablitz) and Anton Englert (Vienna) for their help with cleaning outcrops and sampling fossils. The work was done within the framework of and financially supported by projects of the Federal Government of Lower Austria (K3-F-964/001-2020 and K3-F-964/002-2023, Department Science and Research; headquarters in St. Pölten); the Austrian Academy of Sciences (headquarters in Vienna, project Polzberg Lukeneder 2021) represented by the National Committee for Geo/Hydro Sciences (Earth System Sciences Programme); and the Triassic project "Fenster in die Karnische Krise" of the Society of Friends of the Natural History Museum Vienna (Freunde des Naturhistorischen Museums Wien, CPE Lukeneder 2023). Anton Fürst and Anton Englert (both Vienna) are acknowledged for technical support (thin sections). The manuscript greatly benefited from valuable comments by Dragana Životić (Belgrade) and Alexander Zdravkov (Sofia), the production editor Ashok Arumairaj (Taramani), the in-house series editor Kathleen A. Ritterbush (Utah), the publishing editor Simon Shah-Rohlfs and the production editor Antje Endemann (both Heidelberg).

Author Contribution AL collected and prepared new material, designed research, provided documentation on fossil material. AL basic work, biostratigraphy, lithology, geophysics; preparation of the main manuscript. JW, RS, DM geochemistry, organics.

Funding Open access funding was enabled and the projects financially supported by the Federal Government of Lower Austria (K3-F-964/001-2020 and K3-F-964/002-2023, Department Science and Research; headquarters in St. Pölten); the Austrian Academy of Sciences (headquarters in Vienna, project Polzberg Lukeneder 2021) represented by the National Committee for Geo/Hydro Sciences (Earth System Sciences Programme); the Triassic project "Fenster in die Karnische Krise" of the Society of Friends of the Natural History Museum Vienna (Freunde des Naturhistorischen Museums Wien, CPE Lukeneder 2023); the Natural History Museum Vienna; and the Technical University of Leoben (Chair of Energy Geosciences) (to AL). The author is responsible for the contents of this publication. The funder had no impact on conceptualization, design, data collection, analysis, decision to publish, or preparation of the manuscript.

Competing interests All authors stated that they have no competing interests.

Authors' contributions All authors stated that they have equally contributed to their chapters, read, edited and agreed to submit the final versions of their contributions.

Availability of data and materials All sources of information are provided in the text. Raw data related to the fossil material from Austria are available from the corresponding authors and Alexander Lukeneder upon request.

Additional information The article is part of the Special Volume: *Late Triassic Konservat-Lagerstätten within the Carnian Pluvial Episode. 150 Years of Scientific Excavations in Austria.*

Artificial Intelligence (AI) The chapter was produced without the use of AI.

Language The entire volume uses American English reviewed by a native scientist.

Received 2025.

Published online 2026.

References

Abel O (1906) Fossile Flugfische. Abh K-K Geol RA 56:1–88. https://doi.org/10.5962/bhl.title.61007

Alexander R, Kagi RI, Toh E, van Bronswijk, W. (1988) The use of aromatic hydrocarbons for assessment of thermal histories in sediments. In: The North West Shelf, Australia. Proceedings of the Petroleum Exploration Society of Australia, Perth (P. G. Purcell and R. R. Purcell, eds.), pp 559–562

Alverà G, Dal Corso J, Chu D, Cruciani G, Roghi G, Caggiati M, Song H, Song H, Tian L, Du Y, Haserbek T, Gianolla P (2026) Episodic aridification at the onset of the Carnian Pluvial Episode in Western Tethys. Glob Planet Change 256:105192. https://doi.org/10.1016/j.gloplacha.2025.105192

Andrew C, Hoew P, Paul CRC, Donovam SK (2010) Fatally bitten ammonoids from the lower Lias Group (Lower Jurassic) of Lyme Regis. Dorset Proc Yorks Geol Soc 58(2):81–94. https://doi.org/10.1144/pygs.58.1.276

Andrew C, Howe P, Paul C (2015) Fatally bitten ammonites from septarian concretions of the 'Marston Marble' (Lower Jurassic, Sinemurian), Somerset, UK, and their taphonomy. Proc Geol Assoc 126(1):119–129. https://doi.org/10.1016/j.pgeola.2014.10.006

Aquino Neto FR, Restle A, Connan J, Albrecht P, Ourisson G (1982) Novel tricyclic terpanes (C19, C20) in sediments and petroleums. Tetrahedron Lett 23(19):2027–2030. https://doi.org/10.1016/S0040-4039(00)87251-2

Allison PA (1988) Konservat-Lagerstätten: cause and classification. Paleobiology 14:331–344

Argyriou A, Clauss M, Maxwell EE, Furrer H, Sánchez-Villagra MR (2015) Exceptional preservation reveals gastrointestinal anatomy and evolution in early actinopterygian fishes. Sci Rep 6:18758. https://doi.org/10.1038/srep18758

Arthaber G (1896) Die Cephalopodenfauna der Reiflinger Kalke. Beitr Paläontol Geol Österr-Ungarn Orient 10:1–112. https://doi.org/10.5962/bhl.title.45307

Beardmore S, Furrer H (2016) Taphonomic analysis of *Saurichthys* from two stratigraphic horizons in the Middle Triassic of Monte San Giorgio, Switzerland. Swiss J Geosci 109:1–15. https://doi.org/10.1007/s00015-015-0194-z

Bernardi M, Avanzini M, Bizzarini F (2011) Vertebrate fauna from the San Cassiano Formation (early Carnian) of the Dolomites region. Geo Alp 8:122–127

Berner RA (1984) Sedimentary pyrite formation: an update. Geochim Cosmochim Acta 48:605–615. https://doi.org/10.1016/0016-7037(84)90089-9

Brandner R, Poleschinski W (1986) Stratigraphie und Tektonik am Kalkalpensüdrand zwischen Zirl und Seefeld in Tirol. Jber Mitt Oberrhein Geol Ver 68:67–92. https://doi.org/10.1127/jmogv/68/1986/67

Bray EE, Evans ED (1961) Distribution of n-paraffins as a clue to recognition of source beds. Geochim Cosmochim Acta 22:2–5. https://doi.org/10.1016/0016-7037(61)90069-2

Briggs DEG (2014) Konservat-Lagerstätten 40 years on: the exceptional becomes *mainstream*. In: Laflamme M. Schiffbauer JD, Darroch SAF (eds) Reading and writing of the fossil record: preservational pathways to exceptional fossilization. Paleontological Society Papers, vol 20, pp 1–14. https://doi.org/10.1017/S1089332600002771

Brinkmann W (1982) Die Ichthyosaurier (Reptilia) aus der Grenzbitumenzone (Mitteltrias) des Monte San Giorgio (Tessin, Schweiz) – neue Ergebnisse. Vierteljahresschr Naturforsch Ges Zürich 143(4):165–177

Brinkmann W (1997) Die Ichthyosaurier (Reptilia) aus der Mitteltrias des Monte San Giorgio (Tessin, Schweiz) und von Besano (lombardei, Italien) – der aktuelle Forschungsstand. Vierteljahresschr Naturforsch Ges Zürich 142(2):69–78

Broda K, Marynowski L, Rakociński M, Zatoń M (2019) Coincidence of photic zone euxinia and impoverishment of arthropods in the aftermath of the Frasnian-Famennian biotic crisis. Sci Rep 9:16996. https://doi.org/10.1038/s41598-019-52784-4

Cohen KM, Finney SC, Gibbard PL, Fan, J-X (2013) The ICS international chronostratigraphic chart. Episodes 36:199–204. https://doi.org/10.18814/epiiugs/2013/v36i3/002

Cornford C, Needham, CEJ, Walque L de (1986) Geochemical habitat of North Sea oils. In: Spencer AM (ed) Habitat of hydrocarbons on the norwegian continental shelf, pp 39–54

Cross SRR, Ivanovski N, Duffin CJ, Hildebrandt C, Parker A, Benton MJ (2018) Microvertebrates from the basal Rhaetian Bone Bed (latest Triassic) at Aust Cliff, S.W. England. Proc Geol Assoc 129:635–653. https://doi.org/10.1016/j.pgeola.2018.06.002

Dal Corso J, Bernardi M, Sun Y, Song H, Seyfullah LJ, Preto N, Gianolla P, Ruffell A, Kustatscher E, Roghi G, Merico A, Hohn S, Schmid, AR, Marzoli A, Newton RJ, Wignall PB, Benton MJ (2020) Extinction and dawn of the modern world in the Carnian (Late Triassic). Sci Adv 6(38):eaba0099. https://doi.org/10.1126/sciadv.aba0099

Dal Corso J, Ruffell A, Preto N (2018) The Carnian Pluvial Episode (Late Triassic): new insights into this important time of global environmental and biological change. J Geol Soc Lond 175:986–988. https://doi.org/10.1144/jgs2018-185

Didyk BM, Simoneit, BRT, Brassell SC, EglintonG (1978) Organic geochemical indicators of palaeoenvironmental conditions of sedimentation. Nature 272(5650):216–222. https://doi.org/10.1038/272216a0

Dzik J, Sulej T (2007) A review of the early Late Triassic Krasiejów biota from Silesia, Poland. Palaeontol Pol 64:3–27. https://palaeontologia.pan.pl/PP64/Dzik.pdf

Ekdale AA, Mason TR (1988) Characteristic trace-fossil associations in oxygen-poor sedimentary environments. Geology 16:720–723. https://doi.org/10.1130/0091-7613(1988)016%3c0720:CTFAIO%3e2.3.CO;2

Forchielli A, Pervesler P (2013) Phosphatic cuticle in thylacocephalans: a taphonomic case study of (Arthropoda, Thylacocephala) from the *Fossil-Lagerstätte* Polzberg (Reingraben shales, Carnian, Upper Triassic, Lower Austria). AJES 106:46–61

Furrer HD (2003) Der Monte San Giorgio im Südtessin – vom Berg der Saurier zur Fossil-Lagerstätte internationaler Bedeutung. Neujahrsbl Natur Ges Zürich 206:1–64

Furrer H, Vandelli A (2014) Guide to the Museum of fossils from Monte San Giorgio Meride. Fondazione del Monte San Giorgio, 127 p

Furrer HD (2024) The history of palaeontological research and excavations at Monte San Giorgio. Swiss J Palaeontol 142:18. https://doi.org/10.1186/s13358-024-00314-9

Gelpi E, Schneider H, Mann J, Oró J (1970) Hydrocarbons of geochemical significance in microscopic algae. Phytochem 9(3):603–612. https://doi.org/10.1016/S0031-9422(00)85700-3

Glaessner MF (1931) Eine Crustaceen fauna aus den Lunzer Schichten Niederösterreichs. Jahrb Geol BA 81:467–486

Glaessner MF (1965) Vorkommen fossiler Dekapoden (Crustacea) in Fisch-Schiefern. Senckenb Lethaea 46a:111–122.

de Grande SMB, Aquino Neto FR, Mello MR (1993) Extended tricyclic terpanes in sediments and petroleums. Org Geochem 20(7):1039–1047. https://doi.org/10.1016/0146-6380(93)90112-O

Greenwood PF, Arouri KR, George SC (2000) Tricyclic terpenoid composition of tasmanites kerogen as determined by pyrolysis GC-MS. Geochim Cosmochim Acta 64(7):1249–1263. https://doi.org/10.1016/S0016-7037(99)00326-9

Griffith J (1977) The Upper Triassic fishes from Polzberg bei Lunz. Zool J Linn Soc 60:1–93

Harzhauser M, Lukeneder A, Lukeneder P, Mandic O (2026) Potential modulation of a Late Triassic Pangean Megamonsoon by solar cycles tracing the Carnian Pluvial Episode. In: Lukeneder A (ed) Late Triassic Konservat-Lagerstätten within the Carnian Pluvial Episode: 150 years of scientific excavations in Austria. Top Geobiol 55, Chap 24: 561–583. https://doi.org/10.1007/978-3-032-19777-1_24

Hauke V, Graff R, Wehrung P, Trendel JM, Albrecht P, Riva A et al (1992) Novel triterpene-derived hydrocarbons of the arborane/fernane series in sediments: Part II. Geochim Cosmochim Acta 56(9):3595–3602. https://doi.org/10.1016/0016-7037(92)90405-8

Hoffmann CF, Mackenzie AS, Lewis CA, Maxwell JR, Oudin JL, Durand B, Vandenbroucke M (1984) A biological marker study of coals, shales and oils from the Mahakam Delta, Kalimantan, Indonesia. Chem Geol 42(1–4):1–23. https://doi.org/10.1016/0009-2541(84)90002-0

Hoffmann R, Keupp H (2015) Ammonoid paleopathology. In: Klug C, Korn D, De Baets K, Kruta I, Mapes R (eds) Ammonoid paleobiology: from anatomy to ecology. Top Geobiol 43:877–926. https://doi.org/10.1007/978-94-017-9630-9_21

Hoffmann R, Stevens K, Keupp H, Simonsen S, Schweigert G (2020) Regurgitalites—a winodw into the trophic ecology of fossil cephalopods. J Geol Soc Lond 177:82–102. https://doi.org/10.1144/jgs2019-117

Hornung T (2008) The Carnian Crisis in the Tethys Realm. Dr. Müller, Saarbrücken, 235 p

Hornung T, Brandner R (2005) Biochronostratigraphy of the Reingraben Turnover (Hallstatt Facies Belt): Local black shale events controlled by regional tectonism, climatic change and plate tectonics. Facies 51:460–479. https://doi.org/10.1007/s10347-005-0061-x

Hornung T, Brandner R, Krystyn L, Joachimski MM, Keim L (2007a) Multistratigraphic constraints on the NW Tethyan 'Carnian Crisis'. In: Lucas GS, Spielmann AJ (eds) The global Triassic. New Mex Mus Nat Hist Sci Bull 41:59–67

Hornung T, Krystyn L, Brandner RA (2007b) A Tethys-wide mid-Carnian (Upper Triassic) carbonate productivity crisis: evidence for the Alpine Reingraben Event from Spiti (Indian Himalaya)? J Asian Earth Sci 30(2):285–302. https://doi.org/10.1016/j.jseaes.2006.10.001

Hornung T, Kogan I, Wolf G, Moosleitner G (2019) The Norian fish deposits of Wiestal ("Seefeld Member", Northern Calcareous Alps, Salzburg, Austria)—taxonomy and palaeoenvironmental implications. Austrian J Earth Sci 112(2):125–165. https://doi.org/10.17738/ajes.2019.0008

Huene F (1916) Beiträge zur Kenntnis der Ichthyosaurier im deutschen Muschelkalk. Palaeontogr 62:1–68

Hughes WB, Dzou HAG, Leon IP (1995) The ratios of dibenzothiophene to phenanthrene and pristane to phytane as indicators of depositional environment and lithology of petroleum source rocks. Geochim Cosmochimica Acta 59(17):3581–3598. https://doi.org/10.1016/0016-7037(95)00225-O

Huang W-Y, Meinschein WG (1979) Sterols as ecological indicators. Geochim Cosmochim Acta 915 43:739–745. https://doi.org/10.1016/0016-7037(79)90257-6

Hunt AP, Lucas SG (2012) Descriptive terminology of coprolites and recent feces. In: Hunt AP, Milan J, Lucas, SG, Spielmann JA (eds) Vertebrate Coprolites. New Mex Mus Nat Hist Sci Bull 57:153–160

Hunt AP, Milàn J, Lucas SG, Spielmann JA (2012) Vertebrate coprolites. New Mex Mus Nat Hist Sci Bull 57:387 p

Ji C, Tintori A, Jiang D, Montani R, Confortini F (2020) New Thylacocephala (Crustacea) assemblage from the Spathian (Lower Triassic) of Majiashan (Chaohu, Anhui Province, South China). J Paleontol 95(2):305–319. https://doi.org/10.1017/jpa.2020.92

Jobbins M, Haug C, Klug C (2020) First African thylacocephalans from the Famennian of Morocco and their role in Late Devonian food webs. Sci Rep 10:5129. https://doi.org/10.1038/s41598-020-61770-0

Keupp H (2000). Ammoniten. Paläobiologische Erfolgsspiralen. Jan Thorbecke Verlag, Stuttgart, pp 165

Keupp H (2012) Atlas zur Paläopathologie der Cephalopoden. Berliner Paläobiol Abh 12:392 p

Killops SD, Abbott GD (2022) On the origin of 3-methyl steroids and diasteranes. Chem Geol 607:121014. https://doi.org/10.1016/j.chemgeo.2022.121014

Klompmaker AA, Waljaard NA, Fraaije RHB (2009) Ventral bite marks in Mesozoic ammonoids. Palaeogeogr Palaeoclimatol Palaeoecol 280:245–257. https://doi.org/10.1016/j.palaeo.2009.06.013

Klompmaker AA, Kelley PH, Chattopadhyay D, Clements JC, Huntley JW, Kowalewski M (2019) Predation in the marine fossil record: Studies, data, recognition, environmental factors, and behavior. Earth-Sci Rev 194:472–520. https://doi.org/10.1016/j.earscirev.2019.02.020

Klug C, Schweigert G, Fuchs D, De Baets K (2021a) Distraction sinking and fossilized coleoid predatory behaviour from the German Early Jurassic. Swiss J Palaeontol 140(1):7. https://doi.org/10.1186/s13358-021-00218-y

Klug C, Schweigert G, Tischlinger H, Pochmann H (2021b) Failed prey or peculiar necrolysis? Isolated ammonite soft body from the Late Jurassic of Eichstätt (Germany) with complete digestive tract and male reproductive organs. Swiss J Palaeontol 140(1):3. https://doi.org/10.1186/s13358-020-00215-7

Klug C, Scheyer TM, Klein N, Liu J, Stockar R eds (2024a) Special Issue: 100 years of scientific excavations at UNESCO World Heritage Site Monte San Giorgio and Global Research on Triassic Marine Lagerstätten. Swiss J Palaeontol 143 (Spec Issue), Springer, Basel, 498 p

Klug C, Scheyer TM, Klein N, Liu J, Albisetti D, Furrer H, Stockar R (2024) 100 years of scientific excavations at UNESCO World Heritage Site Monte San Giorgio and global research on Triassic marine Lagerstätten. Swiss J Palaeontol 143:37. https://doi.org/10.1186/s13358-024-00328-3

Korn D, Klug C (2002) Chapter 10: Ammoneae Devonicae. In: Riegraf W (ed) Fossilium catalogus, animalia, vol I. Backhuys, Leiden, 375 p

Kozur HW, Bachmann GH (2010) The middle Carnian Wet Intermezzo of the Stuttgart Formation (Schilfsandstein), Germanic Basin. Palaeogeogr Palaeoclimatol Palaeoecol 290:107–119. https://doi.org/10.1016/j.palaeo.2009.11.004

Kowalewski M (2002) The fossil record of predation: an overview of analytical methods. In: Kowalewski M, Kelley PH (eds) The fossil record of predation. Palentological Society, New Haven. Paleontol Soc Spec Pap 8:3–42. https://doi.org/10.1017/S1089332600001030

Kreuss O (2014) 100 Hieflau. Geologische Karte, 1: 50 000. Geofast, Geosphere Austria. https://doi.org/10.24341/tethys.101

Krystyn L (1978) Eine neue Zonengliederung des mediterranen Unterkarn. Schrift Erdwiss Komm ÖAW 4:37–75

Krystyn L (1991) Die Fossillagerstätten der alpinen Trias. In: Nagel D, Rabeder G (eds) Exkursionen im Jungpaläozoikum und Mesozoikum Österreichs. Österr Paläontol Ges Wien, pp 23–78

Krystyn L, Lein R, Richoz S (2008) Der Gamsstein: Werden und Vergehen einer Wettersteinkalk-Platfrom. Exkursionsführer, Pangeo 2008. J Alp Geol 49:157–172

Kustatscher E, Branz R, Slodownik M, Roghi G, Lukeneder A (2026) The plant fossils and amber of the Carnian (Late Triassic) of Scheiblingbach in Austria. In: Lukeneder A (ed) Late Triassic Konservat-Lagerstätten within the Carnian Pluvial Episode: 150 Years of Scientific Excavations in Austria. Top Geobiol 55, Chap 22:499–525. https://doi.org/10.1007/942978-3-032-19777-1_22

Lahm B (1984) Spumellarienfaunen (Radiolaria) aus den mitteltriassischen Buchensteiner-Schichten von Recoaro (Norditalien) und den obertriassischen Reiflingerkalken von Großreifling (Österreich). Systematik – Stratigraphie. Münchner Geowiss Abh A, Geol Paläontol 1:1–161, Pfeil, München

Lehman JP (1975) A propos de *Ceratodus sturii* Teller, 1891. Bull Mus Nati Hist Nat Paris 375:241–246

Lehman JP (1979) Note sur les Poissons du Trias de Lunz. I. *Thoracopterus* Bronn. Ann Naturhist Mus *Wien* 82:53–66

Liu J, Hu Sx, Rieppel O, Jiang D, Benton MJ, Kelley NP, Aitchison JC, Zhou C, Wen W, Huang J, Xie T, Lv T (2014) A gigantic nothosaur (Reptilia: Sauropterygia) from the Middle Triassic of SW China and its implication for the Triassic biotic recovery. Sci Rep 4:7142. https://doi.org/10.1038/srep07142

Lukeneder A (2015) Ammonoid Habitats and Life History. In: Klug C., Korn D., De Baets K., Kruta I, Mapes, RH (eds) Ammonoid paleobiology: from anatomy to ecology. Top Geobiol 43:689–791. Springer, Dordrecht. https://doi.org/10.1007/978-94-017-9630-9_18

Lukeneder A (2025) 6.6 Rätsel um Fischsaurier von Großreifling. – der verschwunde Kopf. In: Kirchmayer G, Lukeneder A, Riedl WC (eds) UNESCO Global Geopark Steirische Eisenwurzen. Geoparks. Springer, Berlin, pp 109–115

Lukeneder A ed (2026) Late Triassic Konservat-Lagerstätten within the Carnian Pluvial Episode: 150 years of scientific excavations in Austria. Top Geobiol 55:600 p. https://doi.org/10.1007/978-3-032-19777-1

Lukeneder A, Lukeneder P (2021) The Upper Triassic Polzberg palaeobiota from a marine *Konservat-Lagerstätte* deposited during the Carnian Pluvial Episode in Austria. Sci Rep 11:16644. https://doi.org/10.1038/s41598-021-96052-w

Lukeneder A, Lukeneder P (2022a) Taphonomic history and trophic interactions of an ammonoid fauna from the Upper Triassic Polzberg palaeobiota. Sci Rep 12:7455. https://doi.org/10.1038/s41598-022-11496-y

Lukeneder P, Lukeneder A (2022b) Mineralized belemnoid cephalic cartilage from the Late Triassic Polzberg *Konservat-Lagerstätte* (Austria). PLoS ONE 17(4):e0264595. https://doi.org/10.1371/journal.pone.0264595

Lukeneder A, Lukeneder P (2023) New data on the marine Upper Triassic palaeobiota from the Polzberg *Konservat-Lagerstätte* in Austria. Swiss J Palaeontol 142:9. https://doi.org/10.1186/s13358-023-00269-3

Lukeneder A, Lukeneder P (2024) Ammonoid taxonomy of the Carnian Polzberg *Konservat-Lagerstätte* in Austria. Jahrb Geol BA 162(1–4):27–55

Lukeneder A, Tanabe K (2002) *In situ* finds of aptychi in the Barremian of the Alpine Lower Cretaceous (Barremian, Northern Calcareous Alps, Upper Austria). Cretac Res 23:15–24. https://doi.org/10.1006/cres.2001.0303

Lukeneder A, Surmik D, Gorzelak P, Niedźwiedzki R, Brachaniec T, Salamon MA (2020) Bromalites from the Upper Triassic Polzberg section (Austria); insights into trophic interactions and food chains of the Polzberg palaeobiota. Sci Rep 10:20545. https://doi.org/10.1038/s41598-020-77017-x

Lukeneder A, Lukeneder P, Sachsenhofer RF, Roghi G, Rigo M (2024a) Multi-proxy record of the Austrian Upper Triassic Polzberg *Konservat-Lagerstätte* in light of the Carnian Pluvial Episode. Sci Rep 14:11194. https://doi.org/10.1038/s41598-024-60591-9

Lukeneder P, Fuchs D, Lukeneder A (2024b) Morphology, taxonomy and trophic interactions of rostrum-less coleoids from the Late Triassic Polzberg *Konservat-Lagerstätte* (Lower Austria). Swiss J Palaeontol 143:21. https://doi.org/10.1186/s13358-024-00319-4

Lukeneder S, Lukeneder A, Harzhauser M, İslamoğlu Y, Krystyn L, Lein R (2012) A delayed carbonate factory breakdown during the Tethyan-wide Carnian Pluvial Episode along the Cimmerian terranes (Taurus, Turkey). Facies 58:279–296. https://doi.org/10.1007/s10347-011-0279-8

Mackenzie AS, Wolff GA, Maxwell JR (1983) Fatty acids in some biodegraded petroleums. Possible origins and significance. In: Bjoroy M, Albrecht C, Conford C (eds) Advances in organic geochemistry 1981. John Wiley, Chichester, pp 637–649

Mackenzie AS (1984) Applications of Biological Markers in Petroleum Geochemistry. Adv Petrol Geochem 1984:115–214. https://doi.org/10.1016/b978-0-12-032001-1.50008-0

Martin M (1982) Revision von *Tellerodus sturii* (Teller) 1891. Verh Geol BA 1982(2):21–31

Mehl J (1978) Anhäufungen scherbenartiger Fragmente von Ammonitenschalne im süddeutschen Lias und Malm und ihre Deutung als Frassreste. Ber Naturf Ges Freiburg Breisgau 68:75–93

Mojsisovics JAE (1893a) Die Cephalopoden der Hallstätter Kalke. Band I. Abh K-K Geol RA 6:835 p

Mojsisovics JAE (1893b). Die Cephalopoden der Hallstätter Kalke. Band II. Abh K-K Geol RA 6:524 p

Moldowan JM, Seifert Wolfgang K, Gallegos EJ (1983) Identification of an extended series of tricyclic terpanes in petroleum. Geochim Cosmochim Acta 47(8):1531–1534. https://doi.org/10.1016/0016-7037(83)90313-7

Montagna et al. (2018) The Middle Triassic Lagerstätte of Monte San Giorgio reveals the oldest lace bugs (Hemiptera: Tingidae): *Archetingis ladinica* gen. n. sp. n. Riv Ital die Paleont Strat 124:35–44. https://doi.org/10.13130/2039-4942/9623

Mostler H, Scheuring BW (1974) Mikrofloren aus dem Langobard und Cordevol der Nördlichen Kalkalpen und das Problem der Beginns der Keupersedimentation im Germanischen Raum. Geol Paläont Mitt Innsbruck 4(4):1–35

Mueller S, Krystyn L, Kürschner WM (2016) Climate variability during the Carnian Pluvial Phase—a quantitative palynological study of the Carnian sedimentary succession at Lunz am See, Northern Calcareous Alps, Austria. Palaeogeogr Palaeoclimatol Palaeoecol 441:198–211. https://doi.org/10.1016/j.palaeo.2015.06.008

Neenan JM, Klein N, Scheyer TM (2013) European origin of placodont marine reptiles and the evolution of crushing dentition in Placodontia. Nat Commun 4:1621. https://doi.org/10.1038/ncomms2633

Neunlist S, Rohmer M (1985) Novel hopanoids from the methylotrophic bacteria Methylococcus capsulatus and Methylomonas methanica. (22S)-35-aminobacteriohopane-30,31,32,33,34-pentol and (22S)-35-amino-3 beta-methylbacteriohopane-30,31,32,33,34-pentol. The Biochem J 231(3):635–639. https://doi.org/10.1042/bj2310635

Ogg JG (2015) The mysterious Mid-Carnian "Wet Intermezzo" global event. J Earth Sci 26:181–191. https://doi.org/10.1007/s12583-015-0527-x

Olivero EB (2007) Taphonomy of ammonites from the Santonian-Lower Campanian Santa Marta Formation, Antarctica: sedimentological controls on vertically embedded ammonites. Palaios 22:586–597. https://doi.org/10.2110/palo.2005.p05-118r

Peters KE, Moldowan JM, Sundararaman P (1990) Effects of hydrous pyrolysis on biomarker thermal maturity parameters: Monterey Phosphatic and Siliceous members. Org Geochem 15(3):249–265. https://doi.org/10.1016/0146-6380(90)90003-I

Peters KE, Cassa MR (1994) Applied source rock geochemistry. In: Magoon LB, Dow WG (eds) 994 The petroleum system — from source to trap, 60thed. AAPG, pp 93–120. https://doi.org/10.1306/M60585C5

Peters KE, Walters CC, Moldowan, JM (2005) Biomarkers and isotopes in petroleum systems and earth history. In: Peters KE, Walters CC, Moldowan JM (eds) The biomarker guide, 2nd edn. Cambridge University Press, Cambridge, UK. https://doi.org/10.1017/CBO9781107326040

Philp P, Symcox C, Wood M, Nguyen T, Wang H, Kim D (2021) Possible explanations for the predominance of tricyclic terpanes over pentacyclic terpanes in oils and rock extracts. Org Geochem 155:104220. https://doi.org/10.1016/j.orggeochem.2021.104220

Pinna G (1985) Exceptional preservation in the Jurassic of Osteno. Philos Trans R Soc Lond B 311:171–180. https://doi.org/10.1098/rstb.1985.0149

Přikryl T, Košťák M, Mazuch M, Mikuláš R (2012) Evidence for fish predation on a coleoid cephalopod from the Lower Jurassic Posidonia Shale of Germany. Neues Jahrb Geol Paläontol Abh 263:25–33. https://doi.org/10.1127/0077-7749/2012/0206

Radke M, Welte DH (1983) The methylphenanthrene index (MPI). A maturity parameter based on aromatic hydrocarbons. In: Bjorøy M, Albrecht C, Cornford C, de Groot K, Eglinton E, Galimov E, Leythaeuser D, Pelet R, Rullkötter J, Speer G (eds) Advances in organic geochemistry 1981. John Wiley, Chichester, pp 504–512

Reis OM (1900) *Coelacanthus lunzensis* Teller. Jb K-K Geol RA 50(2):187–192

Richter AE (2009) Ammoniten-Gehäuse Mit Bissspuren. Berliner Paläontol Abh 10:297–305

Rigo M, Preto N, Roghi G, Tateo F, Mietto P (2007) A rise in the carbonate compensation depth of western tethys in the carnian (late triassic): deep-water evidence for the carnian pluvial event. Palaeogeogr Palaeoclimatol Palaeoecol 246(2–4):188–205. https://doi.org/10.1016/j.palaeo.2006.09.013

Rigo M, Mazza M, Karádi V, Nicora (2018) A New Upper Triassic conodont biozonation of the Tethyan Realm. In: Tanner LH (ed) The late triassic world: topics in geobiology, vol 46, pp 189–235. https://doi.org/10.1007/978-3-319-68009-5_6

Rubinstein I, Sieskind O, Albrecht P (1975) Rearranged sterenes in a shale: occurrence and simulated formation. J Chem Soc, Perkin Trans 1(19):1833. https://doi.org/10.1039/P19750001833

Ruffell A, Simms MJ, Wignall PB (2015) The carnian humid episode of the late triassic: a review. Geol Mag 153:271–284. https://doi.org/10.1017/S0016756815000424

Rullkötter J, Philip P (1981) Extended hopanes up to C40 in Thornton bitumen. Nature 292(5824):616–618. https://doi.org/10.1038/292616a0

Ruttner A, Schnabel W (1988) Geologische Karte der Republik Österreich. Sheet Ybbsitz 71, 1:50.000. Geologische Bundesanstalt Wien. https://doi.org/10.24341/tethys.6

Savrda CE, Bottjer DJ (1991) Oxygen-related biofacies in marine strata: an overview and update. Geol Soc Lond Spec Publ 58:201–219. https://doi.org/10.1144/gsl.sp.1991.058.01.14

Schädel M, Eldijk T, Winkelhorst H, Reumer JWF, Haug J (2020) Triassic Isopoda—three new species from Central Europe shed light on the early diversity of the group. Bull Geosci 95(2):145–166. https://doi.org/10.3140/bull.geosci.1773

Scheyer TM, Romano C, Jenks J, Bucher H (2014) Early triassic marine biotic recovery: the predators' perspective. PLoS ONE 9(3):e88987. https://doi.org/10.1371/journal.pone.0088987
Schlager W, Schöllnberger W (1974) Das Prinzip stratigraphischer Wenden in der Schichtfolge der Nördlichen Kalkalpen. Mitt Geol Ges Wien 66:165–193
Schultz O (2013) Pisces. In: Piller WE (ed) Austrian academy of sciences. Catal Foss Austr 3:1–576.
Schlumberger M (1890) Seconde note sur les Holothuridées fossiles du Calcaire Grossier. Bull Soc Geol Fr (sér 3) 18:191–206
Seifert WK, Moldowan JM (1980) The effect of thermal stress on source-rock quality as measured by hopane stereochemistry. Phys Chem Earth 12:229–237. https://doi.org/10.1016/0079-1946(79)90107-1
Seifert WK, Moldowan JM (1981) Paleoreconstruction by biological markers. Geochim Cosmochim Acta 45(6):783–794. https://doi.org/10.1016/0016-7037(81)90108-3
Seilacher A (1970) Begriff und Bedeutung der Fossil-Lagerstätten. N Jahrb Geol Paläontol Mh 1970:34–39
Simms MJ, Ruffell AH (2018) The Carnian Pluvial Episode: from discovery, through obscurity, to acceptance. J Geol Soc 175(6):989–992. https://doi.org/10.1144/jgs2018-020
Simms MJ, Ruffell AH (1990) Climatic and biotic change in the late Triassic. J Geol Soc Lond 147:321–327. https://doi.org/10.1144/gsjgs.147.2.0321
Simoneit BRT (1986) Cyclic terpenoids of the geosphere. In: Johns RB (ed) Biological markers in the sedimentary record. Elsevier, Amsterdam, pp 43–99
Simoneit BRT, Leif RN, de Aquino R, Neto F, Almeida AD, Pinto AC, Albrecht P (1990) On the presence of tricyclic terpane hydrocarbons in permian tasmanite algae. Naturwiss 77(8):380–383. https://doi.org/10.1007/BF01135736
Smale MJ (1996) Cephalopods as prey. IV. Fishes. Philos Trans R Soc Lond Biol Sci B 351:1067–1081. https://doi.org/10.1098/rstb.1996.0094
Stur D (1874) Neue Aufschlüsse im Lunzer Sandsteine bei Lunz und ein neuer Fundort von Wengerschiefer im Pölzberg zwischen Lunzersee und Gaming. Verh K-K Geol RA 1:271–273
Stur D (1886) Vorlage des ersten fossilen Schädels von *Ceratodus* aus den ober triadischen Reingrabner Schiefern von Pölzberg nördlich bei Lunz. Verh K-K Geol RA 15:381–383
Tao S, Wang C, Du J, Liu L, Chen Z (2015) Geochemical application of tricyclic and tetracyclic terpanes biomarkers in crude oils of NW China. Mar Petrol Geol 67:460–467. https://doi.org/10.1016/j.marpetgeo.2015.05.030
Teller F (1891) Über den Schädel eines fossilen Dipnoers, *Ceratodus Sturii* nov. spec. aus den Schichten der oberen Trias der Nordalpen. Abh K-K Geol RA 15:1–38
Tintori A (1990) Estherids from the Kalkschieferzone (Triassic) of Lombardy (N. Italy). In: Robba E (ed) Atti del Quarto Simposio di Ecologia e Palecologia delle Comunità' Bentoniche Sorrento, 1988. Mus Reg Sci Nat, Torino, pp 95–105
Tintori A (1992) Fish taphonomy and Triassic anoxic basins from the Alps: a case history. Riv It Paleont Strat 97:393–408
Tollmann A (1976) Analyse des klassischen nordalpinen Mesozoikums. Franz Deuticke, Wien, pp 1–580
Trauth F (1935) Die Aptychen der Trias. Sitzung Akad Wissen, Math-Natur Kl, Abt 1(144):455–483
Trauth F (1948) Geologie des Kalkalpenbereiches der Zweiten Wiener Hochquellenleitung. Abh Geol BAWien 26(1):1–99
van Kaam-Peters HME, Köster J, van der Gaast SJ, Dekker M, de Leeuw JW, Sinninghe DJS (1998) The effect of clay minerals on diasterane/sterane ratios. Geochim Cosmochim Acta 62(17):2923–2929. https://doi.org/10.1016/S0016-7037(98)00191-4
Vullo R (2011) Direct evidence of hybodont shark predation on Late Jurassic ammonites. Naturwiss 98:545–549. https://doi.org/10.1007/s00114-011-0789-9
Wang E, Li M, Ma X, Qian M, Cao T, Li Z, Yang W, Jin Z (2024) Diahopane and diasterane as the proxies for paleoenvironment, hydrocarbon generation condition, and shale oil accumulation. Chem Geol 670:122447. https://doi.org/10.1016/j.chemgeo.2024.122447

Ward PD (1981) Shell sculpture as a defensive adaptation in ammonoids. Paleobiology 7:96–100. https://doi.org/10.1017/S009483730000381X

Wen-Yen H, Meinschein WG (1979) Sterols as ecological indicators. Geochim Cosmochim Acta 43(5):739–745. https://doi.org/10.1016/0016-7037(79)90257-6

Weitz JA, Stojanović KA, Sachsenhofer RF, Misch D, Lukeneder A (2026) Organic petrology and geochemistry of the Scheiblinggraben section: detailed insights into the Carnian Pluvial Episode and comparison to time equivalent profiles. In: Lukeneder A (ed) Late Triassic Konservat-Lagerstätten within the Carnian Pluvial Episode: 150 years of scientific excavations in Austria. Top Geobiol 55, Chap 23:521–560. https://doi.org/10.1007/978-3-032-19777-1_23

Wignall PB, Hallam A (1991) Biofacies, stratigraphic distribution and depositional models of British onshore Jurassic black shales. In: Tyson RV, Pearson TA (eds) Modern and ancient continental shelf anoxia. Geol Soc Lond Spec Publ 58:291–309. https://doi.org/10.1144/gsl.sp.1991.058.01.19

Xiao H, Li M, Nettersheim BJ (2024) Short chain tricyclic terpanes as organic proxies for paleo-depositional conditions. Chem Geol 652:122023. https://doi.org/10.1016/j.chemgeo.2024.122023

Yacobucci MM (2018) Postmortem transport in fossil and modern shelled cephalopods. PeerJ 6:e5909. https://doi.org/10.7717/peerj.5909

Zumberge JE (1987) Prediction of source rock characteristics based on terpane biomarkers in crude oils: a multivariate statistical approach. Geochim Cosmochim Acta 51(6):1625–1637. https://doi.org/10.1016/0016-7037(87)90343-7

Zundel M, Rohmer M (1985) Hopanoids of the methylotrophic bacteria Methylococcus capsulatus and Methylomonas sp. as possible precursors of C 29 and C 30 hopanoid chemical fossils. FEMS Microbiol Lett 28(1):61–64. https://doi.org/10.1111/j.1574-6968.1985.tb00764.x

Chapter 16
C-Isotope Perturbation and Hg Deposition in the Reifling Basin at "Event1" of the Carnian Pluvial Episode (Late Triassic)

Jacopo Dal Corso, Alexander Lukeneder, Greta Alverà, Tastulek Haserbek, Marcello Caggiati, Reinhard F. Sachsenhofer, Johannes A. Weitz, David Misch, Daoliang Chu, Li Tian, Haijun Song, Guido Roghi, and Piero Gianolla

Abstract The Carnian was a pivotal time of environmental change and biological turnover. In this period new faunas radiated and rose to then dominate Mesozoic ecosystems. The long-term changes that ecosystems experienced are linked to discrete short-term events. These events perturbed terrestrial and marine environments, plunging older faunas into crisis and favoring the explosion of animals and

Supplementary Information The online version contains supplementary material available at https://doi.org/10.1007/978-3-032-19777-1_16.

J. Dal Corso (✉) · T. Haserbek · D. Chu · L. Tian · H. Song
State Key Laboratory of Geomicrobiology and Environmental Changes, China University of Geosciences, Wuhan, China
e-mail: j.dalcorso@cug.edu.cn

T. Haserbek
e-mail: tastulek@cug.edu.cn

L. Tian
e-mail: tianlibgeg@cug.edu.cn

H. Song
e-mail: haijunsong@cug.edu.cn

A. Lukeneder
Department of Geology and Paleontology, Natural History Museum Vienna, Vienna, Austria
e-mail: alexander.lukeneder@nhm.at

G. Alverà · P. Gianolla
Department of Physics and Earth Science, University of Ferrara, Ferrara, Italy
e-mail: greta.alvera@unife.it

P. Gianolla
e-mail: piero.gianolla@unife.it

M. Caggiati
Department of Geosciences, University of Padova, Padova, Italy

A. Lukeneder (ed.), *Late Triassic Konservat-Lagerstätten within the Carnian Pluvial Episode*, Topics in Geobiology 55, https://doi.org/10.1007/978-3-032-19777-1_16

plants presenting modern characters. The Carnian events are roughly coeval with the emplacement of Wrangellia Large Igneous Province (LIP), which could release large quantities of volcanic gases, including greenhouse gases such as CO_2, and may have triggered the observed environmental changes. The initial abrupt event of this relatively long (max 2.5 Myr) interval of ecosystem reorganization—known collectively as the Carnian Pluvial Episode (CPE)—is chiefly recorded in the Reifling Basin of the Northern Calcareous Alps, Austria. Here, the Carnian crisis was noted for the first time in the 1970s and named "Reingraben Wende" (Reingraben Turnover) after the Reingraben Formation that represents this first Event of the CPE interval (CPE-Event1) in that area. The Reifling Basin is therefore a reference locality for the study of the Carnian, thanks in part to the spectacular fossil preservation of the Polzberg *Konservat-Lagerstätte.* Despite the increasing number of studies published on the CPE in recent years, the mechanisms of each Event that marked the interval and the timing of the environmental-biological response to it are still not fully understood. The succession of the Reifling Basin offers the opportunity to investigate the CPE-Event1—the original "Reingraben Wende"—in detail to disentangle the global and local processes that drove the environmental change. In this chapter we focus on the record of the negative C-isotope excursion that marks the CPE-Event1, which is evidence of CO_2 injection in the system, and on coeval changes in mercury (Hg) concentrations in the sediments, which indicate changes in the loading of volcanic and/or terrestrial Hg to the Reifling Basin. A revision of published C-isotope records in conjunction with new Hg data contributes to the understanding of global and local phenomena modulating volcanic contribution and terrigenous supply to the Reifling Basin. Furthermore, this study clarifies the relative timing between the onset of the global perturbation and the local environmental response to it.

Keywords Hg deposition · C-isotope perturbation · Carnian Pluvial Episode · Carnian · Late Triassic · Austria

e-mail: marcello.caggiati@unipd.it

R. F. Sachsenhofer · J. A. Weitz · D. Misch
Department Applied Geosciences and Geophysics, Technical University of Leoben, Leoben, Austria
e-mail: reinhard.sachsenhofer@unileoben.ac.at

J. A. Weitz
e-mail: johannes.weitz@unileoben.ac.at

D. Misch
e-mail: David.Misch@unileoben.ac.at

G. Roghi
Institute of Geosciences and Earth Resources, National Research Council (CNR), Padova, Italy
e-mail: guido.roghi@igg.cnr.it

16.1 Introduction

The Carnian (Late Triassic) was punctuated by a series of remarkable environmental and biological events (Dal Corso et al. 2020a; Dal Corso 2025). These events occurred during an interval of global C-cycle perturbations indicated by successive negative C-isotope excursions (NCIE; Fig. 16.1) recorded worldwide in marine and terrestrial Carnian sections, by both organic and carbonate carbon (Dal Corso et al. 2012, 2015, 2018; Mueller et al. 2016a, b; Sun et al. 2016, 2019; Miller et al. 2017; Jin et al. 2019, 2020, 2022a, b, 2023; Shi et al. 2019; Baranyi et al. 2019a; Fu et al. 2020; Lu et al. 2021; Tomimatsu et al. 2021; Rahman et al. 2024; Zhang et al. 2024; Lestari et al. 2024; Haserbek et al. 2025; Jia et al. 2025; Alverà et al. 2026). The C-isotope record suggests multiple, discrete and relatively rapid (ca. 20–150 kyr) emissions of ^{13}C-delpeted CO_2 into the Carnian reservoirs of the exogenic C-cycle (e.g., Dal Corso et al. 2018, 2022, 2024b; Sun et al. 2016; Miller et al. 2017; Dal Corso 2025). Consequently, increases of atmospheric pCO_2 at each CO_2 input could induce global warming as shown by O-isotope records from conodont apatite that indicate 4–7 °C temperature increases in sea-surface waters (Hornung et al. 2007a; Rigo and Joachimski 2010; Trotter et al. 2015; Sun et al. 2016). In general, four events can be clearly distinguished in well-constrained Carnian records based on integrated sedimentological (facies change), biological (fauna and flora turnovers) and geochemical (C-isotope stratigraphy) evidences (Roghi et al. 2010; Dal Corso et al. 2018; Dal Corso 2025; Haserbek et al. 2025). The entire interval including these "Carnian Events" is called Carnian Pluvial Episode (CPE, Fig. 16.1; Simms and Ruffell 1989; Dal Corso et al. 2020a). The CPE had an estimated total duration of ca. 1.2–2.5 Myr (e.g., Miller et al. 2017; Zhang et al. 2024; Dal Corso 2025). Within this relatively long interval, each event was short. According to the current Carnian timescales, stratigraphic framework, and cyclostratigraphic estimates, each NCIE—the global, isochronous and most readily correlative marker of the CPE's Events—had a duration of ca. 40–300 kyr, with the most rapid one being the NCIE-1 (Miller et al. 2017; Zhang et al. 2024; Dal Corso 2025).

The overlap between the age of the CPE and that of emplacement of Wrangellia large igneous province (W-LIP) suggests that volcanic emissions of CO_2 and other gases into the Carnian atmosphere–land–ocean system could have caused the Carnian C-cycle perturbations, the environmental crises and the biological turnovers (e.g., Furin et al. 2006; Dal Corso et al. 2012, 2020a). However, a direct link between W-LIP and the NCIEs is difficult to establish. Existing radioisotope ages of W-LIP are not robust: Ar/Ar dating is affected by widespread alteration, while the multigrain aliquots of zircons analyzed for U/Pb dating were not chemically abraded (Greene et al. 2010; Dal Corso et al. 2020a; Boscaini et al. 2022). Biostratigraphy from strata below and above the basalts allows only to constrain the maximum age span of W-LIP from the late Ladinian to the late Carnian (Greene et al. 2010; Dal Corso et al. 2020a; Boscaini et al. 2022). Magnetostratigraphy suggests W-LIP emplacement lasted for a maximum of ca. 2 Myr (Greene et al. 2010; Boscaini et al. 2022).

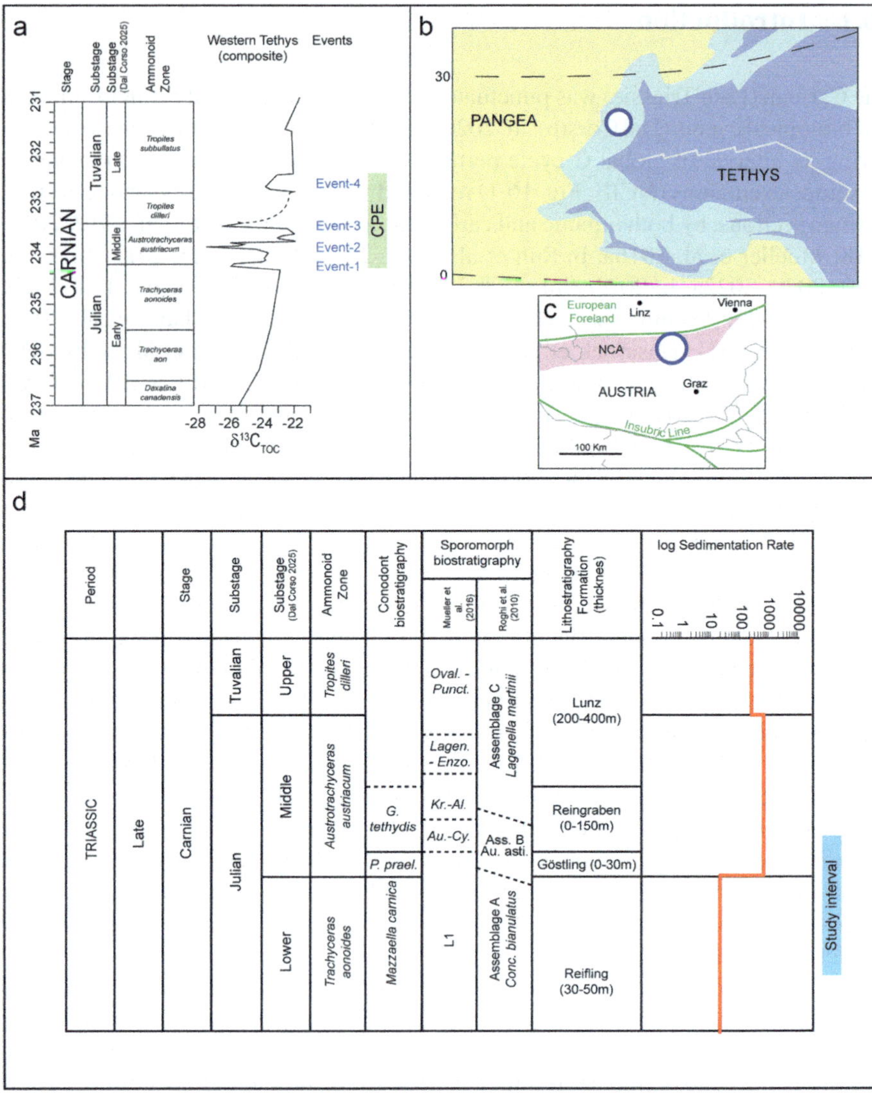

Fig. 16.1 **a** Composite organic C-isotope record from the Western Tethys and the events marking the Carnian Pluvial Episode, CPE (Dal Corso et al. 2018; Dal Corso 2025; Roghi and Lukeneder 2026). **b** Paleogeography of the Western Tethys in the Carnian and location of the Reifling Basin (blue circle) of the Northern Calcareous Alps (NCA; Dal Corso et al. 2018). **c** Map with the locations of the sections of the Reifling Basin (blue circle). **d** Stratigraphic constraints (Roghi et al. 2010; Mueller et al. 2016a, b; Lukeneder et al. 2024), and sedimentation rates (Dal Corso et al. 2024a, b) of the succession of the Reifling Basin (Julian 1–Tuvalian 1). *G. tethydis* = *Gladigondolella tethydis*; *P. prael.* = *Paragondolella praelindae*; *Oval.* = *Ovalipollis*; *Punct.* = *Puncatisporites*; *Lagen.* = *Lagenella*; *Enzo.* = *Enzonalasporites*; *Kr.* = *Kraeuselisporites*; *Al.* = *Alisporites*; *Au.* = *Aulisporites*; *Au. asti.* = *Aulisporites astigmosus*; *Cy.* = *Cycadopites*; *Conc.* = *Concentricisporites*

Traces of volcanic activity can be found in the sedimentary successions recording the NCIEs, thus providing indirect evidence for coeval volcanism. Sedimentary Hg is commonly used as proxy for volcanism (e.g., Grasby et al. 2019), and peaks of Hg concentrations are recorded synchronous to many LIP-related mass extinctions and major climate perturbations of the Phanerozoic (e.g., Percival et al. 2018; Grasby et al. 2019). Hg spikes have also been found in correspondence to the NCIEs of the CPE, both in marine and terrestrial successions, e.g. in Western Tethys (Mazaheri-Johari et al. 2021), Panthalassa (Jin et al. 2023), North and South China (Lu et al. 2021; Zhao et al. 2022) and Northwest China (Haserbek et al. 2025). Furthermore, Hg-isotope analysis indicates a predominance of atmospheric Hg deposition during the spikes, which would support a volcanic source (Zhao et al. 2022; Jin et al. 2023). However, some differences between records (e.g., Mazaheri-Johari et al. 2021) suggest a more complex scenario. The style of LIP emplacement (subaqueous or subaerial) strongly influences the dispersal of Hg (Percival et al. 2018): W-LIP was an oceanic plateau with an extensive subaqueous activity especially in its first phases, and the released volcanic Hg may not have been able to travel too far from the eruption centers at the beginning of the CPE, i.e. at CPE-Event1 (Mazaheri-Johari et al. 2021). Moreover, not only volcanism but also local factors can increase Hg input and drawdown in marine basins as higher terrestrial runoff and seawater redox conditions (e.g., Grasby et al. 2017; Them et al. 2019; Shen et al. 2020; Dal Corso et al. 2020b; Qiu et al. 2022).

The CPE-Event1 (sensu Dal Corso 2025) is chiefly recorded in the Northern Calcareous Alps (NCA), Austria. The Carnian sedimentary succession of the NCA, in particular that of the Reifling Basin (Fig. 16.1), has played a central role for the definition and understanding of the CPE. It is from this region that one of the first accounts of a major event in the Carnian has been given by Schlager and Schöllnberger (1974) by studying the abrupt facies change from the deep-water nodular limestones of the Reifling Formation to the dark shales of the Reingraben Formation (Fig. 16.1). Hence, the event was initially named "Reingraben Wende" (i.e., Reingraben Turning Point; Schlager and Schöllnberger 1974). The concept of a "Reingraben Wende" inspired further research into the stratigraphy, sedimentology, paleontology and geochemistry of this stratigraphic interval in other settings (Hochuli and Frank 2000; Hornung and Brandner 2005; Hornung et al. 2007a, b; Lein et al. 2012), and constituted a major foundation on which the idea of a global "Pluvial Episode" was built (Simms and Ruffell 1989). In the Reifling Basin, the first NCIE of the CPE (NCIE-1) is recorded across the lithological transition from Reifling Formation to the Göstling Formation and then Reingraben Formation (Fig. 16.2; Dal Corso et al. 2015; Mueller et al. 2016b; Lukeneder et al. 2024). Bio- and chemo-stratigraphy constrain the changes in sedimentary facies, the biological turnover, and the local geochemical changes to CPE-Event1 (Dal Corso et al. 2015; Mueller et al. 2016b; Mazaheri-Johari et al. 2021; Lukeneder et al. 2024). CPE-Event1 is therefore equivalent to the original "Reingraben Wende" of Schlager and Schöllnberger (1974).

Among the large amount of data that have been collected in the NCA in the past years, the first sedimentary Hg concentration record across CPE-Event1 comes from a section in the Reifling Basin (Mazaheri-Johari et al. 2021). In the Steinbach (or

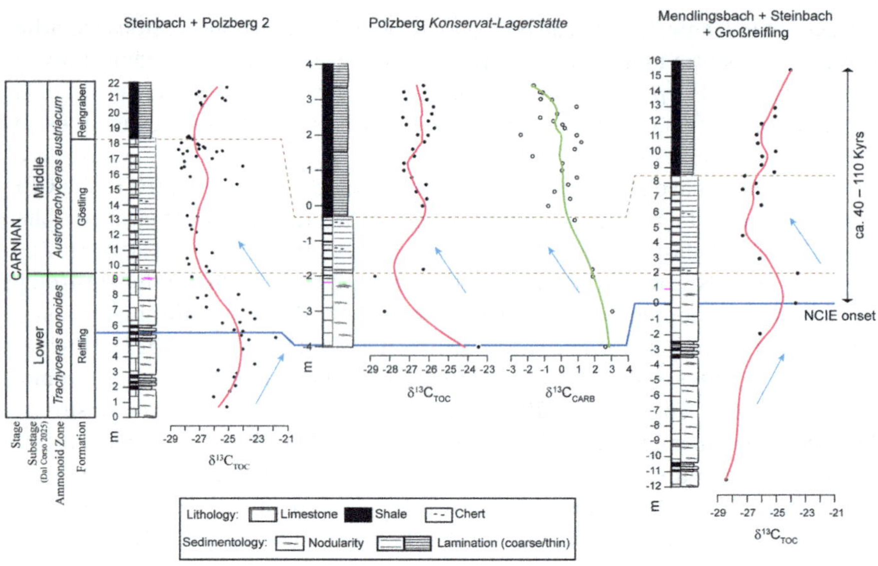

Fig. 16.2 Organic and carbonate C-isotope ($\delta^{13}C_{TOC}$ and $\delta^{13}C_{carb}$) records in sections from the Reifling Basin. Data from Steinbach-Polzberg 2 are from Dal Corso et al. (2015); data from Polzberg *Konservat-Lagerstätte* are from Lukeneder et al. (2024); data from Mendlingsbach + Steinbach + Großreifling are from Mueller et al. (2016a, b). Data are interpolated with a LOESS fitting curve, as in the other following figures. Arrows indicate the major $d^{13}C$ trends

Göstling)—Polzberg 2 section, a major spike in sedimentary Hg is recorded within the topmost Reifling Formation, at the onset of the NCIE-1 (Mazaheri-Johari et al. 2021; Fig. 16.3). In the Reifling Basin—thanks to the high-resolution biostratigraphic framework (e.g., Lukeneder et al. 2024 and references therein), which is the result of decades of detailed stratigraphic, paleontological and sedimentological studies—it is possible to disentangle the fine temporal and causal relationships between global C-cycle changes, volcanism, and local sedimentary controls. The Reifling Basin therefore offers precious insights into the rapid changes that the marine environment and ecosystem underwent during this important interval of the Triassic. Here, we present new Hg concentration data from sections encompassing CPE-Event1 in the succession of the NCA coupled to C-isotope records of the initial CPE global C-cycle disruption. Our study helps disentangling global versus local factors that affected the Reifling Basin during the event at the onset of the CPE.

16.2 Geologic Setting

The study area is located in the eastern Northern Calcareous Alps (NCA), within the Lunz Nappe (part of the Bajuvaric nappes) of the Austroalpine unit (Tollmann 1985; Linzer et al. 1995, 1997; Mandl 2000; Strauss et al. 2023; Fig. 16.1). During the

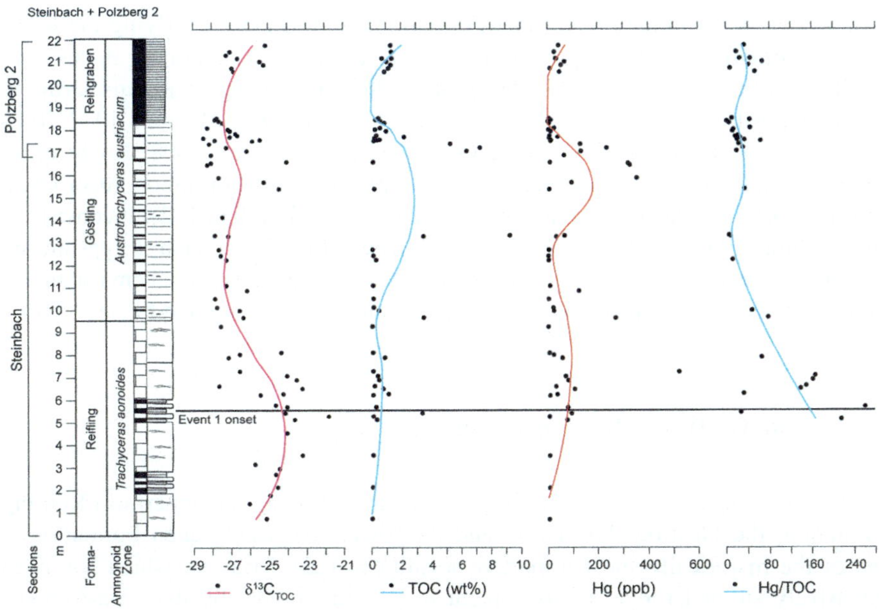

Fig. 16.3 Organic C-isotope ($\delta^{13}C_{TOC}$), total organic carbon, Hg concentrations and Hg/TOC data from Steinbach-Polzberg bis section (Dal Corso et al. 2015; Mazaheri-Johari et al. 2021)

Late Triassic, this region was part of the western Tethys realm (Tollman 1976; Lein 1987; Rüffer and Bechstädt 1999; Mueller et al. 2016b), where extensional tectonics formed isolated carbonate platforms and intraplatform basins. The eastern part of the Reifling Basin, extending between Polzberg and Großreifling, represents one of these depocenters. Here, the marine succession records the progressive infilling of the basin from carbonate platform environments to mixed carbonate–siliciclastic, and ultimately continental systems. The Carnian succession of the Lunz area comprises, in ascending order, the Reifling Formation, the Göstling Formation (which is often considered a member of the Reifling Formation in part of the literature), the Reingraben Formation (Halobienschiefer, Reingrabner Schiefer or *Trachyceras* Schiefer Auct.), the Lunz Formation and the Opponitz Formation (Fig. 16.1; Tollmann 1976; Krystyn 1991; Köppen 1997; Piller et al. 2004; Hornung and Brandner 2005; Roghi et al. 2010; Mueller et al. 2016b; Lukeneder et al. 2024). The Reifling Formation consists mainly of nodular limestones (wackestones to packstones) deposited on the distal slopes of the intraplatform basin. It is overlain by the Göstling Formation, characterized by thin-bedded radiolarian limestones and calciturbidites with intercalated organic-rich layers. Above, the Reingraben Formation represents a dark, finely laminated marl–shale succession with scarce carbonate interbeds and rich fossil content, including ammonoids, halobiids, posidoniids, radiolarians, and exceptional *Konservat-Lagerstätte* assemblages at Polzberg (Krystyn 1991; Forchielli and Pervesler 2013; Lukeneder and Lukeneder 2021; Lukeneder et al. 2024; Lukeneder et al. 2026a this volume). The Lunz Formation, conformably overlying the Reingraben Formation, consists predominantly of sandstones, siltstones, shales, and locally coal-bearing intervals (Sachsenhofer 1987), indicating marginal-marine to deltaic depositional environments (Stur 1874; Verloop 1908; Behrens 1972; Tollmann 1976; Köppen 1997; Hornung et al. 2007a; Pott et al. 2008; Roghi et al. 2010). Then, the Lunz Formation is capped by the Opponitz Formation, recording a return to carbonate deposition during the late Carnian (Tuvalian 1–2). It consists of micritic limestones and dolostones, locally with evaporitic facies such as gypsum and dolomitic breccias, representing shallow-marine to restricted lagoonal environments (Kristan-Tollmann and Hamedani 1973; Tollmann 1976; Dunay and Fisher 1978).

16.3 Biostratigraphic Constraints

The Carnian succession of the Reifling Basin encompassing the uppermost Reifling Formation, the Göstling Formation and the lowermost Reingraben Formation is one of the historically most important localities of the Triassic, which has been extensively studied for its fossil content (e.g., ammonoids, bivalves, gastropods, fishes and sporomorphs) and biostratigraphy (Stur 1874; Teller 1891; Griffith 1977; Krystyn 1978; Mueller et al. 2016b; Lukeneder and Lukeneder 2022, 2021). The basinal Carnian succession of the Reifling intraplatform basin yields a diverse and biostratigraphically significant ammonoid fauna. In the uppermost Reifling Formation, early Carnian (Julian 1, *Trachyceras aonoides* Zone) assemblages are represented by scarce *Trachyceras* spp. and related genera (Lukeneder et al. 2024). The overlying Göstling Formation contains an assemblage dominated by *Austrotrachyceras triadicum*, which defines the *Austrotrachyceras austriacum* Zone (with the *Austrotrachyceras triadicum* Subzone). The Reingraben Formation at Polzberg has yielded abundant and well-preserved specimens, composed mainly of *Austrotrachyceras minor* and *Paratrachyceras haberfellneri*, together with *Sageceras haidingeri*, and *Carnites floridus*, making the section a key regional biostratigraphic reference (Lukeneder and Lukeneder 2022, 2021; Lukeneder et al. 2024). The Lunz Formation lacks ammonoids: the predominance of siliciclastic and coal-bearing deposits reflects marginal-marine to deltaic environments that did not favor the preservation of open-marine cephalopods.

Conodonts were found in the Reifling Formation and in the Göstling Formation (Mostler and Scheuring 1974; Krystyn 1978, 1991; Rüffer and Bechstädt 1999; Lukeneder et al. 2024). There the conodont associations include typical Julian (early–middle Carnian) forms as, e.g., *Paragondolella polygnathiformis*, *Paragondolella auriformis*, *Paragondolella inclinata*, *Gladigondolella tethydis*, *Gladigondolella malayensis malayensis* and *"Mosherella" longnanensis*. According to conodont biostratigraphy (Rigo et al. 2018), Lukeneder et al. (2024; Fig. 16.1) constrained the succession of the Reifling Basin to, from older to younger, the *Mazzaella carnica* Zone (upper Reifling Formation; *Trachyceras aonoides* ammonoid Zone), the *Paragondolella praelindae* Zone (Göstling Formation; *Austrotrachyceras austriacum* ammonoid Zone, *Austrotrachyceras triadicum* Subzone), and the *Gladigondolella tethydis* Zone (Reingraben Formation; *Austrotrachyceras austriacum* ammonoid Zone, *Austrotrachyceras minor* Subzone).

Detailed palynological studies of the succession of the Reifling Basin allowed the definition of ammonoid-calibrated sporomorph assemblages across the Reifling-Göstling-Reingraben Formations. Sporomorphs are in general scarce in the Reifling Formation (Roghi et al. 2010; Mueller et al. 2016b), where a low-diversity assemblage (assemblage A or *Concentricisporites bianulatus* assemblage sensu Roghi et al. 2010, and assemblage L1 sensu Mueller et al. 2016a, b; Fig. 16.1) has been defined. These assemblages are characterized by the presence of *Ovalipollis* spp., *Triadispora* spp., and *Concavisporites* spp. (Roghi et al. 2010), with an increase of *Enzonalasporites vigens* and *Partitisporites* spp. at the top of it (Mueller et al. 2016a, b). Assemblage L1 is found in the *Trachyceras aonoides*–basal *Austrotrachyceras austriacum* ammonoid zones, extending into the Göstling Formation (Mueller et al. 2016a, b; Fig. 16.1). The Reingraben Formation is rich in sporomorphs and shows a high diversity (Roghi et al. 2010; Mueller et al. 2016a, b). At the base of the Reingraben Formation an acme of *Aulisporites astigmosus* and the presence of *Calamospora lunzensis, Distalanulasporites punctus, Kraeuselisporites cooksonae, Leschikisporis aduncus, Tigrisporites halleinis, Duplexisporites* sp. 1 defined the transition to assemblage B or *Aulisporites astigmosus* assemblage (Roghi et al. 2010; Fig. 16.1), within the lower part of the *Austrotrachyceras austriacum* ammonoid Zone. However, new palynological data presented in Roghi and Lukeneder (2026, this volume) show that the *Aulisporites astigmosus* assemblage occurs at least from the base of the Göstling Formation, i.e., its base could coincide with the base of the *Austrotrachyceras austriacum* ammonoid Zone, but *Aulisporites* is also present in the uppermost Reifling Formation (Fig. 16.1). This assemblage has been subdivided into *Aulisporites-Cycadopites* assemblage and *Kraeuselisporites-Alisporites* assemblage by Mueller et al. (2016a, b), the latter being marked by an increase in *Kraeuselisporites cooksonae, Concavisporites* spp. and *Alisporites* spp., and a decrease of *Enzonalasporites vigens, Cycadopites* spp. and *Calamospora lunzensis*. Above, the presence of spores as *Combaculatisporites mesozoicus* and *Raistrickia alpina* and pollen grains as *Lagenella martinii, Equisetosporites chinleanus, Araucariacites* spp., with an increase in *Enzonalasporites vigens, Camerosporites pseudoverrucatus, Lunatisporites* sp. and *Punctatisporites* sp., and a decrease in *Aulisporites astigmosus* and *Aratrisporites* spp. define assemblage C or *Lagenella martini* assemblage (sensu Roghi et al. 2010). This assemblage has been subdivided into *Lagenella-Enzonalasporites* assemblage and *Ovalipollis-Punctatisporites* assemblage by Mueller et al. (2016a, b; Fig. 16.1), the latter being characterised by a decrease of *Cycadopites* spp. and *Densosporites* sp., and an increase of *Aratrisporites* spp., *Punctatisporites* sp., and *Ovalipollis* spp. (Mueller et al. 2016a, b). The *Lagenella martini* assemblage was originally defined in the middle–upper part of the Lunz Formation (Roghi et al. 2010), i.e. upper *Austrotrachyceras austriacum*–lower *Tropites dilleri* ammonoid Zones (Fig. 16.1), but its lower boundary could extend down to the base of the Lunz Formation (Mueller et al. 2016a, b) or even into the Reingraben Formation (Fig. 16.1). Indeed, the above-described sporomorphs assemblages of the NCA can be correlated through the entire western Tethys (e.g., Roghi et al. 2010; Mueller et al. 2016b; Baranyi et al. 2019b; Alverà et al., 2026), and in the Dolomites (Italy) and Transdanubian Range (Hungary) the boundary between the *Aulisporites astigmosus* (assemblage B) and *Lagenella martini* (assemblage C) assemblages is placed in a stratigraphic level within the lower *Austrotrachyceras austriacum* ammonoid Zone, at the end of the positive rebound of the NCIE of CPE-Event1 (Dal Corso et al. 2012, 2018; Baranyi et al. 2019b; Dal Corso 2025), which in the Reifling Basin occurs in the lower part of the Reingraben Formation (Mueller et al. 2016a, b; see discussion below and Fig. 16.2).

16.4 Methods

New Hg data have been collected from Polzberg *Konservat-Lagerstätte* section (Lukeneder et al. 2024; Lukeneder 2026, this volume) and Großau core (Lukeneder et al. 2026, this volume), and coupled with a published Hg record from the composite Steinbach-Polzberg 2 section (Mazaheri-Johari et al. 2021). Polzberg 2 section is located in the same area but is different from the Polzberg *Konservat-Lagerstätte* section (e.g., Lukeneder et al. 2024). The sections cover an interval encompassing the uppermost Reifling Formation, the Göstling Formation and the lowermost Reingraben Formation (Figs. 16.1 and 16.2), which is constrained to the transition between the Julian 1 and the Julian 2, i.e., the *Trachyceras aonoides* to *Astrotrachyceras austriacum* ammonoid Zones (see below), where the onset of the CPE (Event1) has been previously identified and its first C-isotope excursion occurs (Schlager and Schöllnberger 1974; Hornung et al. 2007a; Dal Corso et al. 2015; Mueller et al. 2016b; Lukeneder et al. 2024; Lukeneder 2026, this volume).

Fifty-eight rock samples were cleaned to remove surficial impurities and then powdered using an agate mortar. The concentration of sedimentary Hg was measured using a Lumex RA-915 Mercury Analyzer coupled to a PYRO-915 Pyrolyzer at the State Key Laboratory of Geomicrobiology and Environmental Changes of China University of Geosciences Wuhan. RA-915 M/PYRO-915 has a detection limit for sediments and rocks of 0.5 ppb (https://www.lumexinstruments.com). ~50 mg of powdered sample was weighed in a glass boat, which were then inserted into the instrument. Results were calibrated to soil standard GBW07381 (148 $\pm$ 11 ppbHg). Analytical precision of <8% was assessed using another soil standard (GBW07423; 32 $\pm$ 3 ppbHg), which was analyzed every 10 samples, in duplicate. Total organic carbon (TOC) and total sulfur (TS) contents were measured using an ELTRA Helios CS elemental analyzer. Calibration was done with three standard measurements at the beginning and after every 10 measurements. For the TOC measurements, the samples were decarbonatized with phosphoric acid. All sample measurements were performed in duplicate. TOC, TS, and biomarker data (TAR and CPI) from Polzberg section are from Lukeneder et al. (2024).

16.5 Results

Hg concentration data from Polzberg *Konservat-Lagerstätte* are new and have been plotted together with published TOC, TS and C-isotope data (Lukeneder et al. 2024). Hg, TOC and TS data from Großau (Fig. 16.5; Supplementary material 16.1) are all new. These new records have been coupled and compared to a published record from Steinbach-Polzberg 2 section (Fig. 16.3; Mazaheri-Johari et al. 2021).

Samples from Polzberg *Konservat-Lagerstätte* have Hg concentrations ranging from 1.4 to 113 ppb (Fig. 16.4). The lowest contents of Hg (with a maximum of 5.8 ppb) are found in the Reifling Formation (Fig. 16.4). The single datapoint in the Göstling Formation shows higher Hg concentration (Hg = 45 ppb). Hg remains high in the shales of the Reingraben Formation (Fig. 16.4). In the uppermost part of the section (meters 2–4), Hg measured in the limestone interlayers has lower values then the Hg in the embedding shales (Fig. 16.4). Peak Hg concentrations of >80 ppb, with a maximum of 113 ppb, are measured in the uppermost part of the analyzed interval, i.e. 3 m above the base of the Reingraben Formation (Fig. 16.4). At Polzberg *Konservat-Lagerstätte* Hg is moderately correlated to TOC (R = 0.41), and strongly correlated to TS (R = 0.64; Fig. 16.6).

Samples from Großau have Hg concentrations ranging from 2.5 pbb to 59.5 ppb (Fig. 16.5; Supplementary material 16.1). As at Polzberg *Konservat-Lagerstätte*, very low Hg concentrations (<20 ppb) are found in the Reifling Formation, but Hg concentrations are also comparably low in the Göstling Formation (Fig. 16.5). Hg concentrations in the Reingraben Formation vary between ~30 and ~60 ppb (Fig. 16.5). The highest Hg concentrations (57.7 and 59.5 ppb) are recorded 1.5–2.5 m

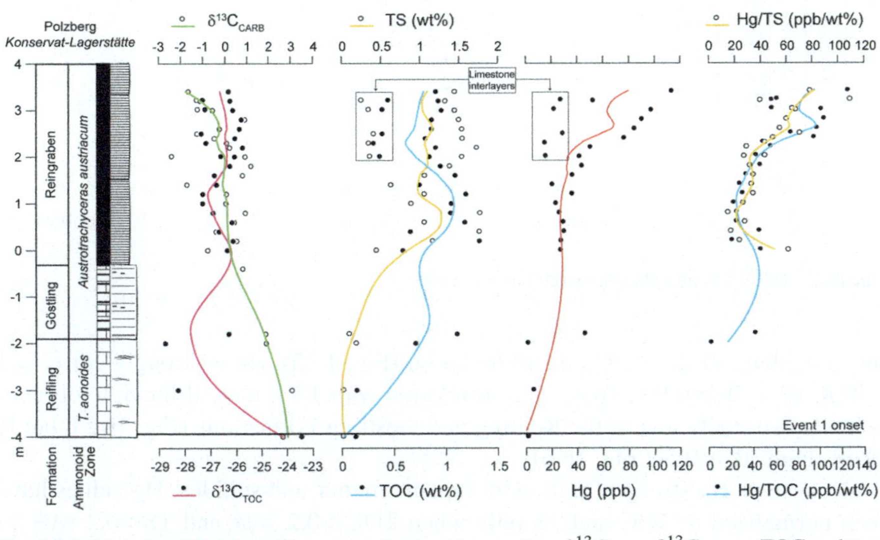

Fig. 16.4 Data from Polzberg *Konservat-Lagerstätte* section. $\delta^{13}C_{TOC}$, $\delta^{13}C_{CARB}$, TOC and total sulfur (TS) are from Lukeneder et al. (2024)

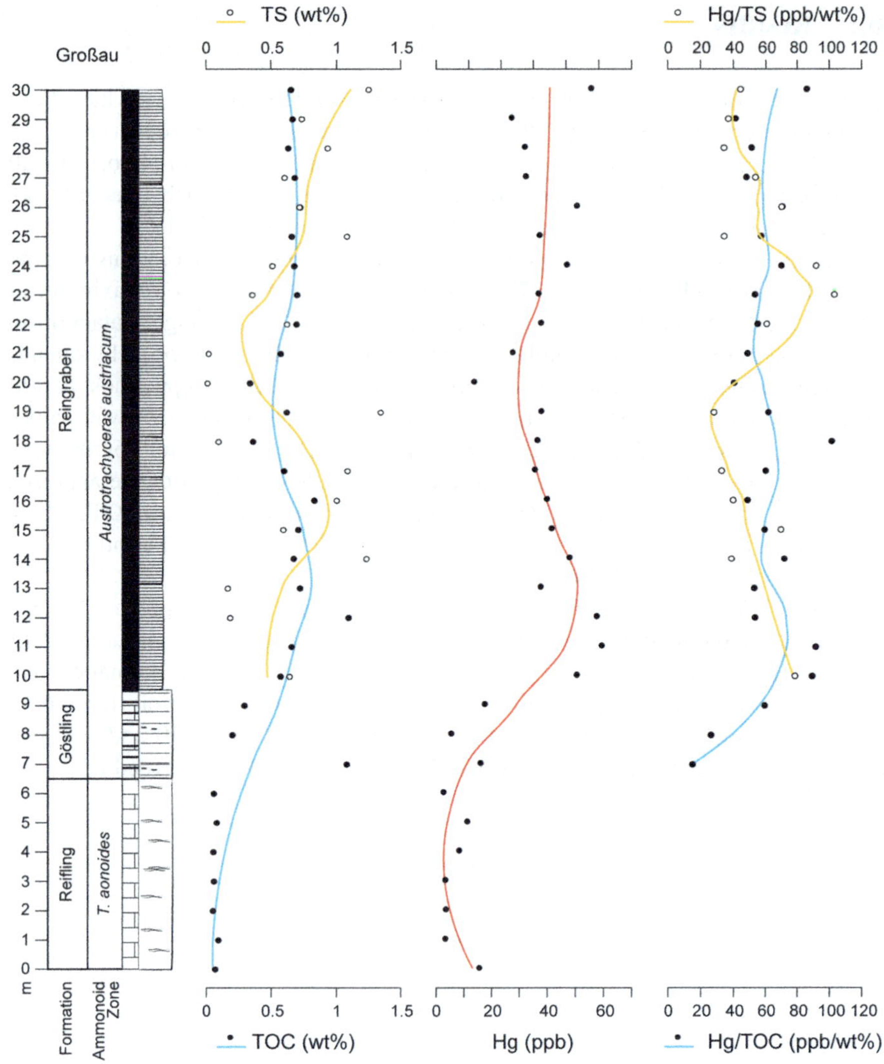

Fig. 16.5 TOC, TS, and Hg data from Großau core

above the base of the Reingraben Formation (Fig. 16.5). Hg is strongly correlated to TOC ($R = 0.76$) (Fig. 16.6). The correlation with TS is more difficult to observe as TS is extremely low in the Reifling and Göstling Formations (Fig. 16.5), but R results high ($R = 0.59$; Fig. 16.6).

Because of Hg strong affinity with organic matter and sulfides, Hg values have been normalized to TOC and TS only when TOC >0.2 wt% and TS >0.2 wt% to minimize biases introduced by low TOC and TS values in estimating environmental enrichment of Hg (Grasby et al. 2019; Shen et al. 2020). Both Hg/TOC and Hg/TS

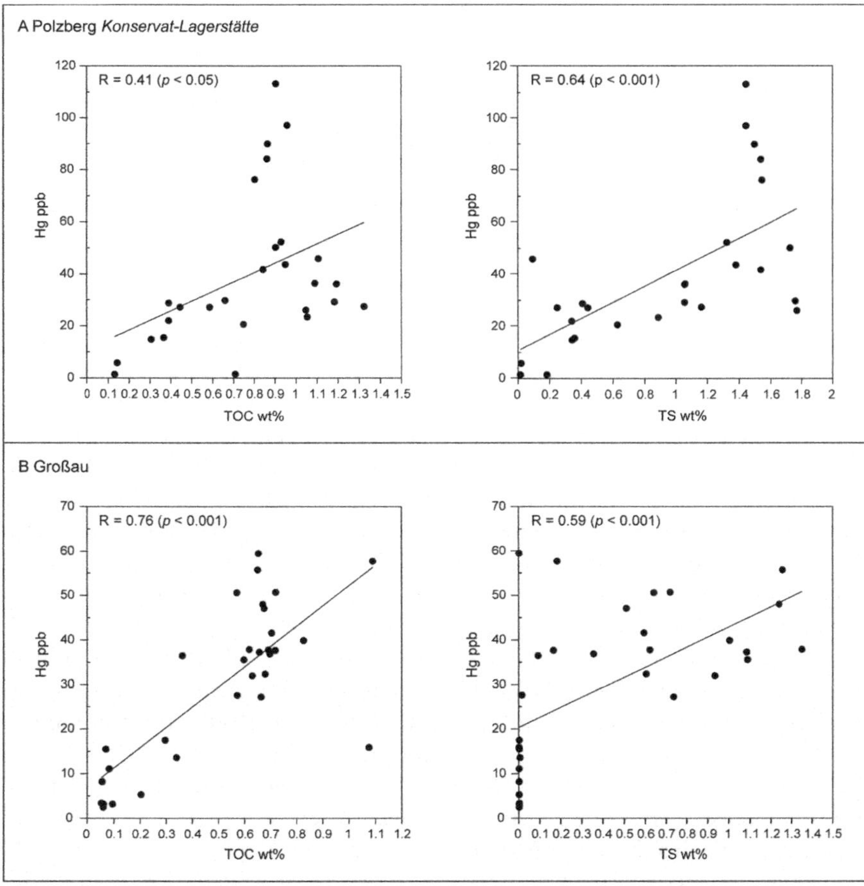

Fig. 16.6 Cross-plots Hg versus TOC and Hg versus TS from Polzberg *Konservat Lagerstätte* section and Großau core

show trends that parallel Hg values, i.e. an upward increase with high ratios in the Reingraben Formation (Figs. 16.4 and 16.5). When normalized to TOC and TS, the differences in Hg concentrations found between the shales and limestone interlayers in the upper part of the Reingraben Formation at Polzberg *Konservat-Lagerstätte* are smoothed (Fig. 16.4). In all considered sections, variations in Hg concentration and Hg/TOC ratio are not associated to lithological changes.

16.6 Discussion

16.6.1 Revision of CPE-Event1 C-Cycle Perturbation as Recorded in the Reifling Basin

The discovery of a major NCIE associated with the onset of the CPE interval showed that the facies and biological changes that were observed in many Carnian sedimentary successions were the local expression of a global C-cycle perturbation (e.g., Dal Corso et al. 2012, 2015, 2018; Sun et al. 2016; Miller et al. 2017). This important step-forward in the comprehension of the CPE was crucial to boost the research on this time interval of the Triassic and to establish the CPE as a major environmental and biological turning point in Earth's history. Before the detection of the NCIE, the idea of a global CPE was questioned and its nature unclear (Visscher et al. 1994; Ruffell et al. 2016; Simms and Ruffell 2018): The link between the NCIE and the emplacement of Wrangellia, which had been previously suggested (Furin et al. 2006), allowed to propose a mechanism of environmental perturbation similar to other LIP-related events of the Phanerozoic (Dal Corso et al. 2012, 2020a).

The events that marked the Carnian are, however, still far from being fully understood and the growing amount of data collected is obviously complicating the initially proposed, rather simple picture. For example, records are still scarce and in many cases with poor age constraint; the environmental changes appear to have been paleogeographically heterogeneous; the response of the marine and terrestrial fauna is unclear; the actual concept of a CPE is becoming questionable (e.g., Dal Corso 2025). Despite the usual complications arising from detailed studies of sedimentary successions, it remains clear that the Carnian exogenic C-cycle was severely perturbed, several times, and that this perturbation is recorded everywhere as successive NCIEs punctuating an interval of the Carnian of ca. 1.2–2.5 Myr named CPE. Four NCIEs are clearly visible in C-isotope records from marine and terrestrial settings, and at different latitudes (Dal Corso et al. 2012, 2015, 2018; Sun et al. 2016, 2019; Mueller et al. 2016a, b; Miller et al. 2017; Jin et al. 2019, 2020, 2022a, b, 2023; Shi et al. 2019; Baranyi et al. 2019a, b; Fu et al. 2020; Lu et al. 2021; Tomimatsu et al. 2021; Lestari et al. 2024; Rahman et al. 2024; Zhang et al. 2024; Haserbek et al. 2025; Jia et al. 2025; Alverà et al. 2026). Biostratigraphy dates the NCIE-1 across the boundary between *Trachyceras aonoides* and *Austrotrachyceras austriacum* ammonoid Zones, NCIE-2 within the *Austrotrachyceras austriacum* ammonoid Zone, NCIE-3 across the boundary between *Austrotrachyceras austriacum* and *Tropites dilleri* ammonoid Zones, and NCIE-4 at the base of *Tropites subbullatus* ammonoid Zone (Fig. 16.1; e.g., Dal Corso et al. 2018). These NCIEs indicate four major injections of ^{13}C-depleted CO_2 into the Carnian atmosphere-land–ocean system, and each of them are linked to discrete environmental changes, hence indicating distinct events within the CPE interval (Fig. 16.1; Dal Corso 2025).

Some of the first records of the NCIE-1 have indeed been produced from sections that are located in the Reifling Basin of the NCA (Dal Corso et al. 2015; Mueller et al. 2016b). In the composite Steinbach + Polzberg 2 section, Dal Corso et al. (2015)

showed a ca. 3–4‰ negative organic C-isotope shift beginning from the topmost Reifling Formation and continuing through the Göstling Formation and basal Reingraben Formation (Fig. 16.2). In this paper, biostratigraphic correlation with other organic C-isotope records from Dolomites (Italy) and the Transdanubian Range (Hungary), allowed to prove synchroneity between the C-cycle perturbation and the environmental change marking the onset of the CPE interval in different settings of the Western Tethys realm (Dal Corso et al. 2015). From a local perspective, the work of Dal Corso et al. (2015) showed that the beginning of the global perturbation slightly preceded the facies changes from limestone to shale in the NCA, which originally defined the "Reingraben Wende" (Schlager and Schöllnberger 1974). Likewise, Mueller et al. (2016a, b) reported an NCIE of similar magnitude in the composite Mendlingsbach + Steinbach + Großreifling section beginning from the topmost Reifling Formation (Fig. 16.2). In this case, thanks to the longer stratigraphic interval analyzed, the end of the NCIE-1 was also captured within the Reingraben Formation (Fig. 16.2; Mueller et al. 2016a, b). The study of Mueller et al. (2016a, b) provided a solid link between this NCIE and a major vegetation change in the western Tethys realm, with a rise of hygrophytic plants. Notably, both records of Dal Corso et al. (2015) and Mueller et al. (2016a, b) show that the NCIE of CPE-Event1 interrupts the positive C-isotope trend that is a prominent feature of the global C-isotope record of the late Ladinian–early Carnian (Fig. 16.2; Korte et al. 2005; Dal Corso et al. 2011, 2022). Lukeneder et al. (2024) presented new C-isotope data from bulk organic matter ($\delta^{13}C_{TOC}$) and carbonates ($\delta^{13}C_{carb}$) measured at Polzberg *Konservat-Lagerstätte* section. Both $\delta^{13}C_{TOC}$ and $\delta^{13}C_{carb}$ display a negative C-isotope shift beginning in the topmost Reifling Formation (Fig. 16.2). As for the $\delta^{13}C_{TOC}$, the magnitude of the $\delta^{13}C_{carb}$ shift is also 3–4‰ (Fig. 16.2). However, $\delta^{13}C_{TOC}$ appears to change faster than $\delta^{13}C_{carb}$, which shows a more gradual variation, but this effect can be related to the very low data resolution in the Reifling Formation and the Göstling Formation (Fig. 16.2). Despite this, the record of Lukeneder et al. (2024) is one of the very few records worldwide that shows, in the same section, a synchronous isotopic change in both the organic and carbonate carbon pools. Alverà et al. (2026) showed that CPE's NCIE-1 is actually composed by at least two main excursions (named NCIE-1α and -1β), and that these two excursions can also be recognized in other records, including the $\delta^{13}C$ curve from the Reifling Basin (NCA).

According to the published cyclostratigraphic estimates, the NCIE of CPE-Event1 lasted for ca. 40–90 kyr (Miller et al. 2017; Zhang et al. 2024). Hence, the sedimentological and biological changes recorded in the Reifling Basin within the interval comprising this NCIE—i.e., the expression of CPE-Event1 (sensu Dal Corso 2025) in this marine basin of the NCA—were geologically very rapid, showing a quick response of the environment to the global C-cycle perturbation.

16.6.2 Mercury Deposition in the Reifling Basin at CPE-Event1

Variations in sedimentary Hg concentrations, Hg/TOC and Hg/TS ratios are not related to lithological changes in the studied sections (Figs. 16.3, 16.4, and 16.5). At the Steinbach-Polzberg 2 section, peak Hg/TOC values found in the nodular limestones of the Reifling Formation at ca. 4–7 m decrease up to the middle part of the laminated mudstones of the Göstling Formation (Fig. 16.3). At the Polzberg *Konservat-Lagerstätte* section, the high-resolution data from the basal 4 m of the Reingraben Formation reveal a gradual increase within the pile of laminated shales (Fig. 16.4). Similarly, at Großau, Hg/TOC values rise from the middle of the Göstling Formation up to 3 m into the Reingraben Formation across the lithological shift from the dark mudstones–grainstones to the shales that differentiate the two formations (Fig. 16.5). Hence, major lithological controls on sedimentary Hg geochemical signals are excluded.

Moreover, significant diagenetic effects on the geochemical proxies from the succession of the Reifling Basin are excluded, as shown in previous studies (Dal Corso et al. 2015; Lukeneder et al. 2024). It has been shown that Hg can be enriched in the sediment by increasing thermal maturation: in mature and overmature successions the measured Hg enrichments relative to immature rocks were up to +132% to + 157% due to re-distribution of mobile and volatile Hg in the sediment (Indraswari et al. 2025). Rock–Eval T_{max} and PI (production index), and biomarker proxies from Polzberg *Konservat-Lagerstätte* section show low thermal maturity (immature to marginal mature; Lukeneder et al. 2024).

High Hg concentrations have been found in connection with the NCIEs of the CPE in different marine and terrestrial geological settings. Mazaheri-Johari et al. (2021) found Hg spikes in coincidence with CPE's Events in marine successions of the Western Tethys, but with some notable differences between the records. In particular, increases in Hg/TOC at CPE's NCIE 1–3 are clearly visible in the more open basins of the Northern Calcareous Alps (Austria) and the Transdanubian Range (Hungary), while in the more restricted Dolomites Basin (Italy) only a Hg/TOC spike at NCIE-3 (CPE-Event3 sensu Dal Corso 2025; Fig. 16.1) is recorded. On the one hand, based on the interpretation of the Hg spikes as of volcanic in origin, the differences between the basins were attributed to the fact that Wrangellia was an oceanic plateau with extensive submarine activity, especially in its initial phases, and the volcanic Hg dispersal was limited by the paleogeography, as observed for other LIPs with a submarine or mixed subaerial/submarine activity (Percival et al. 2018). On the other hand, the basal part of NCIE-1, i.e. the stratigraphic interval where the Hg/TOC spike is recorded in the basins of Austria and Hungary, could be missing in the section of the Dolomites (Dal Corso et al. 2015; Mazaheri-Johari et al. 2021). This second simpler explanation seems most likely, as Hg spikes have been found in the terrestrial (lacustrine) sequences of North China coincident to all NCIEs of the CPE (Lu et al. 2021), and in the Junggar Basin (Nortwest China) the most pronounced Hg/TOC spike is indeed related to the NCIE at CPE-Event1 (Haserbek et al. 2025). If limited marine dispersal of Hg was responsible for the discrepancies between restricted and open marine basins, it is unlikely that the volcanic Hg could reach terrestrial settings that were located far from the sea (Haserbek et al. 2025). Higher Hg loading in marine basins has been also found in successions of South China (Zhao et al. 2022), Tibet (Zhang et al. 2024), and Japan (Jin et al. 2023), although with variable magnitudes and not always clearly recorded during all CPE's events (see compilations in Jin et al. 2023 and Haserbek et al. 2025).

A volcanic origin of the excess Hg deposited during CPE's events appears to be favored in the literature, but other mechanisms could have partially or totally controlled the observed increasing Hg loading to the depositional settings. In particular, the CPE is marked by a remarkable increase in runoff with the delivery of massive amounts of terrestrial material into the basins (Dal Corso et al. 2024b). In some specific depositional settings, terrestrial input was so significant that completely obfuscated the global C-isotope signal in the bulk organic matter (Jia et al. 2025). As soil and vegetation are amongst the largest Hg reservoirs on Earth's surface (Amos et al. 2013; Schuster et al. 2018), erosion or massive oxidation of terrestrial organic matter could indeed result in higher Hg concentrations in lacustrine and marine sediments (e.g., Them et al. 2019; Dal Corso et al. 2020b; Jin et al. 2022a, b). Negative shifts in Δ^{199}Hg associated to the Hg spikes in South China have been interpreted to reflect mixed deposition of both atmospheric and terrestrial Hg (Zhao et al. 2022). The higher Hg concentrations found during NCIE-3 and NCIE-4 in the terrestrial Junggar Basin have been shown to be linked to peaks in transport of material from C3 plants to the lake (Haserbek et al. 2025). In addition to volcanism and runoff, other mechanisms could potentially increase Hg drawdown as the development of euxinic conditions (e.g., Shen et al. 2020), which have been observed in South China during the CPE (Sun et al. 2016; Zhang et al. 2022). In summary, despite the Hg records across the CPE are still geographically limited and sometimes of too low-resolution, available data point to a complex combination of mechanisms controlling Hg deposition during CPE's C-cycle perturbations.

The high-resolution records from the Reifling Basin across CPE-Event1 give important insights into different global and local processes that could govern Hg loading to this marine setting during CPE-Event1. The first straightforward observation is that in all analyzed sections from the Reifling Basin, higher Hg concentrations than background values are recorded during the NCIE (Figs. 16.3, 16.4, and 16.5). Secondly, the Hg spikes do not depend on the amount of TOC and TS in the sediment: although some statistical correlations between Hg versus TOC and Hg versus TS exist (Fig. 16.6), Hg/TOC and Hg/TS show similar trends that are parallel to and consistent with the recorded changes in Hg concentrations through time (Fig. 16.7). Finally, the increase in Hg concentrations recorded in the basal Reingraben Formation run parallel and are highly correlated to biomarkers TAR (terrestrial/aquatic ratio) and CPI (carbon preference index) indices (Fig. 16.8).

TAR is the ratio of long to short chain *n*-alkanes and CPI is the ratio between *n*-alkanes with odd over even carbon-number predominance: both indices are used as proxies for the relative contribution of land-plants to TOC in marine basins (Peters et al. 2005). At Polzberg *Konservat-Lagerstätte* section, both TAR and CPI rise from

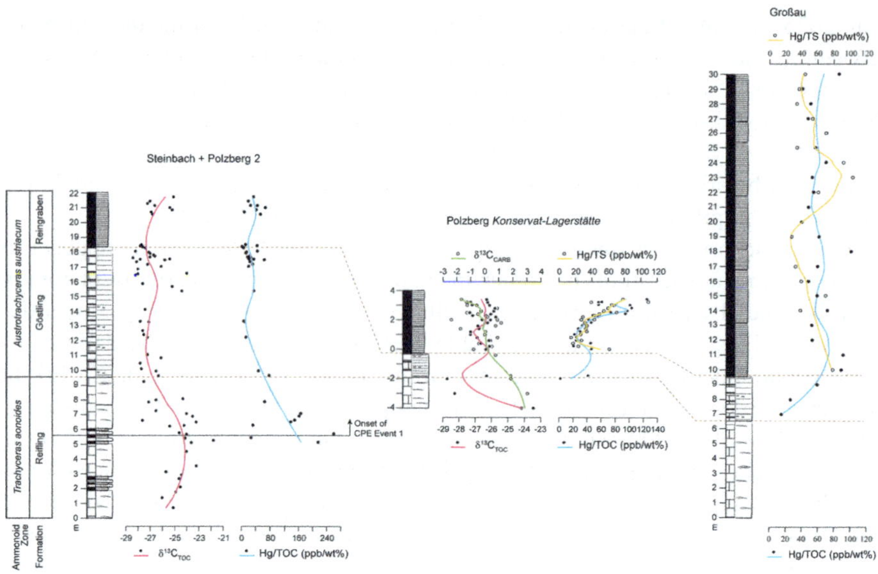

Fig. 16.7 Correlation of C-isotope and Hg (normalized to TOC and TS) records

the top of the Göstling Formation through the first 4 m of the Reingraben Formation (Fig. 16.8), suggesting higher input of land plant material to the Reifling Basin (Lukeneder et al. 2024). Hg and Hg/TOC follow this trend suggesting that increasing Hg loading could be closely linked to increasing terrestrial runoff. As mentioned before, soil and terrestrial vegetation are large reservoirs of Hg on Earth's surface (Amos et al. 2013; Shuster et al. 2018). Massive mobilization of the Hg stored in the terrestrial reservoirs as a consequence of enhanced hydrological cycle, soil erosion and/or massive oxidation of organic matter could produce significant spikes in sedimentary Hg concentrations as those observed, e.g. during the Toarcian Oceanic Anoxic Event (Them et al. 2019; Jin et al. 2022a, b), during the end-Triassic mass extinction (Bos et al. 2024) and during the Permian–Triassic mass extinction (Dal Corso et al. 2020b). Hence, the increase in Hg at the transition between the Göstling and Reingraben Formations is attributed to increasing siliciclastic input, as indicated by integrated geochemical and sedimentological evidence. Interestingly, higher Hg, TAR and CPI are not associated to higher TOC, a feature that has been interpreted as the evidence of either decreasing marine productivity or dilution of marine organic matter due to the high siliciclastic input (Lukeneder et al. 2024). The increase in Hg/TOC in the Göstling – Reingraben Formations can also be seen in the record from Großau section, where Hg/TOC rises on average from ca. 20 ppb/wt% to ca. 60–70 ppb/wt%, as at Polzberg *Konservat-Lagerstätte* (Fig. 16.7). This phenomenon is less pronounced in the Steinbach + Polzberg 2 record, where Hg/TOC at the top of the Göstling Formation is on average 26 ppb/wt% and increases to an average 38 ppb/wt% in the Reingraben Formation (Fig. 16.7). Although this Hg/TOC ratio increase

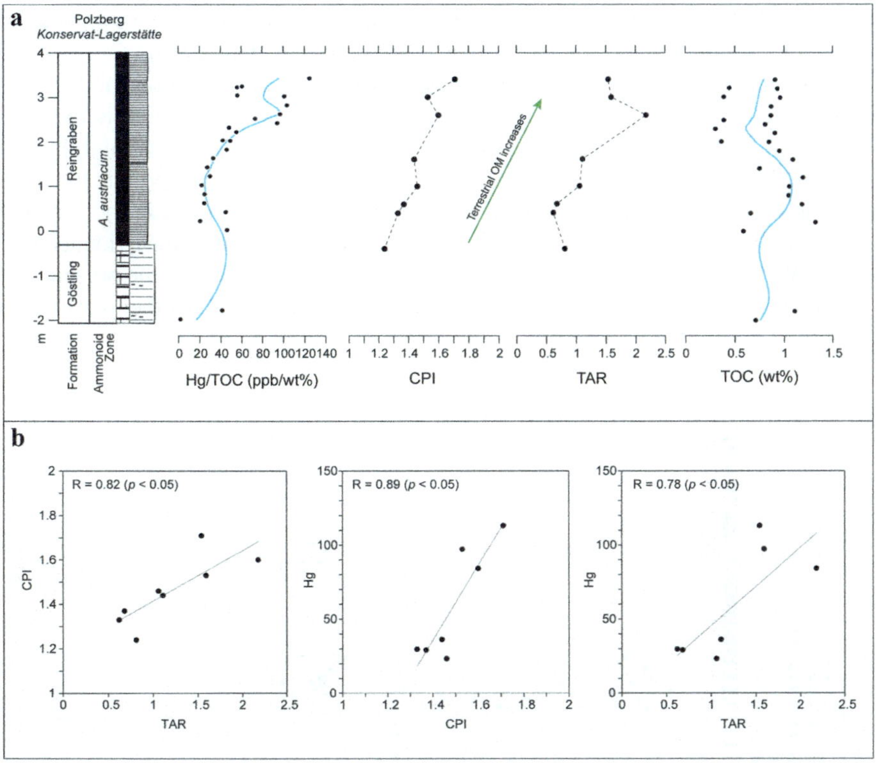

Fig. 16.8 **a** Hg data against biomarker proxies indicating the relative contribution of land-plants to TOC in marine basins. TAR = terrestrial/aquatic ratio; CPI = Carbon Preference Index. Biomarkers are from Lukeneder et al. (2024). **b** Cross-plots showing the correlations between proxies

is recorded in all studied sections in the same stratigraphic interval, the rather small discrepancies between records could be related to sampling biases, or differences between the depositional sites, e.g. a more distal or more proximal position of the section from the source of terrigenous Hg, and different sedimentation rates within the basin. The position of this progressive terrigenous-Hg increase with respect of the $\delta^{13}C_{TOC}$ record is revealing as it occurs in the stratigraphic interval recording the positive rebound of the NCIE (Fig. 16.9). The enhanced hydrological cycle linked to global warming and rising atmospheric pCO_2 levels could overall increase chemical weathering (silicate hydrolysis) and burial of organic matter globally, which are processes that can remove ^{13}C-depleted carbon from the surface system and result in the recorded positive $\delta^{13}C$ rebound, as well as increase sedimentary Hg concentrations in marine basin by higher riverine input and/or higher drawdown in an anoxic environment. In the Dolomites (Western Tethys), in the same stratigraphic interval encompassing the positive rebound of CPE-Event1 NCIE, a general increase of chemical weathering is indeed recorded by increasing kaolinite over chlorite + illite and feldspars, and more riverine input is shown by B/Ga, Ti/Al and Zr/Al data

(Alverà et al. 2026). This follows an interval with episodic aridifications during the pulses of CO_2 into the atmosphere–ocean system (Alverà et al. 2026).

The larger older spike in Hg/TOC at the onset of the NCIE within the Reifling Formation (Fig. 16.2) is recorded only at Steinbach-Polzberg 2, which is the only section analyzed for Hg covering that stratigraphic interval with high-resolution data (Figs. 16.3 and 16.9). The full nature of this spike cannot be evaluated due to a lack of biomarker data or other proxies of terrestrial runoff and erosion in the Reifling Formation of this section. This spike was originally interpreted as the evidence of a pulse of Wrangellia volcanism: the interpretation is substantiated by the presence of large Hg/TOC spikes found within the same stratigraphic interval in other marine and

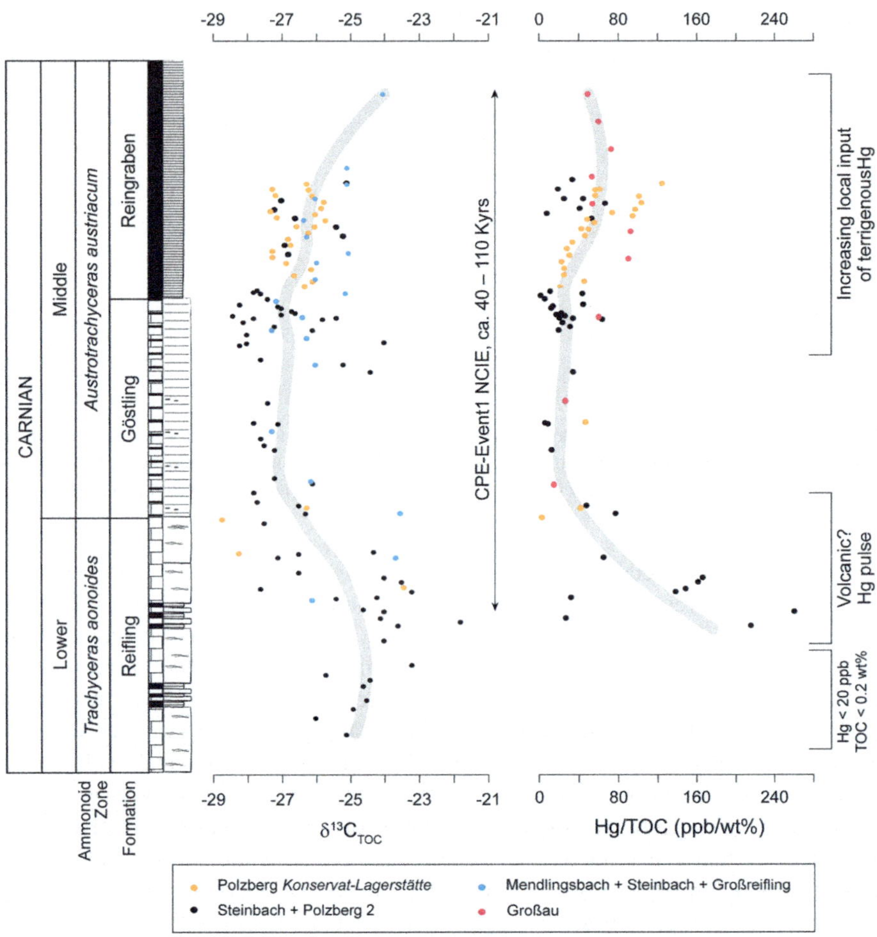

Fig. 16.9 Compilation of all organic C-isotope and Hg data from the Reifling Basin, and interpretation of the main Hg sources to the basin according to the current evidences discussed in the main text. The LOESS fitting is performed on all the compiled data from the studied sections

terrestrial sections worldwide, which in some cases are coupled to $\Delta^{199}Hg$ signatures suggesting a volcanic origin (Mazaheri-Johari et al. 2021; Lu et al. 2021; Zhao et al. 2022; Jin et al. 2023; Zhang et al. 2024; Haserbek et al. 2025). In many records the data resolution is not enough to evaluate the relative timing of Hg/TOC and $\delta^{13}C$ variations, but in the records from Western Tethys the most pronounced Hg/TOC spike is placed at the very onset of the NCIE (Mazaheri-Johari et al. 2021). Moreover, in the Western Tethys this initial Hg/TOC spike occurs before the major facies change from mainly carbonate to siliciclastic sedimentation (Mazaheri-Johari et al. 2021), hence before the increase in input of terrigenous material (Fig. 16.9). In summary, given its widespread occurrence coupled to local sedimentological and geochemical considerations, the first Hg/TOC spike observed in the succession of the Reifling Basin at the onset of the NCIE could likely have been related mainly to LIP volcanism, as previously suggested (Mazaheri-Johari et al. 2021). Interestingly, the relative timing between the registered phenomena could indicate a scenario similar to that suggested for, e.g. the end-Triassic mass extinction, when the Hg injected by the Central Atlantic Magmatic Province first enriched (and poisoned) the terrestrial environments across the Triassic-Jurassic boundary and was later re-mobilized during the Hettangian (Early Jurassic) due to enhancement of the hydrological cycle and erosion (Bos et al. 2024).

Data from the Reifling Basin hence indicate a complex Hg cycling during CPE-Event1. This may include both volcanic and terrestrial Hg inputs, the first coinciding with the beginning of the NCIE, the later occurring during the positive re-bound toward background values (Fig. 16.9). Hg is a highly toxic element for the biota: at the Triassic–Jurassic and Permian–Triassic boundaries, higher sedimentary Hg concentrations are associated to vegetation stress as evidenced by sporomorphs mutagenesis and vegetation collapse (e.g., Bos et al. 2024; Dal Corso et al. 2024a and references therein). The effects of higher Hg deposition to Carnian terrestrial ecosystems should be explored with high-resolution palynological analysis across CPE-Event1 in order to understand the relative timing between the environmental stress and vegetation change. It is notable that during the CPE interval a major turnover in plant communities occurred (Dal Corso et al. 2020a, b). This change, although its timing and tempo are still not precisely defined, resulted in the rise of modern-type floras with the radiation of many new fern and conifer families, as well as the diversification of the Bennettitales (Kustatscher et al. 2018; Dal Corso et al. 2020a; Roghi et al. 2022; Alverà et al. 2026). Exploring the mechanism of such pivotal biological change on land remains of primary importance.

16.7 Conclusions

In this study we presented new sedimentary Hg concentration data, coupled to a revision of the published C-isotope records and other geochemical proxies, measured in sections within the Reifling Basin (Northern Calcareous Alps, Austria) and covering CPE-Event1 (sensu Dal Corso 2025). Data show spikes in Hg loading (expressed as Hg/TOC) to the Reifling Basin coincident with negative C-isotope excursions (NCIEs) in bulk organic matter and carbonates. An initial Hg/TOC peak is recorded at the onset of the NCIE in the uppermost Reifling Formation (uppermost *Trachyceras aonoides* ammonoid Zone), and could have been related to a pulse of the coeval Wrangellia LIP. A further rise in sedimentary Hg concentrations and Hg/TOC is recorded between the uppermost Göstling Formation and basal Reingraben Formation (basal *Austrotrachyceras austriacum* ammonoid Zone) and is related to higher input of land plants material to the basin, as indicated by sedimentology (higher siliciclastic content) and previously published biomarkers. Similar to what was suggested for other LIP-related events, volcanic activity could have initially enriched the Carnian environments in Hg, which could later be re-mobilized when the hydrological cycle and runoff became stronger. In summary, according to the available data, high volcanic CO_2 injection would have perturbed the Carnian C-cycle in the late Julian 1. This perturbation of the C-cycle caused global warming and variations in the hydrological cycle, triggering marked changes in the depositional environments. In the Reifling Basin, the local sedimentological expression of the global perturbations was first an abrupt shift from carbonate deposition (Reifling Formation) to laminated darker limestone (Göstling Formation), followed by a further enhancement of siliciclastic input and demise of carbonate deposition, with the deposition of laminated shales (Reingraben Formation). This higher input of terrigenous material was likely caused by intensification of surface runoff and erosion on land, which was due to the strengthening of the Late Triassic mega-monsoon as a consequence of increasing CO_2 levels in the atmosphere. An enhanced hydrological cycle could increase chemical weathering and/or burial of organic matter globally, removing ^{13}C-depleted CO_2 from the surface system and resulting in the positive $\delta^{13}C$ recovery from the negative shift that is recorded in the stratigraphic interval with higher terrigenous Hg input. This later stage of CPE-Event1 marked the recovery from the C-cycle perturbation. Interestingly, during this phase the Polzberg *Konservat-Lagerstätte* is found in strata of the Reingraben Formation (Lukeneder and Lukeneder 2021). According to available stratigraphic data, CPE-Event1 and the linked environmental changes observed in the Reifling Basin occurred within ca. 40–110 kyrs (Miller et al. 2017; Zhang et al. 2024; Dal Corso 2025).

Acknowledgments The authors are particularly grateful to the owners of the outcrop area around Großreifling, the Austrian Federal Forests Company (Österreichische Bundesforste), the owners of the outcrop area of Polzberg Franziska and Hermann Hofreiter, Eva and Karl Jagersberger (all Gaming), and Felix Pöchhacker (Lunz am See) the owner of the Großau drilling site for sampling and digging permission. The work was done within the framework of and financially supported by projects of the Federal Government of Lower Austria (K3-F-964/001-2020 and K3-F-964/002-2023, Department Science and Research; headquarters in St. Pölten), the Austrian Academy of Sciences (headquarters in Vienna, project Polzberg Lukeneder 2021) represented by the National Committee for Geo/Hydro Sciences (Earth System Sciences Programme), the Triassic project "Fenster in die Karnische Krise" of the society of friends of the Natural History Museum Vienna (Freunde des Naturhistorischen Museums Wien, CPE Lukeneder 2023), and the Natur- und UNESCO Geopark Steirische Eisenwurzen (St Gallen, project Karnische Krise Lukeneder 2024). We thank the reviewer Xin Jin (Chengdu University of Technology) and an anonymous reviewer for the comments that greatly improved the manuscript. The manuscript greatly benefited from valuable comments of the production editor Ashok Arumairaj (Taramani), the in-house series editor Kathleen A. Ritterbush (Utah), the publishing editor Simon Shah-Rohlfs and the production editor Antje Endemann (both Heidelberg).

Author Contribution JDC and AL designed the research. AL collected new material and provided documentation on fossil material. JDC, GA and TH performed Hg analyzes. RS, JW and DM performed TOC and TS analyzes. JDC wrote the original draft of the manuscript and drew the figures, with contributions from AL, PG and GR for the stratigraphy.

Funding Financial support to JDC was provided by the National Natural Science Foundation of China (42172031). Open access funding enabled and projects financially supported by the Federal Government of Lower Austria (K3-F-964/001-2020 and K3-F-964/002-2023, Department Science and Research; headquarters in St. Pölten); the Triassic project "Fenster in die Karnische Krise" of the society of friends of the Natural History Museum Vienna (Freunde des Naturhistorischen Museums Wien, CPE Lukeneder 2023); the Natural History Museum Vienna; and the Technical University of Leoben (Chair of Energy Geosciences). The authors are responsible for the contents of this publication. The funder had no impact on conceptualization, design, data collection, analysis, decision to publish, or preparation of the manuscript.

Competing interests All authors stated that they have no competing interests.

Authors' contributions All authors stated that they have equally contributed to their chapters, read, edited and agreed to submit the final versions of their contributions.

Availability of data and materials All sources of information are provided in the text. Raw data related to the fossil material from Austria are available from the corresponding authors and Alexander Lukeneder upon request.

Additional information This article is part of the Special Volume: *Late Triassic Konservat-Lagerstätten within the Carnian Pluvial Episode. 150 Years of Scientific Excavations in Austria.*

Artificial Intelligence (AI) The chapter was produced without the use of AI.

Language The entire volume uses American English reviewed by a native scientist.

Received 2025.

Published online 2026.

References

Alverà G, Dal Corso J, Chu D, Cruciani G, Roghi G, Caggiati M, Song H-J, Song H-Y, Tian L, Du Y, Haserbek T, Gianolla P (2026) Episodic aridification at the onset of the Carnian Pluvial Episode in Western Tethys. Glob Planet Change 256:105192. https://doi.org/10.1016/j.gloplacha.2025.105192

Amos HM, Jacob DJ, Streets DG, Sunderland EM (2013) Legacy impacts of all-time anthropogenic emissions on the global mercury cycle. Global Biogeochem Cycles 27(2):410–421. https://doi.org/10.1002/gbc.20040

Baranyi V, Miller CS, Ruffell A, Hounslow MW, Kürschner WM (2019a) A continental record of the Carnian Pluvial Episode (CPE) from the Mercia Mudstone Group (UK): palynology and climatic implications. J Geol Soc Lond 176(1):149–166. https://doi.org/10.1144/jgs2017-150

Baranyi V, Rostási Á, Raucsik B, Kürschner WM (2019b) Palynology and weathering proxies reveal climatic fluctuations during the Carnian Pluvial Episode (CPE) (Late Triassic) from marine successions in the Transdanubian Range (western Hungary). Glob Planet Change 177:157–172. https://doi.org/10.1016/j.gloplacha.2019.01.018

Behrens M (1972) Schwermineralverteilungen und Sedimentstrukturen in den Lunzer Schichten (Karn, Trias, Österreich). Jb Geol B-A 116:51–83

Bos R, Zheng W, Lindström S, Sanei H, Waajen I, Fendley IM, Mather TA, Wang Y, Rohovec J, Navrátil T, Sluijs A, van de Schootbrugge B (2024) Climate-forced Hg-remobilization associated with fern mutagenesis in the aftermath of the end-Triassic extinction. Nat Commun 15:3596. https://doi.org/10.1038/s41467-024-47922-0

Boscaini A, Callegaro S, Sun Y, Marzoli A (2022) Late Permian to Late Triassic large igneous provinces: timing, eruptive style and paleoenvironmental perturbations. Front Earth Sci 10:887632. https://doi.org/10.3389/feart.2022.887632

Dal Corso J (2025) Carnian events. Albertiana 49:1–11

Dal Corso J, Preto N, Kustatscher E, Mietto P, Roghi G, Jenkyns HC (2011) Carbon-isotope variability of Triassic amber, as compared with wood and leaves (Southern Alps, Italy). Palaeogeogr Palaeoclimatol Palaeoecol 302:187–193. https://doi.org/10.1016/j.palaeo.2011.01.007

Dal Corso J, Mietto P, Newton RJ, Pancost RD, Preto N, Roghi G, Wignall PB (2012) Discovery of a major negative $\delta^{13}C$ spike in the Carnian (Late Triassic) linked to the eruption of Wrangellia flood basalts. Geology 40(1):79–82. https://doi.org/10.1130/G32473.1

Dal Corso J, Gianolla P, Newton RJ, Franceschi M, Roghi G, Caggiati M, Raucsik B, Budai T, Haas J, Preto N (2015) Carbon isotope records reveal synchronicity between carbon cycle perturbation and the "Carnian Pluvial Event" in the Tethys realm (Late Triassic). Glob Planet Change 127:79–90. https://doi.org/10.1016/j.gloplacha.2015.01.013

Dal Corso J, Gianolla P, Rigo M, Franceschi M, Roghi G, Mietto P, Manfrin S, Raucsik B, Budai T, Jenkyns HC, Reymond CE, Caggiati M, Gattolin G, Breda A, Merico A, Preto N (2018) Multiple negative carbon-isotope excursions during the Carnian Pluvial Episode (Late Triassic). Earth Sci Rev 185:732–750. https://doi.org/10.1016/j.earscirev.2018.07.004

Dal Corso J, Bernardi M, Sun Y, Song H, Seyfullah LJ, Preto N, Gianolla P, Ruffell A, Kustatscher E, Roghi G, Merico A, Hohn S, Schmidt AR, Marzoli A, Newton RJ, Wignall PB, Benton MJ (2020a) Extinction and dawn of the modern world in the Carnian (Late Triassic). Sci Adv 6(38):eaba0099. https://doi.org/10.1126/sciadv.aba0099

Dal Corso J, Mills BJW, Chu D, Newton RJ, Mather TA, Shu W, Wu Y, Tong J, Wignall PB (2020b) Permo-Triassic boundary carbon and mercury cycling linked to terrestrial ecosystem collapse. Nat Commun 11:2962. https://doi.org/10.1038/s41467-020-16725-4

Dal Corso J, Mills BJW, Chu D, Newton RJ, Song H (2022) Background Earth system state amplified Carnian (Late Triassic) environmental changes. Earth Planet Sci Lett 578:117321. https://doi.org/10.1016/j.epsl.2021.117321

Dal Corso J, Newton RJ, Zerkle AL, Chu D, Song H-J, Song H-Y, Tian L, Tong J, Di Rocco T, Claire MW, Mather TA, He T, Gallagher T, Shu W, Wu Y, Bottrell SH, Metcalfe I, Cope HA, Novak M, Jamieson RA, Wignall PB (2024a) Repeated pulses of volcanism drove the end-Permian terrestrial crisis in northwest China. Nat Commun 15:7628. https://doi.org/10.1038/s41467-024-51671-5

Dal Corso J, Sun Y, Kemp DB (2024b) Palaeogeographic heterogeneity of large-amplitude changes in marine sedimentation rates during the Carnian Pluvial Episode (Late Triassic). Glob Planet Change 237:104437. https://doi.org/10.1016/j.gloplacha.2024.104437

Dunay RE, Fisher MJ (1978) The Carnian palynoflora succession in the Northern Calcareous Alps, Lunz am See. Pollen Spores 20:177–187

Forchielli A, Pervesler P (2013) Phosphatic cuticle in thylacocephalans: a taphonomic case study of Austriocaris (Arthropoda, Thylacocephala) from the Fossil-Lagerstätte Polzberg (Reingraben shales, Carnian, Upper triassic, Lower Austria). Austrian J Earth Sci 106(1):46–61

Fu X, Wang J, Wen H, Wang Z, Zeng S, Song C, Chen W, Wan Y (2020) A possible link between the Carnian Pluvial Event, global carbon-cycle perturbation, and volcanism: new data from the Qinghai-Tibet Plateau. Glob Planet Change 194:103300. https://doi.org/10.1016/j.gloplacha.2020.103300

Furin S, Preto N, Rigo M, Roghi G, Gianolla P, Crowley JL, Bowring SA (2006) High-precision U-Pb zircon age from the Triassic of Italy: implications for the Triassic time scale and the Carnian origin of calcareous nannoplankton and dinosaurs. Geology 34(12):1009–1012. https://doi.org/10.1130/G22967A.1

Grasby SE, Shen W, Yin R, Gleason JD, Blum JD, Lepak RF, Hurley JP, Beauchamp B (2017) Isotopic signatures of mercury contamination in latest Permian oceans. Geology 45(1):55–58. https://doi.org/10.1130/G38487.1

Grasby SE, Them TR, Chen Z, Yin R, Ardakani OH (2019) Mercury as a proxy for volcanic emissions in the geologic record. Earth Sci Rev 196:102880. https://doi.org/10.1016/j.earscirev.2019.102880

Greene AR, Scoates JS, Weis D, Katvala EC, Israel S, Nixon GT (2010) The architecture of oceanic plateaus revealed by the volcanic stratigraphy of the accreted Wrangellia oceanic plateau. Geosphere 6(1):47–73. https://doi.org/10.1130/GES00212.1

Griffith J (1977) The Upper Triassic fishes from Polzberg bei Lunz, Austria. Zool J Linn Soc 60

Haserbek T, Tian L, Huang J, Shu W, Liu Y, Hua X, Kemp DB, Roghi G, Alverà G, Gianolla P, Du Y, Song H-Y, Song H-J, Chu D, Cao Y, Shi Y, Dal Corso J (2025) High latitude terrestrial carbon isotope and mercury perturbations across the Late Triassic Carnian Pluvial Episode in Northwestern China. Sci Rep 15:16222. https://doi.org/10.1038/s41598-025-98455-5

Hochuli PA, Frank SM (2000) Palynology (dinoflagellate cysts, spore-pollen) and stratigraphy of the Lower Carnian Raibl Group in the Eastern Swiss Alps. Eclogae Geol Helv 93:429–443

Hornung T, Brandner R (2005) Biochronostratigraphy of the Reingraben Turnover (Hallstatt Facies Belt): local black shale events controlled by regional tectonics, climatic change and plate tectonics. Facies 51:460–479. https://doi.org/10.1007/s10347-005-0061-x

Hornung T, Brandner R, Krystyn L, Joachimski MM, Keim L (2007a) Multistratigraphic constraints on the NW Tethyan "Carnian crisis." N M Mus Nat Hist Sci Bull 41:59–67

Hornung T, Krystyn L, Brandner R (2007b) A Tethys-wide mid-Carnian (Upper Triassic) carbonate productivity crisis: evidence for the Alpine Reingraben Event from Spiti (Indian Himalaya)? J Asian Earth Sci 30(2):285–302. https://doi.org/10.1016/j.jseaes.2006.10.001

Indraswari AO, Frieling J, Idiz E, Mather TA, Dickson AJ, Jenkyns HC, Robinson SA, Fendley IM, Neilson O (2025) The influence of primary depositional conditions on Hg behaviour during early thermal maturation. Chem Geol 690:122884. https://doi.org/10.1016/j.chemgeo.2025.122884

Jia E, Dal Corso J, Wu K, Du Y, Wu Y, Wang F, Dai X, Song H-Y, Chu D, Zhong L, Yuan Z, Chen X, Liu Y, Li Z, Song H-J (2025) Decoupling of $\delta^{13}C_{carb}$ and $\delta^{13}C_{org}$ during the Carnian Pluvial Episode in marine successions of South China. Chem Geol 695:123053. https://doi.org/10.1016/j.chemgeo.2025.123053

Jin X, McRoberts CA, Shi Z, Mietto P, Rigo M, Roghi G, Manfrin S, Franceschi M, Preto N (2019) The aftermath of the CPE and the Carnian-Norian transition in northwestern Sichuan basin, South China. J Geol Soc Lond 176:79–196. https://doi.org/10.1144/jgs2018-104

Jin X, Gianolla P, Shi Z, Franceschi M, Caggiati M, Du Y, Preto N (2020) Synchronized changes in shallow water carbonate production during the Carnian Pluvial Episode (Late Triassic) throughout Tethys. Glob Planet Change 184:103035. https://doi.org/10.1016/j.gloplacha.2019.103035

Jin X, Zhang F, Baranyi V, Kemp DB, Feng XB, Grasby SE, Sun GY, Shi ZQ, Chen WH, Dal Corso J (2022a) Early Jurassic massive release of terrestrial mercury linked to floral crisis. Earth Planet Sci Lett 598:117842. https://doi.org/10.1016/j.epsl.2022.117842

Jin X, Franceschi M, Martini R, Shi Z, Gianolla P, Rigo M, Wall CJ, Schmitz MD, Lu G, Du Y, Huang X, Preto N (2022b) Eustatic sea-level fall and global fluctuations in carbonate production during the Carnian Pluvial Episode. Earth Planet Sci Lett 594:117698. https://doi.org/10.1016/j.epsl.2022.117698

Jin X, Tomimatsu Y, Yin R, Onoue T, Franceschi M, Grasby SE, Du Y, Rigo M (2023) Climax in Wrangellia LIP activity coincident with major Middle Carnian (Late Triassic) climate and biotic changes: mercury isotope evidence from the Panthalassa pelagic domain. Earth Planet Sci Lett 607:118075. https://doi.org/10.1016/j.epsl.2023.118075

Köppen A (1997) Faziesentwicklung in der frühen Obertrias Mitteleuropas — ein sequenzstratigraphischer Vergleich. Gaea Heidelbergensis 2:1–233

Korte C, Kozur HW, Veizer J (2005) $\delta^{13}C$ and $\delta^{18}O$ values of Triassic brachiopods and carbonate rocks as proxies for coeval seawater and palaeotemperature. Palaeogeogr Palaeoclimatol Palaeoecol 226(3–4):287–306. https://doi.org/10.1016/j.palaeo.2005.05.018

Kristan-Tollmann E, Hamedani A (1973) Eine spezifische Mikrofaunen-Vergesellschaftung aus den Opponitzer Schichten des Oberkarn der niederösterreichischen Kalkvoralpen. 143:193–222

Krystyn L (1978) Eine neue Zonengliederung im Alpin-Mediterranen Unterkarn. Schr. Erdwiss. Kommun Österr Akad Wiss 4:37–55

Krystyn L (1991) Die Fossillagerstätten der alpinen Trias. In: Nagel D; Rabeder G (eds) Exkursionen im Jungpaläozoikum und Mesozoikum Österreichs. Exkursionsführer, Österreichische Paläontologische Gesellschaft 1991, Univ Wien, pp 23–78

Kustatscher E, Ash SR, Karasev E, Pott C, Vajda V, Yu J, McLoughlin S (2018) Flora of the Late Triassic. In: Tanner LH (ed) The Late Triassic world. Topics in geobiology, vol 46. Springer, Cham, pp 545–622. https://doi.org/10.1007/978-3-319-68009-5_13

Lein R (1987) Evolution of the Northern Calcareous Alps during Triassic times. In: Flügel H, Faupl P (eds) Geodynamics of the Eastern Alps. Wien Deuticke, pp 85–102

Lein R, Krystyn L, Richoz S, Lieberman H (2012) Middle Triassic platform/basin transition along the Alpine passive continental margin facing the Tethys Ocean—the Gamsstein: the rise and fall of a Wetterstein Limestone Platform (Styria, Austria). J Alpine Geol 54:471–498

Linzer HG, Ratschbacher L, Frisch W (1995) Transpressional collision structures in the upper crust: the fold-thrust belt of the Northern Calcareous Alps. Tectonophysics 242(1–2):41–61. https://doi.org/10.1016/0040-1951(94)00152-Y

Linzer HG, Moser F, Nemes F, Ratschbacher L, Sperner B (1997) Build-up and dismembering of the eastern Northern Calcareous Alps. Tectonophysics 272(2–4):97–124. https://doi.org/10.1016/S0040-1951(96)00254-5

Lestari W, Al-Suwaidi A, Fox CP, Vajda V, Hennhoefer D (2024) Carbon cycle perturbations and environmental change of the middle Permian and Late Triassic Paleo-Antarctic circle. Sci Rep 14:9742. https://doi.org/10.1038/s41598-024-60088-5

Lu J, Zhang P, Dal Corso J, Yang M, Wignall PB, Greene SE, Shao L, Lyu D, Hilton J (2021) Volcanically driven lacustrine ecosystem changes during the Carnian Pluvial Episode (Late Triassic). PNAS 118(40):e2109895118. https://doi.org/10.1073/pnas.2109895118

Lukeneder A (ed) (2026) Late Triassic Konservat-Lagerstätten within the Carnian Pluvial Episode: 150 years of scientific excavations in Austria. Top Geobiol 55:600 p. https://doi.org/10.1007/978-3-032-19777-1

Lukeneder A, Lukeneder P (2021) The Upper Triassic Polzberg palaeobiota from a marine *Konservat-Lagerstätte* deposited during the Carnian Pluvial Episode in Austria. Sci Rep 11:16644. https://doi.org/10.1038/s41598-021-96052-w

Lukeneder A, Lukeneder P (2022) Taphonomic history and trophic interactions of an ammonoid fauna from the Upper Triassic Polzberg palaeobiota. Sci Rep 12:7455. https://doi.org/10.1038/s41598-022-11496-y

Lukeneder A, Lukeneder P, Sachsenhofer RF, Roghi G, Rigo M (2024) Multi-proxy record of the Austrian Upper Triassic Polzberg *Konservat-Lagerstätte* in light of the Carnian Pluvial Episode. Sci Rep 14:11194. https://doi.org/10.1038/s41598-024-60591-9

Lukeneder A, Weitz JA, Sachsenhofer RF, Misch D (2026) Drilling into the Carnian Pluvial Episode. In: Lukeneder A (ed) Late Triassic Konservat-Lagerstätten within the Carnian Pluvial Episode: 150 years of scientific excavations in Austria. Top Geobiol, Chap 15:315–352. https://doi.org/10.1007/978-3-032-19777-1_15

Mandl GW (2000) The Alpine sector of the Tethyan shelf—examples of Triassic to Jurassic sedimentation and deformation from the Northern Calcareous Alps. Mitt Österr Geol Ges 92:61–77

Mazaheri-Johari M, Gianolla P, Mather TA, Frieling J, Chu D, Dal Corso J (2021) Mercury deposition in Western Tethys during the Carnian Pluvial Episode (Late Triassic). Sci Rep 11:17339. https://doi.org/10.1038/s41598-021-96890-8

Miller CS, Peterse F, Da Silva AC, Baranyi V, Reichart GJ, Kürschner WM (2017) Astronomical age constraints and extinction mechanisms of the Late Triassic Carnian crisis. Sci Rep 7:2557. https://doi.org/10.1038/s41598-017-02817-7

Mostler H, Scheuring BW (1974) Mikrofloren aus dem Langobard und Cordevol der Nördlichen Kalkalpen und das Problem des Beginns der Keupersedimentation im Germanischen Raum. Geol Paläontol Mitt 4:1–35

Mueller S, Hounslow MW, Kürschner WM (2016a) Integrated stratigraphy and palaeoclimate history of the Carnian Pluvial Event in the Boreal realm; new data from the upper Triassic Kapp Toscana group in central Spitsbergen (Norway). J Geol Soc Lond 173:186–202. https://doi.org/10.1144/jgs2015-028

Mueller S, Krystyn L, Kürschner WM (2016b) Climate variability during the Carnian Pluvial Phase—a quantitative palynological study of the Carnian sedimentary succession at Lunz am See, Northern Calcareous Alps, Austria. Palaeogeogr Palaeoclimatol Palaeoecol 441(1):198–211. https://doi.org/10.1016/j.palaeo.2015.06.008

Percival LME, Jenkyns HC, Mather TA, Dickson AJ, Batenburg SJ, Ruhl M, Hesselbo SP, Barclay R, Jarvis I, Robinson SA, Woelders L (2018) Does large igneous province volcanism always perturb the mercury cycle? Comparing the records of Oceanic Anoxic Event 2 and the end-Cretaceous to other Mesozoic events. Am J Sci 318:799–860. https://doi.org/10.2475/08.2018.01

Peters KE, Walters CC, Moldowan JM (2005) The biomarker guide, 2nd edn. Cambridge University Press. https://doi.org/10.1017/CBO9780511524868

Piller WE, Egger H, Erhart CW, Gross M, Harzhauser M, Hubmann B, van Husen D, Krenmayr H-G, Krystyn L, Lein R, Lukeneder A, Mandl G, Rögl F, Roetzel R, Rupp C, Schnabel W, Schönlaub HP, Summesberger H, Wagreich M, Wassely G (2004) Die Stratigraphische Tabelle von Österreich 2004 (Sedimentäre Schichtfolgen). Kommission für die paläontologische und stratigraphische Erforschung Österreichs der Österreichischen Akademie der Wissenschaften und Österreichische Stratigraphische Kommission

Pott C, Krings M, Kerp H (2008) The Carnian (Late Triassic) flora from Lunz in Lower Austria: paleoecological considerations. Palaeoworld 17(3–4):172–182. https://doi.org/10.1016/j.palwor.2008.03.001

Qiu Z, Wei H, Tian L, Dal Corso J, Zhang J, Zou C (2022) Different controls on the Hg spikes linked the two pulses of the Late Ordovician mass extinction in South China. Sci Rep 12:5195. https://doi.org/10.1038/s41598-022-08941-3

Rahman NU, Xian B, Fang L, Chen S, Chen P, Ullah Z, Wang P (2024) Volcanically driven terrestrial environmental perturbations during the Carnian Pluvial Episode in the Eastern Tethys. Acta Geol Sin Engl 98(3):753–770. https://doi.org/10.1111/1755-6724.15167

Rigo M, Joachimski MM (2010) Palaeoecology of Late Triassic conodonts: constraints from oxygen isotopes in biogenic apatite. Acta Palaeontol Pol 55(3):471–478. https://doi.org/10.4202/app.2009.0100

Rigo M, Mazza M, Karádi V, Nicora A (2018) New Upper Triassic conodont biozonation of the Tethyan realm. In: Tanner LH (ed) The Late Triassic world. Topics in geobiology, vol 46. Springer, pp 189–235 https://doi.org/10.1007/978-3-319-68009-5_6

Roghi G, Gianolla P, Minarelli L, Pilati C, Preto N (2010) Palynological correlation of Carnian humid pulses throughout western Tethys. Palaeogeogr Palaeoclimatol Palaeoecol 290(1–4):89–106. https://doi.org/10.1016/j.palaeo.2009.11.006

Roghi G, Gianolla P, Kustatscher E, Schmidt AR, Seyfullah LJ (2022) An exceptionally preserved terrestrial record of LIP effects on plants in the Carnian (Upper Triassic) Amber-Bearing Section of the Dolomites, Italy. Front Earth Sci 10:900586. https://doi.org/10.3389/feart.2022.900586

Roghi G, Lukeneder A (2026) Palynological analyses at the onset of the Carnian Pluvial Episode (CPE) in the Reifling Basin (Northern Calcareous Alps, Austria). In: Lukeneder A (ed) Late Triassic Konservat-Lagerstätten within the Carnian Pluvial Episode: 150 years of scientific excavations in Austria topics in geobiology. Top Geobiol 55, Chap 17:383–404. https://doi.org/10.1007/978-3-032-19777-1_17

Ruffell A, Simms MJ, Wignall PB (2016) The Carnian Humid Episode of the late Triassic: a review. Geol Mag 153(2):271–284. https://doi.org/10.1017/S0016756815000424

Rüffer T, Bechstädt T (1999) Triassic sequence stratigraphy in the western part of the Northern Calcareous Alps (Austria). In: de Graciansky P-C; Hardenbol J, Jacquin T, Vail PR (eds) Mesozoic and cenozoic sequence stratigraphy of European Basins. SEPM Spec Publ 60 https://doi.org/10.2110/pec.98.02.0751

Sachsenhofer RF (1987) Fazies und Inkohlung Mesozoischer Kohlen der Alpen Ostösterreichs. Mitt Österr Geol Ges 80:1–45

Schlager W, Schöllnberger W (1974) Das Prinzip stratigraphischer Wenden in der Schichtfolge der Nördlichen Kalkalpen. Mitt Geol Ges Wien 66–67:165–193

Schuster PF, Schaefer KM, Aiken GR, Antweiler RC, Dewild JF, Gryziec JD, Gusmeroli A, Hugelius G, Jafarov E, Krabbenhoft DP, Liu L, Herman-Mercer N, Mu C, Roth DA, Schaefer T, Striegl RG, Wickland KP, Zhang T (2018) Permafrost stores a globally significant amount of mercury. Geophys Res Lett 45(3):1463–1471. https://doi.org/10.1002/2017GL075571

Shen J, Feng Q, Algeo TJ, Liu J, Zhou C, Wei W, Liu J, Them TR, Gill BC, Chen J (2020) Sedimentary host phases of mercury (Hg) and implications for use of Hg as a volcanic proxy. Earth Planet Sci Lett 543:116333. https://doi.org/10.1016/j.epsl.2020.116333

Shi Z, Jin X, Preto N, Rigo M, Du Y, Han L (2019) The Carnian Pluvial Episode at Ma'antang, Jiangyou in Upper Yangtze Block, Southwestern China. J Geol Soc Lond 176:197–207. https://doi.org/10.1144/jgs2018-038

Simms MJ, Ruffell AH (1989) Synchroneity of climatic change and extinctions in the Late Triassic. Geology 17(3):265–268. https://doi.org/10.1130/0091-7613(1989)017%3c0265:SOCCAE%3e2.3.CO;2

Simms MJ, Ruffell AH (2018) The Carnian Pluvial Episode: from discovery, through obscurity, to acceptance. J Geol Soc Lond 175:989–992. https://doi.org/10.1144/jgs2018-020

Strauss P, Granado P, Munoz JA, Bohm K, Schuster R (2023) The Northern Calcareous Alps revisited: formation of a hyperextended margin and mantle exhumation in the Northern Calcareous Alps sector of the Neo-Tethys (Eastern Alps, Austria). Earth Sci Rev 243:104488. https://doi.org/10.1016/j.earscirev.2023.104488

Stur D (1874) Neue Aufschlüsse im Lunzer Sandsteine bei Lunz und ein neuer Fundort von Wengerschiefer im Pölzberg zwischen Lunzersee und Gaming. Verh K-K Geol RA 1:271–273

Sun YD, Wignall PB, Joachimski MM, Bond DPG, Grasby SE, Lai XL, Wang LN, Zhang ZT, Sun S (2016) Climate warming, euxinia and carbon isotope perturbations during the Carnian (Triassic) Crisis in South China. Earth Planet Sci Lett 444:88–100. https://doi.org/10.1016/j.epsl.2016.03.037

Sun YD, Richoz S, Krystyn L, Zhang ZT, Joachimski MM (2019) Perturbations in the carbon cycle during the Carnian Humid Episode: carbonate carbon isotope records from southwestern China and Northern Oman. J Geol Soc Lond 176:167–177. https://doi.org/10.1144/jgs2017-170

Teller F (1891) Über den Schädel eines fossilen Dipnoers, *Ceratodus Sturii* nov. spec. aus den Schichten der oberen Trias der Nordalpen. Abh K K Geol Reichsanst 15:1–38

Them TR, Jagoe CH, Caruthers AH, Gill BC, Grasby SE, Gröcke DR, Yin R, Owens JD (2019) Terrestrial sources as the primary delivery mechanism of mercury to the oceans across the

Toarcian Oceanic Anoxic Event (Early Jurassic). Earth Planet Sci Lett 507:62–72. https://doi.org/10.1016/j.epsl.2018.11.029
Tollmann A (1976) Analyse des klassischen nordalpinen Mesozoikums: Stratigraphie, Fauna und Fazies der Nördlichen Kalkalpen. Franz Deuticke Wien, 580 p
Tollmann A (1985) Geologie von Österreich, Bd. II: Außerzentralalpiner Anteil. Franz Deuticke, Wien, 1–710
Tomimatsu Y, Nozaki T, Sato H, Takaya Y, Kimura JI, Chang Q, Naraoka H, Rigo M, Onoue T (2021) Marine osmium isotope record during the Carnian "pluvial episode" (Late Triassic) in the pelagic Panthalassa Ocean. Glob Planet Change 197:103387. https://doi.org/10.1016/j.gloplacha.2020.103387
Trotter JA, Williams IS, Nicora A, Mazza M, Rigo M (2015) Long-term cycles of Triassic climate change: a new $\delta^{18}O$ record from conodont apatite. Earth Planet Sci Lett 415:165–174. https://doi.org/10.1016/j.epsl.2015.01.038
Verloop JH (1908) Profil der Lunzer Schichten in der Umgebung von Lunz. Z Deutsch Geol Ges 60:81–89
Visscher H, Van Houte M, Brugman WA, Poort RJ (1994) Rejection of a Carnian (Late Triassic) "pluvial event" in Europe. Rev Palaeobot Palynol 83(1–3):217–226. https://doi.org/10.1016/0034-6667(94)90070-1
Zhang Q, Fu X, Wang J, Mansour A, Wei H, Zhang T, Wang M (2024) Orbitally-paced climate change during the Carnian Pluvial Episode. Earth Planet Sci Lett 626:118546. https://doi.org/10.1016/j.epsl.2023.118546
Zhang ZT, Joachimski MM, Grasby SE, Sun YD (2022) Intensive Ocean anoxia and large $\delta^{13}C_{carb}$ perturbations during the Carnian Humid Episode (Late Triassic) in Southwest China. Glob Planet Change 217:103942. https://doi.org/10.1016/j.gloplacha.2022.103942
Zhao H, Grasby SE, Wang X, Zhang L, Liu Y, Chen ZQ, Hu Z, Huang Y (2022) Mercury enrichments during the Carnian Pluvial Event (Late Triassic) in South China. GSA Bull 134(9–10):2709–2720. https://doi.org/10.1130/B36205.1

Chapter 17
Palynological Analyses at the Onset of the Carnian Pluvial Episode (CPE) in the Reifling Basin (Northern Calcareous Alps, Austria)

Guido Roghi and Alexander Lukeneder

Abstract This palynological study describes two continuous sequences from the Reifling Formation, Göstling Formation, and the lower part of the Reingraben Formation. Qualitative and quantitative palynological associations reveal the paleofloristic and paleoenvironmental variations of a Carnian key interval of the Austrian sequences. This study highlights the first paleobotanical associations affected by the onset of the Carnian Pluvial Episode (CPE), during which humidity fluctuated within a general trend of increasing aridity from the top of the Reifling to the lower part of the Reingraben Formation. The Scheiblinggraben and Steinbachgraben sections highlight a definitive reciprocal relationship between vegetation dynamics and the first negative carbon isotope excursion (NCIE) of the CPE.

Keywords Palynology · Carnian Pluvial Episode · Paleoenvironment · Early Carnian · Late Triassic · Austria

17.1 Introduction

Earlier palynological studies focused on the Carnian sequences of the Reifling Basin and the related paleontological finds in the Reifling Formation, Gostling Formation, Reingraben Formation, Lunz Formation and Opponitz Formation. These range from the nineteenth century (Stur 1874; Teller 1891) to more recently (Summesberger and Wagner 1972; Mostler and Scheuring 1974; Griffith 1977; Krystyn 1978, 1991; Lahm 1984; Hornung et al. 2007; Hornung 2008; Mueller et al. 2016). A series of newer

G. Roghi (✉)
Institute of Geosciences and Earth Resources, National Research Council (CNR), Padova, Italy
e-mail: guido.roghi@igg.cnr.it

A. Lukeneder
Department of Geology and Paleontology, Natural History Museum Vienna, Vienna, Austria
e-mail: alexander.lukeneder@nhm.at

A. Lukeneder (ed.), *Late Triassic Konservat-Lagerstätten within the Carnian Pluvial Episode*, Topics in Geobiology 55, https://doi.org/10.1007/978-3-032-19777-1_17

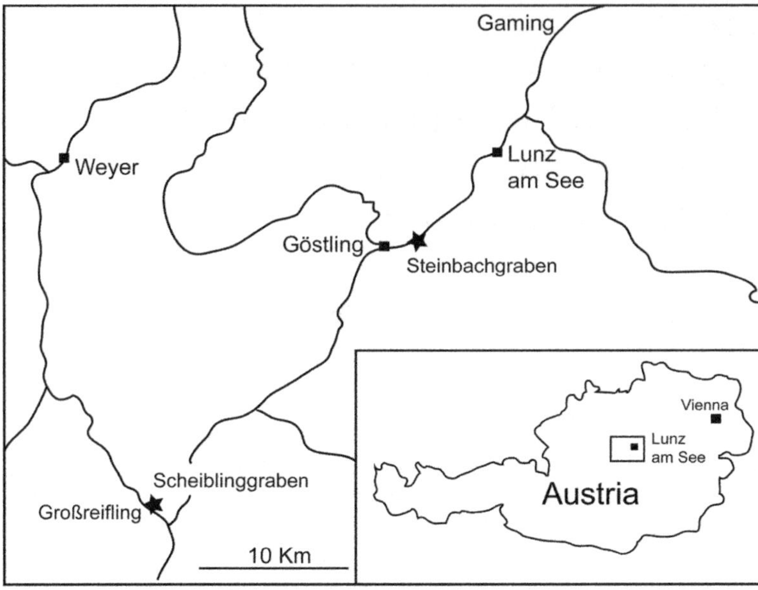

Fig. 17.1 Geographic location of the sampled Steinbachgraben section near Göstling (Lower Austria) and the Scheiblinggraben section near Großreifling (Styria)

excavations have been carried out, yielding important fossil plant and animal assemblages, namely a real Upper Triassic *Konservat-Lagerstätte* located in the Northern Calcarous Alps of Lower Austria (Lukeneder et al. 2020; Lukeneder and Lukeneder 2021, 2022a, b, 2023, 2024; Lukeneder 2026, this volume; Fig. 17.1).

The Reifling Formation, due to its ammonoid content, has been defined as Lower Carnian (Julian 1, *Trachyceras aonoides* Zone, Fig. 17.2). The Göstling Formation and the Reingraben Formation have been attributed to the lower part of the *Austrotrachyceras austriacum* Zone (Julian 2) (Lukeneder et al. 2024; Lukeneder and Lukeneder 2021, 2022a). No ammonites have been found in the overlying Lunz Formation, which is characterized by marginal marine and delta facies. The age of the Lunz Formation was defined based on the age of the above Opponitz Formation, which was attributed to the Tuvalian (Krystyn 1991; Köppen 1997). All these formations therefore constitute an important record for the definition and study of the CPE, as treated in this special volume.

Fossil macroplant assemblages, indicating a continental contribution within the Reifling Basin, have been found at Scheiblingbachgraben and Polzberg (Kustatscher et al. and references therein in this volume). Fossil resin (amber) was also reported in Scheiblingbach (Kustatscher et al., this volume).

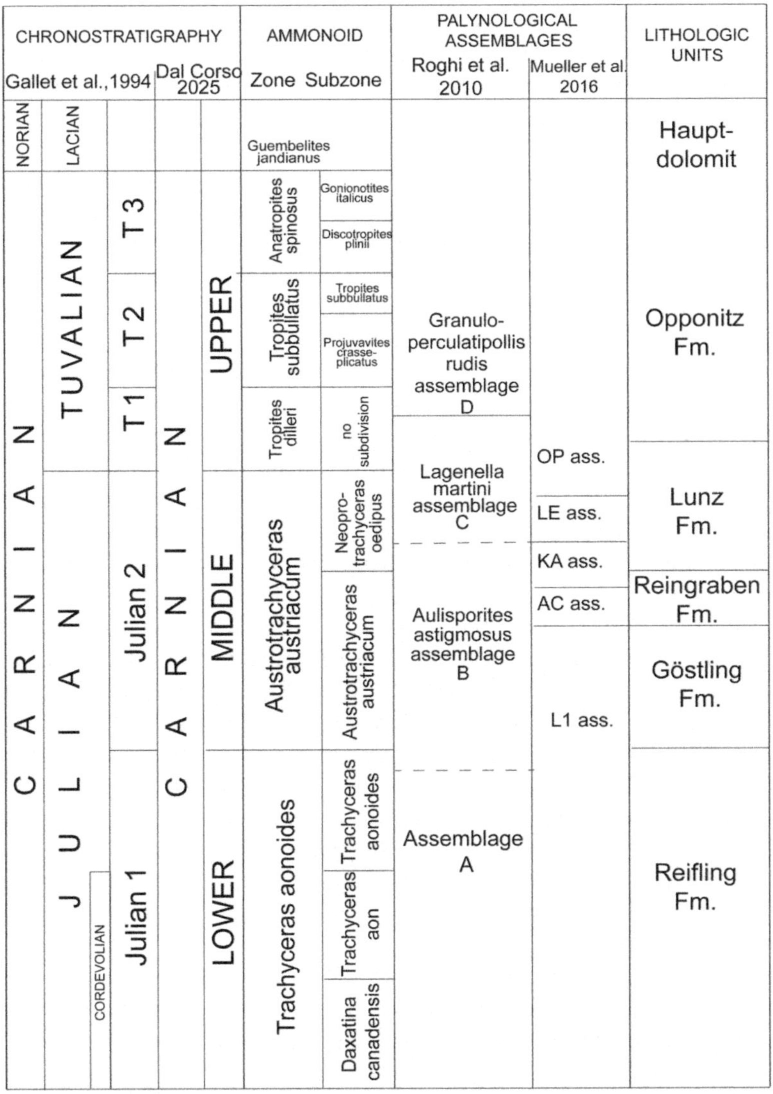

Fig. 17.2 Chronostratigraphy, ammonoid Zone and Subzone and palynological assemblages of the main lithologic units of the Northern Calcareous Alps (Austria); AC ass., Aulisporites-Cycadopites assemblage; KA ass, Kraeuselisporites-Alisporites assemblage; LE ass, Lagenella Enzonalasporites assemblage; OP ass, Ovalipollis-Punctatisporites assemblage

The few palynological works in the area have concentrated on the charcoal sequences referred to the Lunz Formation (Klaus 1960; Bharadwaj and Singh 1963; Dunay and Fisher 1978), where large quantities of fossil plants have also been recovered (Pott et al. 2008; Pott and Lukeneder 2026, this volume). Palynological assemblages from the Carnian ammonoid zones have also been found in the Reifling Basin. Mostler and Scheuring (1974) described a poor palynological assemblage in the uppermost part of the Reifling Formation and Göstling Formation, mainly represented by *Ovalipollis*, *Triadispora* and *Concavisporites* sp., along with many specimens of the marine algae *Tasmanites*. These assemblages of the Reifling and Göstling Formations, with scarce and degraded palynomorphs, were described as assemblages A sensu Roghi et al. (2010) and L1 sensu Mueller et al. (2016). They are correlated with the *Trachyceras aonoides* ammonoid zones and part of the *Austrotrachyceras austriacum*. The overlying Reingraben Formation yielded rich palynological associations, enabling the definition of the assemblage B (*Aulisporites astigmosus* assemblage) corresponding to the *Duplicisporites continuus* assemblage (Roghi 2004; Roghi et al. 2010; Lukeneder et al. 2024). The association is characterized by the presence of azonotriletes, monoletes (*Cycadopites* spp.), *Aulisporites astigmosus*, and a high diversity of monosaccates and bisaccates (Roghi et al. 2010; Lukeneder et al. 2024).

Mueller et al. (2016) divided the Reingraben Formation assemblages into the *Aulisporites-Cycadopites* assemblage and the *Kraeuselisporites-Alisporites* assemblage. Assemblages from the overlying Lunz Formation were defined as assemblage C (*Lagenella martini* assemblage; sensu Roghi et al. 2010). It consists of azonotriletes spores, *Cycadopites*, *Aulisporites astigmosus*, Cavatomonoletes, and the costate forms belonging to *Lagenella* and *Equisetosporites*, along with monosaccates and bisaccates. Mueller et al. (2016) subdivided the assemblages of the Lunz Formation into the *Lagenella-Enzonalasporites* assemblage and the *Ovalipollis-Punctatisporites* assemblage. The boundary between assemblages B and C (*Aulisporites astigmosus* assemblage and *Lagenella martini* assemblage) is apparently located in the lower part of the Lunz Formation (Roghi et al. 2010; Mueller et al. 2016).

This article further investigates the palynological assemblages in the Reifling and Göstling Formations and correlates them with the isotopic curve in order to better define the paleoclimatic aspects at the beginning of this important Late Triassic episode.

17.2 Geologic Setting

The samples used for this palynological study come from the stratigraphic section of the Steinbachgraben, near the village of Göstling, and from the Scheiblinggraben core near Großreifling, both in the Northern Calcareous Alps (Lower Austria, Fig. 17.1). The focus is primarily on the Göstling Formation identified in the Steinbachgraben and Scheiblinggraben, with the aim of defining the qualitative and quantitative palynological record for a better understanding of the early stages of the Carnian Pluvial

Episode (CPE). Eleven samples from Steinbachgraben (J samples) were analyzed. The first three belong to the Reifling Formation, seven to the Göstling Formation, and one from the top belongs to the Reingraben Formation. In the Scheiblinggraben section (SCH samples), six samples from the Göstling Formation and one sample from the base of the overlying Reingraben Formation were analyzed.

17.3 Material and Methods

A total of 22 samples from one section and one core were analyzed. Four samples were sterile. The samples were prepared in the laboratory using HCl (27%) and HF (40%) to remove all carbonates and silica following standard palynological treatment procedures. The residue was sieved (250 and 15 μm), and heavy liquid separation ($ZnCl_2$) was performed where necessary. All slides and remaining sample material are preserved at the Museum of Natural History Vienna.

Quantitative counting was performed when a sufficient number of grains was present (minimum 100 palynomorphs per sample). A Leica DMLB microscope was used. Twenty-eight paleoecological groups were identified based on the relationships of the morphospecies and their paleobotanical affinities. Identifications are based on the following literature: Balme (1995), Baranyi et al. (2017, 2019a, b, 2025), Bonis and Kürschner (2012), Brugman et al. (1993), Clement-Westerhof (1974), Franz et al. (2014), Pacyna et al. (2017), Scibiorski (2024), Visscher and Van der Zwan (1981), Visscher et al. (1994), Zhang et al. (2021).

Annotated Species List (Fig. 17.4)

Spore

LAEVIGATI

Deltoidospora sp.
Concavisporites sp.
Todisporites sp.
Calamospora spp.
Leschikisporis sp.
Trachysporites sp.

APICULATE

Osmundacidites sp.
Uvaesporites sp.
Verrucosisporites sp.
Verrucosisporites morulae (Klaus 1960)
Apiculatisporites sp.
Leptolepidites sp.
Gibeosporites sp.

MURORNATE

Lycopodiacidites sp.
Palaeospongiosporis europeus (Schulz 1965)

CINGULATE?

Annulispora sp.
Striatella seebergensis (Mädler 1964)

AURICULATI

Kyrtomisporis sp.
Camarozonosporites rudis (Leschik in Kräusel et Leschik 1956)

CINGULICAVATI

Densosporites sp.

"PROXIMOCAVATE"

Aulisporites astigmosus Leschik in Kräusel et Leschik 1956 (Klaus 1960)

CAVATOMONOLETES

Aratrisporites sp.

PATINATI

Concentricisporites bianulatus (Neves 1961) Antonescu 1970

ZONOTRILETES

Kraeuselisporites spp.

ALETES

Peroaletes convolutus (Baradwaj and Sigh 1963)

Pollen

MONOSACCITES

Patinasporites densus (Leschik in Kräusel et Leschik 1956)
Vallasporites ignacii (Leschik in Kräusel et Leschik 1956)
Enzonalasporites sp.

DISACCITES

Triadispora spp.
Klausipollenites sp.
Pityosporites sp.
Ovalipollis pseudoalatus (Thiergart 1949) Schuurman 1976

STRIATITI

Lunatisporites acutus (Leschik in Kräusel et Leschik 1956)
Lueckisporites virkkiae (Potonié et Klaus 1954)
Infernopollenites sp.

CIRCUMPOLLES

Duplicisporites granulatus (Leschik in Kräusel et Leschik 1956) Scheuring 1970
Camerosporites secatus (Leschik in Kräusel et Leschik 1956)
Duplicisporites tenebrosa (Scheuring 1978)
Camerosporites pseudoverrucatus (Scheuring 1970)
Paracirculina spp.

MONOCOLPATES

Cycadopites sp.

ALETES

Araucariacites sp.

ACRITARCHA

Micrhystridium (Deflandre 1937) Sarjeant 1967
Baltisphaeridium sp.

GREEN ALGAE

Leiosphaeridia sp.

PRASINOPHTA

Cymatiosphaera sp.
Pterospermella sp.

TASMANITACEAE

Tasmanites sp.

17.4 Results

17.4.1 Spores and Pollen of Terrestrial Plants—Qualitative Analysis

Reifling Formation In Steinbachgraben (Table 17.1), a fossil microflora assemblage with smooth, apiculate, zonate, auriculate, and cingulicavate spores was identified. *Aulispoites astigmosus* was detected in sample J4, directly below the boundary with the Göstling Formation. Taeniate and non-taeniate bisaccates and Circumpolles were found among the pollen. No productive samples belonging to the Reifling Formation were found in Scheiblinggraben (Table 17.2).

Göstling Formation Steinbachgraben (Table 17.1) yielded smooth, apiculate, zonate, cingulicavate, and cavatomonolete spores. The presence of the patinate spore *Concentricisporites bianulatus* has been observed in the lower part of the Göstling Formation, and *Aulispoites astigmosus* is well represented throughout the Göstling Formation. Among the pollen, taeniate and non-taeniate bisaccates, monoletes and Circumpolles are distributed throughout the formation. The monosaccate specimens (*Vallasporites ignacii* and *Patinasporites densus*) appear in the middle part of the Göstling Formation. In Scheiblinggraben (Table 17.2), smooth and apiculate spores, zonate, cavatomonolete spores, and murornate, reticulate and auriculate spores (*Camarozonosporites rudis*) were found. The species *Aulisporites astigmosus* is well represented throughout the Göstling Formation. Among the pollen, notably bisaccate spores were present throughout the formation, both taeniate and non-taeniate, along with Circumpolles, monolete and alete pollen. The monosaccate pollen (*V. ignacii*, *P. densus* and *Enzonalasporites)* appear in the middle part of the Göstling Formation.

Reingraben Formation In Steinbachgraben (Table 17.1), from the very bottom of the formation, smooth, apiculate, and cavatomonolete spores were recorded. *Aulispoites astigmosus* was always present. Among the pollen, monosaccate, bisaccate, non-taeniate, and taeniate, Circumpolles and monolete spores were documented. In Scheiblinggraben (Table 17.2), at the base of Reingraben Formation, smooth spores were found and *Aulisporites astigmosus* was always present. Pollen include monosaccate, bisaccate, both taeniate and non-taeniate, Circumpolles and monolete forms.

The Steinbachgraben and Scheiblinggraben successions (Tables 17.1 and 17.2) have overall proven to be well diversified from a paleobotanical point of view. Based on the affinities of the morphospecies with the parent plants, the presence of lycophytes (*Aratrisporites*, *Kraeuselisporites* and *Camarozonosporites*), sphenophytes (*Calamospora*), and ferns (*Concavisporites*, *Verrucosisporites*, *Apiculatisporites*, *Osmundacidites*) can be highlighted. Among the gymnosperms, many conifers were present, including Araucariaceae (*Araucariacites*), Voltziaceae (*Ovalipollis*, *Triadispora*), Cheirolepidiaceae (Circumpolles), Majonicaceae (*Infernopollenites*, *Lueckisporites*), Bennettitales (*Aulisporites*, *Leschikisporis*) and Cycadales (*Cycadopites*).

Table 17.1 Distribution of palynomorphs in the Steinbachgraben section

Formations	Reifling Fm			Göstling Fm						Reing. Fm	
Position in the section from the base (in metres)	0	10	29	30	31	34	35	35.5	37	38	39
Samples	J1	J2	J4	J5	J6	J7	J8	J9	J10	J11	J12
Spore											
Osmundacidites sp.	×	×		×				×			
Kraeuselisporites sp.	×	×		×		×		×		×	
Concavisporites sp.	×	×		×	×			×		×	
Uvaesporites sp.	×			×							
Kyrtomisporis sp.	×										
Densosporites sp.	×										
Aulisporites astigmosus			×	×		×		×	×	×	×
Aratrisporites sp.				×					×	×	×
Todisporites sp.				×					×		
Concentricisporites bianulatus				×		×					
Leschikisporites sp.										×	×
Verrucosisporites morulae										×	
Calamospora sp.											×
Pollen											
Triadispora spp.	×			×			×				
Ovalipollis pseudoalatus	×			×	×	×	×	×	×	×	×
Duplicisporites granulatus	×		×	×	×		×	×	×	×	×
Camerosporites secatus		×		×							
Lunatisporites acutus		×					×	×			×
Klausipollenites sp.				×	×	×	×				×
Cycadopites sp.				×	×	×		×	×		×
Patinasporites densus						×			×		
Vallasporites ignacii						×	×	×		×	×
Paracirculina tenebrosa								×			
Peroaletes											
Camerosporites pseudoverrucatus										×	

(continued)

Table 17.1 (continued)

Formations	Reifling Fm			Göstling Fm						Reing. Fm	
Position in the section from the base (in metres)	0	10	29	30	31	34	35	35.5	37	38	39
Samples	J1	J2	J4	J5	J6	J7	J8	J9	J10	J11	J12
Lueckisporites virkkiae											×
Amber					×						
Marine elements											
Micrystridium sp.	×	×	×	×	×	×	×	×			
Leiosphaeridia sp.	×	×	×	×		×	×	×	×		
Baltisphaeridium sp.	×	×	×								
Cymatiosphaera sp.	×	×		×	×				×	×	
Tasmanites sp.	×	×	×	×	×	×		×	×	×	×
Pterospermella sp.	×	×									
Foraminiferal linings	×	×	×	×	×						

Amber recovery In sample J6 of the Stainbachgraben section, the palynological slide from the lower part of the Göstling Formation included fossil resin (amber). A small yellow-orange to brown fragment (72 × 40 μm) with the typical conchoidal fracture was recorded.

17.4.2 Palynology of Marine Elements—Qualitative Analysis

Reifling Formation: Acritarchs, algal spores (Chlorophyta, Prasinophyte, and Tasmanaceae), and a foraminiferal lining were present in Steinbachgraben (Table 17.1). No productive samples belonging to the Reifling Formation were found in Scheiblinggraben (Table 17.2).

Göstling Formation: Acritarchs, algal spores (Chlorophyta, Prasinophyte, and Tasmanaceae), and a foraminiferal lining were recorded in Steinbachgraben (Table 17.1). Acritarchs and algal spores (Chlorophyta, Prasinophyte, and Tasmanaceae) were found in Scheiblinggraben (Table 17.2).

Reingraben Formation: Only algal spores (Chlorophyta, Prasinophyte, and Tasmanaceae) were documented in Steinbachgraben, only acritarchs in Scheiblinggraben. Based on the qualitative distribution of aquatic and marine elements, the aquatic component clearly decreased from the upper part of the Reifling Formation to the base of the Reingraben Formation This change is characterized by the disappearance of *Baltisphaeridium* sp. and *Pterospermella* sp. in the Reifling Formation and by the disappearance of *Micrystridium* sp., *Leiosphaeridia* sp. and *Cymatiosphaera*

Table 17.2 Distribution of palynomorphs in the Scheiblinggraben section

Formations	Göstling formation						Rein. Fm
Position in the core log (in cm)	366	415	450	470	610	805	810
Samples	SCH10	SCH15	SCH18	SCH20	SCH33	SCH52	SCH53
Spore							
Calamospora spp.	×	×		×			×
Aulisporites astigmosus	×	×	×	×	×	×	×
Todisporites sp.	×	×					×
Concavisporites sp.	×	×	×	×	×		×
Leschikisporites sp.	×	×	×	×	×		×
Kraeuselisporites spp.	×	×	×	×	×		
Deltoidospora sp.		×		×			
Apiculatisporites sp.		×		×			
Osmundacidites sp.		×		×			
Aratrisporites sp.		×	×	×		×	
Leptolepidites sp.		×					
Uvaesporites sp.		×					
Camarozonosporites rudis		×					
Annulispora sp.		×					
Verrucosisporites sp.		×		×			
Concentricisporites bianulatus		×		×			
Striatella seebergensis		×					
Gibeosporites sp.		×		×			
Lycopodiacidites sp.				×			
Trachysporites sp.				×			
Palaeospongiosporis europeus				×			
Pollen							
Bisaccati ind	×	×	×	×	×	×	×
Triadispora spp.	×	×	×	×	×		×
Klausipollenites sp.	×				×		×
Camerosporites secatus	×	×	×		×	×	
Ovalipollis pseudoalatus			×	×	×	×	×
Duplicisporites granulatus		×	×				×
Paracirculina spp.		×			×		
Cycadopites sp.		×		×	×		×
Araucariacites sp.		×		×	×	×	
Pityosporites sp.		×					

(continued)

Table 17.2 (continued)

Formations	Göstling formation						Rein. Fm
Position in the core log (in cm)	366	415	450	470	610	805	810
Samples	SCH10	SCH15	SCH18	SCH20	SCH33	SCH52	SCH53
Lunatisporites acutus		×					
Enzonalasporites sp.			×	×			×
Patinasporites densus			×	×		×	×
Vallasporites ignacii			×	×			×
Infernopollenites sp.							×
Peroaletes sp.						×	
Marine							
Micrystridium		×					
Leiosphaeridia sp.							
Baltisphaeridium sp.							×
Cymatiosphaera sp.	×						
Tasmanites sp.	×		×				
Pterospermella sp.	×						

sp. in the Göstling Formation. In the upper part of the Reifling Formation and in the first part of the Göstling formation, foraminifera tests linings were present. Although marine palynomorphs reappeared in the overlying Lunz and Opponitz Formations (Roghi et al. 2010; Mueller et al. 2016), the diversity of marine elements clearly declined during the deposition of the Göstling Formation (Tables 17.1 and 17.2).

17.4.3 Definition of Hygro- and Xerophytic Trends

Based on the affinities between morphospecies and parent plants (Table 17.3), a quantitative approach was used to define the most hygrophytic samples and those with a more xerophytic tendency as well as to delineate the related changes throughout the stratigraphic record. The aim was to identify fluctuations from hygro- to xerophytic-dominated intervals in the Göstling Formation (Table 17.4 for Steinbachgraben and Table 17.5 for Scheiblingbachgraben). The results are illustrated in Fig. 17.3, which highlights the ratio of H/(H+X). At both studied locations, a general trend (Ht1 in Fig. 17.3) from more hygrophytic to more xerophytic conditions was evident from the base to the top of the Göstling Formation. Within this trend, a marked fluctuation toward more xerophytic characteristics (α1) is evident in the lower-middle part of the Göstling Formation (Fig. 17.3).

Table 17.3 Affinity of the sporomorph with the producing parent plants. Papers consulted: Balme (1995), Baranyi et al. (2017, 2019a, b), Bonis and Kürschner (2012), Brugman et al. (1993), Clement-Westerhof (1974), Fires (2004), Franz et al. (2014), Pacyna et al. (2017), Scibiorski (2024), Visscher and Van der Zwan (1981), Visscher et al. (1994), Zhang et al. (2021)

Spore	Botanical affinity	Ecology
Aulisporites	*Williamsoniantus keuperianus*, Bennettitalean	Hygrophyte
Todisporites sp.	Ferns, Osmundaceae	Hygrophyte
Aratrisporites	Lycophytes	Hygrophyte
Calamospora	Sphenophytes, Equisetales	Hygrophyte
Deltoidospora sp.	Ferns, Dicksoniaceae and Matoniaceae	Hygrophyte
Concavisporites	Ferns, Dicksoniaceae and Matoniaceae	Hygrophyte
Verrucosisporites sp.	Ferns, Matoniaceae, Dipteridaceae	Hygrophyte
Annulispora sp.	Bryophytes and putative bryophytes	Hygrophyte
Camarozonosporites	Lycophytes	Hygrophyte
Uvaesporites sp.	Lycopsida, Selaginellales	Hygrophyte
Leptolepidites sp.	Lycophytes of uncertain affinity	Hygrophyte
Apiculatisporites sp.	Fern, Marattiales	Hygrophyte
Palaeospongiosporis	Fern	Hygrophyte
Trachysporites sp.	Fern, Osmundaceae	Hygrophyte
Lycopodiacidites sp.	Filicopsida, Ophioglossales	Hygrophyte
Striatella seebergensis	Filicales, Pteridaceae	Hygrophyte
Concentricisporites	Ferns	Hygrophyte
Convolutispora	Ferns, Schizaeaceae	Hygrophyte
Kraeuselisporites	Lycophytes, Lycopodiales	Hygrophyte
Leschikisporites	Lycophytes, Lycopodiales	Hygrophyte
Osmundacidites	Ferns, Osmundaceae	Hygrophyte
Simplicisporites	Lycophytes	Hygrophyte
Trilites	Spore of unknown botanical affinity	Hygrophyte
Pollen		
Araucariacites	Conifer, Araucariaceae (Balme 1995)	Xerophyte
Brachysaccus	Conifer of uncertain affinity (Baranyi et al. 2019a, b)	Xerophyte
Camerosporites	Conifers, Cheirolepidiaceae	Xerophyte
Chordasporites	Conifers, perhaps Voltziaceae	Xerophyte
Duplicisporites	Conifers, Cheirolepidiaceae	Xerophyte
Enzonalasporites	Conifers, Patokaeaceae	Intermediate
Haberkornia	Conifer of uncertain affinity	Xerophyte
Infernopollenites	Conifers, perhaps Majonicaceae	Intermediate
Klausipollenites	Conifers, Voltziaceae	Xerophyte
Lueckisporites	Conifers, Majonicaceae, Voltziaceae	Intermediate
Lunatisporites	Conifers, seed ferns, Peltaspermales	Intermediate

(continued)

Table 17.3 (continued)

Spore	Botanical affinity	Ecology
Ovalipollis	Conifers, Voltziaceae, Cycadophyta	Xerophyte
Paracirculina	Conifers, Cheirolepidiaceae	Xerophyte
Partitisporites	Conifers, Cheirolepidiaceae	Xerophyte
Patinasporites	Conifers, Patokaeaceae	Intermediate
Quadraeculina	Conifer of uncertain affinity	Xerophyte
Samaropollenites	Conifers, Podocarpaceae	Xerophyte
Triadispora	Conifers, Voltziaceae, in situ in *Sertostrobus*, *Darneya*	Xerophyte
Vallasporites	Conifers, Patokaeaceae	Intermediate
Cycadopites sp.	Cycadales, Ginkgoales, Bennettitales	Xerophyte

Tab 17.4 Quantitative dataset for the Steinbachgraben section

		J1	J2	J4	J5	J6	J7	J8	J9	J10	J11	J12
Hygrophytic elements	*Spore tri levigate*	42.0	28.0	40.0	40.0	25.0	14.0	17.0	18.0	11.0	21.0	9.0
	Spore tri ornamentate	5.0	7.0	0.0	12.0	2.0	0.0	6.0	4.0	4.0	4.0	1.0
	Calamospora	0.0	0.0	0.0	0.0	0.0	0.0	0.0	0.0	0.0	2.0	0.0
	Kraeuselisporites	3.0	5.0	0.0	4.0	0	3.0	0	1.0	0	2.0	0.0
	Aratrisporites	0.0	0.0	0.0	1.0	0.0	0.0	0.0	0.0	4.0	1.0	2.0
	Leschikisporites	0.0	0.0	0.0	0.0	0.0	0.0	0.0	0.0	0.0	4.0	4.0
	Aulisporites astigmosus	0.0	0.0	20.0	5.0	0.0	3.0	29.0	20.0	13.0	4.0	9.0
	Cycadopites	0.0	0.0	0.0	3.0	2.0	2.0	0.0	1.0	1.0	0.0	1.0
Xeropytic elements	Bisaccati ind	40.0	40.0	20.0	19.0	51.0	65.0	31.0	41.0	38.0	53.0	42.0
	Bisaccati taeniati ind	2.0	0.0	0.0	0.0	0.0	0.0	1.0	8.0	1.0	0.0	5.0
	Lunatisporites	0.0	0.0	0.0	0.0	0.0	0.0	1.0	0.0	0.0	0.0	4.0
	Infernopollenites	0.0	0.0	0.0	0.0	0.0	0.0	0.0	0.0	0.0	0.0	0.0
	Triadispora	3.0	0.0	0.0	3.0	0.0	0.0	1.0	0.0	1.0	0.0	3.0
	Ovalipollis pseudoalatus	3.0	0.0	0.0	5.0	13.0	6.0	1.0	1.0	1.0	1.0	2.0
	Monosaccati ind	0.0	0.0	0.0	0.0	1.0	5.0	6.0	1.0	10.0	4.0	7.0
	Haberkornia	0.0	0.0	0.0	0.0	0.0	0.0	0.0	0.0	0.0	0.0	0.0
	Patinasporites densus	0.0	0.0	0.0	0.0	0.0	0.0	0.0	0.0	0.0	0.0	0.0
	Vallasporites igniacii	0.0	0.0	0.0	0.0	0.0	0.0	0.0	0.0	0.0	2.0	1.0
	Enzonalasporites	0.0	0.0	0.0	0.0	0.0	0.0	0.0	0.0	0.0	0.0	8.0
	Circumpolles	2.0	20.0	20.0	8.0	6.0	2.0	7.0	5.0	15.0	4.0	2.0

Table 17.5 Quantitative dataset for the Scheiblinggraben section

		SCH10	SCH15	SCH18	SCH20	SCH33	SCH52	SCH53
Hygrophytic elements	*Spore tri levigate*	41.0	32.0	30.0	25.0	24.0	16.0	18.0
	Spore tri ornamentate	3.0	19.0	12.0	19.0	5.0	4.0	1.0
	Calamospora	1.0	2.0	0.0	1.0	0.0	0.0	0.0
	Leschikisporites	6.0	1.0	2.0	4.0	2.0	1.0	2.0
	Kraeuselisporites	1.0	2.0	3.0	1.0	2.0	0.0	0.0
	Aratrisporites	1.0	4.0	2.0	3.0	1.0	2.0	0.0
	Aulisporites astigmosus	25.0	9.0	6.0	8.0	6.0	3.0	5.0
	Camarozonosporites	0.0	1.0	0.0	0.0	0.0	0.0	0.0
	Cycadopites	2.0	4.0	1.0	15.0	1.0	0.0	1.0
Xerophytic elements	Bisaccati ind	16.0	17.0	22.0	11.0	34.0	47.0	50.0
	Bisaccati taeniati ind	1.0	1.0	0.0	1.0	0.0	3.0	4.0
	Lunatisporites	0.0	0.0	0.0	0.0	0.0	0.0	0.0
	Infernopollenites	0.0	0.0	0.0	0.0	0.0	0.0	1.0
	Triadispora	2.0	1.0	0.0	1.0	3.0	0.0	1.0
	Ovalipollis pseudoalatus	0.0	1.0	2.0	1.0	1.0	2.0	1.0
	Monosaccati ind	0.0	1.0	9.0	2.0	4.0	9.0	9.0
	Haberkornia	0.0	0.0	0.0	0.0	0.0	0.0	0.0
	Patinasporites densus	0.0	1.0	1.0	1.0	0.0	2.0	1.0
	Vallasporites igniacii	0.0	0.0	1.0	1.0	0.0	0.0	1.0
	Enzonalasporites	0.0	1.0	1.0	1.0	0.0	0.0	1.0
	Laricoidites	0.0	0.0	0.0	1.0	0.0	2.0	0.0
	Araucariacites	0.0	0.0	0.0	4.0	0.0	1.0	0.0
	Circumpolles	1.0	2.0	8.0	0.0	17.0	8.0	4.0

17.5 Discussion

The analysis of the distribution of the various species in Steinbachgraben and Scheiblinggraben enabled defining for the first time a palynological association for the Göstling Formation containing *Ovalipollis pseudoalatus*, *Camerosporites secatus*, *Duplicisporites granulatus* and *Lunatisporites acutus*. This association, together with *Aulisporites astigmosus*, fits well with assemblage B (*Aulisporites astigmosus* assemblage) of Roghi et al. (2010) and with the *Aulisporites-Cycadopites* Assemblage of Mueller et al. (2016). These assemblages, until now identified only from the base of the Reingraben Formation (Roghi et al. 2010; Mueller et al. 2016), are here lowered at least from the base of the Göstling Formation and could coincide

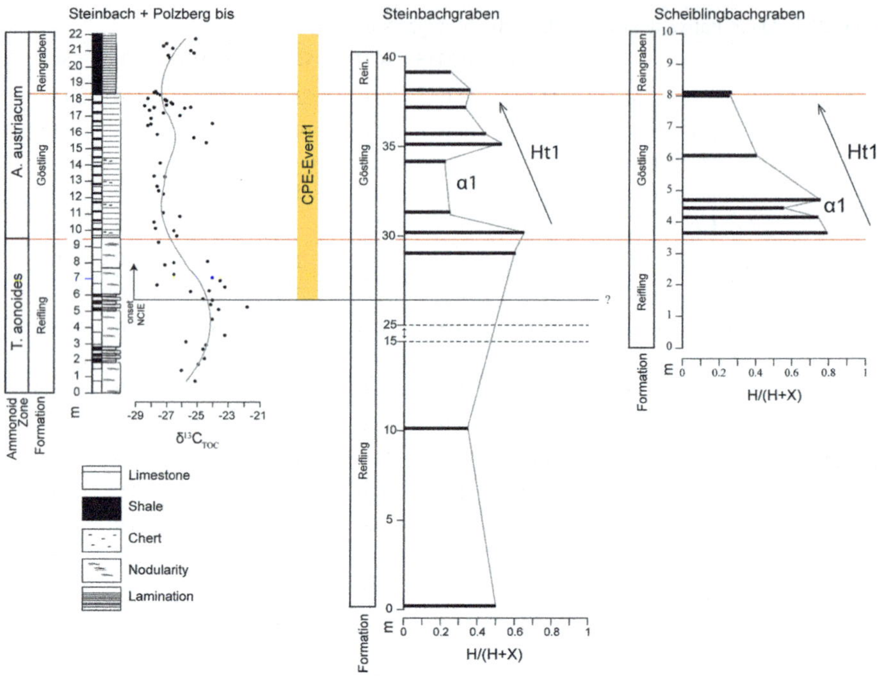

Fig. 17.3 The isotopic curve ($\delta^{13}C_{TOC}$ values) in the Reifling and Göstling Formations, correlated to the ratio of H/(H+X) for the Steinbachgraben and Scheiblingbachgraben sections

with the base of the *Austrotrachyceras austriacum* ammonoid zone (Fig. 17.2). A very poor *Aulisporites astigmosus* assemblage was also recorded at the roof of the Reifling Formation at Steinbachgraben (sample J4, Table 17.1). This finding will need to be verified with more in-depth analyses of the palynological associations in the upper part of the Reifling Formation. The appearance of the monosaccate forms *Patinasporites densus*, *Vallasporites ignacii* and *Enzonalasporites* spp. in the lower part of the Göstling Formation is notable, both in Steinbachgraben (sample J7, Table 17.1) and Scheiblinggraben (sample SCH18, Table 17.2). The appearance of these species, attributed to conifers (Patokaeaceae; cf. Pacyna et al. 2017), occurs precisely in the first arid pulse that characterizes the middle-lower part of the Göstling Formation. Nonetheless, the appearance of these characteristic monosaccate forms is most likely a quantitative fluctuation rather than a qualitative shift because they have already been reported from the beginning of the Carnian (Mietto et al. 2012). These data demonstrate a univocal behavior of the group of monosaccate conifers to the climatic shifts that occurred in this period. This confirms the role of the CPE in the proliferation, migration, extinctions and appearances of new types of conifers that led to the establishment of modern conifers (Dal Corso et al. 2020) (Fig. 17.4).

The subsequent association C (*Lagenella martinii* assemblage) was defined in the lower-middle part of the Lunz Formation (Fig. 17.2), in its Shale and Sandstone

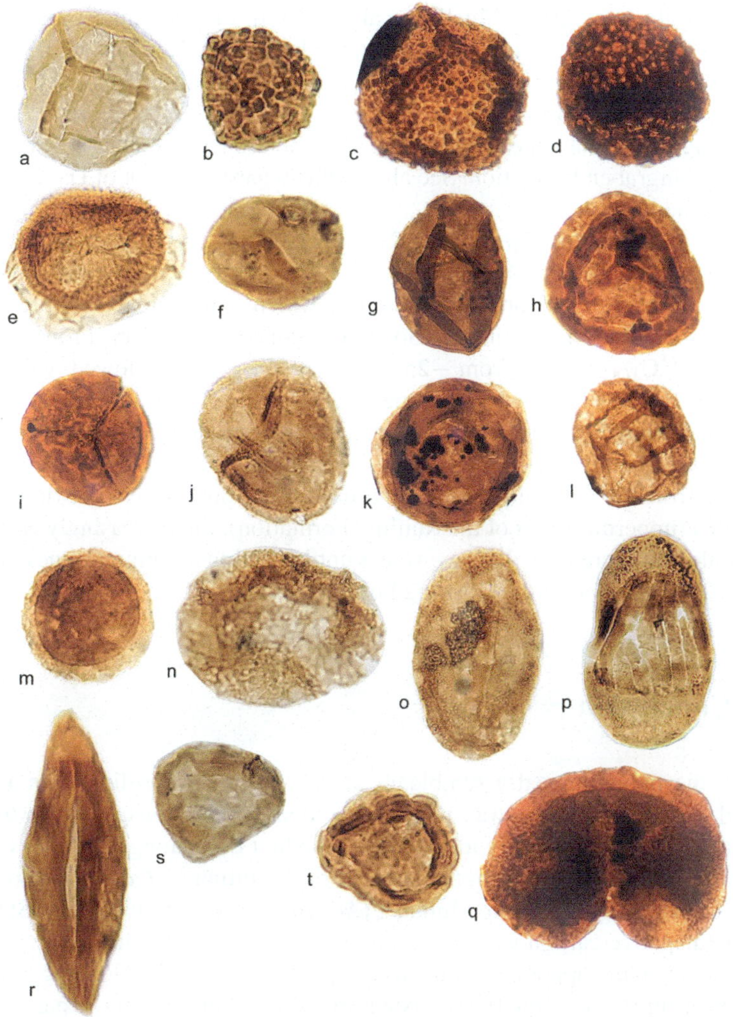

Fig. 17.4 **a** *Calamospora klausii*, SCH53, G64/2, 56 μm diameter, **b** *Uvaesporites* sp., SCH15, F56, 27 μm diameter, **c** *Verrucosisporites* sp., SCH20, S63, 88 μm diameter, **d** *Paleospongiosporis europeus*, SCH20, U47/2, 75 μm diameter, **e** *Kraeuselisporites* sp., SCH15, Q64/2, 61 μm wide, **f** *Aulisporites astigmosus*, SCH10, G62, 38 μm diameter, **g**: *Aulisporites astigmosus*, SCH15, P36/2, 51 μm wide, **h** *Striatella seebergensis*, SCH15, P56/2, 51 μm diameter, **i** *Camarozonosporites rudis*, SCH15, P36/2, 38 μm diameter, **j** *Aratrisporites scabratus*, SCH15, Q49/2, 45 μm diameter, **k** *Peroaletes* sp., SCH20, D46, 56 μm diameter, **l** *Araucariacites* sp., SCH20, K40/4, 68 μm diameter, **m**: *Enzonalasporites* sp., SCH20, L49/2, 36 μm diameter, **n** *Patinasporites densus*, SCH20, 38/3, 70 μm diameter, **o** *Ovalipollis pseudoalatus*, SCH15, R46, 70 μm length, **p** *Lunatisporites* sp., SCH15, N50/4, 72 μm wide, **q** *Pityosporites* sp., SCH15, P46, 60 μm wide, **r** *Cycadopites* sp., SCH15, E58/1, 75 μm high, 25 μm wide, **s** *Duplicisporites granulatus* SCH10, M51/2, 28 μm diameter, **t** *Camerosporites secatus*, SCH15, O63/4, 30 μm diameter

Member (Roghi et al. 2010; Mueller et al. 2016). In the Dolomites, association C (*Lagenella martinii* assemblage) appears in the final part of the first NCIE (negative carbon-isotope excursion; lower part of the Heiligkreuz Formation, Dal Corso et al. 2012), a position that corresponds to the Göstling Formation (see below). This calls for verifying the actual appearance of the aforementioned association, which could fall in the Reingraben Formation. See also in this regard the work of Dal Corso et al. in this volume.

The isotopic curve ($\delta^{13}C_{TOC}$) in the Reifling, Göstling and Reingraben Formation interval was published in Dal Corso et al. (2015). It is partially showed here in Fig. 17.3 and reveals the onset of the CPE, which can be placed approximately 3.5 m below the Reifling Formation-Göstling Formation boundary. From that point, the highest $\delta^{13}C_{TOC}$ values (from -25 to -22 $\delta^{13}C_{TOC}$) are followed, with various fluctuations, by a decreasing trend in the values, with a clear minimum in the lower part of the Göstling Formation. This latter shift can be related to the driest peak/pulse defined by the quantitative palynological diagram ($\alpha 1$ in Fig. 17.3).

Starting from the generally humid conditions at the onset of NCIE1 (initial phase of the CPE) (uppermost part of the Reifling Formation), the increasingly arid conditions parallel the decreasing $d^{13}C_{TOC}$ values until reaching minimum values ($\delta^{13}C_{TOC}$ -27 to -28) at the top of the Göstling Formation.

17.6 Conclusion

Although the palynological assemblages at the top of the Reifling Formation are poor and very degraded, it was possible to identify the *Aulisporites astigmosus* assemblage; it was also clearly identified for the first time along the entire Göstling Formation. A drier climate from the last part of the Julian 1 (*Trachyceras aonoides* Subzone) and the first part of the Julian 2 (*Austrotrachyceras austriacum* Subzone), during the negative carbon isotope excursion, is confirmed by quantitative palynological data from the upper part of the Reifling Formation to the Göstling Formation and the lowermost part of the Reingraben Formation. Thus, changes in palynological assemblages parallel the changes in carbon cycling. The palynological studies of the Göstling Formation initiate the review of the Austrian palynological associations with the aim to locate the various hygrophyte versus xerophyte pulses stratigraphically in relation to the isotope curve.

Acknowledgments The authors are particularly grateful to the owner of the outcrop area—the Austrian Federal Forests (Österreichische Bundesforste)—for sampling and digging permission, especially to Lukas Nöst (forest district Großreifling) and Dominic Asztaller (forest dristrict Großreifling) for ther advice and help during the field campaigns. Jacopo Dal Corso (Wuhan) and Piero Gianolla (Ferrara) are acknowledged for discussion and improved the isotopic dataset. We thank Petra Lukeneder (Gablitz), Veronika Koukal (Vienna) as well as Justine and Umberto Uprimny (Pichlern) for their help with cleaning outcrops and sampling fossils. The work was done within the framework of and financially supported by projects of the Federal Government of Lower Austria (K3-F-964/001-2020 and K3-F-964/002-2023, Department Science and Research; headquarters in

St. Pölten); the Austrian Academy of Sciences (headquarters in Vienna, project Polzberg Lukeneder 2021) represented by the National Committee for Geo/Hydro Sciences (Earth System Sciences Programme); the Triassic project "Fenster in die Karnische Krise" of the Society of Friends of the Natural History Museum Vienna (Freunde des Naturhistorischen Museums Wien, CPE Lukeneder 2023); and the Natur- und UNESCO Geopark Steirische Esienwurzen (St. Gallen, project Karnische Krise Lukeneder 2024). The manuscript greatly benefited from valuable comments by the reviewers Evelyn Kustatscher (Hall in Tyrol) and Elke Schneebeli (Zurich), the production editor Ashok Arumairaj (Taramani), the in-house series editor Kathleen A. Ritterbush (Utah), the publishing editor Simon Shah-Rohlfs and the production editor Antje Endemann (both Heidelberg).

Author Contribution AL collected new material. GR designed the research and provided documentation on fossil material.

Funding Open access funding was enabled and the projects financially supported by the Federal Government of Lower Austria (K3-F-964/001-2020 and K3-F-964/002-2023, Department Science and Research; headquarters in St. Pölten); the Austrian Academy of Sciences (headquarters in Vienna, project Polzberg Lukeneder 2021) represented by the National Committee for Geo/Hydro Sciences (Earth System Sciences Programme); the Natural History Museum Vienna; the Triassic project "Fenster in die Karnische Krise" of the Society of Friends of the Natural History Museum Vienna (Freunde des Naturhistorischen Museums Wien, CPE Lukeneder 2023); and the Natur- und UNESCO Geopark Steirische Esienwurzen (St. Gallen, project Karnische Krise Lukeneder 2024) (to AL). The authors are responsible for the contents of this publication. The funder had no impact on conceptualization, design, data collection, analysis, decision to publish, or preparation of the manuscript.

Competing interests All authors stated that they have no competing interests.

Authors' contributions All authors stated that they have equally contributed to their chapters, read, edited and agreed to submit the final versions of their contributions.

Availability of data and materials All sources of information are provided in the text. Raw data related to the fossil material from Austria are available from the corresponding authors and Alexander Lukeneder upon request.

Additional information The article is part of the Special Volume: *Late Triassic Konservat-Lagerstätten within the Carnian Pluvial Episode. 150 Years of Scientific Excavations in Austria.*

Artificial Intelligence (AI) The chapter was produced without the use of AI.

Language The entire volume uses American English reviewed by a native scientist.

Received 2025.

Published online 2026.

References

Balme BE (1995) Fossil in situ spores and pollen grains: an annotated catalogue. Rev Palaeobot Palynol 87:81–323. https://doi.org/10.1016/0034-6667(95)93235-X

Baranyi V, Rostási A, Raucsik B, Kürschner WM (2019a) Palynology and weathering proxies reveal climatic fluctuations during the Carnian pluvial episode (CPE) (Late Triassic) from marine successions in the Transdanubian Range (western Hungary). Glob Planet Chang 177(2019):157–172

Baranyi V, Rostási Á, Raucsik B, Kürschner WM (2019b) Palynological and X-ray fluorescence (XRF) data of Carnian (Late Triassic) formations from western Hungary. Data Brief 23:103858. https://doi.org/10.1016/j.dib.2019.103858

Baranyi V, Budai T, Karádi V, Rostási Á, Jin X, Kürschner WM, Tóth E (2025) Vegetation and climate record across the Carnian Pluvial episode from the Transdanubian Range, Hungary, Western Tethys. Palaeogeogr Palaeoclimatol Palaeoecol 671:112989. https://doi.org/10.1016/j.palaeo.2025.112989

Baranyi V, Reichgelt T, Olsen PE, Parker WG, Kürschner WM (2017) Norian vegetation history and related environmental changes: new data from the Chinle Formation, Petrified Forest National Park (Arizona, SW USA). GSA Bull 130(5–6):775–795. https://doi.org/10.1130/B31673.1

Barrón E, Gómez JJ, Goy A, Pieren AP (2006) The Triassic-Jurassic boundary in Asturias (Northern Spain): palynological characterisation and facies. Rev Palaeobot Palynol 138:187–208. https://doi.org/10.1016/j.revpalbo.2006.01.002

Bharadwaj DC, Singh HP (1963) An Upper Triassic miospore assemblage from the Coals of Lunz, Austria. Palaeobotany 12:28–44. https://doi.org/10.54991/jop.1963.641

Bonis NR, Kürschner WM (2012) Vegetation history, diversity patterns, and climate change across the Triassic/Jurassic boundary. Paleobiology 38:240–264. https://doi.org/10.1666/09071.1

Brugman WA, Van Bergen PF, Kerp JHF (1993) A quantitative approach to Triassic palynology, the Lettenkeuper of the Germanic Basin as an example. In: Traverse A (ed) Sedimentation of organic particles. Cambridge University Press, Cambridge, pp 409–429

Clement-Westerhof JA (1974) In situ pollen from gymnospermous cones from the Upper Permian of the Italian Alps—a preliminary account. Rev Palaeobot Palynol 17(1–2):63–73. https://doi.org/10.1016/0034-6667(74)90092-X

Dal Corso J, Mietto P, Newton RJ, Pancost RD, Preto N, Roghi G, Wignall PB (2012) Discovery of a major negative $\delta^{13}C$ spike in the Carnian (Late Triassic) linked to the eruption of Wrangellia flood basalts. Geology 40(1):79–82. https://doi.org/10.1130/G32473.1

Dal Corso J, Gianolla P, Newton RJ, Franceschi M, Roghi G, Caggiati M, Raucsik B, Budai T, Haas J, Preto N (2015) Carbon isotope records reveal synchronicity between carbon cycle perturbation and the "Carnian Pluvial Event" in the Tethys realm (Late Triassic). Glob Planet Chang 127:79–90. https://doi.org/10.1016/j.gloplacha.2015.01.013

Dal Corso J, Bernardi M, Sun Y, Song H, Seyfullah LJ, Preto N, Gianolla P, Ruffell A, Kustatscher E, Roghi G, Merico A, Hohn S, Schmidt AR, Marzoli A, Newton RJ, Wignall PB, Benton MJ (2020) Extinction and dawn of the modern world in the Carnian (Late Triassic). Sci Adv 6(38):eaba0099. https://doi.org/10.1126/sciadv.aba00

Dunay RE, Fisher MJ (1978) The Carnian palynofloral succession in the Northern Calcareous Alps. Pollen Spores 20:177–187

Franz M, Nowak K, Berner U, Heunisch C, Bandel K, Röhling HG, Wolfgamm M, (2014) Eustatic and climatic control on the Upper Muschelkalk Sea (late Anisian/Ladinian) in the central European Basin. Glob Planet Chang 135:1–27. https://doi.org/10.1016/j.gloplacha.2015.09.014

Gallet Y, Besse J, Krystyn L, Théveniaut H, Marcoux J (1994) Magnetostratigraphy of the Mayerling section (Austria) and Erenkolu Mezarlik (Turkey) section: improvement of the Carnian (Late Triassic) magnetic polarity time scale. Earth Planet Sci Lett 125(1–4):173–191. https://doi.org/10.1016/0012-821X(94)90214-3

Griffith J (1977) The Upper Triassic fishes from Polzberg bei Lunz. Zool J Linn Soc 60/1:1–93, London. https://doi.org/10.1111/j.1096-3642.1977.tb00834.x

Hornung T (2008) The 'Carnian Crisis' in the Tethys realm—multistratigraphic studies and palaeoclimate constraints. VDM-Verlag, Saarbrücken, pp 1–237. 978-3-639-00035-1

Hornung T, Krystyn L, Brandner R (2007) A Tethys-wide mid-Carnian (Upper Triassic) carbonate productivity crisis: evidence for the Alpine Reingraben Event from Spiti (Indian Himalaya) J Asian Earth Sci 30:285–302. https://doi.org/10.1016/j.jseaes.2006.10.001

Klaus W (1960) Sporen der karnischen Stufe der ostalpinen Trias. Geol Jb BA 5:107–184

Köppen A (1997) Faziesentwicklung in der frühen Obertrias Mitteleuropas—ein sequenzstratigraphischer Vergleich. Gaea Heidelberg, pp 1–233

Krystyn L (1978) Eine neue Zonengliederung im alpin-mediterranen Unterkarn. Schriftreihe der Erdwissenschaftlichen Kommissionen. Österreichische Akademie der Wissenschaften 4:37–75

Krystyn L (1991) Die Fossillagerstätten der alpinen Trias. In: Vasilček W, Krystyn L, Golebiowski R (eds) Exkursionen im Jungpaläozoikum und Mesozoikum Österreichs. Öster Paläontol Ges Wien, pp 23–78

Lahm B (1984) Spumellarienfaunen (Radiolaria) aus den mitteltriassischen Buchensteiner-Schichten von Recoaro (Norditalien) und den obertriassischen Reiflingerkalken von Großreifling (Österreich). Münchner Geowiss Abh A, Geol Paläontol 1:1–161

Lukeneder A (ed) (2026) Late Triassic Konservat-Lagerstätten within the Carnian Pluvial Episode: 150 Years of Scientific Excavations in Austria. Top Geobiol 55:600 p. hhttps://doi.org/10.1007/978-3-032-19777-1

Lukeneder A, Lukeneder P (2021) The Upper Triassic Polzberg palaeobiota from a marine *Konservat-Lagerstätte* deposited during the Carnian Pluvial Episode in Austria. Sci Rep 11:16644. https://doi.org/10.1038/s41598-021-96052-w

Lukeneder A, Lukeneder P (2022a) Taphonomic history and trophic interactions of an ammonoid fauna from the Upper Triassic Polzberg palaeobiota. Sci Rep 12:7455. https://doi.org/10.1038/s41598-022-11496-y

Lukeneder P, Lukeneder A (2022b) Comment on "Triassic coleoid beaks and other structures from the Calcareous Alps revisited" by Doguzhaeva et al. (2022). Acta Palaeontol Pol 67:963–965. https://doi.org/10.4202/app.01016.2022

Lukeneder A, Lukeneder P (2023) New data on the marine Upper Triassic palaeobiota from the Polzberg *Konservat-Lagerstätte* in Austria. Swiss J Palaeont 142:1–18. https://doi.org/10.1186/s13358-023-00269-3

Lukeneder A, Lukeneder P (2024) Ammonite taxonomy of the Carnian Polzberg *Konservat-Lagerstätte* in Austria. Jb GeoSphere Austria (Geol Bundesanstalt) 162:27–55

Lukeneder A, Lukeneder P, Sachsenhofer RF, Roghi G, Rigo M (2024) Multi-proxy record of the Austrian Upper Triassic Polzberg *Konservat-Lagerstätte* in light of the Carnian Pluvial Episode. Sci Rep 14:11194. https://doi.org/10.1038/s41598-024-60591-9

Lukeneder A, Surmik D, Gorzelak P, Niedźwiedzki TB, Salamon MA (2020) Bromalites from the Upper Triassic Polzberg section (Austria); insights into trophic interactions and food chains of the Polzberg palaeobiota. Sci Rep 10:20545. https://doi.org/10.1038/s41598-020-77017-x

Mietto P, Manfrin S, Preto N, Rigo M, Roghi G, Furin S, Gianolla P, Posenato R, Muttoni G, Nicora A, Buratti N, Cirilli S, Spötl C, Ramezani J, Bowring SA (2012) The global boundary stratotype section and point (GSSP) of the Carnian stage (Late Triassic) at Prati di Stuores/Stuores Wiesen Section (Southern Alps, NE Italy). Episodes 35(3):414–430. https://doi.org/10.18814/epiiugs/2012/v35i3/003

Mostler H, Scheuring BW (1974) Mikrofloren aus dem Langobard und Cordevol der Nördlichen Kalkalpen und das Problem des Beginns der Keupersedimentation im Germanischen Raum. Geol Paläontol Mitt 4(4):1–35

Mueller S, Krystyn L, Kürschner WM (2016) Climate variability during the Carnian Pluvial Phase—a quantitative palynological study of the Carnian succession at Lunz am See, Northern Calcareous Alps, Austria. Palaeogeogr Palaeoclimatol Palaeoecol 441(1):198–211. https://doi.org/10.1016/j.palaeo.2015.06.008

Pacyna G, Barbacka M, Zdebska D, Ziaja J, Fijałkowska-Mader A, Bóka K, Sulej T (2017) A new conifer from the Upper Triassic of southern Poland linking the advanced voltzialean type of ovuliferous scale with Brachyphyllum-Pagiophyllum-like leaves. Rev Palaeobot Palynol 245:28–54. https://doi.org/10.1016/j.revpalbo.2017.05.015

Pott C, Lukeneder A (2026) A deltaic swamp forest: coastal wetland vegetation during the climax of the Carnian Pluvial Episode. In: Lukeneder A (ed) Late Triassic Konservat-Lagerstätten within the Carnian pluvial episode: 150 years of scientific excavations. Top Geobiol 55, Chap 18:405–436. https://doi.org/10.1007/978-3-032-19777-1_18

Pott C, Krings M, Kerp H (2008) The Carnian (Late Triassic) flora from Lunz in Lower Austria: paleoecological considerations. Palaeoworld 17(3–4):172–182. https://doi.org/10.1016/j.palwor.2008.03.001

Roghi G (2004) Palynological investigations in the Carnian of the Cave del Prdil area (Julian Alps, NE Italy). Rev Palaeobot Palynol 132(1–2):1–35. https://doi.org/10.1016/j.revpalbo.2004.03.001

Roghi G, Gianolla P, Minarelli L, Pilati C, Preto N (2010) Palynological correlation of Carnian humid pulses throughout western Tethys. Palaeogeogr Palaeoclimatol Palaeoecol 290(1–4):89–106. https://doi.org/10.1016/j.palaeo.2009.11.006

Scheuring B (1974) "Protosaccate" Strukturen, ein weitverbreitetes Pollenmarkmal zur fruhen und mittleren Gymnospermenzeit. Geol Palaontol Mitt 4:1–30

Scibiorski JP (2024) Palynological assemblages from the Middle-Late Triassic Mungaroo Formation, northwestern Australia: palaeobotanical affinities, depositional facies and preferred habitats Lethaia 57(4):1–39 https://doi.org/10.18261/let.57.4.2

Stur D (1874) Neue Aufschlüsse im Lunzer Sandsteine bei Lunz und ein neuer Fundort von Wengerschiefer im Pölzberg zwischen Lunzersee und Gaming. Verh K K Geol Reichsanstalt 1:271–273

Summesberger HK, Wagner L (1972) Der Stratotypus des Anis (Trias). Ann Naturhist Mus Wien 76:515–538

Teller F (1891) Über den Schädel eines fossilen Dipnoers, *Ceratodus Sturii* nov. spec. Aus den Schichten der oberen Trias der Nordalpen. Abh K. K. Geol Reichsanst 15:1–38

Visscher H, Van der Zwan CJ (1981) Palynology of the circum-Mediterranean Triassic phytogeographical and palaeoclimatological implications. Geol Rundsch 70:625–636

Visscher H, Van Houte M, Brugman WA, Poort RJ (1994) Rejection of Carnian (Late Triassic) "pluvial event" in Europe. Rev Palaeobot Palynol 83:217–226

Zhang J, Lenz OK, Wang P, Hornung J (2021) The eco-plant model and its implication on Mesozoic dispersed sporomorphs for Bryophytes, Pteridophytes, and Gymnosperms. Rev Palaeobot Palynol 293:104503. https://doi.org/10.1016/j.revpalbo.2021.104503

Chapter 18
A Deltaic Swamp Forest: Coastal Wetland Vegetation During the Climax of the Carnian Pluvial Episode

Christian Pott and Alexander Lukeneder

Abstract The well-known paleobiota of the *Fossil-Lagerstätte* from the surroundings of Lunz-am-See in Lower Austria were deposited during the early–middle Carnian, within a period commonly referred to as the worldwide Late Triassic Carnian Pluvial Episode (CPE). If vegetation thrives during environmental circumstances that change to wetter and moister conditions with time, the expectation is that the plants will develop or evolve certain adaptations to such conditions, for example to compete for sunlight or cope with heavy rain. Adaptations to wet conditions may be evident on macroscopic scale, but may also remain hidden. The latter can be reflected in characters that are visible only on micro-scale, i.e. on cellular level. Most plant fossils from the Carnian (Upper Triassic) Lunz Formation yield excellently preserved cuticles. An adaptation of this presumed coastal vegetation to moister or wetter conditions, reflecting the influences of their coastal exposition, has repeatedly been emphasized. The environmental conditions that affected the ecosystem are re-evaluated here synoptically based on the composition of the vegetation and the different adaptations of the individual plant species on the macro- and microscopic level. The plant-bearing beds of the Lunz Formation are interpreted as the result of a short-term sea-level lowstand and a meandering river delta system. That system transported terrestrial material from the northeast into a brackish marsh area close to the coast, during the otherwise prevailing marine sedimentation in the Alpine Triassic. The vegetation—if considered as a whole and incorporating all aspects—indicates a humid, moist and swampy, deltaic or near-coastal environment, with autochthonously embedded plant fossils.

Keywords Ecosystem · Environmental conditions · Ecological adaptations · Rainforest · Carnian

---

C. Pott (✉)
LWL-Museum of Natural History, Westphalian State Museum with Planetarium, Münster, Germany
e-mail: christian.pott@lwl.org

A. Lukeneder
Department of Geology and Paleontology, Natural History Museum Vienna, Vienna, Austria
e-mail: alexander.lukeneder@nhm.at

A. Lukeneder (ed.), *Late Triassic Konservat-Lagerstätten within the Carnian Pluvial Episode*, Topics in Geobiology 55, https://doi.org/10.1007/978-3-032-19777-1_18

18.1 Introduction

The well-known paleobiota of the *Fossil-Lagerstätte* from the surroundings of Lunz-am-See in Lower Austria (Pott 2021; Fig. 18.1a) were deposited during the early–middle Carnian, within a period commonly referred to as the worldwide Late Triassic Carnian Pluvial Episode (CPE; Simms and Ruffell 1989, 2018; Hornung et al. 2007; Lukeneder et al. 2012; Ruffell et al. 2015; Dal Corso et al. 2018; Lukeneder and Lukeneder 2021, 2023; Lukeneder et al. 2024; Lukeneder 2026). The CPE featured an enhanced humidification of the Carnian climate, resulting in the deposition of the fossiliferous lower Carnian Reingraben Formation at Scheiblingbach and Polzberg. As noted by Lukeneder and Lukeneder (2021, 2023) and Lukeneder et al. (2020, 2024), such laminated deposits provide excellent conditions for the preservation of entire and soft body fossils, a prerequisite for *Fossil-Lagerstätten.* The paleobiota and fossil assemblages preserved in these deposits reflect a marine environment in the lower Carnian (Julian), yet with intermittent shallow-water influence. This is indicated by the findings of freshwater inhabitants (Lukeneder and Lukeneder 2021, 2023), accompanied by transported and redeposited plant remains, which are accumulated in plant layers. Within the Lunz Nappe in northernmost Styria and Lower Austria, the Reifling Basin (Krystyn 1991; Lukeneder and Lukeneder 2021, 2023; Lukeneder et al. 2024) exposes Upper Triassic (Carnian, Julian) strata, located between Polzberg near Lunz-am-See and Großreifling. The Reifling Basin was a Late Triassic intra-platform basin between Pangea in the northwest and the Meliata Ocean in the southeast. The Meliata Ocean was a small oceanic basin within the Tethyan Realm, separated from the Tethys Ocean by the island chain of the Cimmerian Continent (Hornung and Brandner 2005; Lukeneder et al. 2012).

Enhanced humidification is reflected by a change in the lithology and facies in the Lunz-am-See area. The basal Julian sequence at the Polzberg site is characterized by nodular limestones of the Reifling Formation deposited on the paleoslope of the Reifling Basin. At the base of the Julian *Austrotrachyceras-austriacum* zone, the Reifling Formation is replaced by limestone deposits with organic-rich mudstones of the Göstling Formation and the terrigenous siliciclastic deposits of the Reingraben Formation. Global warming, combined with enhanced humidification during the early Carnian, and the eruption of large amounts of volcanogenic material, probably triggered a climate change during that time interval. The CPE phase was finally peaked by the deposition of the Lunz Formation above the Reingraben Formation with the famous plant fossils of the lower Carnian (Julian 2) Lunz flora (Fig. 18.1b). These fossils characterize the Lunz *Fossil-Lagerstätte* (Krystyn 1991), which can more specifically be defined as a *Konservat-Lagerstätte sensu* Seilacher (1970), Seilacher et al. (1985) and Kimmig and Schiffbauer (2024).

The main sandstone of the Lunz Formation is interpreted as a deltaic deposition which, over time, filled up the up to 150-m-deep Reifling Basin. In contrast, the overlying fossil-rich shales—consisting of dark grey shales intercalated with silt layers, fine-grained sandstone and coal seams up to a few meters thick, and comprising the Lunz swamp paleoenvironments—were formed in coastal lowlands and marshes.

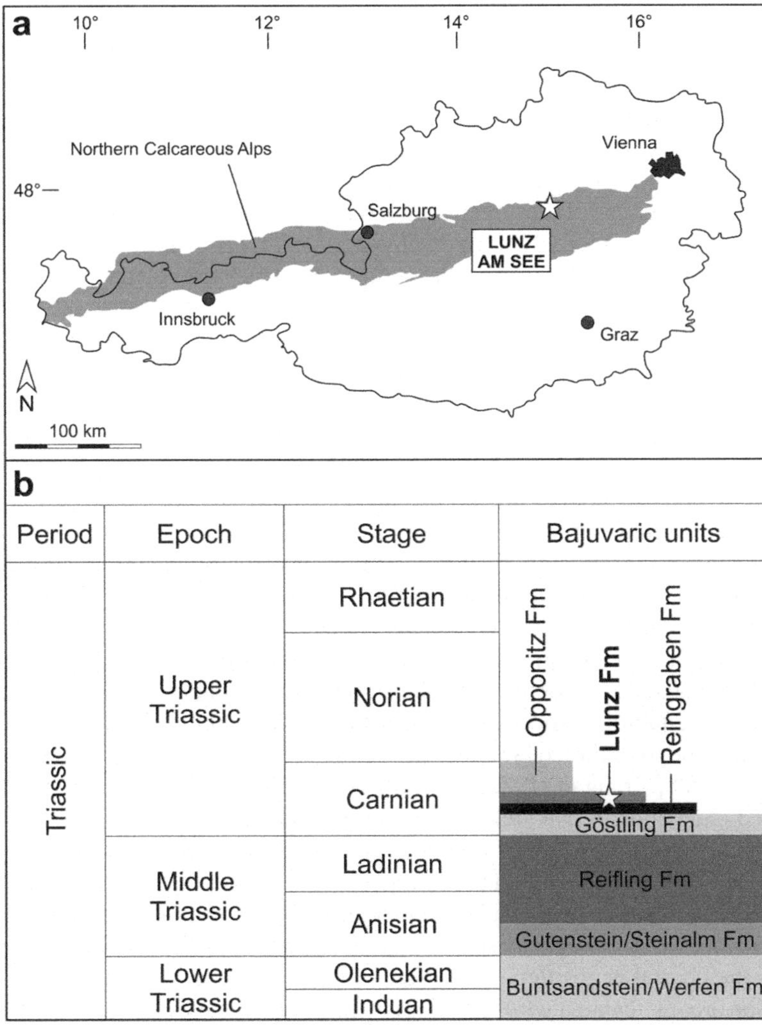

Fig. 18.1 **a** Map of Austria with the position of Lunz-am-See in the Austrian Northern Calcareous Alps (in gray). **b** Stratigraphic position of the Lunz Formation deposited during the early Carnian. White star: position of the Lunz *Konservat-Lagerstätte*. Adapted from Lukeneder and Lukeneder (2021). Fm Formation

The fine-grained sandstones in this complex are interpreted partly as terrigenous deposits and partly as deposits in a very shallow marine or brackish environment (Behrens 1972; Krystyn 1991; Hornung et al. 2007; Meller 2007; Pott et al. 2008b, 2018).

A thriving vegetation during environmental circumstances that change to wetter and moister conditions over time—be it the amount of precipitation such as heavy rain, or wet soil conditions, to name a few—leads to the expectation that those plants develop or evolve certain adaptations to such conditions. This can involve mechanisms to compete for sunlight or cope with heavy rain (e.g., Napp-Zinn 1984; Zotz 2016; Schulze et al. 2019). These strategies can include a layered structure of the forest, anatomical adaptations such as a winding or climbing habit, an epiphytic way of life, or leathery, vertical leaves with robust cuticles and so-called "drip tips", where each leaf apex is extended or pointed to facilitate the quick draining of excess water, which promotes photosynthesis and reduces fungal growth (e.g., Roth 1992; Mabberley 2013; Juniper 2019). In epiphytic plants, roots can absorb nutrients and water from the air and from other plants because wet environments have nutrient-poor soil (Zotz 2016). Leaves commonly are soft, with little fibrous tissue, and may face horizontally towards the sun during rainless periods to capture more light for photosynthesis (Bell 2008). Many plants have large, waxy leaves that are arranged to form funnels that direct water immediately into the root system. Some plants—such as many mosses, lichens and ferns—adapt to shaded places and can inhabit the shady and moist forest floor, live as epiphytes on trees, or shift to a subterranean growth form. Many of these features are, for example, combined and driven to perfection in orchids (e.g., Zotz 2016).

In many cases, adaptations to wet conditions remain hidden, being reflected in characters that are visible on the micro-scale, i.e. on the cellular level only. These include thick, robust cuticles; stomata that are deeply sunken below the level of the epidermal surface or even joined in cavities protected by hairs or papillae; hairy or papillate leaf surfaces; or idiocuticular striae or microreliefs. The latter may have reduced leaf wettability, produced or enhanced the self-cleaning effect of the leaf and/or prevented the formation of a water film on the leaf surface (Napp-Zinn 1984; Roth 1992; Bell 2008; Crang et al. 2019; Schulze et al. 2019). In such cases, examining preserved plant cuticles can yield important information.

The plant fossils from the Carnian (Upper Triassic) Lunz Formation are case in point because most of the taxa are represented by fossils featuring excellently preserved cuticles. Previous studies have repeatedly emphasized an adaptation of this presumed coastal vegetation to moister or wetter conditions, with influences through the coastal exposition of the plants. The coal-producing strata of the Lunz Formation are intercalated with fully marine limestone beds. Consequently, this area may be referred to as a paralic basin (Stremme 1911; Meller 2007; Mueller et al. 2016). As the Alpine Triassic predominantly exhibits a marine character, the plant-bearing beds of the Lunz Formation were interpreted to reflect a short-term sea-level lowstand and a meandering river delta system that transported terrestrial material from the northeast into a brackish marsh area close to the coast (Meller 2007).

Here, we present an overview of the vegetation and the environmental conditions that affected the ecosystem. This summary is based on the composition of the vegetation and on the different adaptations of the individual plants—on the macro- and microscopic level.

18.2 Geologic Setting

The fossils derive from the so-called 'Lunzer Sandstein', which is part of the Lunz Formation (=Lunzer Schichten; Verloop 1908; Fig. 18.1b). The Lunz Formation is subdivided into three members: the lower Shale and Sandstone Member, the Hauptsandstein Member (main sandstone), and the upper Shale and Coal Member (Mueller et al. 2016). The main unit consists generally of sandstone overlain by marine marls that grade upwards into fluvio-lacustrine sandstone, shales and coal. This coal-bearing part is overlain by marls that again are overlain by sandstone. The plant fossils are found in the shales associated with the coal seams (the so-called *Flöznebengestein*). Exact dating of the Lunz Formation remains problematic due to the lack of adequate index fossils. The Lunz Formation is thought to correlate with the upper part of the Reingraben Formation based on the regional and facies-spanning correlation of biostratigraphically well-established sections of the Hallstatt and Reifling intra-platform basins (Hornung and Brandner 2005; Lukeneder and Lukeneder 2021, 2023).

Accordingly, it is most probably late Julian in age, which is supported by palynological studies that indicate a Carnian (Bharadwaj and Singh 1964) and, more precisely, Julian age (Julian 2/II; Dunay and Fisher 1978; Roghi 2004). For the Lunz-am-See area, Mueller et al. (2016) assumed a transition from the Reingraben Formation into the overlying Lunz Formation at the Julian 2/I–Julian 2/II boundary. The upper sub-unit of the Lunzer Schichten (Opponitz Limestone or Opponitzer Schichten; Piller et al. 2004)—nowadays referred to as and separated into the Opponitz Formation (Mueller et al. (2016)—has been dated as Tuvalian by Dunay and Fisher (1978). Julian and Tuvalian are regional chronostratigraphic sub-stages of the Carnian, which correlate to the middle and late Carnian, respectively. More detailed information on the geological setting is provided in Pott (2007) and Pott et al. (2008b, 2017).

18.3 Material and Methods

The flora of the Lunz Formation has been the focus of the scientific work of the first author for almost two decades. The collections studied contain more than 5000 rock specimens from Lunz-am-See in the Northern Calcareous Alps (NCA) of Austria, approximately 100 km west of Vienna (Fig. 18.1a). All studied specimens were collected from former coal mines around Lunz-am-See during the late 19th and early 20th centuries. Unfortunately, precise information on the geographic and stratigraphic occurrence (viz. coal mine) of individual specimens is commonly lacking. Specimens from the Lunz Formation are held in many paleobotanical collections around the world, amongst which the most important are at the Museum of Natural History, Vienna, Austria; the GeoSphere, Vienna, Austria (formerly Geological Survey of Austria); the Swedish Museum of Natural History, Stockholm,

Sweden; the State Museum of Lower Austria (Museum Niederösterreich), St. Pölten, Austria; the Museum of Natural History (Naturalis Biodiversity Center), Leiden, The Netherlands; the Laboratory of Paleobotany and Palynology, University Utrecht, Utrecht, The Netherlands; the State Museum Joanneum (Landesmuseum Joanneum), Graz, Austria; the Senckenberg Museum of Natural History (Senckenberg Naturmuseum), Frankfurt-am-Main, Germany; and the Natural History Museum (Museum für Naturkunde, Leibniz-Institut für Evolutions- und Biodiversitätsforschung) Berlin, Germany.

Hand specimens were commonly photographed with a Nikon D80/Nikkor AF-S Mikro 60 mm 1:2.8G ED system digital camera; in order to enhance contrast, cross-polarization (i.e. polarized light sources together with an analyzing filter in front of the camera lenses) was used. Hand specimens, cuticles and spores were commonly analyzed and photographed with an Olympus SZX10 stereomicroscope and Leitz diaplan and Olympus BX51 light microscopes with mounted digital cameras. For the analysis of cuticles and *in situ* spores, organic material was isolated from the rock surface with a preparation needle. The sediment remains were removed with hydrofluoric acid (40% HF). The residue was macerated using Schulze's reagent (35% HNO_3 with a few crystals of $KClO_3$), bleached with 5–10% potassium hydroxide (KOH) for a few seconds, if necessary, and neutralized in distilled water, and stored in glycerol. Macerated cuticles were washed in distilled water, gently dehydrated in pure glycerin, and subsequently mounted in permanent glycerin-jelly on microscope slides. In order to increase contrast, some cuticles were stained with Sudan III.

In a few cases, scanning electron microscopy (SEM) was used. The specimens were investigated and photographed immediately after sputter coating with gold. SEM micrographs were taken with a Hitachi S-4300 cold field emission scanning electron microscope at 3000 V.

18.4 Results

18.4.1 Short Historical Overview

More than 170 years ago, the first plant fossils were described from the Lunz strata in the region around Lunz-am-See in Lower Austria (Ettingshausen 1851). Since then, numerous fossil sites have been discovered in the Lunz-am-See area and well over 5000 fossils have been recovered. Most of the specimens were collected underground during the mining of the Lunz black coal, and sold (i.e. transported) directly to museums; they were consequently always protected from the elements and are in excellent condition. From around 1860 onwards, when the fossils were increasingly recovered, they became well known far beyond Austria's borders, with many museums in Europe and overseas acquiring specimens for their display collections. As a result, the Lunz flora is one of the most extensive and important floras of the

Alpine Triassic, even though the fossils are spread across many different collections in Europe and around the world.

The fame of the Lunz flora is not only due to the generally excellent preservation of the fossils, but above all due to the various fertile gymnosperm remains (sporophylls, “flowers”, “fruits”, seeds). Most of these belonged to the Bennettitales, an enigmatic group of seed plants that has its origin somewhere in the Early–Middle Triassic (but see Blomenkemper et al. 2021) and became extinct towards the end of the Cretaceous. The Bennettitales from Lunz-am-See are thus among the oldest representatives of this group, whose distribution heyday was in the Early to Middle Jurassic.

Mining of the up to 7-m-thick coal seams around Lunz-am-See began more than 300 years ago. In the 1920s, mining almost ceased due to declining yields. The closure of the last gallery in 1958 marked the end of coal mining here. The fossil-rich outcrops of the Lunz strata are, therefore, no longer accessible today. Nonetheless, with a bit of luck, one can still find beautiful plant fossils on the former mine heaps or in smaller outcrops along streams, smaller rivers and creeks around the village of Lunz-am-See. This material, however, is usually heavily weathered. Nevertheless, significant and scientifically important finds can still be made, as demonstrated by current on-site collection activities (Pott et al. 2017).

The fossils of the Lunz flora have been subject to scientific studies continuously since their discovery (e.g., Stur 1885; Krasser 1909a, 1909b, 1917, 1919; Kräusel 1943, 1948, 1949, 1953; Kräusel and Schaarschmidt 1966; Dobruskina 1998). For almost two decades, the flora of the Lunz Formation has been the research focus of the first author (see References). The respective findings confirm that a very species-rich and diverse flora thrived in the Lunz-am-See region during the Late Triassic. In addition to numerous sphenophytes (Pott et al. 2008a, 2018), herbaceous and tree-like ferns (Pott et al. 2018), cycads (Pott et al. 2007a, 2007b; Pott 2019) and bennettites (Pott et al. 2007c, 2007d, 2007e, 2010 2017; Pott 2016) dominated the “coal swamps” of the Lunz-am-See area (Pott et al. 2008b; Pott and Krings 2010; Pott 2021).

18.4.2 Composition of the Flora from the Lunz Formation

Cryptogams

Local portions of the Lunz flora were rich in sphenophytes (horsetails). The very common *Equisetites arenaceus* (Fig. 18.2a) with its commonly 10–15-cm-thick stems was one of the largest horsetails ever and widespread during the Late Triassic and Jurassic at latitudes around 30° N (Kelber and Van Konijnenburg-van Cittert 1998; Pott et al. 2018). The axes, occasionally proximally producing secondary branches, could grow to a height of 5–10 m and reached diameters of up to 25 cm (Frentzen 1934; Pott et al. 2008a, 2018). Sexual reproduction occurred through spores produced in small, stalked strobili growing on delicate sterile lateral axes in the distal stem area (Kelber and Van Konijnenburg-van Cittert 1998). In addition to *Equisetites arenaceus*, two other, smaller horsetails, viz. *Equisetites conicus* (Fig. 18.2b) and *Neocalamites merianii* (Fig. 18.2c), thrived in the Lunz-am-See region. Both exhibited a herbaceous growth form and probably rarely grew taller than 2–3 m. Axes of *Neocalamites merianii* were covered at regular intervals with whorls of about 60 narrow, up to 20 cm long, individual microphylls (Pott et al. 2008a, 2018). *Neocalamites merianii* reproduced through spores that were formed in terminal sporangia on separate fertile axes (Vladimirovicz 1958; Escapa and Cunéo 2006; Pott et al. 2008a, 2018).

The ferns of the Lunz flora were very species-rich and probably produced a high biomass, which contributed significantly to the formation of the Lunz coal seams. The initially more than 20 taxa (Stur 1885; Krasser 1909a) were mainly assigned to the Marattiaceae, most of whose present-day representatives are large tree-like ferns (Large and Braggins 2004). This certainly also applies to the Lunz ferns. *Asterotheca merianii* (Figs. 18.2d–f), for example, is assumed to have grown up to 15 m tall (Pott et al. 2018). In fact, several up to 10–13-cm-thick branches, associated with *Asterotheca* foliage, were recovered from the Lunz Formation (Pott et al. 2018). In addition, many fossils involve very large fronds and frond portions, suggesting that the species of *Asterotheca*, *Todites*, *Symopteris* and *Danaeopsis* were clearly not small in stature and grew fronds of several meters in length.

A recent extensive study of the Lunz ferns revealed that the original identification of the species and their assignment to different families and orders (Stur 1885; Krasser 1909a) required revision (Pott et al. 2018). According to Pott et al. (2018), three species can be clearly placed in the eusporangiate Marattiales (viz. "*Asterotheca*" *merianii*, *Symopteris lunzensis*, *Danaeopsis lunzensis*; Fig. 18.3a). *Todites linnaeifolius* (Fig. 18.3b) is a representative of the leptosporangiate Osmundales. *Phlebopteris lunzensis* and *Cynepteris reticulata* (Fig. 18.3c) belong to the Matoniaceae, whereas *Dictyophyllum serratum*, *Dictyophyllum lunzensis* (Fig. 18.3e), *Clathropteris reticulata* (Fig. 18.3f) and *Camptopteris lunzensis* (Fig. 18.3d) are representatives of the Dipteridaceae (both leptosporangiate Gleicheniales). *Digitopteris repanda* (Fig. 18.4a) represents a new genus of this family, which was described and erected from the Lunz Formation. *Mertensides bullatus* (Fig. 18.4b) is an enigmatic fern species with a unique combination of characteristics. Its definitive affiliation to a particular order remains to be clarified, but it can probably be classified as Marattiales (Pott et al. 2018). The Lunz flora thus comprises an interesting mixture of eusporangiate and leptosporangiate groups, the former representing a rather primitive group of ferns. The co-occurrence of eusporangiate and leptosporangiate ferns renders the Lunz flora important given the fact that eusporangiate ferns predominated over the more modern, leptosporangiate forms during the Paleozoic, whereas they typically constitute only subordinate components in fern vegetation since the mid-Mesozoic until today.

Seed Plants

The Lunz flora was dominated by numerous gymnospermous seed plants. Recent studies by the first author have shown that although the diversity was not nearly as

Fig. 18.2 Selected cryptogams (sphenophytes and ferns) from the Lunz flora. **a** Main axis of *Equisetites arenaceus*, NHMW 1885/0500/3860. **b** Axis of *Equisetites conicus*, NRM S148253. **c** Leaf-carrying axis of *Neocalamites merianii*, NRM S148335. **d** Frond portion of *Asterotheca merianii*, GBAW 1909/002/0023. **e** Details of fertile and sterile pinnules of *Asterotheca merianii*, GBAW 2010/0102/0017. **f** Sori on the lower surface of fertile pinnules of *Asterotheca merianii*, NHMW 1887/0001/0006. Scale bars 10 mm (a–c), 20 mm (d), 5 mm (e–f). GBAW Geological Survey of Austria (GeoSphere Austria), Vienna, Austria; NHMW Museum of Natural History, Vienna, Austria; NRM Swedish Museum of Natural History, Stockholm, Sweden

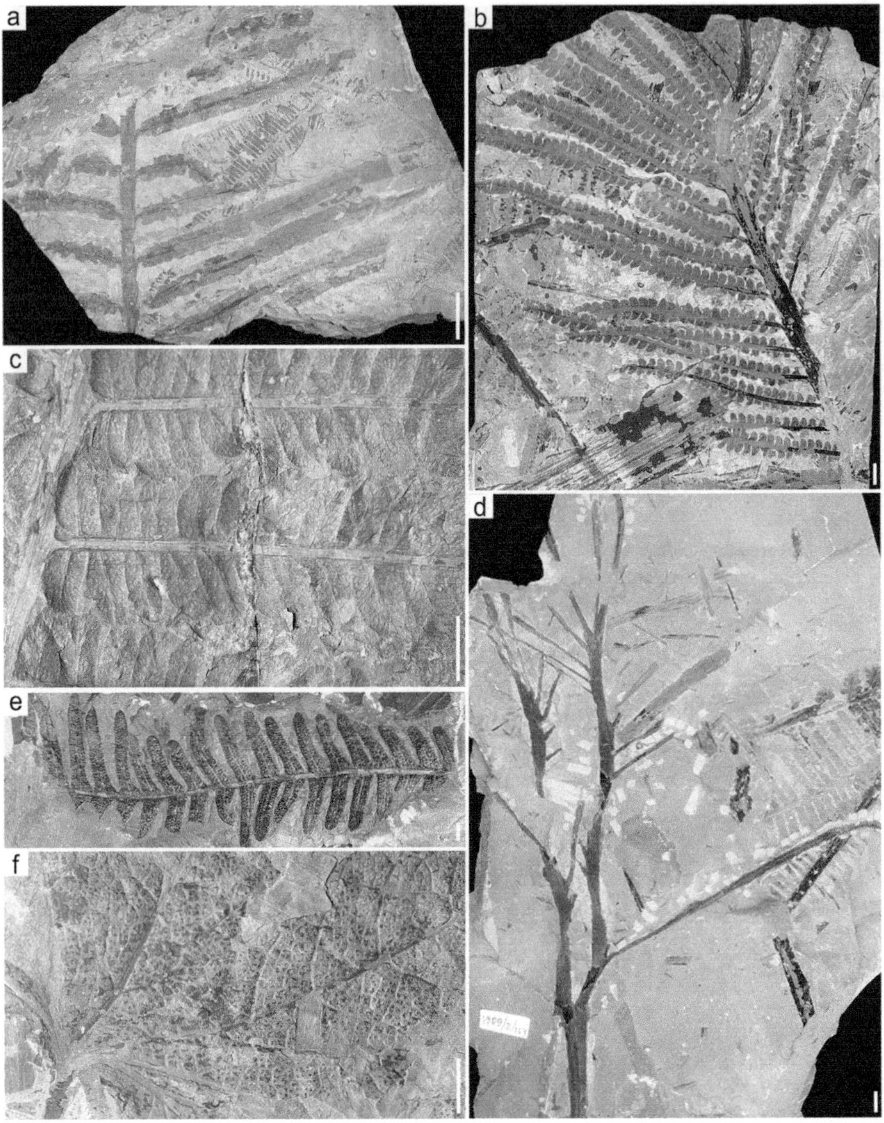

Fig. 18.3 Selected ferns from the Lunz flora. **a** Frond portion of *Danaeopsis marantacea*, NHMW 1889/0006/0006. **b** Frond portion of *Todites linnaeifolius*, NHMW 1887/0001/0004. **c** Detail of a frond portion of *Cynepteris reticulata*, GBAW 1909/002/0551. **d** Almost entire frond of *Camptopteris lunzensis* with the two rachial arms, GBAW 1909/002/0134. **e** Pinna of *Dictyophyllum lunzensis*, GBAW 1909/003/0593. **f** Detail of a frond portion of *Clathropteris reticulata*, GBAW 2010/0102/0022. Scale bars 50 mm (a), 10 mm (b–f). GBAW Geological Survey of Austria (GeoSphere Austria), Vienna, Austria; NHMW Museum of Natural History, Vienna, Austria

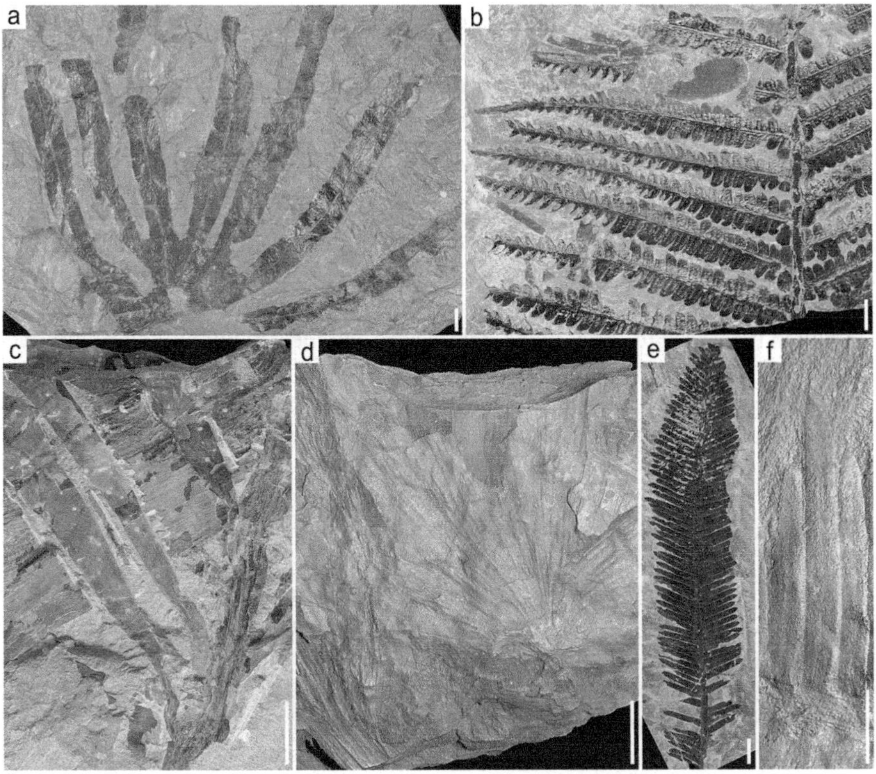

Fig. 18.4 Selected ferns and seed plants from the Lunz flora. **a** Frond of *Digitopteris repanda*, GBAW 1909/003/0112. **b** Frond portion of *Mertensides bullatus*, GBAW 1909/002/0394. **c** Foliate axis of the enigmatic ginkgophyte *Arberophyllum florinii*, NHMW 1889/0006/0008. **d** Possible basal fern frond; fossil originally identified as "*Ginkgoites lunzensis*", GBAW 1942/001/0001. **e** Foliate axis of the conifer *Stachyotaxus lipoldii*, NHMW 1944/0001/0002. **f** Detail of leaflets of *Stachyotaxus lipoldii* with possible "drip tips", NHMW 2006B0008/0006. Scale bars 10 mm (a–e), 5 mm (f). GBAW Geological Survey of Austria (GeoSphere Austria), Vienna, Austria; NHMW Museum of Natural History, Vienna, Austria

high as the first synoptic species lists suggested (Stur 1885; Krasser 1909b), both cycads and the now extinct bennettites were, nevertheless, stand-forming elements of the flora (Pott et al. 2007a, 2007d, 2007e; Pott and Krings 2010). At the same time, ginkgophytes (*Arberophyllum florinii*; Fig. 18.4c; see Dobruskina 1998) and conifers (*Stachyotaxus lipoldii* (Fig. 18.4e–f) are rather underrepresented (Pott et al. 2008b; Pott and Krings 2010). Specimens referred to as "*Ginkgoites lunzensis*" (Kräusel 1920, 1943; Florin 1936) may rather represent basal fern fronds (Fig. 18.4d; see below; Pott, own obs.). Fronds of bennettites and cycads, in contrast, can be found in large numbers on almost every hand specimen.

At first glance, leaves of bennettites and cycads cannot be distinguished from each other. Nonetheless, the two groups have completely different reproductive organs and are therefore probably not closely related (Doyle 2006; Hilton and Bateman 2006; Rudall and Bateman 2007; Crepet and Stevenson 2010; Pott et al. 2012; Pott 2016). While cycads are dioecious plants, in which the micro- and macrosporangiate sporophylls are usually produced in large, terminal cones (strobili) on different plants (Norstog and Nicholls 1997), the micro- and macrosporangiate reproductive organs ("flowers" or "cones") of bennettites are often produced on the same plant or even in a single reproductive organ (Thomas 1915; Harris 1969). These reproductive organs can be surrounded by sterile bracts (Pott et al. 2010, 2017; Pott 2014, 2016). Contrary to general assumptions, bennettites are a rather heterogeneous group and comprise at least two different types (Cycadeoidaceae and Williamsoniaceae) that differ fundamentally in their growth form, plant architecture and the structure of their reproductive organs (Watson and Sincock 1992; Pott 2014, 2016; Pott and McLoughlin 2014; Pott and Axsmith 2015; Pott et al. 2015). The reproductive structures are the most remarkable feature of bennettites, which can most simply be described as flowers. Although these flowers differed in the details of their anatomy from the true flowers of angiosperms, they clearly represented an experiment in plant architecture that preceded but paralleled in architecture some of the adaptations of modern flowering plants.

The leaves of bennettites and cycads can best be distinguished by the architecture of their stomata (Thomas and Bancroft 1913). Stomata ensure the gas exchange that is essential for photosynthesis. They also regulate the plants' water balance by controlling transpiration. The stomata type of the Cycadales is termed haplocheilic (simple or diacytic), that of the Bennettitales is termed syndetocheilic (compound or brachyparacytic). Both types develop through a different ontogeny of the stomata (Florin 1933), which results in the characteristic appearance of the fully developed stomata. The identification of a leaf based on this feature is only possible, however, if the cuticles are preserved. The syndetocheilic type is commonly considered a unique feature of the Bennettitales (Florin 1933; Pott 2016; Pott et al. 2012), but this is currently deemed controversial because syndetocheilic stomata also occur in other plant groups, and therefore do not necessarily represent an autapormorphy of the Bennettitales (cf. Pott et al. 2017).

Bennettites

The two most common plants in the entire Lunz flora were *Pterophyllum filicoides* (Fig. 18.5a) with simply pinnate leaves reaching more than 50 cm in length, and *Pterophyllum brevipenne* (Fig. 18.5b) with leaves up to 25 cm long (Pott et al. 2007e). Almost every hand specimen yields leaves or fragments of one of these two species. Large, incompletely pinnate or entire-margined leaves up to 70 cm long assignable to *Anomozamites lunzensis* (Fig. 18.5c), *Nilssoniopteris haidingeri* (Fig. 18.5d) and *Nilssoniopteris angustior* are also comparatively common. The latter three species, which Krasser (1909a) identified as *Macrotaeniopteris* and placed among marattialean ferns, clearly belong to the Bennettitales based on their epidermal anatomy (Pott et al. 2007d).

Amongst the reproductive organs of clearly bennettitalean nature are the apple-sized *Bennetticarpus wettsteinii* cones (Fig. 18.5e), which were composed of

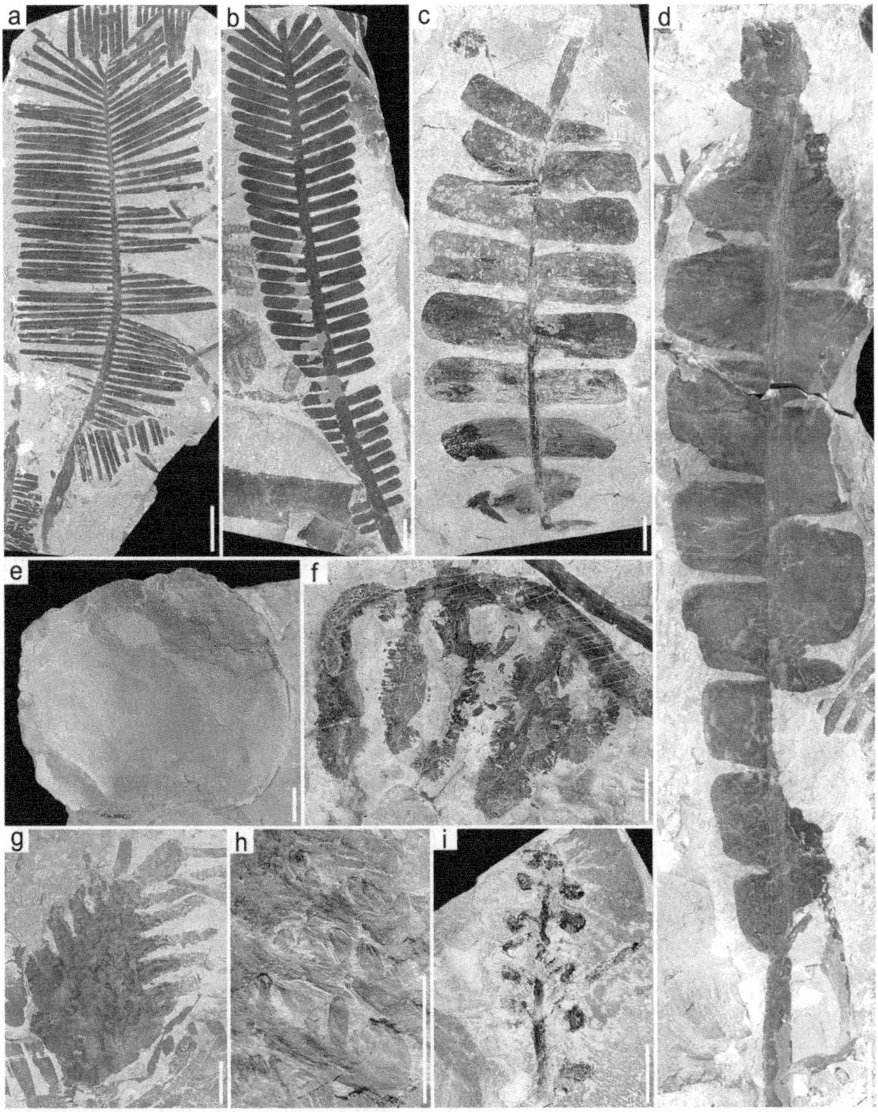

Fig. 18.5 Selected bennettites from the Lunz flora. **a** Leaf of *Pterophyllum filicoides*, NRM S148314. **b** Leaf of *Pterophyllum brevipenne*, NHMW 1884/0500/1209. **c** Leaf of *Anomozamites lunzensis*, NHMW 1888/0001/0018. **d** Leaf of *Nilssoniopteris haidingeri*, NHMW 1884/0500/1203. **e** Seed cone ("fruit") *Bennetticarpus wettsteinii*, NRM S148325. **f** Ovulate organ "*Westersheimia*" *pramelreuthensis*, GBAW 1917/001/0024. **g** Microsporangiate ("pollen") organ *Haitingeria krasseri*, GBAW 1919/002/0001. **h** Detail view of pollen sacks of the microsporangiate organ *Lunzia austriaca*, GBAW 1917/00 1/0012. **i** Enigmatic, presumed caytonialean or corystosperm reproductive structure *Pramelreuthia haberfeldensis*, GBAW 1909/001/0058. Scale bars 20 mm (a, d), 10 mm (b–c, e–i). GBAW Geological Survey of Austria (GeoSphere Austria), Vienna, Austria; NHMW Museum of Natural History, Vienna, Austria; NRM Swedish Museum of Natural History, Stockholm, Sweden

numerous ovules, spirally arranged on a central axis, each bearing a seed (Pott et al. 2010). A pitted outer covering and thick layers of anthracitic coal surrounding the ovules suggest that the cones or "fruits" had a fleshy outer layer (Pott 2014, 2016; Pott and Axsmith 2015). This may have made them attractive to animals, which in turn ensured seed dispersal by consuming the fruits.

Haitingeria krasseri leaves (Fig. 18.5g) are interpreted as pollen organs that carried small pollen sacs on their short pinnae (Pott et al. 2010). They are comparatively common and are always found detached, but often associated with *Nilssoniopteris* foliage. Presumably, they wilted after the mature pollen was released. Sterile scale leaves (*Cycadolepis wettsteinii*) of comparable size and epidermal structure surrounded such a *Haitingeria* flower (Pott et al. 2010) as a form of "tepals". Whether the macrosporangiate inflorescences, which developed into *Bennetticarpus* cones, were produced at the center of such a *Haitingeria* flower is deemed probable (Pott et al. 2010), but recent studies showed that this concept of a hermaphroditic flower may need reconsideration (Pott 2014, 2016).

Westersheimia pramelreuthensis (Fig. 18.5f) is a simply pinnate reproductive organ that produced up to seven pairs of ovule-bearing cones or "catkins" that emerged either hanging or upright from a central axis (Pott 2016). *Westersheimia pramelreuthensis* is always associated with *Pterophyllum filicoides* leaves, and it can be assumed that both organs were produced by the same plant. The pollen-producing organ of this plant is most likely *Lunzia austriaca* (Pott et al. 2017), a cup-shaped flower with radially arranged sterile scale leaves bearing several rows of small pollen sacs. The *Cycadopites*-like pollen of *Lunzia austriaca* (Fig. 18.5h) is present in large numbers in key positions on the cones of *Westersheimia pramelreuthensis*, which can be regarded as solid evidence for conspecificity (Krasser 1917; Pott 2016; Pott et al. 2017).

Some of the reproductive organs attributed by Kräusel (1949) to the Bennettitales are rather cryptic and their affiliation is still unclear. *Pramelreuthia haberfelneri* (Fig. 18.5i) is interpreted by Harris (1964) as the inflorescence of a *Caytonia*-like plant (Caytoniales); an affiliation to other seed ferns such as corystosperms (Umkomasiales) is worthy of discussion (see e.g., Anderson et al. 2019), but sterile foliage attributable to these orders has yet to be reported from the Lunz flora. *Sturiella langeri* is regarded as a small bennettitalean inflorescence (Kräusel 1948), which is similar in structure to *Pramelreuthia*, but this classification is only vaguely justified, especially because the fossil is lost and a detailed examination therefore no longer possible. Pod-like leaves, described as *Leguminanthus siliquosus*, appear similar to *Nilssoniopteris* leaves folded along the midrib; the evidence of pollen sacs on the interior (Kräusel and Schaarschmidt 1966) is very interesting, and the structure of these leaves points to an affiliation with *Nilssoniopteris angustior* (Pott et al. 2007d).

Recently, several specimens of the small microsporophyll of the *Leuthardia* type (Kräusel and Schaarschmidt 1966) have also been recovered from mining heaps in the Lunz coal mining area (Pott and Aschauer, unpublished material). Here as well, a definite attribution requires a more detailed examinationl; Kräusel and Schaarschmidt (1966) assigned *Leuthardia* microsporophylls to the Bennettitales based on the architecture of the stomata.

Cycads

Many of the cycadophyte fronds were assigned to the genus *Nilssonia* based on their epidermal anatomy during the most recent review of these fossils (Pott et al. 2007a). However, the uniformity and systematic affiliation of this genus to Cycadales has recently been questioned again (Pott et al. 2012) and is currently the subject of a morphological and systematic study of the proposed members of the genus (Pott in prep.). The four described *Nilssonia* species of the Lunz flora produced comparatively large to very large fronds (Fig. 18.6a) with long and narrow leaflets (Pott et al. 2007a). Considering the preliminary results of the aforementioned study, they possibly need to be attributed to some other genus clearly affiliated with the Cycadales (Pott et al. 2012). Proper *Nilssonia* leaves are generally characterized by smaller and more slender forms with short, broad leaflets, and should consequently be placed in a separate order (viz. Nilssoniales), probably outside the Cycadales (Nathorst 1909; Schweitzer et al. 2000; Pott et al. 2012; Van Konijnenburg-van Cittert et al. 2017; Pott in prep.). The Nilssoniales were most likely an isolated group of plants besides cycads, bennettites and ginkgo-like plants, and were also outgunned by the explosive Cretaceous dispersal of the angiosperms (Schweitzer et al. 2000; Pott et al. 2012). *Pseudoetenis cornelii* (Fig. 18.6b) is another (genuine) cycad from the Lunz flora (Pott et al. 2007b). Some of the Lunz cycads were quite large (e.g., *Nilssonia neuberi*); in general, however, cycads were less common in the Lunz flora than bennettites.

Reproductive organs of the cycadophytes are abundant and commonly excellently preserved (Krasser 1917, 1919; Kräusel 1943, 1948, 1953; Pott et al. 2010, 2017; Pott

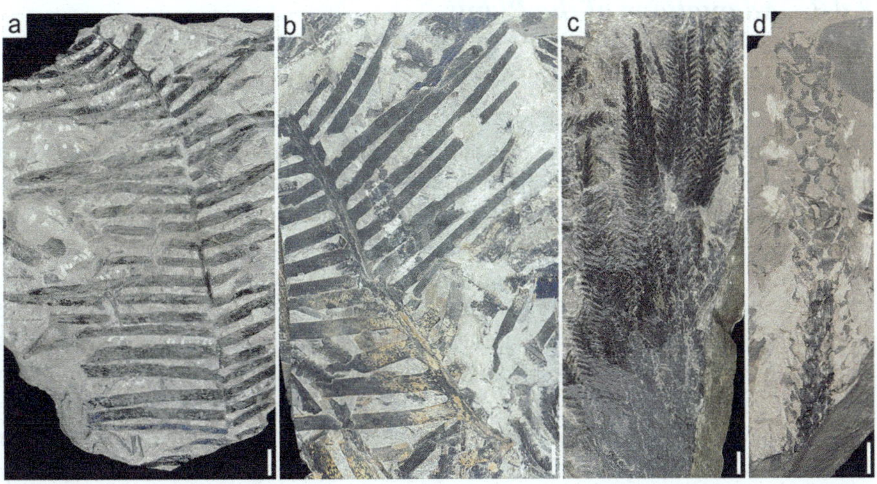

Fig. 18.6 Selected cycads from the Lunz flora. **a** Large leaf of *Nilssonia neuberi*, GBAW 2006/004/0014. **b** Portion of a leaf of *Pseudoctenis cornelii*, NHMW 1877/0001/0013. **c** Several macrosporophylls of *Dioonitocarpidium keuperianum*, GBAW 1917/001/0059. **d** Possible microsporangiate cycad cone (cf. *Androstrobus* sp.), NHMW 1887/0001/0036. Scale bars 50 mm (a), 10 mm (b–d). GBAW Geological Survey of Austria (GeoSphere Austria), Vienna, Austria; NHMW Museum of Natural History, Vienna, Austria

2016, 2019). Most, however, belong to the bennettites (see above). Megasporophylls of the *Dioonitocarpidium* type show a great similarity with those of modern species of the genus *Cycas* (Pott 2019). *Dioonitocarpidium keuperianum* (Fig. 18.6c) is often found in small clusters; *Dioonitocarpidium liliensternii* is sometimes found with attached seeds (Kräusel 1953; Pott 2019). Both cycad megasporophylls and the larger *Nilssonia* leaves (Fig. 18.6a) were probably produced by the same plants. Further fructifications are not yet clearly assignable to cycads, although it is highly unlikely that no microsporangiate strobili are preserved. One specimen (Fig. 18.6d) may represent such a microsporangiate cone of a cycad, preliminarily assignable to the genus *Androstrobus*.

Conifers

Even though the group itself is underrepresented in the Lunz flora, probably due to adaptations to drier hinterland habitats of most species of this group, at least one conifer taxon (i.e. *Stachyotaxus lipoldii*) is present with a number of leaves preserved on several specimens (Pott and Krings 2010). *Stachyotaxus lipoldii* has comparatively large leaves with densely arranged, flat leaflets that form a plane lamina (Fig. 18.4e–f). On some specimens, several leaves occur in the same bedding planes, densely but arbitrarily arranged. This raises the hypothesis that *Stachyotaxus lipoldii* was a deciduous conifer. Leaves appear to have been commonly abscised entirely with all leaflets still attached.

18.4.3 Adaptations of the Plants

Overall, the Lunz flora constitutes a vegetation (Pott 2021) with many moisture-loving plants such as ground ferns, large, tree-like ferns, sphenophytes, ginkgophytes, bennettites and some cycads, but on the other hand comprises conifers and cycads, some of which tend to grow in drier environments today. Recent studies of the cuticles and epidermal anatomies of these plants have provided numerous insights into their ecology and the possible environment or habitats populated by the vegetation in the Lunz-am-See area (summarized in Pott et al. 2008b; Pott 2021).

Most of the plants have robust cuticles and stomata sunken below the level of the epidermis (Figs. 18.7c–e, g), and the leaf surfaces of many plants bear numerous papillae (Figs. 18.7a–c, e–f, h, k). Robust cuticles indicate thick leaves, perhaps even of leathery consistence (Fahn and Cutler 1992; Roth 1992). Sunken stomata help the plants to reduce evaporation, which is otherwise promoted by intense sunlight or wind. Moreover, some species have stomata whose openings are almost completely covered by papillae (Fig. 18.7e). The overhanging papillae are another tool for reducing transpiration. More or less densely arranged papillae on the leaf surfaces of most of the studied species also help to reduce the wind speed on the leaf surface and thus to further reduce transpiration.

Sunken stomata often occur in combination with other xeromorphic characteristics such as leathery leaves, papillae-rich surfaces or robust cuticles, and are generally

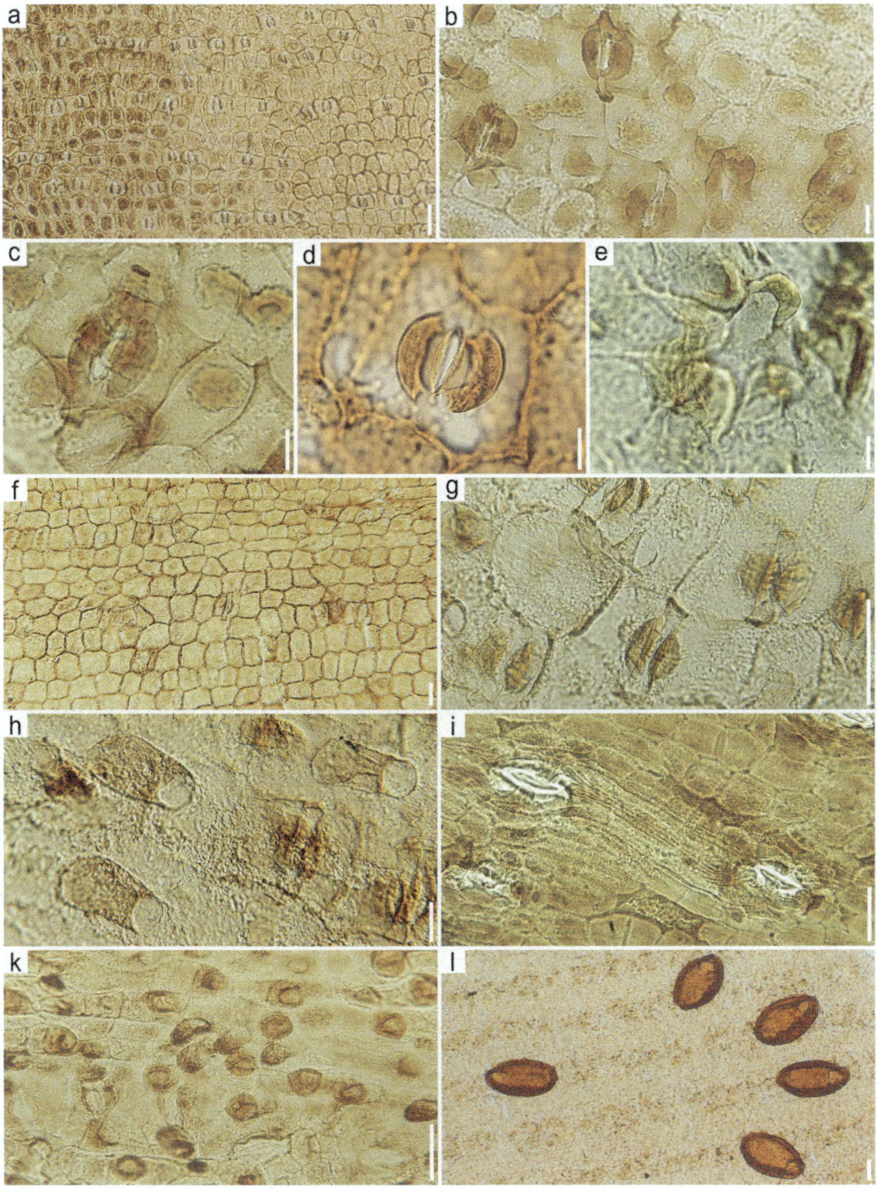
a
b
c
d
e
f
g
h
i
k
l

◄**Fig. 18.7** Cuticles of selected taxa from the Lunz flora. **a** Abaxial leaf surface of *Nilssoniopteris angustior* (Bennettitales), NHMW 1884/0015/0010. **b** Abaxial leaf surface of *Nilssoniopteris haidingeri* (Bennettitales), NHMW 1885/0500/3983/0003. **c** Sunken stoma on the abaxial leaf surface of *Nilssoniopteris haidingeri* (Bennettitales), NHMW 1885/0500/3983/0003. **d** Stoma on the abaxial leaf surface of *Nilssoniopteris angustior* (Bennettitales), NHMW 1884/0015/0012. **e** Stoma covered entirely with papillae on the abaxial leaf surface of *Pseudoctenis cornelii* (Cycadales), NHMW 2007B0002/0005. **f** Abaxial leaf surface of *Anomozamites lunzensis* (Bennettitales), NHMW 1885/0500/4021/0001. **g** Abaxial leaf surface of *Pterophyllum filicoides* (Bennettitales), with stomata and idiocuticular striae, NHMW 1884/0021/0007. **h** Long and hollow papillae on the abaxial leaf surface of *Pterophyllum filicoides* (Bennettitales), GBAW 1909/003/0518/0005; **i** Idiocuticular striae connecting two adjacent stomata on the abaxial leaf surface of *Arberophyllum florinii* (Ginkgoales), NHMW 1886/I/0022/0001. **k** Abaxial leaf surface of *Nilssonia sturii* (Cycadales), densely covered with papillae, GBAW 1909/002/0518/0006. **l** Shells of insect eggs arranged in a circle on the abaxial leaf surface of *Nilssoniopteris haidingeri* (Bennettitales), GBAW 1909/002/0250/0006. Scale bars 100 μm (k), 50 μm (a, f, g, i, j) und 10 μm (b–e, h); GBAW Geological Survey of Austria (GeoSphere Austria), Vienna, Austria; NHMW Museum of Natural History, Vienna, Austria

found on plants that thrive in xeric (dry) conditions (Fahn and Cutler 1992; Bell 2008; Crang et al. 2019). Note, however, that drought does not necessarily have to occur in warm and dry climates. Xeromorphism can also be the result of adaptations to physiological drought (e.g., Fahn and Cutler 1992; Crang et al. 2019). Highly permeable or osmotic soil, sustained wind or lack of soil substrates, a brackish or saline environment, reduced soil pH or an epiphytic lifestyle can also trigger the development of stress-induced adaptations (Crang et al. 2019). Plants growing near areas exposed to the sea or tidal influence are often adapted to salty air and increased salinity in the soil and groundwater (e.g., Roth 1992). Xeromorphic adaptations potentially also reflect seasonal drought. Many orchids, for example, show certain adaptations to xeric conditions even though they live in humid, tropical or subtropical regions. This is because their epiphytic way of life exposes them to periodic drought. The plants are adapted accordingly, regardless of the duration or type and cause of drought. Idiocuticular striae or other surface structures, which are present on the leaves of some Lunz plants (Figs. 18.7g, i), probably help the targeted collection or divertion of atmospheric water (Pott et al. 2007c). Especially the latter may be an adaptation to heavy rainfall by avoiding water build-up on the leaves, which would cause them to break or rot. Too much water can also promote fungal and moss growth on leaves.

18.5 Discussion

The plants from the Lunz flora exhibit many xeromorphic characteristics, but at the same time support the interpretation that the climate in the region around Lunz-am-See was humid during the Carnian (Wing et al. 1992; Kozur and Bachmann 2010; Ogg 2015; Mueller et al. 2016; Pott 2021). The vegetation itself provides several indications of thriving in or having adapted to a humid, moist environment or swampy habitat. Perhaps it even had a rainforest-like appearance, which is reviewed here (Pott 2021; see also the Introduction for the features characterizing a rainforest-like vegetation; see also Napp-Zinn 1984; Bell 2008; Mabberley 2013; Crang et al. 2019; Juniper 2019; Schulze et al. 2019).

18.5.1 Layered Structure of the Forest

A layered structure of the vegetation can be assumed based on the preserved plant species. Herbaceous ferns such as *Phlebopteris lunzensis* and *Cynepteris reticulata* of the Matoniaceae, and *Dictyophyllum serratum, Dictyophyllum lunzensis, Clathropteris reticulata, Camptopteris lunzensis* and *Digitopteris repanda* of the Dipteridaceae (both Gleicheniales) characterized the forest floor and undergrowth layers. Horsetails (*Neocalamites merianii, Equisetites arenaceus* and *Equisetites conicus*) populated the banksides of possibly present water bodies in large monotypic stands (Pott et al. 2018). The undergrowth layer was extended into the understorey layer by bennettite plants (*Pterophyllum, Nilssoniopteris*) and smaller, low-growing cycads (*Nilssonia sturii* and *Nilssonia riegeri*), as well as by ferns forming larger, most likely tree-like plants such as *Asterotheca merianii, Symopteris lunzensis, Danaeopsis lunzensis* of the Marattiales and *Todites linnaeifolius* of the Osmundales (Pott et al. 2007a, 2007d, 2007e, 2018). Larger and taller cycads (*Nilssonia neuberi, Nilssonia lunzensis* and *Pseudoctenis cornelii*) and the few ginkgophyte and conifer species probably extended above this layer to form a possible canopy layer, even if these plants most likely did not grow taller than 25–35 m (Pott et al. 2007a, 2007b).

18.5.2 Winding or Climbing Habit, Epiphytic Way of Life

Unfortunately, none of the plants from the Lunz flora exhibit an unambiguous winding or climbing habit or even an epiphytic way of life. Some *Nilssonia* species have been restored possibly expressing a winding habit (i.e. *Nilssonia schaumburgensis*; Watson and Cusack 2005), but as stated earlier, the *Nilssonia* species from the Lunz flora probably belonged to a different group of plants, unambiguously assignable to the Cycadales (Pott in prep.). The "genuine" *Nilssonia* species formed a separate group of plants affiliated with ginkgophytes (Schweitzer et al. 2000; Pott et al. 2012; Van Konijnenburg-van Cittert et al. 2017; Pott in prep.). Therefore, a climbing or winding habit cannot be assumed for the *Nilssonia* species from the Lunz flora; instead, they probably grew rather tall pachycaul plants as do modern cycads such as *Lepidozamia hopei* or *Dioon spinulosum.*

Several climbing species are also known amongst ferns and fern-like plants, the recent ones mostly represented by tropic and subtropic species, similar to those from periods dating back to the Carboniferous (e.g., Krings et al. 2003a, 2003b; Pott et al. 2026). However, none of the fern species described from the Lunz flora displays features such as hooks and tendrils implying a climbing, scrambling or winding habit (Krings et al. 2003a; Pott et al. 2018).

A hidden hint for the presence of epiphytic ferns may include specimens originally assigned to the ginkgophyte foliage type *Ginkgoites lunzensis* (Fig. 18.4d; Kräusel 1920, 1943; Florin 1936). These leaves are not likely to be interpreted as ginkgophyte foliage, but rather represent basal fern fronds such as those produced by the epiphytic ferns of the genus *Platycerium*. In those plants, such shield-shaped basal fronds serve to protect the fern's roots from damage and desiccation. In some *Platycerium* species, the top margin of these fronds will even grow into an open crown of lobes to catch rainwater, falling forest litter or bird/animal droppings, thus building up their own system of nutrition over many years (Richardson and Walker 2010).

18.5.3 Leathery, Vertical Leaves with Robust Cuticles and "Drip Tips"

In many of the seed plant species reported from the Lunz Formation, robust cuticles and preserved anthracitic coal suggest thick and leathery living leaves. For *Arberophyllum florinii*, one suggestion is that the leathery leaves grew almost vertically on the plants (Kräusel 1943). While this can also be hypothesized for other plants, no substantiating evidence is available. With regard to so-called "drip tips", the leaves of *Arberophyllum florinii* can also be used as evidence: they are extended or pointed, which facilitates the quick draining of excess water. This is enhanced by a micro-structure in the form of idiocuticular striae (Fig. 18.7i) that further support guided runoff of excess water (Pott et al., 2007c). Similar structures are present on the segmented leaves of *Pterophyllum* plants, whose planar architecture (lanceolate–spatulate overall outline with a stiff apical leaflet) may also be regarded as an adaptation to a targeted direction of atmospheric water (Pott et al. 2007e). A similar interpretation may also be forwarded for the large, once pinnate fronds with entire-margined pinnae of the tree-like fern *Danaeopsis lunzensis* (Fig. 18.3a), or for the irregularly segmented or entire-margined, broadly lanceolate leaves of *Anomozamites* and *Nilssoniopteris*, which also formed extended leaf apices or equivalent apical leaflets (Pott et al. 2007d, 2018). If the leaves of the conifer *Stachyotaxus lipoldii* can be compared to those of modern, moisture-living *Sequoia* and *Metasequoia*, then they also probably grew in a similar vertical arrangement on the plants. The conifer *Stachyotaxus lipoldii* also produced comparatively large, laminar leaves with densely arranged planar leaflets forming a lamina that appears entire, where the acute leaflet apices may represent a form of drip tips (Fig. 18.4f).

18.5.4 Large, Waxy Leaves Arranged to Form Funnels that Direct Water Immediately

Besides the above-mentioned idiocuticular striae deemed to guide water towards a certain target, present in a number of species, the whorled arrangement of the leaves of bennettite plants may be seen as a funnel-like arrangement. This is especially the case for the entire-margined leaves of *Nilssoniopteris*, and most ferns. Plants that definitively had a funnel-like leaf arrangement are the above-presented members of the Dipteridaceae and Matoniceae with their umbrella-shaped fronds (viz. *Camptopteris lunzensis, Dictyophyllum serratum, Dictyophyllum lunzensis, Clathropteris reticulata, Phlebopteris lunzensis* and *Digitopteris repanda*; Pott et al. 2018). Large and waxy leaves with possible drip tips were also produced by the bennettites *Nilssoniopteris angustior* and *Nilssoniopteris haidingeri* (Pott et al. 2007d).

18.5.5 Adaptation to Shaded Places

Importantly, many of the herbaceous fern species reported from the Lunz flora (viz. *Camptopteris lunzensis, Dictyophyllum serratum, Dictyophyllum lunzensis, Clathropteris reticulata, Phlebopteris lunzensis* and *Digitopteris repanda*) were members of the shaded forest floor or understory layers, and thus adapted to such shaded places (Van Konijnenburg-van Cittert 2002; Pott et al. 2018). Their relatively large fronds with large tissue areas and very delicate cuticles (if preserved at all) support this view. The leaf tissue area of this morphology increases the sunlight uptake to a maximum and the habitat minimizes protection against UV radiation. Both features are necessary, or eligible, in shaded places only.

18.5.6 Epidermal and Cuticular Features

Beyond the epidermal and cuticular adaptations of the plants noted above (see Results), a few features addressing higher-than-average moisture availability or -presence/-impact are worth mentioning in detail. Microstructures formed by the cuticle (idiocuticular striae) on several leaves of e.g., *Pterophyllum filicoides, Pterophyllum brevipenne* or *Arberophyllum florinii* can be interpreted as an adaptation to increased humidity or heavy seasonal precipitation (Roth 1992; Barthlott and Neinhuis 1997; Pott et al. 2007c, 2007e). These structures presumably reduce the water wettability of the leaves and prevent the formation of a water film on the leaf surface, which can adversely affect the uptake of carbon dioxide. This also optimizes the self-cleaning effect of the leaves so that particles that are harmful to the leaf surface, such as salt crystals or dust particles, are washed off more rapidly and efficiently (Roth

1992; Barthlott and Neinhuis 1997; Koch et al. 2006; Pott et al. 2007c). These adaptations, which minimize wetting and contamination, are prevalent in various plant groups, particularly in aquatic environments or regions with heavy precipitation, enhancing photosynthesis and reducing fungal infections (Neinhuis and Barthlott 1997).

Plant-arthropod interactions may provide additional insights into the growth locality of certain plants. Special damage to the leaves of *Nilssoniopteris angustior* was probably caused by dragonflies depositing their eggs into the leaves (Pott et al. 2008c). These plants must therefore have been growing in the immediate vicinity of stagnant or slow-flowing freshwater so that the larvae would fall into the water immediately after hatching to further develop. Delicate shells of insect eggs on leaves of *Nilssoniopteris haidingeri* (Fig. 18.7l) cannot yet be clearly assigned to any insect group. Ground beetles from the Cupedidae family are possible producers, although dragonflies cannot be completely excluded.

18.5.7 Further Indications of a Vegetation Adapted to Humid, Wet or Moist Environments

Studies of bennettites from other regions (Harris 1932, 1969; Pott and McLoughlin 2009, 2014; Pott 2014, 2026; Pott and Takimoto 2022, 2025) have shown that especially the shrubby *Williamsonia*-like bennettites favored humid, deltaic habitats. "Genuine" members of the Nilssoniales, which are quite similar in their plant architecture to these shrubby Williamsoniaceae, were probably distributed in a comparable, if not the same, habitat (Harris 1964; Pott et al. 2012; Pott and McLoughlin 2014). Especially the former represent dominant portions of the vegetation in the Lunz environment (Pott et al. 2007d, 2007e, 2017; Pott 2016).

Likewise, if the conifer *Stachyotaxus lipoldii* is compared to modern *Sequoia* and *Metasequoia*, it may also have had analogous ecological requirements (Pott and McLoughlin 2014). Accordingly, it would have lived in humid environments like *Metasequoia*, which needs a certain account of precipitation for its growth and is thus commonly found along streams, creeks and in moist gorges. The large, laminar leaves of *Stachyotaxus lipildii* may be regarded as an adaptation to collect moisture from fog as in *Sequoia*, in which coalescence of coastal fog accounts for a considerable part of the trees' water needs.

Circinnate vernation of developing leaves, as it is a common developmental feature in ferns and cycads, has for the first time been demonstrated for bennettites from the Lunz flora (Pott and Krings 2007). However, there is so far no indication that this feature represents an adaptation to moist or humid environments. It is rather regarded as an adaptation to provide protection against invertebrate herbivores through the apposition of two or more layers of leaf (Grubb and Jackson 2007).

Overall, the composition of the Carnian vegetation of the Lunz-am-See area indicates a humid, moist and boggish, deltaic or near-coastal environment. The plant fossils are embedded almost autochthonously. The very large, unaligned and excellently preserved leaves indicate slow-flowing or stagnant freshwaters, along whose immediate shores the plants must have been thriving. Moreover, portions of the flora are characterized by ferns and the general occurrence of large, tree-like ferns and horsetails. These representatives are known to indicate moist soil conditions. Ferns also often colonize large areas in monotypic stands and probably characterized the flora at the edge of a swamp-like landscape (Fig. 18.8), while larger horsetails in particular formed monotypic stands in shallow-water areas. The Lunz environment may be remotely compared with the modern mangrove biome or mangrove forest, a distinct saline woodland or shrubland habitat characterized by depositional coastal environments, where fine sediments often with high organic content collect in areas protected from high-energy wave action.

Paleogeographic reconstructions reveal that the Lunz flora thrived in a coastal lowland region on the northern edge of the Tethys Ocean at around 30° N (Pott et al. 2018; Blakey 2020). The massive coal seams in the region prove that the climate was stable for longer periods of time so that large quantities of plant material could accumulate. The formation of peat (as an educt for the formation of coal) requires an array of special conditions, including a high groundwater table, waterlogged soil, reduced oxygen supply, low pH values and, as a result, a lack of bacterial decay. Such conditions are mainly found in swamps or peat bogs, but also in mangrove areas and many rainforests (e.g., Hutchings and Saenger 1987; Roth 1992).

Macromorphological and epidermal features of the Lunz plants, in combination with geological, sedimentological and paleozoological evidence for a humid and water-dependent environment, suggest a flora that thrived in and at the edge of a coastal, deltaic lowland region. Special structural adaptations such as those of *Arberophyllum florinii* indicate that the plants were also exposed to slight marine or saline influences and sustained wind. Matoniaceae and especially Gleicheniaceae ferns are sometimes also regarded as pioneer plant vegetation, as occurs in coastal areas (Van Konijnenburg-van Cittert, 2002; Pott et al. 2018).

The sedimentological record also corroborates this interpretation. The coal-producing strata of the Lunz Formation are intercalated with fully marine limestone beds. Consequently, this area may be referred to as a paralic basin (Stremme 1911). As the Alpine Triassic is predominantly of a marine character, the plant-bearing beds of the Lunz Formation can be interpreted as the result of a short-term sea-level low-stand and a meandering river delta system that transported terrestrial material from the northeast into a brackish marsh area close to the coast (Meller 2007). Nevertheless, the formation of significant coal seams indicates that the climate was characterized by humid and wet weather for longer periods of time, with at maximum seasonal and short periods of drought, and also by non-marine sedimentation.

Fig. 18.8 Ferns such as *Camptopteris lunzensis* often colonize large areas in monotypic stands and probably characterized the Lunz flora at the edge of a swamp-like landscape. Illustration by Pollyanna van Knorring, Stockholm, Sweden

18.6 Conclusions

During a worldwide phase of enhanced humidification in the Late Triassic, sedimentation of fossiliferous deposits occurred in the Northern Calcareous Alps. The Carnian Pluvial Episode (CPE), characterized by a lithological change from calcareous to siliciclastic in the Reifling intra-platform basin, was initiated at the Julian 2 base. This is reflected by adaptations in the marine and terrestrial facies along the northern Tethyan rim. After normal marine, basinal and limestone deposits of the Reifling

Formation, the CPE started with dark-laminated silicious limestones of the Göstling Formation. This was followed by argillaceous and clay deposits of the Reingraben Formation, which were influenced by enhanced freshwater and terrigenous influx. The final phase of the CPE around the Julian/Tuvalian boundary features enhanced terrigenous influx into the Reifling Basin. Massive sandstones mirror the filling of the Lower Carnian (Julian 2/II), which lasted to the lower/upper Carnian boundary. The so-formed Lunz Formation exhibits hundreds of meters of sandstone, clay and massive coal seams in its upper member. These were subject to extensive mining in the 19th and 20th centuries, yielding excellently preserved plant fossils in the strata accompanying the coal seams. Huge numbers of plant fossil material from the famous Lunz flora are stored in collections of many natural history museums and research institutes (see Material and Methods, Acknowledgements). The plant fossil-bearing intervals of the Lunz Formation were deposited during sea-level low-stands within a meandering river delta system that transported terrestrial material around sea level, from the northeast into a near coastal, brackish marsh area. A coastal wetland vegetation hints at a terrestrial, deltaic swamp forest paleoenvironment during the climax of the CPE in the early Carnian (Julian 2/II).

Plant groups represented in the Lunz flora include horsetails, some as tall as small trees, shrub-like and arborescent ferns, tall cycads, a few enigmatic ginkgophytes and conifers, and—as the dominant plant group—numerous excellently preserved bennettitites. With regard to its composition, the vegetation was—also in individual numbers—dominated by several plant groups and species featuring a (sub)tropical coastal wetland composition—somewhat reminiscent of a mangrove environment. This yielded a characteristic structure and adaptations such as a multi-layered forest. Other features include a winding or climbing habit; an epiphytic way of life; leathery, vertical leaves with robust cuticles and "drip tips"; large, waxy leaves arranged to form funnels that direct water immediately; adaptation to shaded places; and specific plant-arthropod interactions. Others are interpreted to generally have been thriving in such environments., e.g., the bennettites. Cuticular and epidermal features include robust cuticles and stomata sunken below the level of the epidermis, numerous papillae on the leaf surfaces, and idiocuticular striae. These and other surface structures help collect or divert atmospheric water and enhance cleaning effects to avoid accumulation of harmful particles such as dust or salt. Overall, the composition of the vegetation of the Lunz-am-See area indicates a humid, moist and swampy, deltaic or near-coastal environment. Moreover, the plant fossils are embedded autochthonously.

Acknowledgments CP is particularly grateful to the scientific and collection staff at the Museum of Natural History (Vienna, Austria); the Geological Survey of Austria (Vienna, Austria); the Swedish Museum of Natural History (Stockholm, Sweden); the State Museum of Lower Austria (Museum Niederösterreich, St. Pölten, Austria); the Museum of Natural History (Naturalis Biodiversity Center, Leiden, The Netherlands); the Laboratory of Paleobotany and Palynology, University Utrecht (Utrecht, The Netherlands); the State Museum Joanneum (Landesmuseum Joanneum, Graz, Austria); the Senckenberg Museum of Natural History (Senckenberg Naturmuseum, Frankfurt-am-Main, Germany); and the Natural History Museum (Museum für Naturkunde, Leibniz-Institut für Evolutions- und Biodiversitätsforschung, Berlin, Germany). All these institutions made the Lunz material available for study. Hans Kerp and Benjamin Bomfleur at the Palaeobotany Research

Group of the University Münster (Münster, Germany), and Else Marie Friis, Thomas Denk and Steve McLoughlin at the Department of Palaeobiology of the Swedish Museum of Natural History (Stockholm, Sweden), provided the laboratory and research environment along with other facilities, consumables and materials. AL is particularly grateful to the Austrian Federal Forests Company (Österreichische Bundesforste, Purkersdorf, Austria) for sampling and digging permission. The manuscript greatly benefited from valuable comments by Johanna H.A. van Konijnenburg-van Cittert (Utrecht, The Netherlands) and an anonymous reviewer, the production editor Ashok Arumairaj (Taramani), the in-house series editor Kathleen A. Ritterbush (Utah), the publishing editor Simon Shah-Rohlfs and the production editor Antje Endemann (both Heidelberg).

Author Contribution CP designed the research. CP provided documentation of fossil material. CP, AL contributed to writing the paper.

Funding Most of the initial research preceding this manuscript was conducted by CP when he received funding through the German Research Foundation, Bonn, Germany (DFG; grants KR 2125/3-1, KR 2125/3-2), and the Swedish Research Council, Stockholm, Sweden (VR; grant 621-2012-4375). Open access funding was enabled and projects financially supported by the Federal Government of Lower Austria (K3-F-964/001-2020 and K3-F-964/002-2023, Department Science and Research; headquarters in St. Pölten, Austria); the Austrian Academy of Sciences (headquarters in Vienna, project Polzberg Lukeneder 2021) represented by the National Committee for Geo/Hydro Sciences (Earth System Sciences Programme); the Natural History Museum Vienna; and the Triassic project "Fenster in die Karnische Krise" of the Society of Friends of the Natural History Museum Vienna (Freunde des Naturhistorischen Museums Wien, CPE Lukeneder 2023) to AL. The authors are responsible for the contents of this publication. The funders had no impact on conceptualization, design, data collection, analysis, decision to publish, or preparation of the manuscript.

Competing interests All authors stated that they have no competing interests.

Authors' contributions All authors stated that they have equally contributed to their chapters, read, edited and agreed to submit the final versions of their contributions.

Availability of data and materials All sources of information are provided in the text. Raw data related to the fossil material from Austria are available from the corresponding authors and Alexander Lukeneder upon request.

Additional information The article is part of the Special Volume: *Late Triassic Konservat-Lagerstätten within the Carnian Pluvial Episode. 150 Years of Scientific Excavations in Austria.*

Artificial Intelligence (AI) The chapter was produced without the use of AI.

Language The entire volume uses American English reviewed by a native scientist.

Received 2025.

Published online 2026.

References

Anderson HM, Barbacka MK, Bramford MK, Holmes WBK, Anderson JM (2019) *Umkomasia* (megasporophyll): Part 1 of a reassessment of Gondwana Triassic plant genera and a reclassification of some previously attributed. Alcheringa 43:43–70. https://doi.org/10.1080/03115518.2018.1554748

Barthlott W, Neinhuis C (1997) Purity of the Sacred Lotus, or escape from contamination in biological surfaces. Planta 202:1–8. https://doi.org/10.1007/s004250050096

Behrens M (1972) Schwermineralverteilungen und Sedimentstrukturen in den Lunzer Schichten (Karn, Trias, Österreich). Jb Geol BA 116:51–83

Bell AD (2008) Plant form. An illustrated guide to flowering plant morphology. Timber Press, Portland, OR

Bharadwaj DC, Singh HP (1964) An Upper Triassic miospore assemblage from the coals of Lunz, Austria. Palaeobotanist 12:28–44. https://doi.org/10.54991/jop.1963.641

Blakey R (2020) Palaeogeography of North America, Europe and the World. Colorado Plateau Geosystems Inc., Phoenix, AZ. Accessible via https://deeptimemaps.com/

Blomenkemper P, Bäumer R, Backer M, Abu Hamad A, Wang J, Kerp H, Bomfleur B (2021) Bennettitalean leaves from the Permian of equatorial Pangea—The early radiation of an iconic Mesozoic gymnosperm group. Front Earth Sci 9:652699. https://doi.org/10.3389/feart.2021.652699

Crang R, Lyons-Sobaski S, Wise R (2019) Plant anatomy. Springer, Cham. https://doi.org/10.1007/978-3-319-77315-5

Crepet WL, Stevenson DM (2010) The Bennettitales (Cycadeoidales): a preliminary perspective of this arguably enigmatic group. In: Gee CT (ed) Plants in Mesozoic time: morphological innovations, phylogeny, ecosystems. Indiana University Press, Bloomington, pp 215–244

Dal Corso J, Ruffell A, Preto N (2018) The Carnian Pluvial Episode (Late Triassic): new insights into this important time of global environmental and biological change. J Geol Soc Lond 175:986–988. https://doi.org/10.1144/jgs2018-185

Dobruskina IA (1998) Lunz flora in the Austrian Alps—a standard for Carnian floras. Palaeogeogr Palaeoclimatol Palaeoecol 143:307–345. https://doi.org/10.1080/00206818909465973

Doyle JA (2006) Seed ferns and the origin of angiosperms. J Torrey Bot Soc 133:169–209. https://doi.org/10.3159/1095-5674(2006)133[169:SFATOO]2.0.CO;2

Dunay RF, Fisher MJ (1978) The Carnian palynofloral succession in the Northern Calcarous Alp, Lunz-am-See, Austria. Pollen Spores 20:177–187

Escapa I, Cunéo NR (2006) Primer registro de *Neocalamites* (Halle) Vladimirovicz en el Pérmico de Gondwana. Ameghiniana 43:85–92

Fahn A, Cutler DF (1992) Xerophytes. Encyclopedia of Plant Anatomy, XIII, 3. Borntraeger, Berlin

Florin R (1933) Zur Kenntnis der Spaltöffnungsapparate der Bennettitales. Kungl Sv Vetenskapsakad Handl 12:11–31

Florin R (1936) Die fossilen Gingkophyten aus Franz-Joseph-Land nebst Erörterungen über vermeintliche Cordaitales mesozoischen Alters. II Allgemeiner Teil Palaeontographica B 82:1–72

Frentzen K (1934) Über die Schachtelhalmgewächse des Keupers. Aus der Heimat 47:147–152

Grubb PJ, Jackson RV (2007) The adaptive value of young leaves being tightly folded or rolled on monocotyledons in tropical lowland rain forest: an hypothesis in two parts. Plant Ecol 192:317–327. https://doi.org/10.1007/s11258-007-9302-0

Harris TM (1932) The fossil flora of Scoresby Sound East Greenland—Part 3: Caytoniales and Bennettitales. Medd Grønland 85:1–133

Harris TM (1964) The Yorkshire Jurassic Flora III. Trustees BM (NH), London

Harris TM (1969) The Yorkshire Jurassic Flora IV. Trustees BM (NH), London

Hilton J, Bateman RM (2006) Pteridosperms are the backbone of seed-plant phylogeny. J Torrey Bot Soc 133:119–168. https://doi.org/10.3159/1095-5674(2006)133[119:PATBOS]2.0.CO;2

Hornung T, Brandner R (2005) Biochronostratigraphy of the Reingraben turnover (Hallstatt facies belt): local black shale events controlled by regional tectonics, climatic change and plate tectonics. Facies 51:460–479. https://doi.org/10.1007/s10347-005-0061-x

Hornung T, Brandner R, Krystyn L, Joachimsky MM, Keim L (2007) Multistratigraphic constraints on the new Tethyan "Carnian crisis". New Mex Mus Nat Hist Sci Bull 41:59–67

Hutchings P, Saenger P (1987) Ecology of mangroves. University of Queensland Press, St Lucia, QLD

Juniper T (2019) Rainforest—dispatches from Earth's most vital frontlines. Island Press, Washington, DC

Kelber K-P, Van Konijnenburg-van Cittert JHA (1998) *Equisetites arenaceus* from the Upper Triassic of Germany with evidence for reproductive strategies. Rev Palaeobot Palynol 100:1–26. https://doi.org/10.1016/S0034-6667(97)00061-4

Kimmig J, Schiffbauer JD (2024) A modern definition of Fossil-Lagerstätten. Trends Ecol Evol 39:621–624. https://doi.org/10.1016/j.tree.2024.04.004

Koch K, Hartmann D, Schreiber L, Barthlott W, Neinhuis C (2006) Influences of air humidity during the cultivation of plants on wax chemical composition morphology and leaf surface wettability. Environ Experim Botany 56:1–9. https://doi.org/10.1016/j.envexpbot.2004.09.013

Kozur HW, Bachmann GH (2010) The Middle Carnian Wet Intermezzo of the Stuttgart Formation (Schilfsandstein), Germanic Basin. Palaeogeogr Palaeoclimatol Palaeoecol 290:107–119. https://doi.org/10.1016/j.palaeo.2009.11.004

Krasser F (1909a) Die Diagnosen der von Dionysius Stur in der obertriadischen Flora der "Lunzerschichten" als Marattiaceenarten unterschiedenen Farne. Sitzungsber Kaiserl Akad Wiss, Abt 1(118):13–43

Krasser F (1909b) Zur Kenntnis der fossilen Flora der Lunzer Schichten. Jahrb K-K Geol Reichsanst 59:1–26

Krasser F (1917) Studien über die fertile Region der Cycadophyten aus den Lunzer-Schichten: Mikrosporophylle und männliche Zapfen. Denkschr Kaiserl Akad Wiss (Wien). Math-Naturw Kl 94:89–553

Krasser F (1919) Studien über die fertile Region der Cycadophyten aus den Lunzer Schichten: Makrosporophylle. Denkschr Kaiserl Akad Wiss (Wien). Math-Naturw Kl 97:1–32

Kräusel R (1920) Über einige Pflanzen aus dem Keuper von Lunz (Nieder-Österr). Jahrb Preuß Geol Landesanst 41:192–209

Kräusel R (1943) Die Ginkgophyten der Trias von Lunz in Nieder-Österreich und von Neue Welt bei Basel. Palaeontographica B 87:59–93

Kräusel R (1948) *Sturiella langeri* nov. gen., nov. sp., eine Bennettitee aus der Trias von Lunz (Nieder-Österreich). Senckenbergiana 29:141–149

Kräusel R (1949) Koniferen und andere Gymnospermen aus der Trias von Lunz, Nieder-Österreich. Palaeontographica B 89:35–82

Kräusel R (1953) Ein neues *Dioonitocarpidium* aus der Trias von Lunz. Senckenbergiana 34:105–108

Kräusel R, Schaarschmidt F (1966) Die Keuperflora von Neuewelt bei Basel—IV. Pterophyllen und Taeniopteriden. Schweiz Paläontol Abh 84:3–44

Krings M, Grewing A, Taylor TN (2003a) Mariopterid pteridosperms, scrambling and climbing vines of the Late Carboniferous swamp forest: a case study from the Namurian B of Hagen-Vorhalle (Germany). Bot Jahrb Syst 124:427–448. https://doi.org/10.1127/0006-8152/2003/0124-0427

Krings M, Kerp H, Taylor TN, Taylor EL (2003b) How Paleozoic vines and lianas got off the ground: on scrambling and climbing Carboniferous–early Permian pteridosperms. Bot Rev 69:204–224. https://doi.org/10.1663/0006-8101(2003)069[0204:HPVALG]2.0.CO;2

Krystyn L (1991) Die Fossillagerstätten der alpinen Trias. In: Vasicek W, Krystyn L, Golebiowski R (eds) Exkursionen im Jungpaläozoikum und Mesozoikum Österreichs. Österreichische Paläontologische Gesellschaft, Wien, pp 23–78

Large MF, Braggins JE (2004) Tree ferns. Timber Press, Portland, OR, 359 p

Lukeneder A (ed) (2026) Late Triassic Konservat-Lagerstätten within the Carnian Pluvial Episode: 150 Years of Scientific Excavations in Austria. Top Geobiol 55:600 p. https://doi.org/10.1007/978-3-032-19777-1

Lukeneder A, Lukeneder P (2021) The Upper Triassic Polzberg palaeobiota from a marine *Konservat-Lagerstätte* deposited during the Carnian Pluvial Episode in Austria. Sci Rep 11:16644. https://doi.org/10.1038/s41598-021-96052-w

Lukeneder A, Lukeneder P (2023) New data on the marine Upper Triassic palaeobiota from the Polzberg *Konservat-Lagerstätte* in Austria. Swiss J Palaeontol 142:9. https://doi.org/10.1186/s13358-023-00269-3

Lukeneder A, Lukeneder P, Sachsenhofer R, Roghi G, Rigo M (2024) Multi-proxy record of the Austrian Upper Triassic Polzberg *Konservat-Lagerstätte* in light of the Carnian Pluvial Episode. Sci Rep 14:11194. https://doi.org/10.1038/s41598-024-60591-9

Lukeneder A, Surmik D, Gorzelak P, Niedźwiedzki R, Brachaniec T, Salamon MA (2020) Bromalites from the Upper Triassic Polzberg section (Austria); insights into trophic interactions and food chains of the Polzberg palaeobiota. Sci Rep 10:20545. https://doi.org/10.1038/s41598-020-77017-x

Lukeneder S, Lukeneder A, Harzhauser M, İslamoğlu Y, Krystyn L, Lein R (2012) A delayed carbonate factory breakdown during the Tethyan-wide Carnian Pluvial Episode along the Cimmerian terranes (Taurus, Turkey). Facies 58:279–296. https://doi.org/10.1007/s10347-011-0279-8

Mabberley DJ (2013) Tropical rain forest ecology. Chapman and Hall, New York City, NY

Meller B (2007) Lunz am See: Die Lunzer Flora—ein Event mit weltweiter Bedeutung. In: Hofmann T (ed) Wanderungen in die Erdgeschichte 22: Wien, Niederösterreich, Burgenland. Pfeil, München, pp 132–134

Mueller S, Krystyn L, Kürschner WM (2016) Climate variability during the Carnian Pluvial Phase—a quantitative palynological study of the Carnian sedimentary succession at Lunz am See, Northern Calcareous Alps, Austria. Palaeogeogr Palaeoclimatol Palaeoecol 441:198–211. https://doi.org/10.1016/j.palaeo.2015.06.008

Napp-Zinn K (1984) Anatomie des Blattes. II. Blattanatomie der Angiospermen. Encyclopedia of Plant Anatomy, VIII, I. Borntraeger, Berlin

Nathorst AG (1909) Über die Gattung *Nilssonia* Brongn. mit besonderer Berücksichtigung schwedischer Arten. Kungl Sv Vetenskapsakad Handl 43:3–37

Neinhuis C, Barthlott W (1997) Characterization and distribution of water-repellent self-cleaning plant surfaces. Ann Bot 79:667–677. https://doi.org/10.1006/anbo.1997.0400

Norstog KJ, Nicholls TJ (1997) The biology of the cycads. Cornell University Press, Ithaca, NY

Ogg JG (2015) The mysterious mid-Carnian "Wet Intermezzo". J Earth Sci 26:181–191. https://doi.org/10.1007/s12583-015-0527-x

Piller WE, Egger H, Erhart CW, Gross M, Harzhauser M, Hubmann B, Van Husen, D, Krenmayr H-G, Krystyn L, Lein R, Lukeneder A, Mandl GW, Rögl F, Roetzel R, Rupp C, Schnabel W, Schönlaub HP, Summesberger H, Wagreich M, Wessely G (2004) Die stratigraphische Tabelle von Österreich 2004 (sedimentäre Schichtfolgen). Kommission für die paläontologische und stratigraphische Erforschung Österreichs der Österreichischen Akademie der Wissenschaften und Österreichische Stratigraphische Kommission, Vienna

Pott C (2007) Cuticular analysis of gymnosperm foliage from the Carnian (Upper Triassic) of Lunz, Lower Austria. PhD thesis, Universität Münster, IV+274 p

Pott C (2014) A revision of *Wielandiella angustifolia*—a shrub-sized bennettite from the Rhaetian–Hettangian of Scania, Sweden, and Jameson Land, Greenland. Int J Plant Sci 175:467–499. https://doi.org/10.1086/675577

Pott C (2016) *Westersheimia pramelreuthensis* from the Carnian (Upper Triassic) of Lunz, Austria: more evidence for a unitegmic seed coat in early Bennettitales. Int J Plant Sci 177:771–791. https://doi.org/10.1086/688708

Pott C (2019) The cycadalean megasporophyll *Dioonitocarpidium* in the Carnian (Late Triassic) flora of Lunz am See, Austria. Paläontol Zeitschr (PalZ) 93:517–530. https://doi.org/10.1007/s12542-019-00459-w

Pott C (2021) Die fossile Flora der Lunz-Formation (Karnium) von Lunz am See (Niederösterreich). In: Hauschke N, Bachmann GH, Franz M (eds) TRIAS. Aufbruch in das Erdmittelalter, Dr Pfeil, München, pp 447–454

Pott C (2026) When Bennettitales flourished in Westphalia. Abh Westf Mus Naturk 112:187–216

Pott C, Axsmith BJ (2015) *Williamsonia carolinensis* sp. nov. and associated *Eoginkgoites* foliage from the Upper Triassic Pekin Formation: Implications for early evolution in Williamsoniaceae (Bennettitales). Int J Plant Sci 176:174–185. https://doi.org/10.1086/679471

Pott C, Bouchal JM, Choo T, Yousif R, Bomfleur B (2018) Ferns and fern allies from the Carnian (Upper Triassic) of Lunz am See, lower Austria: a melting pot of Mesozoic fern vegetation. Palaeontographica B 297:1–101. https://doi.org/10.1127/palb/2018/0059

Pott C, Fischer T, Aschauer B (2017) *Lunzia austriaca*—a bennettitalean microsporangiate structure with Cycadopites-like *in situ* pollen from the Carnian (Upper Triassic) of Lunz, Austria. Grana 56:321–338. https://doi.org/10.1080/00173134.2017.1282010

Pott C, Kerp H, Krings M (2007a) Morphology and epidermal anatomy of *Nilssonia* (cycadalean foliage) from the Upper Triassic of Lunz (Lower Austria). Rev Palaeobot Palynol 143:197–217. https://doi.org/10.1016/j.revpalbo.2006.07.007

Pott C, Kerp H, Krings M (2007b) *Pseudoctenis cornelii* nov. spec. (cycadalean foliage) from the Carnian (Upper Triassic) of Lunz. Lower Austria. Ann Naturhist Mus Wien 109A:1–17

Pott C, Kerp H, Krings M (2008a) Sphenophytes from the Carnian (Upper Triassic) of Lunz am See (Lower Austria). Jahrb Geol Bundesanst 148:183–199

Pott C, Krings M (2007) First record of circinate vernation in bennettitalean foliage. N Jb Geol Paläont Abh 245:315–321. https://doi.org/10.1127/0077-7749/2007/0245-0315

Pott C, Krings M (2010) Gymnosperm foliage from the Upper Triassic of Lunz, lower Austria: an annotated checklist and identification key. Geo. Alp 7:19–38

Pott C, Krings M, Kerp H (2007c) A surface microrelief on the leaves of *Glossophyllum florinii* (?Ginkgoales) from the Upper Triassic of Lunz, Austria. Bot J Linnean Soc 153:87–95. https://doi.org/10.1111/j.1095-8339.2007.00569.x

Pott C, Krings M, Kerp H (2007d) First record of *Nilssoniopteris* (Gymnospermophyta, Bennettitales) from the Carnian (Upper Triassic) of Lunz, Lower Austria. Palaeontology 50:1299–1318. https://doi.org/10.1111/j.1475-4983.2007.00704.x

Pott C, Krings M, Kerp H (2008b) The Carnian (Late Triassic) flora from Lunz in lower Austria: palaeoecological considerations. Palaeoworld 17:172–182. https://doi.org/10.1016/j.palwor.2008.03.001

Pott C, Krings M, Kerp H, Friis EM (2010) Reconstruction of a bennettitalean flower from the Carnian (Upper Triassic) of Lunz, lower Austria. Rev Palaeobot Palynol 159:94–111. https://doi.org/10.1016/j.revpalbo.2009.11.004

Pott C, Labandeira CC, Krings M, Kerp H (2008c) Fossil insect eggs and ovipositional damage on bennettitalean leaf cuticles from the Carnian (Upper Triassic) of Austria. J Paleont 82:778–789. https://doi.org/10.1666/06-094.1

Pott C, McLoughlin S (2009) Bennettitalean foliage from the Rhaetian-Bajocian (latest Triassic–Middle Jurassic) floras of Scania, southern Sweden. Rev Palaeobot Palynol 158:117–166. https://doi.org/10.1016/j.revpalbo.2009.08.004

Pott C, McLoughlin S (2014) Divaricate growth habit in Williamsoniaceae: unravelling the ecology of a key Mesozoic plant group. Palaeobiodiv Palaeoenviron 94:307–325. https://doi.org/10.1007/s12549-014-0157-9

Pott C, McLoughlin S, Lindström A, Wu SQ, Friis EM (2012) *Baikalophyllum lobatum* and *Rehezamites anisolobus*: two seed plants with "cycadophyte" foliage from the early Cretaceous of Eastern Asia. Int J Plant Sci 173:192–208. https://doi.org/10.1086/663276

Pott C, Schöllmann L, Schwermann AH (2026) Versunkene Welten Nordrhein-Westfalens. 2nd edition. LWL-Museum für Naturkunde, Münster

Pott C, Takimoto H (2022) *Kimuriella* gen. nov. (Bennettitales), a whole-plant bennettite from the Oxfordian (Upper Jurassic) Tochikubo Formation of Shidazawa, Minamisōma, Fukushima Prefecture, Northeast Japan. Paleontol Res 26:158–186. https://doi.org/10.2517/PR200020

Pott C, Takimoto H (2025) *Ohaniella* gen. nov., a new whole-plant bennettite from the Oxfordian (Upper Jurassic) Tochikubo Formation of Shidazawa, Fukushima, northeast Japan. Paleontol Res 29:316–331. https://doi.org/10.2517/prpsj.250010

Pott C, Van Konijnenburg-van Cittert JHA, Kerp H, Krings M (2007e) Revision of the *Pterophyllum* species (Cycadophytina: Bennettitales) in the Carnian (Late Triassic) flora from Lunz, Lower Austria. Rev Palaeobot Palynol 147:3–27. https://doi.org/10.1016/j.revpalbo.2007.03.005

Pott C, Wang X, Zheng X (2015) *Wielandiella villosa* comb. nov. from the middle Jurassic of Daohugou, China: more evidence for divaricate plant architecture in Williamsoniaceae. Botanica Pacifica 4:137–148. https://doi.org/10.17581/bp.2015.04215

Richardson SJ, Walker LR (2010) Nutrient ecology of ferns. In: Mehltreter K, Walker LR, Sharpe JM (eds) Fern ecology. Cambridge University Press, Cambridge, pp 111–139. https://doi.org/10.1017/CBO9780511844898.005

Roghi G (2004) Palynological investigations in the Carnian of the Cave del Predil area (Julian Alps, NE Italy). Rev Palaeobot Palynol 132:1–35. https://doi.org/10.1016/j.revpalbo.2004.03.001

Roth I (1992) Leaf structure: coastal vegetation and mangroves of Venezuela. Encyclopedia of Plant Anatomy, XIV, 2. Borntraeger, Berlin, 172 p

Rudall P, Bateman R (2007) Developmental bases for key innovations in the seed-plant microgametophyte. Trends Plant Sci 12:317–326. https://doi.org/10.1016/j.tplants.2007.06.004

Ruffell A, Simms MJ, Wignall PB (2015) The Carnian humid episode of the Late Triassic: a review. Geol Mag 153:271–284. https://doi.org/10.1017/S0016756815000424

Schulze E-D, Beck E, Buchmann N, Clemens S, Müller-Hohenstein K, Scherer-Lorenzen M (2019) Plant ecology. Springer, Berlin. https://doi.org/10.1007/978-3-662-56233-8

Schweitzer H-J, Kirchner M, Van Konijnenburg-van Cittert JHA (2000) The Rhaeto-Jurassic flora of Iran and Afghanistan. 12. Cycadophyta. II. Nilssoniales. Palaeontographica B 254:1–63. https://doi.org/10.1127/palb/254/2000/1

Seilacher A (1970) Begriff und Bedeutung der Fossil-Lagerstätten. Neues Jahrb Geol Paläontol, Monatshefte 1970:34–39

Seilacher A, Reif W-E, Westphal F (1985) Sedimentological, ecological and temporal patterns of fossil Lagerstätten. Philos Trans R Soc Lond B 311:5–23. https://doi.org/10.1098/rstb.1985.0134

Simms MJ, Ruffell AH (1989) Synchronicity of climatic change in the late Triassic. Geology 17:265–268. https://doi.org/10.1130/0091-7613(1989)017<0265:SOCCAE>2.3.CO;2

Simms MJ, Ruffell AH (2018) The Carnian Pluvial Episode: from discovery, through obscurity, to acceptance. J Geol Soc 175:989–992. https://doi.org/10.1144/jgs2018-020

Stremme H (1911) Über paralische und limnische Kohlenlager und Moore. Int J Earth Sci 2:13–25. https://doi.org/10.1007/BF01798821

Stur D (1885) Die obertriadische Flora der Lunzer-Schichten und des bituminösen Schiefers von Raibl. Denkschr Kaiserl Akad Wiss 3:93–103

Thomas HH (1915) On *Williamsoniella*, a new type of bennettitalean flower. Philos Trans R Soc Lond 207 B: 113–148. https://doi.org/10.1098/rstb.1916.0003

Thomas HH, Bancroft N (1913) On the cuticles of some recent and fossil cycadean fronds. Trans Linnean Soc Lond: Bot 8:55–204. https://doi.org/10.1111/j.1095-8339.1913.tb00284.x

Van Konijnenburg-van Cittert JHA (2002) Ecology of some late Triassic to early Cretaceous ferns in Eurasia. Rev Palaeobot Palynol 119:113–124. https://doi.org/10.1016/S0034-6667(01)00132-4

Van Konijnenburg-van Cittert JHA, Pott C, Cleal C, Zijlstra G (2017) Differentiation of the fossil leaves assigned to *Taeniopteris*, *Nilssoniopteris* and *Nilssonia* with a comparison to similar genera. Rev Palaeobot Palynol 237:100–106. https://doi.org/10.1016/j.revpalbo.2016.11.009

Verloop JH (1908) Profil der Lunzer Schichten in der Umgebung von Lunz. Zeitschr deut geol Gesellsch 60:81–88

Vladimirovicz VP (1958) Découverts des restes de *Neocalamites* avec les strobiles conservés. Dokl Akad Nauk SSSR 122:695–699
Ettingshausen C von (1851) Beiträge zur Flora der Vorwelt. Haidinger's Naturw Abh 4:65–99
Watson J, Cusack HA (2005) Cycadales of the English Wealden. Monogr Palaeontogr Soc Lond 158:1–189. https://doi.org/10.1080/25761900.2022.12131802
Watson J, Sincock CA (1992) Bennettitales of the English Wealden. Monogr Palaeontogr Soc Lond 145:1–228. https://doi.org/10.1080/25761900.2022.12131771
Wing SL, Sues H-D, Tiffney BH, Stucky RK, Weishampel DB, Spicer RA, Jablonski D, Badgley CE, Wilson MVH, Kovach WL (1992) Mesozoic and early Cenozoic terrestrial ecosystems. In: Behrensmeyer AK, Damuth JD, DiMichele WA, Potts R, Sues HD, Wings SL (eds) Terrestrial ecosystems through time: evolutionary paleoecology of terrestrial plants and animals. University of Chicago Press, Chicago, IL, pp 327–416
Zotz G (2016) Plants on plants—the biology of vascular epiphytes. Springer, Heidelberg, 282 p. https://doi.org/10.1007/978-3-319-39237-0

Chapter 19
The Marine Upper Triassic Paleobiota from the Scheiblingbach *Konservat-Lagerstätte* in Austria

Alexander Lukeneder

Abstract This contribution reports new data on the Upper Triassic Scheiblingbach *Konservat-Lagerstätte* in the Northern Calcareous Alps of Styria. New fossil material was obtained and examined from bed-by-bed collections in the silicious Göstling Formation and the well-laminated Reingraben Formation. Over 4,602 new fossils of various marine taxa were collected within the fossiliferous layers from the early Carnian (Julian 2, *Austrotrachyceras austriacum* Zone). The assemblage comprises ammonoids (*Austrotrachyceras*, *Paratrachyceras*, *Carnites*), coleoids (*Mojsisovicsteuthis*, *Phragmoteuthis*), bivalves (*Peribositra*, *Halobia*, div. indet taxa), gastropods (caenogastropods/heterobranchs), thylacocephalan arthropods (*Atropicaris*), branchiopods (*Euestheria*), rare coelacanth remains (scales) and actinopterygian fishes (*Peltopleurus*). Bromalites (coprolites and regurgitalites) produced by piscivorous actinopterygians and durophagous fishes complete the Scheiblingbach paleobiota along with frequent plant remains (*Voltzia,* div. indet plants). The Scheiblingbach paleobiota in the normal marine Reifling Basin characterizes a fluctuating basinal environment with intermittent influx of freshwater, accompanied by the enhanced presence of land plants and colonization by abundant benthic halobiid bivalves. The nektic community is dominated by ammonoids, belemnoids and actinopterygian fish. The prerequisites for preservation – calm and oxygendepleted conditions – prevailed at the seafloor of the Scheiblingbach *Konservat-Lagerstätte* during the deposition of black intermittent layers in the Göstling Formation and in the basal part of the Reingraben Formation. New in situ findings of benthic and nektic taxa highlight the great value of the additional Scheiblingbach paleobiota in comparison to the well-known contemporaneous Polzberg Paleobiota and regarding the autochthonous deposition of the inhabitants within the paleohabitat in a *Konservat-Lagerstätte.* The fauna and flora from Scheiblingbach point to a carbonate platform decline, followed by warmer and wetter conditions during the Carnian Pluvial Episode, triggering the deposition of laminated sediments and forcing inhabitants to adapt.

A. Lukeneder (✉)
Department of Geology and Paleontology, Natural History Museum Vienna, Vienna, Austria
e-mail: alexander.lukeneder@nhm.at

A. Lukeneder (ed.), *Late Triassic Konservat-Lagerstätten within the Carnian Pluvial Episode*, Topics in Geobiology 55, https://doi.org/10.1007/978-3-032-19777-1_19

Keywords *Konservat-Lagerstätte* · Paleobiota · Paleoenvironment · Early Carnian · Late Triassic · Austria

19.1 Introduction

Numerous papers have been published on the comparable Late Triassic paleobiota from the well-known lower Carnian Polzberg *Konservat-Lagerstätte* (Seilacher 1970; Allison 1988; see also Krystyn 1991; Forchielli and Pervesler 2013; Lukeneder 2026, this volume). The latter is equal in age and exhibits similar lithologies, indicating similarities in floral and faunal assemblages. The main difference is the period of sampling: fossils were collected at Polzberg from the 1870s on and from Scheiblingbachgraben (= Scheiblingbach from here on) only since the 1970s. This represents a significant disparity of 100 years of sampling history. The available material mainly increased due to the intensified collecting work by the author since 2022. This includes examining the paleoenvironment and the trophic food web (Lukeneder et al. 2020; Lukeneder and Lukeneder 2021), the taphonomy, trophic history and taxonomy of ammonoids (Lukeneder and Lukeneder 2022, 2024), and new findings of unique mineralized coleoid cartilage (Lukeneder and Lukeneder 2022; Lukeneder et al. 2024). These publications were based on historical and recent paleontological material housed in the official Austrian collections (Natural History Museum Vienna, GeoSphere Austria) and material from private collections (B and K Aschauer).

Fossil remnants from *Konservat-Lagerstätten* form valuable sources of paleobiological information. Excavations involving bed-by-bed collecting were done during several field campaigns (60 days sampling) during four consecutive years starting in September 2022 and extending to August 2025 (and still ongoing). The resulting details on the fauna and flora yielded new insights into the Scheiblingbach paleocommunity of the Northern Calcareous Alps in the north of Styria. They also corroborated the assumptions made by Lukeneder and Lukeneder (2021, 2023) as well as Lukeneder et al. (2024a) for the comparable Polzberg *Konservat-Lagerstätte* located 35 km northeast in Lower Austria. This Special Volume (Lukeneder 2026) provides more details on distinct taxonomic groups (thylacocephalans, bristle worms, conodonts, gastropods, bivalves, fishes, belemnoids, bromalites), the geochemistry, geophysics, lithological aspects and food chains and dietary habits, as well as on the trophic interactions between the inhabitants of Upper Triassic marine ecosystems. Previously unrecorded taxa and fossil assemblage data are presented for the first time and compared to the contemporaneous deposits and fossil remains of the Polzberg *Konservat-Lagerstätte*. These paleobiotas were deposited during the Carnian Julian 2 Ib (*Austrotrachyceras austriacum* Zone, *A. minor* biohorizon), within a period known as the worldwide Upper Triassic Carnian Pluvial Episode (CPE; Lukeneder et al. 2012; Ruffell et al. 2015; Dal Corso et al. 2018; Simms and Ruffell 2018; Lukeneder et al. 2024a; Alverà et al. 2026; Harzhauser et al. 2026; Lukeneder et al. 2026, this volume; Wet Intermezzo, Ogg 2015). The CPE features an enhanced humidification of the Carnian climate, resulting in the deposition of the fossiliferous Reingraben

Formation at Scheiblingbach and Polzberg. As noted by Lukeneder and Lukeneder (2021) and Lukeneder et al. (2020, 2024a), such laminated deposits provide excellent conditions for the preservation of entire and soft body fossils, a prerequisite for *Konservat-Lagerstätten*. The paleobiota and fossil assemblages reflect a marine environment in the lower Carnian (Julian), yet with intermittent shallow-water influence indicated by the findings of freshwater inhabitants such as branchiopods with *Euestheria* (Stur 1874; Lukeneder and Lukeneder 2021; Rusev and Lukeneder 2026, this volume). This resembles Polzberg, which features the lungfish *Tellerodus* (Stur 1886; Teller 1891; Lehman 1975; Martin 1982; Lukeneder and Lukeneder 2021) accompanied by transported and redeposited plant remains, accumulated in plant layers.

The present paper reports for the first time all the recent findings and new taxa of the Scheiblingbach paleobiota, from invertebrates, vertebrates to plants. It also presents the detailed bed-by-bed paleobiota recently collected from a single outcrop (Scheiblingbach) and reveals new details of the lower Carnian food webs. New additional insights into the Late Triassic marine ecosystem of the Reifling Basin are presented and the significance of the assemblage is outlined and compared with the Polzberg *Konservat-Lagerstätte*.

19.2 Material and Methods

The author collected approximately 4,602 fossils (incl. indet) by bed-by-bed collecting during several field campaigns (60 days sampling) between September 2022 and August 2025 (collections are still ongoing) at the Scheiblingbach (= Scheiblingbachgraben) near Großreifling. More specifically, bed-by-bed sampling was possible throughout the Göstling Formation, and the lowermost calcareous part of the Reingraben Formation, whereas in the upper argillaceous part, only distinct layers with abundant bivalves occur in the soft deposits. These abundant new fossil data were compared to the relevant historical material in collections from the comparable Polzberg locality (10,968 fossil remains) in the Natural History Museum Vienna (NHMW), the Geological Survey of Austria (GBA; since 2023 GeoSphere Austria), and the Departments of Earth Sciences at the University of Vienna (DES). Historical excavation campaigns to obtain the Polzberg fossils were organized by the GBA in 1885 and the NHMW in 1909 (Stur 1886; Teller 1891; Griffith 1977).

19.3 Geologic Setting and Lithology

The Scheiblingbachgraben-section (Sch-section) is located on the northern slope of Mount Lärchkogel (1087 m), south of Mount Reiflingberg (712 m), 600 m southwest of the Enns River, 300 m west of the Großreifling center in northern Styria (47°39′52.50"N, 14°42′27.75"E) (geological map, 100 Hieflau, Kreuss 2014;

Fig. 19.1). The northernmost tectonic element of the Northern Calcareous Alps (NCA) in northernmost Styria around Großreifling is the Reifling Unit, part of the Sulzbach Nappe (Fig. 19.1a). Within the Lunz Nappe in northernmost Styria and Lower Austria, the Reifling Basin (Krystyn 1991; Lukeneder and Lukeneder 2021)—an intraplatform basin during the Late Triassic—is located between Polzberg and Großreifling. It exposes Upper Triassic (Carnian, Julian) deposits of the Reifling Basin, which is part of the Lunz Nappe and the southern Sulzbach Nappe within the NCA (Reifling Unit; Lukeneder and Lukeneder 2021; Lukeneder et al. 2024a). During the Late Triassic, the Reifling Basin was an intraplatform basin, which formed between Pangea in the northwest and the Meliata Ocean in the southeast (Fig. 19.1b). The Meliata Ocean was a small oceanic basin within the Tethyan Realm, separated from the Tethys Ocean by the island chain of the Cimmerian continent (Hornung and Brandner 2005; Lukeneder et al. 2012).

The Reifling Formation (0–3.6 m, layers Sch 1–9, Fig. 19.1c) was deposited on the paleoslope of the Reifling Basin. It consists of light gray and gray-green nodular limestones composed of peloidal, filamentous wackestones and packstones with strong bioturbation and a bed-thickness of 5–50 cm. The carbonate content $CaCO_3$ (calcite equivalent percentages) is stable around high values at 85–95% throughout the Formation (Lukeneder et al. 2024a). The Reifling Formation (*Trachyceras aonoides* Zone, *T. aonoides* Subzone) is overlain by the Göstling Formation (3.6–8.35 m, layers Sch 10 to Sch 55, Fig. 19.1c), which represents the *Austrotrachyceras austriacum* Zone (*A. minor* Subzone), indicating the base of Julian 2. The Göstling Formation consists of black, mm- to cm-thick bedded calciturbiditic radiolarian limestones ($CaCO_3$ 60–80%) with organic-rich mudstones. The change from the Reifling Formation to dark, fine-laminated radiolarian wackestones to packstones of the Göstling Formation is abrupt, located between layers Sch 9 and Sch 10. Plant fossils are abundant on the upper part of numerous layers, marking an enhanced terrigenous input. Intense bioturbation is visible in oxidized parts, contrastingly rare or absent in laminated dysoxic to anoxic layers. At Scheiblingbach (Krystyn 1991; Mueller et al. 2016), only about 4.5 m of radiolarian-rich Göstling limestones are present.

After a strongly bioturbated top (surface of layer Sch 55), the overlying terrigeneous siliciclastic deposits of the Reingraben Formation (8.1–32.2 m, layers Sch 56 to Sch 32.2 m; *A. austriacum* Zone, *A. minor* Subzone, Fig. 19.1d) consist of a pronounced alternation of dark gray to black argillaceous marls ($CaCO_3$ 5–20%) intercalated by rare limestone/dolomite layers. No macro-bioturbation is visible. The bivalve *Peribositra* appears in all ontogenetic stages and, in layer-/rock-forming quantities, mostly in articulated butterfly position in lumachelles. The Late Triassic *Konservat-Lagerstätte* at Scheiblingbach comprises invertebrates and vertebrates (fishes). Most of the fossils derive from the lowermost three calcareous meters of the Reingraben Formation, but macrofossils occur throughout the section. The paleobiota shows a nekton-dominated fauna with abundant fish and cephalopods (Lukeneder and Lukeneder 2021, 2023; Lukeneder et al. 2020, 2024a, b). The Reingraben Formation at Scheiblingbach appears with microfossils such as conodonts, ostracods, gastropods and radiolaria in thin sections.

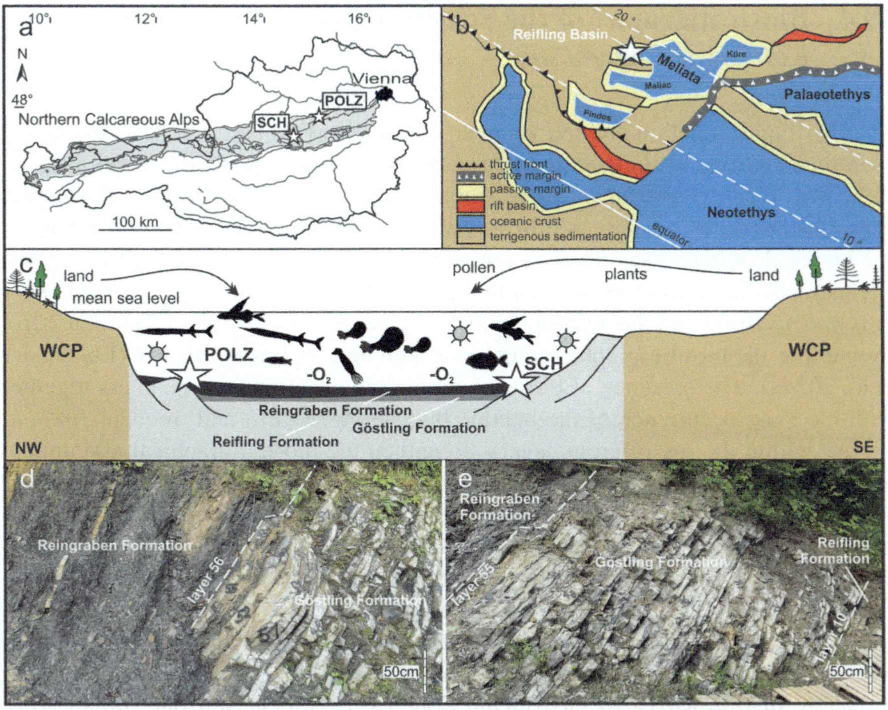

Fig. 19.1 **a** Locality map of the Scheiblingbach area (SCH, Großreifling) in Styria and the Polzberg area (POLZ, Lunz am See) in Lower Austria, with marked position of the Austrian Northern Calcareous Alps (in gray). **b** Paleogeographic position of the Reifling Basin (white asterisk) at the northern rim of the Meliata Ocean. **c** Transect through the Reifling Basin with reconstruction of basinal environments, position of Scheiblingbach and Polzberg (white asterisks), low oxygen conditions (-O_2) and surrounding land with plants. **d** Detailed view on the boundary layers of the Göstling Formation and Reingraben Formation. **e** The top of the light gray Reifling Formation, silicious Göstling Formation (Sch 10 to Sch 55) and overlaying argillaceous Reingraben Formation. NW northwest, SE southeast, WCP Wetterstein Carbonate Platform

Radiolarians and sponge spicules are abundant in the packstones of the Reifling and Göstling Formations, while ammonite and bivalve mudstones are present at the base of the Reingraben Formation. A sudden change from radiolarian wackestones/ packstones (Summesberger and Wagner 1972; Mostler and Scheuring 1974; Lahm 1984; Krystyn 1991) of the Göstling Formation occurs above the boundary to the Reingraben Formation (Tollmann 1976, Fig. 19.1e).

19.4 Biostratigraphy of the Scheiblingbach *Konservat-Lagerstätte*

The Göstling Formation and the younger Reingraben Formation of Scheiblingbach span together over a single ammonoid assemblage, representing the lower Carnian *Austrotrachyceras austriacum* Zone (Julian 2, Fig. 19.2). The Göstling Formation is ascribed to the *A. triadicum* Subzone, whereas the Reingraben deposits belong to the higher *A. minor* Subzone. The lowermost, fossiliferous part of the Reingraben Formation is dated based on the occurrence of *Austrotrachyceras minor*, *Paratrachyceras haberfellneri* and the accompanying *Carnites floridus* as Julian 2 (Julian 2/Ib), resembling the biostratigraphic situation at Polzberg in Lower Austria (Lukeneder et al. 2024a). The presence of biostratigraphic index ammonoid members together with the mass occurrence of the benthic bivalve *Peribositra* and frequent *Halobia rugosa* within the fossiliferous layers strengthen the biostratigraphical assignment to the Julian age (Mueller et al. 2016; Lukeneder and Lukeneder 2021, 2022a; Lukeneder et al. 2024a). Concerning the ammonoid zonation and the specific duration of ammonoid zones in the Late Carnian (Mueller et al. 2016; Dal Corso et al. 2020), the analyzed section encompasses deposits of approximately 500,000 years.

No ammonoids are present in the Reifling Formation, and only trachyceratid specimens with *Austrotrachyceras patroclus*, rare *C. floridus* and *Sirenites* sp. specimens are present in the Göstling Formation. *A. patroclus* is characteristic for the basal *Austrotrachyceras austriacum* Zone (*A. triadicum* Subzone, Julian 2/Ia). Conodonts are restricted to the lowermost part of both the Scheiblingbach and Polzberg sections, namely in deposits of the Reifling and Göstling formations (*Gladigondolella tethydis*, *Gladigondolella malayensis malayensis*, *Gladigondolella* sp. Fig. 19.2). The conodont association collected at the top of the Reifling Formation is

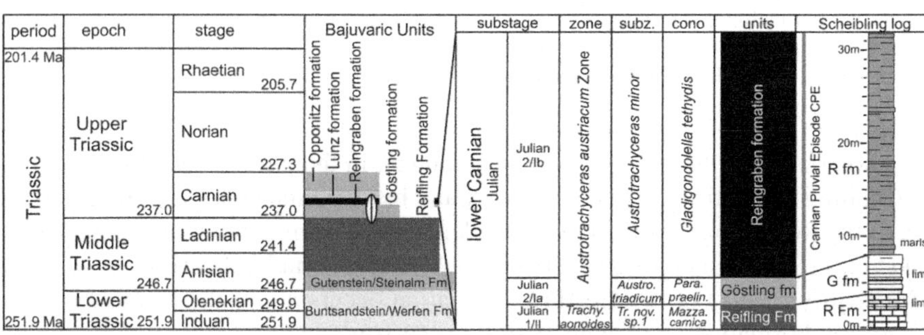

Fig. 19.2 Stratigraphic position of the Scheiblingbach deposits during the lower Carnian. White ellipse: position of the *Konservat-Lagerstätte* Scheiblingbach. Black asterisk and gray bar: stratigraphic age. Adapted after Lukeneder and Lukeneder (2021). Detailed Scheiblingbach section within the lower Carnian Reingraben Formation. Austro *Austrotrachyceras*, Trachy *Trachyceras*, chrono chronostratigraphy, subz subzone, Re Fm Reifling Formation, G fm Göstling Formation, R fm Reingraben Formation, lime limestone, s lime silicious limestone. Adapted after Lukeneder and Lukeneder (2021)

represented by typical Julian forms, including *Paragondolella polygnathiformis*, *P. inclinata*, *P. foliata*, *P. tadpole*, *P. praelindae*, *Paragondolella* aff. *Par. foliata* (sensu Rigo et al. 2007), *Gladigondolella tethydis*, *G. malayensis malayensis* and *"Mosherella" longnanensis*. The presence of *P. polygnathiformis* constrains the age of the samples to Carnian, whereas *P. praelindae* together with *Paragondolella* aff. *Par. foliata* date the top of the Reifling Formation to Julian 2 (Rigo et al. 2007, 2018).

19.5 Results

19.5.1 Bed-By-Bed Collected Paleobiota

Material from the Late Triassic Scheiblingbach *Konservat-Lagerstätte*, with deposits of black, finely laminated Reingraben Formation, is missing in historical collections of the NHMW: Recent excavations (2022–2025) provided 4,602 specimens of various taxonomic groups (n 32, Figs. 19.3 and 19.4). The new findings clearly demonstrate that the majority of the detected fossils derives from the lowermost two calcareous meters of the Reingraben Formation and the topmost meter of the Göstling Formation. Overlying soft argillaceous deposits of the Reingraben deposits were also sampled and bear a restricted fauna, whereby rock-forming masses of bivalves are present in distinct levels (i.e. *Peribositra* coquinas). We sampled in detail one section at Scheiblingbach (locality in Großreifling) comprising a 32-m-long section. That section comprises the top of the Reingraben Formation (Sch 32–Sch 14 m), the basal Reingraben deposits including 600 cm of the fossiliferous *Konservat-Lagerstätte* (14–8.35 m). Below a variegated, extensively borrowed trace fossil level (top of layer Sch 55), the laminated and silicified limestones of the Göstling Formation appear from Sch 55 (8.35 m) to Sch 10 at 3.6 m. The Göstling Formation is underlain by light gray limestones of the Reifling Formation from 3.6–0.0 m (Sch 9 to Sch 1 base). The enormous number of fossils (4,602; all specimens and species) and their quality (entire, fragile morphologies, soft part preservation) yield special insights into the morphology of such otherwise rarely preserved fossil groups and specific taxa. The above number is exclusive of the uncountable masses of rock-forming bivalves (mainly *Peribositra*). *Peribositra* forms meter-thick coquinas in specific levels, hence representing the dominant benthic element at Scheiblingbach. The totality of the Scheiblingbach paleobiota (excl. *Peribositra*) shows a nekton-dominated fauna, with abundant cephalopods (48%; ammonoids, belemnoids) and fishes (13%; mainly actinopterygians, Fig. 19.3).

The basal Reifling Formation (n 4; excl. microfossils) appears mainly with a microfossil content visible in thin sections with abundant radiolaria and filamentous bivalves, numerous foraminifera and rare ammonoids. The stratigraphically following Göstling Formation (n 1,395) shows a change to finely laminated, fossiliferous (radiolaria-dominated) silicious limestones with abundant bivalves (not given in %, not countable numbers; *Peribositra*) in the light gray limestones. Within the

a
b
c
d
e
f
g
h
i
j
k
l
m
n
o
p
q
r
s

◂**Fig. 19.3** Invertebrate and vertebrate members of the lower Carnian Scheiblingbach paleobiota within the argillaceous Reingraben Formation. **a** *Austrotrachyceras minor*, positive, lateral view, note change in adult shell ornamentation, layer Sch 56 top, NHMW 2025/0127/0001. **b** *Paratrachyceras haberfellneri*, positive, lateral view, adult specimens with aragonitic primary shell, layer Sch 56 base, NHMW 2025/0127/0002. **c** *Carnites floridus*, lateral view of a fragmented specimen with narrow radial swellings, note fine growth lines and external keel, NHMW 2025/0127/0003. **d** *Anaptachus lunzensis*, lower jaw of *A. minor*, positive, lateral view, isolated specimen, layer Sch, NHMW 2025/0127/0004. **e** *Phragmoteuthis polzbergensis*, phragmocone with proostracum fragment, layer Sch 56 base, NHMW 2025/0127/0005. **f** *Phragmoteuthis polzbergensis*, cephalic cartilage, carbonized fragments, layer Sch 56 top, NHMW 2022/0214/0002. **g** *Phragmoteuthis polzbergensis*, arm hook, Sch 56 base, NHMW 2025/0127/0006. **h** *Halobia rugosa*, mid-aged specimen, accumulated in butterfly preservation, layer Sch 27.6 m, NHMW 2025/0127/0007. **i** Bivalve indet, single valve, layer Sch 56 base, NHMW 2025/0127/0008. **j** Bivalve indet, single pectinid valve, layer Sch 56 base, NHMW 2025/0127/0009. **k** Gastropod indet, layer Sch 56 top, NHMW 2025/0127/0010. **l** Echinoid indet, regular echinoid, fragment of ambulacral field, layer layer Sch 56 base, NHMW 2025/0127/0011. **m** *Atropicaris* sp., lateral view, layer Sch 58, NHMW 2025/0126/0001. **n** *Euestheria minuta*, positive, lateral view, layer Sch, NHMW 2025/0127/0012. **o** *Platychela trauthi*, lateral view, fragmented specimen, NHMW 2025/0127/0013. **p** Coprolite with fish scales, produced by fish, lateral view of positive near negative mold, layer Sch 56 top, NHMW 2025/0127/0014. **q** Coelacanth, single scale, positive, lateral view, layer Sch 56 top, NHMW 2025/0128/0012. **r** Actinopterygian fish, skull missing, *Elpistoichthys* sp., lateral view, layer 56 top, NHMW 2025/0127/0015. **s** Chondrichthyes, hybodont tooth, layer 56 base, NHMW 2025/0127/0016. Scale bar: 10 mm, except **d**, **g**, **i**, **j**, **o**, **q**, **p** scale bar: 1 mm

black-laminated more marly layers, abundant plant fossils (62%; *Voltzia*, plants spp.), fish scales (19%; no entire specimens), and trachyceratid and carnitid ammonoids (10%) are detectable on layer surfaces (Fig. 19.4). In both the Reifling and the Göstling Formations, distinct layers with intense bioturbation occur. The Reingraben Formation is subdivided in the lower, more calcareous part exhibiting the Scheiblingbach *Konservat-Lagerstätte* (8.35–14.0 m) and the upper, more argillaceous part (14.0–32.0 m) with distinct levels of posidoniid coquinas. Both parts are clearly dominated by posidoniids bivalves (not countable, rock-forming numbers). Besides that bivalves, the basal part (n 2,233) is dominated by austrotrachyceratids (71%) and actinopterygian fishes (10%), accompanied by abundant bromalites (12%). A quite similar picture, with fewer fossil detections (n 66; excl. posidoniids) is notable in the higher part of the Reingraben deposits with dominant ammonoids (61%), actinopterygians (11%) and 23% bromalites (Fig. 19.5).

19.5.2 The Scheiblingbach Konservat-Lagerstätte

Cephalopods dominate the nekton here, with 71.2% ammonoids and 1.0% coleoids, followed by 10.7% fishes (actinopterygians, sarcopterygians, chondrichthyes), accompanied by less frequent undetermined bivalves (0.9%), gastropods (0.3%), arthropods (thylacocephalans, indet. crustaceans; 1 and 2%, respectively), and frequent bromalites with coprolites and regurgitalites (11.6%) as well as allochthonous plant remains (1.3%). The autochthonous macrobenthic community

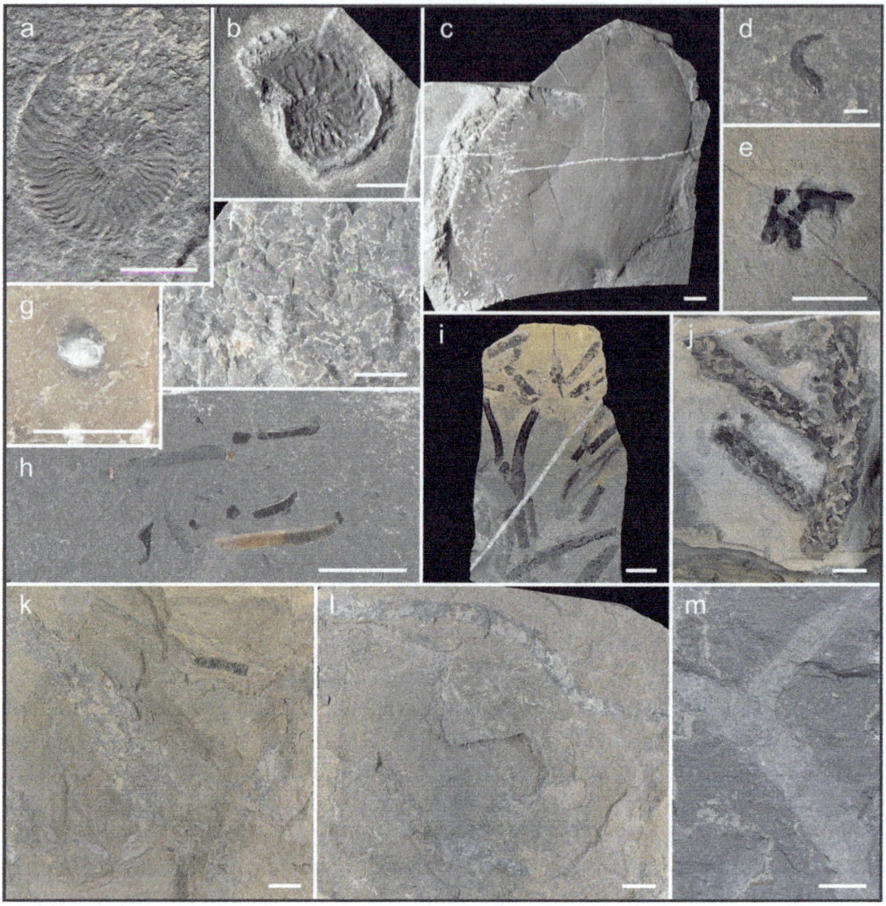

Fig. 19.4 Invertebrate and vertebrate members of the lower Carnian Scheiblingbach paleobiota within the silicious Göstling Formation. **a** *Austrotrachyceras patroclus*, positive, lateral view, layer Sch 51, NHMW 2025/0127/0017. **b** *Sirenites* sp., positive, lateral view, note elongated ridges on ribs, layer Sch 51, NHMW 2025/0127/00018. **c** *Carnites floridus*, fragmented specimen, lateral view, with narrow umbilicus, fine growth lines, layer Sch 52, NHMW 2025/0127/0019. **d** *Mojsisovicsteuthis* sp., curved arm hook, Sch 55, NHMW 2025/0127/0020. **e** *Phragmoteuthis polzbergensis*, cephalic cartilage, carbonized fragments, layer Sch 34, NHMW 2025/0127/0021. **f** *Peribositra*, coquina, single valves, layer Sch 33, NHMW 2025/0127/0022. **g** Crinoid, fragment, layer 6, NHMW 2025/0127/0023. **h** Actinopterygian fish, numerous disseminated fish scales, single scale 15 mm length, layer Sch 18, NHMW 2025/0127/0024. **i** Plant remains, single fragments on layer, layer Sch 52, NHMW 2025/0127/25. **j** *Voltzia* sp., branched morphology, layer Sch 55, NHMW 2025/0127/0026. **k** Trace fossils, branching tube-like burrows, light gray, top of Göstling Formation, layer Sch 55 top, NHMW 2025/0127/0027. **l** Trace fossil, thalassinoid level, top of Göstling Formation, layer Sch 55 top, NHMW 2025/0127/0028. **m** Trace fossil, branched burrow filled with light gray sediment, layer Sch 18, NHMW 2025/0127/0029. Scale bars: 10 mm, except **d** scale bar: 1 mm

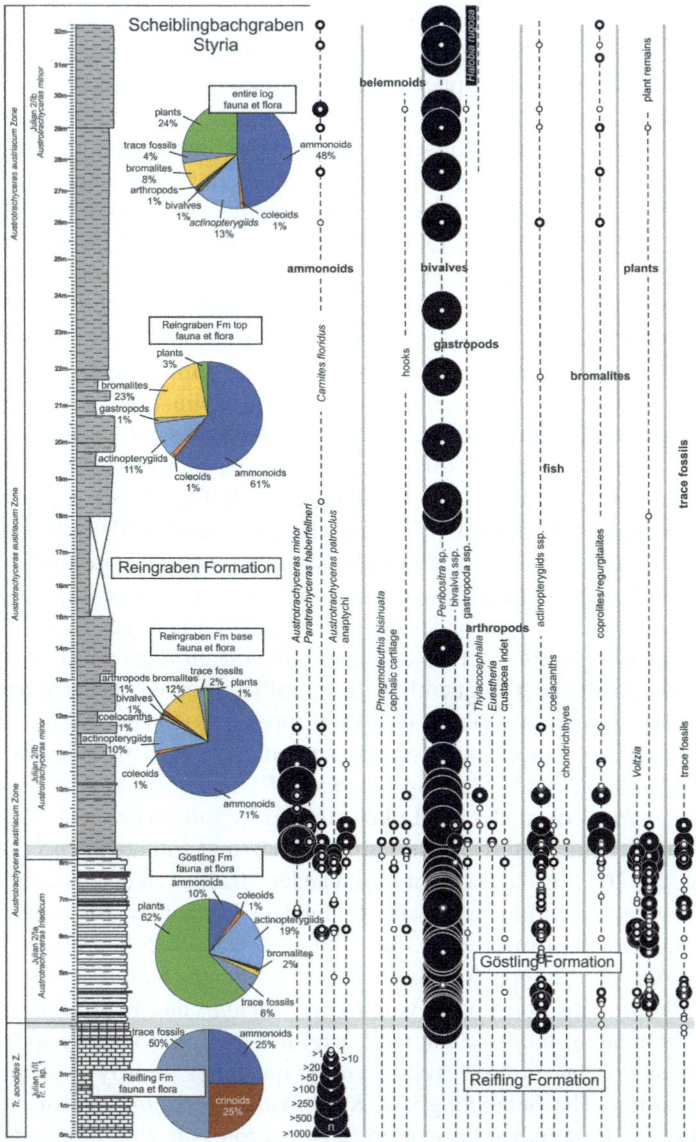

Fig. 19.5 Detailed Scheiblingbach section with indicated sampling layers (sample numbers Sch 1 to Sch 32.2 m) within the lower Carnian Reifling, Göstling and Reingraben Formations. Pie charts for compilation of fossil taxa and composition of Scheiblingbach paleobiota (percentage) throughout the section. Pie chart at the top represents entire fauna (except posidoniids) at Scheiblingbach. Pie charts for each interval: Reingraben Formation top, the Reingraben Formation base, the Göstling Formation, the Reifling Formation. Range and numbers (n) of the occurring fauna and flora with the taxa detected by bed-by-bed sampling. Diameter of black circles increase with quantity from n = 1 to n > 1,000

is dominated by countless posidoniid bivalves, in distinct layers and meter-thick coquina levels. These small (0.1–3.2 cm), thin-shelled clams dominate throughout the section, both with sporadic appearance and in rock-forming accumulations within distinct layers.

The main faunal elements (Figs. 19.3 and X.4) are the benthic bivalves *Halobia rugosa* (n not countable, Fig. 19.3h) along with the ceratitid ammonoids *Austrotrachyceras minor* (n 1,492; = *Trachyceras triadicum* var. minor; Krystyn 1991, Mojsisovics 1893a, b with CLXXXVI = 186, Figs. 1–3, p. 682; see Lukeneder and Lukeneder 2021, 2023, Fig. 19.3a) and rare *Paratrachyceras haberfellneri* (n 10; = *Trachyceras haberfellneri*; Mojsisovics 1893a, b with CLXXXVI = 186, Figs. 6 and 7, CLXXXVII = 187, Figs. 1 and 2, p. 691, Fig. 19.3b). These cephalopods are accompanied by constituents from recent collections with *Carnites floridus* (n 21; Fig. 19.3c), anaptychi of trachyceratids (n 68; partly in situ, Fig. 19.3d), belemnoids (n 22; proostraca and phragmocones n 9, Fig. 19.3e, hooks n 7, cephalic cartilage n 6, Fig. 19.3f and g), halobiid bivalves and undetermined bivalves (n 19; Fig. 19.3i and j), gastropods (n 6; Fig. 19.3k), echinoids (n 1; Fig. 19.3l), arthropods with the thylacocephalan *Atropicaris* (n 15; Fig. 19.3m), euestheriids (n 9; Fig. 19.3n), crustacea indet (n 1; Fig. 19.3o), abundant bromalites (n 258; coprolites and regurgitalites, Fig. 19.3p), coelacanth scales (n 8; Fig. 19.3q), frequent actinopterygian fish (n 231; entire and single scales, Fig. 19.3r), chondrichthyes (n 1; Fig. 19.3s), Coniferopsida with *Voltzia* (n 1) and undetermined plant remains (n 28). *A. minor* and *P. haberfellneri* appear frequently with partially preserved buccal apparatuses of anaptychus-type lower jaws (Trauth 1935) and upper jaws (Lukeneder and Lukeneder 2021, 2022a, 2023). Trace fossils appear sporadically in distinct levels and with different size groups (n 43). Rare specimens of *Phragmoteuthis polzbergensis* and *Mojsisovicsteuthis* (Lukeneder and Lukeneder 2022b; Lukeneder et al. 2024a) occur along with black amorphous cephalic cartilage (n 6) and arm hooks (n 7; Lukeneder and Lukeneder 2021, Lukeneder and Lukeneder 2022b; Lukeneder et al. 2024b). *A. minor* dominates within the lower Reingraben Formation, for example in the *Konservat-Lagerstätte* (lowermost 2 m) with 67% from 8.35 to 14.0 m (Sch 56 to Sch 14 m). Among the cephalopods, *A. minor* dominates with 93% over *P. haberfellneri*, *C. floridus* and belemnoids with *Phragmoteuthis* and *Mojsisovicsteuthis* (all 1%). The dominating faunal element *H. rugosa* (1–32 mm length) appears in all ontogenetic stages and in layer-/rock-forming quantities, mostly in double-valved butterfly position. The Scheiblingbach *Konservat-Lagerstätte* deposits within the lower part of the Reingraben Formation show microfossils with frequent ostracods and gastropods and rare radiolaria in thin sections. The Scheiblingbach collection is housed at the NHMW.

19.6 Discussion

19.6.1 The Scheiblingbach Konservat-Lagerstätte—*Historical Summary and Legacy of the Carnian Pluvial Episode*

The Scheiblingbach locality and the existing formations were noticed by several authors when investigating mainly lithologies and facies changes around the Reifling and Göstling Formations. They described macrofossils (ammonoids, bivalves) and the microfossil content (radiolaria, conodonts; Summesberger and Wagner 1972; Mostler and Scheuring 1974; Lahm 1984; Krystyn 1978). Hornung (2008). Hornung et al. (2007a) suggested that the deposits of the Göstling and the Reingraben Formations from Scheiblingbach were triggered by a humidification during the Carnian Pluvial Episode. The analyses of all these papers end at the base of the Reingraben Formation (Krystyn 1978, 1991).

The Carnian Pluvial/Humid Episode (CPE; Simms and Ruffell 1990; Hornung and Brandner 2005; Kozur and Bachmann 2010; Lukeneder et al. 2012; Ogg 2015, Ruffell et al. 2015; Mueller et al. 2016; Dal Corso et al. 2018, 2020; Lukeneder and Lukeneder 2021) was characterized by an abrupt and increased siliciclastic transport towards the northwestern branch of the Tethys, resulting in the breakdown of the carbonate factory (Hornung and Brandner 2005). The 'Reingraben turning point' is reflected by changes or adaptations of lithologies, facies, paleobiota, biofacies, and evolutionary events. These changes are mirrored in all facies belts of the entire NW Tethyan continental margin (Schlager and Schöllnberger 1974; Hornung and Brandner 2005; Lukeneder and Lukeneder 2021). Global warming and enhanced humidification during the early Carnian, combined with the deposition of large amounts of volcanogenic material, triggered a worldwide climate change (Hornung and Brandner 2005; Hornung et al. 2007a, b; Mueller et al. 2016; Dal Corso et al. 2020; Lukeneder and Lukeneder 2021; Lukeneder et al. 2024; Weitz et al. 2026, this volume). The increasing humidification is reflected by a change in lithology from the basal Julian sequence in the Scheiblingbach and Polzberg area. In the Late Triassic, during the Carnian stage, the Scheiblingbach and Polzberg subbasins and the Reifling Basin were located at the northwestern Tethyan rim in latitudes of 15° N to 30° N (Lukeneder et al. 2012; Mueller et al. 2016; Lukeneder and Lukeneder 2021, Lukeneder et al. 2024). During the lower Carnian the Scheiblingbach locality was located on the southwestern margin of the Reifling Basin. In contrast, the Polzberg site was suggested to be located on the opposite side (distance not to scale) of the basin, to the north of the intraplatform basin (Hornung et al. 2007a; Hornung 2008). Today, the Styrian Scheiblingbach is located 35 km southwest of the Polzberg site (located in Lower Austria). The lowermost nodular limestones, as also seen at the comparable and synchronously deposited Polzberg locality, of the Reifling Formation were deposited on the paleoslope of the Reifling Basin. At the base of Julian 2, the Reifling Formation (*Trachyceras aonoides* Zone, *T. aonoides* Subzone) is replaced by the limestone deposits with organic-rich mudstones of the Göstling Formation in the *Austrotrachyceras austriacum* Zone (*A. austriacum* Subzone, biohorizon of *A. triadicum*). The fauna of the Göstling Formation at Scheiblingbach is less diverse compared to the overlaying Reingraben deposits, with the rare ammonoids *Austrotrachyceras patroclus* (Fig. 19.4a), *Sirenites* sp. (Fig. 19.3b) and *Carnites floridus* (Fig. 19.3b), arm hooks of *Mojsisovicsteuthis* sp. (Fig. 19.4d), cephalic cartilage of *Phragmoteuthis polzbergensis* (Fig. 19.4e), coquina of *Halobia rugosa* (Fig. 19.4f), crinoids, (Fig. 19.4g), frequent disseminated fish scales of actinopterygians, (Fig. 19.4h), abundant plant remains (Fig. 19.4i), frequent specimens of *Voltzia* sp. (Fig. 19.4j), and in contrast to the Reingraben deposits frequent trace fossils (tube-like burrows, Fig. 19.4k–m). On top of the Göstling Formation lies a transitional thalassinoid level (layer Sch 55 top). The Göstling Formation is overlain by deposits of the fossiliferous Reingraben Formation (*A. austriacum* Subzone, biohorizon of *A. minor*).

19.6.2 The Scheiblingbach versus Polzberg Paleobiota

Special conditions such as stagnant environments are required for the formation of *Konservat-Lagerstätten*, which are characterized by their excellent preservation of invertebrate, vertebrate, and plant fossils (Seilacher 1970; Bernardi et al. 2011; Briggs 2014; Lukeneder and Lukeneder 2021, 2023; Lukeneder 2026, this volume). Recent data and fossil collections from Scheiblingbach and coeval Polzberg localities were gained by bed-by-bed sampling within the last decade (Lukeneder and Lukeneder 2023). Comparable but historical material from Polzberg derives from two adits for fossil mining in 1885 and 1909. New findings from Scheiblingbach stem from a single additional, small, 32-m-wide excavation site along a forest road located in the Scheiblingbach ravine at Großreifling (Styria). At both localities, the lowermost two meters of the Reingraben Formation bear a fossiliferous interval comprising *Konservat-Lagerstätten*. The recently collected fossil material Scheiblingbach derived directly from above the top of the Göstling Formation (Fig. 19.5), hence correlatable with the coeval fossil collections from Polzberg (Fig. 19.6). Only the same interval with the *Konservat-Lagerstätten* from the basal Reingraben Formation can be compared because no fossils were collected from the Göstling Formation or from parts higher than 3.3 m in the Reingraben Formation at Polzberg (both intervals not exposed).

Within both paleobiotas (only specimens counted from bed-by-bed collecting) from Scheiblingbach and Polzberg, benthic bivalves with *Peribostra* (uncountable, rock-forming numbers), accompanied by frequent other bivalve taxa (indet) and gastropods, clearly dominate throughout the basal intervals of the Reingraben Formation. Thylacocephalan arthropods are represented only by *Atropicaris* (n 15; 1%) at Scheiblingbach but are more diverse at Polzberg with *Austriocaris carinata* (n 19), *Atropicaris* (n 24), as well as *Paraostenia striata* (n 25) and *Paraostenia* cf. *ambatolokobensis* (n 1). Additional arthropods (3%) at Polzberg include crustaceans with decapods (*Platychela*), branchiopods (*Euestheria*), and isopods (*Obtusotelson*, *Discosalaputium*, Schädel et al. 2020), accompanied by benthic polychaetes (n 4)

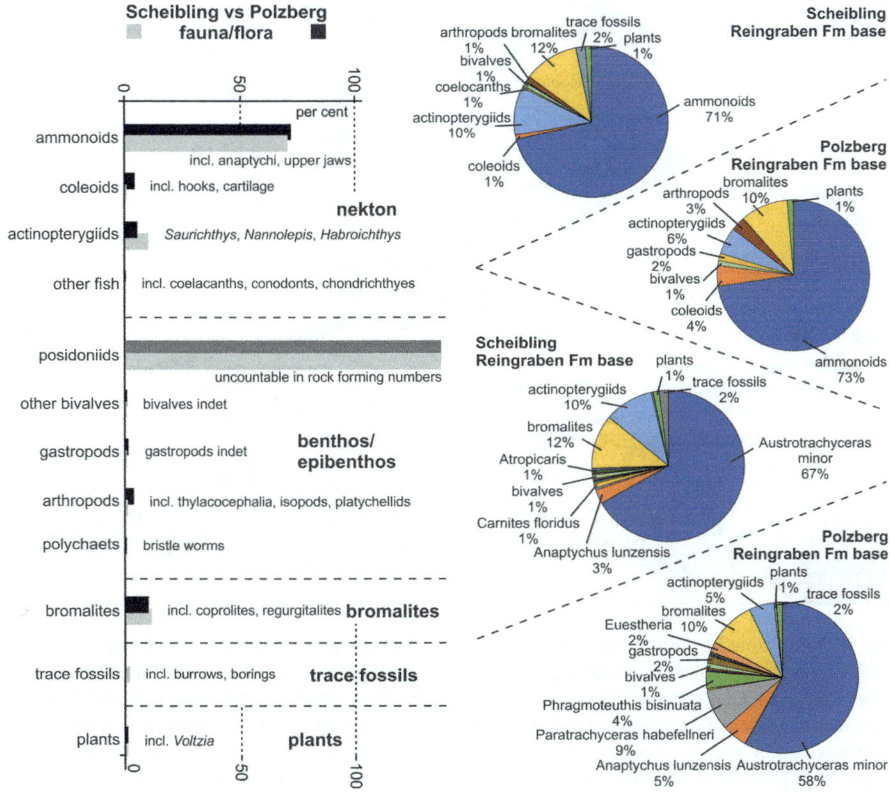

Fig. 19.6 Lower Carnian Scheiblingbach versus Polzberg paleobiota (fauna and flora), assumed mode of life and environment in the Reifling Basin. Mode of life based on direct and indirect evidence from the fossil record of the Scheiblingbach and Polzberg *Konservat-Lagerstätten*. For more details see interpretations in the text. Co-occurring members within both assemblages are dominated by the rock-forming thin-shelled benthic bivalves of *Peribositra*. Nekton is dominated by trachyceratitid ammonoids followed by abundant actinopterygian fishes. Note high numbers in bromalites for both localities. Pie charts (taxonomic group and species level) given for the *Konservat-Lagerstätte* interval at Reingraben Formation base (Scheiblingbach) versus the equivalent basal Reingraben Formation at Polzberg

with *Palaeoaphrodite*, *Machairascolex hofreiteri*; Parry and Lukeneder 2026, this volume). In contrast, at Scheiblingbach arthropods are rare (1%), including a single crustacean specimen (n 1; indet) and rare euestheriids (n 9) forming the benthic members of the marine ecosystem (Fig. 19.7). Trace fossils, as part of the epibenthic and endobenthic community, formed by tube-burrowing crustaceans or worms, are quite frequent (2%) in the basal part of the Reingraben Formation at Scheiblingbach. In contrast, no macro trace fossils were detected at Polzberg.

Nektic members are dominated at both lower Carnian (Julian 2/Ib) localities by ammonoids with the trachyceratids *Austrotrachyceras minor* and *Paratrachyceras haberfellneri*, as well as rare *Carnites floridus* (Fig. 19.6). Bed-by-bed collecting

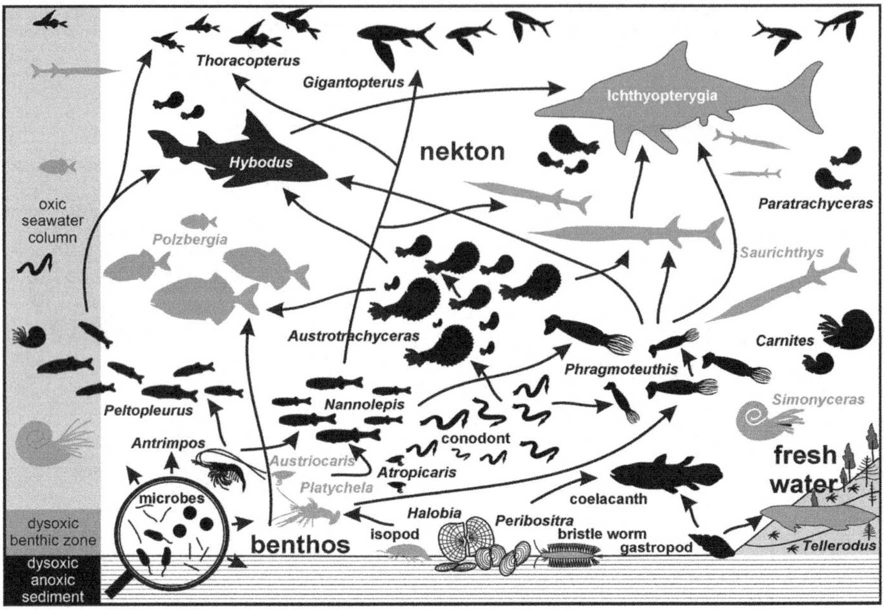

Fig. 19.7 Assumed trophic food web of the lower Carnian Scheiblingbach paleobiota (black) in the Reifling Basin based on direct and indirect evidence from the fossil record of the Scheiblingbach *Konservat-Lagerstätte*. Taxa detected only from the equivalent Polzberg paleobiota are in gray. Direct evidence from bromalites, indirect evidence as generalized interpretation of occurring members within both assemblages. Ichthyopterygia not detected yet, most likely present. Figured members of paleobiota not to scale. Adapted after Lukeneder and Lukeneder (2021)

yielded 1491 specimens of *A. minor* (67%) from Scheiblingbach and 3,075 specimens from Polzberg (58%), 10 specimens of *P. haberfellneri* (0.5%) from Scheiblingbach and 470 specimens from Polzberg (9%), and 21 specimens of *C. floridus* (1%) from Scheiblingbach and 17 specimens from Polzberg (0.3%). Rare *Simonyceras* (n 1) occur only at Polzberg. Frequent nektic coleoids (incl. cephalic cartilage and hooks) complete the cephalopod community: *Phragmoteuthis polzbergensis* and *Mojsisovicsteuthis* sp. are frequent at Scheiblingbach (n 22) and abundant at Polzberg (n 239). Within the layers of the Scheiblingbach *Konservat-Lagerstätte*, nektic actinopterygid fish (10%) appear throughout the section either entire, with fragmented carcasses or as single scales of *Nannolepis elegans*, *Peltopleurus gregarius* and *Thoracopterus*. At Polzberg actinopterygian fishes (6%) were collected with numerous , *Nannolepis elegans* and *Peltopleurus gregarius* along with the flying fish *Thoracopterus polzbergensis* and *Thoracopterus telleri* (Abel 1906; Lehman 1979). *Saurichthys calcaratus* is frequent, *Polzbergia brochatus* rare (n 3; only in historical collections). Other taxa from Polzberg include remnants of several coelacanthid fish (n 5; entire "*Coelacanthus*" *lunzensis*, only historical collection; Reis 1900) and a single specimen (skull) of the lungfish *Tellerodus sturii* (n 1). At Scheiblingbach only coelacanth scales (n 8) have been detected so far. A single heterodont chondrichthyan tooth has been detected from the Scheiblingbach *Konservat-Lagerstätte*, whereas the presence of *Acrodus* as noted by Krystyn (1991) and Lukeneder and Lukeneder (2021, 2023) cannot be confirmed. Conodont clusters, mainly with *Mosherella* specimens, were so far observed only within the Polzberg paleocommunity.

Fossil communities from both the Scheiblingbach and the Polzberg *Konservat-Lagerstätte* show a characteristic preservation of benthic (epifaunal and infaunal) species along with nektic taxa such as fishes, ammonoids and belemnoids (Fig. 19.6). The completeness of the bivalves and their abundant butterfly preservation (still articulated, little fragmentation), ammonoids along with in situ anaptychi (= lower jaws) in or near body chambers of trachyceratid ammonoids, and entire belemnoid specimens with arm-hook crowns point to an autochthonous deposition of the animals or their remains with minimal or no subsequent post-mortem drift or transport. Exceptional soft body preservation (cephalic cartilage, muscle scars), in situ jaws (anaptychi and upper jaws), the completeness of shells and carcasses, and no or restricted bioturbation suggest that both paleobiotas developed in oxygen-depleted basinal waters (Ekdale and Mason 1988; Savrda and Bottjer 1991; Wignall and Hallam 1991; Lukeneder et al. 2020) during calm conditions in the water column and near the seafloor, without major transport or reorientation of the fossil material.

Comparable sites of Triassic age with similar facies and preservational features are rare, but known from Seefeld in Tyrol (Brandner and Poleschinski 1986) or Wiestal in Salzburg (Hornung et al. 2019), both within the Norian Seefeld Formation, and from the Middle Triassic (late Anisian to Ladinian) *Konservat-Lagerstätte* of Monte San Giorgio (Ticino, Switzerland; Brinkmann 1982, 1997; Furrer 2003, 2024; Furrer and Vandelli 2014). The Monte San Giorgio *Lagerstätte* and UNESCO World Heritage Site along with the detected paleobiota are best summarized in the recently published Special Issue by Klug et al. (2024a, b).

Recent findings from field campaigns in 2023 and 2024 encompass thousands of new fossils including well-preserved vertebrate, invertebrate and plant fossils of the Scheiblingbach *Konservat-Lagerstätte*. They mirror the trophic conditions of the paleo-food web at the time of deposition (Seilacher 1970; Briggs 2014; Lukeneder and Lukeneder 2021, 2023; Fig. 19.7). At that time (basal Reingraben Formation), oxygen-depleted bottom water conditions prevailed. Such specific conditions at the sediment–water boundary and within the sediment reduce or prevent the activity of benthic scavengers or bacteria on fragmenting or decomposing shells or carcasses lying on the seafloor (Tintori 1992; Furrer 2003; Beardmore and Furrer 2016; Lukeneder and Lukeneder 2021, 2023; Lukeneder et al. 2024). Additional and frequent findings of bitten shell fragments crushed by nektic predators (Lukeneder and Lukeneder 2022a; Lukeneder et al. 2024a) – along with numerous bromalites (coprolites and regurgitalites, n 258 12%, Polzberg n 534 10%; Hunt and Lucas 2012; Hunt et al. 2012, Lukeneder et al. 2020; Lukeneder and Lukeneder 2021) – shed light on the Scheiblingbach paleobiota. As shown for larger coprolite masses from Polzberg (Lukeneder et al. 2020; Lukeneder and Lukeneder 2022a), entire shells and shell fragments of ammonoids within the bromalite masses are solely from trachyceratid species such as *A. minor* and *P. haberfellneri* (Lukeneder and Lukeneder 2022a). Coleoid fragments with cephalic cartilage and hooks belong exclusively to the belemnoid *Phragmoteuthis polzbergensis* (Lukeneder et al. 2024). Abundant coprolites are clearly dominated by fish scales, followed by crushed and fragmented ammonoid shells and halobiid bivalves, rarely coleoid hooks or carbonized cartilage remains (Lukeneder and Lukeneder 2022b). This suggests that durophagous fishes fed on these cephalopods and fishes (Griffith 1977; Lukeneder et al. 2020; Lukeneder and Lukeneder 2021, 2022a). As shown for the similar paleobioata of the coeval Polzberg *Konservat-Lagerstätte* (Lukeneder et al. 2020; Lukeneder and Lukeneder 2021, 2022a), bottom-dwelling and scavenging arthropods with thylacocephalans (*Austriocaris, Atropicaris, Paraostenia*), crustaceans (decapod *Platychela*, isopods *Obtusotelson, Discosalaputium*), and polychaetes (*Palaeoaphrodite* sp., eunicid polychaete) occurred near or at the seafloor.

A single Mesozoic dipnoan – the lungfish *Tellerodus sturii* (restricted to freshwater) from historical collections (n 1; Stur 1886; Teller 1891; Lukeneder and Lukeneder 2021) – was found only at Polzberg. This specimen is commonly interpreted as the result of post-mortem transport from freshwater ecosystems (Stur 1886; Teller 1891; Lukeneder and Lukeneder 2021, 2023). Conchostracans appear with rare *Euestheria* within the Scheiblingbach paleobiota (n 9, 0.4%) and more frequently in the Polzberg *Konservat-Lagerstätte* (n 111, 2%). Euestheriids typically inhabit freshwater or at least brackish environments (Tintori 1990; Montagna et al. 2018; Lukeneder and Lukeneder 2021, 2023). Both elements – the lungfish *Tellerodus* and the conchostracan shells – indicate a sporadic influx of freshwater or sedimentation from surrounding shallow-water or terrestrial areas into the marginal areas of the restricted Reifling Basin with the Scheiblingbach and Polzberg subbasins (Fig. 19.8a). Plant remains with abundant foliated trunks of the Coniferophyta *Voltzia* support the interpretation of sporadic freshwater influx into the otherwise fully marine basin.

Bromalite contents along with fatal injuries to ammonoids and phragmoteuthids document the prevailing predator–prey relationships among ammonoids (ceratitids), belemnoids (phragmoteuthids) and actinopterygian fishes. This is valid both for the Scheiblingbach and the comparable Polzberg paleobiota within the same intraplatform basin. Bromalites (coprolites, regurgitalites) and ammonoid shell fragment clusters confirm that ammonoid shells were frequently fatally bitten by actinopterygian fish such as *Saurichthys* or belemnoids. The new finding of a hybodont chondrichthyan tooth brings a new predator into the lower Carnian Scheiblingbach paleobiota and trophic food web (Figs. 19.7 and 19.8b). In both marginal basin areas, Scheiblingbach and Polzberg, calm and stagnant environments with oxygen-depleted conditions were hostile to scavengers at the seafloor (Furrer 2003), creating the prerequisites for excellent preservation of entire fish carcasses, in situ buccal masses, cephalic cartilages, muscle scars, ink sacs in the lower Carnian (Julian 2/Ib).

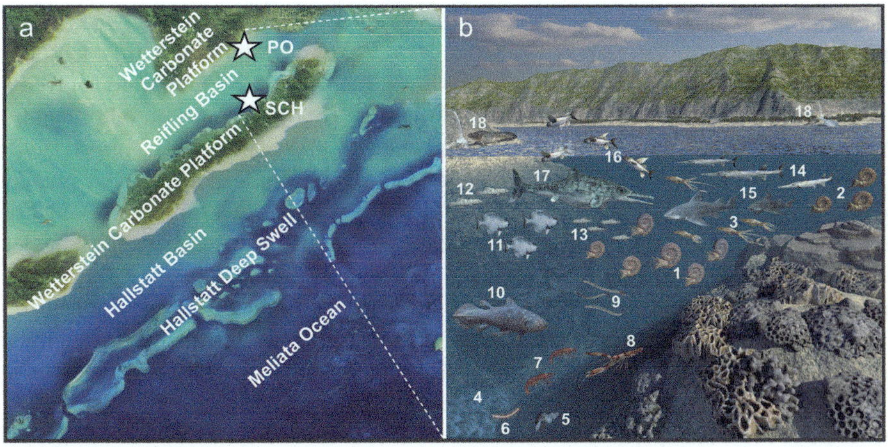

Fig. 19.8 **a** Detailed paleogeography of the Reifling Basin, with the Scheiblingbach (SCH) and Polzberg locality (PO) (white asterisks), with the laminated Reingraben deposits during the Upper Triassic, lower Carnian. **b** Reconstruction of the paleoenvironment and members of the lower Carnian Scheiblingbach and Polzberg paleobiotas. 1 *Austrotrachyceras minor*, 2 *Paratrachyceras haberfellneri*, 3 *Phragmoteuthis polzbergensis* and *Mojsisovicsteuthis*, 4 *Peribositra*, 5 caenogastropod, 6 *Palaeoaphrodite* Po, 7 *Antrimpos* Po, 8 *Platychela trauthi*, 9 conodonts Po, 10 *"Coelacanthus" lunzensis*, 11 *Polzbergia brochatus* Po, 12 *Peltopleurus gregarius*, 13 *Nannolepis elegans*, 14 *Saurichthys calcaratus* Po, 15 *Hybodus*, 16 *Thoracopterus polzbergensis*, 17 potential Ichthyopterygia, 18 *Tellerodus sturii* Po. Po points to taxa found only at Polzberg. Figured members of paleobiota not to scale. **a** Artwork by M Harzhauser, NHMW. **b** Artwork based on Fig. 19.7 by AL, Final artwork by 7reasons; www.7reasons.net; adapted after Lukeneder and Lukeneder (2021). **a** and **b** not to scale

19.6.3 The Paleoenvironment of the Scheiblingbach Konservat-Lagerstätte

The basal Reingraben Formation deposits at Scheiblingbach appear with dark gray, laminated layers. These were formed in a relatively deep marine environment within an intraplatform basin, the Reifling Basin (Lukeneder and Lukeneder 2021, 2023; Fig. 19.8a). This interpretation is inferred from the dominance of benthic posidoniids and a diverse nektic fauna with abundant ammonoids and actinopterygian fishes (Griffith 1977; Forchielli and Pervesler 2013; Schultz 2013; Lukeneder et al. 2020; Lukeneder and Lukeneder 2021, 2023). The bottom water conditions in the entire Reifling Basin during the deposition of the Reingraben limestones and marls were dysoxic to anoxic (Fig. 19.7). This conclusion is based on the preservation of frequent entire fish carcasses with fragile skeletons (Tintori 1992), masses of double-valved posidoniid bivalves, in situ buccal masses (lower jaws) in ammonoids, ammonoid soft parts (carbonization), high levels of organic material in the sediment, the presence of framboidal pyrite, the absence of sessile benthic organisms, and the rareness or even lack of bioturbation (Glaessner 1931, 1965; Lukeneder et al. 2020; Lukeneder and Lukeneder 2021, 2022a, b). The Polzberg subbasin within the Reifling Basin was mainly marine, with ephemeral and limited freshwater input (Griffith 1977; Lukeneder and Lukeneder 2021, 2022a, b). In Scheiblingbach, low energy in the bottom water system, with limited or absent bottom water currents on the seafloor, along with anaerobic to dysaerobic conditions, prevented predators from separating the ammonoid shells from the jaw apparatuses (cf. Tintori 1992; Lukeneder and Tanabe 2002; Lukeneder 2015). This led to the extraordinary preservation of the paleobiota, a situation well known from the lower Carnian Polzberg equivalent. Based on equivalent geochemical data from the Polzberg site (Lukeneder et al. 2024), the oxygen availability was highly variable during the deposition of the lower Carnian basal Reingraben deposits of this basin. Except for the transported freshwater branchiopod *Euestheria* and frequent, occasionally accumulated terrestrial plant material (*Voltzia*), the autochthonous character of the Polzberg deposits is strengthened by the preservation of fragile parts and the extraordinary preservation of in situ buccal (anaptychi) associations within or close to the body chambers of *Austrotrachyceras and Paratrachyceras*. The geochemical results (Lukeneder et al. 2024, see Weitz et al. 2026, this volume), along with the laminated fabric and facies, abundant organic matter and high amounts of sulphur, indicate that the assemblage was deposited under conditions of intermittent oxygen depletion associated with stable water masses. A dynamic environment, controlled by short- and long-term fluctuations in oxygen levels, along with poor circulation of bottom-water currents within an isolated, basin-like region, led to the accumulation of the *Austrotrachyceras* abundance zone in the lowermost part of the Reingraben Formation (encompassing the Scheiblingbach and Polzberg *Konservat-Lagerstätten*).

19.6.4 Trophic Food Web of the Scheibligbach Paleobiota

The great number of recent findings (4602 specimens, excl. posidoniids >> 100,000) and diversity of fossil remains (32 taxonomic groups) enable detailed conclusions on the paleo-trophic levels and food chains (Figs. 19.7 and 19.8b). The marine fauna here is dominated by benthic posidoniids bivalves, nektonic actinopterygian fishes, and ceratitid ammonoids, as already shown for the Polzberg equivalent *Konservat-Lagertstätte* (Lukeneder and Lukeneder 2021, 2023). Pulses of freshwater influx are documented by numerous euestheriids (*Euestheria*), completing the picture of a diverse lower Carnian food web (Fig. 19.7). The new, abundant and high-quality fossil material confirm the data presented earlier on the food web, food chains and predator-prey relations, between fish and cephalopods in this *Konservat-Lagerstätte* by Lukeneder and Lukeneder (2021, 2023).

As these authors noted for the equivalent strata from the Polzberg *Konservat-Lagerstätte*, primary producers are represented on the seafloor mainly by algae and bacteria. Microfilms and microbes were grazed by primary consumers such as gastropods and arthropods on the bottom and filtered by benthic bivalves (Fig. 19.7). Low-level consumers were preyed upon by secondary consumers or nektic predators including actinopterygian fishes, ceratitid ammonoids, phragmoteuthids (*Phragmoteuthis*) and basal coleoids (*Mojsisovicsteuthis*). Smaller fishes and the abundant cephalopods served as prey for larger carnivorous fishes such as *Saurichthys* and hybodonts, and occasionally by coelacanths. Thylacocephalan arthropods, represented by *Atropicaris*, probably either exhibited a scavenging mode of life near the seafloor, as shown for *Ostenocaris* (Pinna 1985; Jobbins et al. 2020), or actively hunted in the water column for conodonts, small cephalopods (Broda et al. 2019) and, as shown for *Concavicaris*, on other small organisms (see also Cross et al. 2018; Broda et al. 2019). These small arthropods, in turn, were hunted by chondrichthyans or other large fishes (Broda et al. 2019; Ji et al. 2020), thus serving as an important food source for numerous fish taxa (Lukeneder and Lukeneder 2023). Valuable information is gained from bromalites, represented by coprolites mostly containing fish scales and regurgitalites with dominant bivalve and ammonoid shell fragments (Lukeneder et al. 2020; Lukeneder and Lukeneder 2023 ane references therein). The regurgitalites here were produced by large durophagous predators, but the cephalopods and arthropods appear to be too small to produce bromalites up to 100 mm in size. The rich ichthyofauna and lack of reptile remains point to predatory fishes as bromalite producers. Krystyn (1991) and Lukeneder et al. (2020) noted the occurrence of the actinopterygian predators *Elpistoichthys*, *Gigantopterus*, *Saurichthys*, *Thoracopterus*, *Nannolepis* and *Peltopleurus*. Griffith (1977) stated that the Upper Triassic ichthyofauna of the equivalent Polzberg *Konservat-Lagerstätte* is characterized by abundant flying fish (Abel 1906), which, according to that author, suggests strong predation pressure in this marine ecosystem. Griffith (1977) and (Lukeneder and Lukeneder 2021, 2023) state that most of the marine fish genera known from the lower Carnian Polzberg site were predatory, as proposed for the Scheiblingbach paleobiota herein. Upper Triassic ichthyosaurs and nothosaurs are known (Arthaber 1896; Huene 1916; Summesberger and Wagner 1972; Dzik and Sulej 2007; Neenan et al. 2013; Liu et al. 2014; Scheyer et al. 2014; Lukeneder 2025) from the Reifling Basin. No evidence is currently available for marine reptiles within either the Scheiblingbach or Polzberg paleobiota (Fig. 19.8b).

19.6.5 Taphonomic Mechanisms

Specific conditions at the seafloor in the constricted Reifling Basin (Lukeneder et al. 2020; Lukeneder and Lukeneder 2021, 2023) hindered or at least reduced alteration processes such as scavenging, decay, disarticulation, abrasion, bioerosion and corrosion. The described unique preservational features highlight the significance of the *Konservat-Lagerstätten* in general and the importance of depositional *Lagerstätten* such as Scheiblingbach and Polzberg.

Black remnants of carbonized cephalic retractor and hyponome retractor muscles were described as black areas within trachyceratitid ammonoids *A. minor* and *P. haberfellneri* from the Polzberg site (Lukeneder and Lukeneder 2022) and were also detected in Scheiblingbach. A specific conservation state of the deposits at both sites is the preservation of in situ buccal masses (upper and lower jaws) described in detail by Trauth (1935) and more recently by Lukeneder and Lukeneder (2022a) for *Anaptychus lunzensis*. Entirely preserved ammonoids that biostratinomically exhibit in situ buccal masses are interpreted as quasi autochthonous faunal elements where the intact shells sank after death of the animal and neither surfaced nor drifted far from their original habitats (Maeda and Seilacher 1996; Korn and Klug 2002; Lukeneder 2015; Yacobucci 2018). The anaptychi mentioned from Scheiblingbach and Polzberg are interpreted as non-mineralized trachyceratid lower jaws (*Anaptychus lunzensis*) and upper jaws of ceratid ammonoids (*A. minor* and *P. haberfellneri*; Lukeneder and Lukeneder 2022a, 2023). This scenario is also strengthened by the faunal (entire fish carcasses, see Tintori 1992), the sedimentological (lamination caused by absence of bioturbation, dark color) and the geochemical data (total organic carbon, pyrite; Lukeneder et al. 2024; Weitz et al. 2026, this volume).

The quantity and type of shell damage in the paleobiota are significant indicators for predator-prey interactions (Kowalewski 2002; Keupp 2012; Hoffmann and Keupp 2015; Lukeneder 2015). Fragmented shells of ammonoids and phragmoteuthids are abundant and bear sharp edges caused by attacks (Lukeneder and Lukeneder 2022a) of actinopterygian fishes or belemnoids. Vertically and subvertically preserved ammonoid shell remains from both *Konservat-Lagerstätten* are exclusively bitten fragments, isolated and separated after being attacked from the rest of the shell on the bedding planes (see also Olivero 2007; Lukeneder and Lukeneder 2022a).

Trophic interactions involve different nektic predators and numerous invertebrate prey species. Cephalopods, mainly the dominant ammonoid genera *Austrotrachyceras* and *Paratrachyceras*, as well as coleoids (*Phragmoteuthis* and *Mojsisovicsteuthis*), are both predators and prey. Generally, predation in the water column can involve several trophic levels as top predators, including ichthyosaurs, mosasaurs, nothosaurs, cartilaginous or bony fish (pycnodonts Ward 1981; holostei Mehl 1978; Richter 2009; teleosts Keupp 2000; semionitids Hoffmann and Keupp 2015). The respective invertebrates include other ammonoids, nautiloids and coleoids (Mehl 1978; Lukeneder 2015; Hoffmann and Keupp 2015; Klug et al. 2021a). Actuopaleontological analyses and observations confirm that modern cephalopods (squids, cuttlefish, octopuses) serve as prey for predatory fish worldwide (Smale 1996). As documented for a variety of Mesozoic ammonoid groups (Klompmaker et al. 2009, 2019) and for ventral bite marks (Andrew et al. 2010, 2015), most attacks were probably caused by coleoids actively preying on the living ammonoids (Lukeneder 2015; Klug et al. 2021b; Lukeneder and Lukeneder 2022a). The Scheiblingbach and Polzberg paleobiotas support this interpretation because numerous caratitid specimens are preserved with ventrally and fatally bitten shells. Phragmoteuthids, in turn, attacked the ventral parts of the ammonoids. Injuries inflicted by extant fish can be sublethal or lethal, in contrast to the always fatal bite attacks by coleoid members indicated by the fossil evidence from equivalent *Konservat-Lagerstätten*. Ammonoid shell clusters, stemming from breakage during bite attacks (Mehl 1978), are frequent in both the Scheiblingbach and Polzberg paleobiotas.

Predatory genera are dominated by actinopterygians within the fish community (Simms and Ruffell 1990; Lukeneder et al. 2020; Lukeneder and Lukeneder 2021). The largest predator of that group is the agile *Saurichthys*, which actively hunted other actinopterygians (Argyriou et al. 2015) and probably also ceratitid ammonoids. In contrast, the belemnoid *Phragmoteuthis* fed on actinopterygians and hunted small and slow austrotrachyceratid ammonoids (Lukeneder and Lukeneder 2022a). Such strategies were reported for Jurassic *Plesioteuthis* and *Trachyteuthis* (Hoffmann et al. 2020), exhibiting stomach contents with ammonoid (i.e. lamellaptychi) and actinopterygian remains. Additional evidence is available for actinopterygian predation on *Phragmoteuthis* (Přikryl et al. 2012) and for predation of coleoids on other coleoids (Klug et al. 2021a). *Saurichthys* is an important predator based on its size, abundance and predatory behavior (Lukeneder et al. 2020; Lukeneder and Lukeneder 2021).

Frequently detected bromalites contain mainly fish scales and fin fragments, all ammonoid size classes from juveniles to adults, and bivalve fragments (posidoniids). Bivalves and ammonoids are either entirely preserved or, in other accumulations, dominated by fragments to shell hash. For the two genera *Austrotrachyceras* and *Paratrachyceras*, the presence of the shell along with anaptychi and upper jaws suggests that the entire animal was swallowed (Lukeneder et al. 2020; Lukeneder and Lukeneder 2021, 2022a). Remnants of cephalic cartilage from *Phragmoteuthis polzbergensis* along with packed ceratitid ammonoids within coprolites are reported from both sites discussed here. Ammonoid shell accumulations consisting of phragmocones or fragments have been described in the literature and interpreted as reflecting predation (Lukeneder 2015; Lukeneder et al. 2020). Research on bromalites, stomach contents and paleopathologies in ammonoids (Hoffmann et al. 2020) led to the conclusion that coleoids (e.g., *Plesioteuthis* and *Trachyteuthis*) were among the key predators of ammonoids in Mesozoic marine ecosystems (Mehl 1978; Klompmaker et al. 2009; Andrew et al. 2010, 2015; Klug et al. 2021a, b).

The excellent preservation of entire organisms along with corresponding bromalites in such specific *Konservat-Lagerstätten* documents trophic interactions (food chains and food web) and predator-prey relationships (Vullo 2011; Přikryl et al. 2012; Lukeneder et al. 2020; Lukeneder and Lukeneder 2021; Klug et al. 2021a, b, Lukeneder and Lukeneder 2022a). Nektic members of the Carnian assemblage such as ammonoids (trachyceratids), coleoids (*Phragmoteuthis* and *Mojsisovicsteuthis,*) or fish (mostly actinopterygians) form the main predators of the Scheiblingbach and Polzberg paleobiotas and paleo-food webs (Lukeneder et al. 2020; Fig. 19.7).

19.7 Conclusions

New excavations with bed-by-bed collecting from the Scheiblingbach *Konservat-Lagerstätte* were done during four field campaigns in 2022 and 2025. Data on thousands of new fossils from the Late Triassic (early Carnian) yielded new details and insights into a new paleocommunity, introducing the Scheiblingbach *Konservat-Lagerstätte*. We show that the Scheiblingbach locality encompasses producers, consumers, as well as small and large predators in the Reifling intraplatform basin from the Carnian Pluvial Episode. This *Konservat-Lagerstätte* is equivalent in age and paleobiota to the well-known Polzberg *Konservat-Lagerstätte* within the same marine basin. The basal sediments of the Reingraben Formation were deposited at an epipelagic to upper mesopelagic position at the seafloor of the NE/SW elongated Reifling Basin, periodically disconnected from oxygenated bottom currents. Low-oxygen ecosystems with limited bottom currents prevail at both the Scheiblingbach and Polzberg sites. Remarkable similarities are detected between the two paleobiotas. The faunal assemblage and specific taxa were found in comparable composition and numbers. Thin-shelled posidoniid bivalves were the dominant (up to rock-forming) epifaunal elements in the lower Carnian deposits of the Reifling Basin, which contains the fossiliferous Scheiblingbach and Polzberg *Konservat-Lagerstätten*. Ceratitid nektic/nektobenthic ammonoids and nektic actinopterygian fishes prevailed in the overlying oxygenated water column. Abundant fishes (actinopterygians) and frequent belemnoids (phragmoteuthids) dominated among nektic predators. Nektic predators produced numerous shell fragments of invertebrate taxa (ammonoids, belemnoids, bivalves). Along with abundant bromalites (coprolites, regurgitalites), this sheds light on the trophic levels and predator-prey relations in the lower Carnian water column. Processes such as fragmentation and decay caused by scavengers and grazers (thylacocephalan arthropods, crustaceans, polychaetes) were minimal on the seafloor at both sites at the time of deposition (Julian2/Ib). The new Scheiblingbach fossil assemblage provides new insights into the Late Triassic food webs of this lower Carnian marine ecosystem, deposited during the CPE. The preservation of fragile skeletons (nektic and benthic taxa), the excellent preservation (entire carcasses, soft body) and rare to missing bioturbation (trace fossils) point to unique paleoenvironmental conditions in the lower Carnian dysoxic bottom water at the marginal edges of the Reifling Basin. The occasional freshwater influx from the surrounding Wetterstein Platform resulted in terrigenous siliciclastic and argillaceous deposition, accompanied by terrestrial plant material and frequent freshwater branchiopods.

Acknowledgments The author is particularly grateful to the owners of the outcrop area, the Austrian Federal Forests Company (Österreichische Bundesforste), for sampling and digging permission. Especially Lukas Nöst (forest district Großreifling) and Dominic Asztaller (forest dristrict Großreifling) for their advice and help during the field campaigns. I thank Petra Lukeneder (Gablitz), Veronika, N.I. Koukal (Vienna) and Justine and Umberto Uprimny (Pichlern) for their help with cleaning outcrops and sampling fossils. The work was done within the framework of and financially supported by projects of the Federal Government of Lower Austria (K3-F-964/001-2020 and K3-F-964/002-2023, Department Science and Research; headquarters in St. Pölten); the Austrian Academy of Sciences (headquarters in Vienna, project Polzberg Lukeneder 2021) represented by the National Committee for Geo/Hydro Sciences (Earth System Sciences Programme); the Triassic project "Fenster in die Karnische Krise" of the Society of Friends of the Natural History Museum Vienna (Freunde des Naturhistorischen Museums Wien, CPE Lukeneder 2023); and the Natur- und UNESCO Geopark Steirische Eisenwurzen (St. Gallen, project Karnische Krise Lukeneder 2024). The manuscript greatly benefited from valuable comments by reviewers Michael Nagl (Vienna) and Petra Lukeneder (Vienna), the production editor Ashok Arumairaj (Taramani), the in-house series editor Kathleen A. Ritterbush (Utah), the publishing editor Simon Shah-Rohlfs and the production editor Antje Endemann (both Heidelberg).

Author Contribution AL collected new material, designed research, provided documentation on fossil material and wrote the paper.

Funding Open access funding enabled and projects financially supported by the Federal Government of Lower Austria (K3-F-964/001-2020 and K3-F-964/002-2023, Department Science and Research; headquarters in St. Pölten); the Triassic project "Fenster in die Karnische Krise" of the Society of Friends of the Natural History Museum Vienna (Freunde des Naturhistorischen Museums Wien, CPE Lukeneder 2023); the Natural History Museum Vienna; and the Natur- und UNESCO Geopark Steirische Eisenwurzen (St. Gallen, project Karnische Krise Lukeneder 2024) (to AL). The author is responsible for the contents of this publication. The funders had no impact on conceptualization, design, data collection, analysis, decision to publish, or preparation of the manuscript.

Competing interests All authors stated that they have no competing interests.

Authors' contributions The author states that he has read, edited and agreed to submit the final version of the contribution.

Availability of data and materials All sources of information are provided in the text. Raw data related to the fossil material from Austria are available from the corresponding authors and Alexander Lukeneder upon request.

Additional information The article is part of the Special Volume: *Late Triassic Konservat-Lagerstätten within the Carnian Pluvial Episode. 150 Years of Scientific Excavations in Austria.*

Artificial Intelligence (AI) The chapter was produced without the use of AI.

Language The entire volume uses American English reviewed by a native scientist.

Received 2025.

Published online 2026.

References

Abel O (1906) Fossile Flugfische. Abh K-K Geol RA 56:1–88. https://doi.org/10.5962/bhl.title.61007

Alverà G, Dal Corso J, Chu D, Cruciani G, Roghi G, Caggiati M, Song H, Song H, Tian L, Du Y, Haserbek T, Gianolla P (2026) Episodic aridification at the onset of the Carnian pluvial episode in Western Tethys. Glob Planet Change 256:105192. https://doi.org/10.1016/j.gloplacha.2025.105192

Andrew C, Hoew P, Paul CRC, Donovam SK (2010) Fatally bitten ammonoids from the lower Lias group (Lower Jurassic) of Lyme Regis, Dorset. Proc Yorks Geol Soc 58(2):81–94. https://doi.org/10.1144/pygs.58.1.276

Andrew C, Howe P, Paul C (2015) Fatally bitten ammonites from septarian concretions of the 'Marston Marble' (Lower Jurassic, Sinemurian), Somerset, UK, and their taphonomy. Proc Geol Assoc 126(1):119–129. https://doi.org/10.1016/j.pgeola.2014.10.006

Allison PA (1988) Konservat-Lagerstätten: cause and classification. Paleobiology 14:331–344

Argyriou A, Clauss M, Maxwell EE, Furrer H, Sánchez-Villagra MR (2015) Exceptional preservation reveals gastrointestinal anatomy and evolution in early actinopterygian fishes. Sci Rep 6:18758. https://doi.org/10.1038/srep18758

Arthaber G (1896) Die Cephalopodenfauna der Reiflinger Kalke. Beitr Paläontol Geol Österr-Ungarn Orient 10:1–112. https://doi.org/10.5962/bhl.title.45307

Beardmore S, Furrer H (2016) Taphonomic analysis of *Saurichthys* from two stratigraphic horizons in the Middle Triassic of Monte San Giorgio, Switzerland. Swiss J Geosci 109:1–16. https://doi.org/10.1007/s00015-015-0194-z

Bernardi M, Avanzini M, Bizzarini F (2011) Vertebrate fauna from the San Cassiano Formation (early Carnian) of the Dolomites region. Geo Alp 8:122–127

Brandner R, Poleschinski W (1986) Stratigraphie und Tektonik am Kalkalpensüdrand zwischen Zirl und Seefeld in Tirol. Jber Mitt Oberrhein Geol Ver 68:67–92. https://doi.org/10.1127/jmogv/68/1986/67

Briggs DEG (2014) Konservat-Lagerstätten 40 years on: the exceptional becomes mainstream. In: Laflamme M, Schiffbauer JD, Darroch SAF (eds) Reading and writing of the fossil record: preservational pathways to exceptional fossilization. Paleontological society papers, vol 20, pp 1–14. https://doi.org/10.1017/S1089332600002771

Brinkmann W (1982) Die Ichthyosaurier (Reptilia) aus der Grenzbitumenzone (Mitteltrias) des Monte San Giorgio (Tessin, Schweiz) – neue Ergebnisse. Vierteljahresschr Naturforsch Ges Zürich 143(4):165–177

Brinkmann W (1997) Die Ichthyosaurier (Reptilia) aus der Mitteltrias des Monte San Giorgio (Tessin, Schweiz) und von Besano (lombardei, Italien) – der aktuelle Forschungsstand. Vierteljahresschr Naturforsch Ges Zürich 142(2):69–78

Broda K, Marynowski L, Rakociński M, Zatoń M (2019) Coincidence of photic zone euxinia and impoverishment of arthropods in the aftermath of the Frasnian-Famennian biotic crisis. Sci Rep 9:16996. https://doi.org/10.1038/s41598-019-52784-4

Cross SRR, Ivanovski N, Duffin CJ, Hildebrandt C, Parker A, Benton MJ (2018) Microvertebrates from the basal Rhaetian Bone Bed (latest Triassic) at Aust Cliff, S.W, England. Proc Geol Assoc 129:635–653. https://doi.org/10.1016/j.pgeola.2018.06.002

Dal Corso J, Bernardi M, Sun Y, Song H, Seyfullah LJ, Preto N, Gianolla P, Ruffell A, Kustatscher E, Roghi G, Merico A, Hohn S, Schmid, AR, Marzoli A, Newton RJ, Wignall PB, Benton MJ (2020) Extinction and dawn of the modern world in the Carnian (Late Triassic). Sci Adv 6(38):eaba0099.https://doi.org/10.1126/sciadv.aba0099

Dal Corso J, Ruffell A, Preto N (2018) The Carnian Pluvial Episode (Late Triassic): new insights into this important time of global environmental and biological change. J Geol Soc Lond 175:986–988. https://doi.org/10.1144/jgs2018-185

Dzik J, Sulej T (2007) A review of the early Late Triassic Krasiejów biota from Silesia, Poland. Palaeontol Pol 64:3–27

Ekdale AA, Mason TR (1988) Characteristic trace-fossil associations in oxygen-poor sedimentary environments. Geology 16:720–723. https://doi.org/10.1130/0091-7613(1988)016%3c0720:CTFAIO%3e2.3.CO;2

Forchielli A, Pervesler P (2013) Phosphatic cuticle in thylacocephalans: a taphonomic case study of (Arthropoda, Thylacocephala) from the *Fossil-Lagerstätte* Polzberg (Reingraben shales, Carnian, Upper Triassic, Lower Austria). AJES 106:46–61

Furrer HD (2003) Der Monte San Giorgio im Südtessin – vom Berg der Saurier zur Fossil-Lagerstätte internationaler Bedeutung. Neujahrsbl Natur Ges Zürich 206:1–64

Furrer H, Vandelli A (2014) Guide to the Museum of fossils from Monte San Giorgio Meride. Fondazione del Monte San Giorgio, 1–127

Furrer HD (2024) The history of palaeontological research and excavations at Monte San Giorgio. Swiss J Palaeontol 142:18. https://doi.org/10.1186/s13358-024-00314-9
Glaessner MF (1931) Eine Crustaceen fauna aus den Lunzer Schichten Niederösterreichs. Jahrb Geol BA 81:467–486
Glaessner MF (1965) Vorkommen fossiler Dekapoden (Crustacea) in Fisch-Schiefern. Senckenb Lethaea 46a:111–122
Griffith J (1977) The Upper Triassic fishes from Polzberg bei Lunz. Zool J Linn Soc 60:1–93
Harzhauser M, Lukeneder A, Lukeneder P, Mandic O (2026) Potentialmodulation of a Late Triassic Pangean Megamonsoon by solar cycles tracing the Carnian Pluvial Episode. In: Lukeneder A (ed) Late Triassic Konservat-Lagerstätten within the Carnian Pluvial Episode: 150 years of scientific excavations in Austria. Top Geobiol 55, Chap 24:xx–xx. https://doi.org/10.1007/978-3-032-19777-1_24
Hoffmann R, Keupp H (2015) Ammonoid paleopathology. In: Klug C, Korn D, De Baets K, Kruta I, Mapes R (eds) Ammonoid paleobiology: from anatomy to ecology. Topics in geobiology, vol 43, pp 877–926. https://doi.org/10.1007/978-94-017-9630-9_21
Hoffmann R, Stevens K, Keupp H, Simonsen S, Schweigert G (2020) Regurgitalites—a winodw into the trophic ecology of fossil cephalopods. J Geol Soc Lond 177:82–102. https://doi.org/10.1144/jgs2019-117
Hornung T (2008) The Carnian Crisis in the Tethyan Realm. Dr. Müller, Saarbrücken, pp 1–235
Hornung T, Brandner R (2005) Biochronostratigraphy of the Reingraben turnover (Hallstatt Facies Belt): local black shale events controlled by regional tectonism, climatic change and plate tectonics. Facies 51:460–479. https://doi.org/10.1007/s10347-005-0061-x
Hornung T, Brandner R, Krystyn L, Joachimski MM, Keim L (2007a) Multistratigraphic constraints on the NW Tethyan 'Carnin Crisis'. In: Lucas GS, Spielmann AJ (eds) The global triassic. New Mexico museum of natural history & science bulletin, vol 41, pp 59–67
Hornung T, Krystyn L, Brandner RA (2007b) A Tethys-wide mid-Carnian (Upper Triassic) carbonate productivity crisis: evidence for the Alpine Reingraben event from Spiti (Indian Himalaya)? J Asian Earth Sci 30(2):285–302. https://doi.org/10.1016/j.jseaes.2006.10.001
Hornung T, Kogan I, Wolf G, Moosleitner G (2019) The Norian fish deposits of Wiestal ("Seefeld Member", Northern Calcareous Alps, Salzburg, Austria)—taxonomy and palaeoenvironmental implications. Austr J Earth Sci 112(2):125–165. https://doi.org/10.17738/ajes.2019.0008
Huene F (1916) Beiträge zur Kenntnis der Ichthyosaurier im deutschen Muschelkalk. Palaeontogr 62:1–68
Hunt AP, Lucas SG (2012) Descriptive terminology of coprolites and recent feces. In: Hunt AP, Milan J, Lucas, SG, Spielmann JA (eds) Vertebrate coprolites. New Mexico Museum of Natural History and Science, vol 57, pp 153–160
Hunt AP, Miln J, Lucas SG, Spielmann JA (2012) Vertebrate coprolites. New Mex Mus Nat Hist Sci Bull 57:1–387
Ji C, Tintori A, Jiang D, Montani R, Confortini F (2020) New Thylacocephala (Crustacea) assemblage from the Spathian (Lower Triassic) of Majiashan (Chaohu, Anhui Province, South China). J Paleontol 95(2):305–319. https://doi.org/10.1017/jpa.2020.92
Jobbins M, Haug C, Klug C (2020) First African thylacocephalans from the Famennian of Morocco and their role in Late Devonian food webs. Sci Rep 10:5129. https://doi.org/10.1038/s41598-020-61770-0
Keupp H (2000). Ammoniten. Paläobiologische Erfolgsspiralen. Jan Thorbecke Verlag, Stuttgart, pp 165
Keupp H (2012) Atlas zur Paläopathologie der Cephalopoden. Berliner Paläobiol Abh 12:1–392
Klompmaker AA, Waljaard NA, Fraaije RHB (2009) Ventral bite marks in Mesozoic ammonoids. Palaeogeogr Palaeoclimatol Palaeoecol 280:245–257. https://doi.org/10.1016/j.palaeo.2009.06.013
Klompmaker AA, Kelley PH, Chattopadhyay D, Clements JC, Huntley JW, Kowalewski M (2019) Predation in the marine fossil record: studies, data, recognition, environmental factors, and behavior. Earth-Sci Rev 194:472–520. https://doi.org/10.1016/j.earscirev.2019.02.020

Klug C, Schweigert G, Fuchs D, De Baets K (2021a) Distraction sinking and fossilized coleoid predatory behaviour from the German Early Jurassic. Swiss J Palaeontol 140(1):7. https://doi.org/10.1186/s13358-021-00218-y

Klug C, Schweigert G, Tischlinger H, Pochmann H (2021b) Failed prey or peculiar necrolysis? Isolated ammonite soft body from the Late Jurassic of Eichstätt (Germany) with complete digestive tract and male reproductive organs. Swiss J Palaeontol 140(1):3. https://doi.org/10.1186/s13358-020-00215-7

Klug C, Scheyer TM, Klein N, Liu J, Stockar R eds (2024a) Special Issue: 100 years of scientific excavations at UNESCO World Heritage Site Monte San Giorgio and Global Research on Triassic Marine Lagerstätten. Swiss J Palaeontol 143 (Spec Issue):1–498, Springer, Basel

Klug C, Scheyer TM, Klein N, Liu J, Albisetti D, Furrer H, Stockar R (2024b) 100 years of scientific excavations at UNESCO World Heritage Site Monte San Giorgio and global research on Triassic marine Lagerstätten. Swiss J Palaeontol 143:37. https://doi.org/10.1186/s13358-024-00328-3

Korn D, Klug C (2002) Chapter 10: Ammoneae devonicae. In: Riegraf W (ed) Fossilium catalogus, animalia I: 1–375. Backhuys, Leiden

Kozur HW, Bachmann GH (2010) The middle Carnian wet intermezzo of the Stuttgart formation (Schilfsandstein), Germanic basin. Palaeogeogr Palaeoclimatol Palaeoecol 290:107–119. https://doi.org/10.1016/j.palaeo.2009.11.004

Kowalewski M (2002) The fossil record of predation: an overview of analytical methods. In: Kowalewski M, Kelley PH (eds) The fossil record of predation. Palentological Society, New Haven. Paleontological Society Papers, vol 8, pp 3–42. https://doi.org/10.1017/S1089332600001030

Kreuss O (2014) 100 Hieflau. Geologische Karte, 1: 50,000. Geofast, Geosphere Austria. https://doi.org/10.24341/tethys.101

Krystyn L (1978) Eine neue Zonengliederung des mediterranen Unterkarn. Schrift Erdwiss Komm ÖAW 4:37–75

Krystyn L (1991) Die Fossillagerstätten der alpinen Trias. In: Nagel D, Rabeder G (eds) Exkursionen im Jungpaläozoikum und Mesozoikum Österreichs. Österr Paläontol Ges Wien, pp 23–78

Lahm B (1984) Spumellarienfaunen (Radiolaria) aus den mitteltriassischen Buchensteiner-Schichten von Recoaro (Norditalien) und den obertriassischen Reiflingerkalken von Großreifling (Österreich). Systematik – Stratigraphie. Münchner Geowiss Abh A, Geol Paläontol 1:1–161, Pfeil, München

Lehman JP (1975) A propos de *Ceratodus sturii* Teller, 1891. Bull Mus Nati Hist Nat Paris 375:241–246

Lehman JP (1979) Note sur les Poissons du Trias de Lunz. I Thoracopterus Bronn. Ann Naturhist Mus Wien 82:53–66

Liu J, Sx H, Rieppel O, Jiang D, Benton MJ, Kelley NP, Aitchison JC, Zhou C, Wen W, Huang J, Xie T, Lv T (2014) A gigantic nothosaur (Reptilia: Sauropterygia) from the Middle Triassic of SW China and its implication for the Triassic biotic recovery. Sci Rep 4:7142. https://doi.org/10.1038/srep07142

Lukeneder A (2015) Ammonoid habitats and life history. In: Klug C, Korn D, De Baets K, Kruta I, Mapes RH (eds) Ammonoid paleobiology: from anatomy to ecology, Top Geobiol 43. Springer, Dordrecht, pp 689–791. https://doi.org/10.1007/978-94-017-9630-9_18

Lukeneder A (2025) 6.6 Rätsel um Fischsaurier von Großreifling. – der verschwunde Kopf. In Kirchmayer G, Lukeneder A, Riedl WC (eds) UNESCO Global Geopark Steirische Eisenwurzen. Geoparks. Springer, Berlin, pp 109–115

Lukeneder A (ed) (2026) Late Triassic Konservat-Lagerstätten within the Carnian Pluvial Episode: 150 years of scientific excavations in Austria. Top Geobiol 55:600 p. https://doi.org/10.1007/978-3-032-19777-1

Lukeneder A, Lukeneder P (2021) The Upper Triassic Polzberg palaeobiota from a marine *Konservat-Lagerstätte* deposited during the Carnian Pluvial Episode in Austria. Sci Rep 11:16644. https://doi.org/10.1038/s41598-021-96052-w

Lukeneder A, Lukeneder P (2022a) Taphonomic history and trophic interactions of an ammonoid fauna from the Upper Triassic Polzberg palaeobiota. Sci Rep 12:7455. https://doi.org/10.1038/s41598-022-11496-y
Lukeneder A, Lukeneder P (2023) New data on the marine Upper Triassic palaeobiota from the Polzberg *Konservat-Lagerstätte* in Austria. Swiss J Palaeontol 142:9. https://doi.org/10.1186/s13358-023-00269-3
Lukeneder A, Lukeneder P (2024) Ammonoid taxonomy of the Carnian Polzberg *Konservat-Lagerstätte* in Austria. Jahrb Geol BA 162(1–4):27–55
Lukeneder A, Tanabe K (2002) *In situ* finds of aptychi in the Barremian of the Alpine Lower Cretaceous (Barremian, Northern Calcareous Alps, Upper Austria). Cretac Res 23:15–24. https://doi.org/10.1006/cres.2001.0303
Lukeneder A, Surmik D, Gorzelak P, Niedźwiedzki R, Brachaniec T, Salamon MA (2020) Bromalites from the Upper Triassic Polzberg section (Austria); insights into trophic interactions and food chains of the Polzberg palaeobiota. Sci Rep 10:20545. https://doi.org/10.1038/s41598-020-77017-x
Lukeneder A, Lukeneder P, Sachsenhofer RF, Roghi G, Rigo M (2024a) Multi-proxy record of the Austrian Upper Triassic Polzberg Konservat-Lagerstätte in light of the Carnian Pluvial episode. Sci Rep 14:11194. https://doi.org/10.1038/s41598-024-60591-9
Lukeneder A, Weitz J, Sachsenhofer RF, Misch D (2026) Drilling into the Carnian Pluvial episode. In: Lukeneder A (ed) Late Triassic Konservat-Lagerstätten within the Carnian Pluvial episode: 150 years of scientific excavations in Austria. Topics in geobiology (This volume)
Lukeneder P, Lukeneder A (2022b) Mineralized belemnoid cephalic cartilage from the Late Triassic Polzberg Konservat-Lagerstätte (Austria). PLoS One 17(4):e0264595. https://doi.org/10.1371/journal.pone.0264595
Lukeneder P, Fuchs D, Lukeneder A (2024b) Morphology, taxonomy and trophic interactions of rostrum-less coleoids from the Late Triassic Polzberg Konservat-Lagerstätte (Lower Austria). Swiss J Palaeontol 143:21 (2024). https://doi.org/10.1186/s13358-024-00319-4
Lukeneder S, Lukeneder A, Harzhauser M, İslamoğlu Y, Krystyn L., Lein R (2012) A delayed carbonate factory breakdown during the Tethyan-wide Carnian Pluvial episode along the Cimmerian terranes (Taurus, Turkey). Facies 58:279–296. https://link.springer.com/article/10.1007/s10347-011-0279-8
Maeda H, Seilacher A (1996) Ammonoid taphonomy. In: Landman NH, Tanabe K, Davis RA (eds) Ammonoid paleobiology. Topics in geobiology, vol 13, pp 543–578. https://doi.org/10.1007/978-1-4757-9153-2_14
Martin M (1982) Revision von *Tellerodus sturii* (Teller) 1891. Verh Geol BA 1982(2):21–31
Mehl J (1978) Anhäufungen scherbenartiger Fragmente von Ammonitenschalne im süddeutschen Lias und Malm und ihre Deutung als Frassreste. Ber Naturf Ges Freiburg Breisgau 68:75–93
Mojsisovics JAE (1893a) Die Cephalopoden der Hallstätter Kalke. Band I. Abh K-K Geol RA 6:1–835 p
Mojsisovics JAE (1893b) Die Cephalopoden der Hallstätter Kalke. Band II. Abh K-K Geol RA 6:1–524 p
Montagna M, Strada L, Dioli P, Tintori A (2018) The Middle Triassic lagerstätte of Monte San Giorgio reveals the oldest lace bugs (Hemiptera: Tingidae): archetingis ladinica gen. n. sp. n. Riv Ital die Paleont Strat 124:35–44. https://doi.org/10.13130/2039-4942/9623
Mostler H, Scheuring BW (1974) Mikrofloren aus dem Langobard und Cordevol der Nördlichen Kalkalpen und das Problem des Beginns der Keupersedimentation im Germanischen Raum. Geol Paläont Mitt Innsbruck 4(4):1–35
Mueller S, Krystyn L, Kürschner WM (2016) Climate variability during the Carnian Pluvial phase—a quantitative palynological study of the Carnian sedimentary succession at Lunz am See, Northern Calcareous Alps, Austria. Palaeogeogr Palaeoclimatol Palaeoecol 441:198–211. https://doi.org/10.1016/j.palaeo.2015.06.008

Neenan JM, Klein N, Scheyer TM (2013) European origin of placodont marine reptiles and the evolution of crushing dentition in Placodontia. Nat Commun 4:1621. https://doi.org/10.1038/ncomms2633

Ogg JG (2015) The mysterious Mid-Carnian "Wet Intermezzo" global event. J Earth Sci 26:181–191. https://doi.org/10.1007/s12583-015-0527-x

Olivero EB (2007) Taphonomy of ammonites from the Santonian-Lower Campanian Santa Marta formation, Antarctica: sedimentological controls on vertically embedded ammonites. Palaios 22:586–597. https://doi.org/10.2110/palo.2005.p05-118r

Parry L, Lukeneder A (2026) Polychaete annelids from the Polzberg Konservat-Lagerstätte. In: Lukeneder A (ed) Late Triassic Konservat-Lagerstätten within the Carnian Pluvial Episode: 150 years of scientific excavations in Austria. Top Geobiol 55; Chap 9:149–170. https://doi.org/10.1007/978-3-032-19777-1_9

Pinna G (1985) Exceptional preservation in the Jurassic of Osteno. Philos Trans R Soc Lond B 311:171–180. https://doi.org/10.1098/rstb.1985.0149

Přikryl T, Košťák M, Mazuch M, Mikuláš R (2012) Evidence for fish predation on a coleoid cephalopod from the Lower Jurassic Posidonia Shale of Germany. Neues Jahrb Geol Paläontol Abh 263:25–33. https://doi.org/10.1127/0077-7749/2012/0206

Reis OM (1900) *Coelacanthus lunzensis* Teller. Jb K-K Geol RA 50(2):187–192

Richter AE (2009) Ammoniten-Gehäuse Mit Bissspuren. Berliner Paläontol Abh 10:297–305

Rigo M, Preto N, Roghi G, Tateo F, Mietto P (2007) A rise in the Carbonate compensation depth of western Tethys in the Carnian (Late Triassic): deep-water evidence for the Carnian Pluvial event. Palaeogeogr Palaeoclimatol Palaeoecol 246(2–4):188–205. https://doi.org/10.1016/j.palaeo.2006.09.013

Rigo M, Mazza M, Karádi V, Nicora A (2018) A new Upper Triassic conodont biozonation of the Tethyan Realm. In: Tanner LH (ed) The Late Triassic world. Top Geobiol 46:189–235. https://doi.org/10.1007/978-3-319-68009-5_6

Ruffell A, Simms MJ, Wignall PB (2015) The Carnian humid episode of the late Triassic: a review. Geol Mag 153:271–284. https://doi.org/10.1017/S0016756815000424

Rusev C, Lukeneder A (2026) Clam shrimp of the Upper Triassic Polzberg *Konservat-Lagerstätten* In: Lukeneder A (ed) Late Triassic Konservat-Lagerstätten within the Carnian Pluvial episode: 150 years of scientific excavations in Austria. Top Geobiol 55, Chap 5:67–85. https://doi.org/10.1007/978-3-032-19777-1_5

Savrda CE, Bottjer DJ (1991) Oxygen-related biofacies in marine strata: an overview and update. Geol Soc London Spec Publ 58:201–219. https://doi.org/10.1144/gsl.sp.1991.058.01.14

Schädel M, Eldijk T, Winkelhorst H, Reumer JWF, Haug J (2020) Triassic Isopoda—three new species from Central Europe shed light on the early diversity of the group. Bull Geosci 95(2):145–166. https://doi.org/10.3140/bull.geosci.1773

Scheyer TM, Romano C, Jenks J, Bucher H (2014) Early Triassic Marine biotic recovery: the predators' perspective. PLoS One 9(3):e88987. https://doi.org/10.1371/journal.pone.0088987

Schlager W, Schöllnberger W (1974) Das Prinzip stratigraphischer Wenden in der Schichtfolge der Nördlichen Kalkalpen. Mitt Geol Ges Wien 66:165–193

Schultz O (2013) Pisces. In: Piller WE (ed) Austrian academy of sciences. Catalogus Fossilium Australia, vol 3, 576 p

Seilacher A (1970) Begriff und Bedeutung der Fossil-Lagerstätten. N Jahrb Geol Paläontol Mh 1970:34–39

Simms MJ, Ruffell AH (2018) The Carnian Pluvial episode: from discovery, through obscurity, to acceptance. J Geol Soc 175(6):989–992. https://doi.org/10.1144/jgs2018-020

Simms MJ, Ruffell AH (1990) Climatic and biotic change in the late Triassic. J Geol Soc London 147:321–327. https://doi.org/10.1144/gsjgs.147.2.0321

Smale MJ (1996) Cephalopods as prey. IV. Fishes. Philos Trans R Soc London Biol Sci B 351:1067–1081. https://doi.org/10.1098/rstb.1996.0094

Stur D (1874) Neue Aufschlüsse im Lunzer Sandsteine bei Lunz und ein neuer Fundort von Wengerschiefer im Pölzberg zwischen Lunzersee und Gaming. Verh K-K Geol RA 1:271–273

Stur D (1886) Vorlage des ersten fossilen Schädels von Ceratodus aus den ober triadischen Reingrabner Schiefern von Pölzberg nördlich bei Lunz. Verh K-K Geol RA 15:381–383. https://www.zobodat.at/pdf/VerhGeolBundesanstalt_1886_0381-0383.pdf
Summesberger H, Wagner L (1972) Der Stratotypus des Anis (Trias). Ann Naturhist Mus Wien 76:515–538
Teller F (1891) Über den Schädel eines fossilen Dipnoers, *Ceratodus Sturii* nov. spec. aus den Schichten der oberen Trias der Nordalpen. Abh K-K Geol RA 15:1–38. https://www.zobodat.at/pdf/AbhGeolBA_15_0001-0038.pdf
Tintori A (1990) Estherids from the Kalkschieferzone (Triassic) of Lombardy (N. Italy). In: Robba E (ed) Atti del Quarto Simposio di Ecologia e Palecologia delle Comunità Bentoniche Sorrento, 1988. Regional Museum of Natural Sciences of Torino, pp 95–105
Tintori A (1992) Fish taphonomy and Triassic anoxic basins from the Alps: a case history. Riv It Paleont Strat 97:393–408
Tollmann A (1976) Analyse des klassischen nordalpinen Mesozoikums. Franz Deuticke, Wien, 580 p
Trauth F (1935) Die Aptychen der Trias. Sitzung Akad Wissen, Math-Natur Kl, Abt 1(144):455–483
Vullo R (2011) Direct evidence of hybodont shark predation on Late Jurassic ammonites. Naturwiss 98:545–549. https://doi.org/10.1007/s00114-011-0789-9
Ward PD (1981) Shell sculpture as a defensive adaptation in ammonoids. Paleobiology 7:96–100. https://doi.org/10.1017/S009483730000381X
Weitz JA, Stojanović KA, Sachsenhofer RF, Misch D, Lukeneder A (2026) Organic petrology and geochemistry of the Scheiblinggraben section: detailed insights into the Carnian Pluvial Episode and comparison to time equivalent profiles. In: Lukeneder A (ed) Late Triassic Konservat-Lagerstätten within the Carnian Pluvial Episode: 150 years of scientific excavations in Austria. Top Geobiol 55, Chap 23:527–566. https://doi.org/10.1007/978-3-032-19777-1_23
Wignall PB, Hallam A (1991) Biofacies, stratigraphic distribution and depositional models of British onshore Jurassic black shales. In: Tyson RV, Pearson TA (eds) Modern and ancient continental shelf anoxia, vol 58. Geological Society, London, Special Publications, pp 291–309. https://doi.org/10.1144/gsl.sp.1991.058.01.19
Yacobucci MM (2018) Postmortem transport in fossil and modern shelled cephalopods. PeerJ 6:e5909. https://doi.org/10.7717/peerj.5909

Chapter 20
New Thylacocephalans from the Upper Triassic Scheiblingbach *Konservat-Lagerstätte*

Thomas Laville and Alexander Lukeneder

Abstract Thylacocephalans are puzzling fossil pancrustaceans known at least from the Silurian to the Cretaceous. In the Triassic, they were an important component of marine faunas, reaching a peak of species diversity and having a widespread distribution. In the Triassic of Austria, thylacocephalans have been known for a century in the Carnian *Polzberg-Lagerstätte*, from which four thylacocephalan taxa have been described. Recent excavations at a new Austrian lower Carnian locality, Scheiblingbach, located about 40 km from the *Polzberg-Lagerstätte*, have uncovered new thylacocephalan specimens. One taxon of Thylacocephala is described from the Scheiblingbach deposits: *Atropicaris*? sp. It represents the second occurrence of thylacocephalans in the Late Triassic of Austria and only the third occurrence of thylacocephalans in the Carnian worldwide. Compared with the *Polzberg-Lagerstätte*, Scheiblingbach displays a low taxonomic diversity of thylacocephalans, which might be the result of a taphonomic bias. In both Polzberg and Scheiblingbach deposits, the specimens of *Atropicaris*? sp. are fragmented and damaged. This might reflect predation by other organisms.

Keywords Arthropoda · Diversity · Early Carnian · Pancrustacea · Predation

20.1 Introduction

Thylacocephala is a puzzling group of fossil pancrustaceans characterized by a shield enveloping most of the body, hypertrophied compound eyes, three pairs of large raptorial appendages and a posterior trunk formed of eight to 22 elongated segments (Schram 2014). Known at least from the Silurian (Haug et al. 2014) to the Cretaceous

T. Laville (✉)
Faculty of Biology, Ludwig-Maximilians-Universität München, Planegg-Martinsried, Germany
e-mail: T.Laville@lmu.de

A. Lukeneder
Department of Geology and Paleontology, Natural History Museum Vienna, Vienna, Austria
e-mail: alexander.lukeneder@nhm.at

A. Lukeneder (ed.), *Late Triassic Konservat-Lagerstätten within the Carnian Pluvial Episode*, Topics in Geobiology 55, https://doi.org/10.1007/978-3-032-19777-1_20

(Charbonnier et al. 2017), thylacocephalans were an important component of Triassic marine faunas, being reported from Asia (Ehiro et al. 2015, 2019; Ji et al. 2017, 2021; Ehiro and Kano 2024), Australia (Haig et al. 2015), Europe (Pinna 1974; Arduini and Brasca 1984; Bürgin et al. 1991), Madagascar (Arduini 1990; Laville et al. 2021) and USA (Charbonnier and Brayard 2019; Laville et al. 2021).

In the Carnian, the fossil record of thylacocephalans is very scarce, being restricted to Slovenia (Laville et al. 2024a) and Austria. In Austria, thylacocephalans have been known for about 100 years. Glaessner (1931) was the first to report Austrian specimens. He described two species, *Austriocaris carinata* Glaessner, 1931 (Fig. 20.1a) and *Austriocaris striata* Glaessner, 1931 (Fig. 20.1c), from the Upper Triassic *Polzberg-Lagerstätte*, which represents an exceptional early Carnian marine assemblage discovered in the Northern Calcareous Alps of Lower Austria (Lukeneder and Lukeneder 2021). He assigned the specimens to Concavicarida, an euarthropod ingroup that was latter reassigned to Thylacocephala (Briggs and Rolfe 1983). More recently, Laville et al. (2024b) undertook a review of the thylacocephalans from the *Polzberg-Lagerstätte*. They reassigned *A. striata* to *Paraostenia* Secretan, 1985 and also reported two additional taxa: *Atropicaris*? sp. (Fig. 20.1b) and *Paraostenia* cf. *ambatolokobensis* (Arduini 1990) (Fig. 20.1d).

Recent excavations at a new locality, Scheiblingbach (Styria), dated from the early Carnian and located about 40 km from the *Polzberg-Lagerstätte*, revealed new

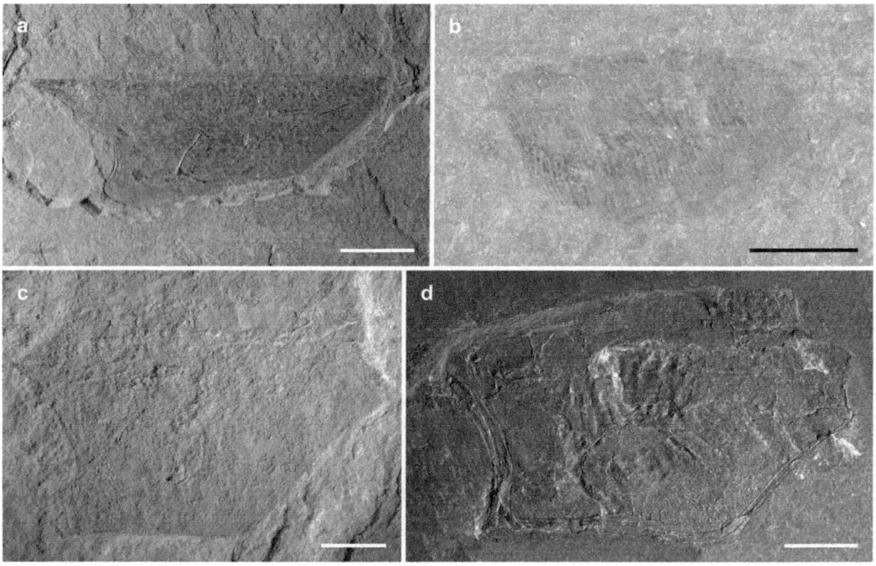

Fig. 20.1 Thylacocephalans from the Polzberg *Lagerstätte*, Austria (early Carnian, Late Triassic). **a** *Austriocaris carinata* Glaessner, 1931 (holotype NHMW 1910/0015/0041). **b** *Atropicaris*? sp. (NHMW 1910/0015/0035). **c** *Paraostenia striata* (Glaessner 1931) (holotype NHMW 1910/0015/0042). **d** *Paraostenia* cf. *ambatolokobensis* (NHMW 2024/0201/0001). Scales: **a**, 20 mm; **b**, 5 mm; **c**, **d**, 4 mm. *Photos* A. Lukeneder

thylacocephalan specimens. We here describe those new specimens and discuss the preservation of these fossils. The similarity of the Scheiblingbach thylacocephalan fauna with that from the *Polzberg-Lagerstätte* is also analyzed.

20.2 Geological Settings

The Scheiblingbachgraben-section (= Scheiblingbach) is located on the northern slope of Mount Lärchkogel (1087 m), south of Mount Reiflingberg (712 m), 600 m southwest of the Enns River, 300 m west of the Großreifling center in northern Styria (47° 39′ 52.50″ N, 14° 42′ 27.75″ E) (geological map, sheet 100 Hieflau GEOFAST, Kreuss 2014; Fig. 20.2). Within the Lunz Nappe in northernmost Styria and Lower Austria, the Reifling Basin (Krystyn 1991; Lukeneder and Lukeneder 2021; Lukeneder 2026a, b, this volume)—an intraplatform basin during the Late Triassic—is located between Polzberg and Großreifling. It exposes Upper Triassic (Carnian, Julian) deposits of the terrigeneous siliciclastic deposits of the Reingraben Formation (8.1 to 32.2 m, layers Sch 56 to Sch 32.2 m; *A. austriacum* Zone, *A. minor* Subzone) consisting of a pronounced alternation of dark gray to black argillaceous marls intercalated by rare limestone/dolomite layers. The Upper Triassic *Konservat-Lagerstätte* at Scheiblingbach comprises invertebrates and vertebrates. Most of the fossils derive from the lowermost three calcareous meters of the Reingraben Formation (for more details see Lukeneder 2026a, b, this volume).

20.3 Material and Methods

The present study is based on 14 thylacocephalan specimens. The basic fossil material is incorporated in the collections of the Naturhistorisches Museum Wien, Austria (NHMW 2012/0288/0001–0014). The material was collected during recent excavations by one of the authors (A.L). Most specimens are compressed laterally, except one specimen which is preserved in butterfly configuration (NHMW 2012/0288/0003; Fig. 20.3a–c), i.e. being dorsally-ventrally compressed. The rock matrix was removed and specimens were mechanically prepared using hammer, chisel and awls, subsequently using various types of pneumatic tools (e.g., air scribes).

Digital high-quality photomicrographs were taken using a Discovery.V20 Stereo Zeiss microscope. The magnifications were ×10, ×20 and ×40 in incident light mode. Data from the AxioCam MRc5 Zeiss were processed and documented using the AxioVision SE64 Rel. 4.9 imaging system at the NHMW. In some photographs, specimens were whitened with ammonium chloride sublimate before photography to enhance details of ornamentation. Plates and line drawings were made with Inkscape 1.2 (GNU).

Institutional abbreviations. NHMW, Naturhistorisches Museum Wien, Austria (Natural History Museum Vienna).

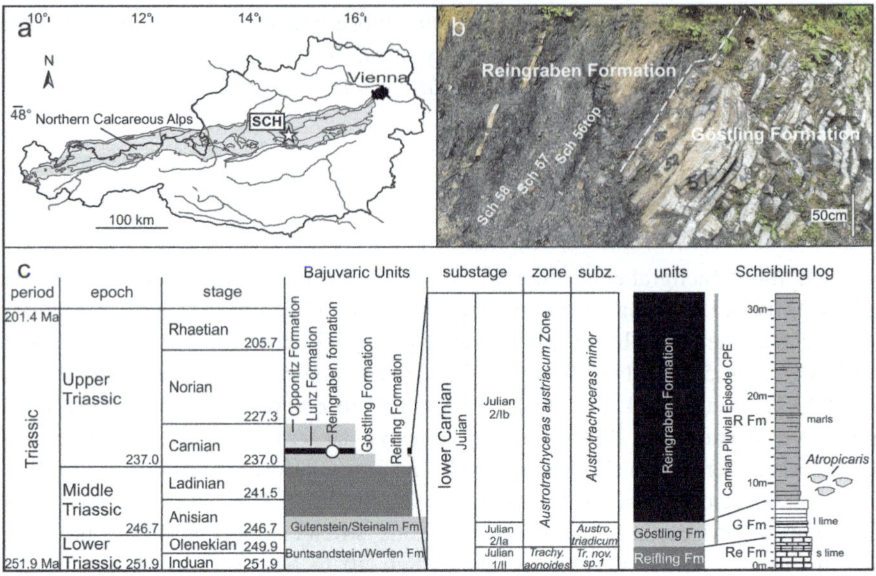

Fig. 20.2 **a** Locality map of the Scheiblingbach area (Sch, Großreifling) in Styria and the Austrian Northern Calcareous Alps (in gray). **b** Outcrop situation at Scheiblingbach with the lower detected Göstling formation and overlying argillaceous, softer Reingraben Formation. Given thylacocephalan layers (Sch 56top, Sch 57, Sch 58) and unit boundaries marked by a white dashed line. **c** Stratigraphic position of the Scheiblingbach deposits during the lower Carnian. White circle: position of the *Konservat-Lagerstätte* Scheiblingbach. Adapted after Lukeneder and Lukeneder (2021). Detailed Scheiblingbach section with indicated sampling layers (*Atropicaris* drawings) within the lower Carnian Reingraben Formation. Austro *Autrotrachyceras*, Trachy *Trachyceras*, chrono chronostratigraphy, subz subzone, Re Fm Reifling Formation, G Fm Göstling Formation, R Fm Reingraben Formation, lime limestone, s lime silicious limestone

20.4 Systematic Paleontology

Euarthropoda Bergström, 1976

Pancrustacea Zrzavý and Štys, 1997

Thylacocephala Pinna et al., 1982

Atropicaris Arduini and Brasca, 1984

Type species Atropicaris rostrata Arduini and Brasca, 1984 by monotypy—Argilliti di Riva di Solto Formation, Sevatian, Norian, Late Triassic; Ponte Giurino, Valle Imagna, Italy.

Species included Atropicaris lintveri Laville et al., 2024a—Anisian, Middle Triassic, Slovenia; *Atropicaris* aff. *lintveri*—Anisian, Middle Triassic, Slovenia (Laville et al. 2024a); *Atropicaris rostrata* Arduini and Brasca, 1984—Late Triassic, Italy;

Fig. 20.3 *Atropicaris*? sp. from Scheiblingbach, Styria, Austria (early Carnian, Late Triassic). **a**–**c** NHMW 2025/0126/0001. **a** Left lateral view; **b** left lateral view (coated with ammonium chloride); **c** close-up of shield ornamentation (coated with ammonium chloride). **d**–**f** NHMW 2025/0126/0003. **d** Dorsal view; **e** dorsal view (coated with ammonium chloride); **f** close-up of shield ornamentation (coated with ammonium chloride). Scales: **a**, **b**, **d**, **e**, 2 mm; **c**, **f**, 1 mm. *Photos* A. Lukeneder

Atropicaris? sp.—Anisian, Middle Triassic, China (Feldmann et al. 2015); *Atropicaris*? sp.—Carnian, Late Triassic, Austria (Laville et al. 2024b); *Atropicaris* sp.—Ladinian, Middle Triassic, Switzerland (Bürgin et al. 1991); *Atropicaris* sp.—Carnian, Late Triassic, Slovenia (Laville et al. 2024a).

Diagnosis (after Laville et al. 2024a) Thylacocephala with a triangular rostrum ending in a spatulate tip, a serrate dorsal midline, a concave posterior margin, a spiny posteroventral corner, and dorsal and ventral intercalary ridges.

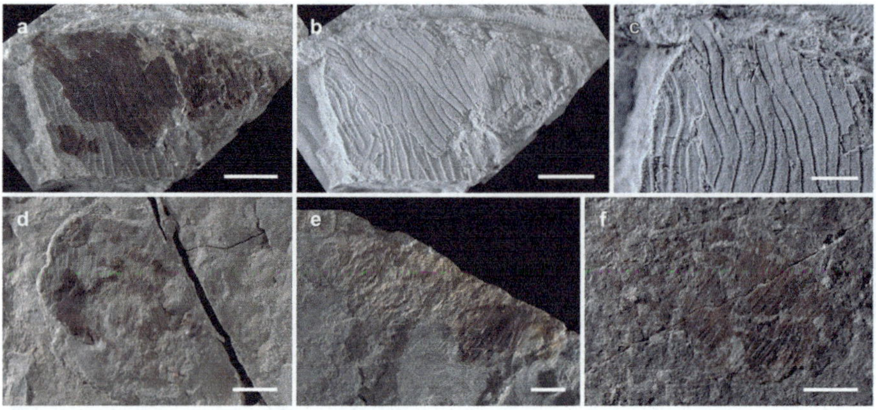

Fig. 20.4 *Atropicaris*? sp. from Scheiblingbach, Styria, Austria (early Carnian, Late Triassic). **a–c** NHMW 2025/0126/0010. **a** Left lateral view; **b** left lateral view (coated with ammonium chloride); **c** close-up of shield ornamentation (coated with ammonium chloride). **d** NHMW 2025/0126/0006. **e** NHMW 2025/0126/0008. **f** NHMW 2025/0126/0014. Scales: **e**, 4 mm; **a**, **b**, **d**, **f**, 2 mm; **c**, 1 mm. *Photos* A. Lukeneder

Atropicaris? sp.

Figures 20.3 and 20.4.

Material 14 specimens: NHMW 2012/0288/0001–0014.

Locality Scheiblingbachgraben (= Scheiblingbach), Styria, Austria.

Age Reingraben Formation, *Austrotrachyceras austriacum* Zone, *Austrotrachycerasminor* Subzone, Julian 2 Ib (regional stage), Carnian, Late Triassic.

Dimensions Due to the deformation and distortion of the specimens, we provide only rough estimates for the length (L_s) and maximum height (H_{max}) of the shield: $L_s >$ 12.8 mm; $H_{max} >$ 7.0 mm.

Description Trapezoidal shield in lateral view, longer than high (Fig. 20.4a, b). Maximal height reached around mid-length. Straight anterior margin slightly inclined posteroventrally (Fig. 20.3). Anterodorsal corner poorly preserved (Fig. 20.3a, b). Convex dorsal midline, poorly preserved (Fig. 20.4a, b). Posterodorsal and posteroventral corners, posterior and ventral margins not preserved. Anteroventral corner rounded. Marginal fold delimited by a shallow groove at least near the ventral margin (Fig. 20.3d, e). Shield covered with at least 48 transverse sigmoid ridges, posteroventrally oriented, including at least 40 primary ridges, 5 dorsal intercalary ridges and 3 ventral intercalary ridges (Fig. 20.3a–c). Appendages, posterior trunk and soft parts not preserved.

Remarks The specimens described here display several similarities with two Triassic taxa, *Atropicaris* and *Ferrecaris* Calzada and Mañé, 1993: a straight anterior margin, slightly tilted posteroventrally, a convex dorsal midline and transverse ridges. Our specimens therefore most likely belong to one of these taxa. In order to assign these specimens either to *Atropicaris* and *Ferrecaris*, it would be necessary to know their rostrum morphology (spatulate tip for *Atropicaris*; sharp, triangular tip for *Ferrecaris*), the ornamentation of their dorsal midline (serrate vs. smooth) and the morphology of their posterior margin (concave surrounded by a posterodorsal and a posteroventral spine vs. straight with rounded posterodorsal and posterodorsal corners). Here, the rostrum and the dorsal midline of the specimens are poorly preserved and the posterior part is missing. Accordingly, it is almost impossible to identify precisely the specimens. One additional character state could help in the identification: only *Atropicaris* apparently bears ventral intercalary ridges. Thus, our specimens most probably belong to *Atropicaris*. Nevertheless, in view of the poor preservation of the type material of *Ferrecaris*, the latter character should be taken with caution: such ridges could potentially be observed if better preserved material was discovered. Based on those comparisons, we leave our specimens in open nomenclature, although probably belonging to *Atropicaris*.

20.5 Discussion

20.5.1 Predation on Thylacocephalans

Within the Scheiblingblach *Konservat-Lagerstätte*, only one taxon has been recovered: ?*Atropicaris*. Similar specimens, also attributed with precaution to *Atropicaris*, have been reported from the equivalent *Polzberg-Lagerstätte* (Laville et al. 2024b). In both deposits, the specimens are poorly preserved, being often crushed and lacking an anterior margin and dorsal midline. A similar state of preservation for *Atropicaris* specimens was previously reported in the Middle Triassic Besano Formation in Italy by Affer and Teruzzi (1999). In this formation, thylacocephalan specimens consist mostly of crushed fragments, some of them being included within coprolites. They have therefore been interpreted as remains from predation. In the Paleozoic, several occurrences of fragmented and damaged thylacocephalans attributed to predation by gnathostomes have also been recorded (Zangerl and Richardson 1963; Briggs and Rolfe 1983; Williams 1990; Broda et al. 2015). Briggs and Rolfe (1983) and Broda et al. (2015) noted in those Paleozoic deposits that predated thylacocephalan remains were often missing anterior and dorsal parts, as in the specimens described here. Based on those previous reports, it is therefore most likely that the fragmented and damaged nature of *Atropicaris* specimens from the Austrian occurrences were the results of predation, although none have been found in bromalites, which are common in the Polzberg *Konservat-Lagerstätte* (Lukeneder et al. 2020). Several predatory taxa are known to occur in the Polzberg locality, including cartilaginous and actinopterygiid fish, cephalopods but also arthropods with the large *Austriocaris* thylacocephalans (Lukeneder and Lukeneder 2021). In the Scheiblingbach material, no bite traces have been discovered on the remains and, as the thylacocephalan remains are not part of bromalites, it is not possible to determine the predator(s).

20.5.2 Diversity of Thylacocephalans in the Early Carnian of Austria

The Scheiblingbach deposits contain a much lower taxonomic diversity of thylacocephalans than the Polzberg *Konservat-Lagerstätte*. Indeed, only one thylacocephalan taxon was found at Scheiblingbach (*Atropicaris*? sp.), whereas four taxa were described in the Polzberg *Konservat-Lagerstätte* (*Austriocaris carinata*, *Atropicaris*? sp., *Paraostenia striata*, *Paraostenia* cf. *ambatolokobensis*). This is surprising, especially because the two deposits are only 40 km apart today. One potential explanation lies in the ecological preferences of the various thylacocephalan taxa. They have been recovered from a variety of marine environments, from proximal environments to the basin itself. Additionally, various lifestyles from benthic to planktonic have been suggested depending on the species. However, similar environmental conditions are thought to have occurred in both Scheiblingbach and Polzberg, making the hypothesis of an ecological differentiation unlikely. Moreover, the lack of data on the soft-parts of Austrian thylacocephalans hinders determining their lifestyle.

The difference in taxonomic composition and diversity might therefore reflect different preservation. In Scheiblingbach, from which about 5000 fossil specimens have been uncovered, the animals are deposited within bivalve lumachelles which are harder components compared to the fine matrix that occurs in some beds in the Polzberg *Konservat-Lagerstätte*. Additionally, taxa such as *Paraostenia* or *Austriocaris* might be much less abundant than the remaining taxa in the localities where they occur. For instance, in the Carnian of Slovenia (Laville et al. 2024a), *Austriocaris* is by far the least abundant taxon. Only three very poorly preserved *Austriocaris* specimens have been discovered there, while 57 specimens of *Atropicaris* and 19 of *Microcaris* Pinna, 1974 have been collected. Nonetheless, hundreds of specimens of *Atropicaris* can be recovered from one locality (e.g., Arduini and Brasca 1984). Therefore, a taphonomic bias most probably affects our perception of the diversity at Scheiblingbach. Finally, for thylacocephalans, similar situations are known from other assemblages such as the Paris Biota (Spathian, Early Triassic). That biota, an exceptionally complex ecosystem dated from immediately after the Smithian/Spathian boundary, was initially described from the Paris Canyon, Idaho, USA (Brayard et al. 2017). Several years later, three new exceptionally diverse assemblages of the same age as the Paris Biota were described: Stewart Canyon and Georgetown from Idaho and NoName from Nevada (Smith et al. 2021). They are taxonomic subsets of the Paris Biota, being less diverse but having most of their taxa in common with the Paris Biota. This is especially true for thylacocephalans: three taxa have been reported from the Paris Biota, but only one or two from the other localities (Laville et al. 2021). Based on this comparison, Scheiblingbach might therefore represent a taxonomic subset of the Polzberg paleobiota, at least for thylacocephalans.

20.6 Conclusion

One taxon of Thylacocephala has been reported from the Scheiblingbach deposit: *Atropicaris*? sp. It represents the second occurrence of thylacocephalans in the Triassic of Austria (Styria) and only the third occurrence of thylacocephalans in the Carnian worldwide. Compared with the Polzberg *Konservat-Lagerstätte* (Lower Austria), the Scheiblingbach *Konservat-Lagerstätte* displays a low thylacocephalan diversity, which might reflect a taphonomic or sampling bias. In both Polzberg and Scheiblingbach deposits, the specimens of *Atropicaris*? sp. are fragmented and damaged. This most probably results from predation by other organisms (actinopterygiids, chondrichthiids, indeterminate pancrustaceans) recovered from those early Carnian localities.

Acknowledgmentss The author is particularly grateful to the owners of the outcrop area, the Austrian Federal Forests Company (Österreichische Bundesforste), for sampling and digging permission. Especially Lukas Nöst (forest district Großreifling) and Dominic Asztaller (forest district Großreifling) are thanked for their advice and help during the field campaigns. I thank Petra Lukeneder (Gablitz), Veronika Koukal (Vienna), and Justine and Umberto Uprimny (Pichlern) for their help with cleaning outcrops and sampling fossils. The work was done within the framework of and financially supported by projects of the Federal Government of Lower Austria (K3-F-964/001-2020 and K3-F-964/002-2023, Department Science and Research; headquarters in St. Pölten); the Austrian Academy of Sciences (headquarters in Vienna, project Polzberg Lukeneder 2021) represented by the National Committee for Geo/Hydro Sciences (Earth System Sciences Programme); the Triassic project "Fenster in die Karnische Krise" of the Society of Friends of the Natural History Museum Vienna (Freunde des Naturhistorischen Museums Wien, CPE Lukeneder 2023); and the Natur- und UNESCO Geopark Steirische Eisenwurzen (St Gallen, project Karnische Krise Lukeneder 2024). The manuscript greatly benefited from valuable comments by the reviewers Florian Braig (Oulu), Masayuki Ehiro (Sendai) and Sylvain Charbonnier (Paris), the production editor Ashok Arumairaj (Taramani), the in-house series editor Kathleen A. Ritterbush (Utah), the publishing editor Simon Shah-Rohlfs and the production editor Antje Endemann (both Heidelberg).

Author Contribution TL Conceptualization. AL collected new material, provided documentation on fossil material. TL, AL interpretation of the fossil material and writing the paper.

Funding Open access funding was enabled and the projects financially supported by the Federal Government of Lower Austria (K3-F-964/001-2020 and K3-F-964/002-2023, Department Science and Research; headquarters in St. Pölten); the Austrian Academy of Sciences (headquarters in Vienna, project Polzberg Lukeneder 2021) represented by the National Committee for Geo/Hydro Sciences (Earth System Sciences Programme); the Triassic project "Fenster in die Karnische Krise" of the Society of Friends of the Natural History Museum Vienna (Freunde des Naturhistorischen Museums Wien, CPE Lukeneder 2023); the Natural History Museum Vienna; and the Natur- und UNESCO Geopark Steirische Esienwurzen (St Gallen, project Karnische Krise Lukeneder 2024) (to AL). The authors are responsible for the contents of this publication. The funders had no impact on conceptualization, design, data collection, analysis, decision to publish, or preparation of the manuscript.

Competing interests All authors stated that they have no competing interests.

Authors' contributions All authors stated that they have equally contributed to their chapters, read, edited and agreed to submit the final versions of their contributions.

Availability of data and materials All sources of information are provided in the text. Raw data related to the fossil material from Austria are available from the corresponding authors and Alexander Lukeneder upon request.

Additional information The article is part of the Special Volume: *Late Triassic Konservat-Lagerstätten within the Carnian Pluvial Episode. 150 Years of Scientific Excavations in Austria.*

Artificial Intelligence (AI) The chapter was produced without the use of AI.

Language The entire volume uses American English reviewed by a native scientist.

Received 2025.

Published online 2026.

References

Affer D, Teruzzi G (1999) Thylacocephalan crustaceans from the Besano Formation, Middle Triassic, N. Italy. Riv Mus Civ Sci Nat "E Caffi" Bergamo 20:5–8

Arduini P, Brasca A (1984) *Atropicaris*: nuovo genere della classe Thylacocephala. Atti Soc Ital Sci Nat Mus Civ Stor Nat Milano 125(1–2):87–93

Arduini P (1990) Studies on Permo-Trias of Madagascar. 4. Thylacocephala from Lower Trias of Madagascar. Atti Soc Ital Sci Nat Mus Civ Stor Nat Milano 131(9):197–204

Brayard A, Krumenacker LJ, Botting JP, Jenks JF, Bylund KG, Fara E, Vennin E, Olivier N, Goudemand N, Saucède T, Charbonnier S, Romano C, Doguzhaeva L, Thuy B, Hautmann M, Stephen DA, Thomazo C, Escarguel G (2017) Unexpected Early Triassic marine ecosystem and the rise of the Modern evolutionary fauna. Sci Adv 3(2):e1602159. https://doi.org/10.1126/sciadv.1602159

Briggs DEG, Rolfe WDI (1983) New Concavicarida (new order: ? Crustacea) from the Upper Devonian of Gogo, Western Australia, and the palaeoecology and affinities of the group. Spec Pap Palaeontol 30:249–276

Broda K, Wolny M, Zatoń M (2015) Palaeobiological significance of damaged and fragmented thylacocephalan carapaces from the Upper Devonian of Poland. Proc Geol Assoc 126(4–5):589–598. https://doi.org/10.1016/j.pgeola.2015.05.005

Bürgin T, Eichenberger U, Furrer H, Tschanz K (1991) The Prosanto Formation—a middle Triassic fossil Lagerstätte of the Silvretta Nappe (Graubunden Canton, Switzerland) with a rich ichthyofauna. Eclogae Geol Helv 84(3):921–990

Calzada S, Mañé R (1993) Primera cita de un Tilacocéfalo (Crustacea) (*Ferrecaris* n. gen. *magransi* n. sp.) en el Ladiniense español. Trab Mus Geol Semin Barc 224:12–16

Charbonnier S, Brayard A (2019) New thylacocephalans from the Early Triassic Paris Biota (Bear Lake County, Idaho, USA). Geobios 54:37–43. https://doi.org/10.1016/j.geobios.2019.04.005

Charbonnier S, Teruzzi G, Audo D, Lasseron M, Haug C, Haug JT (2017) New thylacocephalans from the Cretaceous Lagerstätten of Lebanon. Bull Soc Géol Fr 188(3):19. https://doi.org/10.1051/bsgf/2017176

Ehiro M, Kano H (2024) A new species of *Parisicaris* (Microcarididae, Thylacocephala) from the upper Olenekian (Lower Triassic) Osawa Formation in the South Kitakami Belt, Northeast Japan. Bull Tohoku Univ Mus 23:23–34

Ehiro M, Sasaki O, Kano H, Nemoto J, Kato H (2015) Thylacocephala (Arthropoda) from the Lower Triassic of the South Kitakami Belt, Northeast Japan. Paleontol Res 19(4):269–282. https://doi.org/10.2517/2015PR010

Ehiro M, Sasaki O, Kano H, Nagase T (2019) Additional thylacocephalans (Arthropoda) from the Lower Triassic (upper Olenekian) Osawa Formation of the South Kitakami Belt, Northeast Japan. Palaeoworld 28(3):320–333. https://doi.org/10.1016/j.palwor.2019.03.001

Feldmann RM, Schweitzer CE, Hu S, Huang J, Zhou C, Zhang Q, Wen W, Xie T, Maguire E (2015) Spatial distribution of Crustacea and associated organisms in the Luoping Biota (Anisian, Middle Triassic), Yunnan Province, China: evidence of periodic mass kills. J Paleontol 89(6):1022–1037. https://doi.org/10.1017/jpa.2015.60

Glaessner MF (1931) Eine Crustaceenfauna aus den Lunzer Schichten Niederösterreichs. Jahrb Geol BA 81(3–4):467–486

Haig DW, Martin SK, Mory AJ, McLoughlin S, Backhouse J, Berrell RW, Kear BP, Hall R, Foster CB, Shi GR, Bevan JC (2015) Early Triassic (early Olenekian) life in the interior of East Gondwana: mixed marine–terrestrial biota from the Kockatea Shale, Western Australia. Palaeogeogr Palaeoclimatol Palaeoecol 417:511–533. https://doi.org/10.1016/j.palaeo.2014.10.015

Haug C, Briggs DEG, Mikulic DG, Kluessendorf J, Haug JT (2014) The implications of a Silurian and other thylacocephalan crustaceans for the functional morphology and systematic affinities of the group. BMC Evol Biol 14(1):159. https://doi.org/10.1186/s12862-014-0159-2

Ji C, Tintori A, Jiang D, Motani R, Confortini F (2021) New Thylacocephala (Crustacea) assemblage from the Spathian (Lower Triassic) of Majiashan (Chaohu, Anhui Province, South China). J Paleontol 95(2):305–319. https://doi.org/10.1017/jpa.2020.92

Ji C, Tintori A, Jiang D, Motani R (2017) New species of Thylacocephala (Arthropoda) from the Spathian (Lower Triassic) of Chaohu, Anhui Province of China. PalZ 91. https://doi.org/10.1007/s12542-017-0347-7

Krystyn L (1991) Die Fossillagerstätten der alpinen Trias. In: Vasicek W, Krystyn L, Golebiowski R (eds) Exkursionen im Jungpaläozoikum und Mesozoikum Österreichs. Österreichische Paläontologische Gesellschaft, Wien, pp 24–78

Laville T, Smith CPA, Forel M-B, Brayard A, Charbonnier S (2021) Review of Early Triassic Thylacocephala. Riv Ital Paleontol Stratigr 127(1):73–101. https://doi.org/10.5281/ZENODO.4066340

Laville T, Hitij T, Forel M-B, Gašparič R, Žalohar J, Charbonnier S (2024a) Middle and Late Triassic thylacocephalans from Slovenia: implications for their diversification. Bull Geosci 99(3):219–253. https://doi.org/10.3140/bull.geosci.1907

Laville T, Lukeneder P, Lukeneder A (2024b) Revision of the thylacocephalan biota from the Upper Triassic Polzberg Konservat-Lagerstätte, Austria. Acta Palaeontol Pol 69(4):713–730. https://doi.org/10.4202/app.01180.2024

Lukeneder A (ed) (2026) Late Triassic Konservat-Lagerstätten within the Carnian Pluvial Episode: 150 Years of Scientific Excavations in Austria. Top Geobiol 55:600 p. https://doi.org/10.1007/978-3-032-19777-1

Lukeneder A (2026b) The marine Upper Triassic paleobiota from the Scheiblingbach *Konservat-Lagerstätte* in Austria. In: Lukeneder A (ed) Late Triassic Konservat-Lagerstätten within the Carnian Pluvial Episode: 150 years of scientific excavations in Austria. Top Geobiol 55, Chap 19:437–467. https://doi.org/10.1007/978-3-032-19777-1_19

Lukeneder A, Lukeneder P (2021) The Upper Triassic Polzberg palaeobiota from a marine Konservat-Lagerstätte deposited during the Carnian Pluvial Episode in Austria. Sci Rep 11(1):16644. https://doi.org/10.1038/s41598-021-96052-w

Lukeneder A, Surmik D, Gorzelak P, Niedźwiedzki R, Brachaniec T, Salamon MA (2020) Bromalites from the Upper Triassic Polzberg section (Austria); insights into trophic interactions and

food chains of the Polzberg palaeobiota. Sci Rep 10(1):20545. https://doi.org/10.1038/s41598-020-77017-x
Pinna G (1974) I crostacei della fauna triassica di Cene in Val Seriana (Bergamo). Mem Soc Ital Sci Nat Mus Civ Stor Nat Milano 21(1):5–33
Pinna G, Arduini P, Pesarini C, Teruzzi G (1982) Thylacocephala: una nuova classe di crostacei fossili. Atti Soc Ital Sci Nat Mus Civ Stor Nat Milano 123(4):469–482
Schram FR (2014) Family level classification within Thylacocephala, with comments on their evolution and possible relationships. Crustac 87(3):340–363. https://doi.org/10.1163/15685403-00003289
Secretan S (1985) Conchyliocarida, a class of fossil crustaceans: relationships to Malacostraca and postulated behaviour. Earth Environ Sci Trans R Soc Edinburgh 76(2–3):381–389. https://doi.org/10.1017/S0263593300010592
Smith CPA, Laville T, Fara E, Escarguel G, Olivier N, Vennin E, Goudemand N, Bylund KG, Jenks JF, Stephen DA, Hautmann M, Charbonnier S, Krumenacker LJ, Brayard A (2021) Exceptional fossil assemblages confirm the existence of complex Early Triassic ecosystems during the early Spathian. Sci Rep 11(1):19657. https://doi.org/10.1038/s41598-021-99056-8
Williams ME (1990) Feeding behavior in Cleveland Shale fishes. In: Boucot AJ (ed) Evolutionary paleobiology of behavior and coevolution. Elsevier, Amsterdam, pp 273–287
Zangerl R, Richardson ES (1963) The paleoecological history of two Pennsylvanian black shales. Chicago Natural History Museum, Chicago

Chapter 21
First Hybodontiform Record Reveals Chondrichthyan Role in Carnian Food Webs of the Upper Triassic Reifling Basin, Austria

Sebastian Stumpf, Alexander Lukeneder, and Viola Winkler

Abstract The Upper Triassic Reifling Basin of the Northern Calcareous Alps in Austria has yielded rich marine fossil assemblages, offering rare insights into complex basinal food webs during the Carnian Pluvial Event (CPE), a critical interval of major global environmental upheaval. Despite extensive sampling, chondrichthyan fishes had not been recovered until now. Here we report the first unambiguous chondrichthyan record from the Reifling Basin, represented by a single indeterminate hybodontiform tooth from the lower Carnian of Scheiblingbach. Still largely embedded in matrix, the specimen was examined through micro-CT scanning and digital segmentation. The tooth, indicative of a small- to medium-sized durophagous generalist likely occupying a mid-trophic position, adds to the scarce chondrichthyan record from the Alpine Upper Triassic during the CPE. While the Scheiblingbach tooth bears some resemblance to *Polyacrodus* and *Acrodus*, the deeply unresolved taxonomy and likely polyphyletic nature of these and other historically problematic hybodontiform genera preclude confident identification, exemplifying the persistent difficulty of assigning isolated hybodontiform teeth to specific taxa. Nonetheless, the tooth from Scheiblingbach refines our understanding of Triassic hybodontiform diversity and distribution and highlights the urgent need for future research to overcome long-lasting taxonomic challenges that continue to constrain our perception of early Mesozoic chondrichthyan dynamics.

S. Stumpf (✉) · A. Lukeneder
Department of Geology and Paleontology, Natural History Museum Vienna, Vienna, Austria
e-mail: sebastian.stumpf@nhm.at

A. Lukeneder
e-mail: alexander.lukeneder@nhm.at

S. Stumpf
Department of Paleontology, University of Vienna, Vienna, Austria

V. Winkler
Department of Central Research Laboratories, Natural History Museum Vienna, Vienna, Austria
e-mail: viola.winkler@nhm.at

A. Lukeneder (ed.), *Late Triassic Konservat-Lagerstätten within the Carnian Pluvial Episode*, Topics in Geobiology 55, https://doi.org/10.1007/978-3-032-19777-1_21

Keywords Vertebrata · Chondrichthyes · Hybodontiformes · Carnian Pluvial Episode · Triassic · Austria

21.1 Introduction

Deposited during the Carnian Pluvial Episode (CPE), the marine sedimentary successions of Polzberg and Scheiblingbach, both exposed in the Upper Triassic Reifling Basin of the Northern Calcareous Alps in Austria (Fig. 21.1), have yielded exquisitely preserved fossil assemblages that offer rare glimpses into basinal food webs and ecosystem dynamics during a period of major global environmental upheaval (Griffith 1977; Lukeneder and Lukeneder 2021, 2022, 2023; Lukeneder et al. 2020, 2024). Triggered by extensive volcanic activity, the CPE profoundly affected both marine and terrestrial ecosystems, driving increased precipitation, fluctuating marine salinity, enhanced weathering, global carbon-cycle perturbations, and widespread ecological and biotic turnover (e.g., Bernardi et al. 2018; Dal Corso et al. 2020; Fu et al. 2020; Preto et al. 2010; Rigo and Joachimski 2010; Zhang et al. 2024). Despite these disruptive conditions, the Reifling Basin assemblages document diverse marine communities, and the presence of both invertebrates and vertebrates dominated by nektonic taxa, alongside bromalites (i.e., coprolites and regurgitalites), reflects the persistence of complex, multi-tiered food webs in deeper basinal environments (Lukeneder and Lukeneder 2021, 2022, 2023; Lukeneder 2026a, b, this volume).

Although top predators have not yet been discovered, their presence is inferred from abundant regurgitalites from the lower Carnian of Polzberg (Lukeneder et al. 2020). Notably, extinct hybodontiforms, which were dominant shark-like chondrichthyans throughout much of the early Mesozoic and are regarded the sister group to neoselachians (i.e., sharks and rays; Maisey 1982; Maisey et al. 2004), are hypothesised as the producers of these regurgitalites (Lukeneder et al. 2020). However, direct evidence for hybodontiforms, or chondrichthyans in general, was previously lacking.

This study presents the first unambiguous chondrichthyan record from the Upper Triassic Reifling Basin, represented by an indeterminate hybodontiform tooth from the lower Carnian of Scheiblingbach (Lukeneder 2026a, b). This discovery provides direct evidence of their presence and ecological role in the Reifling Basin during a critical time interval in Earth's history marked by profound global environmental perturbation.

21.2 Material and Methods

The fossil chondrichthyan tooth from the lower Carnian of Scheiblingbach in Styria, Austria, which forms the focus of the present study, is housed and curated at the Natural History Museum Vienna (NHMW) under collection number NHMW-GEO-2025/0127/0016. The specimen originates from terrigenous siliciclastic deposits of the Reingraben Formation (layer Sch 56; *A. austriacum* Zone, *A. minor* Subzone; Lukeneder 2026a, b; Fig. 21.1) and remains largely embedded in matrix, with its lingual face partially exposed. Due to slight damage to these exposed areas and to avoid potential further damage during physical or chemical preparation, the specimen was scanned using microcomputed tomography (micro-CT) and digitally segmented.

The scan was performed using the micro-CT system at the NHMW, Austria (YXLON FF 35 CT, FXE Direct Beam). A current of 160 μA and a voltage of 120 kV were applied to acquire a total of 3600 projection images. The reconstructed image stack had an isotropic voxel size of 6.4 μm and was further processed using Dragonfly 3D World (Version 2024.1, Comet Technologies Canada Inc., Montreal, Canada). The CT dataset is publicly available at: https://doi.org/10.57756/yex8vz.

21.3 Results

21.3.1 Systematic Paleontology

Class **Chondrichthyes** Huxley, 1880

Subclass **Elasmobranchii** Bonaparte, 1838

Cohort **Euselachii** Hay, 1902

Order **Hybodontiformes** Patterson, 1966

Hybodontiformes gen. et sp. indet.

Referred material: NHMW-GEO-2025/0127/0016, one almost complete tooth, still largely embedded in matrix.

Description: The tooth is small, wider than it is high, and slightly convex labially and slightly concave lingually in both occlusal and basal views, measuring 6 mm in mesiodistal width and 2.5 mm in apicobasal height (Fig. 21.2). The crown is robust and exhibits an asymmetrical profile in labiolingual view (Fig. 21.2c, d), suggesting that NHMW-GEO-2025/0127/0016 likely represents a lateral tooth. It bears a low and broad, somewhat pyramidal-shaped main cusp that overhangs the root only slightly both labially and lingually, and is flanked by two very low lateral cusplets on one side and four on the other. A labial peg is absent. A weak transverse crest extends across the main cusp and its adjacent cusplets (Fig. 21.2f). The crown is ornamented

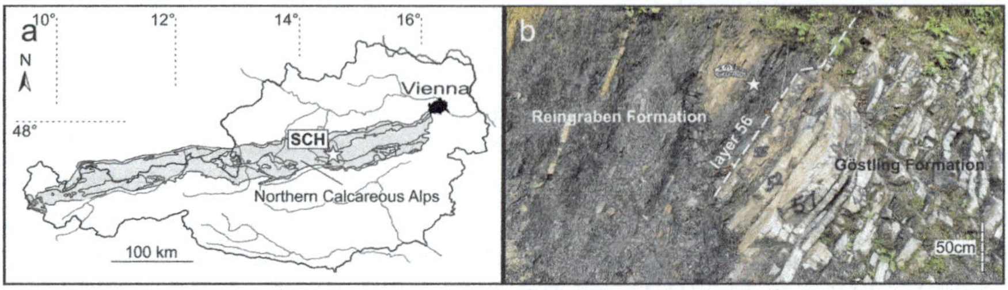

Fig. 21.1 **a** Locality map of the Scheiblingbach area (SCH, Großreifling) in Styria and the Polzberg area (near Lunz am See) in Lower Austria, with marked position of the Austrian Northern Calcareous Alps (in gray). **b** Detailed view on the boundary layers of the Göstling Formation and the Reingraben Formation with indicated hybodontiform layer Sch 56, marked by the white star and tooth drawing

with numerous ridges, which appear rather strongly developed and extend down to the crown shoulder on both the labial and lingual faces, apparently originating from the apices of the main cusp and cusplets. Some of these ridges descend from the transverse crest and appear to bifurcate commonly (Fig. 21.2b–d, f). The labial crown face is incised immediately above the crown-root junction (Fig. 21.2d, h, i).

The root is about as high as the crown, and its labial and lingual faces are perforated by numerous foramina (Fig. 21.1c, d). Lingually, a series of regularly spaced foramina is aligned immediately below the crown-root junction (Fig. 21.2c). Additionally, three larger foramina traverse the root, connecting its labial and lingual faces (Figs. 21.2c–e). The basal face of the root is flat and appears to be largely devoid of foramina (Fig. 21.2g).

As far as can be determined, the dental histology of NHMW-GEO-2025/0127/0016 appears to conform to the osteodont histotype (sensu Jambura et al. 2020). It exhibits a thin layer of enameloid, which is thicker across the main cusp and becomes thinner across the lateral cusplets, encapsulating a core of osteodentine, traversed by irregularly arranged dentinal osteons, with no hollow pulp cavity (Fig. 21.2e). However, the presence of orthodentine cannot be ruled out with certainty based on the available micro-CT data, especially as the orthodentine encapsulating the osteodentine core may be very thin (Jambura et al. 2020, 2021; Maisch and Matzke 2016) and difficult to detect without physical cross-sections.

Discussion of affinities: The overall morphology of NHMW-GEO-2025/0127/0016 from the lower Carnian of Scheiblingbach, combining a low, robust multicuspid crown and a root lacking a specialised device for the articulation of successive teeth of the same tooth file, is inconsistent with referral to Cladodontomorphi (sensu Ginter et al. 2010), but closely resembles hybodontiforms, particularly those traditionally assigned to the genus *Polyacrodus* Jaekel, 1889. *Polyacrodus* was originally established based on isolated teeth from the Anisian–Ladinian of France to include a single species, *P. polycyphus* (Agassiz, 1837), characterised by a stout, high, pyramidal-shaped main cup flanked by up to four pairs of lateral cusplets and bearing a plunging protuberance that extends labially from the base of the main cusp. Initially restricted to Triassic species (Seilacher 1943; Stensiö 1921), the temporal range of *Polyacrodus*

Fig. 21.2 Hybodontiformes gen. et sp. indet., NHMW-GEO-2025/0127/0016, an almost complete tooth from the lower Carnian of Scheiblingbach, Styria, Austria. Photographs: **a** the whole specimen as preserved in lingual view and **b** close-up view of the main cusp; 3D renderings of **c** lingual, **d** labial, **f** occlusal, **g** basal, **h** ?mesial, and **i** ?distal views; **e** virtual section in the frontal plane. Arrows indicate large foramina traversing the root. Abbreviations: en, enameloid; os, osteodentine. Scale bars = 1 mm

has since expanded to encompass records ranging from the Carboniferous to the Late Cretaceous (Cappetta 2012; Ginter et al. 2010). However, the taxonomic content of *Polyacrodus* remains debated and likely represents an unnatural assemblage of unrelated species (Cappetta 2012; Rees 1999), with some authors even suggesting treating *Polyacrodus* as a *nomen dubium* (Rees 2008; Rees and Underwood 2008). This ambiguity is further complicated by the apparent loss of the type material of *P. polycyphus* (Kriwet 2004). Böttcher (2015) referred teeth from the Lower Keuper (Ladinian) of southern Germany to *P. polycyphus*, but despite their larger size, these teeth are readily distinguishable from NHMW-GEO-2025/0127/0016 by their prominent labial protuberance and higher, more acutely pointed main cusp and cusplets.

NHMW-GEO-2025/0127/0016 shows some resemblance to *Acrodus* Agassiz 1837, a species-rich hybodontiform ranging from the Late Carboniferous to the Early Cretaceous (Cappetta 2012; Ginter et al. 2010; Wen et al. 2023). However, as with many historically described hybodontiforms (such as *Polyacrodus*), its taxonomy remains unresolved and in urgent need of revision. *Acrodus* is characterised by low-crowned crushing teeth, with articulated dentitions displaying a gradual transition from short anterior to mesiodistally elongate, sometimes asymmetrical lateral and posterolateral teeth (Day 1864; Duffin 2010; Kuhn 1946; Rieppel 1981; Woodward 1889). The crown usually bears a weak transverse crest from which numerous, densely packed ridges descend down to the crown base. Lateral cusplets, when present, are reduced or vestigial.

NHMW-GEO-2025/0127/0016 resembles posterolateral teeth of *A. anningiae* Agassiz, 1839 from the Lower Jurassic of Dorset, UK (Day 1864; Woodward 1889), as well as a morphologically similar tooth from the Lower Triassic Grippia bonebed of Spitsbergen, Norway, described by Bratvold et al. (2018: Fig. 4B1–4) and referred to *A. spitzbergensis* Hulke, 1873. In contrast to the more bulbous teeth of the *Acrodus* type species, *A. gaillardoti* Agassiz in Alberti, 1834 from the Middle Triassic of Europe, and dentally similar species such as *A. georgii* Mutter, 1998 from the Middle Triassic of Monte San Giorgio, Switzerland, and *A. nobilis* Agassiz, 1838 from the Lower Jurassic of Europe, these teeth exhibit a main cusp that is less laterally expanded and more distinctly pointed. Additionally, they possess more prominent lateral cusplets and a coarser, less densely packed ornamentation. Woodward (1889) suggested that *A. anningiae* might, in part, be based on juvenile specimens of *A. nobilis*, though this remains unconfirmed.

Adding to this ambiguity is a partial skeleton described by Thomson (1982) from the Lower Triassic of Madagascar, which despite possessing teeth reminiscent of *Acrodus*, suggests the presence of a cranio-palatoquadrate articulation that differs markedly from that of other hybodontiforms, including *A. georgii* and likely also *A. nobilis* (cf. Kuhn 1946; Maisey 1982, 1983, 2008; Rieppel 1981). This suggests the independent evolution of similar crushing dentitions among unrelated hybodontiforms, complicating the identification of isolated teeth such as NHMW-GEO-2025/0127/0016.

Teeth reminiscent of *Acrodus* s.l. also characterize the poorly known species *Hybodus medius* Agassiz, 1838 from the Lower Lias of Dorset. However, its referral to the genus *Hybodus* Agassiz, 1837, appears untenable, as the high-crowned, labiolingually compressed, and distinctively multicuspid teeth of the coeval type species *H. reticulatus* Agassiz, 1837 differ markedly from those *H. medius* (cf. Duffin 1993, 2010; Duffin et al. 2023), casting doubt on their close relationship. Likewise, assigning NHMW-GEO-2025/0127/0016 to the genus *Lissodus* Brough, 1935, as defined by Rees and Underwood (2002), also appears to be unjustified due to the absence of a labial peg, the presence of a rather strong ornamentation, more than two pairs of lateral cusplets, and a more robust root.

NHMW-GEO-2025/0127/0016 differs from the problematic elasmobranchs *Acronemus tuberculatus* (Bassani, 1886) from Monte San Giorgio and *Homalodontus*

Mutter et al., 2008, represented by *H. aplopagus* (Mutter et al., 2007) and *H. homalorhizo* (Mutter et al., 2007) from the Lower Triassic of Canada. Both *Acronemus* and *Homalodontus* are among the better-known Mesozoic elasmobranchs, represented by articulated material exhibiting a puzzling combination of morphological features, suggestive of a potential sister group relationship to hybodontiforms and neoselachians, pending further research (Coates and Tietjen 2018; Maisey 2011). Although lateral teeth of *Acronemus* bear some resemblance to NHMW 2025/0127/0016, they clearly differ in lacking lateral cusplets and in possessing a prominent ridge that extends labiolingually across a blunt main cusp, reaching down to the base of the crown (Rieppel 1982). Anterior teeth of *Homalodontus* possess labiolingually narrow crowns with a prominent, acutely pointed main, whereas lateral and posterior teeth lack a main cusp entirely, resulting in a distinctive blade-like morphology that clearly distinguishes them from NHMW-GEO-2025/0127/0016.

Among hybodontiforms, the porous root of NHMW-GEO-2025/0127/0016, with its large foramina connecting the labial and lingual faces, resembles that of the widely distributed European Middle Triassic species *Palaeobates angustissimus* (Agassiz in Geinitz, 1837), as described by Rieppel (1981) based on material from Monte San Giorgio. A similar vascularization pattern is present in the Mesozoic hybodontiform *Strophodus* Agassiz, 1838 (e.g., Rees and Underwood 2008; Rigal and Cuny 2016), as well as in Cretaceous distobatids (e.g., Brito et al. 2025; Landemaine 1991; Neves et al. 2024). However, all of these taxa are readily distinguished from NHMW 2025/0127/0016 by their distinctive teeth, which lack lateral cusplets and suggest a specialised durophagous diet.

NHMW-GEO-2025/0127/0016 is here tentatively interpreted as exhibiting the osteodont tooth histotype. This condition readily distinguishes it from taxa possessing the orthodont tooth histotype, such as *Palaeobates*, in which a hollow pulp cavity is surrounded by orthodentine (Böttcher 2024; Romano and Brinkmann 2010). NHMW-GEO-2025/0127/0016 also differs from *Polyacrodus*, whose dental histology, as described by Jaekel (1889) for *P. polycyphus* is characterised a thick layer of orthodentine surrounding a core of osteodentine and thus conforms to the pseudoosteodont histotype (sensu Jambura et al. 2020); note that some authors have misinterpreted the teeth of *P. polycyphus* as orthodont (e.g., Cappetta 2012; Glickman 1964; Rees 2008; Rees and Underwood 2008).

The dental histology of *Acrodus*, by contrast, remains unresolved. Although its teeth are consistently described as containing a central core of regular osteodentine, the nature of the surrounding tissue remains subject to differing interpretations. Following Radinsky (1961), the surrounding tissue has been interpreted as a very thin layer of orthodentine, whereas others interpreted have regarded it as modified columnar osteodentine characterised by parallel-arranged dentinal osteons (Maisey 1987; Maisey et al. 2004).

In summary, although NHMW-GEO-2025/0127/0016 shares similarities with *Polyacrodus* and *Acrodus*, the unresolved taxonomy and presumably polyphyletic nature of these historically contentious genera prevent confident identification. This highlights the persistent challenges of identifying isolated hybodontiform teeth and underscores the importance of associated skeletal material, which (although rarely preserved) can provide key features for clarifying hybodontiform taxonomy beyond dental traits alone (e.g., Maisey 1987; Stumpf et al. 2021b). NHMW-GEO-2025/0127/0016 is here left in open nomenclature and referred to as Hybodontiformes gen. et sp. indet., pending future taxonomic work that may help clarify its generic affinities.

21.4 Discussion

21.4.1 Tracing Alpine Upper Triassic Chondrichthyans

The single tooth NHMW-GEO-2025/0127/0016 from the lower Carnian of Scheiblingbach, here referred to an indeterminate hybodontiform elasmobranch, constitutes the first unambiguous chondrichthyan record from the Upper Triassic Reifling Basin. Although direct fossil evidence of chondrichthyans had previously been lacking, their presence in the Reifling Basin was inferred from abundant regurgitalites composed primarily of ammonoid shells and rare teuthid remains from the lower Carnian of Polzberg (Lukeneder et al. 2020; Lukeneder and Lukeneder 2022, 2023). The regurgitalites were interpreted by Lukeneder et al. (2020) as having been produced by a large durophagous fish, possibly *Acrodus*. While this genus has not yet been recorded from the Reifling Basin, it remains a plausible candidate given its highly specialised durophagous crushing dentition (Cappetta 2012) and the large body size inferred for some species, such as *A. georgii* (Mutter, 1998) from the Middle Triassic of Monte San Giorgio, which may have reached an adult body length of up to three meters (Kuhn 1946). However, as noted by Lukeneder et al. (2020), any other large Triassic predator inhabiting the NW Tethys margin, such as ichthyosaurs (although not yet recorded from the Reifling Basin) could also have produced these regurgitalites.

The overall morphology of NHMW-GEO-2025/0127/0016 is consistent with a durophagous species likely occupying a mid-trophic position. Comparison with *Lissodus cassangensis* (Teixeira, 1956) from the Lower Triassic of Angola, known from articulated skeletons indicating an adult body length of approximately 20 cm and possessing teeth no wider than 1 mm (Antunes et al. 1990), suggests that NHMW-GEO-2025/0127/0016 derives from a small- to medium-sized species that may have reached an adult length of about a meter. However, as NHMW-GEO-2025/0127/0016 may represent a juvenile, this estimate remains tentative until the discovery of more complete fossil material.

Hybodontiforms appear to be rare in the Alpine Upper Triassic, despite their abundance in the Middle Triassic of Monte San Giorgio (Rieppel 1981, 1982, 2019) and other well-sampled Middle–Late Triassic localities across Europe (e.g., Allard et al. 2015; Böttcher 2015; Brignon 2023; de Lange et al. 2023; Lakin et al. 2016; Manzanares et al. 2016, 2020; Moreau et al. 2021; Pawlak et al. 2022; Pla et al. 2013; Seilacher 1943; Szabó et al. 2019; Weeks et al. 2025). Only few dental and dermal remains have been reported from the Carnian 'Raibl Beds' of the western

Northern Calcareous Alps in Tyrol, Austria (Krainer et al. 2011), and from coeval strata assigned to the San Cassian Formation of the Southern Calcareous Alps in Italy (Bernardi et al. 2011; Bizzarini et al. 2003; Münster 1841; Pindakiewicz et al. 2024). This apparent discrepancy may reflect abiotic or biotic stresses that constrained their distribution, or it may simply result from sampling bias, as small chondrichthyan teeth are easily overlooked.

Münster (1841) established *Hybodus hexagonus* and *H. angulatus* based on dorsal fin spine fragments from the San Cassian Formation; these are here considered *nomina dubia.* Similarly, fragmentary teeth with slender, pointed cusps, reported by Bizzarini et al. (2003) from the San Cassian Formation and referred to *Hybodus* sp., are too incomplete for firm identification and might not even represent hybodontiforms, requiring unambiguous confirmation. Teeth attributed to *Acrodus* sp. by Bernardi et al. (2011) and Pindakiewicz et al. (2024) from the same formation also warrant caution. The finely reticulate, strongly arched teeth described by Bernardi et al. (2011) more closely resemble anterior teeth of *Strophodus* (e.g., Rigal and Cuny 2016; Stumpf et al. 2022) than those of *Acrodus* and are distinct from NHMW-GEO-2025/0127/0016. Likewise, the minute teeth reported by Pindakiewicz et al. (2024), displaying a distinctive reticulate ornamentation, are inconsistent with referral to *Acrodus* or any other known hybodontiform and clearly differ from NHMW-GEO-2025/0127/0016. Teeth assigned to *Palaeobates* sp. have been reported from both the San Cassian Formation (Bizzarini et al. 2003) and the Raibl Beds (Krainer et al. 2011), suggesting a broader, albeit still limited, presence in the Alpine Upper Triassic.

Despite the scarcity of hybodontiforms in the Alpine Upper Triassic and the uncertain generic affinities of many specimens, available evidence suggests different predatory modes, with durophagous generalists appearing to predominate. This pattern mirrors broader Mesozoic trends in the aftermath of the Permian–Triassic mass extinction (PTME), during which marine communities underwent profound and persistent restructuring (Sepkoski 1984), driven largely by escalating selective pressures from increasingly effective predators (Benton 2016; Benton and Wu 2022). By the Middle Triassic, marine communities had regained tiered, functionally complex trophic hierarchies comprising multiple levels from primary consumers to apex predators (Benton et al. 2013; Chen and Benton 2012). This marked increase in ecological complexity reflects a key episode of biotic recovery characterised by enhanced niche differentiation and predator-prey dynamics likely further intensified by the CPE (Benton and Wu 2022; Dal Corso et al. 2020). These developments laid the foundation for the continued rise of modern-style marine ecosystems throughout the Jurassic and Cretaceous.

21.4.2 Mesozoic Hybodontiforms and the Emergence of Modern Chondrichthyan Communities

Although several chondrichthyan lineages survived the PTME, only hybodontiforms and neoselachians flourished in its aftermath, and by the close of the Triassic, most other surviving chondrichthyan groups, such as jalodontiforms and freshwater-dwelling xenacanthiforms, had disappeared (Cappetta 2012; Ivanov et al. 2021). Ultimately, only neoselachians, together with a few holocephalans (the latter never having re-established their Palaeozoic diversity), persisted into the Cenozoic (Friedman and Sallan 2012; Guinot and Cavin 2016).

The impact of the CPE on chondrichthyans remains poorly understood. While global biodiversity data indicate a marked decline of marine bony fishes coinciding with the CPE, chondrichthyan genus-level diversity remains unaffected (Romano et al. 2016). However, this apparent resilience may be misleading due to the considerable uncertainty surrounding the generic affinities of many Triassic species (particularly among hybodontiforms), which likely obscures genuine diversity patterns. It is nevertheless evident that Triassic hybodontiforms were dominated by small- to medium-sized generalist species with clutching- and crushing-type teeth. In contrast, large-bodied hybodontiforms reaching lengths of three meters and equipped with teeth of various shapes, including high-crowned teeth well adapted for piercing and cutting prey, did not became abundant and widespread prior to the Jurassic (e.g., Rees and Underwood 2008; Duffin 2010; Soto et al. 2012; Stumpf et al. 2021a, 2022). The Jurassic marks also the peak of hybodontiform diversity, when they thrived in both marine and continental environments (e.g., Cuny et al. 2009; Duffin and Thies 1997; Klug et al. 2010; Rees and Underwood 2006, 2008; Stumpf et al. 2022, 2023). However, their prominence in open marine habitats had declined by the beginning of the Cretaceous, and by the later part of this period, hybodontiforms had become largely confined to continental waters (Cuny 2012), disappearing entirely during the end-Cretaceous mass extinction. The decline, particularly among open marine forms, and the resulting shift from hybodontiform- to neoselachian-dominated chondrichthyan communities remain unresolved but may have been driven by competitive exclusion by their more advanced chondrichthyan relatives (Maisey 2012; Thies and Reif 1985), which began to rapidly diversify during the Jurassic and went on to become the dominant chondrichthyans of modern marine ecosystems (Guinot and Cavin 2016).

21.5 Conclusions

The single hybodontiform tooth NHMW-GEO-2025/0127/0016 from the lower Carnian of Scheiblingbach provides the first direct evidence for the presence of chondrichthyan fishes in basinal food webs of the Upper Triassic Reifling Basin and adds to the scarce record of chondrichthyans from the Alpine Upper Triassic during the CPE. Prior to this discovery, the presence of chondrichthyans in the Reifling Basin had been inferred only indirectly, based on abundant regurgitalites from the lower Carnian of Polzberg. While the producer of these regurgitalites remains unknown, it has been hypothesised to have been a large durophagous hybodontiform shark-like chondrichthyan. NHMW-GEO-2025/0127/0016, by contrast, is indicative of a small- to medium-sized bodied durophagous generalist, likely occupying a mid-trophic position. Although indeterminate, the specimen refines our understanding of chondrichthyan diversity and distribution in the Alpine Triassic during a time of major global environmental perturbation. Moreover, it highlights the urgent need for future research to address persistent taxonomic uncertainties that continue to constrain our understanding of early Mesozoic chondrichthyan diversity dynamics. Future discoveries from the Reifling Basin and other Alpine Upper Triassic deposits may clarify how chondrichthyans responded to the CPE.

Acknowledgments The authors are particularly grateful to the owners of the outcrop area, the Austrian Federal Forests Company (Österreichische Bundesforste), for sampling and digging permission. Especially Lukas Nöst (forest district Großreifling) and Dominic Asztaller (forest dristrict Großreifling) for their advice and help during the field campaigns. The work was done within the framework of and financially supported by projects of the Federal Government of Lower Austria (K3-F-964/001-2020 and K3-F-964/002-2023, Department Science and Research; headquarters in St. Pölten); the Austrian Academy of Sciences (headquarters in Vienna, project Polzberg Lukeneder 2021) represented by the National Committee for Geo/Hydro Sciences (Earth System Sciences Programme); the Triassic project "Fenster in die Karnische Krise" of the Society of Friends of the Natural History Museum Vienna (Freunde des Naturhistorischen Museums Wien, CPE Lukeneder 2023); the Natural History Museum Vienna; and the Natur- und UNESCO Geopark Steirische Eisenwurzen (St Gallen, project Karnische Krise Lukeneder 2024). We thank the reviewers Christopher Duffin (London) and Jürgen Kriwet (Vienna) for their comments on the manuscript. The manuscript greatly benefited from valuable comments by the production editor Ashok Arumairaj (Taramani), the in-house series editor Kathleen A. Ritterbush (Utah), the publishing editor Simon Shah-Rohlfs and the production editor Antje Endemann (both Heidelberg).

Author Contribution AL collected new material and prepared a figure. SS designed the study, wrote the initial draft of the manuscript, and prepared a figure. VW computed tomography, contributed to data acquisition and figure preparation.

Funding This research was funded in part by the Austrian Science Fund (FWF) [10.55776/PAT2291524] to SS. Open access funding enabled and projects financially supported by the Federal Government of Lower Austria (K3-F-964/001-2020 and K3-F-964/002-2023, Department Science and Research; headquarters in St. Pölten); the Austrian Academy of Sciences (headquarters in Vienna, project Polzberg Lukeneder 2021) represented by the National Committee for Geo/Hydro Sciences (Earth System Sciences Programme); the Triassic project "Fenster in die Karnische Krise" of the Society of Friends of the Natural History Museum Vienna (Freunde des Naturhistorischen Museums Wien, CPE Lukeneder 2023); the Natural History Museum Vienna; and the Natur- und UNESCO Geopark Steirische Eisenwurzen (St Gallen, project Karnische Krise Lukeneder 2024) (to AL). The authors are responsible for the contents of this publication. The funders had no impact on conceptualization, design, data collection, analysis, decision to publish, or preparation of the manuscript.

Competing interests All authors stated that they have no competing interests.

Authors' contributions All authors stated that they have equally contributed to their chapters, read, edited and agreed to submit the final versions of their contributions.

Availability of data and materials All sources of information are provided in the text. Raw data related to the fossil material from Austria are available from the corresponding authors and Alexander Lukeneder upon request.

Additional information The article is part of the Special Volume: *Late Triassic Konservat-Lagerstätten within the Carnian Pluvial Episode. 150 Years of Scientific Excavations in Austria.*

Artificial Intelligence (AI) The chapter was produced without the use of AI.

Language The entire volume uses American English reviewed by a native scientist.

Received 2025.

Published online 2026.

References

Agassiz LJR (1833–1844) Recherches sur les poissons fossils, 5 vols. Petitpierre, Imprimerie de Neuchâtel

von Alberti F (1834) Beitrag zu einer Monographie des bunten Sandsteins, Muschelkalks und Keupers, und die Verbindung dieser Gebilde zu einer Formation. Verlag der J. G. Cotta'schen Buchhandlung, Stuttgart and Tübingen

Allard H, Carpenter SC, Duffin CJ, Benton MJ (2015) Microvertebrates from the classic Rhaetian bone beds of Manor Farm Quarry, near Aust (Bristol, UK). Proc Geol Assoc 126:762–776. https://doi.org/10.1016/j.pgeola.2015.09.002

Antunes MT, Maisey JG, Marques MM, Schaeffer B, Thomson KS (1990) Triassic fishes from the Cassange Depression (R. P. de Angola). Ciências da Terra (UNL), pp 1–64

Benton MJ (2016) The Triassic. Curr Biol 26:R1214–R1218. https://doi.org/10.1016/j.cub.2016.10.060

Bassani F (1886) Sui fossili e sull'età degli schisti bituminosi triasici di Besano in Lombardia. Communicazione preliminare. Atti Soc Ital Sci Nat Mus Civ Stor Nat Milano 29:15–72

Benton MJ, Wu F (2022) Triassic revolution. Front Earth Sci 10 https://doi.org/10.3389/feart.2022.899541

Benton MJ, Zhang Q, Hu S, Chen Z-Q, Wen W, Liu J, Huang J, Zhou C, Xie T, Tong J, Choo B (2013) Exceptional vertebrate biotas from the Triassic of China, and the expansion of marine ecosystems after the Permo-Triassic mass extinction. Earth-Sci Rev 125:199–243. https://doi.org/10.1016/j.earscirev.2013.05.014

Bernardi M, Avanzini M, Bizzarini F (2011) Vertebrate fauna from the San Cassiano Formation (early Carnian) of the Dolomites region. Geo Alp 8:122–127

Bernardi M, Gianolla P, Petti FM, Mietto P, Benton MJ (2018) Dinosaur diversification linked with the Carnian Pluvial Episode. Nat Commun 9:1499. https://doi.org/10.1038/s41467-018-03996-1

Bizzarini F, Prosser G, Prosser F, Prosser I (2003) Osservazioni preliminari sui resti di vertebrati della formazione di S. Cassiano del Bosco di Stuores (Dolomiti Nord-Orientali). Ann Mus Civ Rovereto 17:137–148

Bonaparte CL (1838) Synopsis vertebratorum systematis. Nuovi Ann Sci Nat Bologna 2:105–133

Böttcher R (2015) Fische des Lettenkeupers. In: Hagdorn H, Schoch R, Schweigert G (eds) Der Lettenenkeuper – Ein Fenster in die Zeit vor den Dinosauriern. Plaeodiversity, Sonderband, pp 141–202

Böttcher R (2024) Root resorption during tooth replacement in sharks—a unique character of the Hybodontiformes (Chondrichthyes, Elasmobranchii). Palaeodiversity 17:121–194. https://doi.org/10.18476/pale.v17.a6

Bratvold J, Delsett LL, Hurum JH (2018) Chondrichthyans from the *Grippia* bonebed (Early Triassic) of Marmierfjellet, Spitsbergen. Nor J Geol 98:189–217. https://doi.org/10.17850/njg98-2-03

Brignon A (2023) Les premières découvertes de vertébrés fossiles dans le Trias moyen de la Moselle (nord-est de la France). Bull Soc Hist Nat Ethn Colmar 79:51–88

Brito PM, Veiga IM, Dutheil DB, Bergqvist LP (2025) First occurrence of *Distobatus* Werner, 1989 (Elasmobranchii: Hybodontiformes) in the middle Cretaceous (Albian–Cenomanian) of Brazil: taxonomic and biogeographical implications Cretac Res 171:106119. https://doi.org/10.1016/j.cretres.2025.106119

Brough J (1935) On the structure and relationships of the hybodont sharks. Mem Manch Lit Philos Soc 79:35–49

Cappetta H (2012) Chondrichthyes. Mesozoic and Cenozoic Elasmobranchii: teeth. Handbook of Paleoichthyology 3E. Verlag Dr. Friedrich Pfeil, Munich

Chen ZQ, Benton MJ (2012) The timing and pattern of biotic recovery following the end-Permian mass extinction. Nature Geosci 5:375–383. https://doi.org/10.1038/ngeo1475

Coates MI, Tietjen K (2018) The neurocranium of the Lower Carboniferous shark *Tristychius arcuatus* (Agassiz, 1837). Earth Environ Sci Trans R Soc Edinb 108:19–35. https://doi.org/10.1017/S1755691018000130

Cuny G (2012) Freshwater hybodont sharks in Early Cretaceous ecosystems: a review. In: Godefroit P (ed) Bernissart dinosaurs and Early Cretaceous terrestrial ecosystems. Indiana University Press, Bloomington, pp 519–529

Cuny G, Srisuk P, Khamha S, Suteethorn V, Tong H (2009) A new elasmobranch fauna from the Middle Jurassic of southern Thailand. Geol Soc Lond Spec Publ 315:97–113. https://doi.org/10.1017/S0263593300001346

Dal Corso J, Bernardi M, Sun Y, Song H, Seyfullah LJ, Preto N, Gianolla P, Ruffell A, Kustatscher E, Roghi G, Merico A, Hohn S, Schmidt A, Marzoli A, Newton RJ, Wignall PB, Benton MJ (2020) Extinction and dawn of the modern world in the Carnian (Late Triassic). Sci Adv 6:eaba0099. https://doi.org/10.1126/sciadv.aba0099

Day ECH (1864) III. On *Acrodus Anningiæ*, Agass.; with Remarks upon the Affinities of the Genera *Acrodus* and *Hybodus*. Geol Mag 1:57–65. https://doi.org/10.1017/S001675680011622X

Duffin CJ (1993) Teeth of *Hybodus* (Selachii) from the Early Jurassic of Lyme Regis, Dorset (southern England): Preliminary note. Belg Geol Surv Prof Pap 264:45–52

Duffin CJ (2010). 16. Fishes–Sharks and rays. In: Lord AR, Davis PG (eds) Fossils from the lower lias of the Dorset coast. Palaeontological Association Field Guides to Fossils 13, London, pp 317–340

Duffin CJ, Thies D (1997) Hybodont shark teeth from the Kimmeridgian (Late Jurassic) of northwest Germany. Geol Palaeontol 31:235–256

Duffin CJ, Lauer B, Lauer R (2023) *Hybodus hauffianus* Fraas, 1895 from the Toarcian (Early Jurassic) of Dormettingen, SW Germany. N Jb Geol Paläontol Abh 307:141–154. https://doi.org/10.1127/njgpa/2023/1120

de Lange B, Chenal E, Diependaal HJ, Reumer JWF (2023) Fish remains from the Rhaetian (Late Triassic) of Winterswijk, the Netherlands (Pisces: Chondrichthyes and Actinopterygii). Neth J Geosci 102:e10. https://doi.org/10.1017/njg.2023.10

Friedman M, Sallan LC (2012) Five hundred million years of extinction and recovery: a Phanerozoic survey of large-scale diversity patterns in fishes. Palaeontology 55:707–742. https://doi.org/10.1111/j.1475-4983.2012.01165.x

Fu X, Wang J, Wen H, Wang Z, Zeng S, Song C, Chen W, Wan Y (2020) A possible link between the Carnian Pluvial Event, global carbon-cycle perturbation, and volcanism: new data from the Qinghai-Tibet Plateau. Glob Planet Change 194:103300. https://doi.org/10.1016/j.gloplacha.2020.103300

Geinitz HB (1837) Beitrag zur Kenntnis des thüringer Muschelkalkgebirges. Jena
Ginter M, Hampe O, Duffin CJ (2010) Chondrichthyes. Paleozoic elasmobranchii: teeth. Handbook of paleoichthyology 3D. Verlag Dr. Friedrich Pfeil, Munich
Glickman LS (1964) Class Chondrichthyes, subclass Elasmobranchii. In: Obruchev DV (ed) Fundamentals of paleontology, vol 11. Akad Nauk SSSR, pp 196–237 (in Russian)
Griffith J (1977) The Upper Triassic fishes from Polzberg bei Lunz, Austria. Zool J Linn Soc 60:1–93. https://doi.org/10.1111/j.1096-3642.1977.tb00834.x
Guinot G, Cavin L (2016) 'Fish' (Actinopterygii and Elasmobranchii) diversification patterns through deep time. Biol Rev 91:950–981. https://doi.org/10.1111/brv.12203
Hay OP (1902) Bibliography and catalogue of the fossil vertebrata of North America. Bull US Geol Surv 179:868 p. https://doi.org/10.3133/b179
Hulke JW (1873) Memorandum on some fossil vertebrate remains collected by the Swedish expeditions to Spitzbergen in 1864 and 1868. Bihang till K Svenska Vetensk Akad Handlingar 9:1–11
Huxley TH (1880) On the application of the laws of evolution to the arrangement of the Vertebrata, and more particularly of the Mammalia. Proc Sci Zool Soc London 1880:649–661
Ivanov AO, Duffin CJ, Richter M (2021) Youngest jalodontid shark from the Triassic of Europe and a revision of the Jalodontidae. J Vert Paleontol 41:e1931259. https://doi.org/10.1080/02724634.2021.1931259
Jaekel O (1889) Die Selachier aus dem oberen Muschelkalk Lothringens. Abh Geol. Spezialkarte Els-Lothr 3:273–332
Jambura PL, Türtscher J, Kindlimann R, Metscher B, Pfaff C, Stumpf S, Weber GW, Kriwet J (2020) Evolutionary trajectories of tooth histology patterns in modern sharks (Chondrichthyes, Elasmobranchii). J Anat 236:753–771. https://doi.org/10.1111/joa.13145
Jambura PL, Stumpf S, Kriwet J (2021) Skeletal remains of the oldest known pseudocoracid shark *Pseudocorax kindlimanni* sp. nov. (Chondrichthyes, Lamniformes) from the Late Cretaceous of Lebanon. Cret Res 125:104842. https://doi.org/10.1016/j.cretres.2021.104842
Klug S, Tütken T, Wings O, Pfretzschner H-U, Martin T (2010) A Late Jurassic freshwater shark assemblage (Chondrichthyes, Hybodontiformes) from the southern Junggar Basin, Xinjiang, Northwest China. Palaeobio Palaeoenv 90:241–257. https://doi.org/10.1007/s12549-010-0032-2
Krainer K, Lucas SG, Strasser M (2011) Vertebrate fossils from the Nordalpine Raibl Beds, western Northern Calcareous Alps, Tyrol (Austria). Austrian J Earth Sci 104:97–106
Kriwet J (2004) Late Jurassic selachians (Chondrichthyes: Hybodontiformes, Neoselachii) from Central Portugal. N Jb Geol Paläontol Mh 2004:233–256. https://doi.org/10.1127/njgpm/2004/2004/233
Kuhn E (1946) Über *Acrodus*-Funde aus dem Grenzbitumenhorizont der anisischen Stufe der Trias des Monte San Giorgio (Kanton Tessin). Ecologae Geol Helv 3:662–673
Lakin RJ, Duffin CJ, Hildebrandt C, Benton MJ (2016) The Rhaetian vertebrates of Chipping Sodbury, South Gloucestershire, UK, a comparative study. Proc Geol Assoc 127:40–52. https://doi.org/10.1016/j.pgeola.2016.02.010
Landemaine O (1991) Sélaciens nouveaux du Crétacé supérieur du Sud-Ouest de la France. Quelques apports à la systématique des élasmobranches. Soc Amicale Géol Amateurs 1:1–45
Lukeneder A (ed) (2026a) Late Triassic Konservat-Lagerstätten within the Carnian Pluvial Episode: 150 years of scientific excavations in Austria. Top Geobiol 55:600 p. https://doi.org/10.1007/978-3-032-19777-1
Lukeneder A (2026b) The marine Upper Triassic paleobiota from the Scheiblingbach *Konservat-Lagerstätte* in Austria. In: Lukeneder A (ed) Late Triassic Konservat-Lagerstätten within the Carnian Pluvial Episode: 150 years of scientific excavations in Austria. Top Geobiol 55, Chap 19:437–467. https://doi.org/10.1007/978-3-032-19777-1_19
Lukeneder A, Lukeneder P (2021) The Upper Triassic Polzberg palaeobiota from a marine *Konservat-Lagerstätte* deposited during the Carnian Pluvial Episode in Austria. Sci Rep 11:16644. https://doi.org/10.1038/s41598-021-96052-w

Lukeneder A, Lukeneder P (2022) Taphonomic history and trophic interactions of an ammonoid fauna from the Upper Triassic Polzberg palaeobiota. Sci Rep 12:7455. https://doi.org/10.1038/s41598-022-11496-y

Lukeneder A, Lukeneder P (2023) New data on the marine Upper Triassic palaeobiota from the Polzberg *Konservat-Lagerstätte* in Austria. Swiss J Palaeontol 142:9. https://doi.org/10.1186/s13358-023-00269-3

Lukeneder A, Surmik D, Gorzelak P, Niedźwiedzki R, Brachaniec T, Salamonm MA (2020) Bromalites from the Upper Triassic Polzberg section (Austria); insights into trophic interactions and food chains of the Polzberg palaeobiota. Sci Rep 10:20545. https://doi.org/10.1038/s41598-020-77017-x

Lukeneder A, Lukeneder P, Sachsenhofer RF, Roghi G, Rigo M (2024) Multi-proxy record of the Austrian Upper Triassic Polzberg *Konservat-Lagerstätte* in light of the Carnian Pluvial Episode. Sci Rep 14:11194. https://doi.org/10.1038/s41598-024-60591-9

Maisch MW, Matzke AT (2016) A new hybodontid shark (Chondrichthyes, Hybodontiformes) from the Lower Jurassic Posidonienschiefer Formation of Dotternhausen, SW Germany. N Jb Geol Paläontol Abh 280:241–257. https://doi.org/10.1127/njgpa/2016/0577

Maisey JG (1982) The anatomy and interrelationships of Mesozoic hybodont sharks. Am Mus Novit 2724:1–48

Maisey JG (1983) Cranial anatomy of *Hybodus basanus* Egerton from the Lower Cretaceous of England. Am Mus Novit 2758:1–64

Maisey JG (1987) Cranial anatomy of the lower Jurassic shark *Hybodus reticulatus* (Chondrichthyes: Elasmobranchii), with comments on hybodontid systematics. Am Mus Novi 2878:1–39

Maisey JG (2008) The postorbital palatoquadrate articulation in elasmobranchs. J Morphol 269:1022–1040. https://doi.org/10.1002/jmor.10642

Maisey JG (2011) The braincase of the Middle Triassic shark *Acronemus tuberculatus* (Bassani, 1886). Palaeontology 54:417–428. https://doi.org/10.1111/j.1475-4983.2011.01035.x

Maisey JG (2012) What is an 'elasmobranch'? The impact of palaeontology in understanding elasmobranch phylogeny and evolution. J Fish Biol 80:918–951. https://doi.org/10.1111/j.1095-8649.2012.03245.x

Maisey JG, Naylor GJP, Ward DJ (2004) Mesozoic elasmobranchs, neoselachian phylogeny and the rise of modern elasmobranch diversity. In: Arratia G, Tintori A (eds) Mesozoic Fishes 3. Systematics, paleoenvironments and biodiversity. Verlag Dr. Friedrich Pfeil, Munich, pp 17–56

Manzanares E, Pla C, Martínez-Pérez C, Ferrón H, Botella H (2016) *Lonchidion derenzii*, sp. nov., a new lonchidiid shark (Chondrichthyes, Hybodontiforms) from the Upper Triassic of Spain, with remarks on lonchidiid enameloid. J Vert Paleontol 37:e1253585. https://doi.org/10.1080/02724634.2017.1253585

Manzanares E, Escudero-Mozo MJ, Ferrón H, Martínez-Pérez C, Botella H (2020) Middle Triassic sharks from the Catalan Coastal ranges (NE Spain) and faunal colonization patterns during the westward transgression of Tethys. Palaeogeogr Palaeoclimatol Palaeoecol 539:109489. https://doi.org/10.1016/j.palaeo.2019.109489

Moreau M, Duffin CJ, Hildebrandt C, Hutchinson D, Parker A, Carpenter S, Benton MJ (2021) Microvertebrates from the Rhaetian basal bone bed of Saltford, near Bath, SW England. Proc Geol Assoc 132:174–187. https://doi.org/10.1016/j.pgeola.2020.11.003

Münster GG (1841) Beschreibung und Abbildung der in den Kalkmergelschichten von St. Cassian Gefunden Versteinerungen. Beitr Petrefactenk 4:25–152

Mutter RJ (1998) Zur systematischen Stellung einiger Bezahnungsreste von *Acrodus georgii* sp. nov. (Selachii, Hybodontoidea) aus der Grenzbitumenzone (Mittlere Trias) des Monte San Giorgio (Kanton Tessin Schweiz). Eclogae Geol Helv 91:513–519

Mutter RJ, de Blanger K, Neuman AG (2007) Elasmobranchs from the Lower Triassic Sulphur Mountain Formation near Wapiti Lake (BC, Canada). Zool J Linn Soc 149:309–337. https://doi.org/10.1111/j.1096-3642.2007.00244.x

Mutter RJ, Neuman AG, de Blanger K (2008) *Homalodontus* nom. nov., a replacement name for *Wapitiodus* Mutter, de Blanger and Neuman, 2007 (Homalodontidae nom. nov., ?Hybodontoidea), preoccupied by *Wapitiodus* Orchard, 2005. Zool J Linn Soc 154:419–420. https://doi.org/10.1111/j.1096-3642.2008.00488.x

Neves GS, Medeiros MA, Dutheil DB, Brito PM (2024) First record of *Aegyptobatus* (Hybodontiformes: Distobatidae) in the Cretaceous Alcântara formation (?Albian-Cenomanian) of Maranhão, Northeastern Brazil. Hist Biol 1–7. https://doi.org/10.1080/08912963.2024.2379031

Patterson C (1966) British Wealden sharks. Bull Br Mus Nat Hist 11:251–350

Pawlak W, Rozwalak P, Sulej T (2022) Triassic fish faunas from Miedary (Upper Silesia, Poland) and their implications for understanding paleosalinity. Palaeogeogr Palaeoclimatol Palaeoecol 590:110860. https://doi.org/10.1016/j.palaeo.2022.110860

Pindakiewicz M, Hryniewicz K, Janiszewska K, Kaim A (2024) A microfossil evidence for the composition of fish communities in the Late Triassic of Tethys: examples from Cassian Formation, Italy. Lethaia 57:1–15. https://doi.org/10.18261/let.57.3.7

Pla C, Márquez-Aliaga A, Botella H (2013) The chondrichthyan fauna from the Middle Triassic (Ladinian) of the Iberian Range (Spain). J Vert Paleontol 33:770–785. https://doi.org/10.1080/02724634.2013.748668

Preto N, Kustatscher E, Wignall PB (2010) Triassic climates—state of the art and perspectives. Palaeogeogr Palaeoclimatol Palaeoecol 290:1–10. https://doi.org/10.1016/j.palaeo.2010.03.015

Radinsky L (1961) Tooth histology as a taxonomic criterion for cartilaginous fishes. J Morphol 109:73–92. https://doi.org/10.1002/jmor.1051090106

Rees J (1999) Late Cretaceous hybodont sharks from the Kristianstad Basin, southern Sweden. N Jb Geol Paläontol Mh 1999:257–270

Rees J (2008) Interrelationships of Mesozoic hybodont sharks as indicated by dental morphology—preliminary results. Acta Geol Pol 58:217–221

Rees J, Underwood CJ (2002) The status of the shark genus *Lissodus* Brough, 1935, and the position of nominal *Lissodus* species within the Hybodontoidea (Selachii). J Vert Paleontol 22:471–479. https://doi.org/10.1671/0272-4634(2002)022[0471:TSOTSG]2.0.CO;2

Rees J, Underwood CJ (2006) Hybodont sharks from the Middle Jurassic of the Inner Hebrides, Scotland. Earth Environ Sci Trans R Soc Edinb 96:351–363. https://doi.org/10.1017/S0263593300001346

Rees J, Underwood CJ (2008) Hybodont sharks of the English Bathonian and Callovian (Middle Jurassic). Palaeontology 51:117–147. https://doi.org/10.1111/j.1475-4983.2007.00737.x

Rieppel O (1981) The hybodontiform sharks from the Middle Triassic of Mte. San Giorgio, Switzerland. N Jb Geol Paläontol Abh 16:324–353

Rieppel O (1982) A new genus of shark from the Middle Triassic of Monte San Giorgio, Switzerland. Palaeontology 25:399–412

Rieppel O (2019) Mesozoic sea dragons: Triassic marine life from the ancient tropical lagoon of Monte San Giorgio. Indiana University Press, Bloomington

Rigal S, Cuny G (2016) On the rarity of anterior teeth of *Asteracanthus magnus* (Euselachii: Hybodontiformes). N Jb Geol Paläontol Abh 279:35–41. https://doi.org/10.1127/njgpa/2016/0538

Rigo M, Joachimski MM (2010) Palaeoecology of Late Triassic conodonts: constraints from oxygen isotopes in biogenic apatite. Acta Palaeontol Pol 55:471–478. https://doi.org/10.4202/app.2009.0100

Romano C, Brinkmann W (2010) A new specimen of the hybodont shark *Palaeobates polaris* with three-dimensionally preserved Meckel's cartilage from the Smithian (Early Triassic) of Spitsbergen. J Vert Paleontol 30:1673–1683. https://doi.org/10.1080/02724634.2010.521962

Romano C, Koot MB, Kogan I, Brayard A, Minikh AV, Brinkmann W, Bucher H, Kriwet J (2016) Permian-Triassic Osteichthyes (bony fishes): diversity dynamics and body size evolution. Biol Rev 91:106–147. https://doi.org/10.1111/brv.12161

Seilacher A (1943) Elasmobranchierreste aus dem oberen Muschelkalk und Keuper Württembergs. N Jb Min Geol Paläontol Mh B 1943:256–293

Sepkoski J (1984) A kinetic model of Phanerozoic taxonomic diversity. III. Post-Paleozoic Families and Mass Extinctions. Paleobiology 10:246–267. https://doi.org/10.1017/S0094837300008186

Soto M, Perea D, Toriño P (2012) New remains of *Priohybodus arambourgi* (Hybodontiformes: Hybodontidae) from Late Jurassic-?earliest Cretaceous deposits in Uruguay. Cretac Res 35:118–123. https://doi.org/10.1016/j.cretres.2011.12.001

Stensiö EA (1921) Triassic fishes from Spitsbergen. Part 1. A. Holzhausen, Vienna

Stumpf S, Etches S, Underwood CJ, Kriwet J (2021a) *Durnonovariaodus maiseyi* gen. et sp. nov., a new hybodontiform shark-like chondrichthyan from the Upper Jurassic Kimmeridge Clay Formation of England. PeerJ 9:e11362. https://doi.org/10.7717/peerj.11362

Stumpf S, López-Romero FA, Kindlimann R, Lacombat F, Pohl B, Kriwet J (2021b) A unique hybodontiform skeleton provides novel insights into Mesozoic chondrichthyan life. Pap Palaeontol 7:1479–1505. https://doi.org/10.1002/spp2.1350

Stumpf S, Meng S, Kriwet J (2022) Diversity patterns of Late Jurassic chondrichthyans: new insights from a historically collected hybodontiform tooth assemblage from Poland. Diversity 14:85. https://doi.org/10.3390/d14020085

Stumpf S, Kettler C, Kindlimann R, Cuny G, Kriwet J (2023) The oldest Gondwanan record of the extinct durophagous hybodontiform chondrichthyan, *Strophodus* from the Bajocian of Morocco. Swiss J Palaeontol 142:5. https://doi.org/10.1186/s13358-023-00270-w

Szabó M, Botfalvai G, Ősi A (2019) Taxonomical and palaeoecological investigations of the chondrichthyan and osteichthyan fish remains from the Middle-Late Triassic deposits of the Villány Hills (Southern Hungary). Geobios 75:111–126. https://doi.org/10.1016/j.geobios.2019.10.006

Teixeira C (1956) Sur un Hybodonitdé du Karroo de l'Angola. Rev Faculdade Ciênc Univ Lisboa C 5:135–136

Thies D, Reif W-E (1985) Phylogeny and evolutionary ecology of Mesozoic Neoselachii. N Jb Geol Paläontol Abh 169:333–361

Thomson KS (1982) An early Triassic hybodont shark from northern Madagascar. Postilla 186:1–16

Weeks OJ, Cooper RB, Whiteside DI, Duffin CJ, Copp C, Hildebrandt C, Hutchinson D, Benton MJ (2025) Microvertebrates from a Rhaetian neptunian dyke at Holwell, Somerset: dating the fissures. Proc Geol Assoc 136:101112. https://doi.org/10.1016/j.pgeola.2025.101112

Wen W, Zhang Q, Kriwet J, Hu S, Zhou C, Huang J, Cui X, Min X, Benton MJ (2023) First occurrence of hybodontid teeth in the Luoping Biota (Middle Triassic, Anisian) and recovery of the marine ecosystem after the end-Permian mass extinction. Palaeogeogr Palaeoclimatol Palaeoecol 617:111471. https://doi.org/10.1016/j.palaeo.2023.111471

Woodward AS (1889) Catalogue of the Fossil fishes in the British Museum (Natural History). Part I. British Museum (Natural History), London

Zhang P, Yang M, Lu J, Jiang Z, Vervoort P, Zhou K, Xu X, Chen H, Wang Y, He Z, Bian X, Shao L, Hilton J (2024) Four volcanically driven climatic perturbations led to enhanced continental weathering during the Late Triassic Carnian Pluvial Episode. Earth Planet Sci Lett 626:118517. https://doi.org/10.1016/j.epsl.2023.118517

Chapter 22
The Plant Fossils and Amber of the Carnian (Late Triassic) of the Scheiblinggraben in Austria

Evelyn Kustatscher, Roberta Branz, Miriam Slodownik, Guido Roghi, and Alexander Lukeneder

Abstract The Scheiblinggraben (Northern Calcareous Alps, Austria) provides a well-resolved stratigraphic profile yielding a diverse assemblage of fossils dated to the Carnian Pluvial Episode (Late Triassic). Bed-by-bed excavations yielded 4661 fossils, including 893 plant remains, dominated by conifer shoots, leaves, and wood fragments. Additional taxa include *Elatocladus*, *Pelourdea*, and putative *Radicites* (roots) along with linear leaf fragments with no attributable botanical affinity. The occurrence of amber droplets and resin-filled channels in conifer shoots represents one of the oldest occurrences of resin production in the Alpine Triassic. The plant fossil assemblage is broadly comparable to that of the nearby Polzberg *Konservat-Lagerstätte.* The plant fossil assemblage from Scheiblingbach indicates a stratigraphic variation in the abundance of plant remains that may reflect the relative sedimentation rate and/or transport of organic material from the terrestrial environment into the basin. Palynological data reveal a more diverse assemblage dominated by gymnosperm pollen, with subordinate lycophyte and sphenophyte spores and

E. Kustatscher (✉) · R. Branz
Museum of Nature South Tyrol, Bozen, Italy
e-mail: e.kustatscher@tiroler-landesmuseen.at

E. Kustatscher
Department of Natural History, Tirolean State Museum, Tirol, Austria

Department für Geo– und Umweltwissenschaften, Paläontologie und Geobiologie, Ludwig–Maximilians–Universität and Bayerische Staatssammlung für Paläontologie und Geologie, Munich, Germany

M. Slodownik
Botany Department, Trinity College Dublin, Dublin, Ireland
e-mail: Miriam.Slodownik@tcd.ie

G. Roghi
Institute of Geosciences and Earth Resources—CNR, Padova, Italy
e-mail: guido.roghi@igg.cnr.it

A. Lukeneder
Department of Geology and Paleontology, Natural History Museum Vienna, Vienna, Austria
e-mail: alexander.lukeneder@nhm.at

A. Lukeneder (ed.), *Late Triassic Konservat-Lagerstätten within the Carnian Pluvial Episode*, Topics in Geobiology 55, https://doi.org/10.1007/978-3-032-19777-1_22

only rare ferns and cycadophytes. The composition of the sporomorph assemblage and its attribution to Sporomorph EcoGroups (SEGs) demonstrates a predominance of Lowland and Hinterland/Upland SEG elements, contrasting with the River and Coastal SEG taxa typical of more proximal marine sites such as Polzberg or Lunz am See. This pattern reflects the paleogeographic setting of Scheiblingbach as a distal basinal setting in relation to the coastline, where wind-transported conifer pollen and resin replaced the spore-rich signals of humid Lowland ecosystems. Altogether, these data contribute to a more comprehensive understanding of vegetational responses to climate instability, the development of resin-producing ecosystems, as well as the taphonomic bias in distal marine settings.

Keywords Reingraben Formation · CPE · Amber · Terrestrial ecosystem · Styria

22.1 Introduction

The Scheiblingbachgraben (= Scheiblingbach from here on) in the Northern Calcareous Alps has been a renowned fossil site since the 1970s, when first fossil discoveries were reported from the Reifling, Göstling and Reingraben Formations in the Northern Calcareous Alps (Krystyn 1978, 1991; Summesberger and Wagner 1972; Mostler and Scheuring 1974; Lahm 1984). Early studies primarily focused on lithology, biostratigraphy, and selected fossil groups such as ammonoids, bivalves, radiolarians, and conodonts (Krystyn 1978, 1991; Lahm 1984; Hornung 2008; Hornung et al. 2007). More recently, systematic excavations and bed-by-bed collections have yielded thousands of exceptionally preserved fossils of marine invertebrates, vertebrates, and plants (Lukeneder et al. 2020; Lukeneder and Lukeneder 2021, 2022a, b, 2023, 2024). These new collections significantly expanded our knowledge of the Scheiblingbach biota, enabling a reconstruction of the paleoenvironment and trophic food webs (Lukeneder et al. 2020; Lukeneder and Lukeneder 2021), the taphonomy, and trophic history, as well as the determination of the ammonoids (Lukeneder and Lukeneder 2022a, 2024).

Within the Northern Calcareous Alps, Scheiblingbach is closely linked stratigraphically and paleoenvironmentally to the Polzberg *Konservat-Lagerstätte* in the Northern Calcareous Alps, located about 35 km to the northeast. Both localities represent intraplatform basins of the Reifling Unit and yield fossil assemblages of Julian 2 age (*Austrotrachyceras austriacum* Zone; Krystyn 1991; Mueller et al. 2016; Lukeneder and Lukeneder 2021, 2023; Lukeneder et al. 2024). Together with the renowned Lunz flora from the Lunz Formation (Mueller et al. 2016; Pott and Lukeneder A, this volume and references therein), these deposits form key archives for reconstructing biotic and environmental changes during the Carnian Pluvial Episode (CPE; Simms and Ruffell 1990; Hornung and Brandner 2005; Dal Corso et al. 2018, 2020). Importantly, however, their paleogeographic positions within the Reifling Basin differ: Scheiblingbach was located on the southwestern margin of the basin, whereas Polzberg occupied a more northern, inwards, near-shore setting

(Hornung et al. 2007; Hornung 2008; Lukeneder and Lukeneder 2021, 2023). Lunz, in contrast, was deposited in a deltaic swamp environment (Pott and Lukeneder A, this volume). These differences are reflected in contrasting fossil assemblages, faunal abundances, and preservation styles.

Studying the plant remains from Scheiblingbach is instructive because they complement the well-documented Polzberg and Lunz plant fossil assemblages (Kustatscher et al. 2026; Pott and Lukeneder 2026) and provide further evidence of terrestrial input into the basin. Although plant fossil assemblages are relatively rare and often taphonomically biased in such marine settings, their occurrence within laminated marine deposits alongside marine faunas allows a direct comparison of terrestrial and marine ecosystems during the CPE. Furthermore, the preservation of amber at Scheiblingbach offers insights into terrestrial ecosystems and resin-producing plants not documented from Polzberg. The integration of these plant fossils with sedimentological and paleontological evidence provides a more complete understanding of environmental dynamics, biotic responses, and basin evolution during the CPE.

This paper provides an overview of the plant fossils at Scheiblingbach, their depositional context, and comparisons to the plant fossil assemblage of Polzberg and Lunz. This adds an additional data point to interpret the regional floras of the Northern Calcareous Alps during the Carnian.

22.2 Material and Methods

The Scheiblingbach *Konservat-Lagerstätte* was investigated through a series of high resolution, bed-by-bed excavations conducted under the lead of one of the authors (AL) between 2022 and 2025 in the Scheiblingbachgraben near Großreifling, Styria (47°39′52.50"N, 14°42′27.75"E). The Scheiblingbach excavation point (SCH) is located on the northern slope of Mount Lärchkogel (1087 m), south of Mount Reiflingberg (712 m), 600 m southwest of the Enns River, 300 m west of Großreifling (47°39′52.50"N, 14°42′27.75"E) (geological map, 100 Hieflau, Kreuss 2014; Fig. 22.1). The section exposes the uppermost Reifling Formation, the overlying Göstling Formation, and the basal part of the Reingraben Formation, all of which yielded plant material in addition to abundant marine faunas (Krystyn 1991; Lukeneder and Lukeneder 2021, 2023; Lukeneder et al. 2024). A total of 4661 specimens were collected during 60 days of field campaigns, including both macrofossils and microfossils (Lukeneder et al. 2020; Lukeneder and Lukeneder 2021). Sampling was carried out in a standardized bed-by-bed approach in calcareous intervals, whereas the softer argillaceous parts of the Reingraben Formation were sampled in targeted layers with high fossil density (Lukeneder and Lukeneder 2023).

Plant fossils are preserved as compressions and impressions. They occur either as dispersed fragments within fossiliferous horizons or as concentrated plant layers. The majority of remains consist of conifer shoots, leaves, and wood fragments, with occasionally reproductive organs and amber. Amber occurs both dispersed within

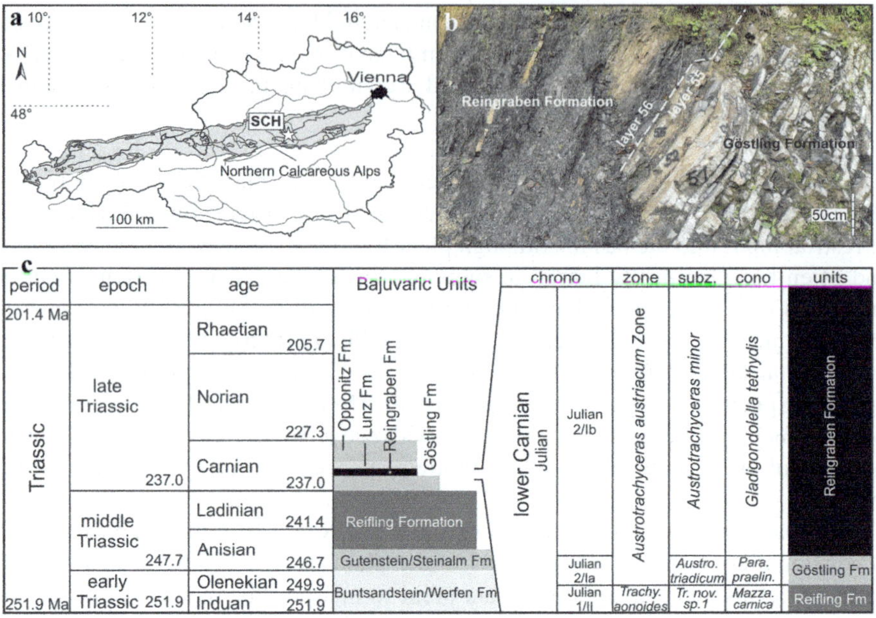

Fig. 22.1 **a** Geographic position of the Scheiblingbach section (SCH). **b** Overview photo of the outcrop yielding the plant fossils. **c** Chrono- and biostratigraphic position of the Scheiblingbach succession

the rock matrix and in situ within the resiniferous channels in the conifer shoots (Lukeneder et al. 2026b, this volume; Harzhauser et al. 2026).

All plant remains were photographed with a NIKON D5200 digital system camera. In situ amber was imaged using fluorescence with a Dino-Lite AM4115T-GFBW hand-held digital microscope with a 480 nm LED light source. UV-fluorescent macro-photographs were taken using a SONY ILCE-7RM2 digital reflex camera and a highpowered Alonefire SV005 10W UV torch with a peak wavelength of 365 nm modified with a 365 nm filter, following the method of Slodownik et al. (2023a, b). The macrofossils were measured from digital photographs using ImageJ1.53a (Schneider et al. 2012).

For the Schulze method (to optain palynomorphs), the organic matter was placed in a glass tube, and a small amount of crystalline potassium chlorate (KCl) was added (an amount slightly exceeding the sample's mass). Then, under a fume hood, 3 ml of 30% nitric acid (HNO3) was added. After each day for the following three days, 1 mL of 67.5% nitric acid was added. On the fifth day, the acid was decanted and the residue washed with deionized water. Then, a few drops of 5% potassium hydroxide (KOH) were added to the residue, and after a few seconds diluted with deionized water to stop the reaction; in case of an instantaneous reaction, water was added immediately. For light microscopy, residues were mounted on glass slides in glycerine jelly and photographed with a Leica DM 2500 LED microscope and a Leica DMC4500 camera (for details see Nowak et al. 2022). Measurements were taken from the images with the Fiji distribution of ImageJ version 1.53c (Schindelin et al. 2012).

The fossil material from Scheiblingbach is curated at the Natural History Museum Vienna (NHMW) and catalogued under collection numbers NHMW 2025/0259/0001-0105, plus two numbers from previous publications NHMW 2025/0127/0025 and 0026 (Lukeneder 2026a, b, overview this volume). Comparative material from Polzberg (NHMW, GeoSphere Austria) was used to evaluate similarities and differences in taphonomy, taxonomic composition, and depositional context (e.g., Kustatscher et al. 2026).

22.3 The Paleobotanical Composition of the Plant Fossil Assemblage

The excavation carried out by AL and colleagues from the NHMW yielded 4661 fossils of various levels of fragmentation and preservation. Of these fossils, 893 were identified as plant fossils, 205 of which were assigned during the excavations to conifers, two to putative roots, and six to straplike plant remains of undeterminable affinity. All other remains are not preserved at a level that permits assignment to any genus (e.g., often only representing fragments of organic material).

Amber (Fig. 22.2a–c)

At least 38 specimens are associated with preserved amber; those with otherwise poor preservation are generally interpreted to belong to conifers because they correspond in overall size and morphology of the shoots and leaves. Some specimens exhibit rather amorphous amber (Fig. 22.2a), while other have intricate resin canals along the axes that branch into the leaves (Fig. 22.2b–c). Most taxa identified from the Scheiblingbach plant fossil assemblage are identical to those described from Polzberg (Kustatscher et al. 2026), to which we refer for more detailed taxonomic descriptions and discussion.

***Radicites* sp. (Fig. 22.2d–e)**

The plant fragments are up to 12.8 mm long and 1.4 mm wide with a monopodial ramification. These are generally assigned to the genus *Radicites*, i.e. root fragments washed into the sediments (e.g., NHMW 2025/0259/0105-106, Fig. 22.2d–e). They are generally attributed to sphenophytes (e.g., Kelber and Hansch 1995; Kustatscher et al. 2008), although it is not completely clear whether all roots belong to this group. They could also belong to other plant groups.

***Elatocladus* sp. (Fig. 22.2f)**

The genus *Elatocladus* is represented by an apical shoot (NHMW 2025/0259/0001, Fig. 22.2F) fragment. The shoot fragment is 65.0 mm long and 22.5 mm wide, with spirally attached lanceolate leaves with a rounded apex. The leaves are 14.2 mm long

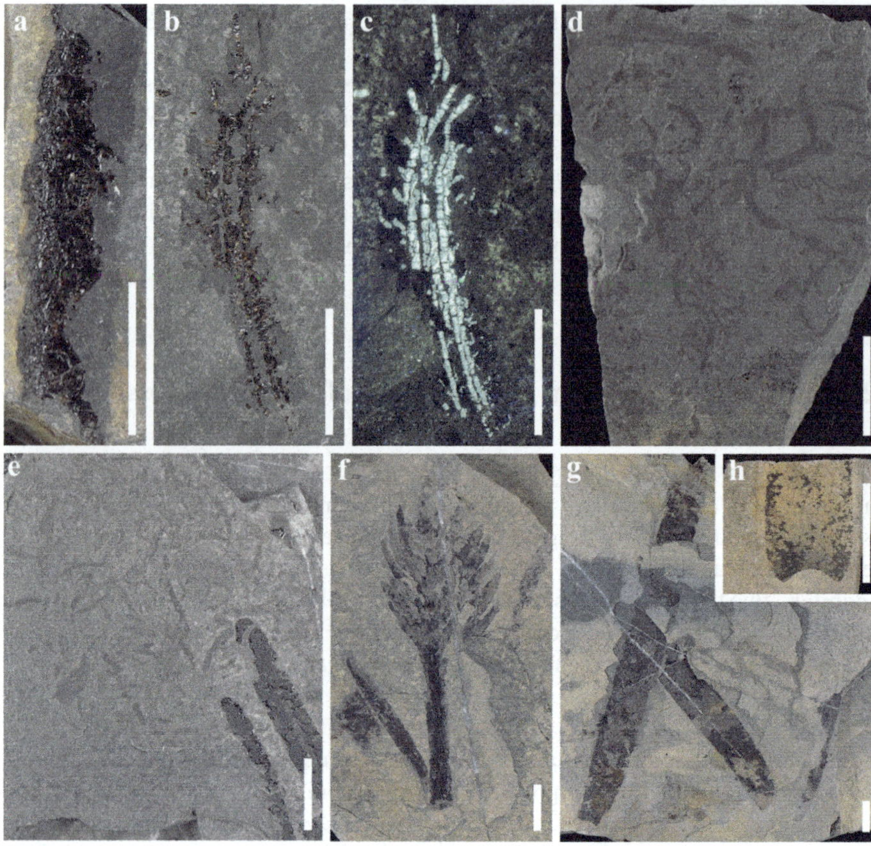

Fig. 22.2 **a** Typical amorphous amber of coniferous origin preserved in the Scheiblingbach site (NHMW 2025/0259/0083). **b** Typical morphologically preserved coniferous resin canals preserved as amber typical for the Scheiblingbach site (NHMW 2025/0259/0098). **c** UV photograph of **b**. **d** *Radicites* sp., roots (NHMW 2025/0259/0105). **e** *Radicites* sp., roots (NHMW 2025/0259/0106). **f** Shoot fragment of *Elatocladus* sp. (NHMW 2025/0259/0001A). **g** *Pelourdea vogesiaca* leaves (NHMW 2025/0259/0015). **h** *Pelourdea vogesiaca* leaf base (NHMW 2025/0259/0039). Scale bars = 10 mm

and 2.0 mm wide and decrease apically in size. At the base of the shoot fragment, a single larger leaf is attached to the axis, 33.4 mm long and 3.7 mm wide. The specimen resembles shoot fragments assigned to *Elatocladus* sp. from the Polzberg assemblage (Kustatscher et al. 2026), but they are narrower and show a stronger apical size reduction. The plant fossil assemblage also contains dispersed leaves. They reach lengths of up to 80 mm and widths of 6.2 mm.

***Pelourdea* sp. (Fig. 22.2g–h)**

The dispersed leaves are up to 58.2 mm long and 9.4 mm wide and show the characteristic crescent-shaped leaf base (NHMW 2025/0259/0001, 2025/0259/0039,

Fig. 22.2g–h). In the Polzberg assemblage, two morphotypes of *Pelourdea* are present, *P. vogesiaca* (wide, lanceolate leaves) and *Pelourdea* sp. (slimmer, curved leaves, Kustatscher et al. 2026, this volume). All fragments in Scheiblingbach likely belong to the small morphotype, *Pelourdea* sp.

***Brachyphyllum* sp. 1 sensu Kustatscher et al.** (2026) **(Fig.** 22.2.3**a–e, g)**

Fig. 22.3 **a** *Brachyphyllum* sp. 1 sensu Kustatscher et al. 2026, Polzberg, this volume, ultimate shoot fragment with resin canals filled with orange-colored amber (NHMW 2025/0259/0025). **b** UV image of A (NHMW 2025/0259/0025). **c** *Brachyphyllum* sp. 1 sensu Kustatscher et al. 2026, Polzberg, this volume, penultimate branch with abundant anatomically preserved resin canals preserved as orange-colored amber (NHMW 2025/0259/0016). **d** UV image of C (NHMW 2025/0259/0016). **e** Dinolite close-up of A showing detail of resin canal (NHMW 2025/0259/0025). **f** *Brachyphyllum* sp. 2 penultimate conifer shoot (NHMW 2025/0127/0025). **g** *Brachyphyllum* sp. 1 sensu Kustatscher et al. 2026, Polzberg in this volume, penultimate branch with preserved cuticle (NHMW 2025/0259/0052). Scale bars a–d, f, g = 10 mm. Scale bar E = 1 mm

The ultimate shoot fragments are up to 44.3 mm long and 6.8 mm wide (e.g., NHMW 2025/0259/016, NHMW 2025/0259/025, Fig. 22.4a–d). The axes of the shoots are covered by helically arranged, imbricate, scale-like leaves. The individual leaves are 8.3 mm long and 3.3 mm wide. The leaf apices are typically obtuse, sometimes acute, and curved towards the axis. Some shoot fragments also preserve amber that follows the resiniferous channels along the branches and bifurcates into the leaves. At least two foliar channels are present, including one abaxial and one adaxial channel that extend into the leaf tip and are up to 0.36 mm wide (e.g., NHMW 2025/0259/0016, 2025/0259/0025, Fig. 22.2.3b, d, e).

Similar shoot types were already described from the Polzberg assemblage as *Brachyphyllum* sp. 1, yet without any traces of amber (Kustatscher et al. 2026, this volume). At Polzberg, however, this type is very rare (three samples only; Kustatscher et al. 2026, this volume), whereas in Scheiblingbach it is one of the more abundant types (at least 23 conifer fragments). Comparable shoots are also known from the Carnian of the Julian Alps, the Karawanken Mountains, and the western Southern Alps (Passoni 1999; Dobruskina 2001).

***Brachyphyllum* sp. 2 sensu Kustatscher et al.** (2026) **(Figs.** 22.3**f and** 22.4)

The penultimate shoot fragments are up to 68.0 mm long and 14.2 mm wide (e.g., NHMW 2025/0127/0025, Fig. 22.3F). The falcate leaves are inserted in a close helix, partly imbricating. The leaves are 7.1 mm long and 3.8 mm wide. The leaf apices are slightly acute and curved towards the axis. The cuticle is moderately thick, with randomly arranged, dicyclic or incomplete dicyclic stomatal complexes. These consist of 6–8 trapezoidal subsidiary cells, lacking papillae (first cycle: average width 16.1 μm, length 25.0 μm; second cycle: average width 17.8 μm, length 38.4 μm). Epidermal cells are isodiametric and lack papillae (average length 38.6 μm, width 29.2 μm). Similar shoots were already described from Polzberg as *Brachyphyllum* sp. 2 (Kustatscher et al. 2026, this volume). Both at Polzberg and Scheiblingbach this type is very rare.

***Brachyphyllum* sp. 3 sensu Kustatscher et al.** (2026) (**Fig. 22.4g–k**)

The ultimate shoot fragments are up to 49.6 mm long and 9.1 mm wide (e.g., NHMW 2025/0259/0044, 2025/0259/070, 2025/0259/0080, Fig. 22.4g–j). The shoots are arranged in a close helix, strongly imbricating, clasping the branch. The shape is approximately triangular with short, pointed apices that are sometimes broken. The leaves are up to 2.7 mm long and 2.1 mm wide. Resin canals are preserved as amber with at least three to four resin canals of up to about 0.2 mm width running along the branch and bifurcating into each leaf, where a single canal continues into the leaf tip (Fig. 22.4F). The cuticle is thick, with randomly arranged monocyclic stomatal complexes consisting of 6–8 trapezoidal subsidiary cells. The stomata are distinct, with larger, more cutinized subsidiary cells compared to the epidermis. Each subsidiary cell has a small papilla extending towards the aperture. Epidermal cells are polygonal to isodiametric, lacking papillae. Some fragments also preserve amber along resin canals or at broken leaf tips (NHMW 2025/0259/0044, Fig. 22.4h). Similar shoots were described from Polzberg as *Brachyphyllum* sp. 3

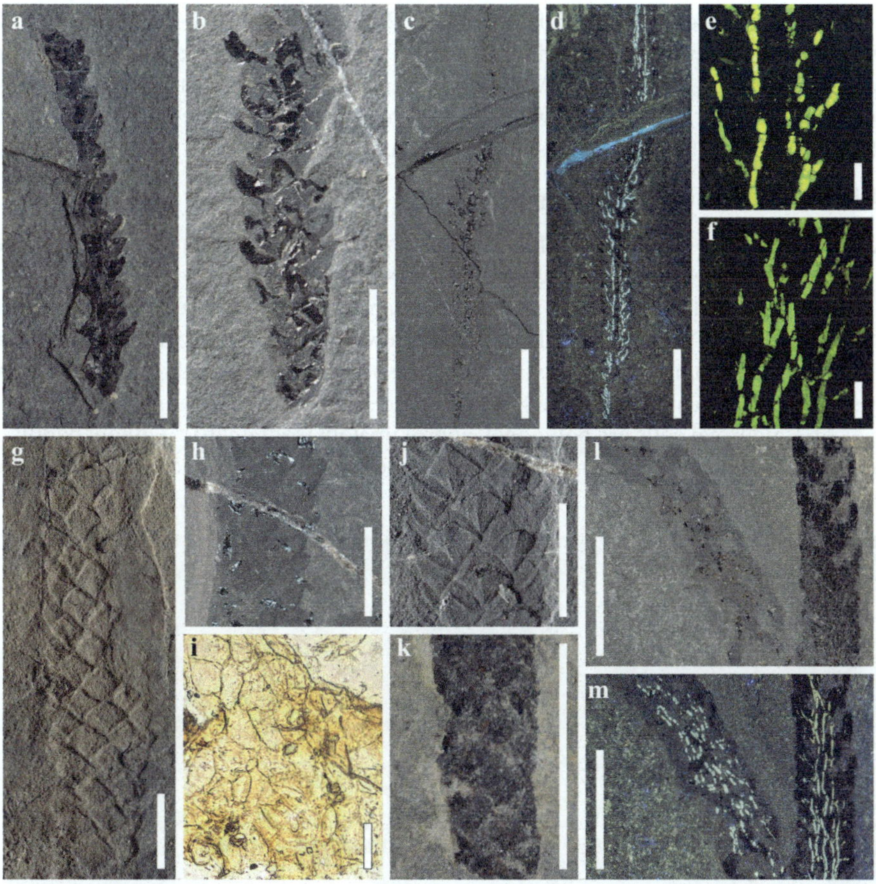

Fig. 22.4 **a** *Pagiophyllum* sp. 1 (NHMW 2025/0259/0011). **b** *Pagiophyllum* sp. 1 (NHMW 2025/0259/0029). **c** *Pagiophyllum* sp. 1, ultimate shoot fragment with amber (NHMW 2025/0259/0063). **d** UV image of C (NHMW 2025/0259/0063). **e** Dinolite image showing detail of resin canals of C (NHMW 2025/0259/063). **f** Detail of *Brachyphyllum* sp. 2 under UV light (NHMW 2025/0127/0025). **g** *Brachyphyllum* sp. 3 (NHMW 2025/0259/0043). **h** *Brachyphyllum* sp. 3 (NHMW 2025/0259/0044). **i** *Brachyphyllum* sp. 3 (NHMW 2025/0259/0069). **j** Detail of *Brachyphyllum* sp. 3 (NHMW 2025/0259/0080). **k** *Brachyphyllum* sp. 3 (NHMW 2025/0259/0094). Scales a–d, g–h, j–k = 10 mm; E–F = 1 mm; i = 50 μm

(Kustatscher et al. 2026, this volume). At Polzberg this type is very rare (one sample), whereas in Scheiblingbach it is the most abundant (at least 32 shoot fragments).

***Pagiophyllum* sp. 1 sensu Kustatscher et al.** (2026) **(Fig.** 22.4**a–e)**

The ultimate shoot fragments are up to 38.9 mm long and 7.4 mm wide (e.g., NHMW 2025/0259/0011, 2025/0259/0029, Fig. 22.4a–b). The shoots are covered by small leaves arising at a wide angle. The leaves are 3.9 mm long and 2.0 mm wide. The leaf apices are slightly acute and curved towards the axis. Some fragments also preserve amber along resin canals (e.g., NHMW 2025/0259/063, Fig. 22.5d–f). Similar shoots and cuticle fragments were described from the Polzberg assemblage as *Pagiophyllum* sp. 1 (Kustatscher et al. 2026). At Polzberg, however, this type is rare (five samples only; Kustatscher et al. 2026), whereas in Scheiblingbach it is relatively abundant (12 out of 70 conifer fragments studied in detail).

Other conifer shoot morphotypes

An additional, an informal conifer shoot morphotype is represented by a 14.0 mm long and 10.2 mm wide shoot fragment (NHMW 2025/0259/0022). Falcate leaves, up to 5.3 mm long and 2.7 mm wide, arise at a wide angle. The apices are acute and curved towards the axis. The cuticle is relatively thin. Few stomata are preserved, represented by monocyclic complexes of six trapezoidal subsidiary cells without papillae (on average 34 × 35 µm). Due to the few preserved stomata, their arrangement remains unclear. Epidermal cells are polygonal, approximately isodiametric, without papillae or wall thickenings, but with relatively thick anticlinal walls (average 44 × 34 µm). In

Fig. 22.5 **a** Mass occurrences of dispersed leaves in parallel orientation (NHMW 2025/0259/099). **b** Mass occurrences of dispersed leaves in random orientation (NHMW 2025/0259/088). **C** Three leaves parallel-oriented, perhaps bifurcating (NHMW 2025/0259/111). **d** Mass of dispersed leaves (NHMW 2025/0259/087). **e** Several leaves arising apparently from a central position (NHMW 2025/0259/090). **f** Bifurcating leaf fragment (NHMW 2025/0259/109). **g** Mass of dispersed leaves, perhaps bifurcating (NHMW 2025/0259/112). Scale bars = 10 mm

macromorphology this shoot fragments resembles *Pagiophyllum* sp. 1 but the cuticle is different with more subsidiary cells in each monocyclic stomata, all subsidiary cells being equally cutinized, and no papillae ornamenting the subsidiary cells. This is why we assigned it to a different morphotype for now.

The second informal conifer shoot morphotype is composed of ultimate shoot fragments, 18.0 mm long and 7.7 mm wide (NHMW 2025/0259/002, 2025/0259/048). The shoots are covered by lanceolate leaves arising at an acute angle. The leaves are 3.2 mm long and 1.7 mm wide. The cuticle is relatively thick, with monocyclic stomatal complexes arranged randomly. These consist of 5–7 trapezoidal subsidiary cells without papillae (average length 35 μm, width 23 μm). The subsidiary cells of the stomata are difficult to distinguish from the epidermis cells due to uniform cutinization and similar cell shapes, except in rare cases where subsidiary cells show thicker anticlinal walls. Epidermal cells are polygonal to isodiametric and lack papillae (average length 35 μm, width 24 μm). Similar cuticles were already described from Polzberg in the poorly preserved fragments assigned to the informal woody branches.

Straplike leaves (Fig. 22.5a–g).

Some strap-like leaf fragments preserved in the plant fossil assemblage (Fig. 22.5a–g). They are up to 100 mm long and less than 10 mm wide. Sometimes they seem to bifurcate. Since cuticle are still missing, it is impossible to determine where these leaf fragments belong to.

22.4 Discussion

22.4.1 Amber

Overall, the Scheiblingbach plant fossils are anatomically well preserved, comprising a mixture of impressions and compressions, and are generally comparable to the preservation of the Polzberg plant fossils. The key distinction is the presence of amber, especially the anatomically preserved amber within the fossils at Scheiblingbach in addition to dispersed amber fragments, both absent in the Polzberg plant fossil assemblage. As the relative composition between the plant fossil assemblages of Scheiblingbach and Polzberg are rather similar, we infer that the presence of amber is due to taphonomic conditions rather than taxonomic factors.

The Carnian Pluvial Episode is famous for its global amber occurrences with remarkable deposits especially from the Dolomites and Julian Alps (Italy; Roghi et al. 2006a, b; Forte et al. 2022). Amber droplets from the Heiligkreuz and Tor formations in the Dolomites (e.g., Rifugio Dibona, Dogna) occur in close association with conifer shoots, suggesting that resin production was mainly driven by stressed conifers growing in coastal and deltaic environments (Roghi et al. 2006a, b; Dal Corso et al. 2011; Seyfullah et al. 2018a, b; Forte et al. 2022). The climatic oscillations and hydrological intensification during the CPE likely promoted resin exudation, as resin secretion in conifers can be a physiological response to environmental stress (stormy events, too much water), disease, or mechanical damage (Langenheim 2003; Wolfe et al. 2009; Seyfullah et al. 2018a, b). The exceptional preservation of amber within marginal-marine sediments indicates rapid burial and low-oxygen conditions, which limited resin degradation (Roghi et al. 2017). The high abundance of amber droplets in paleosols indicates repeated resin exudation under fluctuating environmental conditions, linked to short-lived humid pulses and subsequent drying phases characteristic of the CPE (Gianolla et al. 1998; Seyfullah et al. 2018a, b).

Isotopic analyses of Carnian amber provide additional paleoenvironmental insights. The $\delta^{13}C$ values of amber from Rifugio Dibona show a variability similar to that of modern resins but retain their original isotopic signal, reflecting atmospheric carbon composition during resin secretion (Dal Corso et al. 2011). This allows correlation between terrestrial and marine carbon-isotope records and supports the view that amber can serve as an independent proxy for carbon-cycle perturbations during the CPE. The study of amber within the CPE thus extends beyond paleobotanical significance: it links terrestrial plant physiology, climate/environmental stress such as changing habitat and climate conditions and stormy events, and the carbon cycle, providing a unique geochemical and paleoecological signature. Amber acts both as a signal for climate and environmental stress as well as a physical archive of the atmosphere, recording interactions between terrestrial vegetation and climate change.

22.4.2 Comparison with the Microflora from Scheiblingbach

The Scheiblingbach succession yielded a relatively diverse assemblage of spores and pollen grains (Table 22.1), along with marine palynomorphs such as algal cysts and foraminiferal linings. Lycophyte spores are diverse, with representatives of Isoetales (*Aratrisporites*), Selaginellales (*Kraeuselisporites*), and Lycopodiales (*Camarozonosporites*). Sphenophyte (*Calamospora*) and fern spores (*Concavisporites, Todisporites*) are present but rare. Among the gymnosperms, conifers are the most diverse group, including taxa assigned to Araucariaceae (*Araucariacites*), Cheirolepidiaceae (*Circumpollen*), Patokaeaceae (*Enzonalasporites, Patinasporites, Vallasporites*), and Voltziaceae (*Infernopollenites, Lunatisporites, Ovalipollis, Triadispora*). Additional conifer pollen grains with uncertain affinity include *Haberkornia* and *Ellipsovelatisporites*. Bennettitalean pollen (*Aulisporites, Leschikisporis*) and cycadalean pollen (*Chasmatosporites, Cycadopites*) also contribute to the assemblage.

The palynological assemblage reflects a complex terrestrial ecosystem dominated by a floodplain vegetation, with contributions from river, coastal, and pioneer communities as well as hinterland floras (Table 22.2; compare also with Roghi and Lukeneder 2026, this volume). This reconstruction is consistent with the diverse landscapes inferred from the Lunz plant macrofossil assemblages (Pott and Lukeneder 2026,

Table 22.1 Botanical affinity of the spore and pollen genera of the Scheiblingbach succession (Couper 1958; Townrow 1962; Schulz 1967; Clement-Westerhof 1974; Van Konijnenburg-van Cittert 1978, 1993; Looy et al. 2001; Barrón et al. 2006; Bonis and Kürschner 2012; Costamagna et al. 2018; Baranyi et al. 2019; Zhang et al. 2021)

Spores/pollen genera (Scheiblingbach)	Botanical affinity
Aratrisporites	Lycophytes, Isoetales
Araucariacites	Conifers, Araucariaceae
Aulisporites	Bennettitales
Calamospora	Sphenophytes, Equisetales
Camarozonosporites	Lycophytes, Lycopodiales
Circumpollen	Conifers (Cheirolepidiaceae)
Chasmatosporites	Cycadales
Concavisporites	Ferns, Osmundaceae
Cycadopites	Cycadales
Ellipsovelatisporites	Conifers
Enzonalasporites	Conifers (Patokaeaceae)
Haberkornia	Conifers
Infernopollenites	Conifers, Voltziaceae
Kraeuselisporites	Lycophytes, Selaginellales
Leschikisporis	Bennettitales
Lunatisporites	Conifers, Voltziaceae
Ovalipollis	Conifers, Voltziaceae
Patinasporites	Conifers (Patokaeaceae)
Todisporites	Ferns, Osmundaceae
Triadispora	Conifers, Voltziaceae
Vallasporites	Conifers (Patokaeaceae)

this volume), but contrasts with the comparatively low-diversity macrofossil floras recovered from Polzberg and Scheiblingbach (this study; Kustatscher et al. 2026, this volume). This discrepancy reflects differences in taphonomic and spatial resolution: plant macrofossil assemblages are known to record generally local vegetation. In marine sediments, however, they are strongly shaped by depositional and transport biases, whereas palynological assemblages integrate a broader regional vegetation and paleoclimate signal (e.g., Kustatscher et al. 2010, 2012; Costamagna et al. 2018).

A direct comparison of the Scheiblingbach and Polzberg palynological assemblages (at Polzberg at the stratigraphic level of the major plant fossil bed; see also Roghi and Lukeneder 2026, this volume) reveals marked differences in palynological diversity (Table 22.2). The Polzberg assemblage contains a broader spectrum of taxa, including numerous ferns (*Verrucosisporites*) and conifers (*Classopollis, Duplicisporites, Partitisporites*). The gymnosperms, by contrast, are relatively abundant in Scheiblingbach, represented particularly by *Enzonalasporites*, *Patinasporites*, and

Table 22.2 Attribution of the spore and pollen genera of the Scheiblingbach and Polzberg successions to the SEG (Sporomorph Ecogroups) (Abbink et al. 2004; Kustatscher et al. 2010, 2012; Costamagna et al. 2018)

Genus	SEG group	Polzberg	Scheiblingbach
Aratrisporites	Coastal	x	x
Araucariacites	Coastal	x	x
Kraeuselisporites	Coastal	x	x
Platysaccus	Coastal	x	
Samaropollenites	Coastal	x	
Camerosporites	Pioneer	x	
Circumpollen	Pioneer		x
Classopollis	Pioneer	x	
Duplicisporites	Pioneer	x	
Partitisporites	Pioneer	x	
Alisporites	Lowland	x	
Aulisporites	Lowland	x	x
Brachysaccus	Lowland	x	
Chasmatosporites	Lowland		x
Cycadopites	Lowland	x	x
Distalanulatisporites	Lowland	x	
Enzonalasporites	Lowland	x	x
Haberkornia	Lowland		x
Leschikisporis	Lowland	x	x
Patinasporites	Lowland	x	x
Pseudoenzonalasporites	Lowland	x	
Spiritisporites	Lowland	x	
Anapiculatisporites	River	x	
Calamospora	River	x	x
Camarozonosporites	River		x
Combaculatisporites	River	x	
Concavisporites	River	x	x
Converrucosisporites	River	x	
Densosporites	River	x	
Lunzisporites	River	x	
Lycopodiacidites	River	x	
Punctatisporites	River	x	
Retusotriletes	River	x	
Todisporites	River	x	x
Trachysporites	River	x	

(continued)

Table 22.2 (continued)

Genus	SEG group	Polzberg	Scheiblingbach
Uvaesporites	River	x	
Verrucosisporites	River	x	
Ellipsovelatisporites	Hinterland/Upland	x	x
Infernopollenites	Hinterland/Upland		x
Klausipollenites	Hinterland/Upland	x	
Laricoidites	Hinterland/Upland		x
Lunatisporites	Hinterland/Upland	x	x
Ovalipollis	Hinterland/Upland	x	x
Triadispora	Hinterland/Upland	x	x
Vallasporites	Hinterland/Upland	x	x

Triadispora. This difference can be explained by looking at the paleogeographic setting (Fig. 22.6): the Polzberg assemblage was deposited closer to the continental margin, whereas Scheiblingbach is situated in a more distal intraplatform setting. Further away from the terrestrial environments, the contribution of water-transported spores (particularly ferns and river-associated taxa) diminishes, whereas wind-dispersed conifer pollen becomes relatively enriched (the so-called "Neves effect").

The ecological signal becomes even clearer when viewed through the lens of Sporomorph EcoGroups (SEG; Table 22.2) (e.g., Abbink et al. 2004; Kustatscher et al. 2010, 2012; Costamagna et al. 2018). Most of the taxa absent in the Scheiblingbach record belong to the River SEG (*Converrucosisporites, Uvaesporites*), the Coastal SEG (*Platysaccus, Samaropollenites*), or the Pioneer SEG (*Classopollis, Camerosporites*). By contrast, Scheiblingbach is relatively enriched in taxa of the Lowland SEG (*Cycadopites, Aulisporites, Chasmatosporites, Leschikisporis*) and the Hinterland/Upland SEG (*Triadispora, Infernopollenites, Lunatisporites, Ovalipollis, Vallasporites*). This indicates that the Scheiblingbach assemblage, filtered by long-distance transport and selective taphonomy, preferentially captures the regional pollen dispersal from lowland and upland gymnosperm-dominated vegetation, whereas the more proximal Polzberg record still retains the signature of river, coastal, and pioneer communities.

Thus, the comparison of the Scheiblingbach and Polzberg palynological assemblages underscores the importance of considering taphonomic filters when reconstructing paleoenvironmental and/or paleoclimatic signals from the fossil record. Even within the very similar stratigraphic interval, differences in basin position and transport distance can shift the ecological signal from diverse, spore-rich communities to pollen-dominated assemblages.

Table 22.3 Comparison of the main plant fossil genera in the three main plant fossil assemblages of the Northern Calcareous Alps (Stur, 1874; Pott et al. 2007a, 2007b, 2007c, 2007d, 2007e, 2008, 2010, 2017, 2018; Pott and Krings, 2010, Pott, 2016, 2019; Kustatscher et al. 2026, this volume)

	Scheiblingbach	Polzberg	Lunz
Lycophyta			
Lepacyclotes			x
Sphenophyta			
Equisetites			x
Neocalamites			x
Pteridophyta			
Asterotheca			x
Camptopteris			x
Clathropteris			x
Coniopteris			x
Cynepteris			x
Danaeopsis			x
Dictyophyllum			x
Digitopteris			x
Mertensides			x
Phlebopteris			x
Symopteris			x
Todites			x
Ginkgophyta			
Glossophyllum			x
Arberophyllum			x
Ginkgoites			x
Cycadophyta			
Bennetticarpus			x
Cycadolepis			x
Dioonitocarpidium			x
Haitingeria			x
Lunzia			x
Nilssonia			x
Nilssoniopteris			x
Pseudoctenis			x
Pterophyllum			x
Westersheimia			x
Coniferophyta			
Brachyphyllum	x	x	
Elatocladus	x	x	x
Pagiophyllum	x	x	
Pelourdea	x	x	
Stachytaxus			x

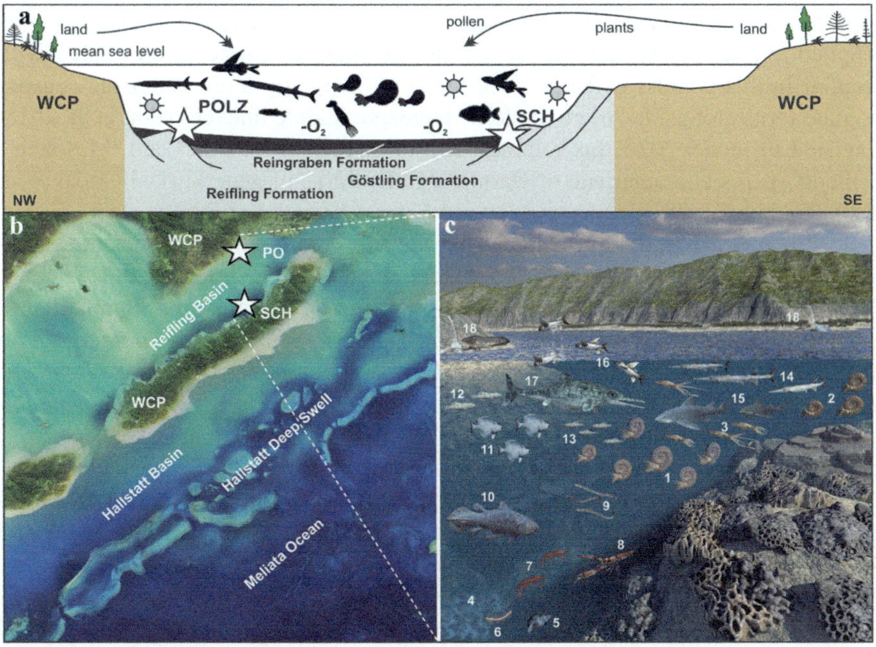

Fig. 22.6 **a** Transect through the Reifling Basin with reconstruction of basinal environments, position of Scheiblingbach (SCH) and Polzberg (POLZ; white asterisks), low oxygen conditions (–O_2) and surrounding land with plants. **b** Detailed paleogeography of the Reifling Basin, with the Scheiblingbach (SCH) and Polzberg locality (PO) (white asterisks) during the Upper Triassic, lower Carnian. **c** Reconstruction of the paleoenvironment and members of the lower Carnian paleobiotas. 1 *Austrotrachyceras*, 2 *Paratrachyceras*, 3 *Phragmoteuthis* and *Mojsisovicsteuthis*, 4 *Peribositra*, 5 caenogastropod, 6 polychaete worm Po, 7 *Antrimpos*, 8 *Platychela*, 9 conodonts, 10 "*Coelacanthus*", 11 *Polzbergia*, 12 *Peltopleurus*, 13 *Nannolepis*, 14 *Saurichthys*, 15 *Hybodus*, 16 *Thoracopterus*, 17 potential Ichthyopterygia, 18 *Tellerodus*. Figured members of paleobiota not to scale. **b** Artwork by M Harzhauser, NHMW. **c** Artwork based on Fig. 7 by AL, final artwork by 7reasons; www.7reasons.net; adapted after Lukeneder and Lukeneder (2021). WCP Wetterstein Carbonate Platform. Hds Hallstatt Deep Swell. **a** and **b** not to scale

22.4.3 Comparison with the Lunz and Polzberg plant fossil assemblages

The most important and best-studied *Fossil-Lagerstätte* of the Northern Calcareous Alps from a plant perspective is the Lunz flora (e.g., Pott 2021; Pott and Lukeneder 2026, this volume, and references therein). The plant fossil assemblage is rich in sphenophytes (*Equisetites, Neocalamites*; Pott et al. 2008a, 2018). Ferns are particularly diverse, including eusporangiate Marattiales (*Asterotheca, Symopteris, Danaeopsis*) and leptosporangiate Osmundales (*Todites*; Pott et al. 2018). Additional taxa such as *Phlebopteris* and *Cynepteris* represent the Matoniaceae, whereas *Camptopteris, Clathropteris, Dictyophyllum,* and *Digitopteris* belong to the Dipteridaceae (leptosporangiate Gleicheniales). Other ferns, such as *Mertensides*, could not yet be assigned to a family but are probably related to the Marattiales (Pott et al. 2018). Peculiar in this context is the co-occurrence of "typically more Paleozoic" eusporangiate ferns alongside "more modern" leptosporangiate lineages (Pott et al. 2018; Pott and Lukeneder 2026, this volume). This mixture of "Paleozoic" and "Mesozoic" elements seems characteristic of Carnian (CPE) floras in general (Dal Corso et al. 2020; Roghi et al. 2022).

Seed plants are important elements of the Lunz flora. Ginkgophytes are represented by *Arberophyllum* (Pott et al. 2007c; Pott and Lukeneder 2026, this volume). The most common plant group in the Lunz assemblage is the Bennettitales, represented both by leaves (*Pterophyllum, Anomozamites, Nilssoniopteris*; Pott et al. 2007d, 2007e, 2008a, b, c; Pott 2021) and reproductive organs (*Bennetticarpus, Haitingeria, Westersheimia, Pramelreuthia, Sturiella, Lunzia, Leguminanthus, Cycadolepis*; Pott 2016, 2021; Pott et al. 2010, 2017; Pott and Lukeneder 2026, this volume). Among the cycads, important leaf genera include *Nilssonia* and *Pseudoctenis* (Pott et al. 2007a, b), and reproductive structures comprise megasporophylls of *Dioonitocarpidium* (Pott 2019). Conifers, by contrast, such as *Elatocladus* and *Voltzia*, are represented mainly by *Stachytaxus* and *Elatocladus lipoldii* (Pott et al. 2008b; Pott and Krings 2010; Pott and Lukeneder 2026, this volume).

In marked contrast, conifers are the dominant plant group in both the Polzberg and Scheiblingbach plant fossil assemblages. This pattern can be linked to different depositional settings. The Lunz flora was preserved autochthonously in a deltaic swamp environment, favoring the accumulation of sphenophytes, ferns, cycads, and bennettitaleans, whereas the Polzberg and Scheiblingbach assemblages were deposited in intraplatform marine basins. Plant remains in the latter settings were likely transported in primarily from coastal, deltaic, and lowland settings rather than from hinterland/upland environments. The transport explains the predominance of conifer shoots and pollen in Polzberg and Scheibling, and the corresponding underrepresentation of ferns and other moisture-dependent groups that characterize the Lunz flora. Interestingly, the composition does not change substantially from the proximal marine setting in Polzberg to the more distal setting of Scheiblingbach. The exception is the decrease of reproductive units, which are present only at Polzberg. Again, a taphonomic bias is suggested here as well.

22.4.4 The paleoenvironment of Scheiblingbach

The Scheiblingbach succession was deposited within the Reifling Basin (Fig. 22.6), an intraplatform basin of the Northern Calcareous Alps, located at the north-western Tethyan margin in latitudes of 15–30° N (e.g., Lukeneder et al. 2012; Lukeneder and Lukeneder 2021). Lithological and paleontological evidence indicates sedimentation under quiet, oxygen-depleted conditions in a distal marine basin (Lukeneder et al. 2024). The plant record, although quantitatively subordinate to the marine fauna, provides crucial information about the surrounding terrestrial ecosystems and the pathways of transport into the basin.

The macrofossil assemblage is dominated by conifers, particularly conifer shoots of the *Brachyphyllum* and *Pagiophyllum* type, accompanied by rare *Elatocladus* and dispersed leaves of *Pelourdea*. Amber is locally preserved, evidencing that conifers in the plant fossil assemblage were resin-producing. Other plant groups such as lycophytes, sphenophytes, and ferns, which are present to abundant in contemporaneous terrestrial assemblages like Lunz (Pott et al. 2018), are conspicuously absent. Since the Scheiblingbach succession was sampled bed-to-bed, the relative distribution within the stratigraphic succession provides additional information. Fossil plants are particularly abundant within certain layers (e.g., Sch 15, Sch 18, 31–34, 49–56 (Lukeneder 2026a, b, this volume). In general, most taxa are present in almost all plant-rich layers throughout the succession, but some taxa are dominant in certain beds, e.g., dispersed straplike-leaves are common throughout beds Sch 29, and 52–53 with fewer other conifer shoot types. It is unclear whether these differences in abundance reflect actual changes in the vegetation on the surrounding landmasses. The selective preservation points rather towards a strong taphonomic filter, favoring the transport and preservation of robust conifer remains that could survive long-distance transport into the basin.

Palynological evidence provides a complementary perspective. The Scheiblingbach assemblage is relatively diverse, but its composition is strongly biased towards gymnosperm pollen (Table 22.1). Lycophyte (*Aratrisporites, Kraeuselisporites, Camarozonosporites*), sphenophyte (*Calamospora*), and fern (*Concavisporites, Todisporites*) spores are relatively rare. Conifer pollen grains are highly diverse, including Araucariaceae (*Araucariacites*), Cheirolepidiaceae (*Circumpollen*), Patokaeaceae (*Enzonalasporites, Patinasporites, Vallasporites*), and Voltziaceae (*Infernopollenites, Lunatisporites, Ovalipollis, Triadispora*). Cycadalean and bennettitalean pollen (*Cycadopites, Chasmatosporites, Aulisporites, Leschikisporis*) are also present. When analyzed in terms of Sporomorph EcoGroups (SEG; Abbink et al. 2004; Kustatscher et al. 2010, 2012; Costamagna et al. 2018), the Scheiblingbach assemblage shows a strong dominance of Lowland and Hinterland/Upland SEG groups, whereas river, coastal, and pioneer elements are underrepresented (Table 22.2). For example, typical River SEG taxa such as *Uvaesporites* are absent, as are Coastal SEG elements like *Platysaccus* and *Samaropollenites*. In contrast, Lowland SEG taxa (*Cycadopites*, *Aulisporites*, *Chasmatosporites*) and Hinterland/Upland SEG pollen grains (*Triadispora*, *Ovalipollis*, *Lunatisporites*) are well represented. This indicates that Scheiblingbach evidences the Neves effect, with a pollen-dominated palynological assemblage even if the flora is dominated at least locally by spore-producing plants.

The comparison with the nearby Polzberg and Lunz macrofossil and Polzberg palynological assemblage highlights this taphonomic and paleogeographic bias (Fig. 22.7a–e). The Lunz plant fossil assemblages (Fig. 22.7a) are autochtonous, almost not transported and fragmented. In Polzberg the macrofossil (Fig. 22.7d) and palynological record still records a wide ecological spectrum, including numerous River and Pioneer SEG taxa (e.g., *Classopollis, Duplicisporites*), but the plant fossils are more fragmented and selected. Scheiblingbach (Fig. 22.7e), in its distal marine environment, yields pollen and an identifiable macrofossil assemblage dominated by conifers due to the fragmented nature of the other elements. The paucity of river/coastal elements can be best explained by the more distal depositional position of Scheiblingbach within the basin. Here, long-distance transport of spores and pollen led to a selective loss of heavier, water-transported spores and an enrichment of wind-dispersed pollen (Neves effect). Thus, the palynological assemblage here overrepresents the conifer forests at a regional scale, whereas typical elements such as horsetails, ferns, cycads, and Bennettitales typical of the lowland, river, and deltaic settings (Pott and Lukeneder 2026, this volume) are underrepresented. Coastal and swamp vegetation is only rarely preserved, not because of its absence in the area, but rather due to a taphonomic filter based on its position in the Reifling Basin.

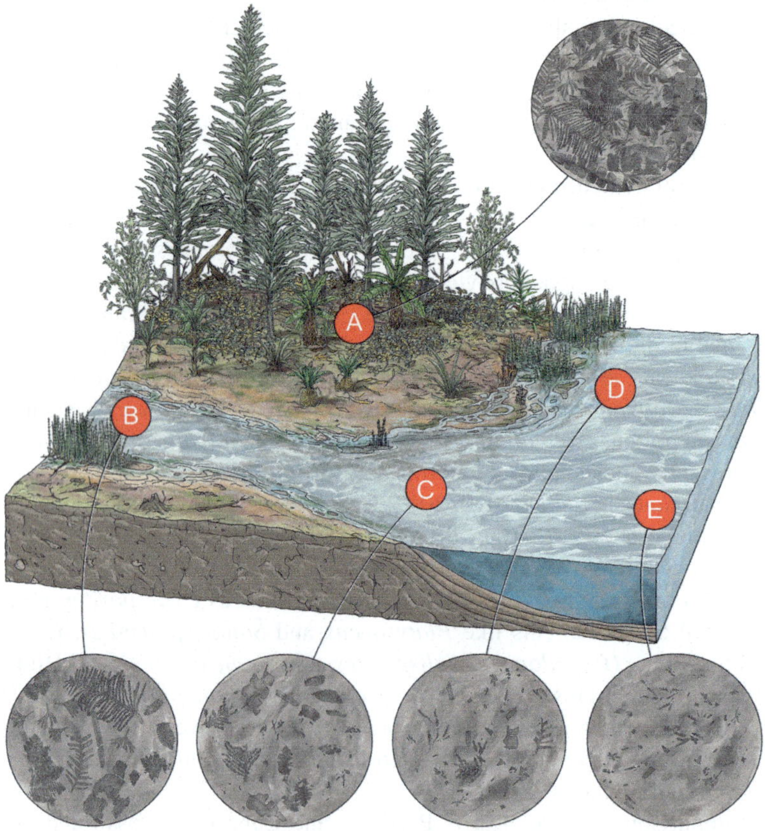

Fig. 22.7 Taphonomic process influencing the plant fossil assemblage composition within a basin from allochtonous swampy plant fossil assemblages (**a**), to poorly sorted and transported plant fossils of a river environment (**b**), to a more sorted and fragmented outer deltaic setting (**c**), to highly fragmented but still identifiable plant fossils of the near-shore marine environment (**d**), and finally the distal marine setting with highly fragmented plant remains (**e**)

22.5 Conclusions

Scheiblingbach is located in a distal position of the Reifling Basin, deposited in a distal marine environment during the Carnian Pluvial Episode. The plant fossil assemblage registers the selective/occasional input of terrestrial remains into an oxygen-depleted marine setting. The plant macrofossil assemblage is dominated by conifers, particularly *Brachyphyllum* and *Elatocladus*, confirming the presence of resin-producing conifers in the Carnian terrestrial vegetation of the Northern Calcareous Alps. Amber preserved both as droplets and within resin channels of conifer shoots provides direct evidence of in situ resin production. Its occurrence coincides with the humid climatic oscillations characteristic of the CPE and supports the interpretation of enhanced physiological stress in conifers linked to fluctuating environmental conditions such as storm events (Seyfullah et al. 2018a, b).

The palynological assemblages are dominated by conifer pollen over fern and lycophyte spores. SEG analysis shows the prevalence of Lowland SEG and Hinterland/Upland SEG taxa, whereas River, Coastal, and Pioneer SEG elements are scarce. This pattern reflects the distal position of Scheiblingbach in the Reifling Basin and thus to long-distance transport and selective taphonomic bias.

The comparison with the Polzberg and Lunz plant fossil assemblages highlights an environmental and taphonomic gradient from humid deltaic systems (Lunz) to nearshore marine basins (Polzberg) and distal marine settings (Scheibling). The consistent dominance of conifers in the two marine sites illustrates their capability of long-distance transport and capacity for resin production under CPE-related climatic stress. The Scheiblingbach plant fossil and amber record therefore represents a conifer-dominated plant fossil assemblage within a humid climate episode. This composition reflects taphonomic selection rather than the original conifer-dominated flora. This once again demonstrates the importance of an integrated geological, sedimentological, paleozoological, paleobotanical, and palynological study for paleoclimatic and paleoenvironmental reconstructions.

Acknowledgments We are particularly grateful to the owners of the outcrop area, the Austrian Federal Forests Company (Österreichische Bundesforste), for sampling and digging permission. Special thanks go to Lukas Nöst (forest district Großreifling) and Dominic Asztaller (forest district Großreifling) for their advice and help during the field campaigns. I thank Petra Lukeneder (Gablitz) and Veronika N.I. Koukal (Vienna) for their help with cleaning outcrops and sampling fossils. The manuscript greatly benefited from valuable comments by the reviewers Christian Pott (Münster) and an anonymous reviewer, the production editor Ashok Arumairaj (Taramani), the in-house series editor Kathleen A. Ritterbush (Utah), the publishing editor Simon Shah-Rohlfs and the production editor Antje Endemann (both Heidelberg).

Author Contribution EK AL designed the research. AL collected new material. EK, AL, MS provided documentation on fossil material. RB carried out the cuticle analyzes and descriptions. GR studied palynologic data. EK, AL, RB, MS contributed to writing the paper. MS, AL produced the figures.

Funding This project (including open access publishing fees) was supported by the Federal Government of Lower Austria (K3-F-964/001-2020 and K3-F-964/002-2023, Department Science and Research; headquarters in St. Pölten); the Austrian Academy of Sciences (headquarters in

Vienna, project Polzberg Lukeneder 2021) represented by the National Committee for Geo/Hydro Sciences (Earth System Sciences Programme); the Triassic project "Fenster in die Karnische Krise" of the Society of Friends of the Natural History Museum Vienna (Freunde des Naturhistorischen Museums Wien, CPE Lukeneder 2023); the Natural History Museum Vienna; and the Natur- und UNESCO Geopark Steirische Esienwurzen (St Gallen, project Karnische Krise Lukeneder 2024) (to AL); and by the Joint Research Project SNF-ST (project WETTEST, CUP H33C23001340003 Dekret 35/2024, Project number 200021L_219219 to EK and Elke Schneebeli). The authors are responsible for the contents of this publication. The funder had no impact on conceptualization, design, data collection, analysis, decision to publish, or preparation of the manuscript.

Competing interests All authors stated that they have no competing interests.

Authors' contributions All authors stated that they have equally contributed to their chapters, read, edited and agreed to submit the final versions of their contributions.

Availability of data and materials All sources of information are provided in the text. Raw data related to the fossil material from Austria are available from the corresponding authors and Alexander Lukeneder upon request.

Additional information The article is part of the Special Volume: *Late Triassic Konservat-Lagerstätten within the Carnian Pluvial Episode. 150 Years of Scientific Excavations in Austria.*

Artificial Intelligence (AI) The chapter was produced without the use of AI.

Language The entire volume uses American English reviewed by a native scientist.

Received 2025.

Published online 2026.

References

Abbink OA, Van Konijnenburg-van Cittert JHA, Van der Zwan CJ, Visscher H (2004) A sporomorph ecogroup model for the Northwest European Jurassic–Lower Cretaceous II: application to an exploration well from the Dutch North Sea. Neth J Geosci 83(2):81–92. https://doi.org/10.1017/S0016774600020059

Baranyi V, Rostási A, Raucsik B, Kürschner WM (2019) Palynology and weathering proxies reveal climatic fluctuations during the Carnian Pluvial Episode (CPE, Late Triassic) from marine successions in the Transdanubian range (western Hungary). Glob Planet Change 177:155–172. https://doi.org/10.1016/j.gloplacha.2019.03.013

Barrón E, Gómez JJ, Goy A, Pieren AP (2006) The Triassic-Jurassic boundary in Asturias (northern Spain): palynological characterization and facies. Rev Palaeobot Palynol 138:187–208. https://doi.org/10.1016/j.revpalbo.2006.01.005

Bonis NR, Kürschner WM (2012) Vegetation history, diversity patterns, and climate change across the Triassic/Jurassic boundary. Paleobiology 38(2):240–264. https://doi.org/10.1666/10026.1

Clement-Westerhof JA (1974) In situ pollen from gymnospermous cones from the Upper Permian of the Italian Alps—a preliminary account. Rev Palaeobot Palynol 17:63–73. https://doi.org/10.1016/0034-6667(74)90092-X

Dal Corso J, Bernardi M, Sun Y, Song H, Seyfullah LJ, Preto N, Gianolla P, Ruffell A, Kustatscher E, Roghi G, Merico A, Hohn S, Schmidt AR, Marzoli A, Newton RJ, Wignall PB, Benton MJ (2020) Extinction and dawn of the modern world in the Carnian (Late Triassic). Sci Adv 6(38):eaba0099. https://doi.org/10.1126/sciadv.aba0099

Costamagna LG, Kustatscher E, Scanu GG, Del Rio M, Pittau P, Van Konijnenburg-van Cittert JHA (2018) A palaeoenvironmental reconstruction of the Middle Jurassic of Sardinia (Italy) based on sedimentological, palynological and palaeobotanical analyses. Palaeobiodiv Palaeoenviron 98:111–138. https://doi.org/10.1007/s12549-017-0306-z

Couper RA (1958) British Mesozoic microspores and pollen grains: a systematic and stratigraphic study. Palaeontographica Abt B 103:75–179

Dal Corso J, Preto N, Mietto P, Roghi G, Jenkyns HC (2011) Carbon-isotope variability of Triassic amber, as compared with wood and leaves (Southern Alps, Italy). Palaeogeogr Palaeoclimatol Palaeoecol 302:187–193

Dal Corso J, Ruffell A, Preto N (2018) The Carnian Pluvial Episode (Late Triassic): new insights into this important time of global environmental and biological change. J Geol Soc London 175:986–988. https://doi.org/10.1144/jgs2018-185

Dobruskina IA (2001) Upper Triassic flora from "Raibl beds" of Julian Alps (Italy) in Karavanke Mts. (Slovenia). Geologija 44:263–290

Forte G, Kustatscher E, Ragazzi E, Roghi G (2022) Amber droplets in the Southern Alps (NE Italy): a link between their occurences and main humid episodes in the Triassic. Riv Ital Paleontol Stratigr 128(1):69–79. https://doi.org/10.54103/2039-4942/15381

Gianolla P, De Zanche V, Mietto P (1998) Triassic sequence stratigraphy in the Southern Alps (Northern Italy): definition of sequences and basin evolution. SEPM Special Publ 60:719–747

Harzhauser M, Lukeneder A, Lukeneder P, Mandic O (2026) Potential modulation of a Late Triassic Pangean Megamonsoon by solar cycles. In: Lukeneder A (ed) Late Triassic Konservat-Lagerstätten within the Carnian Pluvial Episode: 150 years of scientific excavations in Austria. Top Geobiol 55, Chap 24:567–589. https://doi.org/10.1007/978-3-032-19777-1_24

Hornung T, Brandner R (2005) Biochronostratigraphy of the Reingraben Turnover (Hallstatt Facies Belt): local black shale events controlled by regional tectonism, climatic change and plate tectonics. Facies 51:460–479. https://doi.org/10.1007/s10347-005-0061-x

Hornung T, Brandner R, Krystyn L, Joachimski MM, Keim L (2007) Multistratigraphic constraints on the NW Tethyan 'Carnian Crisis'. In: Lucas SG, Spielmann JA (eds) The global triassic. New Mexico museum of natural history & science bulletin, vol 41, pp 59–67

Hornung T (2008) The Carnian Crisis in the Tethyan Realm. 235 p, Dr. Müller, Saarbrücken.

Kelber K-P, Hansch W (1995) Keuperpflanzen. Die Enträtselung einer über 200 Millionen Jahre alten Flora. Museo 11:1–157

Krystyn L (1978) Eine neue Zonengliederung des mediterranen Unterkarn. Schrift Erdwiss Komm ÖAW 4:37–75

Krystyn L (1991) Die Fossillagerstätten der alpinen Trias. In: Vasilček W, Krystyn L, Golebiowski R (eds) Exkursionen im Jungpaläozoikum und Mesozoikum Österreichs. Österreichische Paläontologische Gesellschaft, Wien, pp 23–78

Kustatscher E, Van Konijnenburg-van Cittert JHA (2008) Lycophytes and horsetails from the Triassic flora of Thale (Germany). Neues Jahrb Geol Paläontol Abh 250:65–77

Kustatscher E, Van Konijnenburg-van Cittert JHA, Roghi G (2010) Macrofloras and palynomorphs as proxies for palaeoclimate and palaeoecology: Kühwiesenkopf/Monte Prà della Vacca. Palaeogeogr Palaeoclimatol Palaeoecol 291:71–80. https://doi.org/10.1016/j.palaeo.2009.07.001

Kustatscher E, Heunisch C, Van Konijnenburg-van Cittert JHA (2012) Taphonomical implications of the Ladinian megaflora and palynoflora of Thale (Germany). Palaios 27:753–764. https://doi.org/10.2110/palo.2011.p11-090r

Kustatscher E, Branz R, Slodownik M, Roghi G, Lukeneder A (2026) The plant fossils of the Carnian (Late Triassic) Polzberg *Konservat-Lagerstätte* in Austria. In: Lukeneder A (ed) Late Triassic Konservat-Lagerstätten within the Carnian Pluvial Episode: 150 years of scientific excavations in Austria topics in geobiology. Top Geobiol 55, Chap 12:221–249. https://doi.org/10.1007/978-3-032-19777-1_12

Lahm B (1984) Spumellarienfaunen (Radiolaria) aus den mitteltriassischen Buchensteiner-Schichten von Recoaro (Norditalien) und den obertriassischen Reiflingerkalken von Großreifling (Österreich). Münchner Geowiss Abh A, Geol Paläontol 1:1–161

Langenheim JH (2003) Plant resins: chemistry, evolution, ecology, and ethnobotany. Timber Press, Portland, Cambridge, p 586

Looy CV, Twitchett RJ, Dilcher DL, Van Konijnenburg-van Cittert JHA, Visscher H (2001) Life in the end-Permian dead zone. Proc Natl Acad Sci USA 98:7879–7883. https://doi.org/10.1073/pnas.131218098

Lukeneder A (ed) (2026a) Late Triassic Konservat-Lagerstätten within the Carnian Pluvial Episode: 150 Years of Scientific Excavations in Austria. Top Geobiol 55:600 p. https://doi.org/10.1007/978-3-032-19777-1

Lukeneder A (2026b) The marine Upper Triassic paleobiota from the Scheiblingbach *Konservat-Lagerstätte* in Austria. In: Lukeneder A (ed) Late Triassic Konservat-Lagerstätten within the Carnian Pluvial Episode: 150 years of scientific excavations in Austria. Top Geobiol 55, Chap 19:437–467.https://doi.org/10.1007/978-3-032-19777-1_19

Lukeneder A, Lukeneder P (2021) The Upper Triassic Polzberg palaeobiota from a marine *Konservat-Lagerstätte* deposited during the Carnian Pluvial Episode in Austria. Sci Rep 11:16644. https://doi.org/10.1038/s41598-021-96052-w

Lukeneder A, Lukeneder P (2022a) Taphonomic history and trophic interactions of an ammonoid fauna from the Upper Triassic Polzberg palaeobiota. Sci Rep 12:7455. https://doi.org/10.1038/s41598-022-11496-y

Lukeneder P, Lukeneder A (2022b) Mineralized belemnoid cephalic cartilage from the Late Triassic Polzberg *Konservat-Lagerstätte* (Austria). PLoS One 17(4):e0264595. https://doi.org/10.1371/journal.pone.0264595

Lukeneder A, Lukeneder P (2023) New data on the marine Upper Triassic palaeobiota from the Polzberg Konservat-Lagerstätte in Austria. Swiss J Palaeontol 142:9. https://doi.org/10.1186/s13358-023-00269-3

Lukeneder A, Lukeneder P (2024) Ammonite taxonomy of the Carnian Polzberg *Konservat-Lagerstätte* in Austria. Jb Geol BA Wien 162(1–4):27–55

Lukeneder A, Surmik D, Gorzelak P, Niedźwiedzki R, Brachaniec T, Salamon MA (2020) Bromalites from the Upper Triassic Polzberg section (Austria); insights into trophic interactions and food chains of the Polzberg palaeobiota. Sci Rep 10:20545. https://doi.org/10.1038/s41598-020-77017-x

Lukeneder A, Lukeneder P, Sachsenhofer RF, Roghi G, Rigo M (2024) Multi-proxy record of the Austrian Upper Triassic Polzberg *Konservat-Lagerstätte* in light of the Carnian Pluvial Episode. Sci Rep 14:11194. https://doi.org/10.1038/s41598-024-60591-9

Lukeneder S, Lukeneder A, Harzhauser M, İslamoğlu Y, Krystyn L, Lein R (2012) A delayed carbonate factory breakdown during the Tethyan-wide Carnian Pluvial Episode along the Cimmerian terranes (Taurus, Turkey). Facies 58:279–296

Mostler H, Scheuring BW (1974) Mikrofloren aus dem Langobard und Cordevol der Nördlichen Kalkalpen und das Problem der Beginns der Keupersedimentation im Germanischen Raum. Geol Paläont Mitt Innsbruck 4(4):1–35

Mueller S, Krystyn L, Kürschner WM (2016) Climate variability during the Carnian pluvial phase—a quantitative palynological study of the Carnian sedimentary succession at Lunz am See, Northern Calcareous Alps, Austria. Palaeogeogr Palaeoclimatol Palaeoecol 441:198–211. https://doi.org/10.1016/j.palaeo.2015.06.008

Nowak H, Kustatscher E, Roghi G, Van Konijnenburg-van Cittert JHA (2022) Variability of in situ spores in some leptosporangiate ferns from the Triassic in Italy and Austria. Rev Palaeobot Palynol 296:104544. https://doi.org/10.1016/j.revpalbo.2021.104544

Passoni L (1999) Studio della macroflora e dei palinomorfi del Carnico di Colle di Verano (Monte Pora, Lombardia). PhD–thesis, Università di Milano, 130 p

Pott C (2016) *Westersheimia pramelreuthensis* from the Carnian (Upper Triassic) of Lunz, Austria: more evidence for a unitegmic seed coat in early Bennettitales. Int J Plant Sci 177:771–791. https://doi.org/10.1086/688708

Pott C (2019) The cycadalean megasporophyll *Dioonitocarpidium* in the Carnian (Late Triassic) flora of Lunz am See, Austria. Paläontol Zeitschr (PalZ) 93:517–530. https://doi.org/10.1007/s12542-019-00459-w

Pott C (2021) Die fossile Flora der Lunz-Formation (Karnium) von Lunz am See (Niederösterreich). In: Hauschke N, Bachmann GH, Franz M (eds) TRIAS. Aufbruch in das Erdmittelalter. Dr Pfeil, München, pp 447–454

Pott C, Bouchal JM, Choo T, Yousif R, Bomfleur B (2018) Ferns and fern allies from the Carnian (Upper Triassic) of Lunz am See, Lower Austria: a melting pot of Mesozoic fern vegetation. Palaeontographica B 297:1–101. https://doi.org/10.1127/palb/2018/0059

Pott C, Fischer T, Aschauer B (2017) *Lunzia austriaca*—a bennettitalean microsporangiate structure with *Cycadopites*-like *in situ* pollen from the Carnian (Upper Triassic) of Lunz, Austria. Grana 56:321–338. https://doi.org/10.1080/00173134.2017.1282010

Pott C, Kerp H, Krings M (2007a) Morphology and epidermal anatomy of *Nilssonia* (cycadalean foliage) from the Upper Triassic of Lunz (Lower Austria). Rev Palaeobot Palynol 143:197–217. https://doi.org/10.1016/j.revpalbo.2006.07.007

Pott C, Kerp H, Krings M (2007b) *Pseudoctenis cornelii* nov. spec. (cycadalean foliage) from the Carnian (Upper Triassic) of Lunz, Lower Austria. Ann Naturhist Mus Wien 109A:1–17

Pott C, Kerp H, Krings M (2008a) Sphenophytes from the Carnian (Upper Triassic) of Lunz am see (Lower Austria). Jahrb Geol Bundesanst 148:183–199

Pott C, Krings M (2010) Gymnosperm foliage from the Upper Triassic of Lunz, Lower Austria: an annotated checklist and identification key. Geo. Alp 7:19–38

Pott C, Krings M, Kerp H (2007c) A surface microrelief on the leaves of *Glossophyllum florinii* (?Ginkgoales) from the Upper Triassic of Lunz, Austria. Bot J Linnean Soc 153:87–95. https://doi.org/10.1111/j.1095-8339.2007.00569.x

Pott C, Krings M, Kerp H (2007d) First record of *Nilssoniopteris* (Gymnospermophyta, Bennettitales) from the Carnian (Upper Triassic) of Lunz, Lower Austria. Palaeontology 50:1299–1318. https://doi.org/10.1111/j.1475-4983.2007.00704.x

Pott C, Krings M, Kerp H (2008b) The Carnian (Late Triassic) flora from Lunz in Lower Austria: palaeoecological considerations. Palaeoworld 17:172–182. https://doi.org/10.1016/j.palwor.2008.03.001

Pott C, Krings M, Kerp H, Friis EM (2010) Reconstruction of a bennettitalean flower from the Carnian (Upper Triassic) of Lunz, Lower Austria. Rev Palaeobot Palynol 159:94–111. https://doi.org/10.1016/j.revpalbo.2009.11.004

Pott C, Labandeira CC, Krings M, Kerp H (2008c) Fossil insect eggs and ovipositional damage on bennettitalean leaf cuticles from the Carnian (Upper Triassic) of Austria. J Paleont 82:778–789. https://doi.org/10.1666/06-094.1

Pott C, Lukeneder A (2026) A deltaic swamp forest: coastal wetland vegetation during the climax of the Carnian Pluvial Episode. In: Lukeneder A (ed) Late Triassic Konservat-Lagerstätten within the Carnian Pluvial Episode: 150 years of scientific excavations in Austria. Top Geobiol 55, Chap 18:405–436. https://doi.org/10.1007/978-3-032-19777-1_18

Pott C, Van Konijnenburg-van Cittert JHA, Kerp H, Krings M (2007e) Revision of the *Pterophyllum* species (Cycadophytina: Bennettitales) in the Carnian (Late Triassic) flora from Lunz, Lower Austria. Rev Palaeobot Palynol 147:3–27. https://doi.org/10.1016/j.revpalbo.2007.03.005738

Roghi G, Gianolla P, Kustatscher E, Schmidt AR, Seyfullah LJ (2022) An exceptionally preserved terrestrial record of LIP effects on plants in the Carnian (Upper Triassic) amber-bearing section of the Dolomites, Italy. Front Earth Sci 10:900586. https://doi.org/10.3389/feart.2022.900586

Roghi G, Kustatscher E, Ragazzi E, Giusberti L (2017) Middle Triassic amber associated with voltzialean conifers from the Southern Alps of Italy. Riv Ital Paleontol Stratigr 123(2):193–202

Roghi G, Kustatscher E, Van Konijnenburg-van Cittert JHA (2006) Late Triassic plant from Julian Alps (Italy). Boll Soc Pal Ital 45(I):133–140

Roghi G, Lukeneder A (2026) Palynological analyses at the onset of the Carnian Pluvial Episode (CPE) in the Reifling Basin (Northern Calcareous Alps, Austria). In: Lukeneder A (ed) Late Triassic Konservat-Lagerstätten within the Carnian Pluvial Episode: 150 years of scientific excavations in Austria. Top Geobiol 55, Chap 17:383–404. https://doi.org/10.1007/978-3-032-19777-1_17

Roghi G, Ragazzi E, Gianolla P (2006a) Triassic amber of the Southern Alps (Italy). Palaios 21:143–154

Schindelin J, Arganda-Carreras I, Frise E, Kaynig V, Longair M, Pietzsch T, Preibisch S, Rueden C, Saalfeld S, Schmid B, Tinevez JY, White DJ, Hartenstein V, Eliceiri K, Tomancak P, Cardona A (2012) Fiji: an open-source platform for biological-image analysis. Nat Methods 9(7):676–682. https://doi.org/10.1038/nmeth.2019

Schneider CA, Rasband WS, Eliceiri KW (2012) NIH Image to ImageJ: 25 years of image analysis. Nat Methods 9(7):671–675

Schulz E (1967) Sporenpaläontologische Untersuchungen rätoliassischer Schichten im Zentralteil des Germanischen Beckens. Paläontol Abh B 2:427–633

Seyfullah LJ, Beimforde C, Dal Corso J, Perrichot V, Rikkinen J, Schmidt AR (2018a) Production and preservation of resins—past and present. Biol Rev 93:1684–1714

Seyfullah LJ, Roghi G, Dal Corso J, Schmidt AR (2018b) The Carnian Pluvial Episode and the first global appearance of amber. J Geol Soc 175:1012–1018. https://doi.org/10.1144/jgs2017-143

Simms MJ, Ruffell AH (1990) Climatic and biotic change in the late Triassic. J Geol Soc Lond 147:321–327. https://doi.org/10.1144/gsjgs.147.2.0321

Slodownik MA, Hill RS, McLoughlin S (2023a) Komlopteris: A persistent lineage of post-Triassic corystosperms in Gondwana. Rev Palaeobot Palynol 317:104950. https://doi.org/10.1016/j.revpalbo.2023.104950

Slodownik MA, Escapa I, Mays C, Jordan GJ, Carpenter RJ, Hill RS (2023b) Araucarioides?: A Polar Lineage of Araucariaceae with New Paleogene Fossils from Tasmania, Australia. Int J Plant Sci 184(8):640–58. https://doi.org/10.1086/726183

Stur D (1874) Neue Aufschlüsse im Lunzer Sandsteine bei Lunz und ein neuer Fundort von Wengerschiefer im Pölzberg zwischen Lunzersee und Gaming. Verh K-K Geol RA 1:271–273

Summesberger H, Wagner L (1972) Der Stratotypus des Anis (Trias). Ann Naturhist Mus Wien 76:515–538

Townrow JA (1962) On Pteruchus, a microsporophyll of the Corystospermaceae. Bull Br Mus Nat Hist (Geol) 6:287–320

Van Konijnenburg-van Cittert JHA (1978) Osmundaceous spores in situ from the Jurassic of Yorkshire, England. Rev Palaeobot Palynol 26:125–141. https://doi.org/10.1016/0034-6667(78)90009-X

Van Konijnenburg-van Cittert JHA (1993) A review of the Matoniaceae based on in situ spores. Rev Palaeobot Palynol 78:235–267. https://doi.org/10.1016/0034-6667(93)90012-Y

Wolfe AP, Tappert R, Muehlenbachs K, Boudreau M, McKellar RC, Basinger JF, Garrett A (2009) A new proposal concerning the botanical origin of Baltic amber. Proc. R. Soc. B. Biol Sci 276:3403–3412

Zhang J, Lenz OK, Wang P, Gao Y, Hornung J (2021) The Eco-Plant model and its implication on Mesozoic dispersed sporomorphs for Bryophytes, Pteridophytes, and Gymnosperms. Rev Palaeobot Palynol 293:104503. https://doi.org/10.1016/j.revpalbo.2021.104503

Chapter 23
Organic Petrology and Geochemistry of the Scheiblinggraben Section: Detailed Insights into the Carnian Pluvial Episode and Comparison to Time Equivalent Profiles

Johannes A. Weitz, Ksenija A. Stojanović, Reinhard F. Sachsenhofer, David Misch, and Alexander Lukeneder

Abstract The Carnian Pluvial Episode (CPE) is a time of previously underestimated environmental change, therefore the factors controlling the CPE are not yet fully understood. The sediments of the Göstling Formation represent a transitional lithology, which is characteristic for the intra-platform Reifling Basin and records the shift from carbonate to siliciclastic deposition during the CPE. This transitional interval was investigated at the Scheiblinggraben (Styria, Austria) by means of bulk geochemical, organic petrographical, biomarker and microfacies analysis. Thermal maturity, depositional environment, organic matter source and hydrocarbon potential were assessed and compared to available data from the Großau and Polzberg localities. The depositional environment before the CPE was characterized by normal marine conditions and deposition of almost pure carbonates with virtually

Supplementary Information The online version contains supplementary material available at https://doi.org/10.1007/978-3-032-19777-1_23.

J. A. Weitz · R. F. Sachsenhofer · D. Misch
Department Applied Geosciences and Geophysics, Technical University of Leoben, Leoben, Austria
e-mail: johannes.weitz@unileoben.ac.at

R. F. Sachsenhofer
e-mail: reinhard.sachsenhofer@unileoben.ac.at

D. Misch
e-mail: david.misch@unileoben.ac.at

K. A. Stojanović
Department of Applied Chemistry, University of Belgrade, Belgrade, Serbia
e-mail: ksenija@chem.bg.ac.rs

A. Lukeneder (✉)
Department of Geology and Paleontology, Natural History Museum Vienna, Vienna, Austria
e-mail: alexander.lukeneder@nhm.at

A. Lukeneder (ed.), *Late Triassic Konservat-Lagerstätten within the Carnian Pluvial Episode*, Topics in Geobiology 55, https://doi.org/10.1007/978-3-032-19777-1_23

no organic matter (Reifling Formation). The overlying Göstling Formation comprises 60 siliceous limestone beds, some of them grading into thin dark-laminated layers with partly very high organic matter contents. In the lowermost meter of the Göstling Formation, up to bed 5a, an increase in total organic carbon/sulfur (TOC/S) ratios indicates a change from a normal marine basin to a basin with increased freshwater input. Afterwards, periodic influx of freshwater transported large amounts of siliciclastic material, terrigenous organic matter and nutrients into the basin. Eutrophication favored algal and radiolarian blooms. The high bioproductivity caused oxygen depletion. Thus, both high productivity and enhanced organic matter preservation contributed to the very high TOC contents in the thin, dark interlayers (max. TOC: 11 wt%). The transition to the mudstones of the Reingraben Formation reflects the return to normal marine conditions and an increase in oxygen depletion. Despite that, the amount of organic matter, which is derived mainly from marine organisms, is only moderately high in the first three meters (max. TOC: 1.3 wt%). During deposition of the upper part of the Reingraben Formation, the relative contribution of terrigenous organic matter increased, but the total amount of organic matter decreased. The comparison of geochemical data from the Scheiblinggraben, Polzberg and Großau sections clearly shows a strongly differentiated depositional environment during the deposition of the Göstling Formation, which became more uniform with the start of the Reingraben Formation. The thermal maturity of the three sections varies from immature (Scheiblinggraben: vitrinite reflectance: 0.36–0.43%Rr, T_{max}: 429 °C) to early oil window maturity (Großau: 0.70%Rr; T_{max}: 437 °C). Despite the locally very high TOC contents, the Göstling Formation in the Scheiblinggraben section cannot be considered a potential hydrocarbon source rock.

Keywords Organic petrology · Geochemistry · Carnian Pluvial Episode · Early Carnian · Late Triassic · Austria

23.1 Introduction

During the Carnian (Late Triassic) a major change from carbonate sedimentation associated with the active Wetterstein platform to deeper marine siliciclastic mud- and marlstones occurred in the Northern Calcareous Alps (NCA). This event was referred to as the "Reingrabener Wende" (Reingraben turnover) by Schlager and Schöllnberger (1974). The trigger for the turnover has not yet been conclusively clarified. Tectonic activity has been assumed to be a major factor for the Reingraben turnover (Leuchs and Mosebach 1936; Trümpy 1960; Tollmann 1966; Ogg 2015). It was further realized that time equivalent sections, not only of the Tethyan realm, exhibit similar characteristics. Simms and Ruffell (1989) derived the theory of the Carnian Pluvial Episode (CPE) based on evidence from the Germanic Basin and postulated a link with rift-related volcanism in the Wrangellia Large Igneous Province (LIP) (Preto et al. 2010; Alverà et al. 2026; Lukeneder 2026a, b, this volume), which caused climatic and oceanographic changes. A more detailed analysis of the available literature and the worldwide occurrence of these Carnian perturbations is contained in the comprehensive review by Ogg (2015). The Reifling Basin in the Northern Calcareous Alps, an intra-platform basin surrounded by Wetterstein platforms, is a key reference locality for the study of the CPE.

Our poor understanding of the mechanisms and causes of the Reingraben Event calls for detailed investigations using modern techniques to gain new insights into these mechanisms and their effects. This study establishes high-resolution organic geochemical (e.g., RockEval pyrolysis, GC–MS) and organic petrological records of the Scheiblinggraben section in northeastern Styria (Austria; Fig. 23.1a). The records are combined with microfacies data.

The Scheiblinggraben section is located in the southwestern part of the Reifling Basin and has been of paleontological and biostratigraphic interest since the nineteenth century (Arthaber 1896; Tollmann 1985; Lobitzer et al. 1994). It was selected for the present study because geochemical data from a few samples (Lobitzer et al. 1994; Bruckner-Wein et al. 1996) indicate the presence of organic matter-rich rocks. New paleontological data from the Scheiblinggraben section are reported by Lukeneder (2026a, b, Chap. 19, this volume).

One of the main goals of this study is to identify environmental changes during the CPE at the Scheiblinggraben site. In addition, the newly obtained organic geochemical data is compared with data from the Großau drill core (Lukeneder 2026a, b; see Lukeneder et al. 2026, Chap. 15, this Volume) and the Polzberg section (Lukeneder et al. 2024). This reveals differences within the Reifling Basin and augments the datasets (for location see Fig. 23.1).

23.2 Geologic Setting

The Northern Calcareous Alps (NCA) are the uppermost tectonic unit of the Eastern Alps and belong to the Austroalpine nappe complex. That complex consists predominantly of Mesozoic sedimentary rocks. The Early Triassic was characterized by deposition of shallow shelf siliciclastics (Werfen Formation). From Middle to Late Triassic times the Tethyan shelf was characterised by carbonate sedimentation. Carbonate ramps dominated in the Middle Triassic (Steinalm Formation), and carbonate platforms (Wetterstein, Dachstein) in the Late Triassic (Mandl 2000). The Wetterstein carbonate platforms grew on morphological highs on the shelf that had formed due to Anisian extensional tectonics (Mandl 2000) and bounded the Reifling intraplatform basin. The Wetterstein carbonate platform evolution ended by the Reingraben turnover (=Carnian Pluvial Episode CPE, Carnian Crisis) during the Julian. A major sea level drop here led to subaerial exposure of the platforms. Therefore, on top of the Wetterstein platform and partially on the shallow-water carbonates shed from it (Raming Formation), the change is characterized by a gap with karstification. In the basinal setting, initially characterised by deposition of radiolarian-micrite to radiolarian-filament micrite of the Reifling Formation, the change is marked by the transitional deposition of mixed sediments (calcareous, siliceous and siliciclastic) of

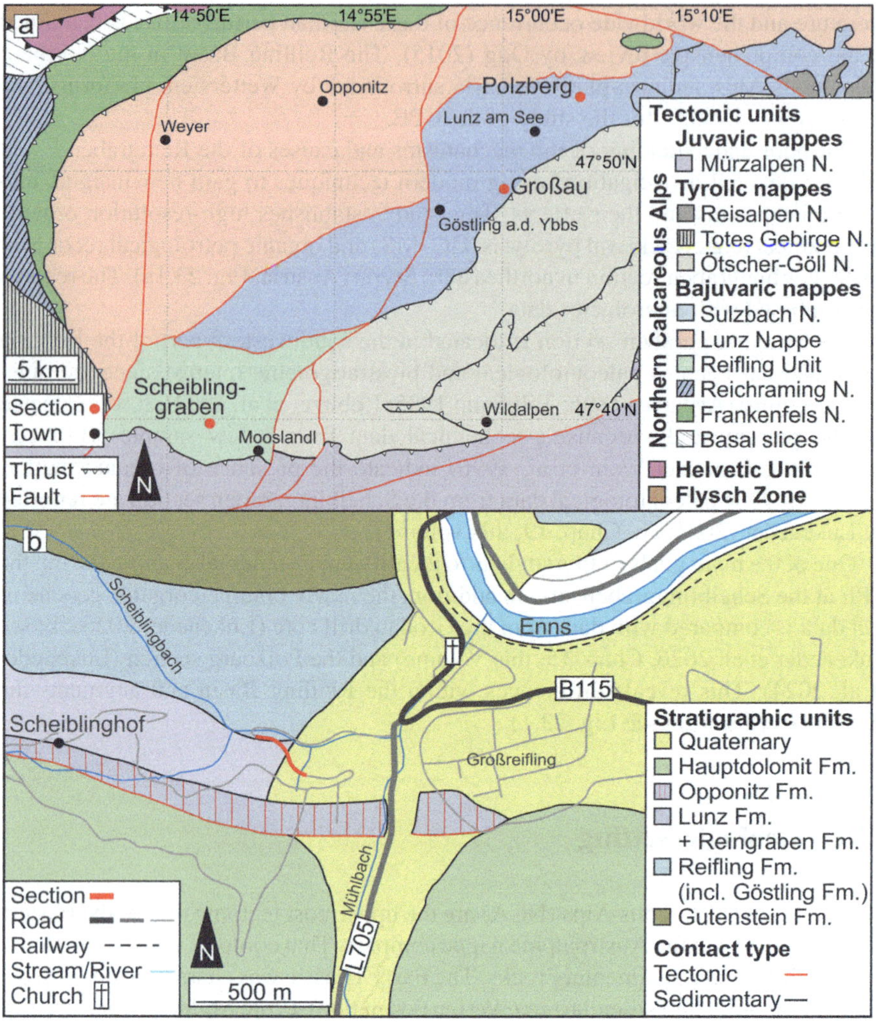

Fig. 23.1 **a** Position of the Scheiblinggraben, Großau and Polzberg sections in the nappe complex of the Northern Calcareous Alps (features from GBA 2021 modified after Mandl 2000). **b** Geologic overview of the Großreifling area and location of the Scheiblinggraben section (red) (Ampferer 1933; GBA 2021)

the Göstling Formation (Schlager and Schöllnberger 1974). The intra-platform basin morphology was filled by siliciclastics, initially the marine, followed by the brackish Lunz Formation (coal-bearing sandstones; Mandl 2000).

The Scheiblinggraben section (= Scheiblingbach) is located in the tectonic Reifling Unit ("Reifling Scholle"), which belongs to the Bajuvaric nappe system of the NCA (Fig. 23.1a). Its Middle Triassic succession includes Partnach marls, siliceous limestones of the Reifling Formation, the mixed carbonate-siliciclastic system of the Göstling Formation and the siliciclastic-dominated system of the Reingraben Formation. This succession is typical for the Carnian in the SW-NE trending Reifling Basin and is the sedimentological manifestation of the Reingraben turnover and the CPE.

23.3 Samples and Methods

23.3.1 Samples

The Scheiblinggraben section is located west of Großreifling, along a forest road connecting Großreifling with the farmhouse "Scheiblinghof" (Fig. 23.1b). The outcrop exposes the upper part of the Reifling Formation, the Göstling Formation and the lower part of the Reingraben Formation (Fig. 23.2).

In April 2024, 94 samples were taken from the Scheiblinggraben outcrop for geochemical analysis. Sampling focused on the approximately 6-m-thick Göstling Formation, but the sample set also includes six samples from the upper 3 m of the Reifling Formation and thirty-two samples from the lower 25 m of the Reingraben Formation. The Göstling Formation is represented by a total of 51 samples. 32 of these samples represent siliceous limestone beds of varying colour (light grey to darker grey, and black). Their sample number corresponds to the bed number (Fig. 23.2). Twenty-five samples from the Göstling Formation represent thinly laminated dark, presumably organic-rich, mudstones. Their sample number corresponds to the number of the underlying bed and includes the letter "a" after the bed number.

Fig. 23.2 Part of the Scheiblinggraben section. The Scheiblinggraben section exposes the upper part of the Reifling Formation (not visible), the Göstling Formation and the Reingraben Formation. Lithologic boundaries and some sample numbers are indicated. The sample numbers refer to the bed numbers. Samples representing thin, dark mudstone interlayers are labelled by adding an "a" to the number of the underlying bed

23.3.2 Thin Sections

Thin sections of rock samples of every bed (1–64) were prepared at the Natural History Museum Vienna. The samples were cut into slabs of 2.5 × 2.5 cm^2. Subsequently, samples were embedded in Araldite epoxy resin where needed, sectioned, mounted on the microscope slides and polished with silicon carbide and aluminium oxide powders to a thickness of about 15–20 μm. Note that it was impossible to prepare thin sections from the clay layers.

23.3.3 Sample Preparation for Organic Geochemistry

To avoid any contamination from soil and modern plant fragments, the samples were washed with water and brushed before further processing. Afterwards the samples were dried at 40 °C for at least 48 h until they reached constant weight. After drying, the samples were crushed with a steel mortar. In a next step, pieces with calcite-filled fissures were removed to determine the original carbonate and organic matter content. The preparation to analytical fineness was done with a mortar grinder. Enough powder was produced from each sample so that all geochemical analyses (bulk geochemistry, biomarker analysis) could be carried out using the same powder.

23.3.4 Bulk Geochemistry

The elemental composition was determined with an ELTRA Helios CS elemental analyzer, burning the rock powder at just below 1350 °C. The machine was calibrated by measuring a standard three times, which was also analyzed after every ten samples. The untreated rock powder was used to determine total carbon (TC [wt%]) and total sulfur content (TS [wt%]). For the determination of the total organic carbon content (TOC [wt%]; Fig. 23.3) the rock powder was treated with concentrated phosphoric acid and hot water. Acid and hot water were added with an intermission of 1 h until no further reaction could be observed. Afterwards at least three washing steps were performed to remove remnants of the phosphorous acid. The total inorganic carbon content (TIC [wt%]) was calculated as the difference between TC and TOC. Assuming that calcite is the only carbonate mineral present, TIC was used to calculate an equivalent calcite content (C_{EQ} [wt%]; $C_{EQ} = \text{TIC} * 8.334$).

Using a Rock Eval 6 instrument, the Rock–Eval parameters (S1 [mg HC/g rock], S2 [mg HC/g rock], T_{max} [°C]) were determined. S1 results from the thermal distillation of already existing hydrocarbons at a constant temperature of 300 °C, whereas S2 results from the cracking of organic matter, thereby generating hydrocarbons, during gradual heating from 300 to 650 °C. The measurements were started with three standards, and after every ten samples another standard was measured. T_{max}

is the temperature at which the maximum hydrocarbons are generated and indicates maturity. The Production index (PI [–]) represents a relation between the already produced hydrocarbons and the total amount of hydrocarbons (PI = S1/(S1 + S2)). The hydrogen index (HI [mg HC/g TOC]) is calculated using the S2 peak and the TOC content (HI = 100 * S2 / TOC; Espitalié et al. 1977; Peters 1986; Langford and Blanc-Valleron 1990; Fig. 23.3). All measurements were conducted in duplicate.

23.3.5 Organic Petrography and Vitrinite Reflectance

Twelve samples were selected for organic petrographical analysis based on their TOC content and position in the succession. The samples were embedded in epoxy resin and cut perpendicular to the layering using a diamond saw with ethylene glycol as lubricant. The cutting surface was covered by epoxy resin before being ground and polished. On these polished blocks, semi-quantitative maceral analysis was performed using reflected white light and blue light excitation. More than 1000 points were counted per sample with a Leica DM 4P microscope at a magnification of 50 times under oil immersion according to the ICCP system (ICCP 1998, 2001; Pickel et al. 2017). Random vitrinite reflectance (%Rr) was measured to determine maturity. Vitrinite reflectance was determined on five driftwood samples from the Göstling Formation. Mean random vitrinite reflectance (%Rr) was measured following established procedures (Taylor et al. 1998; ISO 7404–5 2009). The microscope was calibrated with the N-LASF46A (Rr: 1.311%) synthetic standard.

23.3.6 GC-FID/MS Biomarker Analysis

The bulk geochemical data served as the basis for selecting samples for biomarker analysis. Eleven samples were selected. The extractable organic matter was extracted from 5 to 10 g of rock powder with a Dionex ASE 350 Accelerated Solvent extractor with dichloromethane (DCM) at a temperature of 75 °C and pressure of 100 bar for 1 h. Using a Zymark TurboVap 500, the extract was concentrated to a volume of about 0.5 ml. It was then transferred to a sample tube with a Pasteur pipette and diluted with *n*-hexane to 80:1. The asphaltene fraction was separated by precipitation using a centrifuge. The sample tube was decanted to a beaker and flushed with *n*-hexane, without mobilising the asphaltenes. The obtained asphaltene-free extract was separated into aliphatic, aromatic and polar fractions using a Köhnen-Wilsch medium pressure liquid chromatography (MPLC). Prior to the gas chromatographic analyzes, internal standards were added to the fractions: squalane to the aliphatic and 1,1´-binaphthyl to the aromatic fraction.

The aliphatic hydrocarbon fraction was analyzed with a gas chromatograph (Thermo Scientific Trace GC-Ultra) and a flame ionization detector (GC-FID). The HP-PONA capillary column (inner diameter: 0.20 mm; film thickness: 50 μm) had a length of 50 m. 2-ml of samples were injected at 270 °C, after which the temperature was increased (4 °C/min) to 310 °C (hold time: 35 min).

The aliphatic fraction, additionally to being analyzed with GC-FID, and aromatic hydrocarbon fraction, were analyzed with the gas chromatograph (Thermo Scientific Trace GC-Ultra) in combination with a mass spectrometer (Thermo Fisher ISQ single quadrupole mass spectrometer) (GC–MS). With a scan time of 0.7 ms the spectrometer was operated in the electron ionization mode scanning for mass to charge ratios from 50 to 650 kg/C. A 60-m-long DB-5MS fused capillary column (inner diameter: 0.25 mm; film thickness: 0.25 μm) was used for the measurements. Samples were injected in split mode (ratio: 20, temperature: 260 °C, carrier gas: helium). Initial oven temperature was 40 °C (hold time: 2 min). At 4 °C/min, the oven was heated to 310 °C (hold time: 40 min).

23.4 Results

23.4.1 Lithology and Microfacies

The outcrop at the Scheiblinggraben (Fig. 23.2) exposes the uppermost part of the Reifling Formation, the Göstling Formation, which is about 6 m thick, and the lowermost 25 m of the overlying Reingraben Formation (Fig. 23.3).

Reifling Formation

The exposed part of the Reifling Formation consists of limestone beds, which are up to 3 m thick. The carbonate is predominantly a radiolarian-bivalve packstone with siliceous spicules. The bivalve shells are thin and no preferential bedding is observed. The thin section from bed 3 shows the radiolarian-bivalve packstone. The original fabric is partially recrystallized to blocky calcite, as a component of a monomictic breccia, which has been intensely tectonized. This is evident by the multiple fracture generations with different infilling (calcitic and bituminous) and stylolites (Fig. 23.4a). The carbonate content is generally high (>86 wt%, Fig. 23.3 calcite eq.).

Göstling Formation

The Göstling Formation consists of 60 siliceous limestone beds (beds K-A and 5–55; Fig. 23.2), with a thickness ranging from several centimeters to a few decimeters. Some beds show a gradual transition into thinly laminated dark mudstones (denoted with the addition of an "a" to the bed number). The maximum thickness of the dark mudstones is 5 cm. The dark mudstone interlayers have lower carbonate contents (10–75 wt%) than the siliceous limestones (40–100 wt%). Thin sections showing radiolarian-bivalve packstone, similar to the Reifling Formation, are bioturbated (Fig. 23.4b and c, bioturbation index 2–3). Additionally, rhythmite layers are present, consisting of different alternating microfacies layers. Layers of bivalve packstone,

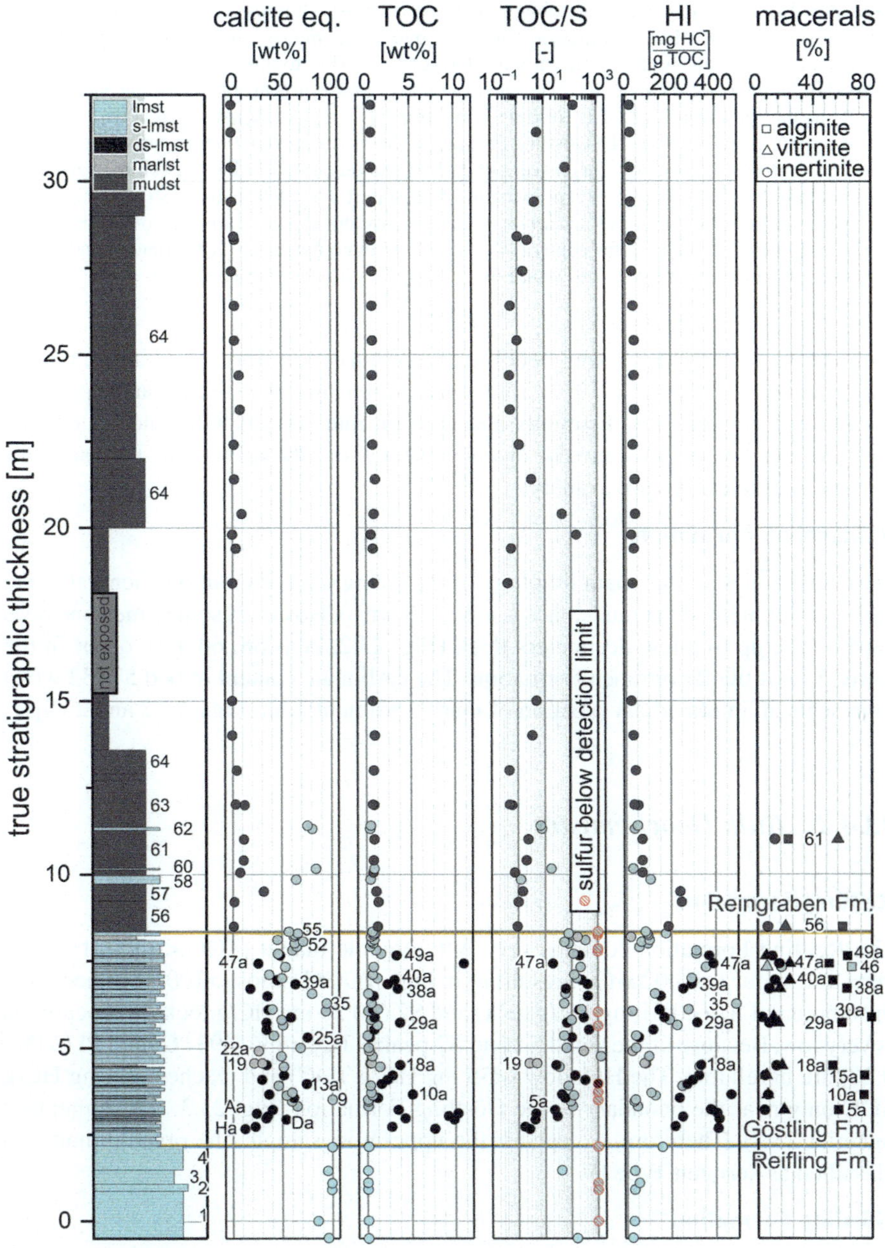
calcite eq.
[wt%]
TOC
[wt%]
TOC/S
[-]
HI
[mg HC/g TOC]
macerals
[%]
true stratigraphic thickness [m]
lmst
s-lmst
ds-lmst
marlst
mudst
not exposed
alginite
vitrinite
inertinite
sulfur below detection limit
Reingraben Fm.
Göstling Fm.
Reifling Fm.

◀**Fig. 23.3** Lithologic log of the Scheiblinggraben section (lmst—limestone, s-lmst—siliceous limestone, ds-lmst—dark siliceous limestone, marlst—marlstone, mudst – mudstone) with carbonate content (calcite eq.) and total organic carbon (TOC) content, TOC to sulfur ratio (TOC/S), hydrogen index (HI, Supplementary material 23.1), and maceral composition (vitrinite, inertinite and alginite; Supplementary material 23.2). The Reifling Formation has a high carbonate content but low TOC, sulfur contents and a low HI. The Göstling Formation shows strong variations in carbonate content. The TOC baseline is generally increased and dark layers reach up to 11 wt%, sulfur is generally low in relation to TOC content, except in the lowermost part. HI reaches up to 489 mg HC/g TOC. The Reingraben Formation has low carbonate contents except in three carbonaceous layers, TOC is generally below 1 wt%, sulfur content is relatively high compared to TOC content, and the HI drops significantly with the first carbonaceous layer above the Göstling Formation

with ostracods, exhibit parallel bedding and bioturbation (Fig. 23.4d and f, bioturbation index ≤ 2, Reineck 1963; Taylor and Goldring 1993). Radiolarian pack- to wackestone contains few filaments and is often interbedded with calci-mudstone. Argillaceous layers show angular changes throughout the formation. Thin sections of dark interlayers are not available.

Reingraben Formation

Dark mudstones are the main lithology of the Reingraben Formation. Generally, the carbonate content of the mudstones is low (<13 wt%). However, some limestone beds (>65 wt%), up to a few decimeters thick (Fig. 23.2, beds 58, 60, 62), occur in the lower 5 m of the Reingraben Formation. The carbonate content of bed 57 (32 wt%) represents an intermediary position of marly sedimentation (Figs. 23.2 and 23.4j).

23.4.2 Bulk Geochemistry

Reifling Formation

The Reifling Formation contains very little organic matter (TOC <0.16 wt%) and very low sulfur (<0.002 wt%), resulting in high TOC/S ratios (>100). Furthermore, the amount of S1 (<0.01 mg HC/g rock) and S2 (<0.26 mg HC/g rock) hydrocarbons is very low. Because of the small S1 and S2 peaks, T_{max} (436–493 °C) and PI (0.00–0.20) are unreliable. The HI is low (<55 mg HC/g TOC) and reaches 158 mg HC/g TOC only near the boundary to the Göstling Formation (Fig. 23.3, Supplementary material 23.1). Therefore, except for the uppermost sample, the organic matter is classified as kerogen Type IV.

Göstling Formation

Depending on the lithotype, TOC contents in the Göstling Formation vary widely (0.08–11 wt%), but most (90% of samples) contents are below 3.5 wt%. The dark interlayers generally exhibit higher TOC (>1.2 wt%), which correlates negatively with carbonate content. Slightly brownish discoloured dark interlayers have lower TOC and deviate from the aforementioned trend, whereas no trend is evident in

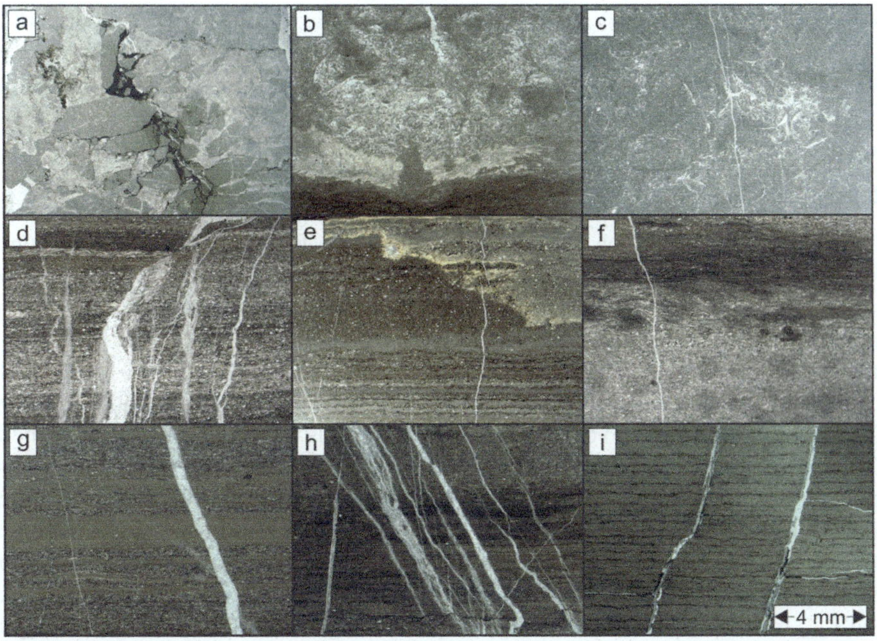

Fig. 23.4 Thin-section photographs from **a** Reifling Formation, **b–h** Göstling Formation, and **i** Reingraben Formation. The scale is the same for all photographs. **a** micro breccia of radiolarian-bivalve (filament) packstone with few sponge spicules, the original fabric is partially recrystallized to blocky calcite, multiple generations of fracture fills (calcite and bitumen) and stylolites (bed 3). **b** argillaceous filament wackestone overlain by bivalve packstone (lumachelle), bivalves have thin micritic envelopes, some peloids and ostracod shells are present; radiolarians are rare, bioturbation is evident by the dark patches consisting of radiolarian wackestone (bed 9). **c** bioturbated radiolarian-filament wacke- to packstone, containing sponge spicules and ostracod shells (bed 23). **d** laminated argillaceous radiolarian-bivalve wackestone with strongly sheared shells and bioclasts, becoming more argillaceous towards the top. Tectonic stress is apparent by extensional fractures, crosscut by shear fractures. Shear fractures are localized in the less competent, more argillaceous layers (bed 29). **e** rhythmite of calci-mudstones and radiolarian packstones, containing filaments and likely ostracod shells. Radiolarian packstones become darker towards the top due to increased argillaceous input, before shifting to calci-mudstones (bed 35). **f** argillaceous bivalve packstone grading into argillaceous mudstone, containing relatively thick shells, radiolarians are sparse, bioturbation is clearly visible (bed 38). **g** deformed rhythmite of argillaceous radiolarian—shell wackestone, calci-mudstones and a thicker calciturbidite layer with bioclasts and oriented thin shells, indicating weak bottom currents (bed 51). **h** argillaceous mudstone changing to argillaceous packstone (bed 51). **i** marlstone with argillaceous interlayers (bed 58). Note that the preparation of thin sections from mudstones was not possible

the siliceous limestones. In the lowest metre of the Göstling Formation, below bed 5a, the sulfur contents are relatively high (average: 1.90 wt%), after which they are generally below 0.05 wt%, except for the marly bed 19 and TOC-rich interlayers (<0.56 wt%). The resulting TOC/S ratios increase up to bed 5a and are high (>20) above, except for the marly bed 19. The amount of S1 (max. 0.73 mg HC/g rock;

90% <0.23 mg HC/g rock) and S2 hydrocarbons (max. 42.72 mg HC/g rock; 90% <12.3 mg HC/g rock) shows a wide range and correlates with TOC (R^2S2-TOC = 0.98). PI is <0.14. T_{max} of samples with S2 >0.2 mg HC/g rock ranges from 423 to 440 °C (average: 429 °C). The maximum HI is 489 mg HC/g TOC, indicating kerogen Type II (Fig. 23.3, Supplementary material 23.1).

Reingraben Formation

The Reingraben Formation consists predominantly of mudstone, while limestone beds are present in the lowermost 5 m (Fig. 23.3). TOC contents are low, both in mudstone and limestone beds, with the exception of the two lowermost mudstones samples (bed 56 and 57, 1.1–1.3 wt% TOC). The behaviour of S1 and S2 hydrocarbons is similar, with slightly higher values in beds 56 and 57 (0.04–0.06 and 2.3–3.3 mg HC/g rock, respectively). PI is below 0.07. HI reaches 250 mg HC/g TOC (kerogen Type II–III) only in the lowermost beds (56 and 57) here. Above bed 57, HI drops below 70 mg HC/g TOC. The sulfur content in Reingraben mudstones is higher than in Göstling samples, but varies strongly (0.0–2.0 wt%). The limestone beds have a low sulfur content (<0.35 wt%). TOC/S ratios are generally low in the lower part of the formation and show trends with upward decreasing values and then increasing values in the uppermost part (Fig. 23.3, Supplementary material 23.1). The T_{max} of samples with S2 >0.2 mg HC/g rock ranges from 423 to 436 °C (average: 430 °C).

23.4.3 Organic Petrography and Vitrinite Reflectance

The semi-quantitative analysis shows a clear dominance of liptinite macerals in the Göstling Formation. Most abundant is lamalginite (50–78 vol% of all macerals, Figs. 23.5a and 23.6a and d), followed by liptodetrinite (5–18 vol%, Fig. 23.6d). Telalginite (<9 vol%, Fig. 23.6a and b) and other liptinites (<7 vol%, Fig. 23.5c), consisting mainly of sporinite and rarely resinite, are of minor importance. Furthermore, vitrinite (6–27 vol%, Figs. 23.5d and 23.6a and c) and inertinite (3–16 vol%, Fig. 23.6c) were identified in significant quantities. The lower sample (bed 56a) from the Reingraben Formation shows the same maceral distribution as the Göstling Formation, but the stratigraphically higher sample (bed 61) exhibits a shift towards an increased amount of vitrinite (56 vol%) at the expense of alginite (20 vol%, macerals, Fig. 23.3, Supplementary material 23.2). Fluorescing bituminous inclusions (exsudatinite) are observed in some calcitic fracture fillings (Fig. 23.5b). Furthermore, pyrite (Figs. 23.5a and 23.6c), oxidised pyrite (Figs. 23.5c and 23.6c) and diagenetic calcite (Fig. 23.5c) are present in the samples. In the Reingraben Formation, in bed 61, microfossils were identified that represent juvenile forms of organisms (Fig. 23.7).

Vitrinite reflectance measurements were conducted on five samples (beds 5a, 15a, 30a, 47a, 52) of the Göstling Formation. Vitrinite reflectance of 5a, 15a and 47a is about 0.42%Rr. Beds 15a and 52 yielded slightly lower values (0.39 and

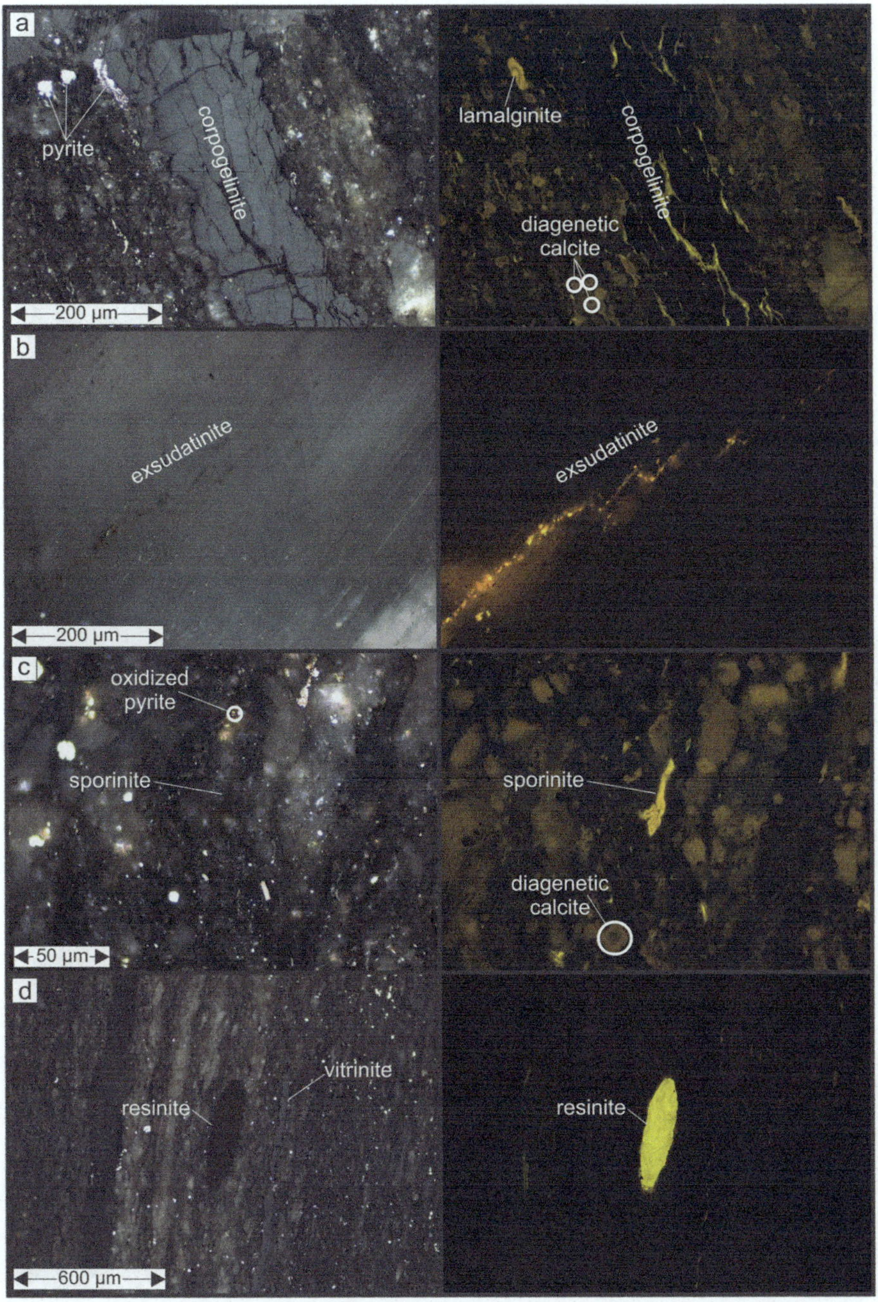

Fig. 23.5 Micro-photographs of polished blocks, in colour under reflected white light (left) and under blue light excitation (right). **a** fragment of plant tissue with corpogelinite, lamalginite, framboidal pyrite and diagenetic calcite (layer 56a). **b** migrated bituminous substance (exsudatinite) in a fracture within a calcite vein (bed 46) **c** sporinite in the center, oxidized framboidal pyrite (reddish reflection) and diagenetic calcite (layer 18a). **d** vitrinite and presumably resinite (layer 38a)

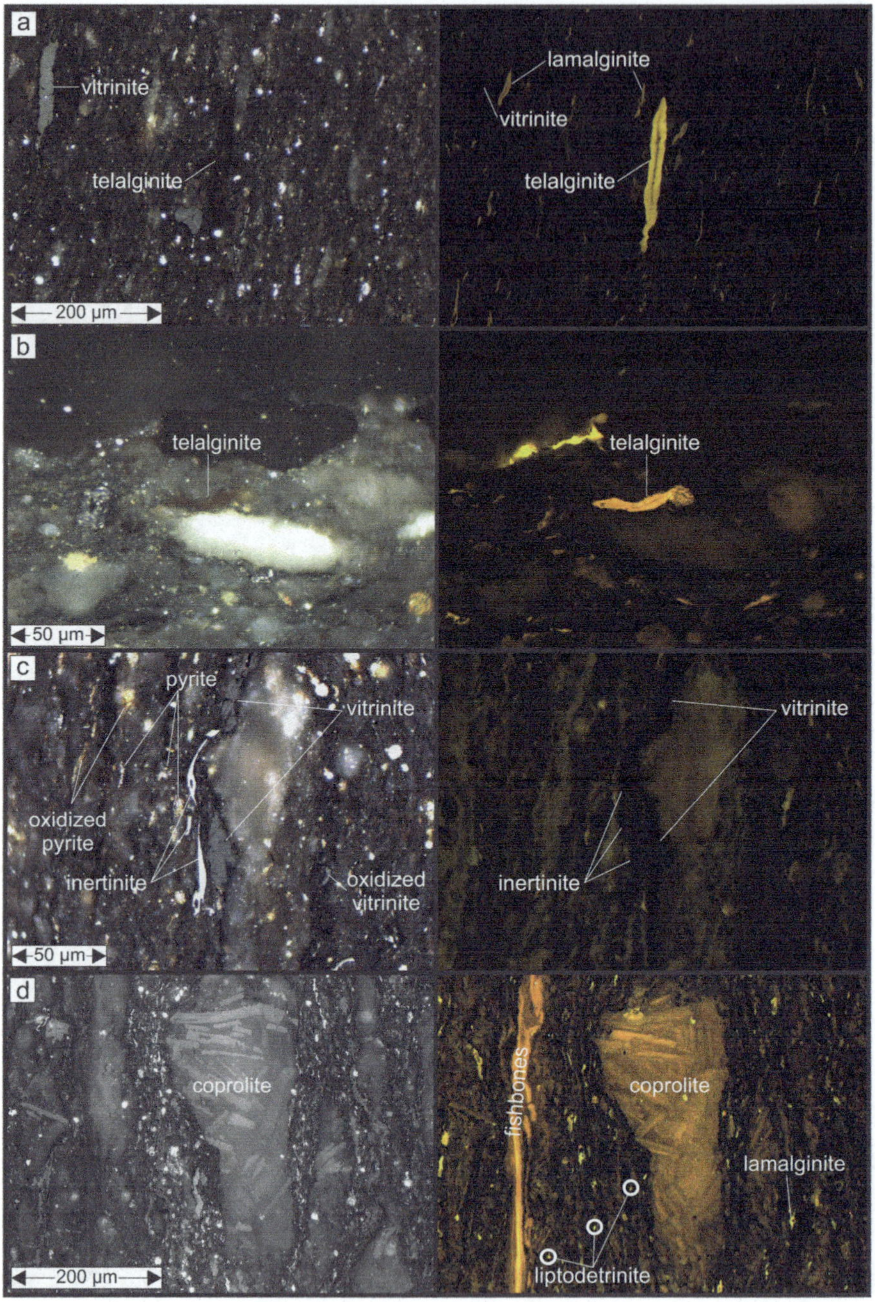

Fig. 23.6 Micro-photographs of polished blocks, in colour under reflected white light (left) and under blue light excitation (right). **a** vitrinite, inertinite, telalginite and lamalginite (layer 38a). **b** telalginite with whorls on the right side (layer 40a). **c** inertinite, vitrinite and oxidized vitrinite, pyrite and its oxidized form (layer 5a). **d** fishbones in a layer and inside a coprolite, lamalginite and liptodetrinite (layer 10a)

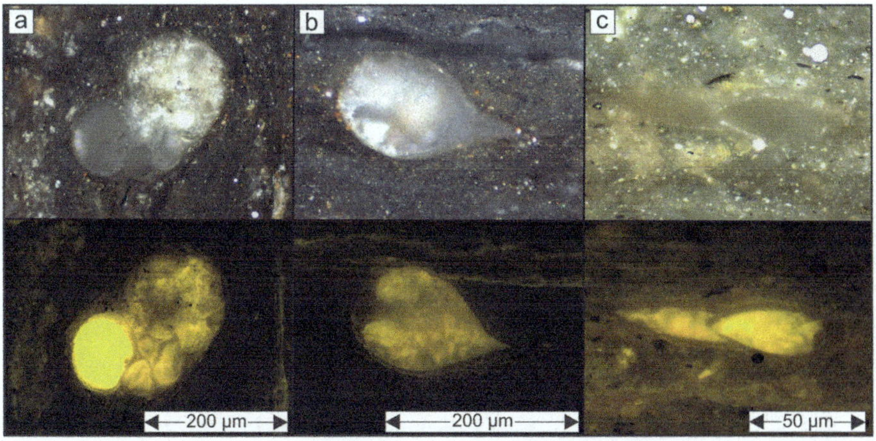

Fig. 23.7 Micro-photographs of polished blocks from the Reingraben Formation (bed 61; top: white light; bottom: blue light excitation). **a** juvenile gastropod, **b** bivalve and **c** foraminifera. Cavities were cemented by diagenetic calcite, which may contain organic matter

0.36%Rr, respectively). In all samples, a second population with higher vitrinite reflectance (>0.6%Rr) is present (Supplementary material 23.2). The higher values were measured in macerals with high heterogeneity (e.g., along the margins of zoned particles, decreasing towards the centre).

23.4.4 Hydrocarbon Composition

The hydrocarbon composition was exclusively determined for thin interlayers from the Göstling Formation with high TOC content (Fig. 23.8), except sample 30a (TOC = 0.84 wt%). Extraction yields varied between 0.2 and 2.6 mg/g rock. The yields correlate well with S1 ($R^2 = 0.92$) and S2 ($R^2 = 0.98$). The relative amount of asphaltenes is decreased in the middle part of the Göstling Formation (≈6%), whereas NSO compounds are increased (≈73%). Aliphatic (8.5–12.9%) and aromatic compounds (5.1–13.0%) show no discernible trend (Supplementary material 23.3). Selected biomarker ratios are plotted versus stratigraphic position in Fig. 23.8.

n-Alkanes and Isoprenoids

n-Alkanes (213–414 μg/g TOC) and isoprenoids (63–255 μg/g TOC) represent the most abundant compound groups present in the rock extract (Supplementary material 23.3). The chain lengths of *n*-alkanes range from $n\text{-}C_{15}$ to $n\text{-}C_{33}$ in all samples (Fig. 23.9). A few samples (10a, 40a and 47a) also show $n\text{-}C_{14}$. Generally, the long-chain *n*-alkanes ($\sum n\text{–}C_{26\text{–}33}/\sum n\text{–}C_{15\text{–}33}$ alkanes: 0.19–0.35) are least abundant. In almost all samples, the abundance of short-chain *n*-alkanes ($\sum n\text{–}C_{15\text{–}20}/\sum n\text{–}C_{15\text{–}33}$

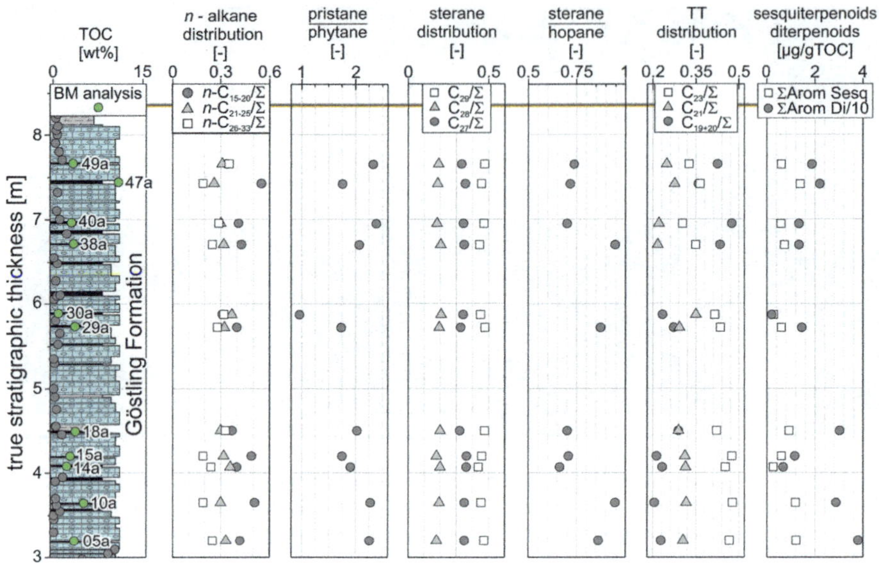

Fig. 23.8 Stratigraphic profile of the Göstling Formation in the Scheiblinggraben section together with the TOC log (Supplementary material 23.1). Samples used for biomarker (BM) analysis (Supplementary material 23.3) are labelled and highlighted by green data points. From left to right the following biomarker ratios are shown: *n*-alkane distribution, pristane/phytane ratio, sterane distribution, sterane/hopane ratio, tricyclic terpane (TT) distribution, content of total aromatic sesquiterpenoids and diterpenoids

alkanes: 0.34–0.55) is the highest, followed by mid-chain homologues ($\sum n$–C_{21-25}/ $\sum n$–C_{15-33} alkanes: 0.26–0.36). The exceptions are samples 18a and 49a, which showed relatively flat distributions of *n*-alkanes with slight dominance of short- and long-chain homologues, and sample 30a in which mid-chain homologues prevailed (*n*-alkane distribution; Fig. 23.8; Supplementary material 23.3). Generally, *n*–C_{17} has the highest peak except in samples 14a, 30a and 49a. However, the abundance of *n*–C_{19} is slightly higher in samples 5a, 14a, 29a, 30a and 49a. The ratio of *n*-alkanes with odd and even carbon numbers shows an increase upwards, observed in the carbon preference index (CPI 1.2–1.5), interrupted by beds 30a and 47a (1.1 and 1.2, respectively, Bray and Evans 1961).

Pristane is more abundant than *n*-C_{17} in all samples, while phytane is more abundant than *n*–C_{18} in samples 5a, 18a, 29a and 30a. The pristane to phytane (Pr/Ph) ratio is generally relatively high (1.7–2.4). Only sample Sch 30a, which has a lower TOC, exhibits a low Pr/Ph ratio of 0.96 (Supplementary material 23.3).

Terpanes

Hopanes are present in all samples. The mass fragmentogram (*m/z* 191) from a typical sample (10a) is shown in Fig. 23.10a. Their concentrations range from 5 to 37 µg/g TOC. The extended homohopanes decrease strongly after C_{31} with increasing carbon number; the typical mass spectrum for C_{35} homohopane is no longer detectable. The

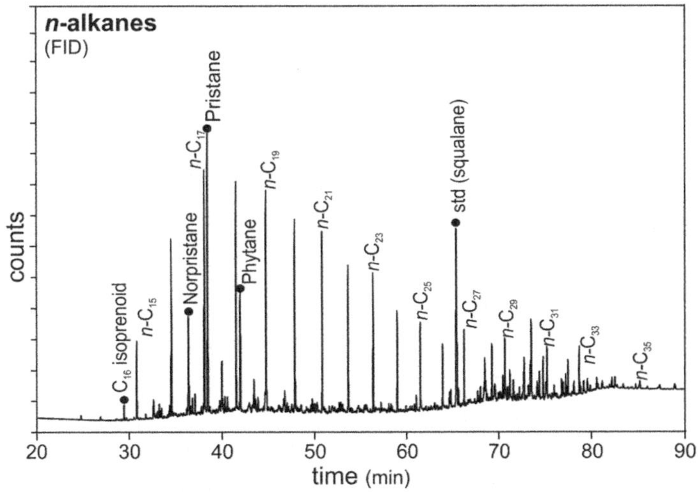

Fig. 23.9 Typical gas-chromatogram of the aliphatic hydrocarbon fraction from layer 10a, analyzed by flame ionization detector (FID). Odd-numbered *n*-alkanes, important isoprenoids and the internal standard (squalane) are labelled

22S/(22S + 22R) isomerization of C_{31} 17α(H), 21β(H)-hopane reaches 0.40 ± 0.04. The sterane to hopane ratio shows two cycles of change, increasing and decreasing. It is generally between 0.66 and 0.95; only sample 30a has a low value of 0.43 (Supplementary material 23.3). Gammacerane could not be identified in any of the samples.

Tricyclic terpanes (TT) were identified up to C_{26} (0.5–1.7 μg/g TOC). In the lower part of the Göstling Formation, TT ratios show the predominance of C_{23} TT (0.42–0.48) over C_{21} TT (0.29–0.35) and C_{19+20} TT (0.21–0.29). In contrast, C_{19+20} TT (0.36–0.48) dominate in the upper part, and the proportions of both C_{23} (0.30–0.36) and C_{21} (0.22–0.28) drop (TT distribution; Fig. 23.8; Supplementary material 23.3). The ratio between C_{25} and C_{26} TT is generally above 1 (1.5–3.4) and increases upwards, except in samples 38a and 40a where the peaks were too small to identify C_{26}. Moreover, C_{24} tetracyclic terpane was identified in all samples (Fig. 23.10a). Norpimarane and norabietane were also detected.

Steranes and Diasteranes

Steranes and diasteranes were identified using mass fragmentograms *m/z* 217 + 218 and *m/z* 259, respectively (Fig. 23.10b). Steranes are present in all samples (5–24 μg/g TOC). Sterane distributions are dominated by C_{29} steranes (C_{29} steranes/ΣC_{27-29} steranes: 0.44–0.56), while C_{28} steranes (C_{28} steranes/ΣC_{27-29} steranes (0.16–0.22)) are least abundant; intermediary is the C_{27} (C_{27} steranes/ΣC_{27-29} steranes: 0.24–0.38) abundance. The distribution is homogeneous over the whole profile (sterane distribution; Fig. 23.8). The C_{29} sterane isomerisation ratios C_{29}ααα20S/C_{29}ααα20(S + R) and C_{29}αββR/(αααR + αββR) are rather uniform in the Göstling Formation, 0.10

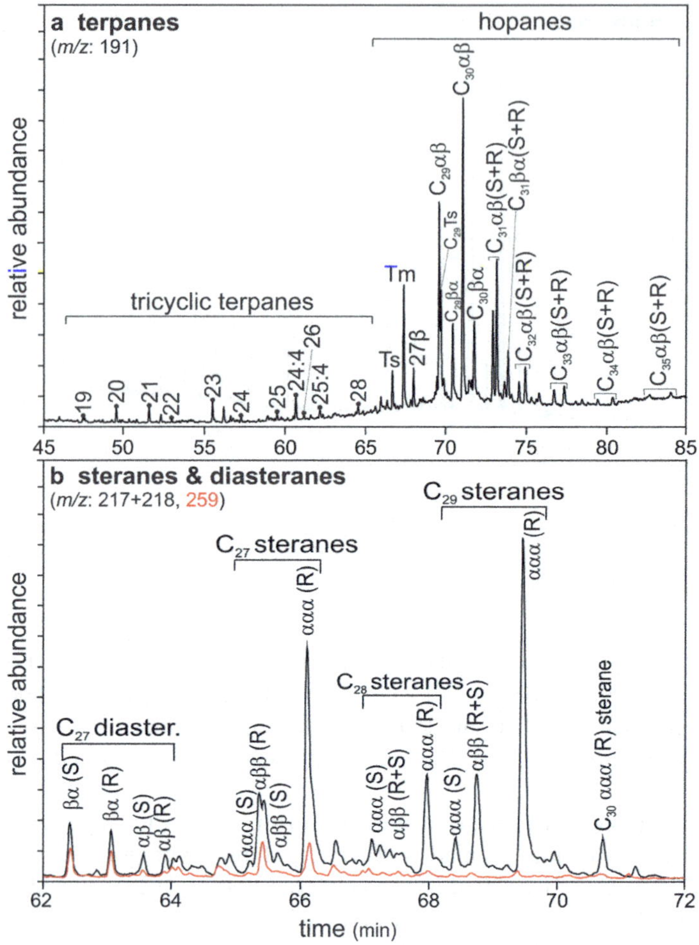

Fig. 23.10 **a** Mass fragmentograms of hopanes and tricyclic terpanes (*m/z*: 191). **b** Mass fragmentograms of steranes (*m/z*: 217 + 218) and diasteranes (*m/z*: 259), both from sample 10a

± 0.02 and 0.25 ± 0.01, respectively (Supplementary material 23.3). C_{27} diasteranes are present in lower abundances than their sterane counterparts. The ratio of C_{27} diasterane to C_{27} sterane varies between 0.17 and 0.30 (Supplementary material 23.3).

Aromatic Compounds

Constituents of the aromatic fractions are compounds of naphthalene, phenanthrene (P) and dibenzothiophene (DBT) types, as well as aromatized steroids (C-ring monoaromatic: MAS; triaromatic: TAS). Ethylnaphthalene (EN) is detectable only in samples 5a and 47a. The dimethylnaphthalenes (DMN) are identifiable in all samples, except sample 30a (Fig. 23.11a). Phenanthrene (P), methylphenanthrenes (MP) and ethyl- + dimethyl-phenanthrenes (EP + DMP) are present in all samples (in sample 30a the peaks are blurred and coeluted). 3- and 2-MP are less abundant than 9- and 1-MP in all samples, whereby 9-MP is more abundant than 1-MP (Fig. 23.11b). DBT/P ranges from 0.14 to 0.85, whereas the MAS/TAS ratio varies between 1.7 and 3.4 (Supplementary material 23.3).

Furthermore, sesquiterpenoids, cadalene and tetrahydrocadalene, as well as the diterpenoid derivatives simonellite, tetrahydroretene and retene are present in all samples (Fig. 23.11). Also, four monoaromatic compounds with fernane skeleton (25-nor-ferna-5,7,9-triene, 25,30-dinor-ferna-5,7,9-triene, 22,25,29,30-tetranor-ferna-5,7,9-triene, des-A-25-nor-ferna-5,7,9-triene) were detected in all samples, using the *m/z* 225. Additionally, fluoranthene, pyrene and methylated pyrenes were found in all samples. Generally, these compounds show a decrease upwards in the first part of the Göstling Formation and reach a minimum at samples 14a–15a, after which they recover and again reach a minimum around sample 47a. The last values are lower than the first (Supplementary material 23.3). Additionally, *n*-alkylbenzenes (*m/z* 92) and benzohopanes cyclised at C-20 (*m/z* 191) were identified but not quantified.

23.5 Discussion

The Scheiblinggraben section is one of the sites in the NCA, alongside the Polzberg and Großau sections, where changes in organic matter characteristics during the CPE were investigated. While the stratigraphic succession of all three sections is similar overall, there are significant differences in the thickness and organic matter content of the Göstling Formation. The following discussion focuses first on the depositional environment and organic matter characteristics of the Göstling Formation in the Scheiblinggraben section. The three sites are then compared to reveal similarities and differences of depositional patterns within the Reifling Basin. Each subheading begins with a brief description of the maturity of organic matter, which can have a major influence on the hydrocarbon composition and the interpretations based on it. In a separate subheading, the hydrocarbon generation potential of the studied stratigraphic interval is discussed.

23.5.1 *The Scheiblinggraben Section*

23.5.1.1 Thermal Maturity

The low average T_{max} value of samples with S2 values above 0.2 mg HC/g TOC (429 °C) suggests that the sediments in the Göstling and Reingraben Formations are immature (Supplementary material 23.3). Vitrinite reflectance measurements of five samples from the Göstling Formation range from 0.36 to 0.43%Rr (Supplementary

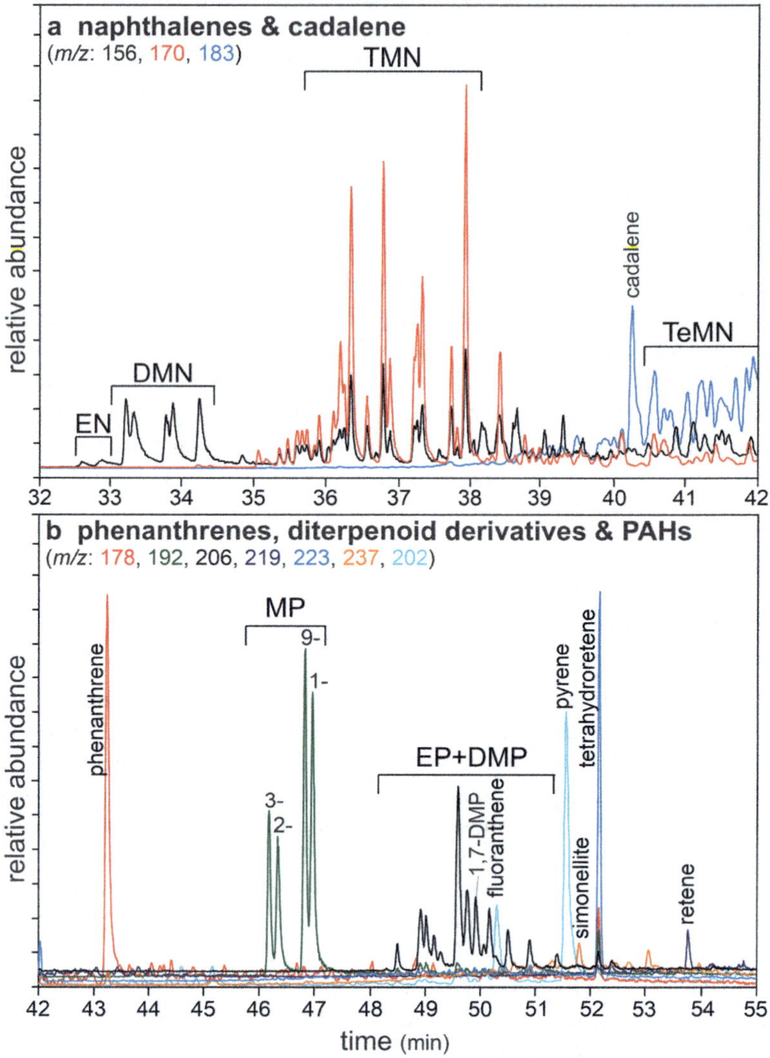

Fig. 23.11 **a** Mass fragmentograms of ethyl- and dimethylnaphthalenes (EN, DMN, *m/z*: 156), trimethylnaphthalenes (TMN, *m/z*: 170), and cadalene and tetramethylnaphthalenes (TeMN, *m/z*:183). **b** Mass fragmentograms of phenanthrene (*m/z*: 178), methylphenanthrenes (MP, *m/z*: 192), ethyl- and dimethylphenanthrenes (EP, DMP, *m/z*: 206), fluoranthene and pyrene (*m/z*: 202), simonellite (*m/z*: 237), tetrahydroretene (*m/z*: 223) and retene (*m/z*: 219). Both from sample 10a

material 23.2). A slightly higher reflectance value (0.53 ± 0.04%Rr) was reported by Sachsenhofer (1987) from coal, 200 m west of the farmhouse "Scheiblinghof" (see Fig. 23.1b), in the overlying Lunz Formation. Our reflectance data are in agreement with previously determined T_{max} values (Peters and Cassa 1994).

Low maturity is also supported by biomarker isomerization ratios (Supplementary material 23.3) including the 22S/22(S + R) C_{31} hopane ratio (0.40 ± 0.04; Ensminger et al. 1977; Seifert and Moldowan 1980; Schoell et al. 1983; Peters et al. 2005), the 20S/20(S + R) C_{29} steranes ratio (0.10 ± 0.02), and the ββ/(αα + ββ) C_{29} steranes ratio (0.25 ± 0.01, Fig. 23.14; Seifert and Moldowan 1981; Mackenzie et al. 1983; Cornford et al. 1986; Peters et al. 2005). The high abundance of monoaromatic steroids (MAS) compared with triaromatic steroids (TAS, MAS/TAS $\geq$ 1.7; Supplementary material 23.3) also indicates low maturity, well below a vitrinite reflectance of 0.65%Rr (Mackenzie 1984; Hoffmann et al. 1984; Peters et al. 1990, 2005). Beyond those arguments, the distribution of methylphenanthrenes with a dominance of 9- and 1-MP over 3- and 2-MP (Fig. 23.11b) also agrees with the previous observations. Note that the average MPI-1 and the calculated vitrinite reflectance (0.36, 0.60%Rc) are too high, due to the fact that MPI-1 does not work for immature samples (Radke and Welte 1983; Peters et al. 2005).

23.5.1.2 Depositional Environment of the Göstling Formation

The Göstling Formation is composed of decimeter-thick siliceous limestone beds separated by thin, dark and often organic matter-rich interlayers. The data on the siliceous limestone beds originate from bulk geochemistry and thin section analysis. Information on the hydrocarbon composition and organic petrography is available for the organic-rich interlayers, but it was not possible to prepare thin sections. Due to the different data sets, siliceous limestones and organic rich layers are discussed separately in the following.

Siliceous Limestone Beds

Siliceous limestone beds form the background sediment. In general, they have a high carbonate content (>40 wt%, average: 61 wt%), a low TOC content (<1.4 wt%) and virtually no sulfur content. This results in very high TOC/S ratios (>50, Fig. 23.3 TOC/S, Fig. 23.15). According to Peters et al. (2005) this is indicative of proximal marine environments with oxidizing conditions. In case of the Göstling Formation the high TOC/S ratios can be linked to strong freshwater influence during deposition, leading to periodically brackish conditions. Thin sections paint a similar picture. The fauna is generally relatively poor: bivalves, ostracods and radiolarians have been identified. Ostracod occurrence, typical biota for low-salinity, reflects the periodic freshwater influx (Flügel 2010). Radiolarian-bivalve packstone, similar to the Reifling Formation, indicates periodically normal marine conditions. Bivalve packstones, with thick shells (likely also from ostracods) often become more argillaceous upwards. They are bedded parallel and contain almost no radiolarians. This indicates turbiditic transport. Bioturbation (bioturbation index $\leq$ 3, Reineck 1963; Taylor and

Goldring 1993) is preferentially found in those layers (Fig. 23.4b, c and f) due to the periodically improved living conditions. Calcimudstones may have formed due to periodic undersaturation with carbonate, which led to increased carbonate dissolution in the aftermath of freshwater input. Radiolarian wackestones containing filaments are indicative of calmer deposition after strong nutrient and oxygen input by freshwater. Sparse bioturbation indicates sub- to dysoxic conditions of the bottom waters.

Organic-Rich Layers

Redox Conditions

Pr/Ph ratios are a widely used indicator of the oxicity of the depositional environment (Didyk et al. 1978) but should not be used without additional constraints (e.g., ten Haven et al. 1987). Relatively high Pr/Ph ratios (1.7–2.4, except sample 30a with a ratio of 0.96) were determined for organic-rich layers in the Göstling Formation (Supplementary material 23.3). In combination with the distributions of C_{31-35} αβ extended hopanes (Rullkötter and Philip 1981), which show an exponentially decreasing trend with increasing number of carbon atoms (Fig. 23.10a), the Pr/Ph ratios indicate initial deposition of organic matter under oxidizing conditions. A low degree of aromatization of fernane derivatives and hopanes, as documented by the presence of monoaromatic compounds (only monoaromatic fernanes and benzohopanes, respectively) suggests oxygen depletion in the water column, but certainly not strictly reducing conditions.

Variations of the Pr/Ph ratio in comparison with the ratio between simonellite (containing 2 aromatic rings) and retene (containing 3 aromatic rings) indicate slight changes in redox conditions during deposition of the studied samples (Supplementary material 23.3). The moderate positive correlation between concentrations of PAHs (fluoranthene + pyrene, Supplementary material 23.3) and inertinite contents ($R^2 = 0.41$; Supplementary material 23.2), as well as the Pr/Ph ratios ($R^2 = 0.43$) denote that these changes were caused by the influx of freshwater. This influx increased the oxygen content in the water column and concurrently transported PAHs, inertinite and terrigenous organic matter from the surrounding areas.

Salinity

A marine depositional environment is proven by abundant marine fossils, although the freshwater influence may have been significant (see above). TOC/S ratios in organic-rich sediments near the base of the Göstling Formation (bed H) are low and reflect normal marine conditions but increase upwards (TOC/S; Fig. 23.3). Above bed 5a, the ratios are very high (>50), suggesting sulphate limitation (e.g., Raiswell and Berner 1984). This is in agreement with the position of the samples in the Pr/Ph versus DBT/P diagram (Hughes et al. 1995; Fig. 23.12). Here, the samples plot in zone 3 (marine shale or lacustrine), which can be linked to a marine environment near a river mouth. Only sample 30a plots in zone 2, which is nonetheless in agreement because it indicates freshwater with a low sulphate content. Similarly, low sulfur kerogens and related oils have been reported in proximal oxidizing environments

(Peters et al. 2005). The absence of gammacerane suggests that water stratification did not play a major role (Sinninghe Damsté et al. 1995). Decreased water salinity is also indicated by relatively abundant C_{21} tricyclic terpane (Zumberge 1987; Tao et al. 2015; Xiao et al. 2024; TT distribution Fig. 23.8).

Organic Matter Sources

The prevalence of alginite (particularly lamalginite; macerals Sect. 23.4.3, Fig. 23.3, Supplementary material 23.2) in all organic matter-rich samples from the Göstling Formation clearly indicates the dominance of marine organic matter. This is further supported by the presence of type II Kerogen (Fig. 23.3 HI, Supplementary material 23.1). Considering the low maturity of the samples (Sect. 23.5.1.1.; Supplementary material 23.1 and 23.3), the predominance of short-chain *n*-alkanes over the mid- and long-chain equivalents (*n*-alkane distribution; Fig. 23.8) provides further evidence for the prevalence of aquatic organic matter (Peters et al. 2005). This is consistent with the results from organic petrography. The high concentrations of C_{17} *n*-alkane are typical for marine organic matter, commonly algae and cyanobacteria. Most cyanobacteria produce 2-methylbacteriohopanepolyols, precursors of 2-methylhopanes, which were not identified in this case. However, a recent comprehensive review of modern cyanobacteria showed that numerous species do not produce these biomarkers (Naafs et al., 2022). The presence of C_{30} $\alpha\alpha\alpha$ (R) sterane (Fig. 23.10b; Neunlist and Rohmer 1985; Zundel and Rohmer 1985), the dominance of the C_{23} homologue in the tricyclic terpane distribution (Moldowan et al. 1983; Zumberge 1987; Simoneit et al. 1990; Grande et al. 1993; Greenwood et al. 2000; Tao et al. 2015; Philp et al. 2021; Fig. 23.8), and the presence of long-chain *n*-alkylnaphthalenes (*m/z* 141; Ellis et al. 1999; Peters et al. 2005) also confirm that

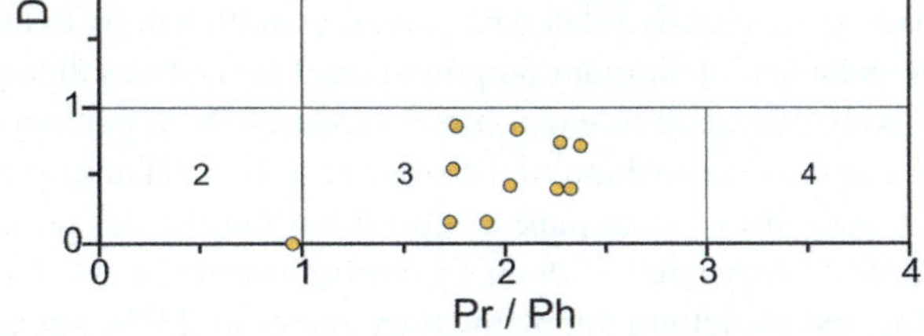

Fig. 23.12 Comparison between the ratio of dibenzothiophene (DBT) to phenanthrene (P) and that of pristane (Pr) to phytane (Ph) (Hughes et al. 1995). Except sample 30a, which plots in zone 2, indicating lacustrine sulphate-poor conditions, all samples plot in zone 3, which indicates a marine shale to lacustrine depositional environment and can be linked to the influence of periodic freshwater influx into the marine depositional realm

marine algae are the main contributor for organic matter. This coincides with abundant mono- and tri-aromatic steroids in the aromatic fractions, and with the C_{31} $\alpha\beta$(R)/C_{30} $\alpha\beta$ ratio >0.25 (Fig. 23.10a, Supplementary material 23.3) throughout the Göstling Formation. The distributions of regular C_{27}–C_{29} $\alpha\alpha\alpha$ (R) steranes are characterized by the prevalence of C_{27} and C_{29} homologues in all samples (Figs. 23.8 and 23.17). Whereas C_{27} sterane is typical for marine organic matter, abundant C_{29} is usually associated with higher plants, although certain marine algae (brown algae) also produce C_{29} sterols (Fleury et al. 1994).

A subordinate contribution of higher land plants is reflected by the presence of odd long-chain *n*-alkanes ($C_{27–33}$) (Peters et al. 2005), which becomes more prominent towards the top of the Göstling Formation (*n*-alkane distribution in Fig. 23.8). The presence of C_{19} and C_{20} tricyclic terpanes also demonstrates the contribution of terrigenous organic matter (Aquino Neto et al. 1982). Their concentrations peak in the middle to upper part of the Göstling Formation, indicating an increase in terrigenous organic matter (TT distribution, Fig. 23.8). This concurs with the presence of aliphatic diterpenoids (norabietane and norpimarane) and aromatic diterpenoid abietane derivatives (simonellite, tetrahydroretene, retene), as well as with the prominent 1,7-DMP and 1-MP in the distribution of methylated phenanthrenes (Simoneit 1986; Alexander et al. 1988; Peters et al. 2005; Fig. 23.11b). Pinaceae (i.e. conifers) are considered as the main source of the mentioned diterpenoids with abietane skeleton (Diefendorf et al. 2019). Norpimarane is mostly related to Pinaceae and Cupressaceae. However, the absence of tetracyclic diterpenoids in all samples argues for the dominant source of norpimarane being Pinaceae. The sesquiterpenoid cadalene, which is widespread in sedimentary organic matter (particularly terrigenous), is also identified in the studied samples (Fig. 23.11a). Although cadalene is a taxonomically non-specific biomarker, it can be produced by Pinaceae (Otto et al. 2007; Lu et al. 2013). The detected abietane type diterpenoids and cadalene (at least partly) were probably derived from Voltziales, which can be discerned even macroscopically in some layers. Indeed, this is reflected in the positive correlation between the contents of total aromatic diterpenoids and abietane skeletons ($\sum$ Arom Di = simonellite + tetrahydroretene + retene) and the contents of total aromatic sesquiterpenoids ($\sum$ Arom Sesq = cadalene + tetrahydrocadalene) ($R^2 = 0.70$) ($\sum$ Arom Di and $\sum$ Arom Sesq; Fig. 23.8, Supplementary material 23.3). Importantly, Voltziales and modern Pinaceae are closely related because they both belong to the class Pinopsida, making the presence of similar compounds and derivatives thereof very likely. Finally, the relatively abundant monoaromatic fernanes (both pentacyclic and tetracyclic) indicate that ferns contribute to the organic matter (Hauke et al. 1992). The increase in the C_{19+20} tricyclic terpane contribution and in the proportion of odd long-chain *n*-alkanes, associated with an increased content of total aromatic diterpenoids with abietane skeleton (Supplementary material 23.3), suggests a relative increase of terrigenous organic matter within the Göstling Formation. Nonetheless, HI does not significantly change in the uppermost samples (Fig. 23.3; Supplementary material 23.1). This result can be attributed to the presence of liptinite group macerals (resinite and sporinite) in the terrestrial organic matter (Figs. 23.5 and

23.6; Supplementary material 23.2) and to the algal blooms caused by the freshwater influx. Along with terrigenous organic matter and siliciclastic material, this influx transported crucial nutrients into the basin. While the contents of diterpenoids increase upwards, the concentrations of fernane derivatives show the opposite trend. This suggests a higher impact of conifers/Pinaceae or its relatives (e.g., detected Voltziales) associated with the decreased contribution by ferns. This is an indication for climate and/or environmental change.

Nutrients and Paleo-Productivity

Our data do not provide direct evidence for the nutrient level. Nonetheless, it is likely that nutrient levels were controlled by the intensity of river runoff. Hence, the occasional inflow of increased amounts of nutrient-rich river water had a three-fold effect: it decreased water salinity, transported land plant material into the Reifling Basin (see above), and also triggered algal and radiolarian blooms in nearshore shallow water. TOC contents, which are very high (often >3 wt%; occasionally >10 wt%; Fig. 23.3), despite the absence of anoxic conditions, show that organic productivity must have been very high during deposition of the organic matter-rich layers. On the other hand, high paleo-productivity created an increased demand for oxygen and caused oxygen-depleted conditions near the sediment–water interface or within the freshly deposited sediment. This favoured organic matter preservation. We therefore speculate that the organic-rich layers indicate episodes of increased river runoff and that the varying TOC contents and redox settings reflect the runoff intensity.

23.5.2 Comparison of Scheiblinggraben, Großau and Polzberg

The geochemical study of the Polzberg section focused on the lower 4 m of the Reingraben Formation (Lukeneder et al. 2024). Accordingly, no information is available for its upper part and only limited data are available for the Reifling and Göstling Formations. At the Scheiblinggraben only the Göstling Formation was evaluated for its hydrocarbon composition. Nonetheless, the organic matter trends recorded in the Scheiblinggraben section, the Großau drill core (Fig. 15.8 in Lukeneder et al. 2026a, b, this volume) and the Polzberg section (Lukeneder et al. 2024, therein Figs. 23.7 and 23.13) yield important information on the similarities and dissimilarities of sedimentological processes at these sites.

Trends in the Scheiblinggraben (Figs. 23.3 and 23.8), Großau (Lukeneder et al. 2026a, b, this volume) and Polzberg (Lukeneder et al. 2024, p. 9, 11) generally show similar evolutions of bulk geochemical data on different scales (Fig. 23.13), reflecting their different positions in the basin. This is further corroborated by thin sections connecting the formations across all locations.

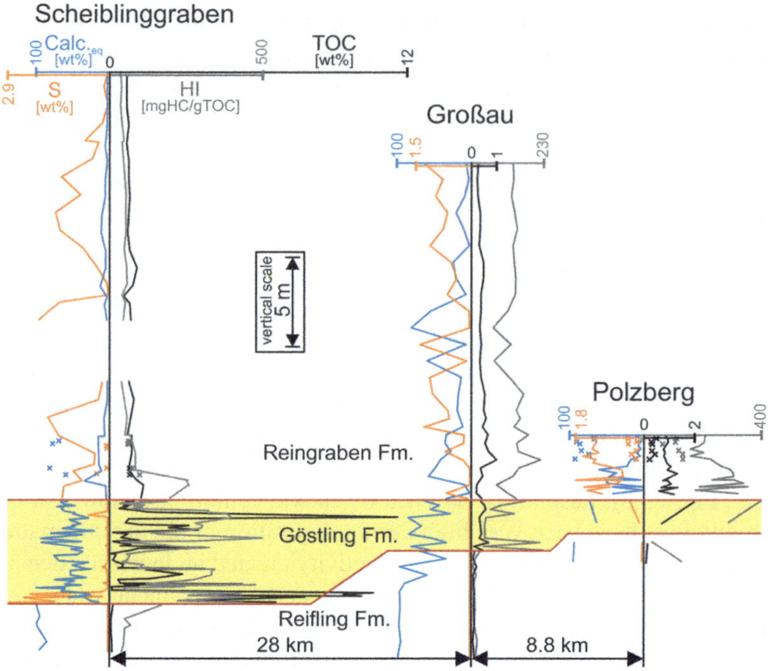

Fig. 23.13 Comparison of bulk geochemical data from Scheiblinggraben, Großau (Lukeneder et al. 2026, Chap. 10, this volume) and Polzberg (Lukeneder et al. 2024), displaying the topmost part of the Reifling Formation, the whole Göstling Formation and part of the Reingraben Formation. Carbonate content (Calc._{eq}), suphur (S), total organic carbon (TOC) and hydrogen index (HI) are shown. All figures are to scale, the distance between the sites is not

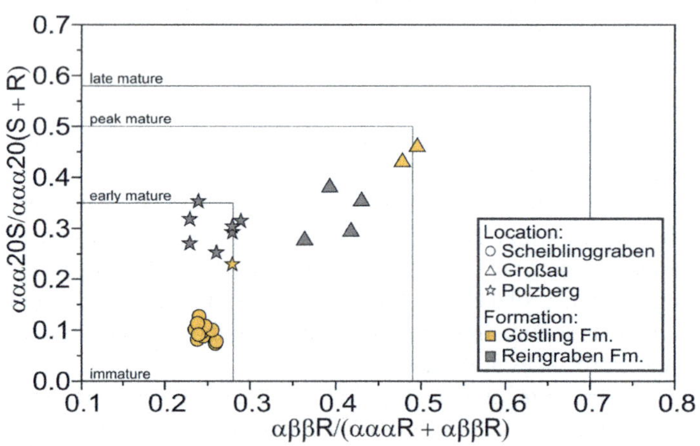

Fig. 23.14 Cross-plot of C_{29} sterane (ααα 20S/ααα (20S + 20R)) and (αββR/(αββR + αααR)) isomerization (Mackenzie and McKenzie 1983; McKenzie et al. 1983; Walters et al. 2012). The Göstling Formation of the Scheiblinggraben succession is least mature. The samples from Polzberg are immature to early mature, and Großau shows early oil window maturity

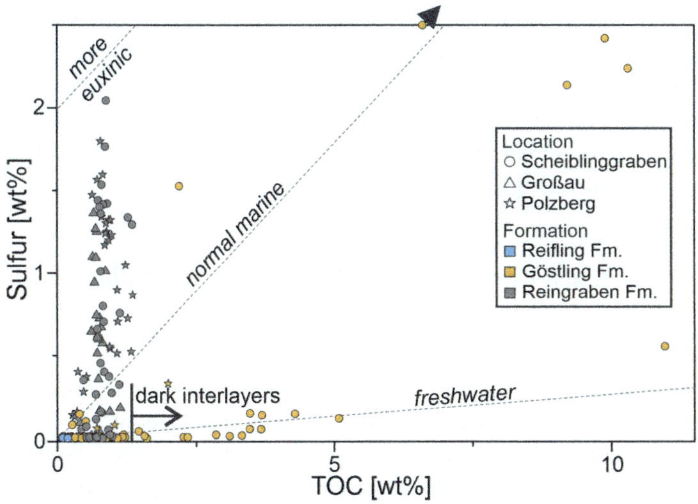

Fig. 23.15 Plot of sulfur against total organic carbon (TOC) content with general trends indicated (Raiswell and Berner 1984; Tada et al. 1999). Low TOC contents in combination with low sulfur do not permit meaningful interpretation (e.g., Reifling Formation). The sulfur content in the Göstling Formation is low in comparison to its TOC content, most evidently in layers from the Scheiblinggraben and one Polzberg sample. This signature is explained by periodic input of river water. In the Reingraben Formation, normal marine to euxinic conditions are suggested. According to the graph, the conditions at the Scheiblinggraben are more euxinic than those at Polzberg and Großau

23.5.2.1 Thermal Maturity

Organic matter in the Scheiblinggraben is immature (see Sect. 5.1.1). In comparison, the thermal maturity of the Großau drill core is significantly higher as shown by vitrinite reflectance (0.7 ± 0.05%Rr) and T_{max} values (437 °C for samples with S2 values >0.2 mg HC/g TOC). The higher maturity of the Großau drill core is also evidenced by higher 20S/20(S + R) (0.38 ± 0.10) and ββ/(αα + ββ) (0.41 ± 0.10) isomerization ratios of C_{29} steranes, which suggests early-to-peak oil window maturity (Seifert and Moldowan 1981; Mackenzie et al. 1983; Cornford et al. 1986; Peters et al. 2005; Fig. 23.14). The C_{31} S/R αβ-hopane isomerization ratio has consistently reached equilibrium (0.58 ± 0.01, Ensminger et al. 1977; Seifert and Moldowan 1980; Schoell et al. 1983; Peters et al. 2005). Generally, the Göstling Formation shows slightly higher 20S/20(S + R) and ββ/(αα + ββ) C_{29} sterane ratios than the Reingraben Formation. The MAS/TAS ratio is 0.39 on average and agrees with the previous statements (Lukeneder A et al. 2026a, b, Chap. 16, this volume).

The thermal maturity at the Polzberg section represents an intermediary position. T_{max} of samples with S2 values >0.2 mg HC/g TOC (431 °C), the C_{29} sterane 20S/20(S + R) (0.29 ± 0.06) and ββ/(αα + ββ) (0.26 ± 0.03) isomerization ratios (Fig. 23.14), the C_{31} hopane isomerization ratio (0.52 ± 0.06) and the MAS/TAS ratio (≈1.7) fall between the respective ranges from the Scheiblinggraben and Großau sections. These data suggest early oil window maturity (Lukeneder et al. 2024, p. 12).

Clearly, the maturity of the sections does not follow a simple E-W trend because Großau lies geographically between the Scheiblinggraben and Polzberg (Fig. 23.1a). The distribution does, however, agree with a maturity map of the overlying Lunz Formation (Sachsenhofer 1987). The reason for the higher maturity of the Großau area remains speculative but may be related to its proximity to the thrust front of the Tyrolian nappe system (Fig. 23.1a).

23.5.2.2 Depositional Environment and Position in the Reifling Basin

Reifling Formation

The Reifling Formation at all three sites has high carbonate contents and consists almost exclusively of carbonates with low organic matter content. The rocks are primarily radiolarian packstones with few sponge spicules. At the Scheiblinggraben, filaments (shells) are found in abundance, whereas at Polzberg and Großau shells are less prevalent. This puts the Scheiblinggraben section into a more proximal position, as compared to the Großau and Polzberg sites.

Göstling Formation

The thickness of the Göstling Formation at the three sites varies significantly (Scheiblinggraben: 6.2 m; Großau: 3.0 m; Polzberg: 2.1 m). TOC contents are very high in organic-rich layers in the Scheiblinggraben section (max. 11 wt%), but significantly lower in the Großau drill core (max. 1.1 wt%). However, the samples with the highest TOC contents occur in the lower and upper part of the Göstling Formation in both sections (Fig. 23.13). A sample with 2.0 wt% from the uppermost part of the Göstling Formation at Polzberg may indicate a similar trend, but the sample number is too low to support this assumption. Interestingly, the TOC contents of siliceous limestone in the Scheiblinggraben section (median 0.77 wt%) are also higher than in the Großau drill core (median 0.30 wt%). This information is useful, as it implies that the lower TOC of the thin organic-rich layers at Großau is not due to a sampling bias (e.g., loss of organic-rich layers during drilling of the Großau borehole).

Major differences exist also in the hydrogen index (HI). At Großau, HI reaches peaks at 160 mg HC/g rock, which classifies the organic matter as kerogen Type III. In contrast, at Scheiblinggraben and Polzberg the maximum HI values (490 and 400 mg HC/g rock, respectively) classify the organic matter as kerogen Type II.

Pr/Ph ratios are relatively high in all sections (Großau $\approx$2.4, Scheiblinggraben $\approx$2.0, Polzberg = 1.3), indicating similar transitional oxidizing to slightly oxygen-depleted redox conditions. A minor influence of land plant input, which can increase Pr/Ph ratios, probably exists (Lukeneder et al. 2026a, b, Chap. 16, this volume; Sect. 5.1.2; Fig. 23.16). Furthermore, the distribution of C_{31-35} $\alpha\beta$ extended homohopanes are similar in all sections (Rullkötter and Philip 1981; Rohmer 1987). Sulfur-limited conditions are suggested by high TOC/S values (Peters et al. 2005) and are further supported by the absence of dibenzothiophenes at Großau and Polzberg and their low concentrations at Scheiblinggraben (low DBT/P ratio).

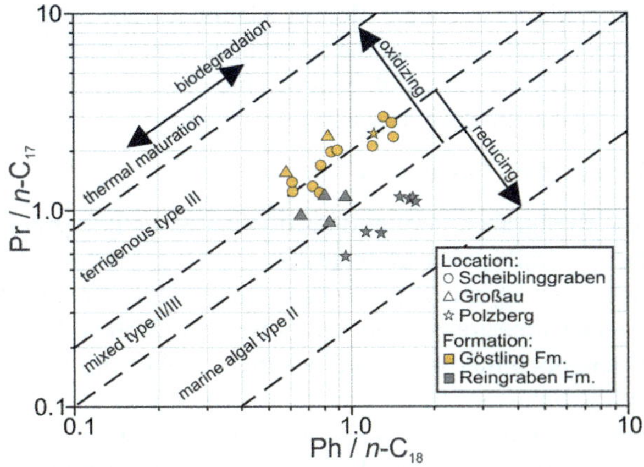

Fig. 23.16 Plot of Pr/n–C_{17} over Ph/n–C_{18} with kerogen type, effects of environmental conditions, maturity and biodegradation (Shanmugam 1985; Peters et al. 2005). The position of the samples from the Göstling and Reingraben formations indicate the relatively more oxidizing conditions during deposition of the Göstling formation

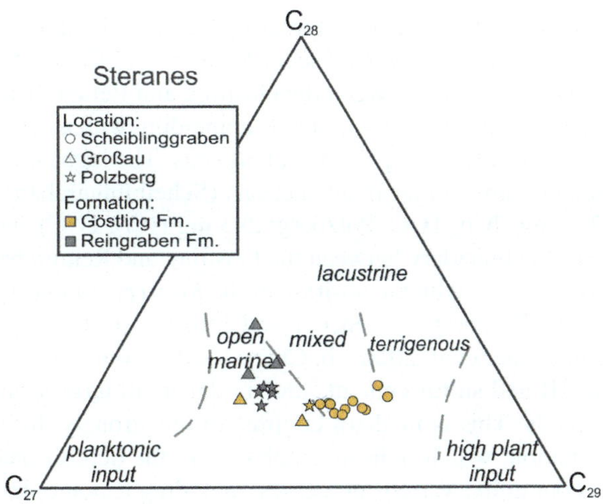

Fig. 23.17 Ternary diagram of the relative contribution of ααα (R) C_{27}, C_{28} and C_{29} steranes. C_{27} sterane is assumed to be of planktonic origin, while C_{29} is probably derived from higher plants. Areas with suggested depositional environment are indicated according to Huang and Meinschein (1979). The samples from the Göstling Formation in the Scheiblinggraben plot into the mixed input field, while the samples of Großau and Polzberg indicate more open marine conditions. All samples from the Reingraben Formation plot into the open marine field

High TOC/S at all sites and abundant C_{21} tricyclic terpane at Großau and Scheiblinggraben (no data from Polzberg, Zumberge 1987; Tao et al. 2015; Xiao et al. 2024) are probably related to periodic freshwater inflow. The river water transported nutrients, sediment and terrigenous organic matter, but only at Scheiblinggraben did this result in massive algal blooms. Gammacerane is absent at Großau and Scheiblinggraben, indicating that the water body was not stratified. The presence of gammacerane in the single Polzberg sample analyzed regarding hydrocarbon composition should be checked critically.

Reingraben Formation

The Reingraben Formation is dominated by fine-grained siliciclastics with varying, but generally low carbonate contents (e.g., <10 wt% at Scheiblinggraben). Distinct limestone layers are present at Scheiblinggraben and Polzberg in the lower 4 m. No distinct limestone layers are present at Großau, but the carbonate contents of the Reingraben mudstones are generally higher (up to 57 wt%) than in other sections. While limestone layers are absent in the lower 4 m of the Großau section, an interval with two maxima in carbonate content (55–80 wt%) was recorded 8–11 m above the base of the Reingraben Formation. This interval may be also present at Scheiblinggraben, albeit at much lower intensity (≈10 wt%). Despite the vertical variations, the carbonate content shows a general decreasing trend in all three sites.

TOC contents are relatively low in all sections (max ≈1.3%) and tend to decrease further upwards. The boundary between the Göstling and Reingraben Formations is characterised by an increase in sulfur content (coinciding with a decrease in TOC/S ratios to normal marine values, Fig. 23.15) at all sites. The highest HI values in the Reingraben Formation are similar in all sections (Scheiblinggraben: 250 mg HC/g TOC, Großau: 230 mg HC/g TOC, Polzberg: 293 mg HC/g TOC) and occur within a few meters above the boundary between the Göstling and Reingraben Formations. This interval also coincides with the position of the *Konservat-Lagerstätten* detected in all three sections (Polzberg: Lukeneder and Lukeneder 2021; Lukeneder et al. 2024, Großau: Lukeneder et al. 2026a, b, Chap. 16, this volume).

Overall, TOC, HI and sulfur contents are similar in all three sections and show similar vertical trends. This is in sharp contrast to the strongly differing values in the Göstling Formation, which indicates a change in the depositional system with increased lateral continuity. Various processes, including tectonic-driven deepening of the basin, erosion resulting in a more uniform basin relief, and change in sediment provenance leading to basin-parallel sediment transport may be responsible for the larger lateral continuity.

Pr/Ph ratios from the Großau drill core (1.3) and the Polzberg Section (0.55) indicate a shift towards more reducing conditions. This shift is also evident by the position of Göstling and Reingraben samples in the cross-plot of Pr/*n*–C_{17} versus Ph/*n*-C_{18} ratios (Shanmugam 1985; Peters et al. 2005; Fig. 23.16). This is in general agreement with decreasing C_{27} diasterane/C_{27} sterane ratios and decreasing TOC/S ratios (Peters et al. 2005; Fig. 23.15). Note, however, that this does not necessarily imply that freshwater input is absent. It may be rather related to a deepening, better connecting the Reifling Basin with the open marine domain. In this case the freshwater input would influence the environment to a lesser degree than in the Göstling Formation.

23.5.2.3 Sources of Organic Matter

The Reifling Formation contains virtually no organic matter. The discussion below therefore starts with the Göstling Formation.

Göstling Formation

As illustrated above, the organic matter in the Scheiblinggraben is dominated by remains of marine organisms, although the relatively high freshwater input from fluviatile systems also introduced significant amounts of terrigenous organic matter (Lukeneder et al. 2024; Lukeneder et al. 2026a, b, Chap. 19, this volume). Based on maceral data, on the concentrations of tricyclic terpanes and on the sterane distributions (Fig. 23.17) at the three sites, the terrigenous organic matter contribution was greatest at Scheiblinggraben, followed by Großau and Polzberg.

Reingraben Formation

Biomarker data from the Reingraben Formation is available only from the Großau and Polzberg sites. Here, the sterane/hopane ratio drastically decreased at the boundary between the Göstling and Reingraben Formations, suggesting increased prokaryotic activity (Peters et al. 2005). Above the boundary, the maceral composition indicates an initial increase in marine organic matter, which is also supported by the high concentration of the C_{23} tricyclic terpane and an increase in C_{27} steranes (Fig. 23.17), which coincides with the *Konservat-Lagerstätten* (Polzberg: Lukeneder and Lukeneder 2021; Lukeneder et al. 2024, Großau: Lukeneder et al. 2026a, b, Chap. 19, this volume). Although biomarker data are not available from the Scheiblinggraben section, HI trends and absolute HI values, which are similar to those in the Großau and Polzberg section (Fig. 23.13), suggest that algal material dominates also in the lower part of the Reingraben Formation at Scheiblinggraben. This interpretation is also supported by high amounts of lamalginite in bed 61. The high amount of marine organic matter is probably related to an ongoing deepening of the Reifling Basin.

The similar HI peaks in all three sections are followed by a concordant decrease in the relative marine organic matter contribution at all sites. Based on HI (Fig. 23.3), the decrease is more pronounced at Scheiblinggraben than at Großau, where both carbonate accumulation and the contribution of marine organic matter remain relatively high. The decrease of marine organic matter coincides with a relative increase of terrigenous organic matter, evident from increasing vitrinite abundance, CPI, and increasing concentrations of total aromatic diterpenoids ($\sum$Arom Di), C_{29} sterane and C_{19+20} tricyclic terpane (Supplementary material 23.3).

23.5.3 *HC Generation Potential*

Hydrocarbon occurrences are known from several locations within the NCA, which according to biomarker and isotope evidence are derived from Mesozoic source rocks (Misch et al. 2017). Prolific source rocks are included in the Seefeld Member of the Upper Triassic Hauptdolomit Formation (e.g., Köster 1989) and in the Lower Jurassic Sachrang Member of the Allgäu Formation (e.g., Neumeister et al. 2015). These organic matter-rich units are known from the western part of the NCA, but a potential source rock from the eastern part of the NCA has not yet been identified. Therefore, the studied stratigraphic interval may also be relevant from a petroleum geological point of view.

The Reifling Formation has very low TOC contents and Rock Eval parameters (blue symbols; Fig. 23.18). Its hydrocarbon generation potential is therefore negligible. The hydrocarbon generation potential of the Reingraben Formation and the siliceous limestones from the Göstling Formation is only slightly better and has to be classified as poor (based on generation potential; S1 + S2) or fair (based on TOC). In contrast, some organic-rich interlayers in the Göstling Formation show a good to very good hydrocarbon generation potential.

The correlation between the genetic potential (S1 + S2) and TOC is similar for all formations and locations (Fig. 23.18). Only samples from the upper part of the Reingraben Formation in the Scheiblinggraben (bed 64) show a major deviation, namely a parallel shift from the general trend, reflecting decreased Rock–Eval S1 and S2 peaks for the observed TOC. This part of the section was previously covered by soil and plants and was uncovered for the project, so the reduced genetic potential might be a result of leaching by organic acids from the overgrowth. An alternative

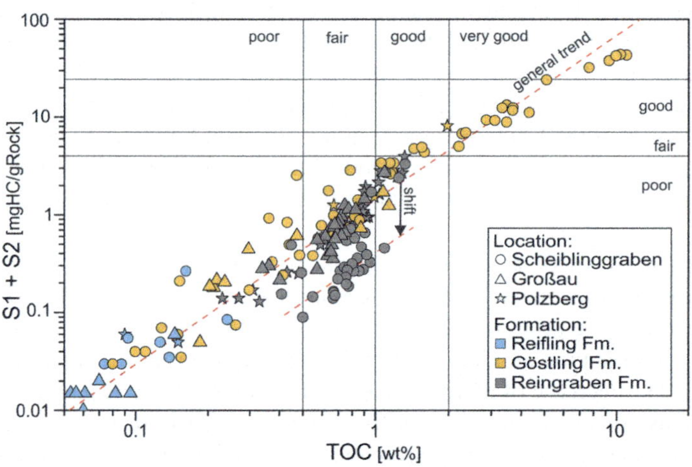

Fig. 23.18 Cross-plot of the genetic potential (S1 + S2) and the total organic carbon (TOC) content. Samples from all locations and formations show good correlation (general trend), except for bed 64 (upper part of the Reingraben Formation) from the Scheiblinggraben section, which has a decreased genetic potential (shift) for the measured TOC content. The Reifling Formation and Reingraben Formation have a poor hydrocarbon generation potential, while points of the Göstling Formation suggest good to very good potential locally; those layers, however, are very thin and do not contribute much overall

explanation may lie in the depositional conditions, such as increased oxygen content or biological degeneration.

The Source Potential Index (SPI = thickness * [S1 + S2] * bulk density/1000; Demaison and Huizinga 1994) was calculated to quantify the source potential of the Göstling and Reingraben Formations. The rock density was assumed to be 2.6 g/ cm^3. For the heterogenous Göstling Formation, the SPI was calculated for each individual organic-rich layer and the results were summed up to obtain an overall potential. The calculations show a SPI of 0.03 t HC/m^2. The low value for the Göstling Formation reflects the very low ("net") thickness of the organic matter-rich layers. For the much thicker Reingraben Formation, the low SPI reflects the low TOC content and the low generation potential (S1 + S2). At Großau and Polzberg the SPI would be even lower. This shows that neither formation can be considered relevant hydrocarbon source rocks. In case of the Göstling Formation, however, a facies characterized by thinner siliceous limestone and thicker organic matter rich layers could be present in (e.g., deeper) parts of the Reifling Basin. Hence, the source rock potential of the Göstling Formation should not be neglected prematurely. Irrespective of the economic significance of the source potential of the studied succession, traces of hydrocarbon migration along fractures are visible in polished blocks and thin-sections.

23.6 Conclusions

Based mainly on information from organic matter, the present study provides new insights into the environmental conditions during the Carnian Pluvial Episode at the Scheiblinggraben. The comparison with data from Großau and Polzberg clarifies the relative positions within the Reifling Basin and augments our understanding of the overall evolution of the depositional system.

During the deposition of the Reifling Formation, carbonatic turbidites were shed from the active Wetterstein platform into the fully marine Reifling Basin. Based on microfacies, the Scheiblinggraben was apparently in a more proximal position than Großau and Polzberg.

The shift to the Göstling Formation is evident in lithology, microfacies, and organic matter contents. The change was caused by increased river runoff, which transported nutrients, siliciclastic material and terrigenous organic matter, particularly from conifers (*Voltzia*) and ferns, into the basin. The nutrient transport resulted in temporary eutrophication of the Reifling Basin and triggered algal and radiolarian blooms. The plankton blooms, in turn, caused episodically oxygen-depleted (but not anoxic) conditions. Both high bioproductivity and improved organic matter preservation resulted in the deposition of thin layers with very high TOC contents. The increased river runoff probably reflects periodically increased precipitation on the continental realm and/or mobilised freshwater from continental water sources (lacustrine systems, groundwater) due to topographic changes. During deposition of the Göstling Formation, the relative contribution of terrigenous organic matter increased, while that of marine organic matter decreased. The thickness of the Göstling Formation, and the amount and type of the organic matter, vary significantly between the three sites (Scheiblinggraben, Großau, Polzberg), suggesting different facies zones within the Reifling Basin.

The transition towards the siliciclastic Reingraben Formation marks a major change in environmental conditions. Remarkably, the conditions at all three sites became similar, i.e. normal marine with an increased relative contribution of marine organic matter in the lower part. This increased lateral homogeneity may be linked to a sea level rise, tectonic subsidence (deepening), a homogenized relief (erosion), or changed sedimentary provenance (basin parallel input). However, the dominance of marine organisms was short-lived, and in the course of the Reingraben evolution a relative increase of terrigenous organic matter (or relative decrease of marine organic matter) occurred. The factors controlling the environmental changes are not entirely clarified, but tectonics, apart from climatological controls, certainly contributed.

Despite the very high TOC contents and the presence of kerogen Type II in the Göstling Formation, this formation cannot be considered a significant hydrocarbon source rock. However, rocks with higher source rock potential may be present in parts of the Reifling Basin. The thermal maturity of the studied sections varies from immature (Scheiblinggraben) to peak oil window maturity (Großau).

Acknowledgments The author is particularly grateful to the owners of the outcrop area, the Austrian Federal Forests Company (Österreichische Bundesforste), for sampling and digging permission. Special thanks go to Lukas Nöst (forest district Grossreifling) and Dominic Asztaller (forest dristrict Grossreifling) for their advice and help during the field campaigns. I thank Petra Lukeneder (Gablitz), Veronika, N.I. Koukal (Vienna) and Justine and Umberto Uprimny (Pichlern) for their help with cleaning outcrops and sampling fossils. The work was done within the framework of and financially supported by projects of the Federal Government of Lower Austria (K3-F-964/001-2020 and K3-F-964/002-2023, Department Science and Research; headquarters in St. Pölten); the Austrian Academy of Sciences (headquarters in Vienna, project Polzberg Lukeneder 2021) represented by the National Committee for Geo/Hydro Sciences (Earth System Sciences Programme); the Triassic project "Fenster in die Karnische Krise" of the Society of Friends of the Natural History Museum Vienna (Freunde des Naturhistorischen Museums Wien, CPE Lukeneder 2023); and the Natur- und UNESCO Geopark Steirische Eisenwurzen (St. Gallen, project Karnische Krise Lukeneder 2024). KS is thankful to the Ministry of Science, Technological Development and Innovation of the Republic of Serbia (contract number 451-03-136/2025-03/200168). The manuscript greatly benefited from valuable comments by the reviewer Dragana Životić (Belgrade) and an anonymous reviewer, the production editor Ashok Arumairaj (Taramani), the in-house series editor Kathleen A. Ritterbush (Utah), the publishing editor Simon Shah-Rohlfs and the production editor Antje Endemann (both Heidelberg).

Author Contribution AL collected new material, designed research, provided documentation on fossil material. JW, KS, RS, DM performed analyses on geochemistry and organics.

Funding Open access funding was enabled and the projects financially supported by the Federal Government of Lower Austria (K3-F-964/001–2020 and K3-F-964/002–2023, Department Science and Research; headquarters in St. Pölten); the Austrian Academy of Sciences (headquarters in Vienna, project Polzberg Lukeneder 2021) represented by the National Committee for Geo/Hydro Sciences (Earth System Sciences Programme); the Triassic project "Fenster in die Karnische Krise" of the Society of Friends of the Natural History Museum Vienna (Freunde des Naturhistorischen Museums Wien, CPE Lukeneder 2023); the Natur- und UNESCO Geopark Steirische Eisenwurzen

(St. Gallen, project Karnische Krise Lukeneder 2024); the Natural History Museum Vienna; and the Technical University of Leoben (Chair of Energy Geosciences) (to AL). The authors are responsible for the contents of this publication. The funders had no impact on conceptualization, design, data collection, analysis, decision to publish, or preparation of the manuscript.

Competing interests All authors stated that they have no competing interests.

Authors' contributions All authors stated that they have equally contributed to their chapters, read, edited and agreed to submit the final versions of their contributions.

Availability of data and materials All sources of information are provided in the text. Raw data related to the fossil material from Austria are available from the corresponding authors and Alexander Lukeneder upon request.

Additional information The article is part of the Special Volume: *Late Triassic Konservat-Lagerstätten within the Carnian Pluvial Episode. 150 Years of Scientific Excavations in Austria.*

Artificial Intelligence (AI) The chapter was produced without the use of AI.

Language The entire volume uses American English reviewed by a native scientist.

Received 2025.

Published online 2026.

References

Alexander R, Kagi RI, Toh E, van Bronswijk W (1988) The use of aromatic hydrocarbons for assessment of thermal histories in sediments. In: Purcell PG, Purcell RR (eds) The North West shelf, Australia: Proc Pet Expl Soc Australia, Perth, pp 559–562

Alverà G, Dal Corso J, Chu D, Cruciani G, Roghi G, Caggiati M, Song H, Song H, Tian L, Du Y, Haserbek T, Gianolla P (2026) Episodic aridification at the onset of the Carnian Pluvial Episode in Western Tethys. Glob Planet Change 256:105192. https://doi.org/10.1016/j.gloplacha.2025.105192

Ampferer O (1933) Admont und Hieflau. Geologische Spezialkarte der Republik Österreich. Assistance of A Bittner, J Stiny, W Hammer, G Hießleitner, M Vacek, KA Redlich et al. GBA (1:75 000)

Aquino Neto FR, Restle A, Connan J, Albrecht P, Ourisson G (1982) Novel tricyclic terpanes (C19, C20) in sediments and petroleums. Tetrahedron Lett 23:2027–2030. https://doi.org/10.1016/S0040-4039(00)87251-2

Arthaber G von (1896) Die Cephalopodenfauna der Reiflinger Kalke. Beitr Paläont Geol Österr-Ungarn Orient 10, I. Abt:112 p

Bray E, Evans E (1961) Distribution of n-paraffins as a clue to recognition of source beds. Geochim Cosmochim Acta 22:2–15. https://doi.org/10.1016/0016-7037(61)90069-2

Bruckner-Wein A, Lobitzer H, Müller P (1996) Organic geochemistry and facies of the Carnian Göstling Beds and Opponitz Formation (Northern Calcareous Alps, Austria). Occ Pap Inst Hung 189:149–157

Cornford C, Needham CEJ, Walque L de (1986) Geochemical habitat of North Sea oils. In: Spencer AM (ed) Habitat of hydrocarbons on the Norwegian continental shelf, pp 39–54.
de Grande S, Aquino Neto FR, Mello MR (1993) Extended tricyclic terpanes in sediments and petroleums. Org Geochem 20:1039–1047. https://doi.org/10.1016/0146-6380(93)90112-O
Demaison G, Huizinga BJ (1994) Genetic classification of petroleum systems using three factors. In: Magoon LB, Dow WG (eds) The petroleum system—from source to trap, 60th ed. AAPG, pp 73–89
Didyk BM, Simoneit BRT, Brassell SC, Eglinton G (1978) Organic geochemical indicators of palaeo-environmental conditions of sedimentation. Nature 272:216–222. https://doi.org/10.1038/272216a0
Diefendorf AF, Leslie AB, Wing SL (2019) A phylogenetic analysis of conifer diterpenoids and their carbon isotopes for chemotaxonomic applications. Org Geochem 127:50–58. https://doi.org/10.1016/j.orggeochem.2018.11.007
Ellis L, Singh RK, Alexander R, Kagi RI (1999) Long-chain alkylnaphthalenes in crude oils and sediments. In: 19th IMOG meeting
Ensminger A, Albrecht P, Ourisson G, Tissot B (1977) Evolution of polycyclic alkanes under the effect of burial (Early Toarcian shales, Paris Basin). In: Adv Org Geochem 1975 (R Campos and J Goni, eds.), pp 45–52
Espitalié J, Laporte JL, Madec M, Marquis F, Leplat P, Paulet J, Boutefeu A (1977) Méthode rapide de caractérisation des roches mètres, de leur potentiel pétrolier et de leur degré d'évolution. Rev Inst Fr Pét 32:23–42. https://doi.org/10.2516/OGST:1977002
Fleury BG, Pereira MV, Da Silva JR, Kaisin M, Teixeira VL, Kelecom A (1994) Sterols from brazilian marine brown algae. Phytochemistry 37:1447–1449. https://doi.org/10.1016/S0031-9422(00)90430-8
Flügel E (2010) Microfacies of carbonate rocks. Analysis, interpretation and application, 2nd ed. Springer Berlin, Heidelberg. https://doi.org/10.1007/978-3-642-03796-2
GBA (2021) GEOFAST—Blatt 100 Hieflau (1:50.000). assistance of I Bayer, M Steinbichler, O Kreuss. https://opac.geologie.ac.at/ais312/dokumente/100_M31_2014.pdf
Greenwood PF, Arouri KR, George SC (2000) Tricyclic terpenoid composition of tasmanites kerogen as determined by pyrolysis GC-MS. Geochim Cosmochim Acta 64:1249–1263. https://doi.org/10.1016/S0016-7037(99)00326-9
Hauke V, Graff R, Wehrung P, Trendel JM, Albrecht P, Riva A et al (1992) Novel triterpene-derived hydrocarbons of the arborane/fernane series in sediments: Part II. Geochim Cosmochim Acta 56:3595–3602. https://doi.org/10.1016/0016-7037(92)90405-8
Haven HL ten, Leeuw JW de, Rullkötter J, Damsté JSS (1987) Restricted utility of the pristane/phytane ratio as a palaeoenvironmental indicator. Nature 330, 641–643. https://doi.org/10.1038/330641a0
Hoffmann CF, Mackenzie AS, Lewis CA, Maxwell JR, Oudin JL, Durand B, Vandenbroucke M (1984) A biological marker study of coals, shales and oils from the Mahakam Delta, Kalimantan, Indonesia. Chem Geol 42:1–23. https://doi.org/10.1016/0009-2541(84)90002-0
Huang W-Y, Meinschein WG (1979) Sterols as ecological indicators. Geochim Cosmochim Acta 43:739–745. https://doi.org/10.1016/0016-7037(79)90257-6
Hughes WB, Holba AG, Dzou LI (1995) The ratios of dibenzothiophene to phenanthrene and pristane to phytane as indicators of depositional environment and lithology of petroleum source rocks. Geochim Cosmochim Acta 59:3581–3598. https://doi.org/10.1016/0016-7037(95)00225-O
ICCP (1998) The new vitrinite classification (ICCP System 1994). Fuel 77:349–358. https://doi.org/10.1016/S0016-2361(98)80024-0
ICCP (2001) The new inertinite classification (ICCP System 1994). Fuel 80:459–471. https://doi.org/10.1016/S0016-2361(00)00102-2
ISO 7404–5 (2009) Methods for the petrographic analysis of coals. Method of determining microscopically the reflectance of vitrinite (ISO/TC 27/SC 5), 3:14 p

Köster J (1989) Organic geochemistry and organic petrology of kerogenous and bituminous intercalations within the "Hauptdolomit" (Triassic, Norian) of the Northern Calcareous Alps. Clausthaler Geow Diss 36

Langford FF, Blanc-Valleron MM (1990) Interpreting rock-eval pyrolysis data using graphs of pyrolizable hydrocarbons versus total organic carbon (1). AAPG Bull 74:799–804. https://doi.org/10.1306/0C9B238F-1710-11D7-8645000102C1865D

Leuchs K, Mosebach R (1936) Die spätladinische Hebung. Zbl Min, Geol Paläont, Abt B, 1936:1–12

Lobitzer H, Solti G, Ravasz C (1994) Fazies und Geochemie ausgewählter Vorkommen bituminöser und kerogenreicher Gesteine Österreichs. Jubiläumsschrift 20 Jahre geologische Zusammenarbeit Österreich-Ungarn. assistance of M Bichler, A Bruckner-Wein, G Császár, L Dosztály, H Egger, M Hètenyi et al. GBA, Wien

Lu Y, Hautevelle Y, Michels R (2013) Determination of the molecular signature of fossil conifers by experimental palaeochemotaxonomy—part 1: The Araucariaceae family. Biogeosciences 10:1943–1962. https://doi.org/10.5194/bg-10-1943-2013

Lukeneder A (ed) (2026a) Late Triassic Konservat-Lagerstätten within the Carnian Pluvial Episode: 150 Years of Scientific Excavations in Austria. Top Geobiol 55:600 p. https://doi.org/10.1007/978-3-032-19777-1

Lukeneder A (2026b) The marine Upper Triassic paleobiota from the Scheiblingbach *Konservat-Lagerstätte* in Austria. In: Lukeneder A (ed) Late Triassic Konservat-Lagerstätten within the Carnian Pluvial Episode: 150 Years of Scientific Excavations in Austria. Top Geobiol 55, Chap 19:437–467. https://doi.org/10.1007/978-3-032-19777-1_19

Lukeneder A, Lukeneder P (2021) The Upper Triassic Polzberg palaeobiota from a marine *Konservat-Lagerstätte* deposited during the Carnian Pluvial Episode in Austria. Sci Rep 11:16644. https://doi.org/10.1038/s41598-021-96052-w

Lukeneder A, Lukeneder P, Sachsenhofer RF, Roghi G, Rigo M (2024) Multi-proxy record of the Austrian Upper Triassic Polzberg *Konservat-Lagerstätte* in light of the Carnian Pluvial Episode. Sci Rep 14:11194. https://doi.org/10.1038/s41598-024-60591-9

Lukeneder A, Weitz JA, Sachsenhofer RF, Misch D (2026) Drilling into the Carnian Pluvial Episode. In: Lukeneder A (ed) Late Triassic Konservat-Lagerstätten within the Carnian Pluvial Episode: 150 years of scientific excavations in Austria. Top Geobiol 55, Chap 15:315–352. https://doi.org/10.1007/978-3-032-19777-1_15

Mackenzie AS (1984) Applications of Bilogical Markers in Petroleum Geochemistry. In: Brooks J, Welte D (eds) Adv Pet Geochem 1984, pp 115–214. https://doi.org/10.1016/b978-0-12-032001-1.50008-0

Mackenzie AS, McKenzie D (1983) Isomerization and aromatization of hydrocarbons in sedimentary basins formed by extension. Geol Mag 120:417–470. https://doi.org/10.1017/S0016756800027461

Mackenzie AS, Wolff GA, Maxwell JR (1983) Fatty acids in some biodegraded petroleums. Possible Origins and Significance, Adv Org Geochem 1981:637–649

Mandl G (2000) The Alpine sector of the Tethys shelf. Examples of Triassic to Jurassic sedimentation and deformation from the Northern Calcareous Alps. Mitt Österr Geol Ges 92:61–77

McKenzie D, Mackenzie AS, Maxwell JR, Sajgó C (1983) Isomerization and aromatization of hydrocarbons in stretched sedimentary basins. Nature 301:504–506. https://doi.org/10.1038/301504a0

Misch D, Leu W, Sachsenhofer RF, Gratzer R, Rupprecht B, Bechtel A (2017) Shallow hydrocarbon indications along the alpine thrust belt and adjacent foreland basin: distribution and implications for petroleum exploration. J Pet Geol 40:341–362. https://doi.org/10.1111/jpg.12684

Moldowan J, Seifert WK, Gallegos EJ (1983) Identification of an extended series of tricyclic terpanes in petroleum. Geochim Cosmochim Acta 47:1531–1534. https://doi.org/10.1016/0016-7037(83)90313-7

Naafs BDA, Bianchini G, Monteiro FM, Sánchez-Baracaldo P (2022) The occurrence of 2-methylhopanoids in modern bacteria and the geological record. Geobiology 20:41–59. https://doi.org/10.1111/gbi.12465

Neumeister S, Gratzer R, Algeo TJ, Bechtel A, Gawlick HJ, Newton RJ, Sachsenhofer RF (2015) Oceanic response to Pliensbachian and Toarcian magmatic events: Implications from an organic-rich basinal succession in the NW Tethys. Glob Planet Change 126:62–83. https://doi.org/10.1016/j.gloplacha.2015.01.007

Neunlist S, Rohmer M (1985) Novel hopanoids from the methylotrophic bacteria Methylococcus capsulatus and Methylomonas methanica. (22S)-35-aminobacteriohopane-30,31,32,33,34-pentol and (22S)-35-amino-3 beta-methylbacteriohopane-30,31,32,33,34-pentol. Biochem J 231:635–639. https://doi.org/10.1042/bj2310635

Ogg JG (2015) The mysterious Mid-Carnian "Wet Intermezzo" global event. J Earth Sci 26:181–191. https://doi.org/10.1007/s12583-015-0527-x

Otto A, Simoneit BRT, Wilde V (2007) Terpenoids as chemosystematic markers in selected fossil and extant species of pine (Pinus, Pinaceae). Bot J Linn Soc 154:129–140. https://doi.org/10.1111/j.1095-8339.2007.00638.x

Peters KE (1986) Guidelines for evaluating petroleum source rock using programmed pyrolysis. AAPG Bulletin 70. https://doi.org/10.1306/94885688-1704-11D7-8645000102C1865D

Peters KE, Cassa MR (1994) Applied source rock geochemistry. In: Magoon LB, Dow WG (eds) The petroleum system—from source to trap, 60th ed. AAPG, pp 93–120

Peters KE., Walters CC., Moldowan, JM (2005) Biomarkers and isotopes in petroleum systems and earth history, 2 ed. Cambridge Univ. Press (The biomarker guide, vol 2), Cambridge, UK:465–1154. https://doi.org/10.1017/CBO9781107326040

Peters KE, Moldowan JM, Sundararaman P (1990) Effects of hydrous pyrolysis on biomarker thermal maturity parameters: Monterey Phosphatic and Siliceous members. Org Geochem 15:249–265. https://doi.org/10.1016/0146-6380(90)90003-I

Philp P, Symcox C, Wood M, Nguyen T, Wang H, Kim D (2021) Possible explanations for the predominance of tricyclic terpanes over pentacyclic terpanes in oils and rock extracts. Org Geochem 155:104220. https://doi.org/10.1016/j.orggeochem.2021.104220

Pickel W, Kus J, Flores D, Kalaitzidis S, Christanis K, Cardott BJ et al (2017) Classification of liptinite—ICCP system 1994. Int J Coal Geol 169:40–61. https://doi.org/10.1016/j.coal.2016.11.004

Preto N, Kustatscher E, Wignall PB (2010) Triassic climates—state of the art and perspectives. Palaeogeogr Palaeoclimatol Palaeoecol 290:1–10. https://doi.org/10.1016/j.palaeo.2010.03.015

Radke M, Welte DH (1983) The methylphenanthrene index (MPI). A maturity parameter based on aromatic hydrocarbons. In: Bjorøy M, Albrecht C, Cornford C (eds) Adv Org Geochem 1981, pp 504–512

Raiswell R, Berner RA (1984) C/S method for distinguishing freshwater from marine sedimentary rocks. Geol 12:365–368

Reineck HE (1963) Sedimentgefüge im Bereich der südlichen Nordsee. Abh Senckenberg Naturf Ges 505:138 p

Rohmer M (1987) The hopanoids, prokaryotic triterpenoids and sterol surrogates. In: Schriner E et al (ed) Surface structures of microorganisms and their interactions with the mammalian host, pp 227–242

Rullkötter J, Philip P (1981) Extended hopanes up to C40 in Thornton bitumen. Nature 292:616–618. https://doi.org/10.1038/292616a0

Sachsenhofer RF (1987) Fazies und Inkohlung mesozoischer Kohlen der Alpen Ostösterreichs. Mitt Österr Geol Ges 80:1–45

Schlager W, Schöllnberger W (1974) Das Prinzip stratigraphischer Wenden in der Schichtfolge der Nördlichen Kalkalpen. Mitt Geol Ges Wien 66/67:165–193

Schoell M, Teschner M, Wehner H, Durand B, Oudin JL (1983) Maturity related biomarker and stable isotope variations and their application to oil/source rock correlation in the Mahakam Delta, Kalimantan. Adv Org Geochem 1981 (M. Bjorøy, C. Albrecht, C. Cornford, et al., eds.), pp 156–163

Seifert WK, Moldowan J (1980) The effect of thermal stress on source-rock quality as measured by hopane stereochemistry. Phys Chem Earth 12:229–237. https://doi.org/10.1016/0079-1946(79)90107-1

Seifert WK, Moldowan J (1981) Paleoreconstruction by biological markers. Geochim Cosmochim Acta 45:783–794. https://doi.org/10.1016/0016-7037(81)90108-3

Shanmugam G (1985) Significance of coniferous rain forests and related organic matter in generating commercial quantities of oil, Gippsland Basin, Australia. AAPG Bull 69:1241–1254. https://doi.org/10.1306/AD462BC3-16F7-11D7-8645000102C1865D

Simms MJ, Ruffell AH (1989) Synchroneity of climatic change and extinctions in the Late Triassic Geol 17:265–268. https://doi.org/10.1130/0091-7613(1989)017%3C0265:SOCCAE%3E2.3.CO;2

Simoneit BRT (1986) Cyclic terpenoids of the geosphere. In: Johns RB (ed) Biological markers in the sedimentary record, pp 43–99

Simoneit BRT, Leif RN, de Aquino R, Neto F, Almeida Azevedo D, Pinto AC, Albrecht P (1990) On the presence of tricyclic terpane hydrocarbons in permian tasmanite algae. Naturwissenschaften 77:380–383. https://doi.org/10.1007/BF01135736

Sinninghe Damsté JS, Kenig F, Koopmans MP, Koster J, Schouten S, Hayes JM, de Leeuw JW (1995) Evidence for gammacerane as an indicator of water column stratification. Geochim Cosmochim Acta 59:1895–1900. https://doi.org/10.1016/0016-7037(95)00073-9

Tada R, Irino T, Koizumi I (1999) Land–ocean linkages over orbital and millennial timescales recorded in Late Quaternary sediments of Japan Sea. Paleooceanogr Paleoclimatol 14(2):236–247. https://doi.org/10.1029/1998PA900016

Tao S, Wang C, Du J, Liu L, Chen Z (2015) Geochemical application of tricyclic and tetracyclic terpanes biomarkers in crude oils of NW China. Mar Pet Geol 67:460–467. https://doi.org/10.1016/j.marpetgeo.2015.05.030

Taylor GH, Teichmüller M, Davis A, Diessel CFK, Littke R, Robert P (1998) Organic petrology. A new handbook incorporating some revised parts of Stach's Textbook of coal petrology. Geol Mag 136. Borntraeger, Berlin, Stuttgart. https://doi.org/10.1017/S0016756899463320

Taylor AM, Goldring R (1993) Description and analysis of bioturbation and ichnofabric. J Geol Soc 150:141–148. https://doi.org/10.1144/gsjgs.150.1.0141

Tollmann A (1985) Geologie von Österreich. Außerzentralalpiner Anteil, 710 p

Tollmann A (1966) Die alpidischen Gebirgsbildungs-Phasen in den Ostalpen und Westkarpaten. Geotektonische Forschungen 21, Schweizerbart Science Publishers, Stuttgart, Germany, 156 p

Trümpy R (1960) Paleotectonic evolution of the central and western alps. Geol Soc America Bull 71:843. https://doi.org/10.1130/0016-7606(1960)71[843:PEOTCA]2.0.CO;2

Walters CC, Lillis PG, Peters KE (2012) Molecular Indicators of Geothermal History. In: Analyzing the Thermal History of Sedimentary Basins: Methods and Case Studies (NB Harris and KE Peters Eds.), SEPM, pp 17–28

Xiao H, Li M, Nettersheim BJ (2024) Short chain tricyclic terpanes as organic proxies for paleo-depositional conditions. Chem Geol 652:122023. https://doi.org/10.1016/j.chemgeo.2024.122023

Zumberge JE (1987) Prediction of source rock characteristics based on terpane biomarkers in crude oils: A multivariate statistical approach. Geochim Cosmochim Acta 51:1625–1637. https://doi.org/10.1016/0016-7037(87)90343-7

Zundel M, Rohmer M (1985) Hopanoids of the methylotrophic bacteria Methylococcus capsulatus and Methylomonas sp. as possible precursors of C 29 and C 30 hopanoid chemical fossils. FEMS Microbiol Lett 28:61–64. https://doi.org/10.1111/j.1574-6968.1985.tb00764.x

Chapter 24
Potential Modulation of a Late Triassic Pangean Megamonsoon by Solar Cycles

Mathias Harzhauser, Alexander Lukeneder, Petra Lukeneder, and Oleg Mandic

Abstract The supercontinent Pangea attained its maximum size with land masses being equally distributed on both hemispheres during the Triassic. The peculiar configuration of continents gave rise to the largest monsoon system of Earth's history, termed Pangean Megamonsoon (PMM). This monsoon system peaked around 234–232 Myr during the Carnian Pluvial Episode, a short-lived period of strongly increased precipitation. We present a section from the Reifling Basin in Austria (Northern Calcareous Alps, central Europe), which formed at that time along the northwestern margin of the Tethys Realm. The record documents a strong modulation of the PMM on millennial and centennial scales over ~215 kyr of early Carnian time. Precessional modulation is expressed in alternations between phases of predominant summer monsoon and phases of strong winter monsoon. Moist landward directed southerlies resulted in increased runoff and strengthened coastal upwelling during summer monsoon. Both mechanisms boosted primary productivity and caused severe bottom water anoxia. Dry northerlies forced downwelling and bottom water ventilation during winter monsoon. Monsoon intensity was also governed by solar cycles. These cycles were not strong enough to initiate downwelling but are expressed by strongly reduced terrestrial input during phases of predominating winter monsoon.

Supplementary Information The online version contains supplementary material available at https://doi.org/10.1007/978-3-032-19777-1_24.

M. Harzhauser (✉) · A. Lukeneder · P. Lukeneder · O. Mandic
Department of Geology and Paleontology, Natural History Museum Vienna, Vienna, Austria
e-mail: mathias.harzhauser@nhm.at

A. Lukeneder
e-mail: alexander.lukeneder@nhm.at

P. Lukeneder
e-mail: petra.lukeneder@nhm.at

O. Mandic
e-mail: oleg.mandic@nhm.at

P. Lukeneder
Institute of Applied Geology, Department of Landscape, Water and Infrastructure, BOKU University, Vienna, Austria

A. Lukeneder (ed.), *Late Triassic Konservat-Lagerstätten within the Carnian Pluvial Episode*, Topics in Geobiology 55, https://doi.org/10.1007/978-3-032-19777-1_24

Keywords Pangea · Tethys · Monsoon · Solar cycles · Precession · Obliquity · Climate

24.1 Introduction

During the Triassic period, the supercontinent Pangea was dominated by a strong monsoon system, often termed 'Pangean Megamonsoon' (Kutzbach and Gallimore 1989; Woods 2005; Ikeda et al. 2017; Zeng et al. 2019; Chu et al. 2020). The Pangean Megamonsoon (PMM) was the largest monsoon system in Earth's history. It became established during the Carboniferous, when Gondwana and Laurasia assembled in the supercontinent Pangea (Soreghan et al. 2002), and peaked during the Triassic, when Pangea reached its maximum size with land masses being equally distributed on both hemispheres (Parrish 1993; Hu et al. 2023; Fig. 24.1a). The peculiar plate configuration caused a considerable seasonal contrast and pressure gradient between the hemispheres, which boosted monsoon effects (Parrish 1993). Along the coasts of the western Tethys, annual precipitation ranged around 1200–1800 mm/yr, whilst the inner parts of Pangea, north and south of the Intertropical Convergence Zone, were arid (Parrish 1993; Hu et al. 2023). During the Jurassic, the rifting of Pangea started and successively terminated the PMM (Francis 2009; Schöllhorn et al. 2020; Hu et al. 2023). The PMM reached its maximum intensity during the Carnian Pluvial Episode (CPE), a geologically short-lived period of strongly increased precipitation from ~234 to 232 my (Dal Corso et al 2020). The massive release of CO_2 during several pulses of the Wrangellia volcanism in Alaska seems to have initiated the CPE (Lu et al. 2021; Zhao et al. 2022; Zhang et al 2023). The CPE was a crucial phase of turnover and diversification for many terrestrial and marine organisms ranging from nannoplankton to dinosaurs (Dal Corso et al 2020; Zhang et al 2023; Bernardi et al 2018). A plethora of papers studied the records and causes of the CPE on regional and global scales (Breda et al. 2009; Kolar-Jurkovšek and Jurkovšek 2010; Arche and López-Gómez 2014; Miller et al 2017; Lukeneder and Lukeneder 2021, 2023; Lukeneder et al. 2024; Alverà et al 2026; Lukeneder 2026a, b, this volume). Similarly, many papers described the impact of the CPE on biota and environments (Sun et al. 2016; Bernardi et al 2018; Dal Corso et al 2020; Rigo et al. 2007, 2012; Zhang et al 2023; Alverà et al 2026). No study, however, focused on high frequency environmental shifts during the CPE and tried to link these to changes in monsoon intensity. Herein, we discuss a high-resolution record from the intraplatform Reifling Basin in Austria (central Europe; Fig. 24.1a, b) spanning about 215 kyr of the CPE. This record exhibits a strong imprint of cyclic modulation of depositional conditions, which are linked to climatic processes, typically occurring during discrete states of monsoon climate.

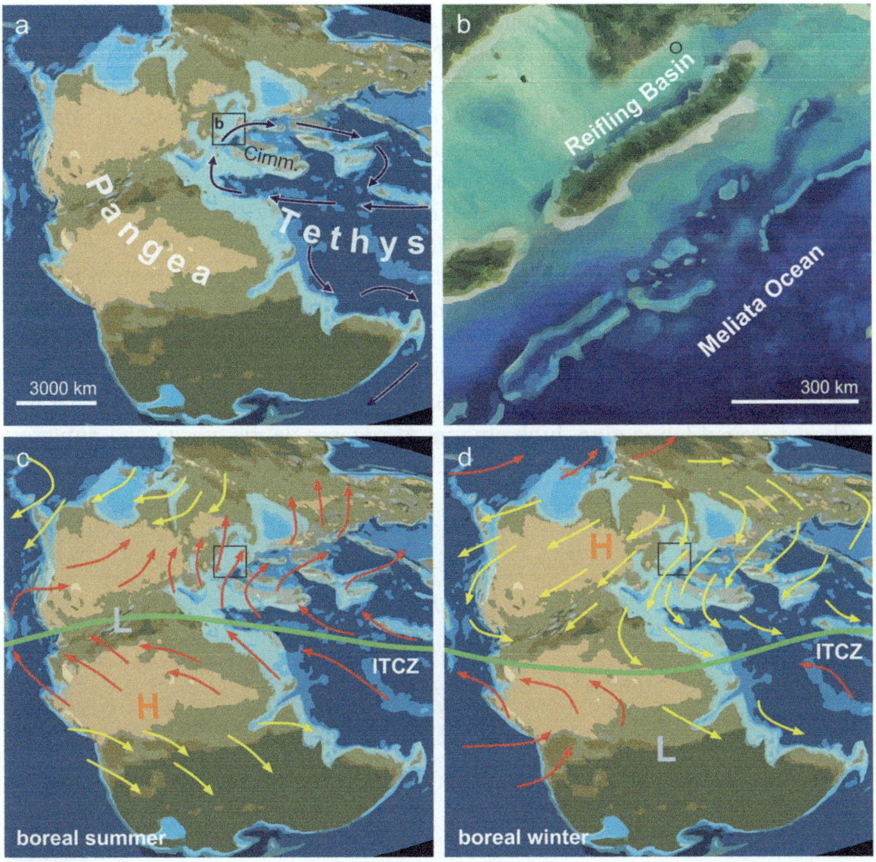

Fig. 24.1 **a** Late Triassic paleogeography and position of the investigation area (black square), modified after Ron Blakey, https://deeptimemaps.com. Ocean currents after Yan and Zhao (2001) and Damborenea et al. (2013). **b** Paleogeography of the Reifling Basin along the northwestern margin of the Meliata Ocean, modified from Lukeneder et al. (2024); circle indicates the position of the SCH-section. **c**, **d** Surface wind patterns of the Pangean mega-monsoon system during the boreal summer (**c**) and the boreal winter (**d**). Wind patterns modified from Winguth and Winguth (2013); Inter Tropical Convergence Zone (ITCZ) after Shozaki and Hasegawa (2022). Red arrows have northern component, yellow arrows indicate southward directions. Cimm, Cimmerian islands

24.2 Geologic and Paleogeographic Setting

Herein, we present a continuous 4-m-thick succession of the Upper Triassic Göstling Formation from the Scheiblingbachgraben (= Scheiblingbach) in Austria. The Scheiblingbachgraben-section (SCH-section) is located on the northern slope of Mount Lärchkogel (1087 m), south of Mount Reiflingberg (712 m), 600 m south-west of the Enns River, 300 m west of the Großreifling center in the north of Styria (47°39′52.50″ N, 14°42′27.75″ E; Fig. 24.2). It exposes Upper Triassic (lower

Carnian, Julian) deposits of the Reifling Basin, which is part of the Lunz Nappe and the southern Sulzbach Nappe within the Northern Calcareous Alps (Reifling Unit, NCA; Lukeneder and Lukeneder 2021, 2023; Lukeneder et al. 2024; Lukeneder 2026a, b, this volume). During the Late Triassic, the Reifling Basin was an intraplatform basin, which formed between Pangea in the northwest and the Meliata Ocean in the southeast (Fig. 24.1b). The Meliata Ocean was a small oceanic basin within the Tethyan Realm, separated from the Tethys Ocean by the island chain of the Cimmerian continent (Neubauer et al. 2019).

At the SCH-section, the Göstling Formation comprises an alternation of three main lithofacies (Figs. 24.3 and 24.4). Lithofacies 1: light grey, bioturbated peloidal limestones with radiolarians and variable amounts of thin bivalve shells and ostracods (Fig. 24.3a–c). Individual beds of this lithofacies attain few centimeters to few decimeters in thickness. Lithofacies 2: dark grey to black silty marly limestones with poorly developed lamination, high amount of organic debris, pyrite and numerous

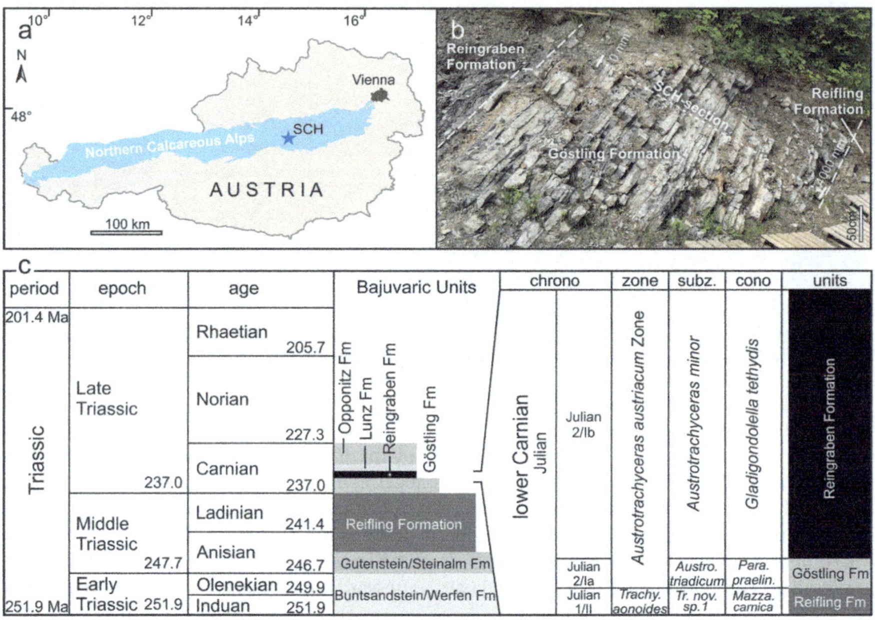

Fig. 24.2 Locality map and stratigraphy of the Upper Triassic Scheiblingbachgraben region. **a** Scheiblingbachgraben area (blue star) within the Northern Calcareous Alps (NCA, Styria). **b** Outcrop situation at Großreifling, basal Reifling Formation, detected Göstling Formation and overlaying Reingraben Formation. Unit boundaries marked by a white dashed lines. **c** Stratigraphic position of the lower Carnian Göstling Formation at Scheiblingbachgraben with the layers comprising the Upper Triassic (lower Carnian) Polzberg paleobiota within the *Austrotrachyceras austriacum* Zone. Column with indicated lithological units, thickness of the SCH section and detected interval with corresponding layers, Sch 2–52. chrono, chronostratigraphy; cono, conodonts; subz., subzone; Fm, Formation; *Austro.*, *Austrotrachyceras*; *Trachy*, *Trachyceras*; *Tr.*, *Trachyceras*; *Para. praelin.*, *Paragondolella praelinda*; *Mazza.*, *Mazzaella*; zone and subzone correspond to ammonoid taxa. Image by AL June 2024. Prepared by AL using CorelDraw X7; www.coreldraw.com

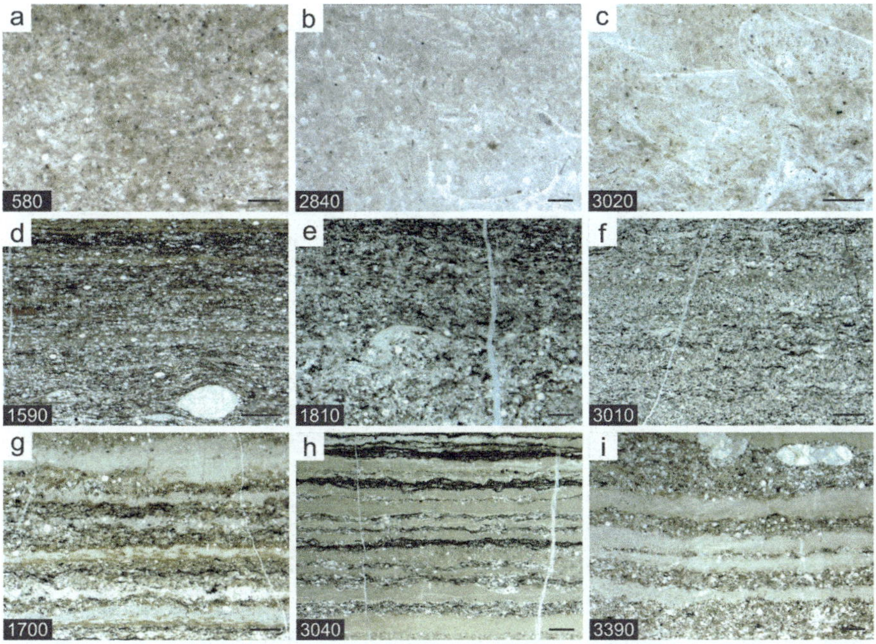

Fig. 24.3 Thin sections of detected lithofacies. **a–c** lithofacies 1, **d–f** lithofacies 2, **g–i** lithofacies 3. Numbers indicate sample position in section in mm in Fig. 24.2. Scale bar: **a–c** 500 μm; **d–i** 1 mm

radiolarians (Fig. 24.3d–f). Individual beds attain few cm in thickness. Lithofacies 3: distinctly laminated units of black silty to marly limestones and light grey cherts, generally lacking bioturbation (Fig. 24.3g–i). The laminated beds may attain few decimeters in thickness, whereas single laminae roughly range around 0.4–1.0 mm. Dark laminae may form radiolarian packstones or may be dominated by organic debris with rare radiolarians. Ammonoids randomly occur in these layers, bivalves are scarce.

Lithofacies 1 reflects oxidized bottom water conditions and oxygen availability within the sediment because the sea bottom became settled by bivalves and ostracods and the lime-mud is intensively bioturbated. Lithofacies 2 indicates anoxic bottom water conditions and high amount of organic input. Anoxic conditions in the sediment allowed pyrite formation by sulphate-reducing bacteria and hampered bioturbation. Nevertheless, lamination is not fully developed or restricted to short intervals without distinct light laminae. The laminated intervals of lithofacies 3 represent rapid fluctuations of terrestrial influx. The lack of bioturbation indicates stable phases of sediment anoxia. Dark laminae are often associated with radiolarian blooms. The scarceness of benthic biota in the light laminae suggests persisting hostile conditions. Biomarker proxies from the close-by Polzberg section (approx. 35 km northeast) document that the organic matter of the Göstling Formation is a mixture of marine and terrestrial material deposited in a dysoxic environment (Lukeneder et al. 2024).

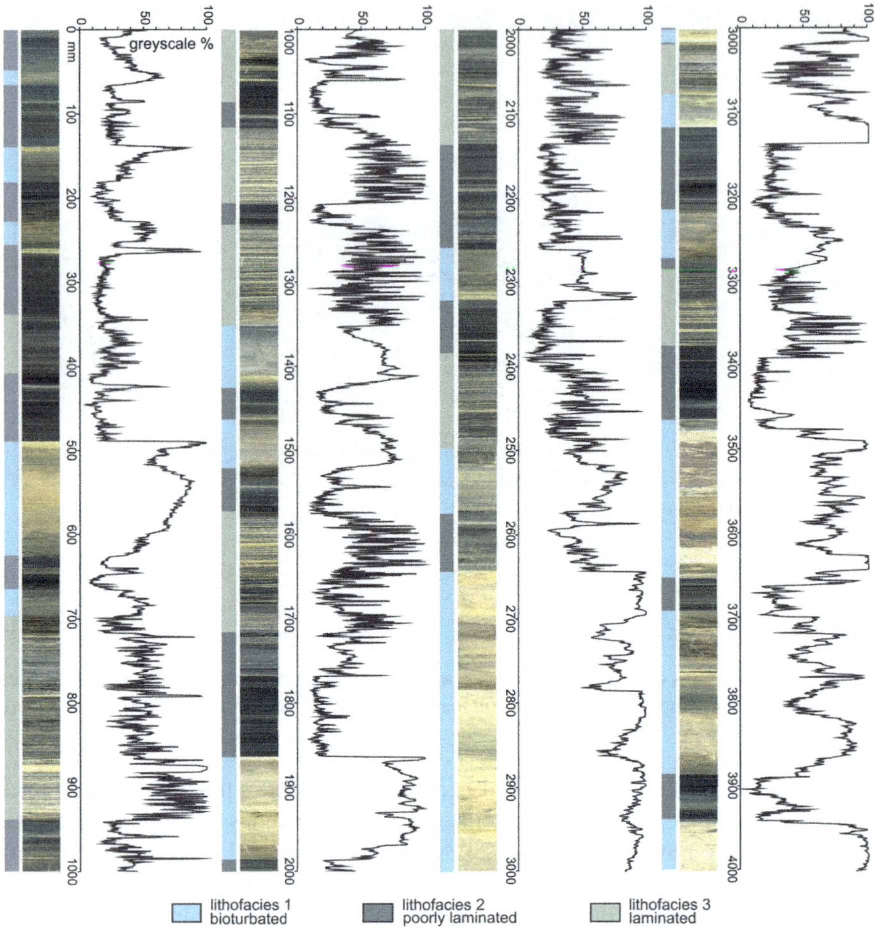

Fig. 24.4 Lithological log of the SCH-section transferred into grey-scale values (in %); prevailing lithofacies are indicated by color bars on the left

Visually, the occurrence of these lithofacies within the SCH-section appears rhythmically and suggests the presence of several cyclicities of different orders. To test for the presence of such cycles, a 4000-mm-thick part of the SCH-section was continuously recovered, cut perpendicularly to the bedding plane, polished, scanned and transformed into grey-scale values (n = 18,900; Fig. 24.4). Spectral analysis was performed on the data set (see Methods).

24.3 Microfacies and Lithology

The Reifling Formation (0 to 3.6 m, beds SCH 1 to SCH 9; Fig. 24.2) was deposited on the paleoslope of the Reifling Basin. It consists of light grey and grey-green nodular limestones composed of peloidal, filamentous wackestone and packstones (standard microfacies types SMF 1, 2 and 3); bed-thickness (0.05–0.5 m). The carbonate content $CaCO_3$ (calcite equivalent percentages) is stable around high values with 85–95% throughout the formation (Lukeneder et al. 2024; Lukeneder 2026a, b, this volume; Weitz et al. 2026, this volume). Strong bioturbation along with rare fish and marine reptile remnants.

The Reifling Formation (*Trachyceras aonoides* Zone, *T. aonoides* Subzone) is overlain by the Göstling Formation (3.6–8.1 m, beds SCH to SCH 52; Fig. 24.2b, c), which represents the *Austrotrachyceras austriacum* Zone (*A. minor* Subzone), indicating the base of Julian 2. The Göstling Formation consists of black, mm- to cm-bedded calciturbiditic radiolarian limestones ($CaCO_3$ 60–80%) with organic-rich mudstones (SMF1, 2 and 3). The change from the Reifling Formation to dark, fine-laminated radiolaria wackestones to packstones of the Göstling Formation is abrupt. On the upper surface of numerous layers plant fossil are abundant, marking an enhanced terrigenous input. Intense bioturbation is visible in oxidized parts, contrastingly rare or absent in laminated dysoxic to anoxic layers. At Scheiblingbachgraben (= Scheiblingbach), only about 4.5 m of radiolarian-rich Göstling limestones can be observed.

The overlying terrigenous siliciclastic deposits of the Reingraben Formation (8.1–32.2 m, beds SCH 53 to SCH 32.2 m; *A. austriacum* Zone, *A. minor* Subzone; Fig. 24.2b, c) consist of a pronounced alternation of dark grey to black argillaceous marls ($CaCO_3$ 5–20%; SMF 3 and 12) intercalated by rare limestone/dolomite layers, no macro-bioturbation is visible. The bivalves *Peribositra* and *Halobia rugosa* appear in all ontogenetic stages and, in layer-/rock-forming quantities, mostly in articulated butterfly position. The Late Triassic *Konservat-Lagerstätte* at Scheiblingbach comprises invertebrates and vertebrates. Most of the fossils derive from the lowermost three calcareous meters of the Reingraben Formation. The paleobiota shows a nekton-dominated fauna with abundant fish and cephalopods (Lukeneder and Lukeneder 2021, 2023). The Reingraben Formation at Scheiblingbach show microfossils such as conodonts, ostracods, gastropods and radiolaria in thin sections.

Microfacies consist mainly of biomicritic limestones, often slightly bioturbated, differentiated into foraminiferal- to radiolarian-dominated wackestones to packstones (Lukeneder et al. 2024). The lower Carnian interval is generally abundant in microfossils. Calcified radiolarians and sponge spicules are abundant in the packstones of both formations, while ammonite and bivalve mudstones appear from the base of the Reingraben Formation. A sudden change from radiolarian wackestones/packstones (Lahm 1984) of the Göstling Formation (Fig. 24.3) occurs above the boundary to the Reingraben Formation (Tollmann 1976). The nomenclature of standard microfacies types SMF are adopted from Flügel (2004) and Wilson (1975).

24.4 Material and Methods

A 4000-mm-thick part of the SCH-section (Fig. 24.2b) was continuously recovered, cut perpendicularly to the bedding plane, polished and scanned with 1200 dpi. A 5-mm-broad composite stripe of the images (120 dpi, 24 bit RGB color TIFF-s; Fig. 24.4), avoiding calcite veins, was converted to the Esri ASCII-format raster file. The resulting dataset comprised 24 colons and 18.900 rows of pixels each given a light-intensity, i.e. grey-scale level where values ranged from 0 to 255 (0 minimum/black to 255 maximum/white). For further analyses, a mean value was calculated for each pixel-row and subsequently transformed into % (0–100). All data points are evenly spaced with a resolution of 0.212 mm (n = 18,900).

Statistical analyses were performed with the software PAST version 4.16c (Hammer et al. 2001). Spectral analysis was applied to detect cyclicities. Raw data were analyzed in Lomb-Scargle periodograms (Fig. 24.5, Supplementary materials 24.1 and 24.2). The data have not been smoothed, to preserve high frequency cycles. Peaks passing the 95 and 99.9% significance levels were considered for further discussion. To test for stability of the detected cycles, we performed spectral analyses on the whole data set and on subsets (0–1000, 1500–2500, 1000–2000, 2500–3500, 2000–3000, 3000–4000 mm). In addition, shorter intervals with distinct lamination have been analyzed to test for high frequency cycles (700–800, 880–930, 1120–1350, 1230–1330, 1600–1700, 1990–2130, 2320–2500, 3300–3390 mm) (Supplementary materials 24.1 and 24.2). The full dataset is too large for REDFIT spectral analysis (Schulz and Mudelsee 2002). To remove noise and to fit data size to the REDFIT module of PAST, the dataset was stepwise reduced (75%, n = 4,737; 25%, n = 594, https://doi.org/10.57756/rmkggz). Frequency oversampling and the number of segments were set to 1 (Monte Carlo simulation) to preserve the maximum of resolution. Resulting models were assessed using the runs value and its 5% acceptance interval. The runs test is a non-parametric test for randomness in a sequence of values such as a time series. The test requires one column of data, which are internally converted to 0 ($x \leq 0$) or 1 ($x > 0$). The test is based on a dichotomy between two values ($x \leq 0$ or $x > 0$). It counts the number of runs and compares this to a theoretical value (Hammer 2021).

All significant peaks (n = 137) derived from Lomb Scargle periodograms were pooled and subjected to Mixture Analysis (Hammer 2021; see https://doi.org/10.57756/rmkggz for data). The solutions (=number of groups) of 20 runs with the best log likelihood values and the lowest Akaike Information Criterion (Akaike 1974) were finally chosen (Fig. 24.6).

Continuous wavelet transforms were calculated to visualize periodicities at different wavelengths and to detect potential non-stationary periodicities through time due to changing sedimentation rates. The 0.05 significance level is shown as black contour lines (Fig. 24.7).

Gaussian filtering was performed with the software AnalySeries 1.1. (Paillard et al. 1996). This is a frequency-selective filtering procedure, which removes unwanted

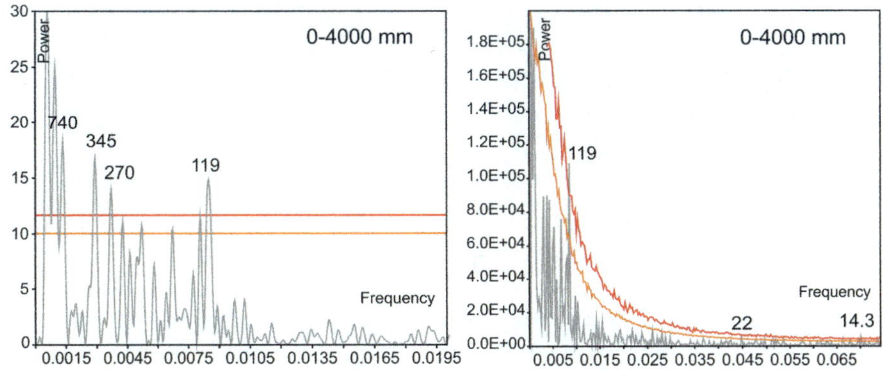

Fig. 24.5 Lomb-Scargle periodogram (left) and REDFIT spectrum (right) based on a reduced dataset (frequencies are transferred into cycle lengths in mm for representative peaks). Red and orange lines represent 99 and 95% confidence intervals. See https://doi.org/10.57756/rmkggz for data

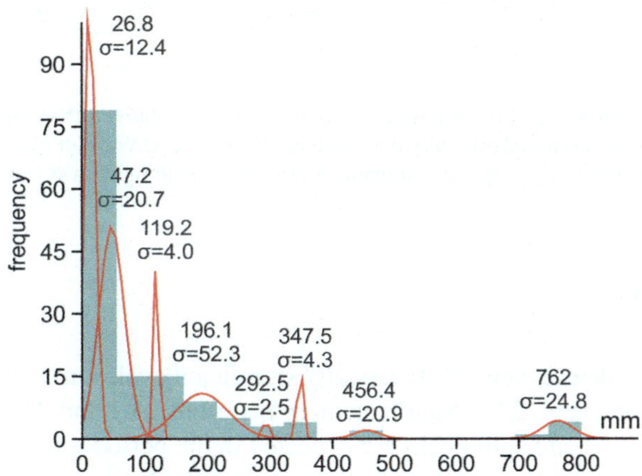

Fig. 24.6 Representative solution of the Mixture Analysis based on pooled peaks from Lomb Scargle periodograms. See https://doi.org/10.57756/rmkggz for data. The Lomb Scargle periodogram of the reduced dataset confirms the low frequency cycles (746, 345, 270, 117 mm). Similarly, the REDFIT analysis confirmed the peak at 199 mm. In addition, REDFIT detected higher frequency cycles at 23, 14.3, and 5.4–4.4 mm. The wavelet transforms indicate the constant presence of the low frequency cycles throughout the record, suggesting no major changes in sedimentation rate (Fig. 24.7). Especially the cycles centered around 119 and 23 mm are very stable in the wavelet spectra

frequency components from the time series. The filter was centered to peaks inferred by spectral analyses and set to a band-width of ±0.0001 (Fig. 24.9).

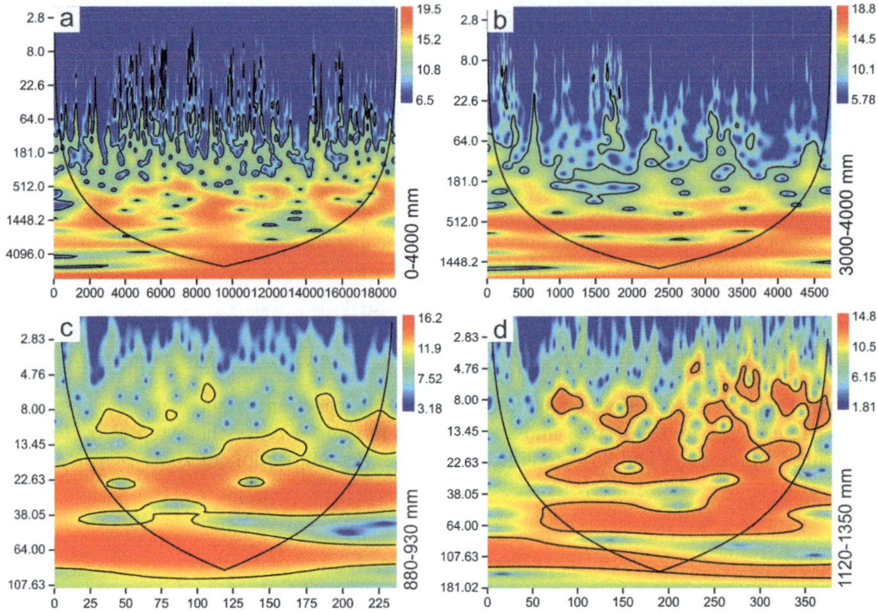

Fig. 24.7 **a** Wavelet spectra of the full dataset and **b** of a representative subset showing a constant and stable signal of the assumed obliquity band around 762 mm. **c–d** Wavelet spectra of two subsets with distinct lamination, revealing the dominant 23–26 mm beat along with at least three additional distinct frequencies

24.5 Results

Spectral analysis detects several statistically significant cycles in the grey-scale data of the SCH-section (Fig. 24.5, Supplementary materials 24.1 and 24.2). Some instability of frequencies and split peaks are expected in the periodograms, due to potential shifts in sedimentation rates and compaction, which not necessarily has been uniform. Therefore, all significant peaks derived from Lomb Scargle periodograms were pooled and subjected to Mixture Analysis, which is a maximum-likelihood method for estimating the parameters of two or more univariate normal distributions (Hammer 2021; Fig. 24.6). This method is an approach to overcome the problem of split peaks in long records (Harzhauser et al. 2023). The Mixture Analysis of the peaks derived from the Lomb-Scargle periodograms indicates the presence of cycles at 762 mm ($\sigma = 24.8$), 456.4 mm ($\sigma = 20.9$), 347.5 mm ($\sigma = 4.3$), 292.5 mm ($\sigma = 2.5$), 119.2 mm ($\sigma = 3.9$), 26.8 mm ($\sigma = 12.4$), 12.2 mm ($\sigma = 8.6$) and 3.9 mm ($\sigma = 1.5$) (https://doi.org/10.57756/rmkggz). Additional peaks at 196.1, 83.0, and 46.9 mm have very large standard deviations between 20 and 50% and are not considered herein.

Due to the shortness of the record, it is impossible to tune the data against the 405-kyr metronome of Laskar (2020) as done by Miller et al. (2017) for coeval deposits in England. Therefore, the detected cyclicities are floating and cannot be tied to any astronomical cycle with certainty. Similarly, it is impossible to apply statistical methods to detect amplitude modulation in the data as done by TimeOpt (Meyers 2015) and eCOCO (Li et al. 2018).

Nevertheless, the phase relation between the detected frequencies allows for testing potential influence of known Milankovitch cycles and solar cycles as summarized by Viaggi (2021). Linking the longest cycle (762 mm) to the 100-kyr eccentricity cycle would result in unrealistically low sedimentation rates and does not reveal meaningful relations of the other peaks with known cycles. Assuming that the long cycle around 762 mm represents the obliquity cycle of 41 kyr derives following depth/time relations: 456 mm = 24.6 kyr, 347 mm = 18.7 kyr, 292 mm = 15.8 kyr, 119 mm = 6.4 kyr, 27 mm = 1.44 kyr, 3.9 mm = 0.209 kyr). Following this age model, the 456 and 347 mm cycles represent the two components of presession at ~23 and ~19 kyr (Ruddiman 2001; Su et al. 2019; Hilgen et al. 2015). The dominant 119 mm cycle correlates with the 6 kyr-Heinrich-Bond cycle (Ruddiman 2001), the 12.4 mm cycle is close to the ~1500 yr cycle (Hasegawa et al. 2022) and the high frequency cycle at 3.9 mm, which is expressed by the bundles of laminae, falls within the range of the ~210 yr De Vries cycle (Stuiver and Braziunas 1993; Wagner et al. 2001). The cycle at 292 mm with a duration of 15.8 kyr might correspond to one of the unnamed cycles with periods around 16.4, 14.9 and 14.5 kyr documented by Hilgen et al. (2015).

This model requires a sedimentation rate of 18.6 mm/kyr, which is nearly identical with sedimentation rates documented by Miller et al. (2017) for coeval Carnian deposits of the Dunscombe Mudstone Fm. in southern England (19.08 mm/kyr). In addition, Ma et al. (2021) documented a sedimentation rate of 10.2 mm/kyr for laminated marine limestone/marlstone couplets from the Cenomanian of North America. These striking parallels support our hypothetical age model. Accordingly, the total section represents 215 kyr of early Carnian time.

24.6 Discussion

Tropical humid conditions with low seasonality prevailed in the tropical rain belt of Pangea, where seasonality was low but seasonality increased distinctly north of the tropics (Landwehrs et al. 2022), which coincides with the paleogeographic position of the studied section. At the junction between Pangea and the Tethys Ocean, the boreal summer correlated with a landward directed south wind system following the northward migration of the Intertropical Convergence Zone (ITCZ; Winguth and Winguth 2013; Shozaki and Hasegawa 2022; Fig. 24.1c). During the boreal winter, the ITCZ shifted southwards on Gondwana. The wind current system across the northwestern Tethys reversed and north and north-easterly winds blew from the continent to the open sea (Winguth and Winguth 2013; Shozaki and Hasegawa 2022; Fig. 24.1d). Consequently, the climate was strongly seasonal in coastal regions, with heavy rains during the boreal summer and dry conditions during the winter season

(Parrish 1993). These states of the PMM (Fig. 24.8) are reflected in the Reifling Basin by distinct lithofacies.

- *PMM summer monsoon* (Fig. 24.8b): the summer monsoon transported moisture from the ocean onto the continent, where high precipitation enhanced transport of terrestrial detritus to the shelf. This phase is reflected in the SCH-section by input of land-plant derived organic carbon in the dark laminae. Increased runoff during the summer monsoon resulted in high nutrient loads in the Reifling Basin and stimulated productivity. In addition, the prevailing south winds enhanced the northwards directed sea surface currents and increased upwelling intensity. Therefore, the wind regime amplified upwelling-derived nutrient flux and additionally fostered primary productivity. Like in modern marine environments, increased primary productivity triggered radiolarian blooms (Nimmergut and Abelmann 2002; Lampitt and Johns 2009; De Wever et al. 2014; Qu et al. 2022). Indeed, the presence of a strong upwelling system along the northwestern margin of the

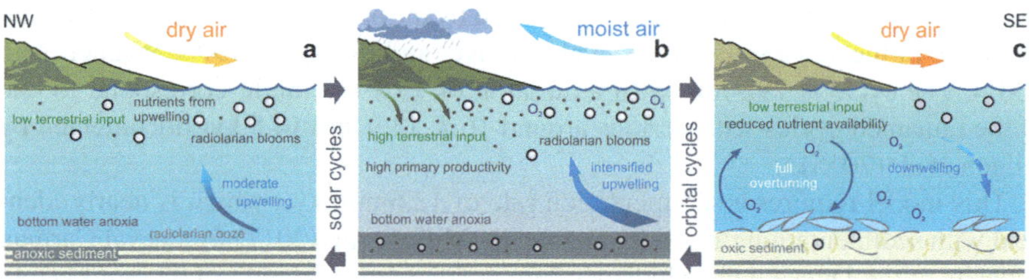

Fig. 24.8 Three alternating monsoon regimes along the Pangea-Tethys transition during the CPE. The alternation between **a** and **b** was governed by solar cycles, whereas shifts between **b** and **c** occurred on Milankovitch scales

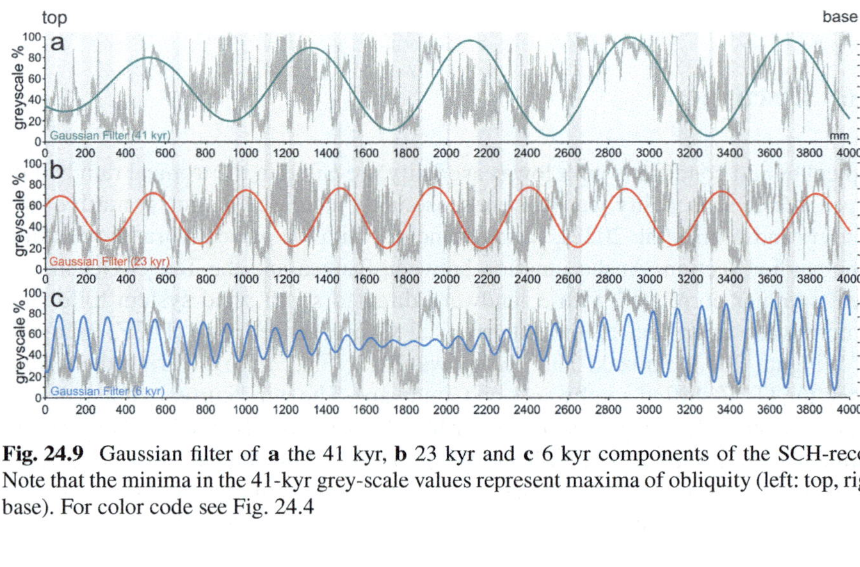

Fig. 24.9 Gaussian filter of **a** the 41 kyr, **b** 23 kyr and **c** 6 kyr components of the SCH-record. Note that the minima in the 41-kyr grey-scale values represent maxima of obliquity (left: top, right: base). For color code see Fig. 24.4

Tethys Ocean is reflected by about 8 °C cooler water masses compared to the open ocean (Rigo et al. 2012) and was supported by prevailing oceanic surface currents (Yan and Zhao 2001; Damborenea et al. 2013).

- Peaking primary productivity and a maximum of terrestrial input resulted in severe bottom water anoxia. Therefore, the dark layers of lithofacies 2 and 3 reflect a prevailing summer monsoon regime.
- *PMM (normal) winter monsoon* (Fig. 24.8a): during the winter-monsoon, dry air was transported from the continent to the sea. Precipitation was low and the input of terrestrial detritus was strongly reduced. Simultaneously, the prevailing northerlies dampened the surface water currents and probably weakened the upwelling system. Still, however, upwelling supplied enough nutrients to maintain a large stock of radiolarians, as reflected by thin layers of radiolarian ooze, which was deposited during these phases. No mixing of the bottom water occurred during these phases and therefore, bottom water conditions remained anoxic as documented by a lack of bioturbation. The light chert laminae in lithofacies 1 reflect this phase in the SCH-section.
- *PMM super winter monsoon* (Fig. 24.8c): the thicker beds of light carbonate of lithofacies 1 indicate fully oxic conditions and require a different mechanism of formation. These beds reflect phases of downwelling as recorded for the modern monsoon system in the Arabian Sea. There, southwesterly winds of the summer monsoon enhance upwelling and primary productivity, leading to a widening of the oxygen minimum zone (Lachkar et al 2018). In contrast, strong seaward directed wind during the winter monsoon season causes downwelling, which ventilates deep waters (DiMarco et al. 2023). Similarly, along the northwestern Tethys coast, it was the seaward directed northerlies of the winter monsoon regime, that periodically promoted longer phases of downwelling, which eventually lead to the oxygenation of the deep water and allowed for settlement by benthic bivalves and ostracods. The comparatively rare layers of lithofacies 1 indicate that the PMM super winter monsoon was a phenomenon, which occurred with a low frequency.

The SCH-record, documents considerable variability of the dominance of these three states of the PMM. Lithofacies 2 and 3 witness high frequency shifts between predominating summer or winter monsoon regimes, whereas lithofacies 1 indicates low frequency interruptions by the super winter monsoon regime. Modulations of the PMM intensity on various orbital scales was already deduced from models and sedimentary records (Whiteside et al. 2011; Winguth and Winguth 2013; Ikeda et al. 2017, 2020; Bahr et al. 2020). Gaussian filter of the 762- and 456-mm cycles, which are interpreted to represent obliquity and precession, suggest a correlation with the alternation between oxic and anoxic parts of the succession (Fig. 24.9, https://doi.org/https://doi.org/10.57756/rmkggz). Phases of bottom water oxygenation and bioturbation appear around maxima of the 762 mm cycle and are most intense when they coincide with maxima of the 456 mm cycle (e.g., 490–610, 2670–3040 mm). Phases with opposing trends (e.g., 762 mm minimum versus 456 mm maximum) are characterized by anoxic phases, often with prominent lamination (e.g., 800–1200, 2000–2270, 3300–3500 mm). Consequently, the grey-scale data can be interpreted as an inverse record of the obliquity band (Fig. 24.9a). Maxima in the 762 mm cycle correspond to obliquity minima, when terrestrial input was slightly reduced and the sediment became oxidized.

Precession extremes had a strong impact on the Triassic monsoon intensity (Whiteside et al. 2011; Winguth and Winguth 2013). Precession maxima correlated with dampened monsoon intensity (Bahr et al. 2020). Similarly, precession minima correspond to increased precipitation in modern monsoon systems (Ziegler et al. 2010; Jalihal et al. 2019; Ding et al. 2021; Armstrong et al. 2023). The mechanism is enforced when precession minima coincide with obliquity maxima (Armstrong et al. 2023). The same relation is documented in the modern Circum-Mediterranean region, where runoff during obliquity maxima is strengthened during precession minima by enhanced insolation seasonality (Tuenter et al. 2005; Marzocchi et al 2015; Bosmans et al. 2020; Armstrong et al. 2023). Similarly, in Eocene marine successions in Dongying (China), the deposition of carbonates occurred during precession maxima, whereas mudstones with high amount of organic carbon deposited during precession minima (Zhang et al. 2022). Assuming that this mechanism governed also the Triassic record, requires that the 456 mm cycle can be read as direct expression of precession with minima around ~300, ~750, ~1200, ~1700, ~2160 and ~2630 mm, all coinciding with anoxic bottom water conditions and a predominant summer monsoon regime (Fig. 24.9b). In contrast, precession maxima tend to correlate with phases of downwelling especially if these fall within obliquity minima (e.g., 550–600, 2800–3000 mm). However, the PMM along the Tethyan coast was not only strongly modulated by precession, as already discussed by Winguth and Winguth (2013) and Bahr et al. (2020), but reflects also the strong imprint of solar cycles, which acted on centennial and millennial scales.

24.6.1 Potential Record of the 6 kyr-Heinrich-Bond Cycle

One of the most prominent sub-Milankovitch cycles in the SCH-record is the ~119 mm cycle, which is interpreted herein as expression of the 6 kyr-Heinrich-Bond cycle (Fig. 24.9c). Given that the De Vries cycle is identically expressed in our Triassic record like in the Cenomanian record described by Ma et al. (2021), it would be tempting to link the dominant ~119 mm cycle to the half-precession cycle, which was shown to have had major influence on the oxygen minimum zone in the Western Interior Seaway during the Cenomanian (Ma et al. 2021). This 10 kyr cycle was also discussed for Triassic continental records of Pangea (Whiteside et al. 2011). Maybe, a reinterpretation of these records will reveal the 6 kyr cycle as well. In any case, interpreting the ~119 mm cycle as expression of the half-precession cycle does not derive meaningful results for the other cycles in the SCH-section. Instead, our age model requires a periodicity of 6.4 kyr for the ~119 mm cycle. Such a ~6 kyr cycle was frequently detected in Pleistocene records and was termed Heinrich-Bond cycle by Viaggi (2021). The cycle was explained by ice sheet dynamics by many authors (Viaggi 2021), which is a highly unlikely mechanism for the Carnian. Alternatively,

Mayewski et al. (1997) interpreted the ~6 kyr cycle to be of solar origin. Xapsos and Burke (2009) identified this cycle in the Holocene sunspot number, supporting its nature as solar cycle. Surprisingly, this cycle seems to be not uncommon in Triassic successions. Wu et al. (2012) documented a comparable cycle of 4.7–5.6 kyr in the early Triassic of South China and a comparable periodicity was proposed by Zühlke et al. (2003) from the Latemar platform in northern Italy (but see Kent et al. 2004). Consequently, Wu et al. (2012) suggested that this sub-Milankovitch cycle was common in the tropical and sub-tropical carbonate platforms of Pangea during the Early and Middle Triassic. In the SCH-record, the Heinrich-Bond cycle is expressed by rapid shifts from light to dark intervals. Therefore, it is an expression of cyclic changes in monsoon intensity.

24.6.2 High Frequency Solar Cycles

Our age model links the 3.9 mm cycle, observed within the bundles of laminae, with a strong 0.2 kyr cyclicity, which corresponds to the quasiperiodic ~210 yr De Vries solar cycle. This cycle is frequently detected in Neogene, Pleistocene and Holocene records (De Vries 1958; Raspopov et al. 2008; Costas et al. 2012; Kern et al. 2012; Steinhilber and Beer 2013; Viaggi 2021; Komitov 2024). The influence of the De Vries cycle on marine sedimentation was proved in oceanographic models by Seidenglanz et al. (2012), who showed that the De Vries cycle penetrates the ocean down to deep-water levels. Mesozoic successions exhibiting the signature of the De Vries cycle are comparatively rare. From terrestrial records, it was documented from latest Triassic tree rings from Brazil (Prestes et al. 2014), from Early Jurassic limestone varves from New Mexico (USA; Dmitriev et al. 2016) and from laminated Aptian lacustrine deposits from Mongolia (Hasegawa et al. 2022) and Brazil (Gomes et al. 2021). Marine examples of the influence of the De Vries cycle were reported from the Cenomanian of the Western Interior Seaway (USA; Ma et al. 2021; Liu et al. 2023). In these records, the De Vries cycle governed also lamination. Sedimentation rate and nature of lamination of the Cenomanian deposits are very similar to those of the herein discussed Triassic record, suggesting comparable forcing mechanisms. Ma et al. (2021) interpreted the alternation of dark and light layers as variations in water-column stratification and redox conditions in response to changes in storm activity and intensity. Liu et al. (2009) observed oscillations in the Asian summer monsoon precipitation during the last millennium, due to changes in the land–sea thermal contrast and hemispheric temperature gradient induced by the De Vries cycle. Similarly, Novello et al. (2016) recognized a 208-year periodicity in the variation of intensity on the South American monsoon, during the last 1500 years. Therefore, the lamination in the herein studied Triassic record indicates a high frequency fluctuation between predominant summer and winter monsoon regimes during the CPE. Consequently, the lamination reflects rapid oscillation in monsoon intensity following the beat of the De Vries cycle, as observed also in modern monsoon systems. Phases of pronounced lamination, in which the De Vries cycle is strongly expressed, are not in phase with obliquity and precession, suggesting that it was decoupled from Milankovitch forcing.

Higher frequency cycles of laminae, constituting the 3.9 mm bundles are not resolved herein. These laminae, however, suggest an additional modulation of monsoon intensity on a decadal scale.

24.7 Conclusions

For the first time, we document a strong modulation of the Pangean Megamonsoon during the Late Triassic on millennial and centennial scales. Major changes in the dominant monsoon regime were bound to Milankovitch scale modulation. This is in line with previous data and models, which assumed a strong PMM modulation by precession (Winguth and Winguth 2013; Bahr et al. 2020). Herein, however, we can link changing monsoon regimes for the first time with distinct responses in marine depositional systems on orbital scales and on the scale of solar cycles. During phases of a predominant summer monsoon regime, primary productivity peaked due the coincidence of enhanced continental runoff and increased nutrient availability by intensified upwelling. This nutrient-saturated system seems to have been at full swing during precession minima, which fell within obliquity maxima. Severe anoxia formed on the seafloor and the oxygen minimum zone widened. In reverse, the Carnian Pluvial Episode was punctuated during phases when obliquity minima coincided with precession maxima. Now, the predominant super winter monsoon regime generated strong seaward directed northerly wind, which caused downwelling, intense bottom water ventilation and settlement of the seafloor by benthic bivalves.

Solar cycles are similarly reflected by shifts of the predominant monsoon regime. In contrast to Milankovitch forcing, the solar cycles did not enhance winter monsoon enough to install downwelling conditions but strongly reduced the input of terrestrial detritus. The northerly winds might have damped upwelling as well, but nutrient availability remained high enough to maintain radiolarian blooms, which formed thin laminae of radiolarian ooze. Solar cycles did not affect the redox conditions on the seafloor of the Reifling Basin as reflected by the continuous lack of bioturbation and the rareness of benthos.

Within the detected hierarchy of climatic cycles, the strong 6 kyr-Heinrich-Bond cycle is an enigmatic feature of assumed solar origin (Xapsos and Burke 2009). Records from the Triassic of China and Italy reveal similar periodicities (Zühlke et al 2003; Wu et al. 2012) and therefore, the 6 kyr cycle might have been an important component of Triassic climate. The highest resolved frequencies in the PMM-record fall in the range of the De Vries solar cycle, which links the PMM to modern monsoon systems (Seidenglanz et al. 2012).

We emphasize that our record cannot be tied to any astronomical cycle with certainty due to its shortness. Consequently, the herein discussed correlation is hypothetical. More high-frequency records, linked to the 405-kyr metronome, will be needed to support or reject the herein proposed hypothesis of monsoon intensity modulation by solar cycles and to test for the presence of the 6 kyr cycle as basic high-frequency beat of Pangea's climate.

Acknowledgments We acknowledge the Austrian Academy of Sciences (headquarters in Vienna) represented by the National Committee for Geo/Hydro Sciences (Polzberg Lukeneder 2021, Earth System Sciences Programme), the Federal Government of Lower Austria (project K3-F-964/001-2020 and K3-F-964/001-2023, Department Science and Research; headquarters in St. Pölten), the UNESCO Geopark Steirische Eisenwurzen and the Natural History Museum Vienna for financial support. We acknowledge the Austrian Federal Forest company (Österreichische Bundesforste) for sampling and digging permission. Lukas Nöst and Dominic Asztaller (both forest dristrict Großreifling) are acknowledged for their advise during field campaigns. Anton Fürst and Anton Englert (both Vienna) are acknowledged for technical support (thin sections). The manuscript greatly benefited from valuable comments by two anonymous reviewers, the production editor Ashok Arumairaj (Taramani), the in-house series editor Kathleen A. Ritterbush (Utah), the publishing editor Simon Shah-Rohlfs and the production editor Antje Endemann (both Heidelberg).

Author Contribution MH Conceptualization. AL, PL fieldwork and data acquisition. OM, MH statistics and analyzes. OM, MH, AL, PL interpretation and writing of the paper.

Funding Open Access funding enabled and organized by Natural History Museum Vienna. The authors acknowledge the financial support by the project ÖAW (Polzberg Lukeneder 2021) and two projects of the Federal Government of Lower Austria (Land NÖ K3-F-964/001-2020, K3-F-964/001-2023); the Natural History Museum Vienna; and the officials of the UNESCO Geopark Steirische Eisenwurzen and by the association of the Freunde des Naturhistorischen Museums Wien (Project Trias Lukeneder). The authors are responsible for the contents of this publication. The funder had no impact on conceptualization, design, data collection, analysis, decision to publish, or preparation of the manuscript.

Competing interests All authors stated that they have no competing interests.

Authors' contributions All authors stated that they have equally contributed to their chapters, read, edited and agreed to submit the final versions of their contributions.

Availability of data and materials All sources of information are provided in the text. Raw data related to the fossil material from Austria are available from the corresponding authors and Alexander Lukeneder upon request.

Additional information The article is part of the Special Volume: *Late Triassic Konservat-Lagerstätten within the Carnian Pluvial Episode. 150 Years of Scientific Excavations in Austria.*

Artificial Intelligence (AI) The chapter was produced without the use of AI.

Language The entire volume uses American English reviewed by a native scientist.

Received 2025.

Published online 2026.

References

Akaike H (1974) A new look at the statistical model identification. IEEE Trans Autom Control 19(6):716–723. https://doi.org/10.1109/TAC.1974.1100705

Alverà G, Dal Corso J, Chu D, Cruciani G, Roghi G, Caggiati M, Song H, Song H, Tian L, Du Y, Haserbek T, Gianolla P (2026) Episodic aridification at the onset of the Carnian Pluvial episode in Western Tethys. Glob Planet Change 256:105192. https://doi.org/10.1016/j.gloplacha.2025.105192

Arche A, López-Gómez J (2014) The Carnian Pluvial Event in Western Europe: new data from Iberia and correlation with the Western Neotethys and Eastern North America–NW Africa regions. Earth-Sci Rev 128:196–231. https://doi.org/10.1016/j.earscirev.2013.10.012

Armstrong E, Tallavaara M, Hopcroft PO, Valdes PJ (2023) North African humid periods over the past 800,000 years. Nat Commun 14:5549. https://doi.org/10.1038/s41467-023-41219-4

Bahr A, Kolber G, Kaboth-Bahr S, Reinhardt L, Friedrich O, Pross J (2020) Mega-monsoon variability during the late Triassic: re-assessing the role of orbital forcing in the deposition of playa sediments in the Germanic Basin. Sedimentology 67(2):951–970. https://doi.org/10.1111/sed.12668

Bernardi M, Gianolla P, Petti FM, Mietto P, Benton MJ (2018) Dinosaur diversification linked with the Carnian Pluvial Episode. Nat Commun 9:1499. https://doi.org/10.1038/s41467-018-03996-1

Bosmans JHC, van der Ent RJ, Haarsma RJ, Drijfhout SS, Hilgen FJ (2020) Precession- and obliquity-induced changes in moisture sources for enhanced precipitation over the Mediterranean Sea. Paleoceanogr Paleoclimatol 35(1):e2019PA003655. https://doi.org/10.1029/2019PA003655

Breda A, Preto N, Roghi G, Furin S, Meneguolo R, Ragazzi E, Fedele P, Gianolla P (2009) The Carnian Pluvial event in the Tofane area (Cortina D'Ampezzo, Dolomites, Italy). Geo Alp 6:80–115. https://www.zobodat.at/pdf/GeoAlp_006_0080-0115.pdf

Chu R, Wu H, Zhu R, Fang Q, Deng S, Cui J, Yang T, Li H, Zhang S (2020) Orbital forcing of Triassic megamonsoon activity documented in lacustrine sediments from Ordos Basin, China. Palaeogeogr Palaeoclimatol Palaeoecol 541:109542. https://doi.org/10.1016/j.palaeo.2019.109542

Costas S, Jerez S, Trigo RM, Goble R, Rebêlo L (2012) Sand invasion along the Portuguese coast forced by westerly shifts during cold climate events. Quat Sci Rev 42:15–28. https://doi.org/10.1016/j.quascirev.2012.03.008

Dal Corso J, Bernardi M, Sun Y, Song H, Seyfullah LJ, Preto N, Gianolla P, Ruffell A, Kustatscher E, Roghi G, Merico A, Hohn S, Schmid, AR, Marzoli A, Newton RJ, Wignall PB, Benton MJ (2020) Extinction and dawn of the modern world in the Carnian (Late Triassic). Sci Adv 6(38):eaba0099. https://doi.org/10.1126/sciadv.aba0099

Damborenea SE, Echevarría J, Ros-Franch S (2013) Southern hemisphere palaeobiogeography of triassic-jurassic marine bivalves. Springer Br Earth Syst Sci 139 pp. https://doi.org/10.1007/978-94-007-5098-2_1

De Vries H (1958) Variation in concentration of radiocarbon with time and location on earth. Kon Ned Akad Wetensch Proc Ser B 61:94–102

De Wever P, O'Dogherty L, Goričan Š (2014) Monsoon as a cause of radiolarite in the Tethyan realm. C R Geosci 346:287–297. https://doi.org/10.1016/j.crte.2014.10.001

DiMarco SF, Wang Z, Chapman P, al-Kharusi L, Belabbassi L, al-Shaqsi H, Stoessel M, Ingle S, Jochens AE, Howard MK (2023) Monsoon-driven seasonal hypoxia along the northern coast of Oman. Front Mar Sci 10:1248005. https://doi.org/10.3389/fmars.2023.1248005

Dmitriev PB, Dergachev VA, Tyasto MI, Blagoveshchenskaya EE (2016) Cycles revealed in Jurassic paleoclimatic data (ca. 200–145 Ma B.P.). Geomagn Aeron 56:914–919. https://doi.org/10.1134/S0016793216070045

Ding Z, Huang G, Liu F, Wu R, Wang P (2021) Responses of global monsoon and seasonal cycle of precipitation to precession and obliquity forcing. Clim Dyn 56:3733–3747. https://doi.org/10.1007/s00382-021-05663-6

Flügel E (2004) Microfacies of carbonate rocks. Analysis, interpretation and application. Springer, Berlin, 976 pp. https://doi.org/10.1007/978-3-662-08726-8

Francis JE (2009) Palaeoclimates of Pangea—geological evidence. Can Soc Petr Geol Spec Publ 17:265–274

Gomes JMP, Rios-Netto AM, Borghi L, Souza Carvalho I, Mendonça Filho JG, Sabaraense LD, Araújo BC (2021) Cyclostratigraphic analysis of the early Cretaceous laminated limestones of the Araripe Basin, NE Brazil: estimating sedimentary depositional rates. J South Am Earth Sci 112(1):103563. https://doi.org/10.1016/j.jsames.2021.103563

Hammer Ø (2021) Paleontological statistics version 4.06. Reference manual. Natural History Museum. University of Oslo, Oslo, 290 pp. https://www.nhm.uio.no/english/research/infrastructure/past/downloads/past4manual.pdf

Hammer Ø, Harper DAT, Ryan PD (2001) PAST: paleontological statistics software package for education and data analysis. Palaeont Elect 4:1–9. https://palaeo-electronica.org/2001_1/past/past.pdf

Harzhauser M, Peresson M, Benold C, Mandic O (2023) Potentially astronomically forced climate sensitivity of Lake Pannon during the Tortonian thermal maximum (Late Miocene, Vienna Basin, Austria). Newsl Stratigr 57(1):89–107. https://doi.org/10.1127/nos/2023/0763

Hasegawa H, Katsuta N, Muraki Y, Heimhofer U, Ichinnorov N, Asahi H, Ando H, Yamamoto K, Murayama M, Ohta T, Yamamoto M, Ikeda M, Ishikawa K, Kuma R, Hasegawa T, Hasebe N, Nishimoto S, Yamaguchi K, Abe F, Tada R, Nakagawa T (2022) Decadal–centennial-scale solar-linked climate variations and millennial-scale internal oscillations during the Early Cretaceous. Sci Rep 12:21894. https://doi.org/10.1038/s41598-022-25815-w

Hilgen FJ, Hinnov LA, Aziz HA, Abels HA, Batenburg S, Bosman JHC, De Boer B, Hüsing SK, Kuiper KF, Lourens LJ, Rivera T, Tuenter E, Van de Wal RSW, Wotzlaw JF, Zeeden C (2015) Stratigraphic continuity and fragmentary sedimentation: the success of cyclostratigraphy as part of integrated stratigraphy. Geol Soc Lond Spec Publ 404:157–197. https://doi.org/10.1144/SP404.12

Hu Y, Li X, Boos WR, Guo J, Lan J, Lin Q, Han J, Zhang J, Bao X, Yuan S, Wei Q, Liu Y, Yang J, Nie J, Guo Z (2023) Emergence of the modern global monsoon from the Pangaea megamonsoon set by palaeogeography. Nat Geosci 16:1041–1046. https://doi.org/10.1038/s41561-023-01288-y

Ikeda M, Ozaki K, Legrand J (2020) Impact of 10-Myr scale monsoon dynamics on Mesozoic climate and ecosystems. Sci Rep 10:11984. https://doi.org/10.1038/s41598-020-68542-w

Ikeda M, Tada R, Ozaki K (2017) Astronomical pacing of the global silica cycle recorded in Mesozoic bedded cherts. Nat Commun 8:15532. https://doi.org/10.1038/ncomms15532

Jalihal C, Bosmans JHC, Srinivasan J, Chakraborty A (2019) The response of tropical precipitation to Earth's precession: the role of energy fluxes and vertical stability. Clim Past 15(2):449–462. https://doi.org/10.5194/cp-15-449-2019

Kolar-Jurkovšek T, Jurkovšek B (2010) New paleontological evidence of the Carnian strata in the Mežica area (Karavanke Mountains, Slovenia): conodont data for the Carnian Pluvial Event. Palaeogeogr Palaeoclimatol Palaeoecol 290:81–88. https://doi.org/10.1016/j.palaeo.2009.06.015

Kent DV, Muttoni G, Brack P (2004) Magnetostratigraphic confirmation of a much faster tempo for sea-level change for the Middle Triassic Latemar platform carbonates. Earth Planet Sci Lett 228(3–4):369–377. https://doi.org/10.1016/j.epsl.2004.10.017

Kern AK, Harzhauser M, Piller WE, Mandic O, Soliman A (2012) Strong evidence for the influence of solar cycles on a Late Miocene lake system revealed by biotic and abiotic proxies. Palaeogeogr Palaeoclimatol Palaeoecol 329–330:124–136. https://doi.org/10.1016/j.palaeo.2012.02.023

Komitov B (2024) About the possible solar nature of the ~200 yr (de Vries/Suess) and ~2000–2500 yr (Hallstadt) cycles and their influences on the earth's climate: the role of solar-triggered tectonic processes in general "sun–climate" relationship. Atmosphere 15:612. https://doi.org/10.3390/atmos15050612

Kutzbach JE, Gallimore RG (1989) Pangaean climates: megamonsoons of the megacontinent. J Geophys Res Atmos 94:3341–3357. https://doi.org/10.1029/JD094iD03p03341

Lachkar Z, Lévy M, Smith S (2018) Intensification and deepening of the Arabian Sea oxygen minimum zone in response to increase in Indian monsoon wind intensity. Biogeosciences 15(1):159–186. https://doi.org/10.5194/bg-15-159-2018

Lahm B (1984) Spumellarienfaunen (Radiolaria) aus den mitteltriassischen Buchensteiner-Schichten von Recoaro (Norditalien) und den obertriasschen Reiflingerkalken von Großreifling (Österreich). Münchner Geowiss Abh A(1):161 p

Lampitt RS, Johns ISD (2009) Radiolaria: major exporters of organic carbon to the deep ocean. Global Biogeochem Cycles 23:1–9. https://doi.org/10.1029/2008GB003221

Landwehrs J, Feulner G, Willeit M, Petri S, Sames B, Wagreich M, Whiteside JH, Olsen PE (2022) Modes of Pangean lake level cyclicity driven by astronomical climate pacing modulated by continental position and pCO_2. PNAS 119(46):e2203818119. https://doi.org/10.1073/pnas.2203818119

Laskar J (2020) Astrochronology. In: Gradstein FM, Ogg JG, Ogg GM (eds) Geological time scale. Elsevier, Amsterdam, pp 139–158. https://doi.org/10.1016/B978-0-12-824360-2.00004-8

Li M, Kump LR, Hinnov LA, Mann ME (2018) Tracking variable sedimentation rates and astronomical forcing in Phanerozoic paleoclimate proxy series with evolutionary correlation coefficients and hypothesis testing. Earth Planet Sci Lett 501:165–179. https://doi.org/10.1016/j.epsl.2018.08.041

Liu J, Wang B, Ding Q, Kuang X, Soon W, Zorita E (2009) Centennial variations of the Global Monsoon precipitation in the last millennium: results from ECHO-G model. J Clim 22:2356–2371. https://doi.org/10.1175/2008JCLI2353.1

Liu Y, Tang W, Zhaou Y, Wang C, Ma C (2023) Centennial to millennial cycles during the Cretaceous Oceanic Anoxic Event II (OAE 2). Quat Sci 43(6):1584–1596. https://doi.org/10.11928/j.issn.1001-7410.2023.06.07

Lu J, Zhang P, Dal Corso J, Yang M, Wignall PB, Greene SE, Shao L, Lyu D, Hilton J (2021) Volcanically driven lacustrine ecosystem changes during the Carnian Pluvial episode (Late Triassic). Earth Atmos Planet Sci 118(40):e2109895118. https://doi.org/10.1073/pnas.2109895118

Lukeneder A (ed) (2026a) Late Triassic Konservat-Lagerstätten within the Carnian Pluvial Episode: 150 Years of Scientific Excavations in Austria. Top Geobiol 55:600 p. https://doi.org/10.1007/978-3-032-19777-1

Lukeneder A (2026b) The marine Upper Triassic paleobiota from the Scheiblingbach *Konservat-Lagerstätte* in Austria. In: Lukeneder A (ed) Late Triassic Konservat-Lagerstätten within the Carnian Pluvial episode: 150 years of scientific excavations in Austria. Top Geobiol 55, Chap 19:437–467. https://doi.org/10.1007/978-3-032-19777-1_19

Lukeneder A, Lukeneder P (2021) The Upper Triassic Polzberg palaeobiota from a marine *Konservat-Lagerstätte* deposited during the Carnian Pluvial episode in Austria. Sci Rep 11:16644. https://doi.org/10.1038/s41598-021-96052-w

Lukeneder A, Lukeneder P (2023) New data on the marine Upper Triassic palaeobiota from the Polzberg *Konservat-Lagerstätte* in Austria. Swiss J Palaeontol 142:9. https://doi.org/10.1186/s13358-023-00269-3

Lukeneder A, Lukeneder P, Sachsenhofer RF, Roghi G, Rigo M (2024) Multi-proxy record of the Austrian Upper Triassic Polzberg *Konservat-Lagerstätte* in light of the Carnian Pluvial episode. Sci Rep 14:11194. https://doi.org/10.1038/s41598-024-60591-9

Ma C, Hinnov LA, Eldrett JS, Meyers SR, Bergman SC, Minisini D, Lutz B (2021) Centennial to millennial variability of greenhouse climate across the mid-Cenomanian event. Geology 50:227–231. https://doi.org/10.1130/G48734.1

Marzocchi A, Lunt DJ, Flecker R, Bradshaw CD, Farnsworth A, Hilgen FJ (2015) Orbital control on late Miocene climate and the North African monsoon: insight from an ensemble of sub-precessional simulations. Clim Past 11(10):1271–1295. https://doi.org/10.5194/cp-11-1271-2015

Mayewski PA, Meeker LD, Twickler MS, Whitlow S, Yang Q, Lyons WB, Prentice M (1997) Major features and forcing of high-latitude norther hemisphere atmospheric circulation using a 110,000 year-long glaciochemical series. J Geophys Res 102(C12):345–366. https://doi.org/10.1029/96JC03365

Meyers SR (2015) The evaluation of eccentricity-related amplitude modulation and bundling in paleoclimate data: an inverse approach for astrochronologic testing and time scale optimization. Paleoceanogr Paleoclimatol 30:1625–1640. https://doi.org/10.1002/2015PA002850

Miller CS, Peterse F, Da Silva AC, Baranyi V, Reichart GJ, Kürschner WM (2017) Astronomical age constraints and extinction mechanisms of the Late Triassic Carnian crisis. Sci Rep 7:2557. https://doi.org/10.1038/s41598-017-02817-7

Neubauer F, Liu Y, Cao S, Yuan S (2019) What is the Austroalpine mega-unit and what are the potential relations to Paleotethys Ocean remnants of southeastern Europe? Geol Carp 70:16–20

Nimmergut A, Abelmann A (2002) Spatial and seasonal changes of radiolarian standing stocks in the Sea of Okhotsk. Deep Sea Res I Oceanogr Res Pap 49(3):463–493. https://doi.org/10.1016/S0967-0637(01)00074-7

Novello V, Vuille M, Cruz FW, Stríkis NM, de Paula MS, Lawrence ER, Cheng H, Karmann I, Jaqueto PF, Trindade RIF, Hartmann GA, Moquet JS (2016) Centennial-scale solar forcing of the South American Monsoon system recorded in stalagmites. Sci Rep 6:24762. https://doi.org/10.1038/srep24762

Paillard D, Labeyrie L, Yiou P (1996) Macintosh program performs time-series analysis. Eos Trans AGU 77(39):379. https://doi.org/10.1029/96EO00259

Parrish JT (1993) Climate of the supercontinent Pangea. J Geol 101:215–233. https://doi.org/10.1086/648217

Prestes A, Rigozo NR, Nordemann DJR, Echer E, Vieira LEA, Souza Echer MP, Wrasse CM, Guarnieri FL (2014) Imprint of climate variability on mesozoic fossil tree rings: evidences of solar activity signals on environmental records around 200 million years ago? Pure Appl Geophys 171:1983–1991. https://doi.org/10.1007/s00024-013-0726-2

Qu H, Wang J, Xu Y, Xu J, Li X (2022) Radiolarian assemblages controlled by ocean production in the western equatorial Pacific. Deep Sea Res I Oceanogr Res Pap 182:103721. https://doi.org/10.1016/j.dsr.2022.103721

Raspopov OM, Dergachev VA, Esper J, Kozyreva OV, Frank D, Ogurtsov M, Kolström T, Shao X (2008) The influence of the de Vries ($\sim$200-year) solar cycle on climate variations: results from the Central Asian Mountains and their global link. Palaeogeogr Palaeoclimatol Palaeoecol 259(1):6–16. https://doi.org/10.1016/j.palaeo.2006.12.017

Rigo M, Preto N, Roghi G, Tateo F, Mietto P (2007) A rise in the carbonate compensation depth of western Tethys in the Carnian (Late Triassic): deep-water evidence for the Carnian Pluvial Event. Palaeogeogr Palaeoclimatol Palaeoecol 246(2–4):188–205. https://doi.org/10.1016/j.palaeo.2006.09.013

Rigo M, Trotter JA, Preto N, Williams IS (2012) Oxygen isotopic evidence for Late Triassic monsoonal upwelling in the northwestern Tethys. Geology 40(6):515–518. https://doi.org/10.1130/G32792.1

Ruddiman WF (2001) Earth's climate: past and future. WH Freeman and Company, New York, 465 p

Schöllhorn I, Adatte T, Van de Schootbrugge B, Houben A, Charbonnier G, Janssen N, Föllmi KB (2020) Climate and environmental response to the break-up of Pangea during the Early Jurassic (Hettangian-Pliensbachian); the Dorset coast (UK) revisited. Glob Planet Change 185:103096. https://doi.org/10.1016/j.gloplacha.2019.103096

Schulz M, Mudelsee M (2002) REDFIT: estimating Red-noise spectra directly from unevenly spaced paleoclimatic time series. Comput Geosci 28(3):421–426. https://doi.org/10.1016/s0098-3004(01)00044-9

Seidenglanz A, Prange M, Varma V, Schulz M (2012) Ocean temperature response to idealized Gleissberg and de Vries solar cycles in a comprehensive climate model. Geophys Res Lett 39(22):L22602. https://doi.org/10.1029/2012GL053624

Shozaki H, Hasegawa H (2022) Development of longitudinal dunes under Pangaean atmospheric circulation. Clim Past 18(7):1529–1539. https://doi.org/10.5194/cp-18-1529-2022

Soreghan MS, Soreghan GS, Hamilton MA (2002) Paleowinds inferred from detrital-zircon geochronology of upper Paleozoic loessite, western equatorial Pangea. Geology 30:695–698. https://doi.org/10.1130/0091-7613(2002)030%3c0695:PIFDZG%3e2.0.CO;2

Steinhilber F, Beer J (2013) Prediction of solar activity for the next 500 years. J Geophys Res Space Phys 118:1861–1867. https://doi.org/10.1002/jgra.50210

Stuiver M, Braziunas TF (1993) Sun, ocean, climate and atmospheric 14CO2, an evaluation of causal and spectral relationships. Holocene 3(4):289–305. https://doi.org/10.1177/095968369300300401

Su Q, Nie J, Luo Z, Li M, Heermance R, Garzione C (2019) Detection of strong precession cycles from the late Pliocene sedimentary records of northeastern Tibetan Plateau. Geochem Geophys Geosyst 20:3901–3912. https://doi.org/10.1029/2019GC008447

Sun YD, Wignall PB, Joachimski MM, Bond DPG, Grasby SE, Lai XL, Wang LN, Zhang ZT, Sun S (2016) Climate warming, euxinia and carbon isotope perturbations during the Carnian (Triassic) Crisis in South China. Earth Planet Sci Lett 444:88–100. https://doi.org/10.1016/j.epsl.2016.03.037

Tollmann A (1976) Analyse des klassischen nordalpinen Mesozoikums. Franz Deuticke, 580 p. https://opac.geologie.ac.at/ais312/dokumente/Tollmann_Klass_noralp_Mesoz_1976_Fulltext.pdf

Tuenter E, Weber SL, Hilgen FJ, Lourens LJ, Ganopolski A (2005) Simulation of climate phase lags in the response to pre-cession and obliquity forcing and the role of vegetation. Clim Dyn 24:279–295. https://doi.org/10.1007/s00382-004-0490-1

Viaggi P (2021) Quantitative impact of astronomical and sun-related cycles on the Pleistocene climate system from Antarctica records. Quat Sci Adv 4:100037. https://doi.org/10.1016/j.qsa.2021.100037

Wagner G, Beer J, Masarik J, Muscheler R, Kubik PW, Mende W, Laj C, Raisbeck Grant M, Yiou F (2001) Presence of the solar de Vries cycle (~205 years) during the last ice age. Geophys Res Lett 28(2):303–306. https://doi.org/10.1029/2000GL006116

Weitz JA, Stojanović KA, Sachsenhofer RF, Misch D, Lukeneder A (2026) Organic Petrology and Geochemistry of the Scheiblinggraben section: detailed insights into the Carnian Pluvial episode and comparison to time equivalent profiles. In: Lukeneder A (ed) Late Triassic Konservat-Lagerstätten within the Carnian Pluvial Episode: 150 years of scientific excavations in Austria. Topics in geobiology. Chapter 23. This volume

Whiteside JH, Grogan DS, Olsen PE, Kent DV (2011) Climatically driven biogeographic provinces of Late Triassic tropical Pangea. Proc Nati Acad Sci USA 108(22):8972–8977. https://doi.org/10.1073/pnas.1102473108

Wilson JL (1975) Carbonate facies in geologic history. Springer, Berlin, 471 pp. https://doi.org/10.1007/978-1-4612-6383-8

Winguth A, Winguth C (2013) Precession-driven monsoon variability at the Permian-Triassic boundary—implications for anoxia and the mass extinction. Glob Planet Change 105:160–170. https://doi.org/10.1016/j.gloplacha.2012.06.006

Woods AD (2005) Paleoceanographic and paleoclimatic context of Early Triassic time. C R Palevol 4:463–472. https://doi.org/10.1016/j.crpv.2005.07.003

Wu H, Zhang S, Feng Q, Jiang G, Li H, Yang T (2012) Milankovitch and sub-Milankovitch cycles of the early Triassic Daye Formation, South China and their geochronological and paleoclimatic implications. Gondwana Res 22(2):748–759. https://doi.org/10.1016/j.gr.2011.12.003

Xapsos MA, Burke EA (2009) Evidence of 6 000-year periodicity in reconstructed sunspot numbers. Sol Phys 257:363–369. https://doi.org/10.1007/s11207-009-9380-3

Yan J, Zhao K (2001) Permo-Triassic paleogeographic, paleoclimatic and paleoceanographic evolutions in eastern Tethys and their coupling. Sci China Ser D Earth Sci 44:968–978. https://doi.org/10.1007/BF02875390

Zeng Z, Zhu H, Yang X, Zeng H, Hu X, Xia C (2019) The Pangaea Megamonsoon records: evidence from the Triassic Mungaroo Formation, northwest shelf of Australia. Gondwana Res 69:1–24. https://doi.org/10.1016/j.gr.2018.11.015

Zhang J, Jiang Z, Liang C, Baars TF, Wang Y, Abels HA (2022) Astronomical forcing of meter-scale organic-rich mudstone–limestone cyclicity in the Eocene Dongying sag, China: implications for shale reservoir exploration. AAPG Bull 106(8):1557–1579. https://doi.org/10.1306/02072220103

Zhang P, Yang M, Lu J, Jiang Z, Zhaou K, Xu X, Wang Y, Wu L, Chen H, Zhu X, Guo Y, Ye H, Shao L, Hilton J (2023) Floral response to the Late Triassic Carnian Pluvial episode. Front Ecol Evol 11:1199121. https://doi.org/10.3389/fevo.2023.1199121

Zhao H, Grasby SE, Wang X, Zhang L, Liu Y, Chen ZQ, Hu Z, Huang Y (2022) Mercury enrichments during the Carnian Pluvial Event (Late Triassic) in South China. GSA Bull 134:2709–2720. https://doi.org/10.1130/B36205.1

Ziegler M, Tuenter E, Lourens LJ (2010) The precession phase of the boreal summer monsoon as viewed from the eastern Mediterranean (ODP Site 968). Quat Sci Rev 29(11–12):1481–1490. https://doi.org/10.1016/j.quascirev.2010.03.011

Zühlke R, Bechstadt T, Mundil R (2003) Sub-Milankovitch and Milankovitch forcing on a model Mesozoic carbonate platform—the Latemar (Middle Triassic, Italy). Terra Nova 15:69–80. https://doi.org/10.1046/j.1365-3121.2003.00366.x

Index

A. Lukeneder (ed.), *Late Triassic Konservat-Lagerstätten within the Carnian Pluvial Episode*, Topics in Geobiology 55, https://doi.org/10.1007/978-3-032-19777-1

Q

R

Symbols

GPSR Compliance

The European Union's (EU) General Product Safety Regulation (GPSR) is a set of rules that requires consumer products to be safe and our obligations to ensure this.

If you have any concerns about our products, you can contact us on ProductSafety@springernature.com

In case Publisher is established outside the EU, the EU authorized representative is:

Springer Nature Customer Service Center GmbH
Europaplatz 3
69115 Heidelberg, Germany

Batch number: 10644391

Printed by Printforce, the Netherlands